T0186883

Pharmacotherapy of Cachexia

Pharmacotherapy of Cachexia

Edited by
Karl G. Hofbauer
Stefan D. Anker
Akio Inui
Janet R. Nicholson

Taylor & Francis
Taylor & Francis Group

Boca Raton   London   New York

A CRC title, part of the Taylor & Francis imprint, a member of the
Taylor & Francis Group, the academic division of T&F Informa plc.

Published in 2006 by
CRC Press
Taylor & Francis Group
6000 Broken Sound Parkway NW, Suite 300
Boca Raton, FL 33487-2742

International Standard Book Number-10: 0-8493-3379-2 (Hardcover)
International Standard Book Number-13: 978-0-8493-3379-8 (Hardcover)
Library of Congress Card Number 2005051075

Library of Congress Cataloging-in-Publication Data

Pharmacotherapy of cachexia / edited by Karl G. Hofbauer.
    p. ; cm.
    Includes bibliographical reverences and index.
    ISBN 0-8493-3379-2 (alk. paper)
    1. Cachexia--Chemotherapy. I. Hofbauer, Karl G. [DNLM: 1. Cachexia--drug therapy. 2. Cachexia--etiology. 3. Cachexia--physiopathology. WB 146 P536 2005]

RB150.C33P43 2005
616'.047--dc22   2005051075

Taylor & Francis Group
is the Academic Division of Informa plc.

Visit the Taylor & Francis Web site at
http://www.taylorandfrancis.com

and the CRC Press Web site at
http://www.crcpress.com

Preface

Cachexia is an underestimated and frequently neglected medical condition. Although it is regarded as an independent risk factor for morbidity and mortality it does not receive sufficient medical attention and treatment. The therapy of cachexia requires a coordinated, flexible, and multidisciplinary approach. Since neither nutritional therapy nor physical exercise is sufficient to treat cachexia, pharmacotherapy plays an important role. A wide variety of drugs with highly diverse mechanisms of action is currently being used but for many of them convincing evidence of therapeutic efficacy is lacking and their benefit/risk ratio is unknown.

The present book was conceived with the aim of providing a comprehensive reference source for health care professionals concerned with the treatment of cachexia. It is organized in six sections; the first one dealing with the pathophysiology of cachexia, the second with the diseases associated with cachexia, and the third with general therapeutic aspects. The subsequent three sections are devoted to pharmacotherapy. A section on drugs in clinical use is followed by one on drugs in research and development and a final section, which is focused on future developments.

Why should a reader interested in cachexia consult a book instead of reading the latest reviews or research articles? The editors strongly believe that a heterogeneous syndrome such as cachexia and its diverse treatment options can be best presented in a single volume where contributions by different authors on various aspects can be found and compared. Such a broad range of chapters allows the reader to find information about various different aspects of the disease and its treatment and to select the most pertinent aspects relating to his specific question by using a single reference source.

We are grateful to the numerous internationally recognized specialists who have contributed chapters on various aspects of cachexia. Through their critical evaluation of pathophysiological mechanisms, clinical features, and therapeutic options this book provides the reader with a valuable source of information on all aspects of this serious and life-threatening syndrome. We would also like to thank the publishers for their expert editorial advice and particularly acknowledge the excellent support of Jny Wittker who was responsible for all administrative and organizational work throughout the preparation of this volume.

Karl G. Hofbauer
Stefan D. Anker
Akio Inui
Janet R. Nicholson

Editors

Karl G. Hofbauer holds the chair for applied pharmacology at the Medical Faculty of the University of Basel, Switzerland. He received his M.D. degree from the University of Vienna, Austria, and specialized in experimental pharmacology at the University of Heidelberg, Germany. He spent more than 20 years in preclinical drug research at Ciba, later Novartis, where he held several research management positions in the area of cardiovascular and metabolic diseases. He has continued to work on these subjects after moving back to academia at the Biozentrum, Basel. His book *Pharmacotherapy of Obesity: Options and Alternatives* published by CRC Press in 2004 is now complemented by the present volume on the pharmacotherapy of cachexia.

Akio Inui is professor of internal medicine in the Department of Behavioral Medicine, Kagoshima University Graduate School of Medical and Dental Sciences. He holds a doctoral degree from Kobe University School of Medicine. Dr. Inui has focused on translational research of peptides in body weight regulation disorders including obesity, cancer cachexia, and eating disorders.

Dr. Inui received the Janssen Award of the American Gastroenterological Association in 2004. He is the editor of *Peptides, Nutrition, International Journal of Oncology, Current Nutrition and Food Science*, and *Recent Patents on CNS Drug Discovery*. His research effort is now directed at elucidating the role of the ghrelin-neuropeptide Y pathway from stomach to brain in appetite and gut motility regulation.

Stefan D. Anker is professor of applied cachexia research in the Charité Campus Virchow-Klinikum in Berlin, Germany (since 2002). He also has an academic appointment at the Department of Clinical Cardiology at the National Heart and Lung Institute (NHLI), Imperial College School of Medicine in London, England.

Dr. Anker studied medicine and obtained his M.D. (1993) from Charité Medical School in Berlin. He went on to earn a Ph.D. (1998) at Imperial College London, based on studies on cachexia in chronic heart failure. As part of his training, Dr. Anker was a cardiology research fellow and a senior research fellow and team leader at the NHLI.

The recipient of numerous clinical research awards and grants, Dr. Anker is also a prodigious writer and has more than 160 original papers, reviews, and editorials to his credit. His published work addresses issues in the pathophysiology and treatment of chronic heart failure, specifically the prognostic importance of weight loss (i.e., cachexia) and the effect of treatments to prevent and reverse cardiac cachexia. Additionally, he is active in immunological research and recently has extended his work to cachexia in other chronic illnesses, like COPD and cancer. As an adjunct to his writing endeavors, Dr. Anker serves on the editorial boards of the *Journal of Cardiac Failure*, *Italian Heart Journal*, *International Journal of Cardiology*, and *Zeitschrift für Kardiologie*. Dr. Anker is member of international steering committees of five clinical trials in heart failure.

Janet R. Nicholson is a post-doctoral research fellow at the Biozentrum, University of Basel, Switzerland. She received her Ph.D. in pharmacology from the University of Cambridge, UK in 1998 and then moved to the USA to complete a 3-year post-doctoral research fellowship at the University of Michigan, where she worked on central mechanisms involved in the regulation of energy balance. In her current position, Dr. Nicholson's work focuses on *in vivo* models of cachexia, and she is actively involved in a drug discovery program aiming to identify novel pharmacological interventions for the treatment of cancer cachexia.

Contributors

Stefan D. Anker
Campus Virchow-Klinikum
Berlin, Germany

Jonas Axelsson
Karolinska Institutet
Karolinska University Hospital at
    Huddinge
Stockholm, Sweden

Waldemar Banasiak
Cardiac Department
Military Hospital
Wroclaw, Poland

Vickie E. Baracos
University of Alberta
Edmonton, Alberta, Canada

Shehzad Basaria
Johns Hopkins University School of
    Medicine
Baltimore, Maryland, USA

Kaspar Berneis
Medical University Clinic
Hospital Bruderholz
Bruderholz, Switzerland

Marijke Brink
University of Basel
Department of Research
University Hospital
Basel, Switzerland

Antonio L.C. Campos
Department of Gastrointestinal
    Surgery
Federal University of Parana
Curitiba, Parana, Brazil

Paul V. Carroll
Guy's & St. Thomas' NHS
    Foundation Trust
London, UK

Tommy Cederholm
Karolinska Institutet
Karolinska University Hospital at
    Huddinge
Stockholm, Sweden

Andrew L. Clark
Castle Hill Hospital
Cottingham, Hull, UK

Ross G. Clark
Tercica Inc.
South San Francisco, California, USA

Andrew J.S. Coats
University of Sydney
Sydney, Australia

Julio C.U. Coelho
Department of Gastrointestinal
    Surgery
Federal University of Parana
Curitiba, Parana, Brazil

Laura A. Coleman
Marshfield Clinic Research
    Foundation
Marshfield, Wisconsin, USA

M. Dahele
Clinical and Surgical Sciences
School of Clinical Sciences and
    Community Health
The University of Edinburgh
Edinburgh, UK

Mellar P. Davis
The Harry R. Horvitz Center for
    Palliative Medicine
Taussig Cancer Center
The Cleveland Clinic Foundation
Cleveland, Ohio, USA

Amie J. Dirks
Wingate University School of
    Pharmacy
Wingate, North Carolina, USA

Adrian S. Dobs
Division of Endocrinology and
    Oncology
Johns Hopkins University School of
    Medicine
Baltimore, Maryland, USA

Wolfram Doehner
Charité Medical School
Berlin, Germany

William J. Evans
University of Arkansas for Medical
    Sciences
Little Rock, Arkansas, USA

Claudine Falconnier
University Hospital
Basel, Switzerland

K.C.H. Fearon
The University of Edinburgh
Edinburgh, UK

David J. Glass
Regeneron Pharmaceuticals
Tarrytown, New York, USA

Carolina G. Goncalves
Department of Surgery
SUNY University
Syracuse, New York, USA

Lindsey Harle
University of Texas Medical Branch
    at Galveston
Galveston, Texas, USA

Olof Heimbürger
Karolinska Institutet
Karolinska University Hospital at
    Huddinge
Stockholm, Sweden

Karl G. Hofbauer
Applied Pharmacology Biozentrum
University of Basel
Basel, Switzerland

Akio Inui
Kagoshima University Graduate
    School of Medical and Dental
    Sciences
Kagoshima, Japan

Ewa A. Jankowska
Military Hospital
Wroclaw, Poland

Aminah Jatoi
Mayo Clinic
Rochester, Minnesota, USA

Matthias John
University Hospital Charité
Berlin, Germany

Contributors

Ulrich Keller
University Hospital
Basel, Switzerland

Charles P. Lambert
University of Arkansas for Medical
    Sciences
Little Rock, Arkansas, USA

Christiaan Leeuwenburgh
University of Florida
College of Medicine
Gainesville, Florida, USA

Bengt Lindholm
Karolinska Institutet
Karolinska University Hospital at
    Huddinge
Stockholm, Sweden

Giovanni Mantovani
Department of Medical Oncology
University of Cagliari
Cagliari, Italy

Daniel L. Marks
Oregon Health & Sciences University
Portland, Oregon, USA

Michael M. Meguid
Department of Surgery
SUNY University
Syracuse, New York, USA

Thomas Meier
Santhera Pharmaceuticals Ltd.
Liestal, Switzerland

John E. Morley
Saint Louis University School
    of Medicine
St. Louis, Missouri, USA

Janet R. Nicholson
Applied Pharmacology Biozentrum
University of Basel
Basel, Switzerland

Smitha Patiyil
Mayo Clinic
Rochester, Minnesota, USA

Simona Perboni
Department of Medical Oncology
University of Cagliari
Cagliari, Italy

Piotr Ponikowski
Cardiac Department
Military Hospital
Wroclaw, Poland

Philip A. Poole-Wilson
National Heart and Lung Institute
Faculty of Medicine
Imperial College
London, UK

Stefanie Possekel
Santhera Pharmaceuticals Ltd.
Liestal, Switzerland

Abdul Rashid Qureshi
Karolinska Institutet
Karolinska University Hospital at
    Huddinge
Stockholm, Sweden

Markus A. Ruegg
Biozentrum
University of Basel
Basel, Switzerland

Annemie M.W.J. Schols
Department of Respiratory Medicine
University of Maastricht
Maastricht, The Netherlands

R.J.E. Skipworth
School of Clinical Sciences and
  Community Health
The University of Edinburgh
Edinburgh, UK

Peter Stenvinkel
Karolinska Institutet
Karolinska University Hospital at
  Huddinge
Stockholm, Sweden

Sabine Strassburg
Department of Cardiology Charité
Campus Virchow-Klinikum
Berlin, Germany

Susumu Suzuki
SUNY University
Syracuse, New York, USA

Yoshiyuki Takimoto
The University of Tokyo
Tokyo, Japan

Ergun Y. Uc
College of Medicine
University of Iowa
Iowa City, Iowa, USA

Katrin Utech
Medical University Clinic
Hospital Bruderholz
Bruderholz, Switzerland

Stephan von Haehling
National Heart & Lung Institute
Department of Clinical Cardiology
Imperial College School of Medicine
London, UK

Contents

Contents

I

Pathophysiology of Cachexia

1 Malnutrition and Cachexia

Carolina G. Goncalves, Antonio L.C. Campos,
Julio C.U. Coelho, Susumu Suzuki, and
Michael M. Meguid

CONTENTS

SUMMARY

Weight loss and cachexia are common features in cancer patients, contributing to decreased quality of life and life expectancy. The mechanisms underlying the onset of malnutrition with tumor growth include anorexia and reduced food intake and an interaction between tumor and host-related mediators, which lead to alterations in carbohydrate, lipid, and protein metabolism. The improvement in the patient's overall nutritional status is a major concern in the overall assessment of cancer patients.

1.1 INTRODUCTION

Cancer is accompanied by weight loss. Malnutrition is a continuum that progresses from an imbalance of nutrient intake in relation to needs and to gross

TABLE 1.1
Factors Interfering with Food Intake

Gastrointestinal obstruction
Pain
Depression
Constipation
Side effects of medication
Neuropeptide and hormonal changes

structural and functional changes. Malnutrition develops when the patient fails to eat sufficiently to meet his or her metabolic needs, thus developing functional and structural changes that lead to body wasting. Anorexia and reduced food intake are frequently encountered in cancer patients,[1] in whom anorexia is usually the presenting symptom. The prevalence of anorexia in cancer patients is difficult to determine, primarily because its presence is often not investigated and the reliability of diagnostic tools is limited. Considering that up to 20% of cancer deaths are due to malnutrition *per se*,[2] it is reasonable to conclude that in many cancer patients starvation contributes to death.

The frequency of cachexia depends on the type of the tumor. In general, patients with haematological malignancies and breast cancer seldom experience substantial weight loss. A high incidence is seen in patients with carcinoma of the pancreas and stomach. Most solid tumors are also associated with a higher frequency of cachexia. At the time of diagnosis, most of the patients with upper gastrointestinal cancers and with lung cancer have substantial weight loss. Cachexia is also more common in children and elderly patients and becomes more pronounced as disease progresses.[3] The incidence of malnutrition is related to the site of the disease and the stage of the disease is also a predictor of weight loss.[4]

1.2 ANOREXIA

During tumor growth, anorexia and reduced food intake are among the major causes leading to malnutrition and eventually cachexia. Food intake is substantially reduced among weight-losing cancer patients.[5] Several factors can influence the reduction in food intake, such as physical obstruction of the gastrointestinal tract, pain, depression, constipation, malabsorption, or side effects of pain management, including the use of opiates, and radiotherapy or chemotherapy (Table 1.1).[6]

Potential mechanisms for cancer anorexia include hormonal and neuropeptide changes. Changes in leptin, serotonin, neuropeptide Y (NPY), and the melanocortin system have been identified as contributory factors in cancer anorexia.

When leptin concentrations are decreased, there is an increase in hypothalamic orexigenic signals that stimulate hunger and increased food intake. Weight loss causes leptin levels to fall in proportion to the loss of body fat and the decrease in leptin concentration stimulates food intake. In contrast to what is seen in starvation, cachectic cancer patients do not experience the increase in food intake promoted by the decrease in leptin. Cytokines may produce long-term inhibition of feeding by stimulating the expression and release of leptin or by mimicking the hypothalamic effect of excessive negative feedback signalling from leptin, leading to the prevention of the normal compensatory mechanisms in the face of both decreased food intake and body weight.[6]

The serotonergic system plays a significant contributory role in the development of anorexia. Evidence in rats indicate a key role for the ventromedial hypothalamic (VMH) serotonergic system in the development of cancer anorexia.[7] Recent data suggest that hypothalamic serotonergic neurotransmission may be critical in linking cytokines to the melanocortin system.[8]

Abnormal hypothalamic NPY activity is one of the mechanisms contributing to the development of anorexia in cancer patients. NPY concentration and release is decreased in anorectic tumor bearing rats, while it is increased in food-restricted rats when compared to control rats fed *ad libitum*.[9] This is confirmed by the decreased levels of NPY protein[10, 11] and its decreased visualization in not only first order but also second order neurons in the hypothalamus.[7, 12] In addition, an up-regulation of α-melanocyte stimulating hormone (α-MSH) occurs in anorectic tumor bearing rats,[13] and a decrease in α-MSH after intracerebroventricular injection of its antagonist results in an increase in food intake in tumor bearing mice.[14] The hypothalamic melanocortin α-MSH derived from pro-opiomelanocortin is an anorexigenic peptide strongly associated with food intake regulation. In cancer cachexia, despite the marked weight-loss, which would normally cause a down-regulation of the α-MSH expression, it is up-regulated.[14]

1.3 PATHOGENESIS OF CACHEXIA

Cachexia is a complex metabolic disorder that involves features of anorexia, anaemia, lipolysis, activation of the acute phase response, and insulin resistance. Several key mediators have been identified as possible etiologic factors of this syndrome.

1.3.1 Tumor-Induced Metabolic Alterations

Host–tumor interactions occur very early during tumor growth, inducing metabolic changes, which are maintained throughout the course of the disease. The presence of a tumor influences the host's metabolic, immunologic, and clinical status. Changes in protein metabolism are profound, but changes also occur in carbohydrate and lipid metabolism (Table 1.2).[15]

Energy expenditure depends on tumor type and tumor volume, ranging from less than 60% to more than 150% of that predicted.[16, 17] The pattern of weight loss seen in cachexia is different from that seen in starvation. The normal response to

TABLE 1.2
Tumor-Induced Metabolic Alterations

Protein metabolism
    Accelerated protein turnover
    Increased skeletal muscle protein degradation
    Decreased skeletal muscle protein synthesis
    Increased hepatic protein synthesis
Carbohydrate metabolism
    Insulin resistance
    Increased glucose turnover
    Impairment of peripheral glucose disposal
    Increased hepatic gluconeogenesis from amino acids and lactate
    Increased glycolysis
Lipid metabolism
    Increased lipolysis
    Decreased lipogenesis
    Decreased activity of lipoprotein lipase

nutrient deprivation produces primarily losses of fat, with relative preservation of lean tissue mass.[18] The cancer patient may lose 75% of muscle protein and 80% of fat with relative preservation of visceral protein. This suggests that other metabolic mechanisms are involved in cancer patients compared with those occurring in starvation.[19]

1.3.1.1 Protein metabolism

Cancer patients have an accelerated total protein turnover as a result of an increase in both protein synthesis and degradation.[20] Skeletal muscle protein synthesis is reduced, while hepatic protein synthesis is increased because of the production of acute phase proteins, such as C-reactive protein and fibrinogen.[21,22] Loss of skeletal muscle accounts for the shorter survival of cachectic cancer patients, with impairment of respiratory muscle function, resulting in the high incidence of hypostatic pneumonia as the terminal event.[23]

    The mechanism for the decrease in protein synthesis in skeletal muscle remains to be fully elucidated, although tumor factors such as proteolysis-inducing factor (PIF)[24] and cytokines such as tumor necrosis factor-α (TNF-α)[25] depress protein synthesis in skeletal muscle of normal animals. However, the increase in protein degradation plays a major in role the atrophy of skeletal muscle in cancer cachexia. There are three pathways for protein degradation: (1) Lysosomal; (2) Ca^{2+}-dependent (calpain); and (3) Ubiquitin–proteasome proteolytic pathway; the latter being the most important mechanism of muscle atrophy in cancer.[26] Both inflammatory cytokines and tumor-secreted factors stimulate the expression of ubiquitin and various subunits of the 26S proteasome.[27] TNF-α and interferon gamma (IFN-γ) have a remarkably high selectivity for the down-regulation of the core myofibrillar proteins myosin heavy chain, suggesting that cancer-induced muscle wasting is derived from the selective down-regulation of specific skeletal muscle gene products.[28]

Tumors are glycolytic and in the absence of adequate nutrient intake the primary source of glucose is muscle.[29] Specific changes in plasma amino acid concentrations have been reported in cancer patients and in tumor-bearing animals.[15,30,31] These changes affect the synthesis of brain neuropeptides by altering the bioavailability of their amino acid precursors.[9] Tryptophan, a serotonin (5-HT) precursor, has a role in the pathogenic mechanisms leading to cancer-related anorexia. We have previously reported a strong relationship between the presence of anorexia and the increase in brain tryptophan availability.[32]

1.3.1.2 Carbohydrate metabolism

Cachectic cancer patients experience relative glucose intolerance and insulin resistance due to tumor related cytokines. There are two major abnormalities in carbohydrate metabolism associated with tumor growth: an increase in glucose turnover and an impairment of peripheral glucose utilization. Gluconeogenesis is increased as a result of tumor related lactate elevation, secondary to the increased tumor glycolysis.[33] The accelerated hepatic gluconeogenesis contributes to cancer patients' protein mass depletion. A 40% increase in hepatic glucose production has been reported in weight-losing cancer patients.[6]

1.3.1.3 Lipid metabolism

Cachectic patients have increased lipid mobilization, decreased lipogenesis, and decreased activity of lipoprotein lipase (LPL).[34] Cytokines inhibit LPL, minimizing the extraction of plasma fatty acids for storage by adipocytes and resulting in a net flux of lipid into the circulation.[35]

Certain tumors synthesize a lipid-mobilizing factor (LMF) identified in a cachexia-inducing murine tumor and in the urine of weight-losing cancer patients.[36] LMF amino acid sequence is identical with the human plasma Zn-α2-glycoprotein (ZAG) and thus LMF and ZAG induce lipolysis in isolated murine adipocytes and cause selective loss of adipose tissue without a change in body water or nonfat carcass mass.[35,36] The induction of lipolysis is mediated by the stimulation of plasma membrane adenylate cyclase, which increases intracellular cyclic AMP, stimulating AMP dependent protein kinase. Glucocorticoids also stimulate ZAG gene expression; therefore the increase in cortisol concentrations that are seen in cachectic patients modulate the increase in lipolysis through the increase in ZAG expression.[37]

The sum of increased tumor glucose utilization, increased gluconeogenesis, and inhibition of fat storage due to the above mentioned mediators in combination with the decrease in food intake associated with anorexia result in the loss of muscle and fat mass that constitutes cancer cachexia.[38]

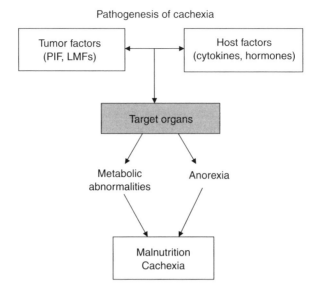

FIGURE 1.1 Both tumor- and host-factors promote complex metabolic alterations and anorexia by acting on different target organs that will ultimately lead to cancer cachexia and malnutrition. Abbreviations: PIF: proteolysis-inducing factor; LMFs: lipid-mobilizing factors.

1.4 MEDIATORS

Endogenous mediators are essential elements in the pathogenesis of anorexia and cachexia (Figure 1.1).[38–40] Cancer cachexia is not merely the result of the competition for ingested nutrients from a tumor. The progression of tumor growth and its interactions with the immune system influences cellular and biological systems both at the site of the tumor and in distant areas of the brain. Malnutrition in cancer patients often develops when the tumor mass is still clinically undetectable or very small. Cancer cachexia is mediated by circulating factors released by the tumor itself or by the host in response to progressive tumor growth and by a combination of both. Cytokines are key mediators in the development of anorexia and cachexia. Several cytokines are involved in the cachectic process, including TNF-α, interleukin-1β (IL-1β), IFN-γ, interleukin-6 (IL-6), and members of the IL-6 super family such as ciliary neurotrophic factor (CNTF) and leukemia inhibitory factor, and their receptors are present in high densities in the hypothalamus.[41]

Tumor necrosis factor-α suppresses LPL activity in adipocytes[42] and its parenteral administration leads to anorexia, weight loss, an acute-phase protein response, protein and fat breakdown, a rise in cortisol and glucagon and a fall in insulin concentration, insulin resistance, anemia, fever, and elevated energy expenditure in both animals[43,44] and humans.[45] However, circulating TNF-α is not often measurable in weight-losing cancer patients[46] and while antibodies to

TNF-α may increase food intake in an animal model of cachexia, they do not ameliorate weight loss in humans.[47] Both IL-1 and TNF-α mediate their anorectic effect through the hypothalamus[48] as well as directly on the gastrointestinal tract, by decreasing gastric emptying.[49] Furthermore, IL-1 and TNF-α administration promote cancer cachexia through increases in resting energy expenditure and also can promote skeletal protein wasting similar to that seen in tumor-bearing animals. Elevated concentrations of IL-6 are found in some weight-losing cancer patients[50] and antibodies to IL-6 will suppress the development of cachexia in animals to some extent.[51]

Ciliary neurotrophic factor produces weight-loss, protein breakdown, fever and an acute-phase protein response in animals.[52] IL-1 also produces anorexia, weight loss, and protein breakdown[53] and the administration of IL-1 antibodies to tumor-bearing animals attenuates some aspects of cachexia.[48] Cortisol and glucagon are among the hormones involved in the cachexia process and high concentrations of both hormones have been documented in weight-losing cancer patients.[54,55]

Cytokines induce a stronger effect on feeding and metabolism when injected directly into the central nervous system (CNS) rather than peripherally. We previously reported that the systemic infusion of soluble TNF-α receptors improved food intake in anorectic tumor-bearing rats.[56] However, other studies have shown that episodic TNF-α administration induces anorexia but does not induce cachexia.[57] We also reported that IL-1 administration in high doses (2 ng) is capable of inducing profound anorexia when it is directly injected into the ventro medial hypothalamic nucleus.[58] IFN-γ inhibits the activity of the enzyme LPL, decreases the rate of synthesis of lipids, and can lead to severe cachexia.[59,60]

Other potential mediators also modulate cancer cachexia. A 24-kDa glycoprotein, known as PIF, has been isolated from the urine of weight-losing cancer patients. This factor produces protein breakdown in animal models.[24] Furthermore, a LMF, identical with the plasma protein ZAG, is produced by tumors, and causes a decrease in body fat by stimulating triglyceride hydrolysis in adipose tissue through a cyclic adenosine monophosphate (AMP)-mediated process by interaction with a beta 3-adrenoreceptor.[36,61] Circulating hormones also have important effects on food intake and body-weight regulation. Ghrelin is a gastrointestinal peptide that stimulates food intake and growth hormone (GH) secretion.[62] Both ghrelin biosynthesis and secretion have been reported to be stimulated in tumor-inoculated cachectic mice.[63] Furthermore, baseline plasma concentrations of ghrelin are elevated in cachectic lung-cancer patients and its synthesis is up-regulated as a compensatory mechanism by chemotherapy-induced anorexia.[64]

We previously reported that plasma leptin concentration is decreased in anorectic methylcholanthrene-induced sarcoma-bearing male rats.[65] However, other studies have shown that leptin concentration is increased during cytokine-induced inflammatory responses in septic patients suggesting that raised leptin concentrations may be related to anorexia.[66,67] Nevertheless, the sum of cumulative animal and human data do not appear to support the view that leptin has a major role in cancer anorexia.[68]

1.5 MALNUTRITION

Cancer anorexia–cachexia syndrome is observed in 80% of patients with advanced-stage cancer. Malnutrition accompanies malignant processes in approximately 50% of patients and eventually leads to severe wasting, which accounts for approximately 30% of cancer related deaths.[69]

The malnourished cancer patient responds poorly to all therapeutic approaches, such as surgery, chemotherapy, and radiotherapy, demonstrating increased morbidity and mortality compared with well-nourished patients.

Cancer patients have their nutritional status affected by several factors, therefore monitoring is essential to prevent or reverse complications resulting from a depleted state. The maintenance of lean body mass above the critical malnutrition thresholds can enhance patient survival and ameliorate quality of life.

Cachexia is expected in cancer patients if an involuntary weight loss of greater than 5% has occurred within a previous 6-month period. A weight loss of 10% or more usually indicates severe depletion, but can be used as a starting criterion for the diagnosis of anorexia–cachexia syndrome in obese patients.[70]

The nutritional status and quality of life of cachectic cancer patients can be improved by an integrated therapeutic strategy, which includes both nutritional counselling and pharmacotherapy.

Although aggressive nutritional support is the initial consideration for the malnourished cancer patient, theoretical concerns regarding stimulation of tumor growth, as well as the lack of well-designed studies documenting benefit, have failed to produce a consensus regarding its therapeutic role. However there are solid data to support its use both pre- and post-operatively for surgery of the gastrointestinal tract in malnourished cancer patients.[71,72]

The nutritional support for cachectic cancer patients includes several approaches: dietary advice to patient and family, as well as information and education of all staff involved in the care of the patient at the hospital and at home are essential. Appetite stimulants and possibly antidepressants are important to relieve symptoms that prevent adequate food intake. Furthermore, enteral and parenteral nutrition is also indicated in selected cases.

The goals of nutritional therapy in cancer patients are to maintain nutritional health and prevent complications of therapy. Nutritional therapy can be used as adjuvant treatment with anticancer therapy or as a long-term administration of nutrients to patients unable to maintain adequate intake during antineoplastic therapy.[73]

Nutritional support should be based on the assessment of the nutritional status, type of tumor, expected response to treatment, and life expectancy.

1.6 CONCLUSION

Cachexia and malnutrition are frequent features in cancer patients. Malnutrition leads to a poor response to all therapeutic modalities with a consequent increase in morbidity and mortality. The mechanisms underlying the onset of malnutrition during the development of a tumor and its growth result from an abnormal host-intermediary metabolism, caused by an interaction of host and tumor-derived substances. Therefore the use of judicious nutritional support in addition to pharmacotherapy that counteract the metabolic effects of cachexia are essential to cancer patients with advanced disease.

ACKNOWLEDGMENTS

This study was supported in part by NIH/DK 003568 and American Institute Cancer Research. A postdoctoral grant from SUNY Upstate Medical University, Hendricks Fund 13230-52 supports CGG.

REFERENCES

1. Sutton, L.M., Demark-Wahnefried, W., and Clipp, E.C. Management of terminal cancer in elderly patients. *Lancet Oncol.*, 4, 149–57, 2003.
2. Daly, J.M., Redmond, H.P., and Gallagher, H. Perioperative nutrition in cancer patients. *J. Parenter. Enteral. Nutr.*, 16, 100S–5S, 1992.
3. Bruera, E. ABC of palliative care. Anorexia, cachexia and nutrition. *Br. Med. J.*, 315, 1219–22, 1997.
4. Meguid, M.M. and Meguid, V. Preoperative identification of the surgical cancer patient in need of postoperative supportive total parenteral nutrition. *Cancer*, 55, 258–62, 1985.
5. Staal-van den Brekel, A.J., Schols, A.M., ten Velde, G.P., Buurman, W.A., and Wouters, E.F. Analysis of the energy balance in lung cancer patients. *Cancer Res.*, 54, 6430–3, 1994.
6. Inui, A. Cancer anorexia–cachexia syndrome: current issues in research and management. *CA Cancer J. Clin.*, 52, 72–91, 2002.
7. Makarenko, I.G., Meguid, M.M., Gatto, L., Goncalves, C.G., Ramos, E.J., Chen, C., and Ugrumov, M.V. Hypothalamic 5-HT$_{1B}$ receptor changes in anorectic tumor bearing rats. *Neurosci. Lett.*, 376, 71–75, 2005.
8. Heisler, L.K., Cowley, M.A., Tecott, L.H., Fan, W., Low, M.J., Smart, J.L., Rubinstein, M., Tatro, J.B., Marcus, J.N., Holstege, H., Lee, C.E., Cone, R.D., and Elmquist, J.K. Activation of central melanocortin pathways by fenfluramine. *Science*, 297, 609–11, 2002.
9. Ramos, E.J., Meguid, M.M., Campos, A.C., and Coelho, J.C. Neuropeptide Y, α-Melanocyte stimulating hormone and monoamines in food intake regulation. *Nutrition.*, 21, 140–52, 2005.
10. Meguid, M.M., Ramos, E.J., Laviano, A., Varma, M., Sato, T., Chen, C., Qi, Y., and Das, U.N. Tumor anorexia: effects on neuropeptide Y and monoamines in paraventricular nucleus. *Peptides*, 25, 261–6, 2004.
11. Ramos, E.J., Suzuki, S., Meguid, M.M., Laviano, A., Sato, T., Chen, C., and Das, U.N. Changes in hypothalamic neuropeptide Y and monoaminergic system in tumor-bearing rats: pre and post tumor resection and at death. *Surgery*, 136, 270–6, 2004.
12. Makarenko, I.G., Meguid, M.M., Gatto, L., Chen, C., and Ugrumov, M.V. Decreased NPY innervation of the hypothalamic nuclei in rats with cancer anorexia. *Brain Res.*, 961, 100–9, 2003.

13. Ramos, E.J., Romanova, I.V., Suzuki, S., Chen, C., Ugrumov, M.V., Sato, T., Meguid, M.M., and Goncalves, C.G. Effects of omega-3 fatty acids on orexigenic and anorexigenic modulators at onset of anorexia. *Brain Res.*, 1046, 157–64, 2005.

14. Marks, D.L., Ling, N., and Cone, R.D. Role of the central melanocortin system in cachexia. *Cancer Res.*, 61, 1432–8, 2001.

15. Muscaritoli, M., Meguid, M.M., Beverly, J.L., Yang, Z.J., Cangiano, C., and Rossi-Fanelli, F. Mechanism of early tumor anorexia. *J. Surg. Res.*, 60, 389–97, 1996.

16. Knox, L.S., Crosby, L.O., Feurer, I.D., Buzby, G.P., Miller, C.L., and Mullen, J.L. Energy expenditure in malnourished cancer patients. *Ann. Surg.*, 197, 152–62, 1983.

17. Fredrix, E.W., Soeters, P.B., Wouters, E.F., Deerenberg, I.M., von Meyenfeldt, M.F., and Saris, W.H. Effect of different tumor types on resting energy expenditure. *Cancer Res.*, 51, 6138–41, 1991.

18. Haslett, P.A. Anticytokine approaches to the treatment of anorexia and cachexia. *Semin. Oncol.*, 25, 53–7, 1998.

19. Heymsfield, S.B., and McManus, C.B. Tissue components of weight loss in cancer patients. *Cancer*, 55, 238–49, 1985.

20. Norton, J.A., Shamberger, R., Stein, T.P., Milne, G.W., and Brennan, M.F. The influence of tumor-bearing on protein metabolism in the rat. *J. Surg. Res.*, 30, 456–62, 1981.

21. Warren, R.S., Jeevanandam, M., and Brennan, M.F. Protein synthesis in the tumor-influenced hepatocyte. *Surgery*, 98, 275–82, 1985.

22. Dworzak, F., et al. Effects of cachexia due to cancer on whole body and skeletal muscle protein turnover. *Cancer*, 82, 42–8, 1998.

23. Inagaki, J., Rodriguez, V., and Bodey, G.P. Proceedings: causes of death in cancer patients. *Cancer*, 33, 568–73, 1974.

24. Lorite, M.J., Cariuk, P., and Tisdale, M.J. Induction of muscle protein degradation by a tumour factor. *Br. J. Cancer*, 76, 1035–40, 1997.

25. Garcia-Martinez, C., Lopez-Soriano, F.J., and Argiles, J.M. Acute treatment with tumour necrosis factor-α induces changes in protein metabolism in rat skeletal muscle. *Mol. Cell. Biochem.*, 125, 11–8, 1993.

26. Bossola, M., Muscaritoli, M., Costelli, P., Grieco, G., Bonelli, G., Pacelli, F., Fanelli, F.R., Doglietto, G.B., and Baccino, F.M. Increased muscle proteasome activity correlates with disease severity in gastric cancer patients. *Ann. Surg.*, 237, 384–9, 2003.

27. Lecker, S.H., Solomon, V., Mitch, W.E., and Goldberg, A.L. Muscle protein breakdown and the critical role of ubiquitin-proteasome pathway in normal and disease states. *J. Nutr.*, 129, 227S–37S, 1999.

28. Acharyya, S., Ladner, K.J., Nelsen, L.L., Damrauer, J., Reiser, P.J., Swoap, S., and Guttridge, D.C. Cancer cachexia is regulated by selective targeting of skeletal muscle gene products. *J. Clin. Intest.*, 114, 370–8, 2004.

29. Norton, J.A., Burt, M.E., and Brennan, M.F. *In vivo* utilization of substrate by human sarcoma-bearing limbs. *Cancer*, 45, 2934–9, 1980.

30. Kubota, A., Meguid, M.M., and Hitch, D.C. Amino acid profiles correlated diagnostically with organ site in three kinds of malignant tumors. *Cancer*, 69, 2348, 1992.

31. Kurzer, M., Janiszewski, J., and Meguid, M.M. Amino acid profiles in tumor-bearing and pair-fed non-tumor bearing malnourished rats. *Cancer*, 62, 1492–6, 1988.

32. Meguid, M.M., Muscaritoli, M., Beverly, J.L., Yang, Z.J., Cangiano, C., and Rossi-Fanelli, F. The early cancer anorexia paradigm: changes in plasma free tryptophan and feeding indexes. *J. Parenter. Enteral Nutr.*, 16, 56S–9S, 1992.

33. Tayek, J.A. A review of cancer cachexia and abnormal glucose metabolism in humans with cancer. *J. Am. Coll. Nutr.*, 11, 445–56, 1992.

34. Tisdale, M.J. Metabolic abnormalities in cachexia and anorexia. *Nutrition*, 16, 1013–4, 2000.

35. Hirai, K., Hussey, H.J., Barber, M.D., Price, S.A., and Tisdale, M.J. Biological evaluation of a lipid-mobilizing factor isolated from the urine of cancer patients. *Cancer Res.*, 58, 2359–65, 1998.

36. Todorov, P.T., McDevitt, T.M., Meyer, D.J., Ueyama, H., Ohkubo, I., and Tisdale, M.J. Purification and characterization of a tumor lipid-mobilizing factor. *Cancer Res.*, 58, 2353–8, 1998.

37. Tisdale, M.J. Tumor-host interactions. *J. Cell Biochem.*, 93, 871–7, 2004.

38. Tisdale, M.J. Cancer anorexia and cachexia. *Nutrition*, 17, 438–42, 2001.

39. Inui, A. Cancer anorexia–cachexia syndrome: are neuropeptides the key? *Cancer Res.*, 59, 4493–501, 1999.

40. Tohgo, A., Kumazawa, E., Akahane, K., Asakawa, A., and Inui, A. Anticancer drugs that induce cancer-associated cachectic syndromes. *Expert Rev. Anticancer Ther.*, 2, 121–9, 2002.

41. Sternberg, E.M. Neural-immune interactions in health and disease. *J. Clin. Invest.*, 100, 2641–7, 1997.

42. Fried, S.K. and Zechner, R. Cachectin/tumor necrosis factor decreases human adipose tissue lipoprotein lipase mRNA levels, synthesis, and activity. *J. Lipid Res.*, 30, 1917–23, 1989.

43. Tracey, K.J., Wei, H., Manogue, K.R., Fong, Y., Hesse, D.G., Nguyen, H.T., Kuo, G.C., Beutler, B., Cotran, R.S., Cerami, A., et al. Cachectin/tumor necrosis factor induces cachexia, anemia, and inflammation. *J. Exp. Med.*, 167, 1211–27, 1988.

44. Mahony, S.M. and Tisdale, M.J. Induction of weight loss and metabolic alterations by human recombinant tumour necrosis factor. *Br. J. Cancer*, 58, 345–9, 1988.

45. Michie, H.R., Spriggs, D.R., Manogue, K.R., Sherman, M.L., Revhaug, A., O'Dwyer, S.T., Arthur, K., Dinarello, C.A., Cerami, A., Wolff, S.M., et al. Tumor necrosis factor and endotoxin induce similar metabolic responses in human beings. *Surgery*, 104, 280–6, 1988.

46. Socher, S.H., Martinez, D., Craig, J.B., Kuhn, J.G., and Oliff, A. Tumor necrosis factor not detectable in patients with clinical cancer cachexia. *J. Natl. Cancer Inst.*, 80, 595–8, 1988.

47. Smith, B.K. and Kluger, M.J. Anti-TNF-α antibodies normalised body temperature and enhanced food intake in tumor-bearing rats. *Am. J. Physiol.*, 265, R615–9, 1993.

48. Turrin, N.P., Ilyin, S.E., Gayle, D.A., Plata-Salaman, C.R., Ramos, E.J., Laviano, A., Das, U.N., Inui, A., and Meguid, M.M. Interleukin-1beta system in anorectic catabolic tumor-bearing rats. *Curr. Opin. Clin. Nutr. Metab. Care*, 7, 419–26, 2004.

49. Bodnar, R.J., Pasternak, G.W., Mann, P.E., Paul, D., Warren, R., and Donner, D.B. Mediation of anorexia by human recombinant tumor necrosis factor through a peripheral action in the rat. *Cancer Res.*, 49, 6280–4, 1989.

50. Scott, H.R., McMillan, D.C., Crilly, A., McArdle, C.S., and Milroy, R. The relationship between weight loss and interleukin 6 in non-small-cell lung cancer. *Br. J. Cancer*, 73, 1560–2, 1996.

51. Strassmann, G., Fong, M., Kenney, J.S., and Jacob, C.O. Evidence for the involvement of interleukin 6 in experimental cancer cachexia. *J. Clin. Invest.*, 89, 1681–4, 1992.

52. Espat, N.J., Auffenberg, T., Rosenberg, J.J., Rogy, M., Martin, D., Fang, C.H., Hasselgren, P.O., Copeland, E.M., and Moldawer, L.L. Ciliary neurotrophic factor is catabolic and shares with IL-6 the capacity to induce an acute phase response. *Am. J. Physiol.*, 271, R185–90, 1996.

53. Hellerstein, M.K., Meydani, S.N., Meydani, M., Wu, K., and Dinarello, C.A. Interleukin-1-induced anorexia in the rat. *Influence of prostaglandins. J. Clin. Invest.*, 84, 228–35, 1989.

54. Schaur, R.J., Fellier, H., Gleispach, H., Fink, E., and Kronberger, L. Tumor host relations. I. Increased plasma cortisol in tumor-bearing humans compared with patients with benign surgical diseases. *J. Cancer Res. Clin. Oncol.*, 93, 281–5, 1979.

55. Knapp, M.L., al-Sheibani, S., Riches, P.G., Hanham, I.W., and Phillips, R.H. Hormonal factors associated with weight loss in patients with advanced breast cancer. *Ann. Clin. Biochem.*, 28, 480–6, 1991.

56. Torelli, G.F., Meguid, M.M., Moldawer, L.L., Edwards, C.K. 3rd, Kim, H.J., Carter, J.L., Laviano, A., and Rossi Fanelli, F. Use of recombinant human soluble TNF receptor in anorectic tumor-bearing rats. *Am. J. Physiol. Regul. Integr. Comp. Physiol.*, 277, R850–5, 1999.

57. Argiles, J.M., Moore-Carrasco, R., Busquets, S., and Lopez-Soriano, F.J. Catabolic mediators as targets for cancer cachexia. *Drug Discov. Today*, 8, 838–44, 2003.

58. Yang, Z.J., Blaha, V., Meguid, M.M., Laviano, A., Oler, A., and Zadak, Z. Interleukin-1alpha injection into ventromedial hypothalamic nucleus of normal rats depresses food intake and increases release of dopamine and serotonin. *Pharmacol. Biochem. Behav.*, 62, 61–5, 1999.

59. Patton, J.S., Shepard, H.M., Wilking, H., Lewis, G., Aggarwal, B.B., Eessalu, T.E., Gavin, L.A., and Grunfeld, C. Interferon and tumor necrosis factor have similar catabolic effect on 3T3 L1 cells. *Proc. Natl Acad. Sci. USA*, 83, 8313–7, 1986.

60. Langstein, H.N., Doherty, G.M., Fraker, D.L., Buresh, C.M., and Norton, J.A. The roles of gama-interferon and tumor necrosis factor in an experimental model of cancer cachexia. *Cancer Res.*, 51, 2302–6, 1991.

61. Russell, S.T., Hirai, K., and Tisdale, M.J. Role of β3-adrenegic receptors in the action of a tumor lipid mobilizing factor. *Br. J. Cancer*, 86, 424–8, 2002.

62. Asakawa, A., Inui, A., Fujimiya, M., Sakamaki, R., Shinfuku, N., Ueta, Y., Meguid, M.M., and Kasuga, M. Stomach regulates energy balance via acylated ghrelin and desaxyl ghrelin. *Gut*, 54, 18–24. 2005.

63. Hanada, T., Toshinai, K., Date, Y., Kajimura, N., Tsukada, T., Hayashi, Y., Kangawa, K., and Nakazato, M. Upregulation of ghrelin expression in cachectic nude mice bearing human melanoma cells. *Metabolism*, 53, 84–8, 2004.

64. Shimizu, Y., Nagaya, N., Isobe, T., Imazu, M., Okumura, H., Hosoda, H., Kojima, M., Kangawa, K., and Kohno, N. Increased plasma ghrelin level in lung cancer cachexia. *Clin. Cancer Res.*, 9, 774–8, 2003.

65. Sato, T., Meguid, M.M., Miyata, G., Chen, C., and Hatakeyama, K. Does leptin really influence cancer anorexia? *Nutrition*, 18, 82–3, 2002.

66. Papathanassoglou, E.D., Moynihan, J.A., Ackerman, M.H., and Mantzoros, C.S. Serum leptin levels are higher but are not independently associated with severity or mortality in the multiple organ dysfunction/systemic inflammatory response syndrome: a matched case control and a longitudinal study. *Clin. Endocrinol.*, 54, 225–33, 2001.

67. Moses, A.G., Dowidar, N., Holloway, B., Waddell, I., Fearon, K.C., and Ross, J.A. Leptin and its relation to weight loss, ob gene expression and the acute-phase response in surgical patients. *Br. J. Surg.*, 88, 588–93, 2001.

68. Laviano, A., Meguid, M.M., Inui, A., Muscaritoli, M., and Rossi-Fanelli, F. Review therapy insight: Cancer anorexia cachexia syndrome when all you can eat is yourself. *Nat. Clin. Pract. Oncol.*, 2, 1–8, 2005.

69. Mantovani, G., Maccio, A., Massa, E., and Madeddu, C. Managing cancer-related anorexia/cachexia. *Drugs*, 61, 499–514, 2001.

70. Palesty, J.A. and Dudrick, S.J. What we have learned about cachexia in gastrointestinal cancer. *Dig. Dis.*, 21, 198–213, 2003.

71. Meguid, M.M., Curtas, M.S., Meguid, V., and Campos, A.C. Effects of preoperative TPN on surgical risk — preliminary status report. *Br. J. Clin. Prac.*, 63, 53–8, 1988.

72. Klein, S., Kinney, J., Jeejeebhoy, K., Alpers, D., Hellerstein, M., Murray, M., and Twomey, P. Nutrition support in clinical practice: review of published data and recommendations for future research directions. National Institutes of Health, American Society for Parenteral and Enteral Nutrition, and American Society for Clinical Nutrition. *J. Parenter. Enteral Nutr.*, 21, 133–56, 1997.

73. Rivadeneira, D.E., Evoy, D., Fahey, T.J. 3rd, Lieberman, M.D., and Daly, J.M. Nutritional support of the cancer patient. *CA Cancer J. Clin.*, 38, 69–80, 1998.

2 Anorexia: Central and Peripheral Mechanisms

Simona Perboni, Giovanni Mantovani,
Akio Inui, and Yoshiyuki Takimoto

CONTENTS

SUMMARY

Anorexia is one of the most common symptoms in advanced cancer and is a frequent cause of discomfort in cancer patients and their families. Pathogenesis of anorexia is multifactorial and involves most of the hypothalamic neuronal signaling pathways modulating energy homeostasis. Cancer anorexia is considered the result of a failure of usual appetite and satiety signals. Loss of appetite can arise from decreased taste and smell of food, dysfunctional hypothalamic signaling pathways and cytokine production. Cytokines appear to play a key role in energy balance through persistent activation of the melanocortin system and inhibition of the neuropeptide Y pathway. The imbalance between anorexigenic and orexigenic peptides leads to suppression of appetite and increasing satiety and satiation associated with a marked weight loss and decline in physical performance. High levels of serotonin appear to contribute to these effects. Despite the increase in the understanding of the molecular and neuronal mechanisms that control body weight homeostasis, few effective therapies are available to improve survival and the quality of life of these patients. The aim of this review is to describe the mechanisms that govern satiety and appetite in cancer patients.

2.1 INTRODUCTION

Anorexia is a persistent and pathological form of satiety that is characterized by a gradual onset and a profound and persistent loss of appetite resulting in a decrease in food intake, which leads to progressive depletion of body energy stores. In cancer patients, the development of anorexia is frequently associated with the presence of cachexia, resulting in the onset of the cancer anorexia–cachexia syndrome.[1] Anorexia and reduced energy intake are clinical conditions clearly interconnected yet different: the former is reported by patients who have lost the desire to eat but can still eat to meet their caloric needs; the latter is an objective measurement of an inappropriate eating behavior, which can be secondary to anorexia as well as dysphagia, psychological discomfort, or economical problems. The pathogenesis of cancer anorexia is considered multifactorial and is the end result of altered central and peripheral neurohormonal signals that govern appetite and satiety.[2] Weight loss is a potent stimulus to food intake in normal humans and animals. Because regulation of hunger is most critical for the survival of a species, a complex circuitry of compensatory and overlapping mechanisms has evolved to protect the individual against deficiency in one or more of these regulators.[3]

The hypothalamus is regarded as the main center of the brain involved in feeding behavior. It receives, integrates, and interprets information from the periphery and other brain regions in order to maintain energy homeostasis. The hypothalamus consists of several nuclei that have extensive connections to regulate energy balance. Hormones, such as leptin and insulin that are involved in the long-term regulation of energy homeostasis, interact in the arcuate nucleus (ARC) of the hypothalamus to indicate the energy present in the body fat stores.[4] The ARC contains at least two distinct groups of neurons controlling energy balance. One set co-expresses neuropeptide Y (NPY) and Agouti-related protein (AgRP), both potent stimulators of food intake while pro-opiomelanocortin (POMC) peptide, the precursor of α-melanocyte stimulating hormone (α-MSH), and cocaine- and amphetamine-regulated transcript (CART), which induce an anorexigenic response, are collocated in an adjacent subset.[5] When circulating leptin levels are suppressed, for example, during fasting, expression of the genes encoding POMC and CART are reduced simultaneously with a marked increase in the genes encoding the orexigenic peptides AgRP and NPY. Therefore, during fasting, an increase in AgRP cooperates in the down-regulation of melanocortin signaling by antagonizing the action of α-MSH, concurrently with an inhibition of POMC gene expression.[6] These two populations send monosynaptic projections to identical target regions within the hypothalamic paraventricular nucleus (PVN) and lateral hypothalamic area (LHA), where signals are integrated and then relayed by independent pathways to regions of the brain governing feeding behavior, energy expenditure, and neuroendocrine functions. It is via this set of integrated responses that depleted body fat stores are gradually replenished. The persistence of anorexia in cancer implies a failure of this adaptive feeding response that is so robust in normal subjects.[7] Anorexia can arise from decreased taste and smell of food, early satiety, and increased brain tryptophan and cytokine production. It involves most of the hypothalamic neuronal signaling pathways modulating energy intake.[8] In this review the mechanisms that govern satiety and appetite in cancer patients will be described.

2.2 SATIETY AND CANCER

The gastrointestinal signals that influence the brain to stop an ongoing meal are collectively called "satiety signals." These inform the central nervous system (CNS) that energy is immediately available from recently ingested foods.[4] Satiety signals provide information about mechanical, for example, stomach stretch and volume, and chemical consequences of food intake, as indicated by cholecystokinin (CCK) release. These signals are conveyed via sensory axons in the vagus and in the sympathetic nerves into the nucleus of the solitary tract (NST) in the brain stem. Within the brain, neuronal circuits integrate information from the NST and several hypothalamic nuclei to determine food intake.[9]

Oral ingestion of food acts as a potent positive stimulus for intake, provided that the food is palatable. However, the processes of satiation (termination of the meal) and satiety (continued inhibition of eating), which regulate the length of an eating episode and dictate the initiation of the next meal, are partly dependent on oral sensation.[10] The stomach has an obvious role in the regulation of food intake. The mechanisms related to stimulation of gastric mechano- and chemoreceptors involved in the capacity and propulsive functions of the stomach are well known. Gastric distention in association with a meal is an adequate stimulus that affects all types of gastrointestinal motor activity.[11] Gastric motility acts as a potent positive stimulus for intake mediated by vagus afferents through the NST.[12] Gastric distention by food may play a major role in eliciting satiety, whereas reduction of hunger feelings after a meal more likely results from an interaction between nutrients and receptors in the upper gut and postabsorptive signals from the liver or brain.[11]

The absorptive phase, posttermination of the meal, is mediated by duodenal release of CCK, which regulates the length of an eating episode by binding to vagal CCK-1 receptors, which reduce the gastrointestinal motility via the NST. An ileal phase involves glucagon-like peptide-1, which induces satiety inhibiting the release of NPY in the hypothalamus and the gastro-intestinal motility in the periphery. These short-term, meal-related signals, are coordinated with the long-term hormonal signals, such as leptin and insulin, which can modulate responses to short-term nutritional inputs to maintain appropriate levels of energy in the fat stores.[4]

It has been suggested that cytokines may play a key role in satiety disturbance. Cytokines may be involved in sensory alterations in the chorda tympani (involved in the transduction of taste), which may lead to changes in taste and in the control of specific food preferences. Patients with end-stage cancer have altered taste thresholds with respect to the bitter modality and these changes are most apparent in those patients with higher concentrations of C-reactive protein, interleukin-1β (IL-1β), interleukin-6 (IL-6), and tumor necrosis factor-α (TNF-α). In these patients the odor threshold was lower than in the healthy subjects.[12] This example of altered gustatory and olfactory sensation illustrates the profound influence that mediators of the inflammatory response can have on dietary intake.[12] Patients treated with chemotherapy frequently present with a decreased taste and smell of food. It has been hypothesized that chemotherapy agents may interfere with the taste-receptor cell turnover in the tongue, leading to important changes in the coding of taste. These changes may be among the causative factors to the development of food-avoidance inducing anorexia.[13] Cancer patients also may experience anorexia secondary to food aversion, which results from the central integration of negative psychological experiences and olfactory and gustatory inputs.[14] Anorectic patients often report early satiety, such that they feel full after ingestion of a small amount of food, indicating lack of gastric accommodation or as a result of gastroparesis and delayed antropyloric transit.[15, 16] Analogous to the effect on food preference, cytokines may sensitize the vagus nerve in the gastric tract, resulting in increased activation of mechanisms mediating sensations of fullness that also contributed to the process of satiety.[12] The altered mechanisms inducing early satiety contribute toward the reduction of the hedonic aspects of food and the psychological pleasantness of having a meal with friends and relatives.[14]

2.3 NEUROHORMONAL SIGNALS THAT GOVERN APPETITE

2.3.1 Cytokines

Inflammatory cytokines released by immune cells have been shown to act in the CNS to control food intake and energy homeostasis.[17] An emerging view is that the anorexia–cachexia syndrome is caused predominantly by cytokines either produced by cancer or released by the immune system as a response to the presence of the cancer.[18] In cancer anorexia cytokines may play a pivotal role in the long-term inhibition of feeding by stimulating the expression and release of leptin and by mimicking the hypothalamic effect of excessive negative feedback signaling from leptin, leading to the prevention of the normal compensatory mechanisms in the face of both decreased food intake and body weight[19] (see Figure 2.1). The inhibition of the NPY/AgRP orexigenic network, persistent stimulation of the POMC anorexigenic pathway, as well as the involvement of other networks including hypothalamic neurotransmitters is a possible explanation of this altered feedback in cancer anorexia.[20]

Chronic administration of inflammatory cytokines, such as TNF-α, IL-1, IL-6, and interferon-γ (IFN-γ), either alone or in combination, are capable of reducing

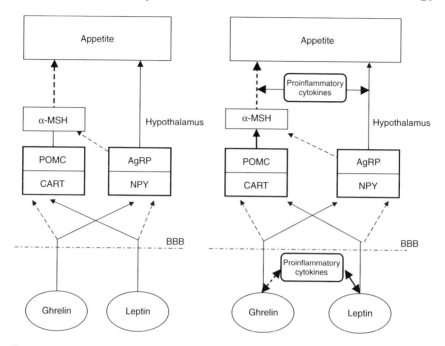

FIGURE 2.1   Simplified model of peptide pathways that regulate appetite. In physiological conditions leptin and ghrelin play an antagonistic role on the neurons of the arcuate hypothalamic nucleus. In cancer, central and peripheral proinflammatory cytokines interact with peptides leading to anorexia. The continuous lines show stimulatory signals while the dashed lines show inhibitory signals. BBB: blood brain barrier, CART: cocaine- and amphetamine-regulated transcript, POMC: proopiomelanocortin peptide, AgRP: Agouti-related protein, NPY: neuropeptide Y, and α-MSH: α-melanocyte stimulating hormone.

food intake and reproducing the distinct features of the cancer anorexia–cachexia syndrome[15,21–23], although the extent to which such effects occur directly in the CNS or indirectly via peripheral tissues is still unclear.[24] IL-1 induces satiety and influences meal size, meal duration, and meal frequency in rats as a result of the activation of IL-1β-induced feeding-regulatory gluco-sensitive neurons in the ventromedial nucleus (VMN) of the hypothalamus.[25,26] IL-1, IL-6, and TNF-α interfere with NPY pathways.[27,28] TNF-α modulates the release of leptin and vice versa. Leptin is also a proinflammatory cytokine and its action modulates the production of other proinflammatory cytokines. All of them can have effects on adipose cell function and insulin resistance. IL-6 is produced in various cells including adipocytes and has been found to regulate leptin levels. IL-6 could also play an additional role, since its receptor as other IL-6 subfamily receptor members (e.g., ciliary neurotrophic factor receptor) share signaling homology with the leptin receptor through gp130, a component of the IL-6 subfamily signal transduction mechanism.[29] Improvements in appetite and weight gain through decreases in cytokine expression after the application of corticosteroids or specific antagonists such as IL-6 antibodies were seen over short periods, but further investigations of the interaction between host and tumor-derived cytokines are necessary to determine their effect on biochemical mechanisms.[30]

Peripheral blood leukocytes of tumor-bearing rats have been shown to secrete greater amounts of IL-1 and TNF activity *in vitro* than cells from healthy controls. Intracellular levels of IL-1, TNF, and IL-6 were not increased in leucocytes of mice implanted with a tumor that did not cause weight loss. Alternatively, several tumor lines that induce the cancer anorexia–cachexia syndrome in rodents constitutively express IL-1, TNF, and IL-6 suggesting a tumor source of anorexigenic cytokines. [31]

In general, serum levels of TNF, IL-1, IL-6, and IFN-γ have not been found to correlate well with weight loss in patients with advanced and terminal cancer.[32] The association of serum cytokines with anorexia is controversial. In one study, megestrol acetate reduced anorexia and improved weight and benefits to appetite were inversely correlated with serum IL-6 levels. In another study, megestrol acetate reduced anorexia independently of serum IL-6 levels.[33,34] It was reported that anorexia in untreated cancer patients did not correlate with circulating cytokines.[35] In the brain, cytokines interact in a paracrine/autocrine way and the local production can be independent from the profiles in the periphery. Cytokines may act in concert with other metabolic pathways to induce anorexia. Increased production and secretion of IL-1 may facilitate tryptophan supply to the brain and IL-1 itself may act on the VMN to increase its neuronal activity and serotonin release. Megesterol acetate may down-regulate the synthesis and release of cytokines and thereby relieve the symptoms of cancer anorexia by mechanisms that reduce the production of cytokines and serotonin. Accordingly, some authors reported up-regulation of IL-1β mRNA in brain regions[36] while others did not confirm this data during the development of anorexia in tumor-bearing models. [24]

2.3.2 Leptin Regulation

Leptin is a hormone that provides information about the status of energy reserves of the organism to the brain, regulating feeding, substrate utilization, energy balance, and the endocrine and immune system.[37] Leptin plays an important role in triggering the adaptive response to starvation since weight loss causes leptin levels to fall in proportion to the loss of body fat.[18] Leptin, a member of the gp 130 family of cytokines, induces a strong T helper-1 lymphocyte response and is regarded as a proinflammatory inducer.[3] Several data suggested a role of leptin in inflammatory diseases. Proinflammatory cytokines up-regulate leptin expression in white adipose tissue and increase plasma leptin levels in hamsters and mice.[38] However, in many common diseases associated with cachexia, such as chronic obstructive pulmonary disease or chronic inflammatory bowel disease, there is an inflammatory status caused by elevated proinflammatory cytokines whereby leptin levels are decreased. In advanced cancer patients the serum leptin levels are not elevated and the lower leptin concentrations seem to be related to body fat mass.[39] In patients with advanced nonsmall-cell lung cancer, serum leptin levels were lower than in controls and more so in the malnourished patients despite an increase in proinflammatory cytokines and acute phase reactants.[40] These data were found also in patients with cancer in different sites.[41-44] In tumor bearing rats, leptin concentrations decreased in plasma and adipose depots 4 days after the tumor cell injections. It has been suggested that leptin secretion in visceral white adipose tissue of tumor bearing rats might be negatively modulated by TNF-α or prostaglandin E2 produced by infiltrating macrophages present in the early stage of cachexia in the adipose tissue. These cytokines may play a role in the local modulation of leptin production in adipose tissue.[37] In tumor-bearing mice, leptin production is decreased while the hypothalamic leptin receptor and NPY expression is increased in response to fat depletion.[45] Taken together these findings suggest that peripheral leptin synthesis is preserved in cancer patients and animals and point to a central dysregulation of the physiological feedback loop.

2.3.3 Ghrelin

Ghrelin is a 28-amino acid acylated polypeptide secreted predominantly from X/A-like enteroendocrine cells of the stomach. It is presently regarded as the only known circulating orexigenic hormone and exerts antagonistic effects on the leptin-induced decrease in food intake through activation of the hypothalamic NPY-Y1 pathway.[46] The orexigenic effect of ghrelin in the hypothalamus could be mediated in part by AgRP. Increased fat mass and hypothalamic expression of AgRP has been shown after the injection of a ghrelin agonist in mice lacking NPY.[47] Ghrelin reduces fat oxidation and increases adiposity.[48] Endogenous ghrelin levels peak before each meal and fall within 1 hour of eating, thus supporting the hypothesis that ghrelin is a hormone that stimulates hunger. Fasting plasma ghrelin levels are inversely related to body mass index and they increase with weight loss induced by caloric restriction.[49] A recent study showed increased levels of active ghrelin in patients with various cancer diagnoses and staging,[50] not just in anorectic lung cancer patients, treated with or without chemotherapy.[51] These data suggest that the increase in ghrelin is inherent to the cachexia syndrome and may represent a compensatory mechanism under catabolic–anabolic imbalance in cancer cachexia.[52] Increased ghrelin levels in cancer patients and animals may still be lower than those in nutritional controls. Despite the increased endogenous ghrelin levels, cancer patients show a "resistance" to the orexigenic effects of this hormone as shown by the lack of increased appetite and the weight loss.[53] In a recent report, sufficient exogenous ghrelin administration appears to overcome any resistance to the appetite-stimulating effects of ghrelin in cancer anorectic patients.[49] It is possible that the ghrelin "resistance" observed in patients with cancer-induced cachexia may be partial, analogous to the insulin resistance state seen in type 2 diabetes mellitus, which is overcome by using high doses of insulin.[50]

In order to understand the mechanisms that regulate ghrelin levels in cancer patients, the relationship between ghrelin and cytokines must be considered. Elevated levels of ghrelin and IL-6 are found in patients with cancer-induced cachexia when compared to individuals with stable body weight with and without cancer.[22,27] Ghrelin administration can antagonize the effect of cytokines on appetite and body weight. Cytokines and ghrelin have opposite effects on body weight and appetite and recent data suggest that cytokines may directly decrease ghrelin production[54] and that ghrelin itself may suppress IL-1β, IL-6, and TNF-α production by human T-lymphocytes and monocytes. Ghrelin functions as a vital counter-regulatory signal in the immune system, controlling not only activation-induced cytokine expression but also leptin-induced expression of the same inflammatory mediators. The reciprocal regulatory effects of these hormones on the expression of IL-1β, IL-6, and TNF-α by immune cells may have widespread implications in management and treatment of anorexia–cachexia syndrome associated with a wide range of inflammatory conditions and cancer.[3]

2.3.4 Melanocortin System

It was reported that aberrant melanocortin signaling might be a contributing factor in anorexia and cachexia.[55,56] The melanocortin system is an important member of the family of catabolic central pathways as supported by solid genetic and pharmacological evidence.[57] This system consists of three endogenous peptide agonists, α-MSH, γ-MSH, and adrenocorticotropic hormone (ACTH), all derived from the precursor POMC peptide, an endogenous peptide antagonist, AgRP, and five subtypes of G protein-coupled receptors. The melanocortin system is involved in numerous functions such as the regulation of appetite and body temperature, memory, behavior, and immunity. Most evidence points toward the hypothalamus as the primary site where the agonist α-MSH and the antagonist AgRP, acting through the type 4 melanocortin receptor (MC4R), interact to influence body weight.[58] AgRP is mainly expressed in the hypothalamic ARC nucleus, co-localized in NPY neurons that project to adjacent hypothalamic areas such as the PVN, LHA, and dorsomedial nucleus. These neurons are a unique subset that is capable of increasing food intake via two different mechanisms, by increasing NPY signaling and by decreasing melanocortin signaling.[6]

Despite marked loss of body weight, which would normally be expected to down-regulate the anorexigenic melanocortin signaling system as a way to conserve energy stores, the melanocortin system remains active during cancer-induced cachexia.[7] α-MSH signaling is increased in tumor-bearing rats and is not decreased by intracerebroventricular ghrelin and NPY infusions. Melanocortin receptor blockade caused a much greater increase of food intake in anorectic rats than did the powerful orexigenic agents, NPY or ghrelin. Treatment with a melanocortin antagonist also reversed some of the negative metabolic consequences of anorexia as determined by serum leptin, ghrelin, insulin, and glucose values.[59] Thus, the central melanocortin receptor blockade by AgRP or other antagonists reversed anorexia and cachexia in animals bearing prostate carcinoma or sarcoma, which suggests a pathogenetic role for this system.[55,56]

The mechanisms contributing to persistent anorexigenic/cachexic activity of the central melanocortin system during cancer are unknown. It has been proposed that the elevations of proinflammatory cytokines might explain this paradox.

The site at which cytokine signaling or tumor products interact with the melano-cortin pathway to decrease food intake remains elusive, but it does not appear to involve changes in hypothalamic POMC or AgRP gene expression.[59]

2.3.5 Serotonin System

Serotonin is a potent anorectic molecule involved in the regulation of food intake and body weight. Food intake leads to increase in serotonin release in the hypothalamus promoting satiation. It has been shown that an increase in tryptophan, a serotonin precursor, occurs in the brains of tumor bearing rats and similar changes in plasma of cancer patients.[24] Intrahypothalamic serotonin concentrations increase in anorectic tumor bearing rats and reverse to normal after tumor-resection, in parallel to improved food intake. Serotonin acts through different types of receptors. Among them, presynaptic serotonin 1A (5-HT1A) receptors, postsynaptic 5-HT1B (1D in humans) receptors, and 5-HT2C receptors are present in the VMN, PVN, and ARC and are responsible for the satiety-enhancing effects of serotonin in the CNS.[60] Changes of 5-HT1B receptor expression are more pronounced in the magnocellular hypothalamic nuclei during the development of cancer anorexia, suggesting that the influence of serotonin on these nuclei may contribute to the decrease in food intake in cancer-related anorexia.[61] Data suggest that serotonin is a critical link between cytokines and neurotransmitters in the CNS (see Figure 2.2). IL-1 increases brain serotoninergic activity[62,63], resulting in a decrease in food intake. Fenfluramine (a serotonin agonist) raises hypothalamic serotonin levels, which activates POMC neurons in the ARC of the hypothalamus, inducing anorexia[64]. Meguid et al.[65] showed an increase of serotonin and a decrease of dopamine concentration in the VMN and a decrease of NPY in the PVN of the hypothalamus. Taken together these findings support that serotonin plays a relevant role in the onset and maintenance of the cancer-related anorexia.

2.3.6 Neuropeptide Y System

Neuropeptide Y is a potent orexigenic neuropeptide. It is synthesized in the neurons of the ARC that project into the PVN, a major integration site for energy homeostasis. NPY evokes feeding, decreases brown adipose tissue thermogenesis, and increases energy storage. Several studies support the hypothesis that the NPY system is altered in cancer anorexia. Previous studies showed a refractory feeding response to NPY when injected into the PVN and recent data showed a decrease in NPY expression in the fibres that innervate the supraoptic nucleus, PVN, suprachiasmatic, and ARC nuclei of the hypothalamus at the onset of anorexia in tumor-bearing rats. The physiological dysfunction of the NPY feeding system in anorectic rats and the decrease in NPY concentration correlate with the degree of anorexia.[66] Integrated data from the literature indicate that there are interactions between the monoamine system and NPY in a tumor-bearing rat model, pointing to a specific and integrated role for these neuromediators in the onset of cancer anorexia. An interesting hypothesis to explain the dysfunctions of

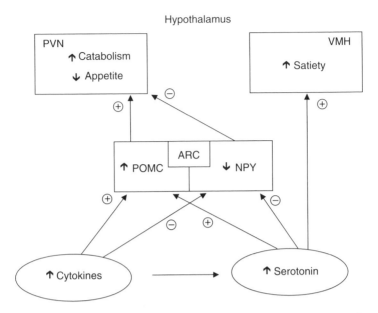

FIGURE 2.2 Proinflammatory cytokines and serotonin interact in the CNS inducing anorexia, satiety, and catabolism in cancer patients. In the hypothalamus increased levels of cytokines play a role activating the anorexigenic pathways and inhibiting the orexigenic pathways. Proinflammatory cytokines seem to increase the serotonin concentrations, which amplify the catabolic effects of cytokines and induce satiety in the ventromedial hypothalamic (VMH) nucleus. POMC: proopiomelanocortin peptide, NPY: neuropeptide Y, ARC: arcuate nucleus of the hypothalamus, and PVN: paraventricular nucleus of the hypothalamus.

the NPY system in cancer anorexia refers to the close relationship between NPY fibres and the hypothalamic neurons that express 5-HT1B receptors in normal rats. NPY and serotonin exert opposing effects on food intake. Recent data showed that NPY and dopamine concentrations decreased simultaneously while serotonin concentration increased in the PVN at the onset of anorexia in tumor-bearing rats.[60] These changes reverted to normal after tumor resection and corresponded to the reversal of anorexia, suggesting the dynamic interaction between monoamines and NPY in the regulation of food intake in normal conditions and cancer anorexia.[65] The hypothalamic NPY system is also one of the key neural pathways disrupted in anorexia induced by IL-1 and other cytokines. IL-1 administered directly into the cerebral ventricles antagonizes NPY-induced feeding in rats at a dose that yields estimated pathophysiological concentrations in the cerebrospinal fluid such as those observed in anorectic tumor-bearing rats.[67–69] Since central administration of cytokines increases hypothalamic serotonin concentrations, it has been speculated that during tumor growth, hypothalamic cytokine expression is increased leading to changes in hypothalamic monoamine concentrations, which in turn activate POMC neurons in the ARC. As a consequence, NPY neurons in the ARC are inhibited and NPY levels in the PVN depressed. Recent data agree with this hypothesis, showing that the inhibition of NPY orexigenic effects is associated with specific changes in serotonin and dopamine concentrations in the PVN, inhibiting food intake and thus promoting the onset of cancer anorexia.[65]

Neuropeptide Y is involved in appetite-stimulating effects of megestrol acetate, one of the drugs actually used in cancer-related anorexia. The mechanisms by which megestrol acetate improves appetite are not well understood. One hypothesis suggests that it is likely to be due to its inhibitory action on proinflammatory cytokines such as IL-1, IL-6, and TNF-α.[70] Another hypothesis suggests that megestrol acetate stimulates NPY synthesis, transport, and release in the hypothalamus.[71] Because NPY and AgRP are produced by the same neurons, agents that affect NPY synthesis and release also influence the synthesis and release of AgRP. Activation of AgRP–NPY neurons acts to promote feeding by two distinct pathways, namely activating NPY receptors and antagonizing MC4R. Studies on the interaction of NPY, melanocortin system, and monoamines would improve the knowledge about appetite mechanisms for new therapeutic approaches to anorexia.[72]

2.4 CONCLUSIONS

The contribution of eating-related disorders to the declining physical, emotional, and social functions of patients with cancer and their families is considerable.[73] There is an urgent need for new and effective therapies for the prevention and treatment of the cachexia–anorexia syndrome. The neurochemical mechanisms responsible for cancer anorexia are still a matter of debate. The data are based on animal studies and the major factor limiting the understanding of anorexia is the lack of knowledge about altered central appetite signals in cancer patients.[13] Multiple mediators are responsible for tumor-associated anorexia. A better understanding of the mechanisms controlling appetite and satiety in cancer-induced anorexia may help with the development of new therapies that could prolong survival and increase quality of life in these patients and their families.

REFERENCES

1. Nelson, K.A., The cancer anorexia–cachexia syndrome, *Semin. Oncol.*, 27, 64, 2000.
2. Rossi-Fanelli, F. and Laviano, A., Cancer anorexia: a model for the understanding and treatment of secondary anorexia, *Int. J. Cardiol.*, 85, 67, 2002.
3. Dixit, V.D. et al., Ghrelin inhibits leptin- and activation-induced proinflammatory cytokine expression by human monocytes and T cells, *J. Clin. Invest.*, 114, 57, 2004.
4. Strader, A.D. and Woods, S.C., Gastrointestinal hormones and food intake, *Gastroenterology*, 128, 75, 2005.
5. Hillebrand, J.J.G., de Wied, D., and Adan, R.A.H., Neuropeptides, food intake and body weight regulation: a hypothalamic focus, *Peptides*, 23, 2283, 2002.

6. Morton, G.J. and Schwartz, M.W., The NPY/AgRP neuron and energy homeostasis, *Int. J. Obes.*, 25, S56, 2001.

7. Lechan, R.M. and Tatro, J.B., Editorial: hypothalamic melanocortin signaling in cachexia, *Endocrinology*, 142, 3288, 2001.

8. Tisdale, M.J., Cancer anorexia and cachexia, *Nutrition*, 17, 442, 2001 (CACS).

9. Woods, S.C., Gastrointestinal satiety signals. I. An overview of gastrointestinal signals that influence food intake, *Am. J. Physiol. Gastrointest. Liver Physiol.*, 286, G7, 2004.

10. Davis, M.P. et al., Appetite and cancer-associated anorexia: a review, *J. Clin. Oncol.*, 22, 1510, 2004.

11. Hellstrom, P.M. and Naslung, E., Interactions between gastric emptying and satiety, with special reference to glucagons-like peptide-1, *Physiol. Behav.*, 74, 735, 2001.

12. Richardson, R.A. and Davidson, H.I.M., Nutritional demands in acute and chronic illness, *Proc. Nutr. Soc.*, 62, 777, 2003.

13. Berteretche, M.V. et al., Decreased taste sensitivity in cancer patients under chemotherapy, *Support Care Cancer*, 12, 571, 2004.

14. Laviano, A. et al., Neurochemical mechanisms for cancer anorexia, *Nutrition*, 18, 100, 2002.

15. Moldawer, L.L., Rogy, M.A., and Lowry, S.F., The role of cytokines in cancer anorexia, *J. Paraenter. Enteral. Nutr.*, 16, 43, 1992.

16. Barber, M.D., Ross, J.A., and Fearon, K.C., Cancer Cahexia, *Surg. Oncol.*, 8, 133, 1999.

17. Dantzen, R., Cytokine induced sickness behavior: mechanisms and implications, *Ann. N.Y. Acad. Sci.*, 933, 222, 2001.

18. Inui, A., Cancer anorexia–cachexia syndrome, *CA. Cancer J. Clin.*, 52, 72, 2002.

19. Wigmore, S.J. et al., Contribution of anorexia and hypermetabolism to weight loss in anicteric patients with pancreatic cancer, *Br. J. Surg.*, 84, 196, 1997.

20. Inui, A., Cancer anorexia–cachexia syndrome: are neuropeptides the key?, *Cancer Res.*, 59, 4493, 1999.

21. Inui, A., Feeding and body-weight regulation by hypothalamic neuropeptides-mediation of the actions of leptin, *Trends Neurosci.*, 22, 62, 1999.

22. Noguchi, Y. et al., Are cytokines possible mediators of cancer cachexia?, *Surg. Today*, 26, 467, 1996.

23. Matthys, P. and Billiau, A., Cytokines and cachexia, *Nutrition*, 13, 763, 1997.

24. Wang, W., Danielsson, A., Svanberg, E., and Lundholm, K., Lack of effects by tricyclic antidepressant and serotonin inhibitors on anorexia in MCG 101 tumor-bearing mice with eicosanoid-related cachexia, *Nutrition*, 19, 47, 2003.

25. Laviano, A. et al., Cracking the riddle of cancer anorexia, *Nutrition*, 12, 706, 1996.

26. Plata-Salaman, C.R., Cytokine-induced anorexia: Behavioral, cellular, and molecular mechanisms, *Ann. N.Y. Acad. Sci.*, 856, 160, 1988.

27. Konsman, J.P. and Dantzer, R., How the immune and nervous systems interact during disease-associated anorexia?, *Nutrition*, 17, 664, 2001.

28. Langhans, W. and Hrupka, B., Interleukins and tumor necrosis factor as inhibitors of food intake, *Neuropeptides*, 33, 415, 1999.

29. Turrin, N.P. et al., Interleukin-1beta system in anorectic catabolic tumor-bearing rats, *Curr. Opin. Clin. Metab. Care*, 7, 419, 2004.

30. Martignoni, M., Kunze, P., and Friess, H., Cancer cachexia, *Mol. Cancer*, 2, 36, 2003.

31. McCarthy, D.O., Inhibitors of prostaglandin synthesis do not improve food intake or body weight of tumor-bearing rats, *Res. Nurs. Health*, 22, 380, 1999.
32. Tisdale, M.J., Cancer cachexia, *Langenbercks Arch. Surg.*, 389, 299, 2004.
33. Mantovani, G. et al., Cytokine involvement in cancer-related anorexia/cachexia: role of megestrol acetate and medroxyprogesterone acetate, *Semin. Oncol.*, 25, 45, 1998.
34. Jatoi, A. et al., Does megastrol acetate down-regulate interleukin-6 in patients with cancer-associated anorexia and weight loss? A North Central Cancer Treatment Group Investigation, *Support Care Cancer*, 10, 71, 2002.
35. Maltoni, M. et al., Serum levels of tumor necrosis factor alpha and other cytokines do not correlate with weight loss and anorexia in cancer patients, *Support Care Cancer*, 5, 130, 1997.
36. Turrin, N.P. et al., Interleukin-1beta system in anorectic catabolic tumor-bearing rats, *Curr. Opin. Clin. Metab. Care*, 7, 419, 2004.
37. Machado, A.P., Costa Rosa, L.F.P.B., and Seelaender, M.C.L., Adipose tissue in Walker 256 tumour-induced cachexia: possible association between decreased leptin concentration and mononuclear cell infiltration, *Cell Tissue Res.*, 318, 53, 2004.
38. Lugarini, F. et al., Acute and chronic administration of immunomodulators induces anorexia in zucher rats, *Physiol. Behav.*, 84, 165, 2005.
39. Somasundar, P. et al., Leptin is a growth factor in cancer, *J. Surg. Res.*, 116, 337, 2004.
40. Aleman, M.R. et al., Leptin role in advanced lung cancer. A mediator of the acute phase response or a marker of the status of nutrition?, *Cytokine*, 19, 21, 2002.
41. Simons, J.P. et al., Plasma concentration of total leptin and human lung-cancer-associated cachexia, *Clin. Sci. (Lond.)*, 93, 273, 1997.
42. Brown, D.R., Berkowitz, D.E., and Breslow, M.J., Weight loss is not associated with hyperleptinemia in humans with pancreatic cancer, *J. Clin. Endocrinol. Metab.*, 86, 162, 2001.
43. Wallace, A.M., Sattar, N., and McMillan, D.C., Effect of weight loss and the inflammatory response on leptin concentrations in gastrointestinal cancer patients, *Clin. Cancer Res.*, 4, 2977, 1998.
44. Mantovani, G. et al., Serum levels of leptin and proinflammatory cytokines in patients with advanced-stage cancer at different sites, *J. Mol. Med.*, 78, 554, 2000.
45. Bing, C. et al., Cachexia in MAC16 adenocarcinoma: suppression of hunger despite normal regulation of leptin, insulin and hypothalamic neuropeptide Y, *J. Neurochem.*, 79, 1004, 2001.
46. Inui, A., Ghrelin: an orexigenic and somatotrophic signal from the stomach, *Nat. Rev. Neurosci.*, 2, 551, 2001.
47. Tschop, M. et al., GH-Releasing Peptide-2 increases fat mass in mice lacking NPY: indication for a crucial mediating role of hypothalamic Agouti-related protein, *Endocrinology*, 143, 558, 2004.
48. Plata-Salaman, C.R., Central nervous system mechanisms contributing to the cachexia–anorexia syndrome, *Nutrition*, 16, 1009, 2000.
49. Neary, N.M. et al., Ghrelin increases energy intake in cancer patients with impaired appetite: acute, randomized, placebo-controlled trial, *J. Clin. Endocrinol. Metab.*, 89, 2832, 2004.
50. Garcia, J.M. et al., Active ghrelin levels and active/total ghrelin ratio in cancer-induced cachexia, *J. Clin. Endocrinol. Metab.* 90, 2920, 2005.

51. Shimizu, Y. et al., Increased plasma ghrelin level in lung cancer cachexia, *Clin. Cancer Res.*, 9, 774, 2003.

52. Hanada, T. et al., Upregulation of ghrelin expression in cachectic nude mice bearing human melanoma cells, *Metabolism*, 53, 84, 2004.

53. Zigman, J.M. and Elmquist, J.K., Minireview: from anorexia to obesity — The yin and yang of body weight control, *Endocrinology*, 144, 3749, 2003.

54. Asakawa, A. et al., Ghrelin is an appetite-stimulatory signal from stomach with structural resemblance to motilin, *Gastroenterology*, 120, 337, 2001.

55. Marks, D.L., Ling, N., and Cone, R.D., Role of the central melanocortin system in cachexia, *Cancer Res.*, 61, 1432, 2001.

56. Wisse, B.E. et al., Reversal of cancer anorexia by blockade of central melanocortin receptors in rats, *Endocrinology*, 142, 3292, 2001.

57. Inui, A. et al., Ghrelin, appetite, and gastric motility: the emerging role of the stomach as an endocrine organ, *FASEB J.*, 18, 439, 2004.

58. Foster, A.C. et al., Body weight regulation by selective MC4 receptor agonists and antagonists, *Ann. N.Y. Acad. Sci.*, 994, 103, 2003.

59. Wisse, B.E., Schwartz, M.W., and Cummings, D.E., Melanocortin signaling and anorexia in chronic disease states, *Ann. N.Y. Acad. Sci.*, 994, 275, 2003.

60. Ramos, E.J. et al., Cancer anorexia–cachexia syndrome: cytokines and neuropeptides, *Curr. Opin. Clin. Nutr. Care*, 7, 427, 2004.

61. Makarenko, I.G. et al., Hypothalamic 5-HT1B-receptor changes in anorectic bearing rats, *Neurosci. Lett.*, 376, 71, 2005.

62. Shintani, F. et al., Interleukin-1beta augments release of norepinephrine, dopamine and serotonin in the rat anterior hypothalamus, *J. Neurosci.*, 7, 3574, 1993.

63. Laviano, A. et al., Effects of intra-VMN mianserin and IL-1ra on meal number in anorectic tumour-bearing rats, *J. Invest. Med.*, 48, 40, 2000.

64. Heisler, L.K. et al., Activation of central melanocortin pathways by fenfluramine, *Science*, 297, 609, 2002.

65. Meguid, M.M., et al., Tumor anorexia: effects on neuropeptide Y and monoamines in paraventricular nucleus, *Peptides*, 25, 261, 2004.

66. Makarenko, I.G. et al., Decreased NPY innervation of the hypothalamic nuclei in rats with cancer anorexia, *Brain Res.*, 961, 100, 2003.

67. Sonti, G., Llyin, S.E., and Plata-Salaman, C.R., Neuropeptide Y blocks and reveres interleukin-1β-induced anorexia in rats, *Peptides*, 17, 517, 1996.

68. Gayle, D., Llyin, S.E., and Plata-Salaman, C.R., Central nervous system IL-1β system and neuropeptide Y mRNA during IL-1β — induced anorexia in rats, *Brain Res. Bull.*, 44, 311, 1997.

69. Sonti, G., Llyin, S.E., and Plata-Salaman, C.R., Anorexia induced by cytokine interactions at pathophysiological concentrations, *Am. J. Physiol.*, 270, 1349, 1996.

70. Lopez, A.P. et al., Systematic review of megestrol acetate in the treatment of anorexia-cachexia syndrome, *J. Pain Symptom Manage.*, 27, 360, 2004.

71. y McCarthy, H.D. et al., Megestrol acetate stimulates food intake and water intake in the rat: effects on regional hypothalamic neuropeptide Y concentrations, *Eur. J. Pharmacol.*, 256, 99, 1994.

72. Ramos, E.J.B. et al., Neuropeptide Y, α-melanocyte-stimulating-hormone, and monoamines in food intake regulation, *Nutrition*, 21, 269, 2005.

73. Strasser, F., Eating-related disorders in patients with advanced cancer, *Support Care Cancer*, 11, 11, 2003.

3 Skeletal Muscle Atrophy and Hypertrophy

David J. Glass

CONTENTS

SUMMARY

Skeletal muscle responds to increases in load, such as induced by weight-bearing exercise, by inducing a compensatory increase in mass, which is called "hypertrophy." In the adult animal, hypertrophy occurs as a result of an increase in the size, as opposed to the number, of preexisting skeletal muscle fibers. Administration of insulin-like growth factor 1 (IGF1), a secreted protein growth factor, is sufficient to induce skeletal muscle hypertrophy; further, there is evidence that exercise induces hypertrophy by inducing the muscle to produce increased levels of IGF1. Over the past few years, the signaling pathways downstream of IGF1 receptor activation that mediate hypertrophy have been established. These pathways cooperate to induce an increase in protein synthesis, and are discussed in detail in this chapter. The converse of skeletal muscle hypertrophy is referred to as "atrophy." Skeletal muscle atrophy occurs in multiple clinical settings, including: treatment with glucocorticoids, cancer, denervation, immobilization, prolonged bed rest, congestive heart failure, renal failure, AIDS, and sepsis. Atrophy occurs as a result of an increase in the rate of protein breakdown — which occurs via the stimulation of ATP-dependent ubiquitin-mediated proteolysis. Atrophy-induced proteolysis requires two particular E3 ubiquitin ligases, MAFbx (Muscle Atrophy F-box Protein; also known as Atrogin-1) and MuRF1 (Muscle Ring Finger protein 1). Studies with genetically altered mice have demonstrated that MAFbx and MuRF1 are required for muscle atrophy, and that MuRF1 is transcriptionally up-regulated by the IKK (IκB kinase complex)/NF-κB (nuclear factor-kappaB) proinflammatory pathway. NF-kappaB activation via IKK is sufficient to induce skeletal muscle atrophy.

Finally, it has been demonstrated that the transcriptional up-regulation of *MuRF1* and *MAFbx* can be blocked by administration of IGF1. This last finding demonstrates that in addition to increasing protein synthesis the IGF1-induced hypertrophy pathways also dominantly regulate the induction of key mediators of skeletal muscle atrophy.

3.1 CLINICAL SYNDROMES

There are a variety of clinical settings in which there is a significant loss of skeletal muscle mass, or muscle "atrophy" (Glass, 2003). Perhaps the best-known cause of atrophy is a syndrome known as "cachexia," characterized by a significant anorexia and disproportionate muscle wasting. Diseases that can induce cachexia include AIDS and cancer. Further, in acute settings such as sepsis, severe burns, renal failure, and congestive heart disease, a significant loss of skeletal mass can be induced. It is now recognized that a common set of molecular markers are activated in skeletal muscle during many of these different settings of atrophy, suggesting that a common pathway is activated during atrophy, no matter what the cause. These markers, MuRF1 and MAFbx are discussed later in this chapter. Other clinical settings that can induce skeletal muscle atrophy include muscle inactivity, such as that caused by prolonged bed rest, or when a limb is put into a cast; denervation, such as after a spinal injury or in diseases such as Spinal Muscular Atrophy; and prolonged treatment with glucocorticoids — or illnesses that cause an increase in cortisol.

The muscle loss associated with aging, sarcopenia, may be distinct from other atrophy conditions (Conboy et al., 2005). With aging, there is a general loss in the integrity of muscle satellite cells — cells responsible for regenerating skeletal muscle fibers — that may be responsible for the gradual loss of muscle integrity. However, it has been shown that agents that can induce skeletal muscle hypertrophy, such as IGF1, can be helpful in counteracting sarcopenia (Musaro et al., 2001).

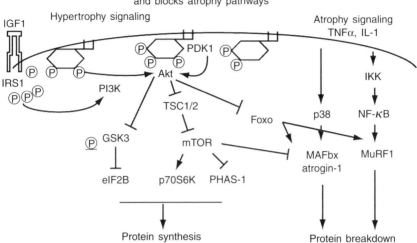

FIGURE 3.1   Left Side. Hypertrophy. Weight-bearing exercise leads to increased expression of Insulin-Like Growth Factor 1 (IGF1). The IGF1 signaling pathways relevant to hypertrophy are presented. Abbreviations: IRS1, Insulin Receptor Substrate-1; PI3K, Phosphatidylinositol-3 Kinase; PDK1, 3-Phosphoinositide-Dependent Kinase 1; GSK3, Glycogen Synthase Kinase 3; mTOR, Mammalian Target of Rapamycin; eIF2B, Eukaryotic Initation Factor 2B; PHAS1, 4E-BP or 4E-binding protein. Right side. Atrophy. Multiple different perturbations can induce skeletal muscle atrophy. Those that have been shown to cause the up-regulation of MuRF1 and MAFbx/Atrogin are illustrated. Abbreviations: TNFα, Tumor Necrosis Factor alpha; IL-1 Interleukin-1; IKK, IκB kinase; NF-κB, Nuclear Factor-κB; MuRF1, Muscle RING Finger Protein1; MAFbx — Muscle Atrophy F-Box protein. Cross-talk. The Akt pathway blocks the transcriptional up-regulation of MuRF1 and MAFbx by phosphorylating and thereby inhibiting the FOXO family of transcription factors.

3.2 PROTEIN SYNTHESIS PATHWAYS DOWNSTREAM OF INSULIN-LIKE GROWTH FACTOR 1 (IGF1)

Weight-bearing exercise leads to an increase in muscle expression of IGF1 (DeVol et al., 1990), which is sufficient to induce hypertrophy of skeletal muscle (Vandenburgh et al., 1991). Transgenic mice in which IGF1 is overexpressed in skeletal muscle undergo hypertrophy (Coleman et al., 1995; Musaro et al., 2001). IGF1 causes the increase in muscle mass by stimulating the Phosphatidylinositol-3 Kinase (PI3K)/Akt1 (also known as PKB, for Protein Kinase B) pathway, resulting in the downstream activation of targets which are required for protein synthesis (Bodine et al., 2001b; Rommel et al., 2001) (Figure 3.1). Recently, it was demonstrated that activation of Akt1 was sufficient to induce skeletal muscle hypertrophy *in vivo*; transgenic mice, in which a mutant, constitutively active, form of Akt1 was conditionally expressed in adult skeletal muscle, underwent rapid hypertrophy when expression of Akt1 was triggered (Lai et al., 2004).

3.3 HYPERTROPHY MEDIATORS DOWNSTREAM OF PI3K AND AKT: THE mTOR AND GSK3 PATHWAYS

Experiments in *Drosophila* delineated a pathway downstream of PI3K and Akt1 that can control cell size (Figure 3.1). Genetic loss of IRS-1 (insulin receptor substrate-1; Bohni et al., 1999), PI3K (Leevers et al., 1996), the *Drosophila* homologue of the mammalian target of rapamycin (mTOR, also known as FRAP or RAFT-1) (Zhang et al., 2000), as well as of p70 S6 Kinase (Montagne et al., 1999) (p70S6K) all resulted in decreases in cell size in the *Drosophila* wing (Figure 3.1). Akt1 induces the activation of mTOR via inhibition of the Tsc1/Tsc2 complex (Gao et al., 2002; Inoki et al., 2002; Tee et al., 2002), and both Akt1 phosphorylation (Bodine et al., 2001b) and mTOR activation are increased during muscle hypertrophy (Reynolds et al., 2002). While IGF1 activates mTOR and p70S6K downstream of PI3K/Akt1 activation, amino acids can activate mTOR directly causing a subsequent stimulation of p70S6K activity (Burnett et al., 1998; Hara et al., 1998). Thus mTOR appears to have an important and central function in integrating a variety of growth signals, from simple nutritional stimulation to activation by protein growth factors, resulting in protein synthesis.

Rapamycin is a chemical that binds mTOR and inhibits its function (Pallafacchina et al., 2002). This reagent has been useful in elucidating the Akt/mTOR/p70S6K pathway. Rapamycin, when complexed with a protein called FK506-binding protein (FKBP12), disrupts activation of mTOR (Pallafacchina et al., 2002). *In vitro*, when applied to myotube cultures, rapamycin blocks activation of p70S6K downstream of either activated Akt1 or IGF1 stimulation (Pallafacchina et al., 2002; Rommel et al., 2001; Rommel et al., 1999). However, rapamycin does not completely block IGF1-mediated hypertrophy *in vitro*, which may suggest that other pathways downstream of Akt1 — but independent of mTOR — play a role in some settings of hypertrophy. *In vivo* treatment with rapamycin entirely blocked load-induced hypertrophy, and inhibited the activation of p70S6K normally observed in this hypertrophy model (Bodine et al., 2001b). Treatment with rapamycin during load-induced hypertrophy did not block activation of Akt1, again demonstrating that Akt1 is upstream of mTOR, and that p70S6K activation requires the activation of mTOR (Bodine et al., 2001b). Experiments with rapamycin thus provide pharmacologic evidence for the activation of a linear Akt1/mTOR/p70S6K pathway during hypertrophy (Figure 3.1).

Genetic support for a linear Akt1/mTOR/p70S6K pathway came recently from reports that demonstrated that the tuberous sclerosis complex-1 and -2 proteins (Tsc1 and Tsc2) can inhibit mTOR. Akt1 phosphorylates Tsc2, thereby activating mTOR at least in part by disrupting the Tsc1–Tsc2 complex (Inoki et al., 2002). Furthermore, in insulin or serum-stimulated cells, activation of p70S6K is inhibited by expression of the Tsc1–Tsc2 complex (Inoki et al., 2002; Tee et al., 2002). This finding demonstrates that introduction of a genetic inhibitor of mTOR, downstream of Akt1, inhibits the activation of p70S6K, adding genetic evidence for an Akt1/mTOR/p70S6K pathway. Since PDK1 has been shown to phosphorylate p70S6 kinase directly (Pullen et al., 1998), one might have assumed that activation of mTOR was dispensable in some settings of p70S6K activation. That might still be the case; alternatively, it may be that p70S6K must first be primed by another kinase, such as mTOR, before being activated by PDK1 (3-Phosphoinosetide Dependent Protein Kinase-1) (Hannan et al., 2003; Saitoh et al., 2002) (Figure 3.1). Activation of p70S6 kinase is necessary for muscle fibers to achieve normal size, since skeletal muscle cells are smaller when they are null for the gene (Ohanna et al., 2005).

In addition to stimulating p70S6 kinase-mediated protein translation, activation of mTOR inhibits PHAS-1 (also known as 4E-BP), which is a negative regulator of the protein initiation factor eIF-4E (Hara et al., 1997) (Figure 3.1). It has recently been shown that PHAS-1 can directly bind a protein called raptor, which also binds mTOR (Hara et al., 1997; Kim et al., 2000). Mutations in PHAS-1, which inhibit interaction with raptor, also inhibit mTOR-mediated phosphorylation of PHAS-1 (Choi et al., 2003). Finally, overexpression of raptor can enhance the phosphorylation of PHAS-1 by mTOR *in vitro* (Choi et al., 2003; Schalm et al., 2003). mTOR binds PHAS-1 by a TOR signaling (TOS) motif; this same motif is found in p70S6K (Schalm et al., 2003).

Raptor's binding to mTOR can be enhanced by a protein called GβL (G-protein β-subunit-like protein, pronounced "gable") (Kim do et al., 2003). Upon binding of GβL to mTOR, p70S6K activation is enhanced (Kim do et al., 2003). However, GβL has not yet been assessed in skeletal muscle and expression studies indicate that it is not strongly expressed in that tissue (Rodgers et al., 2001).

In summary, mTOR can increase protein synthesis by modulating two distinct pathways, the p70S6K pathway and the PHAS-1 pathway (Figure 3.1). Blockade of PHAS-1 might be a potential route to increasing protein synthesis and therefore hypertrophy. Stimulation of p70S6K might be a second route, since muscle cells are smaller when p70S6K is absent (Ohanna et al., 2005). Interestingly, the mTOR pathway may also be important for signaling to eIF2B (eukaryotic Initiation Factor 2B), a translation initiation factor (Kubica et al., 2005). When rats were subjected to exercise, eIF2Bε protein expression increased — however this increase was blocked by the mTOR inhibitor rapamycin (Kubica et al., 2005). This finding seemingly reconnects distinct Akt signaling pathways, the Akt/GSK3/eIF2B pathway, and the Akt/mTOR/p70S6K pathway (Kubica et al., 2005).

3.4 A DISTINCT HYPERTROPHY MEDIATOR DOWNSTREAM OF PI3K AND AKT: GSK3β

Glycogen synthase kinase 3 beta, GSK3β, is a distinct substrate of Akt1 that has been shown to modulate hypertrophy. GSK3β activity is inhibited by Akt1 phosphorylation (Cross et al., 1995). Expression of a dominant negative, kinase inactive form of GSK3β induces dramatic hypertrophy in skeletal myotubes (Rommel et al., 2001). In cardiac hypertrophy, GSK3β phosphorylation is also evident (Hardt and Sadoshima, 2002) and expression of a dominant negative form of GSK3β can induce cardiac hypertrophy (Hardt and Sadoshima, 2002). Cardiac hypertrophy was also shown to proceed via a PI3K dependent process, linking GSK3β to the

PI3K/Akt1 pathway in the heart (Haq et al., 2000). In addition to inducing Akt activation (Bodine et al., 2001b) exercise inhibits GSK3β (Lajoie et al., 2004; Markuns et al., 1999; Sakamoto et al., 2004). A variety of agents that inhibit GSK3β were shown to induce skeletal muscle cell hypertrophy (Rochat et al., 2004). GSK3β blocks protein translation initiated by the eIF2B protein (Hardt and Sadoshima, 2002). Therefore GSK3β inhibition may induce hypertrophy by stimulating protein synthesis independent of the mTOR pathway.

3.5 Satellite Cells: Role in Hypertrophy?

Satellite cells are mononuclear cells that neighbor skeletal muscle fibers. Satellite cells are stimulated upon injury, whereupon they proliferate and fuse to preexisting fibers, aiding in repair. Satellite cells are also stimulated and found to be required for load-induced skeletal muscle hypertrophy (Rosenblatt et al., 1994). The requirement for satellite cell proliferation and fusion in some instances of hypertrophy somewhat blurs the hypertrophy/proliferation distinction, suggesting that strategies to induce proliferation, survival, and fusion of satellite cells may be a distinct way to induce hypertrophy. Indeed, this may be the mechanism by which the TGFβ family member myostatin exerts its negative effect on skeletal muscle — via the inhibition of differentiation (Patel and Amthor, 2005). However, it is not clear if activation of satellite cells can be separated from activation of the Akt1 pathway, since Akt1 is also necessary for cell survival and proliferation in many mitosis-competent cells (Matsui et al., 2003).

3.6 Clenbuterol and Other Agonists of the Gα-s Coupled Receptors

Agonists of the β2-adrenergic receptor, such as clenbuterol have been shown to induce skeletal muscle hypertrophy, and to blunt atrophy (Hinkle et al., 2004). Most of the β2-adrenergic receptor signaling studies have focused on activation of cAMP pathways, mediated by the Gα_s protein. However, it has also been reported that the Gβ/γ proteins activate the PI3K/Akt1 pathway (Crespo et al., 1994), raising the possibility that just as weight-bearing exercise and IGF1 seem to converge on the PI3K/Akt1 pathway so as to mediate muscle hypertrophy; other hypertrophy-promoting agents may also converge on this critical pathway.

Activation of other receptors on skeletal muscles that are coupled in a manner similar to that of the β2-adrenergic receptor also induces hypertrophy. For example, it was recently demonstrated that the CRF2 receptor could induce skeletal muscle hypertrophy and that the CRF2 receptor was similarly coupled to Gα_s (Hinkle et al., 2004).

3.7 Atrophy via Induction of Ubiquitin Ligase Pathways

One might ask whether skeletal muscle atrophy is the converse of skeletal muscle hypertrophy. Indeed, there are sets of genes that are inversely regulated during atrophy as opposed to hypertrophy, which would seem to indicate that at least some processes are opposing during the two physiologic settings (Latres et al., 2005; Sacheck et al., 2004). During skeletal muscle atrophy, an entirely distinct process is stimulated: a dramatic increase in protein degradation and turnover. Furthermore, unique transcriptional pathways are activated (Haddad et al., 2003; Jagoe et al., 2002) and many of these are not inversely regulated during hypertrophy (Latres et al., 2005; Sacheck et al., 2004). This stimulation of proteolysis was shown to occur at least in part due to an activation of the ubiquitin–proteasome pathway (Jagoe et al., 2002).

Ubiquitin is a short peptide, which can be conjugated to specific protein substrates. A chain-reaction may then ensue: a second ubiquitin peptide is ligated to the first, and a third to the second; in this way a chain of polyubiquitin is built onto the substrate and this ubiquitin chain targets the substrate to a structure called the proteasome, where the substrate is proteolyzed into small peptides (Jagoe et al., 2002).

The addition of ubiquitin to a protein substrate has come to be recognized as an exquisitely modulated process. This process requires three distinct enzymatic components, an E1 ubiquitin-activating enzyme, an E2 ubiquitin-conjugating enzyme, and an E3 ubiquitin-ligating enzyme. The E3 ubiquitin ligases are the components that confer substrate specificity. Several hundred distinct E3s have already been identified and it is likely that each modulates the ubiquitination of a distinct set of substrates. Thus, the regulation of ubiquitination appears to be a coordinate signaling pathway, analogous to phosphorylation, in which key pathways might be activated by the enhanced proteolysis of a key inhibitor protein, or in which pathways might be inactivated via the degradation of an activating enzyme.

The involvement of the ubiquitin–proteasome pathway in skeletal muscle atrophy had been well established: rates of protein breakdown increase during atrophy; inhibition of the proteasome blocks these increases (Tawa et al., 1997); the amount of polyubiquitin conjugation per total protein measured increases during atrophy (Lecker et al., 1999); mRNA levels of genes, which encode distinct components of the ubiquitin pathway, increase during atrophy (Jagoe et al., 2002) (Figure 3.1).

Differential expression screening studies, designed to identify markers of the atrophy process, identified two genes whose expression increased significantly in multiple models of skeletal muscle atrophy: *MuRF1* (Bodine et al., 2001a) (for *Muscle Ring Finger1*) and *MAFbx* (Bodine et al., 2001a) (for *Muscle Atrophy F-box*; also called *Atrogin-1* (Gomes et al., 2001). Both of these genes were shown to encode E3 ubiquitin ligases (Bodine et al., 2001a). Expression of *MuRF1* and *MAFbx* is stimulated when the nerve innervating a muscle is cut, thus resulting in paralysis and severe atrophy; these genes are also up-regulated by simple immobilization of the muscle, or by treatment with a glucocorticoid, which causes muscle cachexia (Bodine et al., 2001a). In all, 13 distinct models of skeletal muscle atrophy have shown to result in an increase of MAFbx/atrogin and MuRF1 (Bodine et al., 2001; Dehoux et al., 2003; Deruisseau et al., 2004; Gomes et al., 2001; Li et al., 2003). In the case of sepsis-induced atrophy both MuRF1 and

MAFbx were up-regulated several-fold and this up-regulation could be blocked by a pharmacologic inhibitor of glucocorticoids (Wray et al., 2003), suggesting a distinct mechanism by which sepsis induces atrophy, that is, through activation of glucocorticoid signaling.

MuRF1 encodes a protein that contains three domains: a RING-Finger domain (Borden and Freemont, 1996), which is required for ubiquitin ligase activity (Kamura et al., 1999); a "B-box," whose function is unclear, and a "coiled-coil domain," which may be required for the formation of heterodimers between MuRF1 and a related protein, MuRF2 (Centner et al., 2001). Proteins that have these three domains have been called "RBCC" proteins (for RING, B-BOX, Coiled-Coil domain) (Saurin et al., 1996), or "TRIM" proteins (for tripartite motif) (Reymond et al., 2001). MuRF1 has been demonstrated to have ubiquitin ligase activity, which depends on the presence of the RING domain for that activity (Bodine et al., 2001a). As for substrates, it was recently shown that MuRF1 could induce the ubiquitination of the cardiac form of Troponin I (Kedar et al., 2004), indicating that MuRF1 may act by degrading components of the contractile apparatus. Also suggesting that possibility is that MuRF1 has been shown to bind to the myofibrillar protein titin, at the M line (Centner et al., 2001; McElhinny et al., 2002; Pizon et al., 2002). Overexpression of MuRF1 results in the disruption of the subdomain of titin that binds MuRF1, suggesting that MuRF1 may play a role in titin turnover (McElhinny et al., 2002). MuRF1 has also been demonstrated to be in the nucleus and indications that it interacts with transcription-regulating elements such as GMEB-1 (Glucocorticoid Modulatory Element Bound protein-1) suggest a potential role for MuRF1 in modulating transcription (McElhinny et al., 2002). This has not yet been demonstrated, however.

MAFbx/Atrogin-1 contains an F-box domain, a characteristic motif seen in a family of E3 ubiquitin ligases called SCFs (for Skp1, Cullin, F-box) (Jackson and Eldridge, 2002). F-box containing E3 ligases usually bind a substrate only after that substrate has first been posttranslationally modified, for example by phosphorylation (Jackson and Eldridge, 2002). This suggests the possibility of a signaling pathway in which a potential substrate is first phosphorylated as a response to an atrophy-induced stimulus, and then degraded via MAFbx. Recently, substrates have been suggested for MAFbx, including MyoD (Tintignac et al., 2005). However, it has not yet been demonstrated if MyoD is ubiquitinated by MAFbx in skeletal muscle or during atrophy conditions.

MuRF1$^{-/-}$ and MAFbx$^{-/-}$ mice appear phenotypically normal. However, under atrophy conditions, significantly less muscle mass is lost in either MuRF1$^{-/-}$ or MAFbx$^{-/-}$ animals in comparison to control littermates (Bodine et al., 2001a). This finding demonstrated for the first time that inhibition of discrete ubiquitin ligases could moderate the amount of muscle lost after an atrophy-inducing stimulus. Therefore MuRF1 or MAFbx may be attractive targets for pharmacologic intervention. They may also serve as early markers of skeletal muscle atrophy, aiding in the diagnosis of muscle disease.

There are probably additional E3s that play important roles in skeletal muscle atrophy. For example, a protein called E3-alphaII was shown to be enriched in skeletal muscle, and its expression is positively regulated by proinflammatory cytokines, which can induce skeletal muscle atrophy. During progression of cancer cachexia, *E3alpha-II* mRNA levels were induced, accompanied by an increase in protein ubiquitination (Kwak et al., 2004). Whether loss of E3-alpha-II in any way perturbs the course of muscle atrophy remains to be established; however, it may be important downstream of breakdown of the contractile apparatus, because it has been implicated as an "N-end rule" ubiquitin ligase, which would presumably be necessary to break down contractile apparatus proteins clipped by other proteases (Kwak et al., 2004).

3.8 Akt Inhibition of FOXO Transcription Factors Blocks Up-Regulation of MuRF1 and MAFbx

Studies of differentiated myotube cultures demonstrated that treatment of myotubes with the cachectic glucocorticoid dexamethasone promotes enhanced protein breakdown and increased expression of genes broadly involved in the ubiquitin–proteasome proteolytic pathway (Du et al., 2000; Hong and Forsberg, 1995; Wang et al., 1998). More recent studies showed that *in vitro* treatment of myotubes with dexamethasone induces atrophy, accompanied by the specific increased expression of *MAFbx* and *MuRF1* (Sandri et al., 2004; Stitt et al., 2004). The up-regulation of *MAFbx* and *MuRF1* was antagonized by simultaneous treatment with IGF1 (Sandri et al., 2004; Stitt et al., 2004), acting through the PI3K/Akt pathway (Sandri et al., 2004; Stitt et al., 2004); this finding demonstrated a novel role for Akt — in addition to stimulating skeletal muscle hypertrophy, Akt stimulation could dominantly inhibit the induction of atrophy signaling (Figure 3.1). Similarly, *MuRF1* and *MAFbx* were activated in a separate model of atrophy, diabetes, and here too IGF1 blocked the transcriptional up-regulation (Lee et al., 2004). Genetic activation of Akt was shown to be sufficient to block the atrophy-associated increases in *MAFbx* and *MuRF1* transcription (Stitt et al., 2004). The mechanism by which Akt inhibited *MAFbx* and *MuRF1* up-regulation was demonstrated to involve the FOXO family of transcription factors (Sandri et al., 2004; Stitt et al., 2004). In myotubes, FOXO transcription factors are excluded from the nucleus when phosphorylated by Akt, and translocate to the nucleus upon dephosphorylation. The translocation and activity of FOXO transcription factors is required for up-regulation of MuRF1 and MAFbx — in the case of FOXO3, activation was demonstrated to be sufficient to induce atrophy (Sandri et al., 2004), a finding that was subsequently supported by the transgenic expression of FOXO1, which resulted in an atrophic phenotype (Kamei et al., 2004).

3.9 Cachectic Triggers of Atrophy

Several cytokines have been shown to induce muscle wasting, most notably TNFα (Tumor Necrosis Factor α, a proinflammatory secreted cytokine that was originally called "cachectin" (Argiles and Lopez-Soriano, 1998; Beutler et al., 1985;

Tisdale, 2000). TNF-α levels are elevated in the circulation of patients with cancer cachexia, contributing to negative nitrogen balance (Argiles and Lopez-Soriano, 1998). TNF-α binding to its receptor induces the activation of the Rel/NF-κB (NF-κB) family of transcription factors (von Haehling et al., 2002). NF-κB activation was shown to be required for cytokine-induced loss of skeletal muscle proteins (Ladner et al., 2003).

Since NF-κB in muscle is activated by disuse (Hunter et al., 2002) or sepsis (Penner et al., 2001), it might play a role in the pathogenesis of these conditions. Consistent with a requisite role of NF-κB in atrophy, *in vitro* blockade inhibits protein loss in C2C12 myotubes (Li and Reid, 2000). Also consistent with a role for a requisite role of NF-κB in atrophy, treatment with TNF-α, which induces NF-κB signaling, attenuates insulin-stimulated protein synthesis (Williamson et al., 2005). In cells of the immune and inflammatory systems, NF-κB was demonstrated to be a central integration site for proinflammatory signals and a regulator of related target genes. NF-κB activation in these cells directly increases the production of cytokines, demonstrating the presence of a "feed-forward" system. Studies in cultured C2C12 muscle cells suggested that TNF-α inhibits myocyte differentiation through NF-κB activation (Guttridge et al., 2000; Ladner et al., 2003; Langen et al., 2004). The TNF-α inhibition of myogenesis was postulated to inhibit the ability of muscle precursor, or "satellite," cells to be recruited into the muscle, further enhancing the atrophic effect (Guttridge et al., 2000). A recent paper demonstrated the loss and dysfunction of satellite cells during atrophy, although no link to NF-κB signaling was made in that paper (Mitchell and Pavlath, 2004).

Activation of NF-κB is controlled by the IKK. Upon phosphorylation by IKK, IκB is ubiquitinated and targeted to the proteasome for degradation, resulting in the activation of NF-κB (Yaron et al., 1998). In a recent study, the NF-κB pathway was activated or inhibited selectively in the skeletal muscle of transgenic mice, through the muscle-specific expression of either constitutively active IKKβ or a dominant inhibitory form of IκBα (Cai et al., 2004). These mice were referred to as MIKK (for muscle-specific expression of IKK, which results in activation of the NF-κB in the transgenic animals,) or MISR (for muscle-specific "super-repressor," since expression of a dominant negative form of IκB represses NF-κB activity) (Cai et al., 2004).

Activation of the NF-κB pathway was demonstrated to be sufficient to induce significant atrophy, as measured by increases in *in vivo* amino acid excretion and tyrosine turnover in isolated muscles (Cai et al., 2004). The expression of the ubiquitin ligase *MuRF1*, but not *MAFbx*, was up-regulated in the MIKK skeletal muscle; this finding provided the first functional dissection of *MuRF1* and *MAFbx* signaling (Cai et al., 2004). A distinct *in vitro* study demonstrated that overexpression of IκBα could block loss of myosin in TNF-α treated C2C12 myotubes (Ladner et al., 2003).

When MIKK mice were crossed into a MuRF1 null background, there was a significant reduction in muscle loss, demonstrating that transcriptional activation of *MuRF1* by NF-κB is a requisite step in NF-κB induced atrophy. Given this data, a linear IKKβ/NF-κB/MuRF1 signaling pathway was proposed (Figure 3.1);

atrophic stimuli cause the activation of NF-κB, which thereby induces atrophy in part through the transcriptional up-regulation of MuRF1 (Cai et al., 2004).

A distinct study, using a knockout of the p105/p50 NF-κB1 gene, demonstrated that there was less atrophy in NF-κB1($-/-$) mice, as well as a blockade of the fiber-type switching associated with atrophy (Hunter and Kandarian, 2004). Also, it was shown that induction of proteasome expression could be attenuated by inhibiting NF-κB activity (Wyke et al., 2004), giving further credence to the notion that inhibition of NF-κB signaling may be a fruitful mechanism to ameliorate skeletal muscle atrophy.

It was noteworthy that MAFbx/Atrogin-1 was not perturbed upon NF-κB activation. That finding demonstrated that *MAFbx* up-regulation was not required for NF-κB induced muscle loss. However, given the multiple physiological settings that do result in *MAFbx/atrogin* up-regulation, the implication was that a second, parallel pathway, distinct from NF-κB, is usually induced during atrophy (Figure 3.1).

The trigger for up-regulation of *MAFbx* has recently been determined to be p38 (Li et al., 2005). It was shown that TNF-α acts to stimulate expression of *MAFbx*, and that this up-regulation could be blocked by pharmacologic inhibitors of p38 (Li et al., 2005) (Figure 3.1). Whether or not p38 has any cross-effect on MuRF1 up-regulation, or on the NF-κB pathway, has yet to be determined.

3.10 OTHER POTENTIAL MEDIATORS OF ATROPHY: MYOSTATIN

The development of a strain of cows that yielded excessive amounts of beef led to the isolation of myostatin (Grobet et al., 1997; McPherron et al., 1997). The "double-muscled" cows were shown to have a mutation in *myostatin* (Grobet et al., 1997; Kambadur et al., 1997; McPherron and Lee, 1997). Mice engineered to be myostatin null similarly have a large increase in muscle mass relative to wild-type littermates (McPherron and Lee, 1997). However, when muscle obtained from Myostatin$^{-/-}$ animals was analyzed, it was shown to be larger as a result of an increase in the number of muscle fibers and not as a result of hypertrophy. Therefore, it was thought that myostatin acts in a way that is distinct from the activation of atrophy or the inhibition of hypertrophy; that it simply blocks the proliferation of muscle precursors, thus decreasing muscle mass. However, a more recent experiment has complicated matters; when adult animals are given an inhibitory antibody to myostatin, they undergo what appears to be muscle hypertrophy (Whittemore et al., 2003). Also, mice given myostatin undergo muscle atrophy (Zimmers et al., 2002). Whether myostatin is acting directly on preexisting muscle fibers or whether these data uncover a requisite role for satellite cell proliferation and fusion in the maintenance of normal muscle mass remains to be seen. But the possibility exists that blockade of the myostatin pathway may be now be a distinct route to inhibiting skeletal muscle atrophy. Inhibition of myostatin has already been shown to be beneficial for dystrophic muscle (Bogdanovich et al., 2002); however, this might be a unique circumstance due to the rapid turnover of muscle and the subsequent need for satellite cell proliferation and differentiation in the dystrophic animal.

3.11 CONCLUSION

A considerable amount of recent progress has been made in the understanding of the signaling pathways that mediate skeletal muscle hypertrophy and atrophy. These findings help to give hope that novel drug targets may be found to block skeletal muscle atrophy and the gradual loss of strength seen even in normal aging. The lack of approved drugs for skeletal muscle disease helps to highlight the need for continued research in this area.

ACKNOWLEDGMENTS

Sincere apologies to scientific colleagues whose work was omitted from this review due to space constraints.

REFERENCES

Argiles, J.M. and Lopez–Soriano, F.J. (1998). Catabolic proinflammatory cytokines. *Curr. Opin. Clin. Nutr. Metab. Care.* 1, 245–251.

Beutler, B., Mahoney, J., Le Trang, N., Pekala, P., and Cerami, A. (1985). Purification of cachectin, a lipoprotein lipase-suppressing hormone secreted by endotoxin-induced RAW 264.7 cells. *J. Exp. Med.* 161, 984–995.

Bodine, S.C., Latres, E., Baumhueter, S., Lai, V.K., Nunez, L., Clarke, B.A., Poueymirou, W.T., Panaro, F.J., Na, E., Dharmarajan, K., et al. (2001a). Identification of ubiquitin ligases required for skeletal muscle atrophy. *Science* 294, 1704–1708.

Bodine, S.C., Stitt, T.N., Gonzalez, M., Kline, W.O., Stover, G.L., Bauerlein, R., Zlotchenko, E., Scrimgeour, A., Lawrence, J.C., Glass, D.J., and Yancopoulos, G.D. (2001b). Akt/mTOR pathway is a crucial regulator of skeletal muscle hypertrophy and can prevent muscle atrophy *in vivo. Nat. Cell Biol.* 3, 1014–1019.

Bogdanovich, S., Krag, T.O., Barton, E.R., Morris, L.D., Whittemore, L.A., Ahima, R.S., and Khurana, T.S. (2002). Functional improvement of dystrophic muscle by myostatin blockade. *Nature* 420, 418–421.

Bohni, R., Riesgo-Escovar, J., Oldham, S., Brogiolo, W., Stocker, H., Andruss, B.F., Beckingham, K., and Hafen, E. (1999). Autonomous control of cell and organ size by CHICO, a Drosophila homolog of vertebrate IRS1-4. *Cell* 97, 865–875.

Borden, K.L. and Freemont, P.S. (1996). The RING finger domain: a recent example of a sequence-structure family. *Curr. Opin. Struct. Biol.* 6, 396–401.

Burnett, P.E., Barrow, R.K., Cohen, N.A., Snyder, S.H., and Sabatini, D.M. (1998). RAFT1 phosphorylation of the translational regulators p70 S6 kinase and 4E-BP1. *Proc. Natl Acad. Sci. USA* 95, 1432–1437.

Cai, D., Frantz, J.D., Tawa, N.E., Jr., Melendez, P.A., Lidov, H.G.W., Hasselgren, P.O., Frontera, W.R., Lee, J., Glass, D.G., and Shoelson, S.E. (2004). IKKbeta/NF-kappaB activation causes severe muscle wasting in mice. *Cell* 119, 285–298.

Centner, T., Yano, J., Kimura, E., McElhinny, A.S., Pelin, K., Witt, C.C., Bang, M.L., Trombitas, K., Granzier, H., Gregorio, C.C., et al. (2001). Identification of muscle specific ring finger proteins as potential regulators of the titin kinase domain. *J. Mol. Biol.* 306, 717–726.

Choi, K.M., McMahon, L.P., and Lawrence., J.C., Jr. (2003). Two motifs in the translational repressor PHAS-I required for efficient phosphorylation by mTOR and recognition by raptor. *J. Biol. Chem.*, 278, 19667–19673.

Coleman, M.E., DeMayo, F., Yin, K.C., Lee, H.M., Geske, R., Montgomery, C., and Schwartz, R.J. (1995). Myogenic vector expression of insulin-like growth factor I stimulates muscle cell differentiation and myofiber hypertrophy in transgenic mice. *J. Biol. Chem.* 270, 12109–12116.

Conboy, I.M., Conboy, M.J., Wagers, A.J., Girma, E.R., Weissman, I.L., and Rando, T.A. (2005). Rejuvenation of aged progenitor cells by exposure to a young systemic environment. *Nature* 433, 760.

Crespo, P., Xu, N., Simonds, W.F., and Gutkind, J.S. (1994). Ras-dependent activation of MAP kinase pathway mediated by G-protein beta gamma subunits. *Nature* 369, 418–420.

Cross, D.A., Alessi, D.R., Cohen, P., Andjelkovich, M., and Hemmings, B.A. (1995). Inhibition of glycogen synthase kinase-3 by insulin mediated by protein kinase B. *Nature* 378, 785–789.

Dehoux, M.J.M., van Beneden, R.P., Fernandez-Celemin, L., Lause, P.L., and Thissen, J.-P.M. (2003). Induction of MafBx and Murf ubiquitin ligase mRNAs in rat skeletal muscle after LPS injection. *FEBS Lett.* 544, 214–217.

Deruisseau, K.C., Kavazis, A.N., Deering, M.A., Falk, D.J., Van Gammeren, D., Yimlamai, T., Ordway, G.A., and Powers, S.K. (2004). Mechanical ventilation induces alterations of the ubiquitin-proteasome pathway in the diaphragm. *J. Appl. Physiol.* 98, 1314–1321.

DeVol, D.L., Rotwein, P., Sadow, J.L., Novakofski, J., and Bechtel, P.J. (1990). Activation of insulin-like growth factor gene expression during work-induced skeletal muscle growth. *Am. J. Physiol.* 259, E89–E95.

Du, J., Mitch, W.E., Wang, X., and Price, S.R. (2000). Glucocorticoids induce proteasome C3 subunit expression in L6 muscle cells by opposing the suppression of its transcription by NF-kappa B. *J. Biol. Chem.* 275, 19661–19666.

Gao, X., Zhang, Y., Arrazola, P., Hino, O., Kobayashi, T., Yeung, R.S., Ru, B., and Pan, D. (2002). Tsc tumour suppressor proteins antagonize amino-acid#150; TOR signalling. *Nat. Cell Biol.* 4, 699–704.

Glass, D.J. (2003). Molecular mechanism modulating muscle mass. *Trends Mol. Med.* 9, 344–350.

Gomes, M.D., Lecker, S.H., Jagoe, R.T., Navon, A., and Goldberg, A.L. (2001). Atrogin-1, a muscle-specific F-box protein highly expressed during muscle atrophy. *Proc. Natl Acad. Sci. USA* 98, 14440–14445.

Grobet, L., Poncelet, D., Royo, L.J., Brouwers, B., Pirottin, D., Michaux, C., Menissier, F., Zanotti, M., Dunner, S., and Georges, M. (1998). Molecular definition of an allelic series of mutations disrupting the myostatin function and causing double-muscling in cattle. *Mamm. Genome.* 9, 210–213.

Guttridge, D.C., Mayo, M.W., Madrid, L.V., Wang, C.Y., and Baldwin, A.S., Jr. (2000). NF-kappaB-induced loss of MyoD messenger RNA: possible role in muscle decay and cachexia. *Science* 289, 2363–2366.

Haddad, F., Roy, R.R., Zhong, H., Edgerton, V.R., and Baldwin, K.M. (2003). Atrophy responses to muscle inactivity II: molecular markers of protein deficits. *J. Appl. Physiol.* 25, 25.

Hannan, K.M., Thomas, G., and Pearson, R.B. (2003). Activation of S6K1 (p70 ribosomal protein S6 kinase 1) requires an initial calcium-dependent priming event involving formation of a high-molecular-mass signalling complex. *Biochem. J.* 370, 469–477.

Haq, S., Choukroun, G., Kang, Z.B., Ranu, H., Matsui, T., Rosenzweig, A., Molkentin, J.D., Alessandrini, A., Woodgett, J., Hajjar, R., et al. (2000). Glycogen synthase kinase-3{beta} is a negative regulator of cardiomyocyte hypertrophy. *J. Cell Biol.* 151, 117–130.

Hara, K., Yonezawa, K., Kozlowski, M.T., Sugimoto, T., Andrabi, K., Weng, Q.P., Kasuga, M., Nishimoto, I., and Avruch, J. (1997). Regulation of eIF-4E BP1 phosphorylation by mTOR. *J. Biol. Chem.* 272, 26457–26463.

Hara, K., Yonezawa, K., Weng, Q.P., Kozlowski, M.T., Belham, C., and Avruch, J. (1998). Amino acid sufficiency and mTOR regulate p70 S6 kinase and eIF-4E BP1 through a common effector mechanism (published erratum appears in *J. Biol. Chem.* 273, 22160). *J. Biol. Chem.* 273, 14484–14494.

Hardt, S.E. and Sadoshima, J. (2002). Glycogen synthase kinase-3beta: a novel regulator of cardiac hypertrophy and development. *Circ. Res.* 90, 1055–1063.

Hinkle, R.T., Donnelly, E., Cody, D.B., Bauer, M.B., Sheldon, R.J., and Isfort, R.J. (2004). Corticotropin releasing factor 2 receptor agonists reduce the denervation-induced loss of rat skeletal muscle mass and force and increase non-atrophying skeletal muscle mass and force. *J. Muscle Res. Cell Motil.* 25, 539–547.

Hong, D.H. and Forsberg, N.E. (1995). Effects of dexamethasone on protein degradation and protease gene expression in rat L8 myotube cultures. *Mol. Cell Endocrinol.* 108, 199–209.

Hunter, R.B., and Kandarian, S.C. (2004). Disruption of either the Nfkb1 or the Bcl3 gene inhibits skeletal muscle atrophy. *J. Clin. Invest.* 114, 1504–1511.

Hunter, R.B., Stevenson, E., Koncarevic, A., Mitchell-Felton, H., Essig, D.A., and Kandarian, S.C. (2002). Activation of an alternative NF-kappaB pathway in skeletal muscle during disuse atrophy. *FASEB J.* 16, 529–538.

Inoki, K., Li, Y., Zhu, T., Wu, J., and Guan, K.L. (2002). TSC2 is phosphorylated and inhibited by Akt and suppresses mTOR signalling. *Nat. Cell Biol.* 4, 648–657.

Jackson, P.K., and Eldridge, A.G. (2002). The SCF ubiquitin ligase: an extended look. *Mol. Cell 9*, 923–925.

Jagoe, R.T., Lecker, S.H., Gomes, M., and Goldberg, A.L. (2002). Patterns of gene expression in atrophying skeletal muscles: response to food deprivation. *FASEB J.* 16, 1697–1712.

Kamei, Y., Miura, S., Suzuk, M., Kai, Y., Mizukami, J., Taniguchi, T., Mochida, K., Hata, T., Matsuda, J., Aburatani, H., et al. (2004). Skeletal muscle FOXO1 (FKHR) transgenic mice have less skeletal muscle mass, down-regulated Type I (slow twitch/red muscle) fiber genes, and impaired glycemic control. *J. Biol. Chem.* 279, 41114–411123.

Kambadur, R., Sharma, M., Smith, T.P., and Bass, J.J. (1997). Mutations in myostatin (GDF8) in double-muscled Belgian Blue and Piedmontese cattle. *Genome Res.* 7, 910–916.

Kamura, T., Koepp, D.M., Conrad, M.N., Skowyra, D., Moreland, R.J., Iliopoulos, O., Lane, W.S., Kaelin, W.G., Jr., Elledge, S. J., Conaway, R.C., et al. (1999). Rbx1, a component of the VHL tumor suppressor complex and SCF ubiquitin ligase. *Science* 284, 657–661.

Kedar, V., McDonough, H., Arya, R., Li, H.-H., Rockman, H.A., and Patterson, C. (2004). Muscle-specific RING finger 1 is a bona fide ubiquitin ligase that degrades cardiac troponin I. *Proc. Natl Acad. Sci. USA* 101, 18135–18140.

Kim do, H., Sarbassov dos, D., Ali, S.M., Latek, R.R., Guntur, K.V., Erdjument–Bromage, H., Tempst, P., and Sabatini, D.M. (2003). GbetaL, a positive regulator of the rapamycin-sensitive pathway required for the nutrient-sensitive interaction between raptor and mTOR. *Mol. Cell* 11, 895–904.

Kim, S., Jung, Y., Kim, D., Koh, H., and Chung, J. (2000). Extracellular zinc activates p70 S6 kinase through the phosphatidylinositol 3-kinase signaling pathway. *J. Biol. Chem.* 275, 25979–25984.

Kubica, N., Bolster, D.R., Farrell, P.A., Kimball, S.R., and Jefferson, L.S. (2005). Resistance exercise increases muscle protein synthesis and translation of eukaryotic initiation factor 2b{epsilon} mRNA in a mammalian target of rapamycin-dependent manner. *J. Biol. Chem.* 280, 7570–7580.

Kwak, K.S., Zhou, X., Solomon, V., Baracos, V.E., Davis, J., Bannon, A.W., Boyle, W.J., Lacey, D.L., and Han, H.Q. (2004). Regulation of protein catabolism by muscle-specific and cytokine-inducible ubiquitin ligase E3{alpha}-II during cancer cachexia. *Cancer Res.* 64, 8193–8198.

Ladner, K.J., Caligiuri, M.A., and Guttridge, D.C. (2003). Tumor necrosis factor-regulated biphasic activation of NF-kappa B is required for cytokine-induced loss of skeletal muscle gene products. *J. Biol. Chem.* 278, 2294–2303.

Lai, K.-M., Gonzalez, M., Poueymirou, W.T., Kline, W.O., Na, E., Zlotchenko, E., Stitt, T.N., Economides, A., Yancopoulos, G.D., and Glass, D.J. (2004). Conditional activation of akt in adult skeletal muscle induces rapid hypertrophy. *Mol. Cell Biol.* 24, 9295–9304.

Lajoie, C., Calderone, A., Trudeau, F., Lavoie, N., Massicotte, G., Gagnon, S., and Beliveau, L. (2004). Exercise training attenuated the PKB and GSK-3 dephosphorylation in the myocardium of ZDF rats. *J. Appl. Physiol.* 96, 1606–1612.

Langen, R.C.J., Van Der Velden, J.L.J., Schols, A.M.W.J., Kelders, M.C.J.M., Wouters, E.F.M., and Janssen-Heininger, Y.M.W. (2004). Tumor necrosis factor-alpha inhibits myogenic differentiation through MyoD protein destabilization. *FASEB J.* 18, 227–237.

Latres, E., Amini, A.R., Amini, A.A., Griffiths, J., Martin, F. J., Wei, Y., Lin, H.C., Yancopoulos, G.D., and Glass, D.J. (2005). Insulin-like growth factor-1 (IGF1) inversely regulates atrophy-induced genes via the phosphatidylinositol 3-kinase/Akt/mammalian target of rapamycin (PI3K/Akt/mTOR) pathway. *J. Biol. Chem.* 280, 2737–2744.

Lecker, S.H., Solomon, V., Price, S.R., Kwon, Y.T., Mitch, W. E., and Goldberg, A.L. (1999). Ubiquitin conjugation by the N-end rule pathway and mRNAs for its components increase in muscles of diabetic rats. *J. Clin. Invest.* 104, 1411–1420.

Lee, S.W., Dai, G., Hu, Z., Wang, X., Du, J., and Mitch, W.E. (2004). Regulation of muscle protein degradation: coordinated control of apoptotic and ubiquitin-proteasome systems by phosphatidylinositol 3 kinase. *J. Am. Soc. Nephrol.* 15, 1537–1545.

Leevers, S.J., Weinkove, D., MacDougall, L.K., Hafen, E., and Waterfield, M.D. (1996). The Drosophila phosphoinositide 3-kinase Dp110 promotes cell growth. *Embo. J.* 15, 6584–6594.

Li, Y.-P. and Reid, M.B. (2000). NF-kappa B mediates the protein loss induced by TNF-alpha in differentiated skeletal muscle myotubes. *Am. J. Physiol. Regul. Integr. Comp. Physiol.* 279, R1165–1170.

Li, Y.-P., Chen, Y., Li, A.S., and Reid, M.B. (2003). Hydrogen peroxide stimulates ubiquitin conjugating activity and expression of genes for specific E2 and E3 proteins in skeletal muscle myotubes. *Am. J. Physiol. Cell. Physiol.* 00129.02003.

Li, Y.-P., Chen, Y., John, J., Moylan, J., Jin, B., Mann, D.L., and Reid, M.B. (2005). TNF-{alpha} acts via p38 MAPK to stimulate expression of the ubiquitin ligase atrogin1/MAFbx in skeletal muscle. *FASEB J.* 19, 362–370.

Markuns, J.F., Wojtaszewski, J.F., and Goodyear, L.J. (1999). Insulin and exercise decrease glycogen synthase kinase-3 activity by different mechanisms in rat skeletal muscle. *J. Biol. Chem.* 274, 24896–24900.

Matsui, T., Nagoshi, T., and Rosenzweig, A. (2003). Akt and PI 3-kinase signaling in cardiomyocyte hypertrophy and survival. *Cell Cycle* 2, 220–223.

McElhinny, A.S., Kakinuma, K., Sorimachi, H., Labeit, S., and Gregorio, C.C. (2002). Muscle-specific RING finger-1 interacts with titin to regulate sarcomeric M-line and thick filament structure and may have nuclear functions via its interaction with glucocorticoid modulatory element binding protein-1. *J. Cell Biol.* 157, 125–136.

McPherron, A.C. and Lee, S.J. (1997). Double muscling in cattle due to mutations in the myostatin gene. *Proc. Natl Acad. Sci. USA* 94, 12457–12461.

McPherron, A.C., Lawler, A.M., and Lee, S.J. (1997). Regulation of skeletal muscle mass in mice by a new TGF-beta superfamily member. *Nature* 387, 83–90.

Mitchell, P.O. and Pavlath, G.K. (2004). Skeletal muscle atrophy leads to loss and dysfunction of muscle precursor cells. *Am. J. Physiol. Cell Physiol.* 287, C1753–1762.

Montagne, J., Stewart, M.J., Stocker, H., Hafen, E., Kozma, S. C., and Thomas, G. (1999). Drosophila S6 kinase: a regulator of cell size. *Science* 285, 2126–2129.

Musaro, A., McCullagh, K., Paul, A., Houghton, L., Dobrowolny, G., Molinaro, M., Barton, E.R., Sweeney, H.L., and Rosenthal, N. (2001). Localized IGF-1 transgene expression sustains hypertrophy and regeneration in senescent skeletal muscle. *Nat. Genet.* 27, 195–200.

Ohanna, M., Sobering, A.K., Lapointe, T., Lorenzo, L., Praud, C., Petroulakis, E., Sonenberg, N., Kelly, P.A., Sotiropoulos, A., and Pende, M. (2005). Atrophy of S6K1−/− skeletal muscle cells reveals distinct mTOR effectors for cell cycle and size control. *Nat. Cell Biol.* 7, 286.

Pallafacchina, G., Calabria, E., Serrano, A.L., Kalhovde, J.M., and Schiaffino, S. (2002). A protein kinase B-dependent and rapamycin-sensitive pathway controls skeletal muscle growth but not fiber type specification. *Proc. Natl Acad. Sci. USA* 25, 25.

Patel, K. and Amthor, H. (2005). The function of myostatin and strategies of myostatin blockade-new hope for therapies aimed at promoting growth of skeletal muscle. *Neuromuscul. Disord.* 15, 117–126.

Penner, C.G., Gang, G., Wray, C., Fischer, J.E., and Hasselgren, P.O. (2001). The transcription factors NF-kappab and AP-1 are differentially regulated in skeletal muscle during sepsis. *Biochem. Biophys. Res. Commun.* 281, 1331–1336.

Peterson, R.T., Desai, B.N., Hardwick, J.S., and Schreiber, S.L. (1999). Protein phosphatase 2A interacts with the 70-kDa S6 kinase and is activated by inhibition of FKBP12-rapamycinassociated protein. *Proc. Natl Acad. Sci. USA* 96, 4438–4442.

Pizon, V., Iakovenko, A., Van Der Ven, P.F., Kelly, R., Fatu, C., Furst, D.O., Karsenti, E., and Gautel, M. (2002). Transient association of titin and myosin with microtubules in nascent myofibrils directed by the MURF2 RING-finger protein. 115, 4469–4482.

Pullen, N., Dennis, P.B., Andjelkovic, M., Dufner, A., Kozma, S. C., Hemmings, B.A., and Thomas, G. (1998). Phosphorylation and activation of p70s6k by PDK1. *Science* 279, 707–710.

Reymond, A., Meroni, G., Fantozzi, A., Merla, G., Cairo, S., Luzi, L., Riganelli, D., Zanaria, E., Messali, S., Cainarca, S., et al. (2001). The tripartite motif family identifies cell compartments. *EMBO J.* 20, 2140–2151.

Reynolds, T.H. t., Bodine, S.C., and Lawrence, J.C., Jr. (2002). Control of Ser2448 phosphorylation in the mammalian target of rapamycin by insulin and skeletal muscle load. *J. Biol. Chem.* 277, 17657–17662.

Rochat, A., Fernandez, A., Vandromme, M., Moles, J.-P., Bouschet, T., Carnac, G., and Lamb, N.J.C. (2004). Insulin and Wnt1 pathways cooperate to induce reserve cell activation in differentiation and myotube hypertrophy. *Mol. Biol. Cell* 15, 4544–4555.

Rodgers, B.D., Levine, M.A., Bernier, M., and Montrose-Rafizadeh, C. (2001). Insulin regulation of a novel WD-40 repeat protein in adipocytes. *J. Endocrinol.* 168, 325–332.

Rommel, C., Bodine, S.C., Clarke, B.A., Rossman, R., Nunez, L., Stitt, T.N., Yancopoulos, G.D., and Glass, D.J. (2001). Mediation of IGF1-induced skeletal myotube hypertrophy by PI(3)K/Akt/mTOR and PI(3)K/Akt/GSK3 pathways. *Nat. Cell Biol.* 3, 1009–1013.

Rommel, C., Clarke, B.A., Zimmermann, S., Nunez, L., Rossman, R., Reid, K., Moelling, K., Yancopoulos, G.D., and Glass, D.J. (1999). Differentiation stage-specific inhibition of the raf-MEK-ERK pathway by Akt. *Science* 286, 1738–1741.

Rosenblatt, J.D., Yong, D., and Parry, D.J. (1994). Satellite cell activity is required for hypertrophy of overloaded adult rat muscle. *Muscle Nerve* 17, 608–613.

Sacheck, J.M., Ohtsuka, A., McLary, S.C., and Goldberg, A.L. (2004). IGF1 stimulates muscle growth by suppressing protein breakdown and expression of atrophy-related ubiquitin-ligases, atrogin-1 and MuRF1. *Am. J. Physiol. Endocrinol. Metab.* 287, E591–E601.

Saitoh, M., Pullen, N., Brennan, P., Cantrell, D., Dennis, P.B., and Thomas, G. (2002). Regulation of an Activated S6 Kinase 1 Variant Reveals a Novel Mammalian Target of Rapamycin Phosphorylation Site. *J. Biol. Chem.* 277, 20104–20112.

Sakamoto, K., Arnolds, D.E., Ekberg, I., Thorell, A., and Goodyear, L.J. (2004). Exercise regulates Akt and glycogen synthase kinase-3 activities in human skeletal muscle. *Biochem. Biophys. Res. Commun.* 319, 419–425.

Sandri, M., Sandri, C., Gilbert, A., Skurk, C., Calabria, E., Picard, A., Walsh, K., Schiaffino, S., Lecker, S.H., and Goldberg, A.L. (2004). Foxo transcription factors induce the atrophy-related ubiquitin ligase atrogin-1 and cause skeletal muscle atrophy. *Cell* 117, 399–412.

Saurin, A.J., Borden, K.L., Boddy, M.N., and Freemont, P.S. (1996). Does this have a familiar RING? *Trends Biochem. Sci.* 21, 208–214.

Schalm, S.S., Fingar, D.C., Sabatini, D.M., and Blenis, J. (2003). TOS motif-mediated raptor binding regulates 4E-BP1 multisite phosphorylation and function. *Curr. Biol.* 13, 797–806.

Stitt, T.N., Drujan, D., Clarke, B.A., Panaro, F.J., Timofeyva, Y., Kline, W.O., Gonzalez, M., Yancopoulos, G.D., and Glass, D.G. (2004). The IGF1/PI3K/Akt pathway prevents expression of muscle atrophy-induced ubiquitin ligases by inhibiting FOXO transcription factors. *Mol. Cell* 14, 395–403.

Tawa, N.E., Jr., Odessey, R., and Goldberg, A.L. (1997). Inhibitors of the proteasome reduce the accelerated proteolysis in atrophying rat skeletal muscles. *J. Clin. Invest.* 100, 197–203.

Tee, A.R., Fingar, D.C., Manning, B.D., Kwiatkowski, D.J., Cantley, L.C., and Blenis, J. (2002). Tuberous sclerosis complex-1 and -2 gene products function together to inhibit mammalian target of rapamycin (mTOR)-mediated downstream signaling. *Proc. Natl Acad. Sci. USA*, 99, 13571-13576.

Tintignac, L.A., Lagirand, J., Batonnet, S., Sirri, V., Leibovitch, M.P., and Leibovitch, S.A. (2005). Degradation of MyoD mediated by the SCF (MAFbx) ubiquitin ligase. *J. Biol. Chem.* 280, 2847–2856.

Tisdale, M.J. (2000). Biomedicine. Protein loss in cancer cachexia. *Science* 289, 2293–2294.

Vandenburgh, H.H., Karlisch, P., Shansky, J., and Feldstein, R. (1991). Insulin and IGF-I induce pronounced hypertrophy of skeletal myofibers in tissue culture. *Am. J. Physiol.* 260, C475–484.

von Haehling, S., Genth-Zotz, S., Anker, S.D., and Volk, H.D. (2002). Cachexia: a therapeutic approach beyond cytokine antagonism. *Int. J. Cardiol.* 85, 173–183.

Wang, D. and Sul, H.S. (1998). Insulin stimulation of the fatty acid synthase promoter is mediated by the phosphatidylinositol 3-kinase pathway. Involvement of protein kinase B/Akt. *J. Biol. Chem.* 273, 25420–25426.

Whittemore, L.A., Song, K., Li, X., Aghajanian, J., Davies, M., Girgenrath, S., Hill, J.J., Jalenak, M., Kelley, P., Knight, A., et al. (2003). Inhibition of myostatin in adult mice increases skeletal muscle mass and strength. *Biochem. Biophys. Res. Commun.* 300, 965–971.

Williamson, D.L., Kimball, S.R., and Jefferson, L.S. (2005). Acute treatment with TNF-{alpha} attenuates insulin-stimulated protein synthesis in cultures of C2C12 myotubes through a MEK1-sensitive mechanism. *Am. J. Physiol. Endocrinol. Metab.*

Wyke, S.M., Russel, S.T., and Tisdale, M.J. (2004). Induction of proteasome expression in skeletal muscle is attenuated by inhibitors of NF-kappaB activation. *Br. J. Cancer* 91, 1742–1750.

Yaron, A., Hatzubai, A., Davis, M., Lavon, I., Amit, S., Manning, A.M., Andersen, J.S., Mann, M., Mercurio, F., and Ben-Neriah, Y. (1998). Identification of the receptor component of the IkappaBalpha-ubiquitin ligase. *Nature* 396, 590–594.

Zhang, H., Stallock, J.P., Ng, J.C., Reinhard, C., and Neufeld, T.P. (2000). Regulation of cellular growth by the drosophila target of rapamycin dTOR [In Process Citation]. *Genes Dev.* 14, 2712–2724.

Zimmers, T.A., Davies, M.V., Koniaris, L.G., Haynes, P., Esquela, A.F., Tomkinson, K.N., McPherron, A.C., Wolfman, N.M., and Lee, S.J. (2002). Induction of cachexia in mice by systemically administered myostatin. *Science* 296, 1486–1488.

4 Skeletal Muscle Apoptosis in Cachexia and Aging

Amie J. Dirks and Christiaan Leeuwenburgh

Contents

Abbreviations

AIF   apoptosis-inducing factor
AP-1   activating protein-1
Apaf-1   apoptosis protease activating factor-1
ARC   apoptosis repressor with caspase-associated recruitment domain
Bad   Bcl-2 antagonist of apoptosis
Bak   Bcl-2 homologous antagonist/killer
Bax   Bcl-2 associated protein X
Bcl-2   B-cell lymphoma-2
$Bcl-X_L$ B-cell lymphoma X long
Bid   BH3 interacting domain death agonist
cIAP-1, 2   cellular inhibitor of apoptosis protein-1, 2

cFlip   FADD-like interleukin-1 beta converting enzyme
              inhibitory protein
dATP   deoxyadenosine triphosphate
FADD   fas associated death domain
IL-1 6, 15   interleukin-1, 6, 15
NF-κB nuclear factor-kappa B
Omi/HtrA2   high-temperature requirement A2
RIP1   receptor interacting protein-1
Smac/Diablo   second mitochondrial activator of caspases/direct IAP binding
              protein with low pI
TNF-α tumor necrosis factor-alpha
TNFR1, 2   tumor necrosis factor receptor 1, 2
TRADD   TNF receptor associated death domain
TRAF2   TNF receptor associated factor 2
TUNEL   terminal dUTP nick end labeling
XIAP   X-linked inhibitor of apoptosis protein

SUMMARY

Apoptosis, programmed cell death, is an essential process in development and homeostasis of the human body. Altered regulation of apoptosis contributes to the pathophysiology of a plethora of disease states leading to either an abnormal resistance or susceptibility to cell death. An aberrant increase in the rate of apoptosis in skeletal muscle is associated with several conditions and diseases, including cachexia and aging, which contributes to the wasting of muscle mass. Apoptosis that occurs in cachexia may likely be induced by elevated levels of tumor necrosis factor-α (TNF-α), which is a common feature among the cachexia-inducing diseases. In comparison, apoptosis that occurs with normal aging in skeletal muscle is likely due to mitochondrial dysfunction or sarcoplasmic reticulum stress. Elucidating the signaling pathways leading to apoptosis associated with cachexia and aging will lead to targeted therapies to prevent muscle wasting.

4.1 INTRODUCTION

Apoptosis, programmed cell death, is an evolutionary conserved process by which individual cells of a multicellular organism commit suicide. Apoptosis plays an essential role in regulating cell number both during the developmental and postdevelopmental stages of the mammalian lifespan. For example, in the development of the central nervous system (CNS) half of the neurons produced during neurogenesis die via apoptosis before CNS maturation.[1] In the mature adult, apoptosis is responsible for the death and shedding of cells in the epidermis and gastrointestinal tract. Furthermore, death of white blood cells via apoptosis after the immune response is necessary to maintain immunocompetency. Apoptosis is also responsible for ridding the body of damaged or infected cells that may be harmful to the organism. Cells containing damaged DNA or dysfunctional mitochondria or cells infected with bacteria or viruses die via apoptosis. Hence, apoptosis is an essential conserved process required for the development and homeostasis of an organism.

Although apoptosis is important in maintaining health, excessive or inadequate apoptosis can contribute to disease pathophysiology. Excessive apoptosis has been implicated in the progression of many age-related diseases such as Alzheimer's, Parkinson's, cardiovascular disease such as atherosclerosis, and diabetes mellitus. For example, loss of cholinergic neurons in the basal forebrain and loss of dopaminergic neurons in the basal ganglia via apoptosis results in the characteristic features and progression of Alzheimer's and Parkinson's disease, respectively.[2-6] Apoptosis of vascular smooth muscle cells plays a part in vascular calcification in atherosclerosis and apoptosis of pancreatic beta cells contributes to the lack of insulin secretion in diabetes mellitus.[7] Excessive apoptosis also contributes to tissue damage and the decline of organ function in ischemic conditions such as stroke and myocardial infarction.[8,9] Inadequate apoptosis is known to be a classical hallmark of tumor formation and progression in all cancers. Inadequate clearance of apoptotic cells is thought to be critical in the development of autoimmune diseases such as systemic lupus erythematosus and rheumatoid arthritis.[10] As the focus of this chapter, apoptosis also may play a role in muscle wasting of cachectic patients with diseases such as cancer, chronic heart failure, and chronic obstructive pulmonary disease and also contributes to the loss of muscle during normal aging.

4.2 APOPTOTIC SIGNALING

Apoptosis is executed by specific cellular signaling pathways and is therefore characterized by specific biochemical and morphological events. Some of these identifying features of apoptosis include chromatin condensation and DNA fragmentation into mono- and oligonucleosomes, cellular shrinkage, maintenance of organelle membrane integrity, and membrane blebbing forming apoptotic bodies, which are engulfed by macrophages or neighboring cells. Death via apoptosis ensures death of a single cell without causing an inflammatory response and therefore is not disruptive to surrounding tissues.

Apoptosis is mediated by activation of a variety of cysteine proteases, known as caspases. Caspases normally exist in an inactivated state called procaspases but can be activated by proteolytic cleavage and subsequent heterodimerization. Initiation of apoptosis involves activation of a caspase cascade in which "initiator" caspases (i.e., caspase-8, caspase-9, and caspase-12) first become activated and then cleave and activate "effector" caspases (i.e., caspase-3, caspase-6, and caspase-7). The effector caspases carry out the proteolytic events that result in cellular breakdown and demise. There are 14 known mammalian caspases (i.e., caspase-1 to caspase-14), which participate in the apoptotic process depending on the stimulus and respective signaling pathway that is activated and the cell type undergoing apoptosis. The two major pathways extensively described include the mitochondrion-mediated and receptor-mediated apoptotic signaling (Figure 4.1). However, other pathways, such as the sarcoplasmic reticulum-mediated pathway and lysosome-mediated pathways do exist.[11, 12]

4.2.1 Mitochondrion-Mediated Signaling

Mitochondria play a central role in initiating apoptosis. Upon stimulation, mitochondria can release cytochrome c into the cytosol, which forms a complex, known as the apoptosome, with procaspase-9, Apaf-1, and dATP. Once the apoptosome is formed, procaspase-9 can cleave and activate itself. The active enzyme, caspase-9, can cleave and activate effector caspases such as procaspase-3, which leads to the typical morphological features of apoptosis. This process is highly regulated at a number of levels. First, cytochrome c release from the mitochondria is regulated. The Bcl-2 family of proteins was the first described to affect the release of cytochrome c. This family consists of a number of proteins, which are antiapoptotic or pro-apoptotic. For example, Bcl-2 and Bcl-X_L protect against cytochrome c release and are therefore antiapoptotic while Bax, Bak, Bad, and Bid favor cytochrome c release and are therefore pro-apoptotic. The ratio and interaction of the Bcl-2 family antiapoptotic and pro-apoptotic proteins determines the fate of cytochrome c release from the mitochondria. Often the Bcl-2/Bax ratio is used as an indicator of apoptotic potential where a high ratio protects against apoptosis and a low ratio favors apoptosis. Apoptosis repressor with caspase-associated recruitment domain (ARC) is another protein that regulates cytochrome c release. Upon stimulation, ARC translocates from the cytosol to the mitochondrial membrane and prevents cytochrome c release.[13] Recent data show that ARC may protect against apoptosis by binding to Bax and interfering with its activation, which would ultimately prevent cytochrome c release.[14] A second level of regulation is the inhibition of apoptosome formation by various heat-shock proteins (HSP's). HSP 70 and 90 can associate with Apaf-1 to prevent the recruitment and activation of procaspase-9 to the apoptosome.[15, 16] A third level of regulation involves the inhibition of caspases by the IAP's. The IAP's (i.e., XIAP, cIAP-1, and cIAP-2) can bind to cleaved and activated caspase-9 and -3 to inhibit their enzyme activity and to prevent apoptosis. Finally, the mitochondria can release additional proteins, along with cytochrome c, to relieve the inhibition exerted by the IAPs so indeed apoptosis can be executed. These proteins include Smac/Diablo and Omi/HtrA2.[17–19]

Mitochondria can also release pro-apoptotic proteins that are not involved in the activation of the caspase cascade. Mitochondria can release AIF and endonuclease G (EndoG), which translocate to the nucleus to induce chromatin condensation and DNA fragmentation in a caspase independent manner.[20, 21]

In summary, mitochondria play a central role in the execution of apoptosis (see Figure 4.1). Mitochondria can release proteins that function to activate the caspase cascade or proteins that can directly induce chromatin condensation and DNA fragmentation in a caspase independent manner. These processes are regulated at multiple levels in order to maintain precise control over cell death and survival

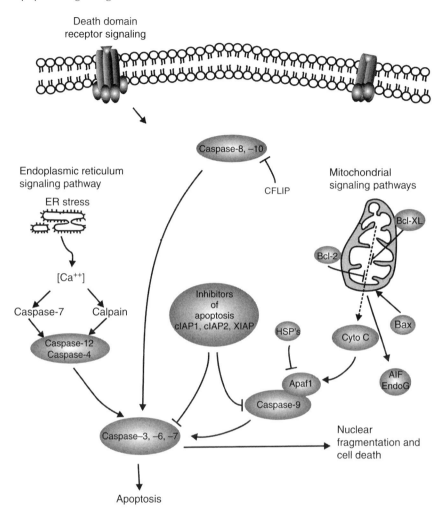

FIGURE 4.1 Schematic representation of potential pathways to induce skeletal muscle apoptosis. Mitochondrion-mediated signaling is initiated by the release of cytochrome c from the mitochondrion. Cytochrome c, Apaf-1, dATP, and inactive caspase-9 forms a complex, called the apoptosome, which results in activation of caspase-9. Caspase-9 then activates caspase-3, which leads to the morphological changes associated with apoptosis. The Bcl-2 proteins regulate the release of cytochrome c. IAPs regulate the activity and/or activation of caspase-9 and caspase-3. HSPs can inhibit formation of the apoptosome. The mitochondria can also release AIF and EndoG, which induce apoptosis in a caspase independent manner by causing DNA fragmentation. Receptor-mediated signaling is initiated by the binding of TNF-α to TNFR1, which leads to the activation of caspase-8. Caspase-8 cleaves and activates caspase-3 leading to apoptosis. cFLIP can inhibit the activation of caspase-8. Apoptosis induced by ER stress leads to the activation of caspase-12/4, which inactivates caspase-3.

when challenged by a changing environment such as occurs *in vivo* with disease states and aging.

4.2.2 Receptor-Mediated Signaling

Cytokines can induce apoptosis in some cell types via their interaction with specific receptors of the tumor necrosis factor receptor (TNFR) superfamily. TNF-α is a cytokine that can elicit a broad spectrum of responses. Exposure of cells to TNF-α most commonly causes activation of NF-κB and AP-1 resulting in the expression of genes involved in cell survival and acute and chronic inflammatory responses.[22] However, TNF-α is capable of inducing apoptosis. TNF-α signals via two membrane receptors: TNFR1 and TNFR2. These receptors are homologous in their extracellular domains, but are structurally different in their cytoplasmic domains. TNFR1 contains a death domain, whereas TNFR2 does not. TNFR1 mediates signaling for both apoptosis and cell survival while TNFR2 mostly transduces signals favoring cell survival. Ligand binding to TNFR1 can induce apoptosis in an effector cell via the activation of procaspase-8, which cleaves and activates procaspase-3 initiating the caspase cascade. Alternatively, binding of TNF-α to TNFR1 can induce a pro-inflammatory/antiapoptotic response mediated through the transcription factor NF-κB. Recent evidence suggests that cytokines can increase the levels of antiapoptotic proteins in mitotic cells. For example, a recent paper by Micheau and Tschopp,[23] shows that the stimulation of NF-κB and subsequent transcriptional activity of NF-κB determines the cells' fate. Specifically, cells accommodating defective NF-κB signals (resulting in low quantities of antiapoptotic proteins) undergo TNF-α-induced apoptotic elimination. The presence and recruitment of adaptor proteins to the cytoplasmic domain of TNFR1 determine the outcome: caspase activation resulting in apoptosis or NF-κB activation resulting in cell survival. Adaptor proteins include TRADD, TRAF2, RIP1, and FADD. To activate NF-κB thereby promoting cell survival, TRADD associates with TNFR1 and is able to recruit RIP1 and TRAF2 to form a TNFR1–TRADD–RIP1–TRAF2 complex. This leads to activation of NF-κB, with consequent expression of inflammatory and antiapoptotic proteins. To induce apoptosis upon ligand binding to TNFR1, TRADD can instead recruit RIP1 and FADD forming a TNFR1–TRADD–RIP1–FADD complex leading to procaspase-8 activation. Apoptosis mediated via TNFR1 can be inhibited by cFLIP, which interferes with the activation of procaspase-8. See Figure 4.2.

Once apoptosis is initiated via caspase-8, the release of cytochrome c from the mitochondria and activation of the mitochondrion-mediated signaling may occur, but is downstream from caspase-8 activation. Active caspase-8 cleaves Bid, which then stimulates Bax and Bak activity resulting in cytochrome c release. Some cell types require activation of the mitochondrion-mediated signaling via Bid to execute apoptosis and others do not.

Tumor necrosis factor-α can induce apoptosis by more than one pathway, although the most widely accepted pathway involves TRADD, FADD, and caspase-8.[22] Evidence for an alternative pathway comes from the residual apoptotic

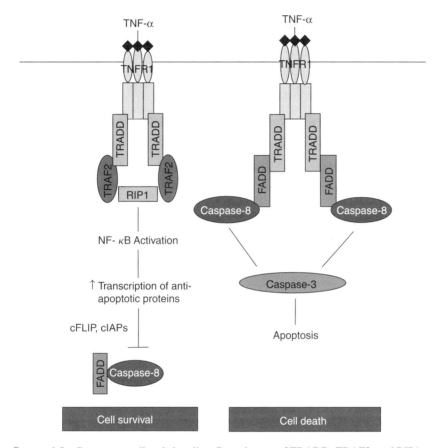

FIGURE 4.2 Receptor-mediated signaling. Recruitment of TRADD, TRAF2, and RIP1 to TNFR1 leads to NF-κB activation. NF-κB activation results in transcription of antiapoptotic proteins, such as cFLIP and cIAP's, which favors cell survival. Recruitment of TRADD, FADD, caspase-8 to TNFR1 leads to the activation of caspase-3 thereby resulting in cell death.

response observed when FADD deficient cells are exposed to TNF-α.[24] Activation of the TNFR1 receptor can also lead to the up-regulation of inducible nitric oxide synthase (iNOS) resulting in the excessive production of nitric oxide (NO$^\bullet$), which can lead to apoptosis.

4.3 APOPTOSIS IN SKELETAL MUSCLE

The role of apoptosis has been extensively described during development and aging and under a plethora of conditions and diseases in various cell types and organ systems such as the nervous system and immune system. However, relatively little is known about the role of apoptosis in skeletal muscle physiology and pathophysiology. In part, this is due to the fact that skeletal muscle is a postmitotic tissue (fully differentiated and multinucleated) and therefore many scientists believed that myocytes would not have an active programmed cell-death mechanism. Although progress is advancing, much is still to be elucidated to understand the importance of apoptosis in skeletal muscle.

4.3.1 Skeletal Muscle Apoptosis in Cachexia

Apoptosis has been implicated as a mechanism contributing to the wasting of muscle mass during cachectic conditions. Cachexia has been associated with diseases such as cancer, chronic heart failure, and chronic obstructive pulmonary disorder among others and is a strong predictor of mortality.[23,25–29] For example, cachexia occurs in most cancer patients before death and is responsible for up to ~25% of cancer deaths.[30,31] Muscle wasting and protein depletion can lead to general frailty and exertional fatigue, respiratory failure, cardiac dysfunction, and a compromised immune system.[30,32–34] In the case of cancer, cachectic patients are more susceptible to the toxic side effects of chemotherapy.[32] This effect can interfere with appropriate dosing of patients and therefore lead to insufficient treatment to prevent tumor growth. Prevention of cachexia would allow treatment with higher doses of medicine appropriate to slow tumor growth. This is just one example, among many, of the benefits of ongoing investigation into the mechanisms causing muscle wasting in the patients with diseases such as cancer. Activation of myocyte apoptosis has been shown to be a mechanism contributing to cachexia.

Apoptosis has been reported in muscle of experimental models of cachexia and of cachectic patients. A common indicator of apoptosis is fragmentation of nuclear DNA into mono- and oligonucleosomes. TUNEL staining is a histochemical technique that labels fragmented double-stranded DNA within the nucleus, which can then be observed under the microscope. The percentage of TUNEL-positive nuclei in muscle of patients with cachexia correlates with the degree of muscle wasting[35,36] and has been reported to be as high as 57% in patients with chronic obstructive pulmonary disorder.[26] In comparison, age-matched healthy control subjects revealed 3.8% of TUNEL-positive nuclei. In experimental models of cachexia-inducing diseases, such as chronic heart failure, the number of TUNEL-positive nuclei in wasting muscle of rodents range from 40 to 189 nuclei/mm^3 compared to 1 to 7 nuclei/mm^3 in muscle from healthy control animals.[37–39] TUNEL staining is considered to be the "gold" standard in apoptosis detection. This method is advantageous in determining the amount of apoptotic nuclei in muscle tissue because it allows the discrimination between the varying cell types found in muscle tissue. Visual inspection under the microscope can determine that the TUNEL-positive nuclei are myocyte nuclei rather than the nuclei of resident satellite cells, phagocytes, fibroblasts, or white blood cells. However, this method is not without its limitations, one being that DNA undergoing repair can show false positive results. For this reason, many studies include more than one method of determining the occurrence of apoptosis. Another common method of detecting apoptotic DNA fragmentation is the appearance of "DNA ladder-ing" when the isolated DNA is subjected to agarose gel electrophoresis. The DNA fragments separate in the gel every 200 nucleotides, indicating the pres-ence of mono- and oligonucleosomes. Using this method, in experimental models of cancer data support a significant increase in the amount of DNA laddering in muscle of cachectic rodents.[40,41] This method also has its experimental disad-vantages in that it cannot be determined if the DNA isolated is exclusively from myocyte nuclei. Furthermore, it is not possible to quantify the number of nuclei or number of cells undergoing apoptosis. Often the TUNEL staining and DNA lad-dering techniques will be used together to determine the best estimate of myocyte death via apoptosis that may be contributing to the muscle wasting of cachexia.

Alterations in the components of the apoptotic machinery, such as caspases and members of the Bcl-2 protein family, are also analyzed to evaluate the role of apop-tosis in cachexia. The activity of various caspases is increased in cachectic skeletal muscle from rodents with experimentally induced chronic heart failure and tumor burden.[37–39,42] Several authors have reported significant increases in caspase-3 and -9 activities in the tibialis anterior muscle from rats that had chronic heart failure (CHF)[37–39] compared to healthy controls. Tisdale et al. compared activity of cas-pases in the gastrocnemius muscle from mice implanted with cachexia-inducing tumors (MAC 16) and from mice implanted with tumors that are known not to induce cachexia (MAC 13).[42] In the mice bearing the MAC 16 tumors, the activities of caspase-1, -3, -6, -8, and -9 in the cachectic muscle were significantly elevated compared to the caspase activities in muscle of mice bearing the MAC 13 tumors. However, the authors did not find nucleosomal DNA fragmentation via agarose gel electrophoresis to confirm the execution of apoptosis due to the activation of caspases. The authors suggested that the DNA could have been degraded as a result of sample preparation and therefore may explain why DNA appeared fragmented in all samples and without the nucleosomal pattern. Consistent with elevated cas-pase activity in cachectic muscle, the Bax/Bcl-2 ratio also increases in cachectic muscle in experimental animals.[37,39,40] An increase in the Bax/Bcl-2 ratio would favor cytochrome c release and activation of the mitochondrion-mediated signal-ing pathway. In support, cytosolic cytochrome c has been found to be elevated in cachectic muscle.[39]

In summary, the data support the view that apoptosis does contribute to the wasting of skeletal muscle in cachexia-inducing diseases. Cachectic muscle is associated with increased DNA fragmentation characteristic of apoptosis, as indic-ated by the presence of TUNEL-positive nuclei and DNA laddering, increased activities of various caspases, and an increase in the Bax/Bcl-2 ratio.

4.3.1.1 Causes of apoptosis in cachexia

It is thought that the cause of apoptosis in cachectic muscle is an alteration in the growth factor to cytokine ratio in the extracellular milieu of muscle tissue. Circulating levels of pro-inflammatory cytokines are elevated in patients with cachexia-inducing diseases such as chronic heart failure, cancer, and chronic obstructive pulmonary disease.[43,44] The cytokines thought to be involved with the muscle wasting effects of cachexia are TNF-α, IL-6, and IL-1. Of those, TNF-α has been the focus of attention as to its role in inducing apoptosis of skeletal muscle thereby contributing to muscle wasting in cachectic states.

Animals infused with TNF-α exhibit significant muscle wasting with nucleosomal DNA fragmentation suggesting that apoptotic mechanisms are activated.[45] The apoptosis-inducing effects of TNF-α have also been demonstrated in tumor burden TNFR1 knockout mice.[45] The TNFR1 knockout mice implanted with cachexia-inducing tumors exhibited significantly less nucleosomal DNA fragmentation and less muscle wasting in the gastrocnemius muscle compared to wild-type mice with tumor burden. These findings suggest that TNF-α mediated signaling via TNFR1 does in large part contribute to the induction of apoptosis although not entirely since markers of apoptosis were not completely abolished. Hence, TNF-α must also induce apoptosis via alternative receptors in skeletal muscle, albeit to a much lesser extent, or other cytokines signal via their cognate receptors contribute to apoptosis.

The mechanism by which TNF-α induces apoptosis in conditions of cachexia has not been clearly defined. Several studies have shown that TNF-α-induced apoptosis in skeletal muscle is associated with the activation of procaspase-8,[46] iNOS up-regulation,[47] and the increased production of sphingosine.[38] Degenerating myofibers normally are regenerated by the proliferation and differentiation of myoblasts, which fuse with the existing myofibers. Chronic exposure of skeletal muscle to TNF-α can cause apoptosis in both myoblasts and myofibers and therefore the degenerating and atrophying myofibers are unable to regenerate. Myoblasts exposed to chronically elevated TNF-α undergo apoptosis and show procaspase-8 cleavage.[46] It appears that the initial effects of TNF-α on myoblasts are cleavage and activation of procaspase-8 as well as activation of NF-κB, which favors survival and counteracts the apoptotic signal. Continued exposure to TNF-α leads to the inactivation of NF-κB, possibly due to cleavage by the active caspases, which alters the balance of the survival and death signals resulting in apoptosis of the myoblast.[46] Inhibition of NF-κB, via a dominant inhibitory IκBα-containing adenoviral vector, accelerates the apoptotic response to TNF-α.[46] Caspase-8 inhibition significantly reduces TNF-α-induced apoptosis. The data suggest that in the condition of cachexia, which is characterized by a chronically elevated cytokine milieu, TNF-α exposure results in the activation of caspase-8 and inactivation of NF-κB causing apoptosis of myoblasts thereby decreasing the potential for myofiber regeneration or hypertrophy.

It is important to point out that wasting of differentiated myotubes in cachectic conditions is shown to be mediated via TNF-α-induced activation of NF-κB[48] suggesting that inhibition of NF-κB would prevent muscle wasting in cachexia. In differentiated myofibers, TNF-α-induced apoptosis may be mediated by iNOS up-regulation and sphingosine production.[38,47] iNOS is under the regulation of NF-κB, therefore it is not surprising that iNOS expression is found to be elevated approximately two times in muscle samples taken by biopsy from chronic obstructive pulmonary disease (COPD) patients with cachexia.[47] iNOS up-regulation can lead to the excessive production of nitric oxide, as much as thousand times increase, resulting in protein degradation and apoptosis.[49] Excessive nitric oxide (NO) production can result in the formation of peroxynitrite (ONOO$^-$) via its interaction with superoxide anion. ONOO$^-$ is a highly reactive molecule that can nitrate proteins and lipids within the cell rendering them inactive and can enhance their degradation by the ubiquitin–proteasome system.[50–52] Although NO production in cachectic muscle has not been directly measured, protein nitrotyrosination is elevated.[53] Therefore, apoptosis in differentiated myofibers may be, in part, due to excessive NO production as a result of chronic TNF-α-induced NF-κB activation.

In addition to NF-κB activation, TNF-α stimulation can also result in the production of sphingolipids, such as sphingosine. The binding of TNF-α to TNFR1 leads to the catabolism of sphingomyelin resulting in the generation of sphingosine. Dalla Libera and colleagues have found that apoptosis of skeletal myofibers in experimental CHF is associated with increased levels of TNF-α and sphingosine.[38] They report a significant correlation of $r^2 = 0.74$ between serum levels of TNF-α and TUNEL-positive myocytes and a correlation of $r^2 = 0.87$ between serum levels of sphingosine and TUNEL-positive myocytes.[38] Furthermore, it was demonstrated *in vitro* that sphingosine can directly induce apoptosis in cultured myofibers taken from CHF rats.[38] Sphingosine may induce apoptosis via the production of ceramide that has been shown to induce apoptosis via activation of the mitochondrion-mediated signaling pathway.[54]

In the case of cancer, proteolysis inducing factor (PIF) has been found to contribute to muscle wasting and apoptosis.[55] PIF is a sulfated glycoprotein that was isolated from the urine of cancer patients.[56] PIF has been shown to decrease the rate of protein synthesis and also increase the rate of degradation by activation of the ubiquitin–proteasome system in myotubes. Furthermore, PIF can induce apoptosis in myotubes presumably via the release of free arachidonic acid and activation of the mitochondrion-mediated signaling pathway. Treatment of myotubes with PIF results in an eightfold increase in the release of arachidonic acid, increased Bax expression, release of cytochrome c, and activation of caspases.[55] Pretreatment of myotubes with eicosapentaenoic acid attenuated the apoptotic effects of PIF preventing the release of arachidonic acid and activation of the mitochondrial mediated signaling. Indeed, the use of fish oil, which contains eicosapentaenoic acid, has been shown to be effective in pancreatic cancer patients to attenuate muscle wasting and to improve quality of life.[57]

4.3.1.2 Experimental treatments for prevention of skeletal muscle apoptosis in cachexia

Other studies looking at possible treatments to prevent cachexia-related apoptosis involve use of L-carnitine, IL-15, and growth hormone. L-carnitine is well known for its metabolic effects in fatty acid oxidation. However, L-carnitine has been shown to have alternative roles in regulating gene expression and activity of caspases.[58] L-carnitine has been shown to prevent apoptosis in cardiac myocytes treated with doxorubicin by preventing sphingomyelin hydrolysis and generation of ceramide.[59] Treatment of rats with L-carnitine attenuated the amount of skeletal muscle apoptosis occurring with experimental heart failure with a decrease in TNF-α and sphingosine levels.[37] Bcl-2 levels were elevated and activated caspases-9 and -3 were reduced in the skeletal muscle of CHF rats treated with L-carnitine compared to untreated CHF rodents. Furthermore, *in vitro* studies demonstrate that L-carnitine treatment can prevent staurosporine-induced apoptosis in skeletal myotubes.[37] These data suggest that L-carnitine may prevent apoptosis by blocking the mitochondrion mediated signaling pathway or acts further upstream by preventing ceramide production.

IL-15 acts as an anabolic cytokine in skeletal muscle. IL-15 has the ability to increase total myofibrillar protein content in muscle cells by decreasing protein degradation.[45] IL-15 decreases the activity of the ubiquitin–proteasome system in skeletal muscle of tumor-bearing rodents resulting in an increase in protein content.[45] Furthermore IL-15 treatment also prevents apoptosis in cachectic skeletal muscle.[40] Daily treatments of IL-15 given to tumor-bearing rats completely attenuated the DNA fragmentation associated with apoptosis. IL-15 may have prevented apoptosis by altering TNF-α signaling since the expression of TNFR1 and TNFR2 were both reduced to control levels.[40] Furthermore, protein levels of iNOS were significantly reduced compared to untreated tumor-bearing rats and were similar to control levels. These data suggest that IL-15 may prevent cachexia-related apoptosis by decreasing TNF-α-induced expression of iNOS thereby decreasing NO production and apoptosis.

Treatment of CHF rodents with growth hormone prevents apoptosis in skeletal muscle.[39] Daily treatment of CHF rats with a high dose of growth hormone ($1.0\,mg\,kg^{-1}\,day^{-1}$) for two weeks attenuated the increase in Bax/Bcl-2 ratio, cytochrome c release, and increased caspase-9 activity normally seen in CHF rodents without treatment.[39] It is believed that the effects may be mediated through insulin-like growth factor-1 (IGF-1), which has been shown to activate cell survival pathways via NF-κB and PI3 kinase signaling pathways.[39] Of note, a similar treatment plan using a low dose of growth hormone ($0.2\,mg\,kg^{-1}\,day^{-1}$) did not have the same beneficial effects. The authors suggest that this may be due to the fact that growth hormone resistance is a common characteristic in patients with CHF.[39] Despite the beneficial effects, the use of growth hormone as a treatment for cachexia may not be a clinically viable option due to its dangerous side effects, such as the onset of diabetes mellitus.

4.3.2 Skeletal Muscle Apoptosis in Normal Aging

Aging individuals lose muscle mass at an approximate rate of 1 to 2% per year past the age of 50.[33] The loss of muscle mass with age, known as sarcopenia, is due to both a decrease in the number of fibers as well as atrophy of the remaining fibers. It is estimated that individuals lose $\sim$30% of their muscle fibers by the age of 80 years.[60] Recent evidence has supported apoptosis as a mechanism contributing to the loss of myofibers with age.[61-64] Aged skeletal muscle is associated with mitochondrial dysfunction and sarcoplasmic reticulum (SR) stress, both of which may be a stimulus for the induction of apoptosis in aging muscle.

Recently, we found a significant increase in the protein levels of caspase-3[62] and a correlation between the apoptotic index and the muscle weight of the gastrocnemius in rats aged 6 months vs. 26 months, suggesting that apoptosis does indeed play a role in sarcopenia. Alway et al.[63,64] have shown that Bax increases and Bcl-2 decreases with age in the plantaris muscle of rats, providing a mechanism to explain increases in apoptosis. Caspase-9 activity also increases with age in the plantaris muscle.[63,64] It was also demonstrated that aged muscle contains elevated levels of AIF and Apaf-1, which suggests that the apoptotic potential significantly increases with age.[62] Since the aforementioned apoptotic proteins play a role in the mitochondrion-mediated pathway, these data suggest that the mitochondria may play a role in apoptotic processes with age. Although some data support the role of mitochondria in the initiation of apoptosis in aged muscle, some contradictory data exist. We have shown that cytosolic cytochrome c levels decrease in the gastrocnemius muscle of aged rats compared to young, consistent with an increase in the mitochondrial content of ARC, which would inhibit cytochrome c release.[62] These contradictory data suggest that the role of the mitochondrion-mediated apoptosis in aging muscle may be muscle-specific since these studies were done in different types of muscle fibers.

In skeletal muscle myocytes, the maintenance of optimal calcium levels for proper contractility is provided by the SR. Recently, it has been shown that increased intracellular Ca^{2+} concentrations and endoplasmic reticulum (ER) stress can contribute significantly to increased susceptibility to apoptosis via activation of procaspase-12, a caspase localized on the cytoplasmic side of the SR.[65,66] Furthermore, it is known that aged skeletal muscle is characterized by SR dysfunction.[67,68] Therefore, it seems plausible to hypothesize that apoptosis in aging muscle may be initiated by SR-mediated mechanisms involving procaspase-12. Activation of procaspase-12 leads to apoptosis via the activation of procaspase-3. We showed significant increases of 350% in procaspase-12 content, the inactive form of the enzyme, in the gastrocnemius from 26-month old Fisher-344 rats compared with 12-month old rats.[62] Interestingly, Rao et al. showed that induction of ER stress was a specific signal to induce expression of procaspase-12.[69] This may suggest that the apoptotic potential, due to SR stress, increases with age in skeletal muscle.

In 2002, the gene for the human procaspase-12 was cloned.[70] Unlike rodent procaspase-12, the gene is interrupted by a frame shift and premature stop codon, which results in a truncated protein, and also has an amino acid substitution in the critical site for caspase activity.[70] These data suggest that the apoptotic function of procaspase-12 in response to ER stress is lost in humans. However, Hitomi et al. provide data to support that procaspase-4 may have the same role in humans as procaspase-12 has in rodents and other species.[71] They report that in some human cell lines apoptosis induced by ER stress is initiated by the activation of procaspase-4, which is localized to the ER membrane. Cells with decreased procaspase-4 expression, produced by RNA interference techniques, are more resistant to apoptosis induced by ER stress than are normal cells.[71] Thus, if SR stress contributes to myocyte loss with age in humans, it likely involves procaspase-4.

In summary, mitochondrial dysfunction and SR stress may play a key role in activating apoptotic processes in aged skeletal muscle thereby contributing to the loss of myocytes and muscle mass. This is a rather novel area of investigation that will require more focus to elucidate specific apoptotic signaling pathways contributing to sarcopenia. It is likely that several pathways of apoptosis are involved in the pathology of sarcopenia.

4.3.3 Exercise and Skeletal Muscle Apoptosis

Exercise improves muscle function, increases muscle mass and strength, and decreases fatigability. Resistance training has been shown to be an effective measure against cachexia and sarcopenia.[72–78] The effects of resistance training on muscle mass and strength involve enhanced motor unit recruitment, improved excitation–contraction (E–C) coupling and calcium handling, and an increase in the synthesis of contractile proteins. Recently, it has also been shown that resistance training can reduce oxidative stress, a known factor contributing to cellular aging and sarcopenia.[79] Six months of low-intensity or high-intensity resistance training three times per week attenuated lipid peroxidation levels in men ranging in age between 60 and 83 years.[79] However, despite oxidative stress being a potent stimulus for apoptosis, it is not known whether resistance training reduces the rate of skeletal muscle apoptosis or affects proteins involved in determining the apoptotic potential.

Aerobic exercise has been shown to decrease muscle fatigability, improve aerobic capacity, and attenuate the catabolic process in cachexia.[80] Aerobic training decreases local cytokine production and decreases iNOS expression possibly contributing to the mechanism by which exercise slows the catabolic process.[80] In conditions of cachexia, it is not known how exercise training affects skeletal muscle apoptosis. However, the effects of aerobic exercise on apoptosis in non-diseased states has been described.[81–84] The current data describe the ability of acute exercise to increase apoptosis in skeletal muscle,[82–85] while exercise training decreases the apoptotic potential.[81] Strenuous acute exercise, such as wheel running, enhanced the number of apoptotic nuclei and decreased the Bcl-2/Bax ratio.[83,84] In contrast, eight weeks of treadmill training increased the BCL-2/Bax ratio and decreased Apaf-1 in the soleus muscle.[81] Furthermore, exercise training increases expression of heat-shock protein 70 (HSP70),[81] which has been shown to play a role as an antiapoptotic protein by inhibiting apoptosome formation[86,87] and also by acting as an antagonist of AIF.[88] Exercise training was also shown to attenuate the loss of skeletal muscle nuclei due to apoptosis after hindlimb suspension.[89] In summary, aerobic exercise training decreases the apoptotic potential in non-disease states. Although the effects of aerobic training on apoptosis in cachectic muscle have not been investigated, it seems possible that the apoptotic potential would be reduced due to the decreased production of local cytokines and iNOS expression. However, the effects of resistance training on apoptosis in cachectic and noncachectic muscles are unknown.

4.4 PERSPECTIVE

The significance of apoptosis in skeletal muscle is somewhat obscure. Even though there is evidence that apoptosis does indeed occur in skeletal muscle, it is difficult to quantify its contribution to overall muscle loss during aging or with pathologic conditions. In most cell types, apoptosis of the nucleus equates to apoptosis and death of the cell. Since skeletal muscle is multinucleated, apoptosis of a nucleus may not equate to apoptosis of the entire muscle fiber. Indeed, when fibers atrophy the nucleus to fiber area ratio remains the same, which means that nuclei are lost by an unknown mechanism. Allen and colleagues[89] hypothesize that the nuclear loss is due to a mechanism involving apoptosis. It is not known under what conditions or how many nuclei must be lost before the entire fiber undergoes "classical" apoptosis.

Another consideration when studying skeletal muscle apoptosis is the fiber type composition of the muscle. Although with limited published data, it appears that type II muscle fibers may be more susceptible to apoptosis compared to type I muscle fibers.[20, 35] Therefore, muscles composed of a higher percentage of type II fibers, such as the superficial vastus lateralis (SVL), would be affected to a much greater extent compared to muscles with primarily type I fibers, such as the soleus.[20, 35] Data comparing apoptotic signaling in the type I soleus and the type II SVL muscle suggests a more pronounced stimulation of cell death signaling in the type II muscle which would concur with the greater loss of muscle mass demonstrated with age in this particular muscle.[90, 91] Previous research has reported a tendency for type II muscles to be more prone to develop apoptosis than type I muscles.[20, 35] These researchers explored the skeletal muscle myopathy linked with muscle bulk loss in a model of congestive heart failure. In the tibialis anterior muscle, composed primarily of type II fibers, the magnitude of apoptosis detected in the tissue mirrored the increase in circulating TNF-α and was accompanied by muscle atrophy.[35] At variance, however, was the degree of apoptosis and muscle atrophy detected in the soleus. Despite, the levels of TNF-α being equivalent to those reported in the sister study[35] the investigators did not observe any degree of muscle atrophy in the soleus, and noted the near threefold reduction in myocyte apoptosis compared to that determined in the type II fiber muscle.[20] Recent, yet unpublished data, from our laboratory show an increase in apoptosis in the SVL with age while undetectable levels occurred in the soleus. On the basis of the current findings, type II muscles possess a predisposition to undergo apoptosis to a greater extent than type I muscles.

4.5 CONCLUSION

Apoptosis is a cell suicide program that plays a role in development and in the maintenance of health. It is also a process that contributes to disease progression and occurs more frequently in various organs as we age, which may contribute to the loss of function. In particular, apoptosis occurs in skeletal muscle of cachectic patients suffering from a variety of diseases such as cancer, chronic heart failure, chronic obstructive pulmonary disease, and others. Common among these diseases is the chronic elevation of circulating cytokine levels, which is thought to be a major inducer of skeletal muscle apoptosis and wasting. TNF-α has been given the most attention and is thought to be the major cytokine responsible for apoptosis of myocytes. TNF-α may induce apoptosis via the activation of caspase-8 or by the induction of iNOS expression and production of NO$^{\cdot}$. Apoptosis also occurs in aging muscle and may contribute to muscle wasting in the elderly. It has been shown that the apoptotic potential of myocytes increases with age, however it is not known what signaling pathways are activated to induce apoptosis. Aging muscle has been characterized by having mitochondrial dysfunction and ER dysfunction, which may be stimuli in aging muscle to induce apoptosis. Recent research supports this hypothesis.

Although experiments in animal models have shown some promise in the prevention of apoptosis, currently it is not known what pharmacological treatments may inhibit the occurrence of apoptosis in cachectic patients or aging individuals. Exercise training has been shown to be effective in attenuating cachexia in disease patients and muscle loss in aging individuals. However, the effects of resistance exercise training on specific apoptosis pathways in these populations have not been investigated. Experiments have shown that exercise training decreases apoptosis and the apoptotic potential in the skeletal muscle of healthy young rodents. It is likely that exercise may have the same effect in cachectic or aging muscle. It is quite likely that further research may lead to effective treatments in the prevention of apoptosis thereby reducing muscle wasting in patients and in the elderly. The ultimate outcome would be an enhanced quality of life and a reduction in morbidity due to excessive muscle loss.

REFERENCES

1. Lossi, L. and Merighi, A. *In vivo* cellular and molecular mechanisms of neuronal apoptosis in the mammalian CNS. *Prog. Neurobiol.* 69, 287–312 (2003).
2. Maruyama, W., Youdim, M.B., and Naoi, M. Antiapoptotic properties of rasagiline, N-propargylamine-1(R)-aminoindan, and its optical (S)-isomer, TV1022. *Ann. N.Y. Acad. Sci.* 939, 320–9 (2001).
3. Lev, N., Melamed, E., and Offen, D. Apoptosis and Parkinson's disease. *Prog. Neuropsychopharmacol. Biol. Psychiatry* 27, 245–50 (2003).
4. Hirsch, E.C. et al. The role of glial reaction and inflammation in Parkinson's disease. *Ann. N. Y. Acad. Sci.* 991, 214–28 (2003).
5. Waldmeier, P.C. Prospects for antiapoptotic drug therapy of neurodegenerative diseases. *Prog. Neuropsychopharmacol. Biol. Psychiatry* 27, 303–21 (2003).
6. Tatton, W.G. and Chalmers-Redman, R.M. Modulation of gene expression rather than monoamine oxidase inhibition: (−)-deprenyl-related compounds in controlling neurodegeneration. *Neurology* 47, S171–83 (1996).

7. Trion, A. and van der Laarse, A. Vascular smooth muscle cells and calcification in atherosclerosis. *Am. Heart J.* 147, 808–14 (2004).

8. Thatte, U. and Dahanukar, S. Apoptosis: clinical relevance and pharmacological manipulation. *Drugs* 54, 511–32 (1997).

9. Wallace, J.L., Ignarro, L.J., and Fiorucci, S. Potential cardioprotective actions of no-releasing aspirin. *Nat. Rev. Drug Discov.* 1, 375–82 (2002).

10. Cline, A.M. and Radic, M.Z. Apoptosis, subcellular particles, and autoimmunity. *Clin. Immunol.* 112, 175–82 (2004).

11. Bursch, W. The autophagosomal–lysosomal compartment in programmed cell death. *Cell Death Differ.* 8, 569–81 (2001).

12. Leeuwenburgh, C. Role of apoptosis in sarcopenia. *J. Gerontol. A Biol. Sci. Med. Sci.* 58, 999–1001 (2003).

13. Ekhterae, D. et al. ARC inhibits cytochrome c release from mitochondria and protects against hypoxia-induced apoptosis in heart-derived H9c2 cells. *Circ. Res.* 85, e70–7 (1999).

14. Gustafsson, A.B., Tsai, J.G., Logue, S.E., Crow, M.T., and Gottlieb, R.A. Apoptosis repressor with caspase recruitment domain protects against cell death by interfering with Bax activation. *J. Biol. Chem.* 279, 21233–8 (2004).

15. Beere, H.M. et al. Heat-shock protein 70 inhibits apoptosis by preventing recruitment of procaspase-9 to the Apaf-1 apoptosome. *Nat. Cell Biol.* 2, 469–75 (2000).

16. Pandey, P. et al. Negative regulation of cytochrome c-mediated oligomerization of Apaf-1 and activation of procaspase-9 by heat shock protein 90. *EMBO J.* 19, 4310–22 (2000).

17. Yang, Q.H., Church-Hajduk, R., Ren, J., Newton, M.L., and Du, C. Omi/HtrA2 catalytic cleavage of inhibitor of apoptosis (IAP) irreversibly inactivates IAPs and facilitates caspase activity in apoptosis. *Genes. Dev.* 17, 1487–96 (2003).

18. Vaux, D.L. and Silke, J. HtrA2/Omi, a sheep in wolf's clothing. *Cell* 115, 251–3 (2003).

19. Suzuki, Y. et al. A serine protease, HtrA2, is released from the mitochondria and interacts with XIAP, inducing cell death. *Mol. Cell* 8, 613–21 (2001).

20. Susin, S.A. et al. Molecular characterization of mitochondrial apoptosis-inducing factor. *Nature* 397, 441–6 (1999).

21. Li, L.Y., Luo, X., and Wang, X. Endonuclease G is an apoptotic DNase when released from mitochondria. *Nature* 412, 95–9 (2001).

22. Baud, V. and Karin, M. Signal transduction by tumor necrosis factor and its relatives. *Trends Cell Biol.* 11, 372–7 (2001).

23. Micheau, O. and Tschopp, J. Induction of TNF receptor I-mediated apoptosis via two sequential signaling complexes. *Cell* 114, 181–90 (2003).

24. Yeh, W.C. et al. FADD: essential for embryo development and signaling from some, but not all, inducers of apoptosis. *Science* 279, 1954–8 (1998).

25. Anker, S.D. et al. Wasting as independent risk factor for mortality in chronic heart failure. *Lancet* 349, 1050–3 (1997).

26. Lewis, M.I. Apoptosis as a potential mechanism of muscle cachexia in chronic obstructive pulmonary disease. *Am. J. Respir. Crit. Care Med.* 166, 434–6 (2002).

27. Schols, A.M. Pulmonary cachexia. *Int. J. Cardiol.* 85, 101–10 (2002).

28. Berry, J.K. and Baum, C. Reversal of chronic obstructive pulmonary disease-associated weight loss: are there pharmacological treatment options? *Drugs* 64, 1041–52 (2004).

29. Wouters, E.F., Creutzberg, E.C., and Schols, A.M. Systemic effects in COPD. *Chest* 121, 127S–30S (2002).
30. Dworzak, F., Ferrari, P., Gavazzi, C., Maiorana, C., and Bozzetti, F. Effects of cachexia due to cancer on whole body and skeletal muscle protein turnover. *Cancer* 82, 42–8 (1998).
31. Warren, S. The immediate cause of death in cancer. *Am. J. Med. Sci.* 184, 610–13 (1932).
32. Andreyev, H.J., Norman, A.R., Oates, J., and Cunningham, D. Why do patients with weight loss have a worse outcome when undergoing chemotherapy for gastrointestinal malignancies? *Eur. J. Cancer* 34, 503–9 (1998).
33. Marcell, T.J. Sarcopenia: causes, consequences, and preventions. *J. Gerontol. A Biol. Sci. Med. Sci.* 58, M911–6 (2003).
34. Drott, C., Ekman, L., Holm, S., Waldenstrom, A., and Lundholm, K. Effects of tumor-load and malnutrition on myocardial function in the isolated working rat heart. *J. Mol. Cell Cardiol.* 18, 1165–76 (1986).
35. Vescovo, G. et al. Apoptosis of skeletal muscle myofibers and interstitial cells in experimental heart failure. *J. Mol. Cell Cardiol.* 30, 2449–59 (1998).
36. Vescovo, G. et al. Apoptosis in the skeletal muscle of patients with heart failure: investigation of clinical and biochemical changes. *Heart* 84, 431–7 (2000).
37. Vescovo, G. et al. L-Carnitine: a potential treatment for blocking apoptosis and preventing skeletal muscle myopathy in heart failure. *Am. J. Physiol. Cell Physiol.* 283, C802–10 (2002).
38. Dalla Libera, L. et al. Apoptosis in the skeletal muscle of rats with heart failure is associated with increased serum levels of TNF-alpha and sphingosine. *J. Mol. Cell Cardiol.* 33, 1871–8 (2001).
39. Dalla Libera, L. et al. Beneficial effects of GH/IGF-1 on skeletal muscle atrophy and function in experimental heart failure. *Am. J. Physiol. Cell Physiol.* 286, C138–44 (2004).
40. Figueras, M. et al. Interleukin-15 is able to suppress the increased DNA fragmentation associated with muscle wasting in tumour-bearing rats. *FEBS Lett.* 569, 201–6 (2004).
41. van Royen, M. et al. DNA fragmentation occurs in skeletal muscle during tumor growth: A link with cancer cachexia? *Biochem. Biophys. Res. Commun.* 270, 533–37 (2000).
42. Belizario, J.E., Lorite, M.J., and Tisdale, M.J. Cleavage of caspases-1, -3, -6, -8 and -9 substrates by proteases in skeletal muscles from mice undergoing cancer cachexia. *Br. J. Cancer* 84, 1135–40 (2001).
43. Sharma, R. and Anker, S.D. Cytokines, apoptosis and cachexia: the potential for TNF antagonism. *Int. J. Cardiol* 85, 161–71 (2002).
44. Sharma, R., Al-Nasser, F.O., and Anker, S.D. The importance of tumor necrosis factor and lipoproteins in the pathogenesis of chronic heart failure. *Heart Fail. Monit.* 2, 42–7 (2001).
45. Carbo, N. et al. TNF-alpha is involved in activating DNA fragmentation in skeletal muscle. *Br. J. Cancer* 86, 1012–6 (2002).
46. Stewart, C.E., Newcomb, P.V., and Holly, J.M. Multifaceted roles of TNF-alpha in myoblast destruction: a multitude of signal transduction pathways. *J. Cell Physiol.* 198, 237–47 (2004).
47. Agusti, A., Morla, M., Sauleda, J., Saus, C., and Busquets, X. NF-kappaB activation and iNOS upregulation in skeletal muscle of patients with COPD and low body weight. *Thorax* 59, 483–7 (2004).

48. Li, Y.P. and Reid, M.B. NF-kappaB mediates the protein loss induced by TNF-alpha in differentiated skeletal muscle myotubes. *Am. J. Physiol. Regul. Integr. Comp. Physiol.* 279, R1165–70 (2000).

49. Brune, B., von Knethen, A., and Sandau, K.B. Nitric oxide and its role in apoptosis. *Eur. J. Pharmacol.* 351, 261–72 (1998).

50. Beckman, J.S., Viera, L., Estevez, A.G., and Teng, R. Nitric oxide and peroxynitrite in the perinatal period. *Semin. Perinatol.* 24, 37–41 (2000).

51. Beckman, J.S. Protein tyrosine nitration and peroxynitrite. *FASEB J.* 16, 1144 (2002).

52. Jatoi, A., Cleary, M.P., Tee, C.M., and Nguyen, P.L. Weight gain does not preclude increased ubiquitin conjugation in skeletal muscle: an exploratory study in tumor-bearing mice. *Ann. Nutr. Metab.* 45, 116–20 (2001).

53. Barreiro, E., Gea, J., Corominas, J.M., and Hussain, S.N. Nitric oxide synthases and protein oxidation in the quadriceps femoris of patients with chronic obstructive pulmonary disease. *Am. J. Respir. Cell Mol. Biol.* 29, 771–8 (2003).

54. Sandri, M. Apoptotic signaling in skeletal muscle fibers during atrophy. *Curr. Opin. Clin. Nutr. Metab. Care* 5, 249–53 (2002).

55. Smith, H.J. and Tisdale, M.J. Induction of apoptosis by a cachectic-factor in murine myotubes and inhibition by eicosapentaenoic acid. *Apoptosis* 8, 161–9 (2003).

56. Todorov, P. et al. Characterization of a cancer cachectic factor. *Nature* 379, 739–42 (1996).

57. Barber, M.D., Ross, J.A., Voss, A.C., Tisdale, M.J., and Fearon, K.C. The effect of an oral nutritional supplement enriched with fish oil on weight-loss in patients with pancreatic cancer. *Br. J. Cancer* 81, 80–6 (1999).

58. Mutomba, M.C. et al. Regulation of the activity of caspases by L-carnitine and palmitoylcarnitine. *FEBS Lett.* 478, 19–25 (2000).

59. Andrieu-Abadie, N. et al. L-carnitine prevents doxorubicin-induced apoptosis of cardiac myocytes: role of inhibition of ceramide generation. *FASEB J.* 13, 1501–10 (1999).

60. Lexell, J. Human aging, muscle mass, and fiber type composition. *J. Gerontol A Biol. Sci. Med. Sci.* 50, 11–6 (1995).

61. Dirks, A. and Leeuwenburgh, C. Apoptosis in skeletal muscle with aging. *Am. J. Physiol. Regul. Integr. Comp. Physiol.* 282, R519–27 (2002).

62. Dirks, A.J. and Leeuwenburgh, C. Aging and lifelong calorie restriction result in adaptations of skeletal muscle apoptosis repressor, apoptosis-inducing factor, X-linked inhibitor of apoptosis, caspase-3, and caspase-12. *Free Radic. Biol. Med.* 36, 27–39 (2004).

63. Alway, S.E., Degens, H., Krishnamurthy, G., and Smith, C.A. Potential role for Id myogenic repressors in apoptosis and attenuation of hypertrophy in muscles of aged rats. *Am. J. Physiol. Cell. Physiol.* 283, C66–76 (2002).

64. Alway, S.E., Martyn, J.K., Ouyang, J., Chaudhrai, A., and Murlasits, Z.S. Id2 expression during apoptosis and satellite cell activation in unloaded and loaded quail skeletal muscles. *Am. J. Physiol. Regul. Integr. Comp. Physiol.* 284, R540–9 (2003).

65. Nakagawa, T. and Yuan, J. Cross-talk between two cysteine protease families. Activation of caspase-12 by calpain in apoptosis. *J. Cell Biol.* 150, 887–94 (2000).

66. Nakagawa, T. et al. Caspase-12 mediates endoplasmic-reticulum-specific apoptosis and cytotoxicity by amyloid-beta. *Nature* 403, 98–103 (2000).

67. Narayanan, N., Jones, D.L., Xu, A., and Yu, J.C. Effects of aging on sarcoplasmic reticulum function and contraction duration in skeletal muscles of the rat. *Am. J. Physiol.* 271, C1032–40 (1996).

68. Payne, A.M., Dodd, S.L., and Leeuwenburgh, C. Life-long calorie restriction in Fischer 344 rats attenuates age-related loss in skeletal muscle-specific force and reduces extracellular space. *J. Appl. Physiol.* 95, 2554–62 (2003).

69. Rao, R.V. et al. Coupling endoplasmic reticulum stress to the cell death program. Mechanism of caspase activation. *J. Biol. Chem.* 276, 33869–74 (2001).

70. Fischer, H., Koenig, U., Eckhart, L., and Tschachler, E. Human caspase 12 has acquired deleterious mutations. *Biochem. Biophys. Res. Commun.* 293, 722–6 (2002).

71. Hitomi, J. et al. Involvement of caspase-4 in endoplasmic reticulum stress-induced apoptosis and Abeta-induced cell death. *J. Cell Biol.* 165, 347–56 (2004).

72. al-Majid, S. and McCarthy, D.O. Cancer-induced fatigue and skeletal muscle wasting: the role of exercise. *Biol. Res. Nurs.* 2, 186–97 (2001).

73. al-Majid, S. and McCarthy, D.O. Resistance exercise training attenuates wasting of the extensor digitorum longus muscle in mice bearing the colon-26 adenocarcinoma. *Biol. Res. Nurs.* 2, 155–66 (2001).

74. Walsmith, J. and Roubenoff, R. Cachexia in rheumatoid arthritis. *Int. J. Cardiol.* 85, 89–99 (2002).

75. McCartney, N., Hicks, A.L., Martin, J., and Webber, C.E. A longitudinal trial of weight training in the elderly: continued improvements in year 2. *J. Gerontol. A Biol. Sci. Med. Sci.* 51, B425–33 (1996).

76. McCartney, N. and McKelvie, R.S. The role of resistance training in patients with cardiac disease. *J. Cardiovasc. Risk* 3, 160–6 (1996).

77. Nelson, M.E. et al. Effects of high-intensity strength training on multiple risk factors for osteoporotic fractures. A randomized controlled trial. *JAMA* 272, 1909–14 (1994).

78. Vincent, K.R. et al. Resistance exercise and physical performance in adults aged 60 to 83. *J. Am. Geriatr. Soc.* 50, 1100–7 (2002).

79. Vincent, K.R., Vincent, H.K., Braith, R.W., Lennon, S.L., and Lowenthal, D.T. Resistance exercise training attenuates exercise-induced lipid peroxidation in the elderly. *Eur. J. Appl. Physiol.* 87, 416–23 (2002).

80. Schulze, P.C., Gielen, S., Schuler, G., and Hambrecht, R. Chronic heart failure and skeletal–muscle catabolism: effects of exercise training. *Int. J. Cardiol.* 85, 141–9 (2002).

81. Siu, P.M., Bryner, R.W., Martyn, J.K., and Alway, S.E. Apoptotic adaptations from exercise training in skeletal and cardiac muscles. *FASEB J.* 18, 1150–2 (2004).

82. Carraro, U. and Franceschi, C. Apoptosis of skeletal and cardiac muscles and physical exercise. *Aging (Milano)* 9, 19–34 (1997).

83. Podhorska-Okolow, M. et al. Apoptosis of myofibres and satellite cells: exercise-induced damage in skeletal muscle of the mouse. *Neuropathol. Appl. Neurobiol.* 24, 518–31 (1998).

84. Sandri, M. et al. Apoptosis, DNA damage and ubiquitin expression in normal and mdx muscle fibers after exercise. *FEBS Lett.* 373, 291–5 (1995).

85. Arslan, S., Erdem, S., Sivri, A., Hascelik, Z., and Tan, E. Exercise-induced apoptosis of rat skeletal muscle and the effect of meloxicam. *Rheumatol. Int.* 21, 133–6 (2002).

86. Kobayashi, Y. et al. Chaperones Hsp70 and Hsp40 suppress aggregate formation and apoptosis in cultured neuronal cells expressing truncated androgen receptor protein with expanded polyglutamine tract. *J. Biol. Chem.* 275, 8772–8 (2000).

87. Saleh, A., Srinivasula, S.M., Balkir, L., Robbins, P.D., and Alnemri, E.S. Negative regulation of the Apaf-1 apoptosome by Hsp70. *Nat. Cell Biol.* 2, 476–83 (2000).

88. Cande, C., et al. Apoptosis-inducing factor (AIF): a novel caspase-independent death effector released from mitochondria. *Biochimie.* 84, 215–22 (2002).

89. Allen, D.L., et al. Apoptosis: a mechanism contributing to remodeling of skeletal muscle in response to hindlimb unweighting. *Am. J. Physiol.* 273, C579–87 (1997).

90. Alnaqeeb, M.A. and Goldspink, G. Changes in fibre type, number and diameter in developing and ageing skeletal muscle. *J. Anat.* 153, 31–45 (1987).

91. Holloszy, J.O., Chen, M., Cartee, G.D., and Young, J.C. Skeletal muscle atrophy in old rats: differential changes in the three fiber types. *Mech. Ageing Dev.* 60, 199–213 (1991).

5 Neurohormonal Factors

Andrew L. Clark

CONTENTS

SUMMARY

Many of the body's hormone systems have the maintenance of normal energy balance as part of their role. Many specific endocrine diseases such as thyrotoxicosis, diabetes mellitus, and Addison's disease cause profound changes in body weight. Thyroxine regulates basal metabolism and excess causes an increase in basal metabolic rate with weight loss. Growth hormone, insulin, and some adrenocorticosteroids are anabolic factors, the deficiency of which causes weight loss.

The sympathetic nervous system and the renin–angiotensin system are primarily concerned with control of blood pressure and autonomic reflexes, but their stimulation can also lead to weight loss.

In patients with cachexia several neurohormonal mechanisms may contribute to weight loss. Prolonged sympathetic and renin-angiotensin system activation are common, particularly in chronic heart failure. Hormonal changes are also common, and a noteworthy feature in cachexia is resistance to the anabolic effects of hormones. Thus insulin resistance develops, as does growth hormone resistance, with a rise in the ratio of growth hormone to its mediator, insulin-like growth factor. The result is a general shift from anabolic metabolic processes to predominantly catabolic processes, contributing to the progression of cachexia.

5.1 INTRODUCTION

For most normal humans, long-term weight oscillates only a little between narrow bounds. In the intact organism, weight loss can only occur if there is greater energy expenditure than consumption. In many chronic disease states associated with cachexia, there is evidence of decreased energy intake, but also for increased energy expenditure. In this chapter, I will discuss the evidence that neurohormonal factors are involved in the development of cachexia: abnormal patterns of neurohormonal activation could potentially contribute to cachexia by affecting appetite or intestinal absorption, but also, crucially, by affecting metabolic rate and increasing energy consumption.

It is somewhat artificial to separate neurohormones from cytokine and immune activation, but these will be considered in another chapter: the effects of neurohormones on appetite control and energy metabolism have been considered in an earlier chapter.

5.2 LESSON FROM GENERAL ENDOCRINOLOGY

Long-term energy balance in the intact adult human is modulated by the influence of many hormone systems. These systems serve gradually to modulate metabolic processes to maintain a constant internal milieu for metabolizing tissues in contrast to neural mechanisms, which allow rapid responses to environmental challenge. Several endocrine pathologies are associated with weight loss and cachexia and these conditions give insight into how normal hormone activity contributes to the control of energy balance; and how abnormal hormonal activity might contribute to cachexia in other disease states (Table 5.1).

5.2.1 Thyrotoxicosis

Thyrotoxicosis can be considered as the paradigm of the potential for neurohormonal factors to cause cachexia. The hypermetabolic state induced by excessive exposure to thyroid hormones is usually due to excessive production by the thyroid gland; the commonest form being autoimmune or Graves' disease. However, the effects of excess thyroxine are recognized by potential dieters, and factitious thyrotoxicosis is not rare.

The typical features of thyrotoxicosis are well known: fast pulse, palpitations tremor, eyelid lag, warm moist skin, and weight loss. Diarrhea may also contribute

TABLE 5.1
Possible Neurohormonal
Contributors to the
Catabolic State

Thyroxine
Growth hormone — IGF-1
Steroid metabolism
Insulin
Sympathetic nervous system
Renin–angiotensin system

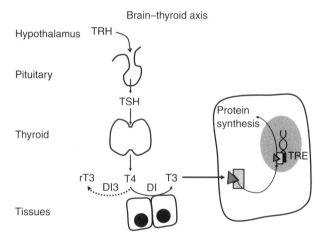

FIGURE 5.1   The brain–thyroid axis. TRH is thyrotropin releasing hormone. TSH is thyroid stimulating hormone. DI is deiodinase. DI3 is deiodinase type 3. rT3 is reverse T3. TRE is thyroid response element on DNA which is the target for the T3-receptor complex. Genes with TREs are triggered to be transcribed by binding.

to weight loss. Heat intolerance is almost pathognomonic. The clinical features can be less striking, particularly in the elderly, who may present with only atrial fibrillation, lethargy, or weight loss.

Thyroxine (T4; so called because each molecule has four iodine atoms) is produced in response to stimulation of the thyroid by thyroid stimulating hormone (TSH) from the pituitary, itself in turn stimulated by hypothalamic production of thyrotropin releasing hormone (see Figure 5.1). In the peripheral tissues, deiodinase removes one iodine from each T4 molecule to leave the metabolically active form, tri-iodothyronine, T3.

Thyroxine can also be deiodinated to the metabolically inactive rT3. There are three different deiodinases with different activities. The DI3 enzyme converts T4 to rT3 (and T3 to inactive T2). Some biological control happens at this level: in thyrotoxicosis, expression of DI3 is upregulated, and in hypothyroidism, downregulated.[1]

Tri-iodothyronine diffuses to the nucleus where it binds to specific receptors. The T3-receptor complex binds to response elements on DNA,[2] causing changes in the transcription of multiple genes. Genes affected include structural proteins, such as myosin heavy chain,[3] and other functional proteins such as Na, K-ATPase,[4] and G-proteins.[5] There are additional interactions, such as an increase in the amount of insulin-like growth factor-1.[6] The overall effect is an increase in basal metabolic rate, an effect that has been known for many years.[7,8] The mechanisms underlying the increase in basal metabolic rate remain incompletely understood, but include the promotion of futile metabolic cycles,[9,10] and uncoupling of mitochondrial respiration from ATP synthesis.[11,12] This uncoupling seems to be mediated through increased expression of uncoupling protein-3.[13]

It is easy to see why this direct increase in the basal metabolic rate will lead to weight loss,[14,15] and therapies for obesity have been suggested based on selective stimulation of thyroid receptors.[16]

In contrast to hyperthyroidism, where the main loss of weight is from the muscle compartment, hypothyroidism induces gain in weight, predominantly of fat tissue.[17] However, it is important to note that an adequate level of thyroxine is required for the function of other (anabolic) hormones. Thus, for example, for growth hormone to have its maximum effect, the individual has to be euthyroid.[18-20] In this sense, thyroxine is a facultative anabolic hormone.

5.2.2 Growth Hormone Deficiency

Growth hormone deficiency in adults was long thought not to have any importance, but there is now good evidence that growth hormone deficient adults may benefit from replacement therapy.[21,22] Growth hormone deficiency may be associated with a specific cardiomyopathy.[23]

Growth hormone is released in circadian rhythm in adults (being highest during sleep) and in response to stress and exercise. Its regulation is complex (see Figure 5.2).[24,25] Growth hormone-releasing hormone (GH-RH) released in the hypothalamus causes growth hormone release from the anterior pituitary. The effect of GH-RH is antagonized by somatostatin.

In the circulation, around 50% of the growth hormone is bound to a specific binding protein, which appears to be the same moiety as cellular growth hormone receptor, presumably shed into the circulation.[26] There are some direct effects of growth hormone itself, but most of its effects are mediated through the production of insulin-like growth factor (previously known as somatomedin) released predominantly by the liver.

Insulin-like growth factor (principally IGF-1) is, in turn, highly protein bound in the circulation to a variety of IGF-binding proteins,[27] and mediates its effects through binding to cell surface receptors (mainly the type 1 receptor subtype). It has some cross-reactivity with the insulin receptor. IGF-1 has many effects,

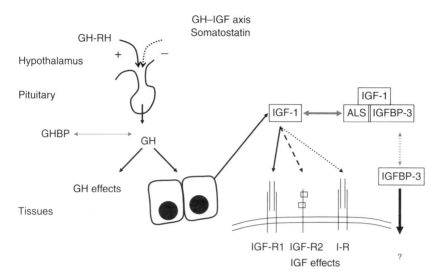

FIGURE 5.2 The growth hormone — insulin like growth factor-1 (IGF-1) axis. GH-RH is growth hormone releasing hormone. GHBP is growth hormone binding protein — presumed to be identical to growth hormone receptor shed into the circulation. Most IGF-1 circulates in a ternary complex bound to IGFBP-3 and acid labile subunit (ALS). There are many other IGF binding proteins (not shown). The two IGF receptors (IGF-R1 and -R2) and the insulin receptor (I-R) are shown. There is some evidence to suggest that IGFBP-3 may have some direct biological effects itself.

including an increase in the uptake of glucose and amino acids by muscle cells, increased protein synthesis[28] and suppression of protein degradation.[29,30] In addition, IGF-1 can cause cultured skeletal muscle cells to divide and differentiate,[31] and antagonizes the catabolic effects of glucocorticoids.[32]

Reduction in IGF-1 may have additional adverse consequences; tumor-necrosis factor α induced cell killing is prevented by IGF-1 alone[33] and expression of IGF-1 may protect cells, including cardiac myocytes, from apoptosis.[34–36]

Adult growth hormone deficiency is usually associated with weight gain rather than cachexia, but its importance in the present context is that growth hormone deficient adults have alterations in body composition. While fat mass increases,[37] there is decreased skeletal muscle mass and bone mineral density with concomitant increase in the risk of osteoporotic fracture.[38]

Treatment of growth hormone deficiency improves the distribution of body mass with a reduction in central body fat and an increase in muscle mass.[39,40] In addition, bony mineral density improves after prolonged treatment.[41,42]

5.2.3 Addison's Disease and Steroid Hormones

The commonest cause of adrenal insufficiency in industrialized societies is autoimmune destruction of the adrenal cortex.[43] Weight loss and anorexia are common presenting features, and advanced weight loss can be very severe in patients with a prolonged clinical course.[44]

The hypothalamus regulates adrenocorticotrophin (ACTH) release through the production of corticotrophin releasing hormone.[45] ACTH in turn causes cortisol production in the adrenal glands via both an increase in transport of the hormone precursor, cholesterol, into mitochondria and an increase in cortisol synthetic enzyme transcription.[46]

Cortisol is the principal catabolic steroid. It mediates its effects through binding to the glucocorticoid receptor, which subsequently dimerises. The dimerised receptor then diffuses into the nucleus where it binds to DNA.[47] DNA binding is to glucocorticoid response elements.[48] This binding can have an antiinflammatory effect, for example, by upregulating IκBα (I kappa B alpha inhibitory protein) which sequesters NF-κB (nuclear factor kappa B), preventing its proinflammatory effects.[49] NF-κB is activated by inflammatory stimuli and is a ubiquitous nuclear transcription factor which promotes the expression of many inflammatory genes.[50,51] Cortisol is also a catabolic hormone, promoting activation of gluconeogenesis[52] and protein catabolism and lipolysis.[53] The protein catabolism seems to be mediated by activation of the ubiquitin pathway.[54,55] Consequently, patients with Cushing's disease have profound muscle wasting and osteoporosis.[56,57] The muscle wasting is the dominant cause of loss of body cell mass.[58]

Endogenous anabolic steroids are principally thought of as androgenic, but their anabolic effects are important in women as well as men. Dehydroepiandrosterone (DHEA) is the major anabolic steroid precursor. Although its intrinsic activity is low, it is converted to androstenedione and thence testosterone. Anabolic steroids mediate their effects partly by blocking cortisol's binding to its receptor,[59] partly by increasing protein synthesis[60] and encouraging a positive nitrogen balance.[61,62] Hypogonadism in the adult male leads to loss of muscle bulk. That these phenomena accompany normal aging has led to a large literature on the possible beneficial effects on muscle bulk of testosterone supplementation in older men.[63]

5.2.4 Diabetes Mellitus

Deficiency of insulin causes hyperglycaemia and diabetes mellitus, but insulin is also, of course, a vitally important anabolic hormone. Young patients with insulin-dependent diabetes have an absolute deficiency of insulin, and as well as developing thirst and polyuria, tend to be thin and have rapid weight loss.[64] Starvation diets, while staving off the onset of ketoacidosis, only worsened the weight loss.

Insulin is usually thought of and discussed in terms of its effects on glucose metabolism. The insulin receptor is an insulin-regulated tyrosine kinase.[65,66]

On binding to its receptor, insulin triggers an increase in the number of cell-surface glucose transporter molecules, promotes glycogen storage and inhibits glycogen breakdown. However, insulin's nonglucose effects are at least as important; insulin is an anticatabolic hormone, preventing lipolysis in adipocytes, as well as increasing lipid storage,[67] and preventing adipocyte apoptosis.[68,69] It also increases cellular uptake of amino acids and inhibits protein breakdown.[70–72] Insulin has other anabolic effects, such as promoting osteoblast activity.[73]

All these anabolic effects are lost in the young patient with insulin-dependent diabetes: weight loss is an inevitable consequence.

Glucagonoma, a rare tumor arising from α cells in the pancreas, is associated with marked cachexia.[74,75] In health, the principal role of glucagon is to increase glycogenolysis and promote gluconeogenesis to raise blood glucose concentration. Glucagon also enhances lipolysis and prevents fatty acid uptake by the liver.

5.2.5 Panhypopituitarism

In panhypopituitarism, weight loss and frank cachexia, sometimes known as Simmond's disease, can occur. The pattern of symptoms depends upon the aetiology of the hypopituitarism and pattern of hormone loss. Adrenocorticotrophic hormone deficiency is particularly associated with weight loss, while thyroid-stimulating hormone deficiency is associated with weight gain.

5.2.6 Conclusion

Many endocrine pathologies are associated with weight loss or altered body composition. An excess of hormone production can be responsible as in thyrotoxicosis and Cushing's disease; or a deficiency can be responsible, as in diabetes and Addison's disease. We will see later that a third possibility, that of *hormone resistance*, may contribute to cachexia.

5.3 Neurohormonal Systems and Weight Control

5.3.1 Sympathetic Nervous System

The autonomic nervous system functions to provide immediate and short-term responses to environmental stresses. The dominant role of the autonomic nervous system is to mediate visceral reflexes and in the case of the sympathetic nervous system to provide an immediate response to environmental stress by mobilizing the body's resources to permit immediate strenuous physical exercise.

The general effect of mass sympathetic stimulation is catabolic, led by glycogenolysis, both through a direct effect and via inhibition of insulin secretion and promotion of glucagon secretion.[76] Sympathetic stimulation also causes lipolysis.[77] The overall effect is of mobilizing energy resources for immediate consumption by exercising skeletal muscle.

The physiological role of the autonomic nervous system in the long-term control of energy metabolism is less clear. Activation of the sympathetic nervous system causes an increase in basal metabolic rate[78,79] and heat production.[80] By contrast, beta adrenoceptor antagonism causes a fall in basal metabolic rate,[81] an effect that may underlie the increase in weight seen in some patients on long-term β-blocker therapy.[82,83] Evidence that sympathetic activity is important in determining longer term energy balance comes from observations of Pima Indians who have extremely high rates of obesity. Muscle sympathetic nerve activity correlates with basal metabolic rate in white people, but not in Pima Indians.[84,85] and Pima Indians have lower basal muscle sympathetic activity than Caucasians.[85]

The traditional understanding of the sympathetic nervous system is that it is the "flight or fight" system, and mass sympathetic discharge fulfils this role. However, much sympathetic activity is discrete.[86] The effect of sympathetic discharge on lipolysis can vary depending upon where the fat is: the lipolytic response is lower in subcutaneous gluteal fat, and higher in omental fat.[87,88]

Sympathetic stimulation can also effect substrate use in metabolism. The respiratory quotient, RQ, is the rate of carbon dioxide production relative to oxygen consumption. An RQ of 1 represents exclusive carbohydrate metabolism. Fat oxidation has a lower RQ (around 0.7) as fat is more highly reduced than carbohydrate. A high resting RQ predisposes to weight gain[89,90] and there is an inverse relation between 24 h RQ and basal muscle sympathetic nerve activity (see Figure 5.3).[91] These findings suggest that sympathetic activation causes a shift toward preferential lipid oxidation, that may in turn be related to weight control.

Catecholamines also have profound effects on protein metabolism. Whereas the effects of catecholamines on carbohydrate and lipid metabolism are to promote

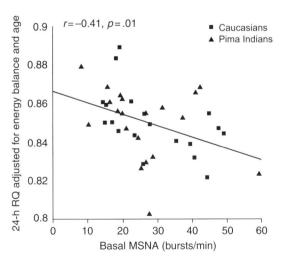

FIGURE 5.3   The relation between basal muscle sympathetic neurone activity (MSNA) and the 24 h average respiratory quotient (carbon dioxide production divided by oxygen consumption; RQ). The higher the sympathetic activity, the lower the RQ, which by inference suggests a higher proportion of fatty acid oxidation relative to carbohydrate (Taken from Snitker S., et al., *J. Clin. Endocrinol. Metab.* 1998; 83: 3977–9. With permission.).

catabolism, catecholamines have an *anabolic* effect on skeletal muscle[92,93] and *inhibit* proteolysis.[94] Removal of the adrenal glands results in an increase in proteolysis, at least in rats.[95] The effect seems to be mediated by circulating catecholamines released from the adrenal medulla, rather than through sympathetic nerve endings, with catecholamines providing a tonic restraining effect on muscle proteolysis by inhibiting calcium dependent proteolytic pathways.[95,96]

5.3.2 Renin–Angiotensin System

The physiological role of the renin–angiotensin system is to maintain blood pressure. Renin is released from the kidney in response to low arterial pressure, and this acts on circulating angiotensinogen to generate angiotensin I (Ang I) (Figure 5.4). In turn, Ang I is converted into Ang II by the angiotensin converting enzyme (ACE). Ang II, acting through its specific receptors, is an extremely potent vasoconstrictor, and also promotes sodium and water retention by the kidneys, both directly and through stimulating aldosterone release by the adrenal glands.

In a rat model, high levels of Ang II cause profound weight loss, independent of vasoconstriction, at least in part through suppression of IGF-1.[97,98] This weight loss is to some extent due to loss of skeletal muscle through excess muscle breakdown (and not impaired synthesis).[98] Ang II can also cause lipolysis.[99]

Whether these mechanisms are important in controlling body weight and energy metabolism is not clear. In disease states, Ang II may be important. The levels seen in rat models[97,98] are similar to those seen in heart failure.[100,101] Some evidence of the importance of the renin–angiotensin system involvement in weight control comes from studies in heart failure where treatment with the angiotensin

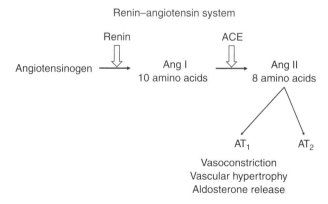

FIGURE 5.4 The renin–angiotensin system. Renin, an enzyme released from the kidney in response to a fall in renal perfusion, acts on angiotensinogen to produce angiotensin (Ang) I. Angiotensin converting enzyme (ACE) converts Ang I to Ang II. Ang II is the active moiety, and works through angiotensin receptors. The physiological role of the AT_2 receptor is not clear.

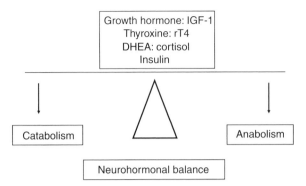

FIGURE 5.5   Catabolic/anabolic balance in health. IGF-1 is insulin-like growth factor-1;
rT4 is reverse T4; DHEA is dehydroepiandrosterone.

converting enzyme inhibitor, enalapril, resulted in a reduction in the risk of weight
loss.[102, 103]

5.4   NEUROHORMONAL PATTERNS IN HEALTH

In highlighting abnormalities of energy balance and tissue distribution that occur
in endocrine disease, a picture of neurohormonal balance in health emerges
(Figure 5.5). There is constant cycling and recycling of metabolites and cellular fuels: neurohormonal systems have evolved to deal with the environmental
loads that unbalance the system. Thus feeding induces rapid production of insulin
and suppression of glucagon, the need for exercise causes catabolic steroid release
and sympathetic nervous system activation for rapid supply of energy sources to
skeletal muscle.

On top of the short-term responses, hormonal systems are concerned with
growth and maturation of the individual and regulating the internal economy of
the body.

Many of the neurohormonal systems interact: for example, the response to
infused cathecholamines is increased in patients with mild hyperthyroidism.[104]
Angiotensin II and the sympathetic nervous system enhance each others' actions.[105]
The cells of the body are constantly bathed in a sea of small molecules regulating
their actions with the net effect of maintaining of the *status quo*. In the next
section, we will turn to the possible effects of the neurohormonal systems under
consideration on cachexia in nonendocrine disease.

5.5   CACHEXIA IN SYSTEMIC DISEASE

Cachexia is a common phenomenon in the natural history of many chronic diseases.
It is so common that little attention is paid to it outside cancer studies. However,

it does seem strange that cachexia develops in, for example, chronic heart failure: what has the heart to do with long-term weight control?

A substantial body of evidence has accumulated to demonstrate that chronic conditions are associated with an increase in basal metabolic rate (BMR). This is seen, for example, in chronic heart failure[106–108] (in which the BMR can account for 70% of daily energy expenditure[109]); chronic lung disease[110–112]; rheumatoid arthritis[113,114]; AIDS[115,116]; and burns.[117,118] Given the relation between neuro-hormonal balance and the basal metabolic rate, it seems reasonable to suspect that there may be some relation between the development of cachexia and the increase in BMR and neurohormonal balance. In normal subjects, infusions of catabolic hormones (hydrocortisone, glucagon, and adrenaline) induce hypergly-cemia, hyper-insulinemia, insulin resistance, negative nitrogen balance, which are precisely the changes seen in the cachexia syndrome.[119,120]

5.5.1 Heart Failure

Patients with chronic heart failure may develop cachexia, but even in the early stages of the disease, patients with heart failure lose muscle bulk,[121] as a con-sequence of both the disease process and physical inactivity.[122] Reduced bone mass is common in heart failure,[123] perhaps related to low levels of serum vitamin D and secondary hyperparathyroidism.[124] How does this wasting process come about?

Sympathetic and renin–angiotensin system activation: Chronic heart fail-ure is *par excellence* the condition characterized by neurohormonal activation. There is activation of the sympathetic nervous system[125,126] and renin–angiotensin system.[127,128] This neurohormonal activation is presumed to occur in response to a fall in tissue perfusion in an attempt to maintain blood pressure and nutrient supply to the tissues. Hyper-reninaemia results in elevated circulating concentrations of stress hormones such as angiotensin II, aldosterone, and noradrenaline.[128–130]

Neurohormonal activation follows myocardial damage.[131,132] The increase in plasma levels of the hormones further increase with the severity of heart failure,[132] and they have important prognostic significance in both asymptomatic left ventricular dysfunction[133–135] and overt heart failure.[131,136]

Thyroid function: Patients with chronic heart failure frequently have abnormal thyroid hormone metabolism, with the so-called sick euthyroid syndrome.[137,138] T4 levels are low with an increase in rT3 levels in the presence of normal levels of TSH. How this happens is unclear, but may involve an inhibitor of de-iodination from T4 to T3.[139] The degree of abnormality seems to be related to the degree of left ventricular dysfunction,[137] and there is some evidence to suggest that treatment with T3 might improve cardiac performance.[140,141]

Insulin resistance: As noted above, insulin is a major anabolic hormone. The insulin resistance syndrome commonly researched and discussed in cardiovascular medicine is that associated with obesity. However, we have shown that insulin resistance is common in heart failure patients (see Figure 5.6).[142,143] This insulin

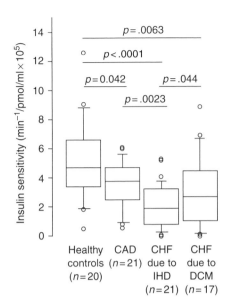

FIGURE 5.6 Insulin sensitivity in patients with chronic heart failure. CAD is coronary artery disease; CHF is chronic heart failure; IHD is ischaemic heart disease; and DCM is dilated cardiomyopathy (Taken from Swan J.W., et al., *J. Am. Coll. Cardiol.* 1997; 30: 527–32. With permission.).

resistance is characterized by high circulating levels of insulin, and normal fasting glucose. Insulin levels appear to be higher in heart failure patients with cachexia than in those without,[144] and body mass index is related to the degree of insulin resistance.[145] However, insulin resistance is less clearly related to the presence of cachexia when defined as an active process of weight loss.[146]

Insulin resistance may be implicated in other forms of cachexia. For example, in tumor bearing rats, cachexia is staved off by exercise: the improvement in skeletal muscle protein balance is associated with increased insulin sensitivity.[147] Insulin resistance is particularly associated with pancreatic cancer even in the absence of pancreatic destruction.[148] In this condition, the insulin resistance is related to pancreatic production of islet amyloid polypeptide, a hormonal factor secreted from the pancreatic beta cells, that reduces insulin sensitivity.[149] Cachexia in liver cirrhosis is associated with decreased insulin sensitivity and high levels of circulating insulin,[150] as well as with low levels of IGFs.

The origin of insulin resistance is unclear. There is evidence to suggest that tumor necrosis factor α (TNFα) might induce insulin resistance in a variety of conditions including obesity[151–153] and cancer.[154] In mice, inhibition of TNFα promotes insulin sensitivity.[151,155] The mechanism appears to be through TNFα induced phosphorylation of the intracellular insulin receptor substrate, thus reducing insulin action.[156]

Steroid metabolism: The catabolic steroid cortisol can be grossly elevated in untreated patients with chronic heart failure.[128] In treated patients, cortisol was

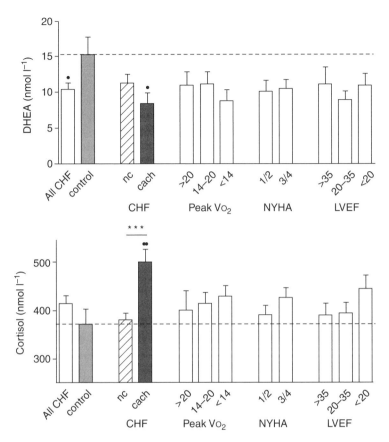

FIGURE 5.7   DHEA and cortisol levels in patients with chronic heart failure. Patients are divided in different ways. nc is noncachectic, cach is cachectic. Peak V_{O_2} is peak oxygen consumption, measured in $ml\,kg^{-1}\,min^{-1}$. NYHA is New York Heart Association symptom classification, and LVEF is left ventricular ejection fraction. Note that the presence of cachexia is the best discriminator between low and normal cortisol levels (Taken from Anker S.D., et al., *Circulation* 1997; 96: 526–34. With permission.).

only elevated in those with cachexia, where cachexia was defined as an active process of weight loss.[144] In this study of 37 noncachectic and 16 cachectic patients with chronic heart failure and no peripheral oedema, cortisol levels were normal in patients without cachexia, and were a third higher than normal in those with cachexia.

By contrast, the anabolic steroid precursor, DHEA was low in both groups of heart failure patients, more strikingly so in the cachectic group (see Figure 5.7).[144] We have also described a rise in the *ratio* of cortisol to DHEA in chronic heart failure patients.[157] The ratio correlates negatively with body mass index (BMI) (the higher the BMI, the lower the ratio of catabolic to anabolic steroid). In a study of 55 patients with chronic heart failure, we again found an increase in the cortisol : DHEA ratio, and the ratio correlated with reduced muscle and fat tissue content and reduced bone mass.[158] Other forms of cachexia are also associated with alteration in cortisol: DHEA ratio, such as that associated with tuberculosis.[159] Cortisol levels are high in weight losing cancer patients.[160, 161]

We have not found any relation between sex hormone levels and cachexia in heart failure.[157] Others have shown that testosterone is lower in patients with lung cancer who develop weight loss,[162] and in male patients with other forms of cancer.[163] Decreased levels of testosterone have also been seen in patients with chronic lung disease.[164]

As with insulin resistance, there is a relation between the cortisol: DHEA ratio and levels of circulating TNFα.[157] Testosterone production can be inhibited by TNFα production.[165, 166]

Growth hormone — IGF-1 axis: Growth hormone levels can also be strikingly elevated in untreated heart failure patients.[128] However, it should be borne in mind that in health, most growth hormone secretion is nocturnal, and others have reported decreased growth hormone secretion by studying average nocturnal levels.[167] Single day time readings may be of little value, so what follows should be interpreted with caution. We have shown, however, that overnight GH production was raised in heart failure,[168] and relates to day time measures.

High levels of growth hormone can coexist with inappropriately normal levels of IGF-1,[144] suggesting that growth hormone resistance may be present (Figure 5.8; Table 5.2). We found further evidence of growth hormone resistance in a group of 52 heart failure patients.[169] Those with low levels of IGF-1 had very much higher levels of growth hormone, and this implied growth hormone resistance was associated with decreased muscle bulk and strength. Interestingly, the high growth hormone: IGF-1 was associated with an increased cortisol: DHEA ratio, increased catecholamine levels, and TNFα level.[169]

Growth hormone binding protein itself may have a role. Low levels suggest low levels of GH receptor (as the binding protein is shed cellular receptor). Growth hormone binding protein (GHBP) appears to be low in those heart failure patients with cachexia,[168] and high levels of GHBP predict a beneficial response to exogenous growth hormone when used for treatment of heart failure.[170] In addition, while GH levels are high in cachexia with low GHBP, IGF-1 and IGFBP-3 were also low.

However, other investigators have found that *free* IGF-1 levels may be high in heart failure. The explanation is that the major IGF carrier protein, IGF-binding protein (IGFBP)-3 is low; although total IGF-1 is also low, the free hormone may be elevated.[171] The low total IGF-1 was not seen in patients receiving angiotensin converting enzyme inhibitors, suggesting that the renin–angiotensin system may have a role in suppressing IGF-1.[171] This is supported by the finding that in a rat model of cachexia, the fall in IGF-1 apparently induced by Ang II infusion was blocked by the Ang receptor blocker, losartan.[97] However, patients in this study[171] had presented acute heart failure, and did not have cachexia. Others have reported low IGF-1 levels in association with normal IGFBP-3.[172]

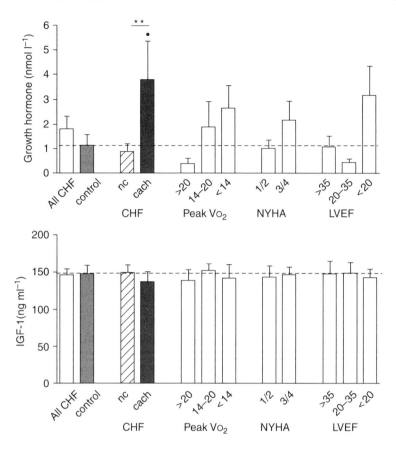

FIGURE 5.8   Growth hormone and IGF-1 levels in patients with chronic heart failure. Abbreviations as in Figure 5.6 (Taken from Anker S.D., et al., *Circulation* 1997; 96: 526–34. With permission.).

TABLE 5.2
Possible Sites of GH Resistance

Decreased GFBP	Reflects decreased GH receptors
Reduced IGF-1 production in response to GH	Abnormal response to high GH level
Reduced IGFBP-3	Less circulating IGF-1
Abnormal IGFBP-3	Less circulating IGF-1
Increased IGFBP-1	Inhibitory IGFBP

Other cachexia syndromes are associated with decreased IGF-1, such as burns.[173] Other conditions associated with growth hormone resistance include cirrhosis,[174] HIV infection,[175] chronic pulmonary disease,[176] and cancer.[177]

That altered IGF-1 metabolism may be related to cachexia is supported by the observation that its administration to rats with various cachexia syndromes attenuates protein loss.[178,179] IGF-1 administration can reduce protein oxidation in human burns patients.[180] However, the protein sparing effect of IGF-1 is not seen in the Ang II infusion model,[98] suggesting that there may be an interaction between Ang II and IGF-1 at tissue level. Resistance to IGF-1 effects is also implied by the inability of patients with renal failure to respond to IGF-1.[181]

A further point at which IGF-1 activity can be modulated is via the IGFBPs. The most important in health is IGFBP-3, and most IGF-1 circulates in a ternary complex with IGFBP-3 and an acid-labile subunit.[182] In the rat Ang II infusion model, IGFBP-3 is reduced and IGFBP-2 increased by Ang II.[97] In humans, IGFBP-3 falls in some wasting conditions.[183] Furthermore, the IGFBP-3 molecule itself can be abnormal in cachexia, thus binding less IGF-1.[184] IGFBP-1 inhibits IGF-1 activity[185] and is over-expressed in some wasting conditions.[186]

The origin of IGF resistance, at whatever level, is uncertain, but as noted above, activation of the renin–angiotensin system is a prime suspect. Additionally, there is an interaction between the GH-IGF-1 axis and TNFα. TNFα interferes with the GH-stimulated production of IGF-1[187] and can block IGF-1 stimulated protein synthesis.[188]

5.6 NEUROHORMONAL BALANCE AND CACHEXIA: CONCLUDING REMARKS

Cells are bathed in the interstitial fluid and are bombarded with multiple, often conflicting, signals from neurotransmitters and hormones. The signals are in balance: in response to stress from the environment, the balance may shift one way or the other, but then returns to equilibrium. For example, acute stress, such as the need to run away, leads to a transient switch to catabolism to mobilize resources which reverts to balance when the stress is removed; following a heavy meal and rest, anabolism temporarily predominates.

The need we have to measure things scientifically leads to (partial) neglect of the obvious fact that in this model of cellular life, there are multiple simultaneous influences, and many of the neurohormones interact with one another. Thus, for example, activation of the renin–angiotensin system leads to GH/IGF imbalance; DHEA and IGF-1 interact with each other to enhance their activity. T4 is required for optimal behavior of other hormone systems.

When a catabolic illness starts, there is usually an initial stress to which the body responds by mobilizing (catabolic) resources. The stress, however, is not removed, and so the catabolic state persists. Rather than a loss of anabolic hormones, there is the development of resistance to anabolic hormones. There is no discrete loss of any one system that is "the cause" of catabolism; rather there is a general shift away from the anabolic to the catabolic (Figure 5.9).

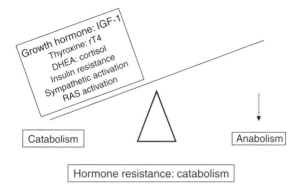

FIGURE 5.9 Hormonal imbalance in cachexia. IGF-1 is insulin-like growth factor, rT4 is reverse T4 and DHEA is dehydroepiandrosterone, RAS is reninangiotensin system. Compare with Figure 5.5.

The interaction with the immune system is of great importance. TNFα in particular is associated with abnormalities of many of the catabolic: anabolic pairs, and the role of the cytokine systems in the establishment of cachexia will be considered in other chapter.

REFERENCES

1. Tu H.M., Legradi G., Bartha T., Salvatore D., Lechan R.M., and Larsen P.R. Regional expression of the type 3 iodothyronine deiodinase messenger ribonucleic acid in the rat central nervous system and its regulation by thyroid hormone. *Endocrinology* 1999; 140: 784–90.
2. Koenig R.J. Thyroid hormone receptor coactivators and corepressors. *Thyroid* 1998; 8: 703–13.
3. Gustafson T.A., Markham B.E., Bahl J.J., and Morkin E. Thyroid hormone regulates expression of a transfected alpha-myosin heavy-chain fusion gene in fetal heart cells. *Proc. Natl Acad. Sci. USA* 1987; 84: 3122–6.
4. Horowitz B., Hensley C.B., Quintero M., Azuma K.K., Putnam D., and McDonough A.A. Differential regulation of Na,K-ATPase alpha 1, alpha 2, and beta subunit mRNA and protein levels by thyroid hormone. *J. Biol. Chem.* 1990; 265: 14308–14.
5. Rapiejko P.J., Watkins D.C., Ros M., and Malbon C.C. Thyroid hormones regulate G-protein beta-subunit mRNA expression *in vivo. J. Biol. Chem.* 1989; 264: 16183–9.
6. Kupfer J.M. and Rubin S.A. Differential regulation of insulin-like growth factor I by growth hormone and thyroid hormone in the heart of juvenile hypophysectomized rats. *J. Mol. Cell. Cardiol.* 1992; 24: 631–9.
7. Magnus-Levi A. Uber den respiratorischen Gaswechsel unter dem Einfluss des Thyroidea sowie unter verschiedenen pathologishen Zuständen. *Berl Klin Wochenschr* 1895; 34: 650.
8. Winkler A.W., Criscuolo J., and Lavietes P.H. Quantitative relationship between basal metabolic rate and thyroid dosage in patients with true myxedema. *J. Clin. Invest.* 1943; 22: 531–4.

9. Shulman G.I., Ladenson P.W., Wolfe M.H., Ridgway E.C., and Wolfe R.R. Substrate cycling between gluconeogenesis and glycolysis in euthyroid, hypothyroid, and hyperthyroid man. *J. Clin. Invest.* 1985; 76: 757–64.

10. Silva J.E. Thyroid hormone control of thermogenesis and energy balance. *Thyroid* 1995; 5: 481–92.

11. Freake H.C. and Oppenheimer J.H. Thermogenesis and thyroid function. *Annu. Rev. Nutr.* 1995; 15: 263–91.

12. Lebon V., Dufour S., Petersen K.F., Ren J., Jucker B.M., Slezak L.A., Cline G.W., Rothman D.L., and Shulman G.I. Effect of triiodothyronine on mitochondrial energy coupling in human skeletal muscle. *J. Clin. Invest.* 2001; 108: 733–7.

13. Gong D.W., He Y., Karas M., and Reitman M. Uncoupling protein-3 is a mediator of thermogenesis regulated by thyroid hormone, beta3-adrenergic agonists, and leptin. *J. Biol. Chem.* 1997; 272: 24129–32.

14. Bratusch-Marrain P., Schmid P., Waldhausl W., and Schlick W. Specific weight loss in hyperthyroidism. *Horm. Metab. Res.* 1978; 10: 412–5.

15. Hoogwerf B.J. and Nuttall F.Q. Long-term weight regulation in treated hyperthyroid and hypothyroid subjects. *Am. J. Med.* 1984; 76: 963–70.

16. Grover G.J., Mellstrom K., Ye L., Malm J., Li Y.L., Bladh L.G., Sleph P.G., Smith M.A., George R., Vennstrom B., Mookhtiar K., Horvath R., Speelman J., Egan D., and Baxter J.D. Selective thyroid hormone receptor-beta activation: a strategy for reduction of weight, cholesterol, and lipoprotein (a) with reduced cardiovascular liability. *Proc. Natl Acad. Sci. USA* 2003; 100: 10067–72.

17. Seppel T., Kosel A., and Schlaghecke R. Bioelectrical impedance assessment of body composition in thyroid disease. *Eur. J. Endocrinol.* 1997; 136: 493–8.

18. Svanberg E., Healey J., and Mascarenhas D. Anabolic effects of rhIGF-I/IGFBP-3 in vivo are influenced by thyroid status. *Eur. J. Clin. Invest.* 2001; 31: 329–36.

19. Samuels M.H., Wierman M.E., Wang C., and Ridgway E.C. The effect of altered thyroid status on pituitary hormone messenger ribonucleic acid concentrations in the rat. *Endocrinology* 1989; 124: 2277–82.

20. Nanto-Salonen K., Muller H.L., Hoffman A.R., Vu T.H., and Rosenfeld R.G. Mechanisms of thyroid hormone action on the insulin-like growth factor system: all thyroid hormone effects are not growth hormone mediated. *Endocrinology* 1993; 132: 781–8.

21. De Boer H., Blok G.-J., and Van der Keen E.A. Clinical aspects of growth hormone deficiency in adults. *Endocrine Rev.* 1995; 16: 63–86.

22. Bengtsson B.Å., Eden S., Lönn L, Kvist H., Stockland A., Lindstedt G., Bosaeus I., Tolli J., Sjostrom L., and Isaksson O.G. Treatment of adults with growth hormone deficiency with recombinant human growth hormone. *J. Clin. Endocrinol. Metab.* 1993; 76: 309–17.

23. Frustaci A., Gentiloni N., Corsello S.M., Caldarulo M., and Russo M.A. Reversible dilated cardiomyopathy due to growth hormone deficiency. *Chest* 1992; 102: 326–7.

24. Matthews L.S., Norstadt G., and Palmirez R.D. Regulation of insulin-like growth factor-1 gene expression by growth hormone. *Proc. Natl Acad. Sci. USA* 1986; 83: 9343–7.

25. Underwood L.E. and Van Wyk J.J. Normal and aberrant growth. In: Wilson J.D., Foster D.W., Editors. *Williams Textbook of Endocrinology*. Philadelphia: WB Saunders 1992; pp. 1079–138.

26. Baumann G. Growth hormone binding protein 2001. *J. Pediatr. Endocrinol. Metab.* 2001; 14: 355–75.
27. Murphy L.J. Insulin-like growth factor-binding proteins: functional diversity or redundancy? *J. Mol. Endocrinol.* 1998; 21: 97–107.
28. Russell-Jones D.L., Umpleby A.M., Hennessy T.R., Bowes S.B., Shojaee-Moradie F., Hopkins K.D., Jackson N.C., Kelly J.M., Jones R.H., and Sonksen P.H. Use of a leucine clamp to demonstrate that IGF-I actively stimulates protein synthesis in normal humans. *Am. J. Physiol.* 1994; 267: E591–8.
29. Fryburg D.A., Jahn L.A., Hill S.A., Oliveras D.M., and Barrett E.J. Insulin and insulin-like growth factor-I enhance human skeletal muscle protein anabolism during hyperaminoacidemia by different mechanisms. *J. Clin. Invest.* 1995; 96: 1722–9.
30. Ding H., Gao X.-L., Hirschberg R., Vadgama J.V., and Kopple J.D. Impaired actions of insulin-like growth factor 1 on protein synthesis and degradation in skeletal muscle of rats with chronic renal failure. *J. Clin. Invest.* 1996; 97: 1064–75.
31. Husmann I., Soulet L., Gautron J., Martelly I., and Barritault D. Growth factors in skeletal muscle regeneration. *Cytokine Growth Factor Rev.* 1996; 7: 249–58.
32. Kanda F., Takatani K., Okuda S., Matsushita T., and Chihara K. Preventive effects of insulin-like growth factor-I on steroid-induced muscle atrophy. *Muscle Nerve* 1999; 22: 213–7.
33. Wu Y., Tewari M., Cui S., and Rubin R. Activation of the insulin-like growth factor-I receptor inhibits tumour necrosis factor-induced cell death. *J. Cell. Physiol.* 1996; 168: 499–509.
34. Wang L., Ma W., Markovich R., Chen J.W., and Wang P.H. Regulation of cardiomyocyte apoptotic signaling by insulin-like growth factor I. *Circ. Res.* 1998; 83: 516–22.
35. Resnicoff M., Abraham D., Yutanawiboonchai W., Rotman H.L., Kajstura J., Rubin R., Zoltick P., and Baserga R. The insulin-like growth factor receptor protects tumor cells from apoptosis in vivo. *Cancer Res.* 1995; 55: 2463–9.
36. Resnicoff M., Burgaud J.L., Rotman H.L., Abraham D., and Baserga R. Correlation between apoptosis, tumorigenesis, and levels of insulin-factor I receptors. *Cancer Res.* 1995; 55: 3739–41.
37. Rosen T., Bosaeus I., Tolli J., Lindstedt G., and Bengtsson B.A. Increased body fat mass and decreased extracellular fluid volume in adults with growth hormone deficiency. *Clin. Endocrinol.* 1993; 38: 63–71.
38. Wuster C., Abs R., Bengtsson B.A., Bennmarker H., Feldt-Rasmussen U., Hernberg-Stahl E., Monson J.P., Westberg B., and Wilton P; KIMS Study Group and the KIMS International Board. Pharmacia & Upjohn International Metabolic Database. The influence of GH deficiency, GH replacement therapy and other aspects of hypopituitarism on fracture rate and bone mineral density. *J. Bone Miner. Res.* 2001; 16: 398–405.
39. Amato G., Carella C., Fazio S., La Montagne G., Cittadini A., Sabatini D., Marciano-Mone C., Sacca L., and Bellastella A. Body composition, bone metabolism and heart structure and function in growth hormone (GH) deficient adults before and after GH replacement therapy at low doses. *J. Clin. Endocrinol. Metab.* 1993; 77: 1671–6.
40. Weaver J.U., Monson J.P., Noonan K., John W.G., Edwards A., Evans K.A., and Cunningham J.C. The effect of low dose recombinant human growth hormone

replacement on regional fat distribution and cardiovascular risk factors in hypopituitary adults. *J. Clin. Endocrinol. Metab.* 1995; 80: 153–9.

41. Weaver J.U., Monson J.P., Noonan K., Price C., Edwards A., Evans K.A., James I., and Cunningham J. The effects of low dose recombinant human growth hormone replacement on indices of bone remodelling and bone mineral density in hypopituitary growth hormone deficient adults. *Endocrinol. Metab.* 1996; 3: 55–61.

42. Johansson G., Rosén T., Bosaeus I., Sjoström L., and Bengtsson B.-Å. Two years of growth hormone GH treatment increases bone mineral content and density in hypopituitary patients with adult onset GH deficiency. *J. Clin. Endocrinol. Metab.* 1996; 81: 2865–73.

43. Carey R.M. The changing clinical spectrum of adrenal insufficiency. *Ann. Intern. Med.* 1997; 127: 1103–5.

44. Blaustein S.A., Golden N.H., and Shenker I.R. Addison's disease mimicking anorexia nervosa. *Clin. Pediatr. (Phila.)* 1998; 37: 631–2.

45. Orth D.N. Cushing's syndrome. *N. Engl. J. Med.* 1995; 332: 791–803.

46. Waterman M.R. and Bischof L.J. Cytochromes P450 12: diversity of ACTH (cAMP)-dependent transcription of bovine steroid hydroxylase genes. *FASEB J.* 1997; 11: 419–27.

47. Karin M. New twists in gene regulation by glucocorticoid receptor: is DNA binding dispensable? *Cell* 1998; 93: 487–90.

48. Beato M. Gene regulation by steroid hormones. *Cell* 56: 335–44.

49. Auphan N., Didonato J.A., Rosette C., Helmberg A., and Karin M. Immunosuppression by glucocorticosteroids: inhibition of NF-κB activity through induction of IκBα synthesis. *Science* 1995; 270: 286–90.

50. Lenardo M.J. and Baltimore D. NF kappa B : a pleiotropic mediator of inducible and tissue specific gene control. *Cell* 1989; 58: 227–9.

51. Barnes P.J. and Karin M. Nuclear factor-κB — a pivotal transcription factor in chronic inflammatory disease. *N. Engl. J. Med.* 1997; 336: 1066–71.

52. Khani S. and Tayek J.A. Cortisol increases gluconeogenesis in humans: its role in the metabolic syndrome. *Clin. Sci.* 2001; 101: 739–47.

53. Brillon D.J., Zheng B., Campbell R.G., and Matthews D.E. Effect of cortisol on energy expenditure and amino acid metabolism in humans. *Am. J. Physiol.* 1995; 268: E501–13.

54. Auclair D., Garrel D.R., Chaouki Zerouala A., and Ferland L.H. Activation of the ubiquitin pathway in rat skeletal muscle by catabolic doses of glucocorticoids. *Am. J. Physiol.* 1997; 272: C1007–16.

55. Tiao G., Fagan J., Roegner V., Lieberman M., Wang J.J., Fischer J.E., and Hasselgren P.O. Energy-ubiquitin-dependent muscle proteolysis during sepsis in rats is regulated by glucocorticoids. *J. Clin. Invest.* 1996; 97: 339–48.

56. LaPier T.K. Glucocorticoid-induced muscle atrophy. The role of exercise in treatment and prevention. *J. Cardiopulm. Rehabil.* 1997; 17: 76–84.

57. Hickson R.C., Czerwinski S.M., Falduto M.T., and Young A.P. Glucocorticoid antagonism by exercise and androgenic-anabolic steroids. *Med. Sci. Sports Exerc.* 1990; 22: 331–40.

58. Pirlich M., Biering H., Gerl H., Ventz M., Schmidt B., Ertl S., and Lochs H. Loss of body cell mass in Cushing's syndrome: effect of treatment. *J. Clin. Endocrinol. Metab.* 2002; 87: 1078–84.

59. Gribbin H. and Flavell Matts S. Mode of action and use of anabolic steroids. *Br. J. Clin. Pract.* 1976; 30: 3–9.

60. Bhasin S., Storer T.W., Berman N., Yarasheski K.E., Clevenger B., Phillips J., Lee W.P., Bunnell T.J., and Casaburi R. Testosterone replacement increases fat-free mass and muscle size in hypogonadal men. *J. Clin. Endocrinol. Metab.* 1997; 82: 407–13.

61. Martinez J.A., Buttery P.J., and Pearson J.T. The mode of action of anabolic agents: the effect of testosterone on muscle protein metabolism in the female rat. *Br. J. Nutr.* 1984; 52: 515–21.

62. Michelsen C.B., Askanazi J., Kinney J.M., Gump F.E., and Elwyn D.H. Effect of an anabolic steroid on nitrogen balance and amino acid patterns after total hip replacement. *J. Trauma* 1982; 22: 410–13.

63. Gruenewald D.A. and Matsumoto A.M. Testosterone supplementation therapy for older men: potential benefits and risks. *J. Am. Geriatr. Soc.* 2003; 51: 101–15.

64. Himsworth H.P. Diabetes melitus. Its differentiation into insulin-sensitive and insulin-insensitive types. *Lancet* 1936; I: 127–30.

65. Kasuga M., Karlsson F.A., and Kahn C.R. Insulin stimulates the phosphorylation of the 95,000-dalton subunit of its own receptor. *Science* 1982; 215: 185–7.

66. Petruzzelli L.M., Gangaly S., Smith C.R., Cobb M.H., Rubin C.S., and Rosen O.M. Insulin activates a tyrosine specific protein kinase in extracts of 3T3-LI adipocytes and human placenta. *Proc. Natl Acad. Sci. USA* 1982; 79: 6792–6.

67. Meisner H. and Carter J. Jr. Regulation of lipolysis in adipose tissue. *Horiz. Biochem. Biophys.* 1977; 4: 91–103.

68. Prins J.B. and O'Rahilly S. Regulation of adipose cell number in man. *Clin. Sci.* 1997; 92: 3–11.

69. Geloen A., Collet A.J., Guay G., and Bukowiecki L.J. Insulin stimulates in vivo cell proliferation in white adipose tissue. *Am. J. Physiol.* 1989; 256: C190–6.

70. Fukagawa N.K., Minaker K.L., Rowe J.W., Goodman M.N., Matthews D.E., Bier D.M., and Young V.R. Insulin-mediated reduction of whole-body protein breakdown. Dose–response effects on leucine metabolism in postabsorptive men. *J. Clin. Invest.* 1985; 76: 2306–11.

71. Flakoll P.J., Kulaylat M., Frexes-Steed M., Hourani H., Brown L.L., Hill J.O., and Abumrad N.N. Amino acids augment insulin's suppression of whole body proteolysis. *Am. J. Physiol.* 1989; 257: E839–47.

72. Bonadonna R.C., Saccomani M.P., Cobelli C., and DeFronzo R.A. Effect of insulin on system A amino acid transport in human skeletal muscle. *J. Clin. Invest.* 1993; 91: 514–21.

73. Hill P.A., Tumber A., and Meikle M.C. Multiple extracellular signals promote osteoblast survival and apoptosis. *Endocrinology* 1997; 138: 3849–58.

74. Frankton S. and Bloom S.R. Glucagonomas. *Ballières Clin. Gastroenterol.* 1996; 10: 697–707.

75. Chastain M.A. The glucagonoma syndrome: a review of its features and discussion of new perspectives. *Am. J. Med. Sci.* 2001; 321: 306–20.

76. Webber J. and Macdonald I.A. Metabolic actions of catecholamines in man. *Baillieres Clin. Endocrinol. Metab.* 1993; 7: 393–413.

77. Lafontan M. and Berlan M. Fat cell adrenergic receptors and the control of white and brown fat cell function. *J. Lipid Res.* 1993; 34: 1057–91.

78. Staten M.A., Matthews D.E., Cryer P.E., and Bier D.M. Physiological increments in epinephrine stimulate metabolic rate in humans. *Am. J. Physiol.* 1987; 253: E322–30.

79. Simonsen L., Bulow J., Madsen J., and Christensen, N.J. Thermogenic response to epinephrine in the forearm and abdominal subcutaneous adipose tissue. *Am. J. Physiol.* 1992; 263: E850–5.

80. Blaak E.E., van Baak M.A., Kempen K.P., and Saris W.H. Role of alpha- and beta-adrenoceptors in the sympathetically mediated thermogenesis. *Am. J. Physiol.* 1993; 264: E11–17.

81. Monroe M.B., Seals D.R., Shapiro L.F., Bell C., Johnson D., and Parker Jones P. Direct evidence for tonic sympathetic support for resting metabolic rate in healthy adult humans. *Am. J. Physiol.* 2001; 280: E740–4.

82. Rossner S., Taylor C.L., Byington R.P., and Furberg C.D. Long term propranolol treatment and changes in body weight after myocardial infarction. *Br. Med. J.* 1990; 300: 902–3.

83. Sharma A.M., Pischon T., Hardt S., Kunz I., and Luft F.C. Hypothesis. Beta-adrenergic receptor blockers and weight gain: a systematic analysis. *Hypertension* 2001; 37: 250–4.

84. Saad M.F., Alger S.A., Zurlo F., Young J.B., Bogardus C., and Ravussin E. Ethnic differences in sympathetic nervous system-mediated energy expenditure. *Am. J. Physiol.* 1991; 261: E789–94.

85. Spraul M., Ravussin E., Fontvieille A.M., Rising R., Larson D.E., and Anderson E.A. Reduced sympathetic nervous activity. A potential mechanism predisposing to body weight gain. *J. Clin. Invest.* 1993; 92: 1730–5.

86. Muntzel M.S., Morgan D.A., Mark A.L., and Johnson A.K. Intracerebroventricular insulin produces nonuniform regional increases in sympathetic nerve activity. *Am. J. Physiol.* 1994; 267: R1350–5.

87. Mauriège P., Despres J.P., Prud'homme D., Pouliot M.C., Marcotte M., Tremblay A., and Bouchard C. Regional variation in adipose tissue lipolysis in lean and obese men. *J. Lipid. Res.* 1991; 32: 1625–33.

88. Mauriège P., Prud'homme D., Lemieux S., Tremblay A., and Despres J.P. Regional differences in adipose tissue lipolysis from lean and obese women: existence of postreceptor alterations. *Am. J. Physiol.* 1995; 269: E341–50.

89. Seidell J.C., Muller D.C., Sorkin J.D., and Andres R. Fasting respiratory exchange ratio and resting metabolic rate as predictors of weight gain: the Baltimore Longitudinal Study on Aging. *Int. J. Obes. Relat. Metab. Disord.* 1992; 16: 667–74.

90. Zurlo F., Lillioja S., Esposito-Del Puente A., Nyomba B.L., Raz I., Saad M.F., Swinburn B.A., Knowler W.C., Bogardus C., and Ravussin E. Low ratio of fat to carbohydrate oxidation as predictor of weight gain: study of 24-h RQ. *Am. J. Physiol.* 1990; 259: E650–7.

91. Snitker S., Tataranni P.A., and Ravussin E. Respiratory quotient is inversely associated with muscle sympathetic nerve activity. *J. Clin. Endocrinol. Metab.* 1998; 83: 3977–9.

92. Kim Y.S. and Sainz R.D. β-Adrenergic agonists and hypertrophy of skeletal muscles. *Life Sci.* 1991; 50: 397–407.

93. Reeds P.J., Hay S.M., Dorwood P.M., and Palmer R.M. Stimulation of muscle growth by clenbuterol: lack of effect on muscle protein biosynthesis. *Br. J. Nutr.* 1986; 56: 249–58.

94. Shamoon H., Jacob R., and Sherwin R.S. Epinephrine-induced hypoaminoacidemia in normal and diabetic human subjects. Effect of beta blockade. *Diabetes* 1980; 29: 875–81.

95. Navegantes L.C., Resano N.M., Migliorini R.H., and Kettelhut I.C. Catecholamines inhibit Ca(2+)-dependent proteolysis in rat skeletal muscle through beta(2)-adrenoceptors and cAMP. *Am. J. Physiol. Endocrinol. Metab.* 2001; 281: E449–54.

96. Navegantes L.C., Resano N.M., Migliorini R.H., and Kettelhut I.C. Effect of guanethidine-induced adrenergic blockade on the different proteolytic systems in rat skeletal muscle. *Am. J. Physiol.* 1999; 277: E883–9.

97. Brink M., Wellen J., and Delafontaine P. Angiotensin II causes weight loss and decreases circulating insulin-like growth factor I in rats through a pressor-independent mechanism. *J. Clin. Invest.* 1996; 97: 2509–16.

98. Brink M., Price S.R., Chrast J., Bailey J.L., Anwar A., Mitch W.E., and Delafontaine P. Angiotensin II induces skeletal muscle wasting through enhanced protein degradation and down-regulates autocrine insulin-like growth factor I. *Endocrinology* 2001; 142: 1489–96.

99. Cassis L.A., Marshall D.E., Fettinger M.J., Rosenbluth B., and Lodder R.A. Mechanisms contributing to angiotensin II regulation of body weight. *Am. J. Physiol.* 1998; 274: E867–76.

100. Pedersen E.B., Danielsen H., Jensen T., Madsen M., Sorensen S.S., and Thomsen O.O. Angiotensin II, aldosterone and arginine vasopressin in plasma in congestive heart failure. *Eur J. Clin. Invest.* 1986; 16: 56–60.

101. Staroukine M., Devriendt J., Decoodt P., and Verniory A. Relationships between plasma epinephrine, norepinephrine, dopamine and angiotensin II concentrations, renin activity, hemodynamic state and prognosis in acute heart failure. *Acta Cardiol.* 1984; 39: 131–8.

102. Adigun A.Q. and Ajayi A.A. The effects of enalapril-digoxin-diuretic combination therapy on nutritional and anthropometric indices in chronic congestive heart failure: preliminary findings in cardiac cachexia. *Eur. J. Heart Fail.* 2001; 3: 359–63.

103. Anker S.D., Negassa A., Coats A.J., Afzal R., Poole-Wilson P.A., Cohn J.N., and Yusuf S. Prognostic importance of weight loss in chronic heart failure and the effect of treatment with angiotensin-converting-enzyme inhibitors: an observational study. *Lancet* 2003; 361: 1077–83.

104. Gelfand R.A., Hutchinson-Williams K.A., Bonde A.A., Castellino P., and Sherwin R.S. Catabolic effects of thyroid hormone excess: the contribution of adrenergic activity to hypermetabolism and protein breakdown. *Metabolism* 1987; 36: 562–9.

105. Zimmerman B., Syberte E., and Wong P. Interactions between sympathetic and renin–angiotensin system. *Hypertension* 1984; 2: 581–7.

106. Poehlman E.T., Scheffers J., Gottlieb S.S., Fisher M.L., and Vaitekevicius P. Increased metabolic rate in patients with congestive heart failure. *Ann. Intern. Med.* 1994; 121: 860–2.

107. Riley M., Elborn J.S., McKane W.R., Bell N., Stanford C.F., and Nicholls. Resting energy expenditure in chronic cardiac failure. *Clin. Sci.* 1991; 80: 633–9.

108. Lommi J., Kupari M., and Yki-Jarvinen. Free fatty acid kinetics and oxidation in congestive heart failure. *Am. J. Cardiol.* 1998; 81: 45–50.

109. Obiesan T., Toth M.J., and Kendall D. Energy expenditure and symptom severity in men with heart failure. *Am. J. Cardiol.* 1996; 77: 1250–2.

110. Schols A.M., Fredrix E.W., Soeters P.B., Westerterp K.R., and Wouters E.F. Resting energy expenditure in patients with chronic obstructive pulmonary disease. *Am. J. Clin. Nutr.* 1991; 54: 983–7.

111. Congleton J. and Muers M.F. Resting energy expenditure in cryptogenic fibrosing alveolitis. *Eur. Respi. J.* 1997; 10: 2744–8.

112. Creutzberg E.C., Schols A.M.W.J., Bothmer-Quaedvlieg F.C.M., and Wouters E.F.M. Prevalence of an elevated resting energy expenditure in patients with chronic obstructive pulmonary disease in relation to body composition and lung function. *Eur. J. Clin. Nutr.* 1998; 52: 396–401.

113. Roubenoff R., Roubenoff R.A., Ward L.M., Holland S.M., and Hellmann D.B. Rheumatoid cachexia: depletion of lean body mass in rheumatoid arthritis. Possible association with tumor necrosis factor. *J. Rheumatol.* 1992; 19: 1505–10.

114. Roubenoff R., Roubenoff R.A., Cannon J.G., Kehayias J.J., Zhuang H., Dawson-Hughes B., Dinarello C.A., and Rosenberg I.H. Rheumatoid cachexia: cytokine-driven hypermetabolism accompanying reduced body cell mass in chronic inflammation. *J. Clin. Invest.* 1994; 93: 2379–86.

115. Mulligan K., Tai V.W., and Schambelan M. Energy expenditure in human immunodeficiency virus infection. *N. Engl. J. Med.* 1997; 336: 70–71.

116. Grunfeld C., Pang M., Shimizu L., Shigenaga J.K., Jensen P., and Feingold K.R. Resting energy expenditure caloric intake and short-term weight change in human immunodeficiency virus infection and the acquired immunodeficiency syndrome. *Am. J. Clin. Nutr.* 1995; 55: 455–60.

117. Matsuda T., Clark N., Hariyani G.D., Bryant R.S., Hanumadass M.L., and Kagan R.J. The effect of burn wound size on resting energy expenditure. *J. Trauma* 1987; 27: 115–8.

118. Milner E.A., Cioffi W.G., Mason A.D., McManus W.F., and Pruitt B.A. Jr. A longitudinal study of resting energy expenditure in thermally injured patients. *J. Trauma* 1994; 37: 167–70.

119. Bessey P.Q., Watters J.M., Aoki T.T., and Wilmore D.W. Combined hormonal infusion simulates the metabolic response to injury. *Ann. Surg.* 1984; 200: 264–81.

120. Watters J.M., Bessey P.Q., Dinarello C.A., Wolff S.M., and Wilmore D.W. Both inflammatory and endocrine mediators stimulate host responses to sepsis. *Arch. Surg.* 1986; 121: 179–90.

121. Mancini D.M., Walter G., Reichnek N., Lenkinski R., McCully K.K., Mullen J.L., and Wilson J.R. Contribution of skeletal muscle atrophy to exercise intolerance and altered muscle metabolism in heart failure. *Circulation* 1992; 85: 1364–73.

122. Simonini A., Long C.E., Dudley G.A., Yue P., McElhinny J., and Massie B.M. Heart failure in rats causes changes in skeletal muscle morphology and gene expression that are not explained by reduced activity. *Circ. Res.* 1996; 79: 128–36.

123. Anker S.D., Clark A.L., Teixeira M.M., Hellewell P.G., and Coats A.J.S. Loss of bone mineral in patients with cachexia due to chronic heart failure. *Am. J. Cardiol.* 1999: 83: 612–15.

124. Shane E., Mancini D., Aaronson K., Silverberg S.J., Seibel M.J., Addesso V., and McMahon D.J. Bone mass, vitamin D deficiency, and hyperparathyroidism in congestive heart failure. *Am. J. Med.* 1997; 103: 197–207.

125. Thomas J.A. and Marks B.H. Plasma norepinephrine in congestive heart failure. *Am. J. Cardiol.* 1978; 41: 233–43.

126. Leimbach W.N., Wallin G., Victor R.G., Aylward Pe., Sundolf G., and Mark A.L. Direct evidence from intraneural recordings of increased central sympathetic outflow in patients with heart failure. *Circulation* 1986; 73: 913–19.

127. Francis G.S., Goldsmith S.R., Levine T.B., Olivari M.T., and Cohn J.N. The neurohumoral axis in congestive heart failure. *Ann. Intern. Med.* 1984; 101: 370–7.

128. Anand I.S., Ferrari R., Kalra G.S., Wahi P.L., Poole-Wilson P.A., and Harris P.C. Edema of cardiac origin. Studies of body water and sodium, renal function, hemo-dynamic indexes, and plasma hormones in untreated congestive cardiac failure. *Circulation* 1989; 80: 299–305.

129. Pederson E.B., Danielson H., Jensen T., Madsen M., and Sorenson S.S., and Thomsen O.O. Angiotensin II, aldosterone and arginine vasopressin in plasma in congestive heart failure. *Eur J. Clin. Invest.* 1986; 16: 56–60.

130. Staroukine M., Devriendt J., Decoodt P., and Verniory A. Relationships between plasma epinephrine, norepinephrine, dopamine, and angiotensin II concentrations, renin activity, hemodynamic state and prognosis in acute heart failure. *Acta Cardiol.* 1984; 39: 131–8.

131. Rouleau J.L., de Champlain J., Klein M., Bichet D., Moye L., and Packer M. Activation of neurohormonal systems in post-infarction left ventricular dysfunction. *J. Am. Coll. Cardiol.* 1993; 22: 390–8.

132. Francis G.S., Benedict C., Johnstone D.E., Kirlin P.C., Nicklas J., Liang C.S., Kubo S.H., Rudin-Toretsky E., and Yusuf S. Comparison of neuroendocrine activ-ation in patients with left ventricular dysfunction with and without congestive heart failure. *Circulation* 1990; 82: 1724–9.

133. Rouleau J.L., Packer M., Moye L., de Champlain J., Bichet D., and Klein M. Prognostic significance of neurohormonal activation in patients with an acute myocardial infarction: effect of captopril. *J. Am. Coll. Cardiol.* 1994; 24: 583–9.

134. Benedict C.R., Shelton B., Johnstone D.E., Francis G., Greenberg B., Konstam M., Probstfield J.L., and Yusuf S., for the SOLVD Investigators. Prognostic signficance of plasma norepinephrine in patients with asymptomatic left ventricular dysfunction (SOLVD). *Circulation* 1990; 82: 1724–9.

135. Vantrimpont P., Rouleau J.L., Ciampi A., Harel F., de Champlain J., Bichet D., Moye L.A., and Pfeffer M. Two-year time course and significance of neurohumoral activation in the Survival and Ventricular Enlargement (SAVE) Study. *Eur. Heart J.* 1998; 19: 1552–63.

136. Swedberg K., Eneroth P., Kjekshus J., and Wilhelmsen L., for the CONSENSUS Trial Study Group. Hormones regulating cardiovascular function in patients with severe congestive heart failure and their relation to mortality. *Circulation* 1990; 82: 1730–6.

137. Opasich C., Pacini F., Ambrosino N., Riccardi P.G., Febo O., Ferrari R., Cobelli F., and Tavazzi L. Sick euthyroid syndrome in patients with moderate-to-severe chronic heart failure. *Eur. Heart J.* 1996; 17: 1860–6.

138. Hamilton M.A., Stevenson L.W., Luu M., and Walden J.A. Altered thyroid hormone metabolism in advanced heart failure. *J. Am. Coll. Cardiol.* 1990; 16: 91–5.

139. Chopra I.J., Huang T.S., Beredo A., Solomon D.H., and Chua Teco G.N. Serum thyroid hormone binding inhibitor in nonthyroidal illnesses. *Metabolism* 1986; 35: 152–9.

140. Moruzzi P., Doria E., and Agostoni P.G. Medium-term effectiveness of L-thyroxine treatment in idiopathic dilated cardiomyopathy. *Am. J. Med.* 1996; 101: 461–7.

141. Hamilton M.A., Stevenson L.W., Fonarow G.C., Steimle A., Goldhaber J.I., Child J.S., Chopra I.J., Moriguchi J.D., and Hage A. Safety and hemodynamic effects of intravenous triiodothyronine in advanced congestive heart failure. *Am. J. Cardiol.* 1998; 81: 443–7.

142. Swan J.W., Walton C., Godsland I.F., Clark A.L., Coats A.J., and Oliver M.F. Insulin resistance in chronic heart failure. *Eur. Heart J.* 1994; 15: 1528–32.

143. Swan J.W., Anker S.D., Walton C., Godsland I.F., Clark A.L., Leyva F., Stevenson J.C., and Coats A.J. Insulin resistance in chronic heart failure: relation to severity and aetiology of heart failure. *J. Am. Coll. Cardiol.* 1997: 30: 527–32.

144. Anker S.D., Chua T.P., Ponikowski P., Harrington D., Swan J.W., Kox W.J., Poole-Wilson P.A., and Coats A.J. Hormonal changes and catabolic/anabolic imbalance in chronic heart failure and their importance for cardiac cachexia. *Circulation* 1997; 96: 526–34.

145. Sabelis L.W., Senden P.J., Zonderland M.L., van de Wiel A., Wielders J.P., Huisveld I.A., van Haeften T.W., and Mosterd W.L. Determinants of insulin sensitivity in chronic heart failure. *Eur. J. Heart Fail.* 2003; 5: 759–65.

146. Doehner W., Pflaum C.D., Rauchhaus M., Godsland I.F., Egerer K., Cicoira M., Florea V.G., Sharma R., Bolger A.P., Coats A.J., Anker S.D., and Strasburger C.J. Leptin, insulin sensitivity and growth hormone binding protein in chronic heart failure with and without cardiac cachexia. *Eur. J. Endocrinol.* 2001; 145: 727–35.

147. Daneryd P., Hafstrom L., Svanberg E., and Karlberg I. Insulin sensitivity, hormonal levels and skeletal muscle protein metabolism in tumour-bearing exercising rats. *Eur. J. Cancer* 1995; 31A: 97–103.

148. Permert J., Adrian T.E., Jacobsson P., Jorfelt L., Fruin A.B., and Larsson J. Is profound peripheral insulin resistance in patients with pancreatic cancer caused by a tumor-associated factor? *Am. J. Surg.* 1993; 165: 61–7.

149. Permert J., Larsson J., Westernark G.T., Herrington M.K., Christmanson L., Pour P.M., Westernark P., and Adrian T.E. Islet amyloid polypeptide in patients with pancreatic cancer and diabetes. *N. Engl. J. Med.* 1994; 330: 313–8.

150. Nolte W., Hartmann H., and Ramadori G. Glucose metabolism and liver cirrhosis. *Exp. Clin. Endocrinol. Diabetes* 1995; 103: 63–74.

151. Hotamisligil G.S., Shargill N.S., and Spiegelman B.M. Adipose expression of tumor necrosis factor-alpha: direct role in obesity-linked insulin resistance. *Science* 1993; 259: 87–91.

152. Hofmann C., Lorenz K., Braithwaite S.S., Colca J.R., Palazuk B.J., Hotamisligil G.S., and Spiegelman B.M. Altered gene expression for tumor necrosis factor-a and its receptors during drug and dietary modulation of insulin resistance. *Endocrinology* 1994; 134: 264–70.

153. Hotamisligil G.S. and Spiegelman B.M. Tumor necrosis factor α: a key component of the obesity-diabetes link. *Diabetes* 1994; 43: 1271–8.

154. McCall J.L., Tuckey J.A., and Parry B.R. Serum tumour necrosis factor alpha and insulin resistance in gastrointestinal cancer. *Br. J. Surg.* 1992; 79: 1361–3.

155. Hotamisligil G.S., Budavari A., Murray D., and Spiegelman B.M. Reduced tyrosine kinase activity of the insulin receptor in obesity-diabetes. Central role of tumor necrosis factor-α. *J. Clin. Invest.* 1994; 94: 1543–9.

156. Kanety H., Feinstein R., Papai M.Z., Hemi R., and Karasik A. Tumor necrosis factor a-induced phosphorylation of insulin receptor substrate-1 (IRS-1). *J. Biol. Chem.* 1995; 270: 23780–4.

157. Anker S.D., Clark A.L., Kemp M., Salsbury C., Teixeira M.M., Hellewell P.G., and Coats A.J. Tumour necrosis factor and steroid metabolism in chronic heart failure: possible relation to muscle wasting. *J. Am. Coll. Cardiol.* 1997; 30: 997–1001.

158. Anker S.D., Ponikowski P.P., Clark A.L., Leyva F., Rauchhaus M., Kemp M., Teixeira M.M., Hellewell P.G., Hooper J., Poole-Wilson P.A., and Coats A.J. Cytokines and neurohormones relating to body composition alterations in the wasting syndrome of chronic heart failure. *Eur. Heart J.* 1999; 20: 683–93.

159. Rook G.A.W., Honour J., Kon O.M., Wilkinson R.J., Davidson R., and Shaw R.J. Urinary adrenal steroid metabolites in tuberculosis — a new clue to pathogenesis? *Q. J. Med.* 1996; 89: 333–41.

160. Burt M.E., Aoki T.T., Gorschboth C.M., and Brennan M.F. Peripheral tissue metabolism in cancer-bearing man. *Ann. Surg.* 1983; 198: 685–91.

161. Knapp M.L., Al-Sheibani S., Riches P.G., Hanham I.W.F., and Phillips R.H. Hormonal factors associated with weight loss in patients with advanced breast cancer. *Ann. Clin. Biochem.* 1991; 28: 480–6.

162. Simons J.P., Schols A.M., Buurman W.A., and Wouters E.F. Weight loss and low body cell mass in males with lung cancer: relationship with systemic inflammation, acute-phase response, resting energy expenditure, and catabolic and anabolic hormones. *Clin. Sci.* 1999; 97: 215–23.

163. Todd B.D. Pancreatic carcinoma and low serum testosterone; a correlation secondary to cancer cachexia? *Eur. J. Surg. Oncol.* 1988; 14: 199–202.

164. Kamischke A., Kemper D.E., Castel M.A., Luthke M., Rolf C., Behre H.M., Magnussen H., and Nieschlag E. Testosterone levels in men with chronic obstructive pulmonary disease with or without glucocorticoid therapy. *Eur. Respir. J.* 1998; 11: 41–5.

165. Mealy K., Robinson B., Millette C.F., Majzoub J., and Wilmore D.W. The testicular effects of tumor necrosis factor. *Ann. Surg.* 1990; 211: 470–5.

166. van der Poll T., Romijn J.A., Endert E., and Sauerwein H.P. Effects of tumor necrosis factor on the hypothalamic-pituitary-testicular axis in healthy men. *Metab. Clin. Exp.* 1993; 42: 303–7.

167. Giustina A., Lorusso R., Borghetti V., Bugari G., Misitano V., and Alfieri O. Impaired spontaneous growth hormone secretion in patients with severe dilated cardiomyopathy. *Am. Heart J.* 1996; 131: 620–2.

168. Anker S.D., Volterrani M., Pflaum C.D., Strasburger C.J., Osterziel K.J., Doehner W., Ranke M.B., Poole-Wilson P.A., Giustina A., Dietz R., and Coats A.J. Acquired growth hormone resistance in patients with chronic heart failure: implications for therapy with growth hormone. *J. Am. Coll. Cardiol.* 2001; 38: 443–52.

169. Niebauer J., Pflaum C.-D., Clark A.L., Strasburger C.J., Hooper J., Poole-Wilson P.A., Coats A.J., and Anker S.D. Deficient insulin-like growth factor-I in chronic heart failure predicts altered body composition, anabolic deficiency, cytokine and neurohormonal activation. *J. Am. Coll. Cardiol.* 1998; 32: 393–7.

170. Osterziel K.J., Strohm O., Schuler J., Friedrich M., Hänlein D., Willenbrock R., Anker S.D., Poole-Wilson P.A., Ranke M.B., and Dietz R. Randomised, double

blind, placebo-controlled trial of human recombinant growth hormone in patients with chronic heart failure due to dilated cardiomyopathy. *Lancet* 1998; 351: 1233–7.

171. Anwar A., Gaspoz J.M., Pampallona S., Zahid A.A., Sigaud P., Pichard C., Brink M., and Delafontaine P. Effect of congestive heart failure on the insulin-like growth factor-1 system. *Am. J. Cardiol.* 2002; 90: 1402–5.

172. Broglio F., Fubini A., Morello M., Arvat E., Aimaretti G., Gianotti L., Boghen M.F., Deghenghi R., Mangiardi L., and Ghigo E. Activity of GH/IGF-1 axis in patients with dilated cardiomyopathy. *Clin. Endocrinol.* 1999; 50: 417–30.

173. Jeffries M.K. and Vance M.L. Growth hormone and cortisol secretion in patients with burn injury. *J. Burn Care Rehabil.* 1992; 13: 391–5.

174. Crawford D.H., Shepherd R.W., Halliday J.W., Cooksley G.W., Golding S.D., Cheng W.S., et al. Body composition in nonalcoholic cirrhosis: the effect of disease etiology and severity on nutritional compartments. *Gastroenterology* 1994; 106: 1611–7.

175. Scevola D., Di Matteo A., Uberti F., Minoia G., Poletti F., and Faga A. Reversal of cachexia in patients treated with potent antiretroviral therapy. *AIDS Read* 2000; 10: 365–9.

176. Braun S.R., Keim N.L., Dixon R.M., Clagnaz P., Anderegg A., and Shrago E.S. The prevalence and determinants of nutritional changes in chronic obstructive pulmonary disease. *Chest* 1984; 86: 558–63.

177. Jenkins R.C. and Ross R.J.M. Acquired growth hormone resistance in catabolic states. *Baillières Clin. Endocrinol. Metab.* 1996; 10: 411–9.

178. Tomas F.M., Knowles S.E., Owens P.C., Read L.C., Chandler C.S., Gargosky S.E., and Ballard F.J. Increased weight gain, nitrogen retention and muscle protein synthesis following treatment of diabetic rats with insulin-like growth factor (IGF)-I and des(1-3)IGF-I. *Biochem. J.* 1991; 276: 547–54.

179. Fang C.H., Li B.G., Wang J.J., Fischer J.E., and Hasselgren P.O. Treatment of burned rats with insulin-like growth factor I inhibits the catabolic response in skeletal muscle. *Am. J. Physiol.* 1998; 275: R1091–8.

180. Cioffi W.G., Gore D.C., Rue L.W., Carrougher G., Guler H.P., McManus W.F., and Pruitt B.A. Insulin-like growth factor-I lowers protein oxidation in patients with thermal injury. *Ann. Surg.* 1994; 220: 310–19.

181. Hirschberg R., Kopple J., Lipsett P., Benjamin E., Minei J., Albertson T., Munger M., Metzler M., Zaloga G., Murray M., Lowry S., Conger J., McKeown W., O'shea M., Baughman R., Wood K., Haupt M., Kaiser R., Simms H., Warnock D., Summer W., Hintz R., Myers B., Haenftling K., Capra W., et al. Multicenter clinical trial of recombinant human insulin-like growth factor I in patients with acute renal failure. *Kidney Int.* 1999; 55: 2423–32.

182. Baxter R.C., Martin J.L., and Beniac V.A. High molecular weight insulin-like growth factor binding protein complex. Purification and properties of the acid-labile subunit from human serum. *J. Biol. Chem.* 1989; 264: 11843–8.

183. Gelato M.C. and Frost R.A. IGFBP-3. Functional and structural implications in aging and wasting syndromes. *Endocrine* 1997; 7: 81–5.

184. Lee C.Y. and Rechler M.M. Proteolysis of insulin-like growth factor (IGF)-binding protein-3 (IGFBP-3) in 150-kilodalton IGFBP complexes by a cation-dependent protease activity in adult rat serum promotes the release of bound IGF-I. *Endocrinology* 1996; 137: 2051–8.

185. Frost R.A. and Lang C.H. Differential effects of insulin-like growth factor I (IGF-I) and IGF-binding protein-1 on protein metabolism in human skeletal muscle cells. *Endocrinology* 1999; 140: 3962–70.

186. Mynarcik D.C., Frost R.A., Lang C.H., DeCristofaro K., McNurlan M.A., Garlick P.J., Steigbigel R.T., Fuhrer J., Ahnn S., and Gelato M.C. Insulin-like growth factor system in patients with HIV infection: effect of exogenous growth hormone administration. *J. Acquir. Immune Defic. Syndr.* 1999; 22: 49–55.

187. Wolf M., Bohm S., Brand M., and Kreymann G. Proinflammatory cytokines interleukin 1 beta and tumor necrosis factor alpha inhibit growth hormone stimulation of insulin-like growth factor I synthesis and growth hormone receptor mRNA levels in cultured rat liver cells. *Eur. J. Endocrinol.* 1996; 135: 729–37.

188. Frost R.A., Lang C.H., and Gelato M.C. Transient exposure of human myoblasts to tumor necrosis factor-alpha inhibits serum and insulin-like growth factor-I stimulated protein synthesis. *Endocrinology* 1997; 138: 4153–9.

6 Cytokines and the Pathophysiology of Skeletal Muscle Atrophy

Vickie E. Baracos

CONTENTS

SUMMARY

Just over 20 years have elapsed since our 1983 report that a cytokine, interleukin-1 has a direct action to increase protein catabolism in skeletal muscle.[1] This observation created a key link in our understanding of the parallel occurrence of inflammation, infection, injury, and skeletal muscle wasting. Research on cytokines and skeletal muscle continues apace, with about 50 new citations appearing per year on various aspects of cytokine expression and action in skeletal muscle tissue. The particular emphasis of this chapter is the most recent evidence in our developing an understanding of muscle cytokines, and the regulation of catabolic processes. Skeletal muscle expresses and responds to a series of catabolic, pro-inflammatory cytokines which elicit induction of proteolytic pathways and suppress the anabolic actions of growth hormone, insulin and insulin-like growth factor (IGF)-1. The role of a group of anabolic, myotrophic cytokines that would appear to counter these catabolic effects is emerging.

6.1 INTRODUCTION: CYTOKINES OF SKELETAL MUSCLE

6.1.1 Catabolic, Pro-Inflammatory Cytokines

Over the last 20 years, our understanding of the role of pro-inflammatory cytokines as cachectins has evolved and these are now believed to be the key factors underlying various forms of wasting. The three main elements of the wasting effect are those originally described, namely activation of proteolysis[1–9]; as well as inhibition of growth hormone (GH) and insulin-like growth factor (IGF)-1 signaling,[10–17] and induction of insulin resistance,[18–24] which has subsequently been described. Of the pro-inflammatory cytokines, tumor necrosis factor (TNF)-α interleukin (Il)-1β and Il-6, as well as interferon (IFN)-γ are currently thought to be the principal catabolic actors in skeletal muscle. Il-1 was the first cytokine reported to act on skeletal muscle[1] and the description of TNF-α as a cachectin followed closely thereafter.[25] Both of these cytokines have been extensively studied in skeletal muscle in the same way as Il-6.[19, 20, 26–36]

The effects of cytokines such as TNF-α are both catabolic and antianabolic and, in addition, to a powerful induction of proteolysis, TNF-α plus IFN-γ strongly reduces myosin expression.[3] TNF-α interferes with cell cycle exit and represses the accumulation of transcripts encoding muscle-specific genes in differentiating C2C12 myoblasts.[5] The muscle regulatory factor MyoD, which induces muscle-specific transcription is degraded by the ubiquitin–proteasome pathway in response to TNF-α.[5] Muscle and myotubes express IFN-γ receptors[37] and this factor induces proteases including cathepsins B and L in myotubes.[38]

The actions of these cytokines on skeletal muscle are simultaneous and synergistic. Multiple cytokines are locally expressed, as are multiple cytokine receptors, upon stimulation. The pro-inflammatory cytokines show the feature of potentiation of each other's actions and a variety of potent synergies among these factors in their actions on muscle have been demonstrated (i.e., TNF-α + IFN-γ) for protein catabolism;[3] TNF-α + IFN-γ + endotoxin (lipopolysaccharide (LPS)) for cytokine receptor induction[37] and TNF-α + IFN-γ + LPS for nitric oxide synthesis and insulin resistance.[39]

6.1.1.1 The cytokine axis of skeletal muscle

A particularly fascinating aspect of our current understanding is the local production of cytokines and cytokine response elements within muscle cells and tissue, quite independently of any systemic inflammatory response (Figure 6.1). It is an important initial clarification that isolated myocytes and myotubes in the absence of any other cell type have been shown to respond to various stimuli to produce cytokines and respond to them through locally expressed receptor proteins.[37] C2C12 cells,[5, 12, 17, 40–45] L6 cells,[24] as well as human primary muscle

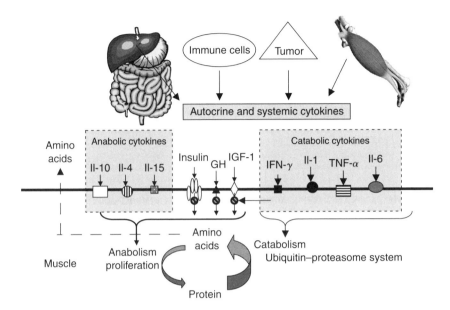

FIGURE 6.1   Cytokine axis of skeletal muscle. Cytokines acting on muscle originate systemically in various tissues, or locally in muscle cells or mononuclear phagocytes. These respectively, display a catabolic character and promote protein degradation as well as receptor resistance to GH, insulin, and IGF-1. These actions are countered by a group of anti-inflammatory cytokines expressing anabolic and pro-proliferative effects.

cultures[14] have been investigated. In differentiated skeletal muscle tissue, there are many other resident cell types that could be the source of cytokine production, and during certain types of inflammation and injury, both resident and recruited mononuclear phagocytes are potentially a significant source of multiple types of pro-inflammatory mediators.[46] Il-8, a chemokine, is locally expressed in skeletal muscle[47] and is a chemo-attractant for neutrophils.[48]

Thus, within the composite tissue that is called muscle, the muscle cells are one of the several direct sources of cytokine production. In reviewing the potential sources of cytokine signals to muscle, it is to be noted that systemic levels of cytokines vary and that sites of inflammation (or tumors[33]) distant from muscle can signal to this tissue via the systemic circulation of cytokines.

Lang and Frost[49,50] suggest that the scope of the muscle cytokine axis qualifies its inclusion as a component of the innate immune system, along with phagocytes, such as neutrophils and macrophages, its classically identified cellular components. Recent data from many laboratories clearly show that skeletal muscle is capable of recognizing and responding to invading pathogens and the catabolic hormonal milieu. Initial studies indicated the presence of toll-like receptor (TLR)-2 and -4 in both skeletal muscle *in vivo* and cultured C2C12 myocytes.[49,50] TLRs act as sentinels or monitors for the presence of pathogens.[51,52] TLR4, for example, recognizes LPS from Gram-negative bacteria such as *Escherichia coli* and LPS stimulates numerous effectors including early- and late-phase cytokines and the generation of nitric oxide by nitric oxide synthase-2. These findings, coupled with the earlier demonstration of TNF-α, Il-6, Il-1, and IFN-γ receptors in muscle,[37] provide convincing evidence for the afferent limb of the innate immune system in muscle. In addition, the efferent limb of the system is also present in myocytes as evidenced by the synthesis and secretion of various inflammatory mediators. For example, after exposure to LPS, skeletal muscle and myocytes demonstrate a dose- and time-dependent increase in the expression of cytokines, such as TNF-α, Il-1, and Il-6.[49,50]

6.2 METHODS AND APPROACHES IN MUSCLE CYTOKINE RESEARCH

A variety of new methodological approaches has allowed the further discrimination of which cytokines are expressed in skeletal muscle and the nature of their actions.

In contrast to a large literature on cultured cells and tissues from experimental animals, cytokine responses of human muscle were less well known. This tissue is relatively inaccessible, not usually biopsied, and the biopsy procedure is inevitably uncomfortable. Information from biopsies is becoming more prevalent, with over a dozen new citations in the last 2 years on cytokines and cytokine gene expression in human muscle.[6,8,27,30,53–60] Biopsies have been done on otherwise healthy individuals pre- and post-exercise[54,55,58–60] and in muscle injury induced by eccentric contraction.[27,30] There is additionally some information on muscle cytokine expression after ischemia-reperfusion injury,[47,53] in cachexia of chronic heart failure,[56] in end-stage renal disease,[6] dystrophy, myopathy[61,62] and in response to hyperinsulinemic, euglycemic clamp.[57]

Biopsied tissue, from humans and animals, has in turn been subjected to analysis by gene expression microarray, and there are a small number of recent additions to this area.[7,46,47,63,64] Microarray is a powerful means both of profiling the gene expression response of skeletal muscle to treatment with TNF-α or other cytokines,[64] as well as to probe cytokine gene expression in response to physiological factors such as age,[7] insulin,[63] or pathologies such as ischemia-reperfusion injury,[47] type 2 diabetes,[63] or muscle injury.[46] Some of these changes are very dramatic, for example, the largest increases in gene expression detected using oligonucleotide array analysis of 12,625 genes were for Il-6 (20-fold), Il-8 (18-fold), and TNF type II receptor (13-fold) in skeletal muscles harvested from patients who underwent coronary artery bypass grafting before and after cardiopulmonary bypass and cardioplegic arrest.[47] It seems likely that this concerted response results in a strong potentiation of inflammation locally within the muscle tissue during ischemic injury.

In terms of implicating cytokines as having a causal role in regulation of muscle metabolism, a number of ways of specifically modulating cytokine production or action are at hand, including gene knockout, monoclonal antibodies, soluble receptors, and drugs such as pentoxyfylline that block cytokine production. Knockout approaches directed toward TNF receptor subtypes as well as Il-6,[29,35,65,66] were most recently used as an attempt to identify more clearly the role of these cytokines in obesity, insulin resistance, and in response to exercise. Knockout of TNF type I receptors were earlier been used to establish the key role of TNF in muscle protein catabolism in mice bearing the Lewis lung carcinoma.[66] A variety of monoclonal antibodies toward different cytokines were used for a similar purpose, including antiTNF-α (infliximab, Remicade), antiIFN-γ, and CNTO 328, a human-mouse chimeric monoclonal antibody to Il-6. Infliximab has been used in preliminary studies in attempts to alter the course of insulin resistance in morbidly obese humans; however, this approach was ineffective.[21] In a pilot study of two patients with myositis, infliximab appeared to provide subjective and objective improvement.[67] Remicade protected dystrophic skeletal muscle from necrosis in an animal model.[61] The possible action of CNTO 328 to inhibit tumor-induced cachexia was tested in two human tumor-induced cachexia models in nude mice,[33] human melanoma cells, or human prostate tumor cells. Since CNTO 328 blocks human Il-6 but not mouse Il-6, the ability of this antibody to block or partially reverse weight loss in these models suggest that tumor cell-secreted (i.e., human) Il-6 directly contributes to body weight loss. By neutralizing mono-clonal antibodies and of mice congenitally deficient in IFN-γ, Madihally[68] has shown that in burned animals, the absence of IFN-γ virtually abolished skeletal muscle hypercatabolism.

While these results highlight the potential of anticytokine treatments in cach-exia therapy, many questions remain unanswered. In the case where TNF or another cytokine is produced in muscle and acts in an autocrine fashion within the tissue, it remains unclear whether these antibodies can actually penetrate sufficiently to act. It would also appear crucial to be able to inhibit early-phase cytokine production and it remains unknown to what extent the inhibition of cytokine pro-duction after inflammation and catabolic responses that are fully engaged will be effective.

Polymorphisms in various cytokines or cytokine receptors can account for at least part of the variation in muscle mass and response to various physiological and pathological situations.[30,69] Dennis et al.[30] have provided the first evidence that several single nucleotide polymorphisms (SNP) that were previously shown to alter Il-1 activity are associated with altered inflammatory response in human muscle to acute resistance exercise. Five polymorphic sites in the Il-1 gene cluster: Il-1α ($+4845$), Il-1β ($+3954$), Il-1β (-511), Il-1β (-3737), and Il-1RN ($+2018$), were evaluated before and after a standard bout of resistance leg exercise. Vastus lateralis biopsies were obtained pre-, and at 24-, and 72-h postexercise. Inflammatory responses were significantly associated with specific haplotype patterns and were also influenced by individual SNP. Subjects with SNP at Il-1β ($+3954$) or at Il-1β (-3737) had a 2-fold higher median induction of several inflammatory markers. The Il-1RN ($+2018$) SNP maximized the response specifically within these groups and was associated with increased macrophage recruitment. There are numerous described polymorphisms in either coding or promoter regions of TNF-α, Il-6, Il-10, and IFN-γ; however, to date, their implications for skeletal muscle responses are unknown.

6.3 THE SCOPE OF CYTOKINE INVOLVEMENT IN SKELETAL MUSCLE REGULATION

6.3.1 Catabolic Mediation: Induction of Proteolytic Systems

Muscle wasting is primarily mediated by the activation of the ubiquitin–proteasome system, which is responsible for degrading the bulk intracellular proteins. Using transcriptional screening techniques, specific changes in gene expression have been identified that appear to be characteristics of muscular wasting processes. Concurrent expression of inflammatory cytokines and multiple elements of proteolytic systems is a consistent theme in gene expression microarray studies in sarcopenia of aging[7] and ischemia-reperfusion injury.[47] Causal associations between these phenomena have been pursued using TNF-α and Il-6 as agonists in specific model systems. Of particular interest is the atrophy-related and cytokine-inducible expression of a number of elements of this proteolytic system.[4,9]

E3 or ubiquitin ligases are highly specific regulators of the ubiquitin–proteasome pathway, which target proteins for proteolytic breakdown by conferring substrate specificity to the ubiquitination reaction.[3] Three muscle-specific ubiquitin ligases have been reported to date. Kwak et al[2] recently reported the full-length cloning of E3α-II, a novel "N-end rule" ubiquitin ligase, which is highly enriched in skeletal muscle, and its expression is regulated by pro-inflammatory cytokines, as treatment of myotubes with TNF-α or Il-6, elicited marked increases in E3α-II, and ubiquitin conjugation activity in parallel.

Tumor necrosis factor-α also increases ubiquitin-conjugating activity in skeletal muscle by up-regulating UbcH2, an ubiquitin carrier protein.[9] UbcH2 is constitutively expressed by human skeletal muscles, murine limb muscle, and cultured myotubes, and TNF-α stimulates UbcH2 expression. In extracts from TNF-α treated myotubes, ubiquitin-conjugating activity is limited by UbcH2 availability when UbcH2 activity is inhibited by an antiserum to UbcH2 or a dominant negative mutant of UbcH2 and is enhanced by wild-type UbcH2. Thus, UbcH2 up-regulation is a novel response to TNF-α signaling in skeletal muscle that appears to be essential for the increased ubiquitin conjugation induced by this cytokine.[9]

6.3.2 Inhibition of GH and/or IGF-1 Signaling

Growth hormone and IGF-1 are potent regulators of muscle mass in health and disease. This somatomedin axis is markedly deranged in various catabolic conditions in which circulating and tissue levels of inflammatory cytokines are elevated.[10] Frost and Lang[11] have reviewed the role of secondary changes in the GH and IGF-1 systems in the loss of muscle protein during expression of pro-inflammatory cytokines. In animal models of cachexia and muscle atrophy, increased cytokine expression and decreased IGF-1 expression seem to appear together in a defined sequence. In rats, endotoxin increased TNF-α mRNA in tibialis muscle (7-fold after 1 h) and this was followed 12 h later by a fall in IGF-1 mRNA (-73%).

In an animal model of cachexia of chronic heart failure,[15] the local expression of IGF-1 in skeletal muscle was reduced by 39% and this correlated inversely with local expression of Il-1β. Muscle cells *in vitro* allow the chance to explore the direct relationship between these events. Myocytes cultured with TNF-α are GH resistant.[11] In C2C12 myotubes, TNF-α strongly suppressed levels of IGF-1 mRNA (-73%) and protein (-47%) within 72 h.[70] TNF-α decreases IGF-1 mRNA expression in C2C12 myoblasts via a Jun N-terminal kinase pathway.[17] Ceramide, a second messenger in both TNF-α and Il-1β receptor signaling pathways, is a key downstream mediator that leads to IGF-1 resistance.[12] TNF-α acts on muscle cells to induce a state of IGF-1 receptor resistance.[16] TNF-α inhibits IGF-1-stimulated protein synthesis in primary porcine myoblasts and in C2C12 myoblasts, where as little as 0.01 ng/ml TNF-α significantly inhibits protein synthesis induced by IGF-1. TNF-α reduces by $\sim$50% IGF-1-stimulated tyrosine phosphorylation of two of the major downstream receptor docking molecules, insulin receptor substrate (IRS)-1 and IRS-2. Very low physiological concentrations of TNF-α thus interfere with both protein synthesis and muscle cell development by inducing a state of IGF-1 receptor resistance.[16]

Collectively, these changes act to limit IGF-1 availability in muscle, which disturbs protein balance and results in the loss of protein stores in catabolic and inflammatory conditions.

6.3.3 Cytokine-Induced Insulin Resistance

From cell or organ systems *in vitro*, it seems clear that cytokines inhibit insulin actions, mainly on glucose and fatty acid metabolism. Although the presence of insulin resistance of protein metabolism has not been explored, it may reasonably be assumed that this would be present, given the global suppression of upstream insulin signal transduction.[16] For example, in primary neonatal rat myotubes, TNF-α caused insulin resistance on glucose uptake and glucose transporter (GLUT) 4 translocation by impairing insulin stimulation of insulin receptor (IR) and IRS-1 and IRS-2 tyrosine phosphorylation, IRS-associated phosphatidylinositol 3-kinase activation, and Akt phosphorylation.[23, 24] Bruce and Dyck[20] used a pulse-chase technique to determine the effect of Il-6 and TNF-α on fatty acid metabolism in isolated rat soleus muscle. Il-6 increased exogenous and endogenous fatty acid oxidation by $\sim$50%. When Il-6 and insulin were both present, Il-6 attenuated insulin's suppressive effect on fatty acid oxidation. Furthermore, in the presence of insulin, Il-6 reduced the esterification of fatty acid to triacylglycerol. These results demonstrate that Il-6 plays a role in regulating fat metabolism in muscle, increasing rates of oxidation, and attenuating insulin's lipogenic effects.

Kim et al.[22] studied the effects of Il-6 treatment on whole-body insulin action and glucose metabolism *in vivo* during hyperinsulinemic–euglycemic clamps in conscious mice. Pretreatment with Il-6 acutely reduced insulin-stimulated glucose uptake in skeletal muscle, and this was associated with defects in insulin-stimulated IRS-1-associated PI 3-kinase activity and increases in fatty acyl-CoA levels in skeletal muscle. The situation in human muscle is less clear. In quadriceps femoris muscle of type 2 diabetic patients, TNF-α levels are about 4-fold higher than in matched nondiabetic controls[18]; however, when infliximab was used in attempts to alter the course of insulin resistance in morbidly obese humans, this approach was ineffective.[21]

6.4 ANABOLIC MYOTROPHIC CYTOKINES

The actions of the catabolic pro-inflammatory cytokines appear to be countered by the anabolic actions of Ils-4, -10, and -15. These three predominantly anti-inflammatory cytokines are anabolic and myotrophic. Il-4, -10, and -15 are known to promote myogenesis, as well as, to act to suppress the direct actions of Il-6 and TNF-α to promote protein catabolism, apoptosis, and insulin resistance.

Interleukin-10 is expressed in skeletal muscle[36,71,72] and is physiologically regulated with age,[36] resistive loading,[71] and aerobic exercise.[72] Il-10 suppresses the activation of muscle protein degradation in an animal model of cancer cachexia, in which TNF-α as been shown as the primary inducer of protein catabolism.[66] The counter-regulatory actions of Il-10 on Il-6-induced insulin resistance comprise a nice example of the reciprocal actions of these factors. Kim et al.[22] studied the effects of Il-6 on insulin action and glucose metabolism *in vivo* during hyperinsulinemic–euglycemic clamps in conscious mice. Acute Il-6 treatment reduced insulin-stimulated glucose uptake in skeletal muscle; however, concurrent treatment with Il-10 protected skeletal muscle from Il-6 and lipid-induced defects in insulin action and signaling activity, and these effects were associated with decreases in intramuscular fatty acyl-CoA levels. Il-10 also prevented Il-6-induced defects in hepatic insulin action.

Interleukin-4 is expressed in skeletal muscle[71,73] and Pavlath and coworkers[73] have demonstrated a novel role for Il-4 in regulating the fusion of myoblasts with differentiated myotubes. The transcription factor NFATc2 controls myoblast fusion at a specific stage of myogenesis after the initial formation of a myotube and is necessary for further cell growth. By examining genes regulated by NFATc2 in muscle, these authors identified Il-4 as a molecular signal that controls myoblast fusion with myotubes. Muscle cells that lack Il-4 or the Il-4α receptor subunit form normally but are reduced in size and have a reduced number of nuclei. Il-4 is expressed by a subset of muscle cells in fusing muscle cultures and requires the Il-4α receptor subunit on myoblasts to promote fusion and growth.

Interleukin-15 is expressed in skeletal muscle[54,72] and is able to reduce protein loss and apoptosis related to muscle wasting during cancer cachexia in experimental animals.[74,75] Administration of Il-15 to rats bearing the Yoshida AH-130 ascites hepatoma (which induces a TNF-α-dependent cachectic response) resulted in a significant reduction of muscle wasting and muscle protein catabolism. In addition, Il-15 completely reversed the increased DNA fragmentation observed in skeletal muscle of tumor-bearing animals. Il-15 resulted in a considerable decrease in both type 1 and type 2 TNF-α receptors, suggesting a direct counter-regulatory effect at the level of TNF-α signaling.

Although Il-15 expression was not induced in muscle by an hour of bicycle ergometry[54] or a 3 h run at 75% of VO_2 max,[72] a SNP in exon 7 of Il-15 receptor α was associated with muscle hypertrophy in response to a resistance training program in humans and accounted for 7.1% of the variation in this response in regression modeling.[69] A polymorphism in exon 4 was independently associated with muscle hypertrophy and accounted for an additional 3.5% of the variation. These results suggest that Il-15 is an important mediator of muscle mass response to resistance exercise training in humans. Related work shows that Il-15 increases myosin accretion in human skeletal myogenic cultures.[76] In primary human skeletal myogenic cells, accretion of a major myofibrillar protein, myosin heavy chain (MHC), was induced and Il-15 was most effective at stimulating MHC accretion when added to cultures after differentiation of myoblasts had occurred.

6.5 IMPLICATIONS FOR THE PHARMACOTHERAPY OF CACHEXIA

Our increasing understanding of the cytokine axis of skeletal muscle provides us with a series of potential targets for anticachexia therapy. One available approach would be to employ monoclonal antibodies against specific cytokines, as several of these agents have been developed keeping in mind the other applications of anticancer therapy. It seems that the efficacy of this highly specific monotherapy directed at a single cytokine will depend on the degree to which the cytokine in question is causally associated with the wasting syndrome. As different cancer cachexias in animals seem to involve different distinctive cytokines (i.e., TNF-α but not Il-6, or vice versa), there is considerable uncertainty about the general applicability of such an approach. Alternatively, interventions that more broadly inhibit the production and action of the pro-inflammatory cytokines in muscle as well as systemically may be more likely to achieve down-regulation of catabolic effects. The relative efficacy, side effect profile, and cost of different genres of anti-inflammatory therapies in this regard remain to be systematically tested.

The potential therapeutic use of Ils-4, -10 and -15 for muscle atrophy is also unknown. This would depend on the variety of systemic effects caused by these factors in the context of different diseases, and in cancer the potential for these factors to alter the proliferation of tumor cells or their response to antineoplastic therapy would require consideration.

REFERENCES

1. Baracos V.E., Rodemann H.P., Dinarello C.A., and Goldberg A.L. Stimulation of muscle protein degradation and prostaglandin E_2 release by leukocytic pyrogen (interleukin-1). A mechanism for the increased degradation of muscle proteins during fever. *N. Engl. J. Med.* 1983; 308: 553–8.
2. Kwak K.S., Zhou X., Solomon V., Baracos V.E., Davis J., Bannon A.W., Boyle W.J., Lacey D.L., and Han H.Q. Regulation of protein catabolism by muscle-specific and cytokine-inducible ubiquitin ligase E3α-II during cancer cachexia. *Cancer Res.* 2004; 64: 8193–8.

3. Acharyya S., Ladner K.J., Nelsen L.L., Damrauer J., Reiser P.J., Swoap S., and Guttridge D.C. Cancer cachexia is regulated by selective targeting of skeletal muscle gene products. *J. Clin. Invest.* 2004; 114: 370–8

4. Spate U. and Schulze P.C. Proinflammatory cytokines and skeletal muscle. *Curr. Opin. Clin. Nutr. Metab. Care* 2004; 7: 265–9.

5. Langen R.C., Van Der Velden J.L., Schols A.M., Kelders M.C., Wouters E.F., and Janssen-Heininger Y.M. Tumor necrosis factor α inhibits myogenic differentiation through MyoD protein destabilization. *FASEB J.* 2004; 18: 227–37.

6. Raj D.S., Shah H., Shah V.O., Ferrando A., Bankhurst A., Wolfe R., and Zager P.G. Markers of inflammation, proteolysis, and apoptosis in ESRD. *Am. J. Kidney Dis.* 2003; 42: 1212–20.

7. Pattison J.S., Folk L.C., Madsen R.W., Childs T.E., and Booth F.W. Transcriptional profiling identifies extensive downregulation of extracellular matrix gene expression in sarcopenic rat soleus muscle. *Physiol. Genomics.* 2003; 15: 34–43.

8. Welle S., Brooks A.I., Delehanty J.M., Needler N., and Thornton C.A. Gene expression profile of aging in human muscle. *Physiol. Genomics* 2003; 14: 149–59.

9. Li Y.P., Lecker S.H., Chen Y., Waddell I.D., Goldberg A.L., and Reid M.B. TNF-α increases ubiquitin-conjugating activity in skeletal muscle by up-regulating UbcH2/E220k. *FASEB J.* 2003; 17: 1048–57.

10. Lang C.H., Hong-Brown L., and Frost R.A. Cytokine inhibition of JAK-STAT signaling: a new mechanism of growth hormone resistance. *Pediatr. Nephrol.* 2005; 20: 306–12.

11. Frost R.A. and Lang C.H. Alteration of somatotropic function by proinflammatory cytokines. *J. Anim. Sci.* 2004; 82: E100–9.

12. Strle K., Broussard S.R., McCusker R.H., Shen W.H., Johnson R.W., Freund G.G., Dantzer R., and Kelley K.W. Proinflammatory cytokine impairment of insulin-like growth factor I-induced protein synthesis in skeletal muscle myoblasts requires ceramide. *Endocrinology* 2004; 145: 4592–602.

13. Tuomisto T.T., Rissanen T.T., Vajanto I., Korkeela A., Rutanen J., and Yla-Herttuala S. HIF-VEGF-VEGFR-2, TNF-α and IGF pathways are upregulated in critical human skeletal muscle ischemia as studied with DNA array. *Atherosclerosis* 2004; 174: 111–20.

14. Foulstone E.J., Huser C., Crown A.L., Holly J.M., and Stewart C.E. Differential signalling mechanisms predisposing primary human skeletal muscle cells to altered proliferation and differentiation: roles of IGF-1 and TNF-α. *Exp. Cell. Res.* 2004; 294: 223–35.

15. Schulze P.C., Gielen S., Adams V., Linke A., Mobius-Winkler S., Erbs S., Kratzsch J., Hambrecht R., and Schuler G. Muscular levels of proinflammatory cytokines correlate with a reduced expression of insulin like growth factor-I in chronic heart failure. *Basic Res. Cardiol.* 2003; 98: 267–74.

16. Broussard S.R., McCusker R.H., Novakofski J.E., Strle K., Shen W.H., Johnson R.W., Freund G.G., Dantzer R., and Kelley K.W. Cytokine-hormone interactions: tumor necrosis factor α impairs biologic activity and downstream activation signals of the insulin-like growth factor I receptor in myoblasts. *Endocrinology* 2003; 144: 2988–96.

17. Frost R.A., Nystrom G.J., and Lang C.H. Tumor necrosis factor-α decreases insulin-like growth factor-I messenger ribonucleic acid expression in C2C12 myoblasts via a Jun N-terminal kinase pathway. *Endocrinology* 2003; 144: 1770–9.

18. Torres S.H., De Sanctis J.B., de L Briceno M., Hernandez N., and Finol H.J. Inflammation and nitric oxide production in skeletal muscle of type 2 diabetic patients. *J. Endocrinol.* 2004; 181: 419–27.

19. Carey A.L. and Febbraio M.A. Interleukin-6 and insulin sensitivity: friend or foe? *Diabetologia* 2004; 47: 1135–42.

20. Bruce C.R. and Dyck D.J. Cytokine regulation of skeletal muscle fatty acid metabolism: effect of interleukin-6 and tumor necrosis factor-α. *Am. J. Physiol. Endocrinol. Metab.* 2004; 287: E616–21.

21. Di Rocco P., Manco M., Rosa G., Greco A.V., and Mingrone G. Lowered tumor necrosis factor receptors, but not increased insulin sensitivity, with infliximab. *Obes. Res.* 2004; 12: 734–9.

22. Kim H.J., Higashimori T., Park S.Y., Choi H., Dong J., Kim Y.J., Noh H.L., Cho Y.R., Cline G., Kim Y.B., and Kim J.K. Differential effects of interleukin-6 and -10 on skeletal muscle and liver insulin action *in vivo*. *Diabetes* 2004; 53: 1060–7.

23. de Alvaro C., Teruel T., Hernandez R., and Lorenzo M. Tumor necrosis factor alpha produces insulin resistance in skeletal muscle by activation of inhibitor κB kinase in a p38 MAPK-dependent manner. *J. Biol. Chem.* 2004; 279: 17070–8.

24. Yamaguchi K., Higashiura K., Ura N., Murakami H., Hyakukoku M., Furuhashi M., and Shimamoto K. The effect of tumor necrosis factor-α on tissue specificity and selectivity to insulin signaling. *Hypertens. Res.* 2003; 26: 389–96.

25. Beutler B. and Cerami A. Cachectin and tumour necrosis factor as two sides of the same biological coin. *Nature* 1986; 320: 584–8.

26. Lappas M., Yee K., Permezel M., and Rice G.E. Sulphasalazine and BAY 11-7082 interfere with the NF-κB and IKKβ pathway to regulate the release of pro-inflammatory cytokines from human adipose tissue and skeletal muscle, *in vitro*. *Endocrinology* 2005; 146: 1491–7.

27. Hamada K., Vannier E., Sacheck J.M., Witsell A.L., and Roubenoff R. Senescence of human skeletal muscle impairs the local inflammatory cytokine response to acute eccentric exercise. *FASEB J.* 2005; 19: 264–6.

28. Haddad F., Zaldivar F.P., Cooper D.M., and Adams G.R. IL-6 induced skeletal muscle atrophy. *J. Appl. Physiol.* 2005; 98: 911–7.

29. Keller C., Keller P., Giralt M., Hidalgo J., and Pedersen B.K. Exercise normalises overexpression of TNF-α in knockout mice. *Biochem. Biophys. Res. Commun.* 2004; 321: 179–82

30. Dennis R.A., Trappe T.A., Simpson P., Carroll C., Huang B.E., Nagarajan R., Bearden E., Gurley C., Duff G.W., Evans W.J., Kornman K., and Peterson C.A. Interleukin-1 polymorphisms are associated with the inflammatory response in human muscle to acute resistance exercise. *J. Physiol.* 2004; 560: 617–26.

31. Rieusset J., Bouzakri K., Chevillotte E., Ricard N., Jacquet D., Bastard J.P., Laville M., and Vidal H. Suppressor of cytokine signaling 3 expression and insulin resistance in skeletal muscle of obese and type 2 diabetic patients. *Diabetes* 2004; 53: 2232–41.

32. Tomas E., Kelly M., Xiang X., Tsao T.S., Keller C., Keller P., Luo Z., Lodish H., Saha A.K., Unger R., and Ruderman N.B. Metabolic and hormonal interactions between muscle and adipose tissue. *Proc. Nutr. Soc.* 2004; 63: 381–5.

33. Zaki M.H., Nemeth J.A., and Trikha M. CNTO 328, a monoclonal antibody to IL-6, inhibits human tumor-induced cachexia in nude mice. *Int. J. Cancer* 2004; 111: 592–5.

34. Febbraio M.A., Hiscock N., Sacchetti M., Fischer C.P., and Pedersen B.K. Interleukin-6 is a novel factor mediating glucose homeostasis during skeletal muscle contraction. *Diabetes* 2004; 53: 1643–8.

35. Kelly M., Keller C., Avilucea P.R., Keller P., Luo Z., Xiang X., Giralt M., Hidalgo J., Saha A.K., Pedersen B.K., and Ruderman N.B. AMPK activity is diminished in tissues of Il-6 knockout mice: the effect of exercise. *Biochem. Biophys. Res. Commun.* 2004; 320: 449–54.

36. Hacham M., White R.M., Argov S., Segal S., and Apte R.N. Interleukin-6 and interleukin-10 are expressed in organs of normal young and old mice. *Eur. Cytokine. Netw.* 2004, 15: 37–46.

37. Zhang Y., Pilon G., Marette A., and Baracos V.E. Cytokines and endotoxin induce cytokine receptors in skeletal muscle. *Am. J. Phys.* 2000; 279: E196–205

38. Gallardo E., de Andres I., and Illa I. Cathepsins are upregulated by IFN-γ/STAT1 in human muscle culture: a possible active factor in dermatomyositis. *J. Neuropathol. Exp. Neurol.* 2001; 60: 847–55.

39. Bedard S., Marcotte B., and Marette A. Cytokines modulate glucose transport in skeletal muscle by inducing the expression of inducible nitric oxide synthase. *Biochem. J.* 1997; 325: 487–93

40. Delaigle A.M., Jonas J.C., Bauche I.B., Cornu O., and Brichard S.M. Induction of adiponectin in skeletal muscle by inflammatory cytokines: *in vivo* and *in vitro* studies. *Endocrinology* 2004; 145: 5589–97.

41. Carballo-Jane E., Pandit S., Santoro J.C., Freund C., Luell S., Harris G., Forrest M.J., and Sitlani A. Skeletal muscle: a dual system to measure glucocorticoid-dependent transactivation and transrepression of gene regulation. *J. Steroid Biochem. Mol. Biol.* 2004; 88: 191–201.

42. Baeza-Raja B. and Munoz-Canoves P. p38 MAPK-induced nuclear factor-κB activity is required for skeletal muscle differentiation: role of interleukin-6. *Mol. Biol. Cell.* 2004; 15: 2013–26.

43. Frost R.A., Nystrom G.J., and Lang C.H. Epinephrine stimulates Il-6 expression in skeletal muscle and C2C12 myoblasts: role of c-Jun N.H.2-terminal kinase and histone deacetylase activity. *Am. J. Physiol. Endocrinol Metab.* 2004; 286: E809–17. [Epub 2004 Jan. 13.]

44. Frost R.A., Nystrom G.J., and Lang C.H. Lipopolysaccharide and proinflammatory cytokines stimulate interleukin-6 expression in C2C12 myoblasts: role of the Jun NH2-terminal kinase. *Am. J. Physiol. Regul. Integr. Comp. Physiol.* 2003; 285: R1153-[64].

45. Luo G., Hershko D.D., Robb B.W., Wray C.J., and Hasselgren P.O. Il-1β stimulates Il-6 production in cultured skeletal muscle cells through activation of MAP kinase signaling pathway and NF-κB. *Am. J. Physiol. Regul. Integr. Comp. Physiol.* 2003; 284: R1249–54.

46. Summan M., McKinstry M, Warren G.L., Hulderman T., Mishra D., Brumbaugh K., Luster M.I., and Simeonova P.P. Inflammatory mediators and skeletal muscle injury: a DNA microarray analysis. *J. Interferon Cytokine Res.* 2003; 23: 237–45.

47. Ruel M., Bianchi C., Khan T.A., Xu S., Liddicoat J.R., Voisine P., Araujo E., Lyon H., Kohane I.S., Libermann T.A., and Sellke F.W. Gene expression profile after cardiopulmonary bypass and cardioplegic arrest. *J. Thorac. Cardiovasc. Surg.* 2003; 126: 1521–30.

48. Tsivitse S.K., Mylona E., Peterson J.M., Gunning W.T., and Pizza F.X. Mechanical loading and injury induce human myotubes to release neutrophil chemoattractants. *Am. J. Physiol. Cell Physiol.* 2005; 288: C721–9.

49. Lang C.H., Silvis C., Deshpande N., Nystrom G., and Frost R.A. Endotoxin stimulates *in vivo* expression of inflammatory cytokines tumor necrosis factor α, interleukin-1β, -6, and high-mobility-group protein-1 in skeletal muscle. *Shock* 2003; 19: 538–46.

50. Frost R.A. and Lang C.H. Skeletal muscle cytokines: regulation by pathogen associated molecules and catabolic hormones. *Curr. Opin. Clin. Nutr. Metab. Care* 2005; 8: 255–63.

51. Iwasaki A. and Medzhitov R. Toll-like receptor control of the adaptive immune responses. *Nat. Immunol.* 2004, **5**: 987–95

52. Cook D.N., Pisetsky D.S., and Schwartz D.A. Toll-like receptors in the pathogenesis of human disease. *Nat. Immunol.* 2004, **5**: 975–9.

53. Adembri C., Kastamoniti E., Bertolozzi L., Vanni S., Dorigo W., Coppo M., Pratesi C., De Gaudio A.R., Gensini G.F., and Modesti P.A. Pulmonary injury follows systemic inflammatory reaction in infrarenal aortic surgery. *Crit. Care Med.* 2004; 32: 1170–7.

54. Chan M.H., Carey A.L., Watt M.J., and Febbraio M.A. Cytokine gene expression in human skeletal muscle during concentric contraction: evidence that Il-8, like Il-6, is influenced by glycogen availability. *Am. J. Physiol. Regul. Integr. Comp. Physiol.* 2004; 287: R322–7.

55. Hiscock N., Chan M.H., Bisucci T., Darby I.A., and Febbraio M.A. Skeletal myocytes are a source of interleukin-6 mRNA expression and protein release during contraction: evidence of fiber type specificity. *FASEB J.* 2004; 18: 992–4.

56. Krankel N., Adams V., Gielen S., Linke A., Erbs S., Schuler G., and Hambrecht R. Differential gene expression in skeletal muscle after induction of heart failure: impact of cytokines on protein phosphatase 2A expression. *Mol. Genet. Metab.* 2003; 80: 262–71.

57. Krogh-Madsen R., Plomgaard P., Keller P., Keller C., and Pedersen B.K. Insulin stimulates interleukin-6 and tumor necrosis factor-α gene expression in human subcutaneous adipose tissue. *Am. J. Physiol. Endocrinol. Metab.* 2004; 286: E234–8.

58. Keller P., Keller C., Carey A.L., Jauffred S., Fischer C.P., Steensberg A., and Pedersen B.K. Interleukin-6 production by contracting human skeletal muscle: autocrine regulation by IL-6. *Biochem. Biophys. Res. Commun.* 2003; 310: 550–4.

59. Penkowa M., Keller C., Keller P., Jauffred S., and Pedersen B.K. Immunohistochemical detection of interleukin-6 in human skeletal muscle fibers following exercise. *FASEB J.* 2003; 17: 2166–8.

60. Pedersen B.K., Steensberg A., Keller P., Keller C., Fischer C., Hiscock N., van Hall G., Plomgaard P., and Febbraio M.A. Muscle-derived interleukin-6: lipolytic, anti-inflammatory and immune regulatory effects. *Pflugers Arch.* 2003; 446: 9–16.

61. Grounds M.D. and Torrisi J. Anti-TNF-α (Remicade) therapy protects dystrophic skeletal muscle from necrosis. *FASEB J.* 2004; 18: 676–82.

62. Kuru S., Inukai A., Kato T., Liang Y., Kimura S., and Sobue G. Expression of tumor necrosis factor-α in regenerating muscle fibers in inflammatory and noninflammatory myopathies. *Acta. Neuropathol. (Berl.)* 2003; 105: 217–24.

63. Hansen L., Gaster M., Oakeley E.J., Brusgaard K., Damsgaard Nielsen E.M., Beck-Nielsen H., Pedersen O., and Hemmings B.A. Expression profiling of insulin action in human myotubes: induction of inflammatory and pro-angiogenic pathways in relationship with glycogen synthesis and type 2 diabetes. *Biochem. Biophys. Res. Commun.* 2004; 323: 685–95.

64. Alon T., Friedman J.M., and Socci N.D. Cytokine-induced patterns of gene expression in skeletal muscle tissue. *J. Biol. Chem.* 2003; 278: 32324–34.

65. Faldt J., Wernstedt I., Fitzgerald S.M., Wallenius K., Bergstrom G., and Jansson J.O. Reduced exercise endurance in interleukin-6-deficient mice. *Endocrinology* 2004; 145: 2680–6.

66. Llovera M., Garcia-Martinez C., Lopez-Soriano J., Carbo N., Agell N, Lopez-Soriano F.J., and Argiles J.M. Role of TNF receptor 1 in protein turnover during cancer cachexia using gene knockout mice. *Mol. Cell. Endocrinol.* 1998; 142: 183–9.

67. Hengstman G.J., van den Hoogen F.H., Barrera P., Netea M.G., Pieterse A., van de Putte L.B., and van Engelen B.G. Successful treatment of dermatomyositis and polymyositis with anti-tumor-necrosis-factor-α: preliminary observations. *Eur. Neurol.* 2003; 50: 10–5.

68. Madihally S.V., Toner M., Yarmush M.L., and Mitchell R.N. Interferon γ modulates trauma-induced muscle wasting and immune dysfunction. *Ann. Surg.* 2002; 236: 649–57.

69. Riechman S.E., Balasekaran G., Roth S.M., and Ferrell R.E. Association of interleukin-15 protein and interleukin-15 receptor genetic variation with resistance exercise training responses. *J. Appl. Physiol.* 2004; 97: 2214–9.

70. Fernandez-Celemin L., Pasko N., Blomart V., and Thissen J.P. Inhibition of muscle insulin-like growth factor I expression by tumor necrosis factor-α. *Am. J. Physiol. Endocrinol. Metab.* 2002; 283: E1279–90.

71. Vassilakopoulos T., Divangahi M., Rallis G., Kishta O., Petrof B., Comtois A., and Hussain S.N. Differential cytokine gene expression in the diaphragm in response to strenuous resistive breathing. *Am. J. Respir. Crit. Care Med.* 2004; 170: 154–61.

72. Nieman D.C., Davis J.M., Henson D.A., Walberg-Rankin J., Shute M., Dumke C.L., Utter A.C., Vinci D.M., Carson J.A., Brown A., Lee W.J., McAnulty S.R., and McAnulty L.S. Carbohydrate ingestion influences skeletal muscle cytokine mRNA and plasma cytokine levels after a 3-h run. *J. Appl. Physiol.* 2003; 94: 1917–25.

73. Horsley V., Jansen K.M., Mills S.T., and Pavlath G.K. Il-4 acts as a myoblast recruitment factor during mammalian muscle growth. *Cell* 2003; 113: 483–94.

74. Figueras M., Busquets S., Carbo N., Barreiro E., Almendro V., Argiles J.M., and Lopez-Soriano F.J. Interleukin-15 is able to suppress the increased DNA fragmentation associated with muscle wasting in tumour-bearing rats. *FEBS Lett.* 2004; 569: 201–6.

75. Carbo N., Lopez-Soriano J., Costelli P., Busquets S., Alvarez B, Baccino F.M., Quinn L.S., Lopez-Soriano F.J., and Argiles J.M. Interleukin-15 antagonizes muscle protein waste in tumour-bearing rats. *Br. J. Cancer* 2000; 83: 526–31.

76. Furmanczyk P.S. and Quinn L.S. Interleukin-15 increases myosin accretion in human skeletal myogenic cultures. *Cell. Biol. Int.* 2003; 27: 845–51.

II

Diseases Associated with Cachexia

7 Diseases Associated with Cachexia: Cancer

R.J.E. Skipworth, M. Dahele, and K.C.H. Fearon

CONTENTS

SUMMARY

Cachexia is a chronic wasting syndrome involving loss of both adipose tissue and lean body mass (LBM), and which is difficult to reverse with conventional nutritional support. Indeed, the translation of the word *cachexia* from its Greek origin is "bad condition." A patient with the syndrome typically demonstrates anorexia, early satiety, severe weight loss, weakness, anemia, and oedema.[1]

Cachexia occurs in up to one-half of all patients diagnosed with cancer.[2] It results from a complex interaction of responses between the tumor and the host, the outcome of which is dependent on both tumor phenotype and host genotype. Mediator pathways include activation of the pro-inflammatory cytokine network, production of tumor-specific cachectic factors, and stimulation of the neuroendocrine stress response.

Cachexia represents a significant physical and mental burden for patients with cancer. Cachectic individuals suffer from decreased functional status, reduced response to conventional antineoplastic therapy,[3,4] and shortened survival.

7.1 THE DEFINITION OF CANCER CACHEXIA

Cachexia is a clinical syndrome that is difficult to define.[5] The complex, multi-factorial origin of cachexia precludes a uniform pathophysiological definition. The difficulty in defining cachexia was highlighted at a workshop held in 1997 entitled 'Clinical Trials for the Treatment of Secondary Wasting and Cachexia: Selection of Appropriate End-Points.[6] Various authors have attempted to reach a concise definition for cachexia. MacDonald,[7] for example, proposed — "*a wasting syndrome involving loss of muscle and fat directly caused by tumour factors, or indirectly caused by an aberrant host response to tumour presence.*" This definition, however, does not indicate the depth of complexity of cachexia or the expected prognosis of the cachectic patient.

Cachexia is not a state that exists independent of time. It represents a developing condition that passes through early, middle, and late phases (Figure 7.1). Advanced cachexia is easily diagnosed in patients with severe non-volitional weight loss, muscle wasting, anorexia, early satiety, reduced physical function, fatigue, anemia, and oedema. However, these features are non-specific and cannot be used to identify a patient in the early stages of cachexia. Specific features of cachexia are more difficult to define. In contrast to simple starvation, cachectic patients seem to undergo selective depletion of their skeletal muscle mass with relative preservation of visceral protein mass.[8] Preservation of the viscera might be explained partly by the stimulation of a hepatic acute phase protein response (APR) to the tumor. However, recent evidence has suggested a specific targeted loss of myosin heavy chains (MyHC) by pro-inflammatory cytokines to account for the marked loss of skeletal muscle in this condition.[9] A further unique feature of cachexia is that there appears to be a suboptimal response to conventional nutritional support. Patients may undergo repletion of fat stores but restoration of LBM is much more difficult to achieve.[10]

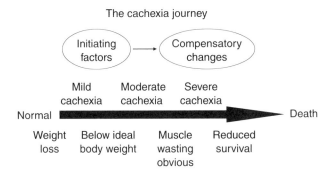

FIGURE 7.1 The chronological phases of cancer cachexia. In the early stages, it is worth noting that obese patients may be suffering from mild cachexia despite their body weight being above the ideal range.

The lack of strict diagnostic criteria for cancer cachexia hinders clinical studies, both at a mechanistic level, and in terms of the introduction of effective therapy via clinical trials. In the context of clinical trials, the absence of standardized inclusion criteria has been further compounded by the lack of structured endpoints. In determining an appropriate future definition of cancer cachexia, it will be important to identify those components of cachexia that demonstrate prognostic power and influence on an important domain of the patient's quality of life (QoL). Furthermore, it should be an aim to quantify these components by degree of severity so that cachectic study populations can be stratified. In identifying subgroups of cachectic patients with different dominant features, we may be able to commence targeted therapy. For example, the efficacy of systemic inflammatory response/APR modulators may be strongly affected by the level of activation of such pathways.

7.2 THE INCIDENCE OF CACHEXIA IN CANCER PATIENTS

The incidence of cachexia in cancer patients is usually determined by assessing their degree of weight loss. However, weight loss may be compromised as a direct index of wasting in those patients who develop fluid retention. Although marked loss of LBM (primarily skeletal muscle) is a characteristic of cachexia[11] and distinguishes the syndrome from simple starvation, LBM includes the extracellular water space and therefore suffers from the same problem. Recently, cross-sectional imaging of muscle mass and visceral fat (using computerized tomography [CT] or magnetic resonance [MRI]) was proposed as a direct clinical measure that could be used to assess cachexia.[12] However, this method measures tissue (e.g., muscle) volume rather than tissue content (e.g., protein) and is therefore open to misinterpretation.

In 1980, Dewys and the Eastern Cooperative Oncology Group[13] performed one of the most comprehensive clinical studies to examine the prognostic potential of weight loss in cancer patients who had not yet received chemotherapy. The incidence and severity of weight loss in the previous 6 months was recorded in a range of different tumor types. In the order of decreasing association with weight loss, the cancer types were ranked thus:

- Gastric cancer (85%)
- Pancreatic cancer (83%)
- Non-small cell lung cancer (NSCLC) (61%)
- Small cell lung cancer (SCLC) (57%)
- Prostate cancer (56%)
- Colon cancer (54%)
- Unfavorable non-Hodgkin's lymphoma (48%)
- Sarcoma (40%)
- Acute nonlymphocytic leukemia (39%)
- Breast cancer (36%)
- Favourable non-Hodgkin's lymphoma (31%)

The *incidence* of weight loss in each disease also correlated with *severity* of weight loss. It is worth noting, that unlike the rest of the data, weight loss in pancreatic cancer patients was only analyzed over the preceding 2 months indicating the severity of cachexia in this particular disease.

In a study focused on patients with unresectable pancreatic cancer, 85% of patients were shown to have lost a median of 14% of their pre-illness stable weight at diagnosis.[14] This weight loss increased to a median of 25% at or near the time of death (median inter-assessment period = 27 weeks). Over this period, LBM fell from 43.4 to 40.1 kg and adipose mass decreased from 12.5 to 9.6 kg.

Other than patients with upper gastrointestinal (GI) malignancy, patients with lung cancer are the major group known to sustain most significant weight loss. Weight loss at presentation was reported as observed in 59%, 58%, and 76% of patients with SCLC, unresectable NSCLC, and mesothelioma, respectively.[3]

At the commencement of radiotherapy, 57% of head and neck cancer patients were observed to have lost weight,[15] with a mean weight loss of 6.5 kg, equating to approximately 10% of body weight. However, whether head and neck cancer patients are truly cachectic or are predominantly suffering from semi-starvation is open to question since most patients will respond readily to enteral nutritional support.

The choice of weight loss as a marker of cachexia is understandable as weight loss is easily definable, simple to measure, and clinically related to patient outcome. However, it is possible that weight loss is an insensitive marker of the physiological dysfunction driving cachexia. By the time a patient manifests significant weight loss, the cachectic process is firmly established and is likely to be more difficult to slow or reverse. There is therefore a need for practical early markers of the cachectic process in cancer patients. Abnormalities in resting energy expenditure (REE) that precede weight loss have now been identified in certain groups of cancer patients and may represent one such marker. For example, patients with nonmetastatic NSCLC exhibit an increase in REE of 140 ± 35 kcal per day (after adjusting for differences in LBM) despite less than 25% having significant (>5%) weight loss.[16] These results suggest that intervention trials that target hypermetabolism in cancer cachexia could be initiated prior to documented weight loss.

7.3 THE PATHOPHYSIOLOGY OF CANCER CACHEXIA: TUMOR VS. HOST

Cancer cachexia appears to be the result of a variety of interactions between the host and the tumor, the nature of which is incompletely understood[17–19] (Figure 7.2). The complexity of these interactions may partly explain why patients with macroscopically similar tumors can demonstrate considerable variation in their tendency to develop cachexia. The host response often includes activation of the acute phase response/systemic inflammatory response and the neuroendocrine stress response. Pro-inflammatory cytokines (e.g., tumor necrosis factor [TNF]-α, interleukins [IL]-1, IL-6, IL-8, IL-12, and interferon [IFN]-γ) may be produced by tumor cells, host inflammatory cells, or a combination of the two.[20] The tumor's influence includes amplification of the pro-inflammatory/stress response and direct expression of pro-cachectic factors that have direct catabolic effects, such as proteolysis-inducing factor (PIF)[21] and lipid mobilizing factor (LMF).[22] The relative importance of different pathways and mediators in individual patients or tumor types remains unclear.

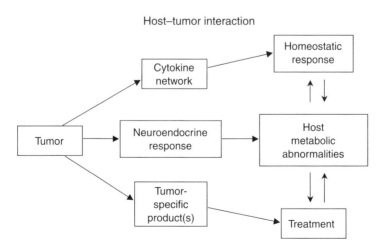

FIGURE 7.2 Mediator pathways implicated in the genesis of cancer cachexia. Different pathways contribute to a variable extent depending, in part, on host genotype and tumor phenotype.

7.4 TUMOR-SPECIFIC PRODUCTS

7.4.1 Proteolysis-Inducing Factor

The induction of cancer cachexia in parabiotic rats provided early evidence that cachexia is mediated by circulating factors.[23] PIF appears to be one such factor. It was first identified as a glycosylated polypeptide using an antibody cloned from splenocytes of mice transplanted with the MAC16 adenocarcinoma.[21] This antibody binds to a sulfated glycoprotein (24,000 kD), which is also present in the urine of cachectic human cancer patients. PIF has since been identified in the urine of weight-losing patients with carcinoma of the pancreas, breast, ovary, lung, colon, rectum, and liver.[24] Patients with pancreatic cancer excreting PIF in the urine have a significantly greater total weight loss and rate of weight loss than patients whose urine does not contain PIF.[25] If PIF is isolated from human urine and injected into mice, it also induces muscle wasting in these creatures.[24]

Proteolysis-inducing factor transcripts appear to be absent or minimally present in normal tissues, with the exception of sweat glands,[26] the pons, and the paracentral gyrus of the cerebral cortex.[27] In contrast, PIF mRNA has been demonstrated in human GI cancer,[28] prostate cancer,[29] and melanoma.[30] Studies in human breast cancer cells have suggested that PIF shares sequence homology with dermcidin (DCD)[27], an antibiotic peptide normally secreted by sweat glands[26], and with YP-30, a neural survival factor.[27] Expression of YP-30 is induced by oxidative stress[27] and promotes cell growth and survival in breast cancer cells *in vitro*.[27] Tumor implantation studies in mice with MCF7 clones expressing human PIF also suggest that PIF may have a survival-promoting component.[31]

Proteolysis-inducing factor is capable of inducing skeletal muscle breakdown, both *in vitro* and *in vivo*, and is thought to do so via activation of the ubiquitin–proteasome pathway[11] (Figure 7.3). In this pathway, proteins are tagged for degradation by the attachment of a polyubiquitin chain, which is recognized by the 26S proteasome, a large multi-subunit catalytic complex. PIF-induced protein degradation, both in gastrocnemius muscle *in vivo* and murine fibroblasts *in vitro*, is associated with increased expression of both mRNA and protein for the ubiquitin conjugating protein and the proteasome α and β subunits.[32] Upregulated expression of the ubiquitin–proteasome proteolytic pathway in murine myotubes is associated with PIF-induced activation of the transcription factor nuclear factor kappa B (NF-κB),[33] possibly through activation of protein kinase C.[34] PIF has also been shown to activate NF-κB and STAT3 in hepatocytes resulting in increased production of IL-6, IL-8, and C-reactive protein (CRP), and decreased production of transferrin.[35] Thus, PIF and Il-6 may be common mediators of both cachexia and systemic inflammation.

There are, however, unresolved issues concerning the biology of PIF. It appears to be transcribed from a gene that, depending on the extent of transcription, may result in other factors such as dermcidin or YP-30 being produced. Moreover, the biological activity of PIF (in terms of the induction of skeletal muscle proteolysis) appears to be dependent on the protein being glycosylated. Stable forced expression

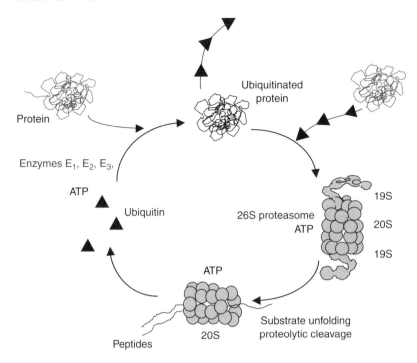

FIGURE 7.3 The ubiquitin–proteasome pathway is a complex multistep process, which requires ATP and results in the tagging of specific proteins with ubiquitin for degradation in the proteasome.

of the human gene homologue of PIF in multiple murine and human cell lines produces a non-glycosylated secreted protein that does not induce murine cachexia *in vivo*.[31] These findings suggest potential cross-species differences in the protein sequence and posttranslational processing of PIF. However, the presence of at least PIF core peptide in some normal tissues suggests that pro-cachectic tumors must produce a certain glycosylated variant of PIF. Full characterization of PIF to account for such specificity is awaited.

7.4.2 Lipid Mobilizing Factor

Lipid mobilizing factor is produced by cachexia-inducing tumors and is involved in the specific mobilization of adipose tissue, with increased oxidation of released fatty acids, possibly via induction of uncoupling protein (UCP) expression.[36] It has been shown that LMF isolated from the MAC16 murine tumor, or from patients with cancer cachexia, stimulates lipolysis directly through interaction with adenylate cyclase in a GTP-dependent process.[22] This effect may also be produced by the interaction of LMF with the β_3-adrenoreceptor.[37]

Mitochondrial UCPs transport protons across the inner mitochondrial membrane, not linked to ATP synthesis. They constitute a potential energy sink and thus the over-expression of UCPs may be another aetiological factor in cachexia. LMF has been shown to increase both mRNA and protein levels of UCP-1, -2, and -3 in brown adipose tissue (BAT) and UCP-2 in murine skeletal muscle and liver.[36] LMF is thought to be the cause of lipid mobilization and the decrease in plasma leptin observed in the MAC16 colon adenocarcinoma mouse model. In this model, UCP-1 mRNA levels are elevated in BAT, and UCP-2 and UCP-3 levels are increased in skeletal muscle.[38] Transgenic mice over-expressing UCP-3 in skeletal muscle are hyperphagic and weigh less than wild-type littermates, largely through a reduction in adipose tissue mass.[39] Increased levels of UCP-2 and UCP-3 mRNA expression are also demonstrated in the gastrocnemius and soleus muscles of cachectic rats bearing the Yoshida AH-130 ascites hepatoma.[40] This increase in mRNA content is associated with a twofold increase in fatty acid, triglyceride, and cholesterol levels. Reduction of hyperlipemia with nicotinic acid also reduces UCP-3 expression in soleus, but not in gastrocnemius. This suggests that circulating fatty acids may be involved in the regulation of UCP-3 gene expression in aerobic muscles during cancer cachexia. Cachexia in the Yoshida AH-130 ascites hepatoma model is thought to be mediated by TNF-α, and a single IV injection of TNF-α administered to these rats does cause a significant increase in skeletal muscle UCP-2 and -3.[41] These results suggest that LMF or TNF-α may be responsible for the elevation of UCP-3 mRNA seen in the skeletal muscle of cachectic cancer patients, possibly through elevation of serum lipid levels.

Despite these observations, the role of UCPs in body weight regulation has been questioned, and evidence in human subjects is scant.[42] However, one supportive study has demonstrated that the skeletal muscle of patients with gastric adenocarcinoma and weight loss demonstrates a fivefold increase in mRNA levels for the mitochondrial UCP-3 compared with controls and cancer patients who have not lost weight.[43]

7.4.3 Cytokines

A variety of human cancer cell lines have been shown to produce both pro-inflammatory and anti-inflammatory cytokines.[44,45] However, cytokines released by tumor cells are generally not detectable in the circulation and probably act only locally to promote inflammation and activate host inflammatory cells passing through the tumor. These activated host cells then release their own cytokine cascade, which initiates the APR. In cancer cases where serum levels of cytokines (e.g., TNF-α) are elevated, these levels generally correlate with the stage of the disease, reflecting tumor size and metastasis.[46]

Cytokines, such as IL-6, one of the key inducers of the APR in humans, are incapable of inducing the full cachexia syndrome on their own. Patients receiving treatment with IL-6 as part of antineoplastic trials report side-effects of fatigue and flu-like symptoms but only a proportion develop weight loss.[47] Studies using incubated rat skeletal muscle have also clearly demonstrated that IL-6 has no direct effect on muscle proteolysis[48] (although recent data has shown that MyHC protein expression is decreased in the colon-26 adenocarcinoma mouse,[9] a largely IL-6-dependent model[49]). Therefore, it seems likely that a combination of mediators work together to induce cachexia.

Other putative cytokine mediators of cachexia include TNF, IFN-γ, IL-1, leukaemia inhibitory factor (LIF), and transforming growth factor (TGF)-β. Chinese hamster ovary (CHO) cells transfected with the human TNF gene produce cachexia when implanted into nude mice.[50] A similar result is also found if the same mice are transfected with CHO cells constitutively producing IFN-γ[51]. An anti-IFN-γ monoclonal antibody is able to reverse the wasting syndrome associated with the murine Lewis lung carcinoma.[52] TNF and IFN-γ work cooperatively to down-regulate transcription of the MyHC gene *in vitro* and *in vivo*.

Some cytokines may be potential repressors of cachexia. IL-4, IL-10,[53] and IL-13 all demonstrate anti-inflammatory, and hence anticachectic, activity.[54] The final wasting status of the cachectic patient presumably depends on the balance between pro-inflammatory and anti-inflammatory cytokines.

7.5 HOST RESPONSE MECHANISMS

7.5.1 The Acute Phase Response/Systemic Inflammation

The APR is frequently identified in weight-losing cancer patients and is now recognized as an independent adverse prognostic factor. The presence of such a response is associated with accelerated weight loss and shortened survival duration in patients with several different forms of cancer (e.g., pancreatic,[55] renal,[56] lung,[57] and gastroesophageal cancer[58]).

The mechanism by which the APR is related to weight loss and survival is unknown, but it has been suggested that acute phase protein production represents a sink for amino acids that contributes to the loss of skeletal muscle[59] (Figure 7.4). The APR is also associated with anorexia and reduced food intake, reduced muscle protein synthesis, increased catabolism, and reduced voluntary activity (leading to secondary muscle wasting).

These responses are probably initiated and modulated by pro-inflammatory cytokines released from either the tumor or the host monocyte/macrophage cell system. Target organ changes are then mediated through secondary messengers (e.g., eicosanoids, arachidonic acid, platelet activating factor, and nitric oxide). The main cytokine influencing the APR in humans is thought to be IL-6. It has been shown that peripheral blood mononuclear cells (PBMC) from cancer patients induce a hepatic APR via an IL-6-dependent mechanism.[60,61] Both TNF and IL-1 are capable of inducing IL-6 production from tumor and host cells.[62]

To utilize the APR not only as a prognostic variable but also as a target for therapy, it is vital to understand the mechanisms underlying the development of this response and how these translate into accelerated demise. Recent pilot data from estimation of cytokine mRNA in gastroesophageal cancer tissue have shown that tumor cytokine gene expression correlates weakly with the serum acute

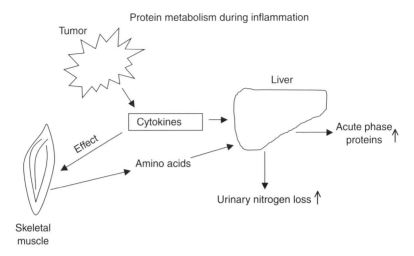

FIGURE 7.4   Pro-inflammatory cytokines may induce muscle wasting either directly or indirectly (via the acute phase response). During an acute phase response, increased synthesis of hepatic export proteins (such as fibrinogen) puts an added demand on the body's labile amino acid reserves, which is met, in part, by the breakdown of skeletal muscle.

phase proteins haptoglobin and alpha-1 antichymotrypsin, but not CRP.[63] One interpretation of these findings is that gastroesophageal cancer is associated with a range of expression of pro-inflammatory cytokine genes but that this expression does not appear to influence circulatory acute phase protein levels directly. Thus, the modulation of the systemic inflammatory response may not only be dependent on the tumor but also on the "normal" tissues of the host.

It has been postulated that the interplay between tumor cells, host lymphocytes, and tumor-associated macrophages (TAMs) results in alternatively activated monocytes giving rise to a microenvironment in which cell-mediated immune responses and apoptosis are reduced, while angiogenesis is promoted.[64–66] This predominantly Th-2 type microenvironment may favor tumor progression,[67,68] which, in turn, by virtue of tumor cell phenotype (production of pro-inflammatory mediators) and local tissue effects/destruction, may paradoxically promote systemic inflammation. However, little is known about the predominant cytokine profile of tumor cells and macrophages within a cancer mass or the surrounding "normal" tissue. The final mediator pathway of systemic inflammation probably includes monocytes activated within the tumor mass. Such cells can then re-enter the circulation and release cytokines at distant target organs.[69]

The relationship between host cytokine genotype and survival in human cancer has previously been demonstrated. For example, a polymorphism of the human IL-1β gene has been shown to be associated with reduced survival of patients with pancreatic cancer.[70] However, no such correlation could be demonstrated between two polymorphisms of the human TNF gene and survival.[71] These studies have focused on pro-inflammatory cytokines but it stands to reason that similar future studies concentrating on anti-inflammatory cytokine genotype may also demonstrate influence on patient survival. Inflammation may be an inevitable consequence of cancer but the effect of this inflammation on the host may be directly related to the ability of the host to regulate such inflammation.

7.5.2 Increased Catabolism and Anabolic Failure in the Host

One view of cancer-associated weight loss is that it reflects increased catabolism and inadequate anabolic activity. The energy expenditure (and therefore energy demand) of a cachectic patient is largely governed by the patient's REE and the energy expended on physical activity (PA). Cancer patients may lose weight despite a normal food intake suggesting that resting energy requirements are increased (i.e., catabolism is increased). However, increased,[72] normal,[73] and reduced REE[74] have all been described. The REE of cancer patients may be comparable to normal controls even when it can be shown that whole-body protein turnover rate is significantly elevated.[75] It has been suggested that different tumor types (and stage of disease) may be associated with different effects on REE; lung and pancreatic cancer inducing an increased REE,[76,77] while gastric and colorectal cancer having little effect.[78]

Mediators of catabolism in cachexia include endogenous corticosteroids, PIF, LMF, and pro-inflammatory cytokines. Inadequate anabolic activity might reflect hypogonadism or reduced insulin production. Alternatively, it may reflect an impaired response to anabolic mediators such as anabolic-androgenic steroids, growth hormone, and insulin, or an inadequate supply of energy and macro/micronutrients. Weight-losing cancer patients frequently exhibit relative glucose intolerance, and resistance to insulin and growth hormone.[17-19]

7.6 Other Causes of Cachexia in Cancer

7.6.1 Inadequate Nutritional Intake

Patients with advanced cancer may have inadequate nutritional intake and fail to appropriately increase their intake in response to increased resting energy demands.[79] Intake may be reduced by 'primary' or 'secondary' mechanisms. Primary mechanisms represent a constitutive part of the cachexia syndrome and manifest as anorexia or early satiety. In this situation, anorexia is defined as "a reduction in food intake caused primarily by reduced appetite," as opposed to the more literal definition of "not eating." Anorexia is common in cancer patients, with an incidence of 15 to 40% at diagnosis.[80] However, it does not fully explain a patient's weight loss as measured food intake fails to correlate with the degree of malnutrition.

The pathogenesis of 'primary' anorexia/early satiety and the control of human appetite are incompletely understood. At the present time, cytokines, (including Il-1 β, Il-6, and TNF), central neurotransmitters (e.g., serotonin), feeding-stimulating neuropeptides (e.g., neuropeptide Y [NPY]), hormones (e.g., leptin and ghrelin), and tumor-derived compounds[81] have all been implicated.[82] However, two peptide systems in particular appear to be strongly influential in the control of feeding behavior. These are the orexigenic NPY and the anorexigenic pro-opiomelanocortin (POMC) systems. Both originate in the hypothalamic arcuate nucleus and extend projections widely over the brain.[83] They have been shown to be intricately linked with each other and to operate in parallel. POMC neurons are the source of the potent melanocortin neuropeptides, such as α-melanocyte-stimulating hormone (α-MSH), which via interaction with the central melanocortin-3 and 4 receptors (MC3/4-R), induce an anorectic state in the individual. Studies in knockout mice have demonstrated that POMC-derived α-MSH may be post-transcriptionally regulated by neuronal basic helix-loop-helix transcription factors.[84]

Other putative mediators of human appetite may well operate through communication with the POMC and NPY systems. For example, leptin is known to increase the frequency of action potentials in POMC neurons by two different mechanisms[83]: depolarization through a nonspecific cation channel; and reduced inhibition by local orexigenic NPY/gamma aminobutyric acid (GABA) neurons. Furthermore, downstream melanocortin peptides have an auto-inhibitory effect on this circuit.[83]

In cancer cachexia, leptin levels are markedly low because of weight loss and therefore melanocortin signaling should be reduced. However, studies show that the POMC system still appears to be important in the development of cancer-associated anorexia. For example, when administered into the third cerebral ventricle of Lobund–Wistar rats (an anorexic animal model of prostate cancer), the melanocortin receptor antagonist SHU9119 completely reverts anorexia and produces weight gain comparable to that observed in non-tumor bearing control animals.[85] However, intracerebroventricular (icv) injections of NPY and ghrelin were unable to elicit the same response.

The role of cytokines in cancer anorexia may well be effected through influence on both the NPY and POMC systems. In particular, IL-1 has been associated with the induction of anorexia[86] by blocking NPY-induced feeding. NPY levels are reduced in anorectic tumor-bearing rats,[87] and a correlation between reduced food intake and brain IL-1 is also demonstrated in the same animals. The mechanism involved in the attenuation of NPY activity by cytokines may involve inhibition of NPY synthesis, inhibition of neuronal firing rates, or an attenuation of its postsynaptic effects.[88] Studies in rats have also shown that the anorexic, but not pyrogenic actions, of IL-1β are modulated by central MC-3/4 receptors,[89] and that the carboxyl terminal dipeptide of α-MSH, Lys-Pro, transiently antagonizes anorexia induced by IL-1β.[90]

Cytokines may further mediate anorexia via an increase in corticotropin-releasing factor (CRF), a central neurotransmitter that suppresses the function of glucose-sensitive neurons and food intake. The neuroanatomical site of action of this pathway is likely to be the bed nucleus of the stria terminalis.[91] Surprisingly, CRF-induced anorexia is not influenced by MC-4 R blockage,[92] despite the fact that

α-MSH can antagonize the anorexia induced by CRF,[93] suggesting the existence of a feedback loop.

Other proposed neurotransmitter mediators of cancer-associated anorexia include serotonin. Changes in the circulating levels of free tryptophan induce changes in brain serotonin concentration and, consequently, changes in food intake.[94]

The decreased PA of cachectic cancer patients may also have the potential to reduce appetite by improving the sensitivity of the physiological satiety signaling system, by adjusting macronutrient preferences or food choices, and by altering the hedonic response to food.[95]

Secondary mechanisms of reduced nutritional intake include physiological and mental problems associated with the physical presence of tumor, for example, mechanical gut obstruction, dysphagia, nausea, constipation, depression, GI fungal infection, and side-effects of treatment. Secondary problems should be proactively sought by medical staff and appropriately managed. If they are not appropriately managed, not only will the patient's physical status deteriorate, but the inanition, depression, and fatigue that cachectic patients suffer will also be further compounded by social isolation as they are unable to take part in the normal social patterns of eating. Medical staff should always be alert to the risk of deteriorating nutritional status when patients are hospitalized.[96,97]

7.6.2 Causes of Cachexia Related to Antineoplastic Therapy

Many cancer interventions will exacerbate already reduced energy and nutrient intake. Surgical patients may be fasted for prolonged periods peri-operatively, and both chemotherapy and radiotherapy can induce side-effects such as anorexia, nausea, vomiting, mucositis, taste change, or lethargy.[98] Symptoms will depend on the nature and course of the chemotherapeutic drugs being used and the location, volume, and dose of radiotherapy. Some cytotoxic drugs may even generate their own cachexia-like side-effects.[99] For example, antitubulin taxanes induce greater loss of body weight in tumor-bearing mice than in healthy mice, even when the agents significantly reduce tumor growth. The complex interaction between nutrition, cachexia, and chemotherapy still requires elucidation.[100–102]

7.7 CLINICAL ASSESSMENT OF CACHEXIA

7.7.1 Weight Loss and Body Composition

Changes in body composition can be assessed with methods of varying complexity. Simple and inexpensive measures include body weight, body mass index (BMI), and bedside anthropometry using age-adjusted reference values, for example, mid-arm circumference, arm muscle circumference and triceps skin-fold thickness. Bio-impedance analysis (BIA) is a safe, quick, and practical field-methodology that allows calculation of total body water (TBW), extracellular fluid (ECF) volume, intracellular fluid (ICF) volume, LBM, and fat mass. Dual-energy x-ray absorptiometry (DEXA), *in vitro* neutron activation and tritiated water analysis, total body potassium, and underwater weighing are all more specialized techniques. The role of body composition analysis in wasting has been reviewed by Koch et al.[103]

It is important to note that how weight loss is defined may influence the conclusions reached by clinical studies of cachexia. One such trial prospectively studied eight weight-loss-related variables in patients with NSCLC.[104] The authors found that *total* weight loss was the best predictor of prognosis and that it was a superior measure to *rate* of weight loss.

7.7.2 Nutritional Status

Although the routine formal assessment of functional nutritional status is uncommon in medicine, practical approaches to cancer nutrition should be highlighted and encouraged.[105] There are a number of objective assessment tools available that can be used to guide therapy.[106] The "mini-nutritional assessment" (MNA) was initially developed as an inexpensive assessment tool to identify elderly persons (≥65 years) who were at a risk of malnutrition. It consists of 18 questions in 4 categories relating to simple anthropometry, general assessment, dietary assessment, and subjective assessment. It has been shown to correlate with weight loss and CRP in patients with advanced cancer receiving palliative chemotherapy.[107] The "subjective global assessment" (SGA) is a clinical tool that combines subjective and objective information from patient history and examination and then categorizes patients according to nutritional state. It is effective in detecting cancer patients at risk of malnutrition.[108]

The prognostic inflammatory nutritional index (PINI) was originally developed for the assessment of nutritional status and prognosis in critically ill patients. In developing the tool, discriminant analysis of blood markers, including those pertaining to nutritional status, was used to select acute phase reactants (orosmucoid, CRP) and visceral proteins (albumin, prealbumin). These were then combined in a formula to yield a score that was able to stratify critically ill patients according to their risk of complication and mortality. The authors suggested that it could be used to assess most pathological conditions, and it has been shown that the PINI is indeed highly abnormal in cancer patients.[109, 110] In these patients, high PINI scores correlate with elevated serum IL-6 levels.[110]

7.7.3 Functional Status

Poor physical function in cachexia may relate to many factors including loss of body mass, reduced substrate supply (food intake), and reduced volitional effort (fatigue/depression), all of which can be related, at least in part, to the effects of systemic inflammation. The loss of physical reserve can lead to a devastating loss of independence. Furthermore, it may also exacerbate functional decline as inactivity is linked to muscle atrophy.

Functional status can be assessed subjectively or objectively. Subjective assessment is usually performed using scores of performance status (PS). Most PS tools rely on physician assessment of subjective variables including patient independence and their ability to work and be active. Examples of such tools include the Karnofsky performance score (KPS), the World Health Organisation (WHO) score, and the Eastern Cooperative Oncology Group (ECOG) scores. These scores may be subject to bias but general agreement has been demonstrated between oncologist and patient-assessed ECOG scores.[111]

Palliative therapy (i.e., chemotherapy and radiotherapy) is heavily influenced by estimates of patient PS. It is used as a study outcome measure, in the assessment of patient suitability for therapy, and in the prediction of benefit to be derived from the said therapy. PS tools have proven value as prognostic indicators in cancer patients[112] and as clinical trial inclusion criteria.[113] Attempts have also been made to validate accurately the scores as global markers of functional status.[114] However, despite growing interest in the functional assessment of cancer patients,[115] it remains unclear exactly what PS scores measure and how they relate to true levels of patient activity. It has been suggested that PS suffers from being too narrow a tool,[116] and that conventional PS scores may be less informative in older patients.

Objective assessment of a patient's functional status can be performed using either simple techniques, or more involved technologies that accurately measure a patient's physical activity level (PAL). Simple techniques involve direct measurement of muscle power (e.g., grip strength, treadmill testing, and dynamic evaluation of leg extensor power). Leg strength correlates well with self-reported functional status in elderly women,[117] and improvements in lower extremity strength are associated with gains in chair rise performance, gait speed, mobility, and "mobility confidence" in the elderly.[118] Handgrip dynamometry is a readily measured tool, which is an accurate prognostic indicator in surgical patients.[119] The Simmonds Functional Assessment (SFA) combines a self-reported questionnaire, which collects information on symptoms and function, and a panel of nine physical tasks, including tying a belt, putting on a sock, and walking tests.

The gold standard methodology for determining PAL involves combined measurement of total energy expenditure (TEE) by the doubly labeled water technique and REE by indirect calorimetry. PAL represents estimated energy requirements as a multiple of the basal metabolic rate (BMR) and is derived from the ratio of TEE to REE (Figure 7.5). PA is a significant and variable component of daily energy expenditure in free-living individuals. Although REE may be increased in some wasted cancer patients, TEE may actually fall, suggesting that they modulate energy demand by reducing PA.[120]

7.7.4 Quality of Life

The palliative therapy of advanced cancer patients is often administered to improve symptoms, functional status, and QoL. This is because objective tumor responses

Measurement of physical activity

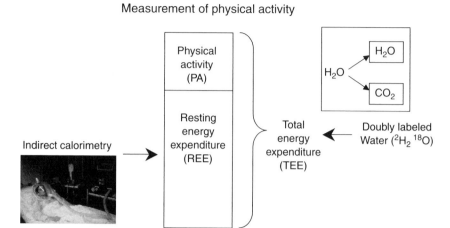

Physical activity level (PAL) = TEE/REE

FIGURE 7.5 Total energy expenditure (TEE) is composed of resting energy expenditure (REE) and physical activity (PA). The level of PA performed by an individual can be expressed as the ratio of TEE/REE and is known as the physical activity level (PAL).

vary between only 20 and 70%, and often there is no survival advantage. For these reasons, research into QoL assessment has undergone a significant expansion in clinical oncology over recent years.[114] QoL is a multidomain phenomenon and subject to observer variation; physician, patient, and partner perception of a patient's QoL can differ.[121] When asked what factors are important to QoL, patients group their responses into five distinct domains — their own state, the quality of their palliative care, their physical environment, relationships, and outlook.[122] No commonly used single instrument at this time addresses all these issues. Instead, most tools use researcher-defined domains such as physical, functional, social, and emotional well-being,[123] and there is a concern that in doing so they may reflect the views of health professionals rather than patients. Indeed, the emphasis placed on health by many QoL tools may be questionable in patients with advanced cancer.[124] The measurement, analysis, and interpretation of QoL data is challenging,[125] and the complexity of the tools has resulted in selection bias secondary to patient dropout in some QoL studies.[126] Currently, the commonest tools for specific use in cancer patients include the EORTC QLQ C30 questionnaire and the Functional Assessment of Chronic Illness Therapy (FACIT) questionnaires. The latter group of questionnaires has a wide range of tools for use in different forms of cancer and cancer therapy, including one for the specific assessment of anorexia and cachexia therapy (Functional Assessment of Anorexia/Cachexia Therapy [FAACT]). Unfortunately, common cancer QoL tools are not interchangeable and are unlikely to be directly comparable.[127] Therefore, there have been recent calls for an international consensus on how best to assess QoL in oncology.[128]

7.8 Cancer Cachexia as an Independent Risk Factor in Patient Prognosis and Survival

Pretreatment weight loss has long been identified as an important prognostic factor in oncology.[13] The objective clinical consequences of cachexia are profound. Reduced survival, impaired response to anticancer therapy, impaired immunity, lower performance status, increased symptomatology, reduced PA, and impaired QoL are all associated with cancer cachexia. A number of studies also report the impact of cachexia with additional stratification according to the presence or absence of systemic inflammation (determined by CRP).

7.8.1 The Effect of Cachexia on Functional Status and QoL

Due to the poor prognosis of patients with lung and upper GI cancer, and the associated high incidence of cachexia, it is these diseases that were studied most extensively. For example, in a 1-month study of 24 patients with lung cancer, the KPS and 2 patient self-reported tools were used to show that physical function deteriorated in those not receiving chemotherapy.[129] Thirty-three per cent of patients experienced difficulty walking one block or more, 79% complained of serious fatigue, and 44% had difficulty with household chores. Only 21% were completely satisfied with their level of activity.

In a series of studies on patients with advanced GI cancer,[130–132] patients with more than 5% weight loss had a higher CRP, lower albumin, lower anthropometry measures, poorer appetite, and lower QoL scores. Both PS and CRP were associated with survival independent of weight loss. The authors suggested that a loss or gain of more than 2.5 kg over 6 to 8 weeks was required to produce a significant change in PS in weight-losing patients with GI cancer.

The objective physical burden that cachectic cancer patients experience was demonstrated by Moses et al[120] who showed that in 24 hypermetabolic, weight-losing pancreatic cancer patients TEE was significantly reduced due to a lower PAL. Measured PAL (mean 1.24 [SD 0.04]) was much lower than that recorded in healthy adults of similar age (mean 1.62 [SD 0.28]).[133] It is entirely plausible that levels of activity as low as this may exacerbate the muscle wasting seen in cachexia.[134] These levels are comparable to those observed in spinal cord injury patients living at home[135] or in patients with cerebral palsy.[136] Over an 8-week period, it was possible to increase TEE and PAL in the pancreatic cancer patients with a specialized nutritional supplement containing fish-oils but not with an isocaloric, isonitrogenous control supplement.[120]

7.8.2 The Effect of Cachexia on Patient Response to Palliative Therapy

A retrospective review of 1555 patients receiving chemotherapy for GI malignancy determined that patients with weight loss received lower initial drug doses but developed more frequent and severe dose-limiting toxicity.[4] On average, they received 1 month less treatment and weight loss correlated with reduced failure-free and overall survival. Patients who stopped losing weight had an improved overall survival. Weight loss at presentation was an independent prognostic variable. The authors concluded that these results provided a rationale for randomized nutritional intervention studies in this population.

Similar results have been found in studies of lung cancer. Patient with weight loss and NSCLC or mesothelioma failed more frequently to complete at least three cycles of chemotherapy.[3] Anemia also occurred more frequently as a drug-associated toxicity in NSCLC patients with weight loss. In NSCLC patients receiving chemotherapy, the Voorrips Physical Activity Questionnaire (PAQ), another measure of PS, has been shown to be predictive of WHO grade 3 or higher toxicity, but not 90-day disease progression.[137]

7.8.3 The Effect of Cachexia on Survival

Weight loss and KPS have long been known to be predictive of survival. In the ECOG's paper of 1980,[13] weight loss prior to chemotherapy was noted to be a significant prognostic factor for survival in a wide range of malignancies. Within each of the 12 cancer types studied (measurable gastric, nonmeasurable gastric, pancreas, NSCLC, SCLC, prostate, colon, unfavorable nonHodgkin's lymphoma, sarcoma, acute nonlymphocytic leukemia, breast, and favorable non-Hodgkin's lymphoma), survival was shorter in the patients who had experienced weight loss compared with patients who had not. For 9 of 12 comparisons this difference reached statistical significance. For sarcoma, unfavorable non-Hodgkin's lymphoma, colon, and prostate cancer, the median survival was approximately twice as long in the patients with no weight loss as in the patients with weight loss. When the data were analyzed by weight loss categories, the greatest difference in survival was noted between the no weight loss and the 0 to 5% weight loss categories. This suggests that the *incidence* of cachexia is more important as a marker of survival than the *severity* of the cachectic process at presentation. Weight loss also correlated with deterioration in PS in 11 of 12 studies (9 of 12 studies reached statistical significance).

The presence of an acute phase response (as demonstrated by raised serum CRP) has been shown to be a significant indicator of patient survival in a number of types of cancer. For example, in patients with gastric cancer, those with an inflammatory response exhibit a severely shortened median survival time (9 weeks vs. 53 weeks, $p < .001$).[138] Other diseases in which this association has also been demonstrated include esophageal cancer,[58, 139] pancreatic cancer,[55] colorectal cancer,[140] NSCLC,[57] metastatic breast cancer,[141] and advanced renal cancer.[142] However, whether the prognostic significance of inflammation is independent of the stage of the disease has not yet been determined. Moreover, the mechanism linking inflammation with shortened survival is also obscure. Systemic inflammation may simply reflect a more aggressive tumor phenotype, or may in itself be a mediator of accelerated systemic decline (cachexia), thereby shortening survival. Further analysis of patient genotype may be important in resolving these issues.[70,71]

7.9 CHALLENGES IN TREATING CANCER CACHEXIA

Many single anticancer cachexia therapies have met with little or no success. It seems likely that a combination of therapies, perhaps targeted at patient-specific pathophysiological abnormalities, will be required. There is need for consensus on research definitions of the various phases of cachexia, and also a need for criteria of therapeutic success. For example, does weight gain mean longer survival, better response to palliative therapy, or improved PS and QoL? Which body compartments should be targeted? How should different anticachexia interventions (e.g., pharmacological[143] and exercise[144,145] therapy) be integrated with conventional anticancer therapies and each other? These are questions that will take time to answer but as with any trials of medical intervention, the guiding principle must always be "Primum Non Nocere" (first do no harm).

REFERENCES

1. Tisdale M.J., Cachexia in cancer patients, *Nat. Rev. Cancer*, 2, 862, 2002.
2. Palesty J.A. and Dudrick S.J. What we have learned about cachexia in gastrointestinal cancer, *Dig. Dis.*, 21, 198, 2003.
3. Ross J.P., Do patients with weight loss have a worse outcome when undergoing chemotherapy for lung cancers? *Br. J. Cancer*, 90, 1905, 2004.
4. Andreyev H.J. et al., Why do patients with weight loss have a worse outcome when undergoing chemotherapy for gastrointestinal malignancies? *Eur. J. Cancer*, 34, 503, 1998.
5. Argiles J.M., Busquets S., and Lopez-Soriano F.J., Cytokines in the pathogenesis of cancer cachexia, *Curr. Opin. Clin. Nutr. Metab. Care*, 6, 401, 2003.
6. Raiten D.J. and Talbot J.M. Clinical Trials for the Treatment of Secondary Wasting and Cachexia: selection of Appropriate Endpoints, *J. Nutr.*, 1S, 1, 1999.
7. MacDonald N. et al., Understanding and managing cancer cachexia, *J. Am. Coll. Surg.*, 197, 143, 2003.
8. Fearon K.C. and Preston T., Body composition in cancer cachexia, *Infusionstherapie*, 17, 63, 1990.
9. Acharrya S. et al., Cancer cachexia is regulated by selective targeting of skeletal muscle gene products, *J. Clin. Invest.*, 114, 370, 2004.
10. Nixon D.W. and Lawson D.H., Nutritional support of the cancer patient, *Hosp. Formul.*, 8, 616, 1983.
11. Tisdale M.J., Loss of skeletal muscle in cancer: biochemical mechanisms, *Front. Biosci.*, 6, D164, 2001.
12. Ogiwara H. et al., Diminished visceral adipose tissue in cancer cachexia, *J. Surg. Oncol.*, 57, 129, 1994.
13. Dewys W.D. et al., Prognostic effect of weight loss prior to chemotherapy in cancer patients, Eastern Cooperative Oncology Group, *Am. J. Med.*, 69, 491, 1980.
14. Wigmore S.J. et al., Changes in nutritional status associated with unresectable pancreatic cancer, *Br. J. Cancer*, 75, 106, 1997.

15. Lees J. et al., Incidence of weight loss in head and neck cancer patients on commencing radiotherapy treatment at a regional oncology centre, *Eur. J. Cancer Care (Engl.)*, 8, 133, 1999.

16. Jatoi A. et al., Do patients with nonmetastatic non-small cell lung cancer demonstrate altered resting energy expenditure? *Ann. Thorac. Surg.*, 72, 348, 2001.

17. Heber D., Byerley L.O., and Tchekmedyian N.S., Hormonal and metabolic abnormalities in the malnourished cancer patient: effect on tumor host interaction, *JPEN J. Parenter. Enteral Nutr.*, 16, 60S, 1992.

18. Crown A.L. et al., What is the role of the insulin-like growth factor system in the pathophysiology of cancer cachexia, and how is it regulated? *Clin. Endocrinol. (Oxf).*, 56, 723, 2002.

19. Brink M., Anwar A., and Delafontaine P., Neurohormonal factors in the development of catabolic/anabolic imbalance and cachexia, *Int. J. Cardiol.*, 85, 111, 2002.

20. Strassman G., Masui Y., Chizzonite R., and Fong M. Mechanisms of experimental cancer cachexia. Local involvement of IL-1 in colon-26 tumor, *J. Immunol.* 150, 2341–2345, 1993.

21. Todorov P. et al., Characterisation of a cancer cachectic factor, *Nature*, 379, 739, 1996.

22. Hirai K. et al., Biological evaluation of a lipid-mobilizing factor isolated from the urine of cancer patients, *Cancer Res.*, 58, 2359, 1998.

23. Norton J.A. et al., Parabiotic transfer of cancer/anorexia in male rats, *Cancer Res.*, 45, 5547, 1985.

24. Cariuk P., Induction of cachexia in mice by a product isolated from the urine of cachectic cancer patients, *Br. J. Cancer*, 76, 606, 1997.

25. Wigmore S.J. et al., Effect of oral eicosapentaenoic acid on patients with weight loss in patients with pancreatic cancer, *Nutr. Cancer*, 36, 177, 2000.

26. Schittek B. et al., Dermicidin: a novel antibiotic peptide secreted by sweat glands, *Nat. Immunol.*, 2, 1133, 2001.

27. Porter D. et al., A neural survival factor is a candidate oncogene in breast cancer, *Proc. Natl Acad. Sci. USA*, 100, 10931, 2003.

28. Cabal-Mazano R. et al., Proteolysis-inducing factor is expressed in tumours of patients with gastrointestinal cancers and correlates with weight loss, *Br. J. Cancer*, 84, 1599, 2001.

29. Wang Z. et al., Expression of the human cachexia-associated protein (HCAP) in prostate cancer and in a prostate cancer animal model of cachexia, *Int. J. Cancer*, 105, 123, 2003.

30. Todorov P.T., Field W.N., and Tisdale M.J., Role of a proteolysis-inducing factor (PIF) in cachexia induced by a human melanoma (G361), *Br. J. Cancer*, 80, 1734, 1999.

31. Monitto C.L. et al., Characterization of a human homologue of proteolysis-inducing factor and its role in cancer cachexia, *Clin. Cancer Res.*, 10, 5862, 2004.

32. Lorite et al., Activation of ATP-ubiquitin-dependent proteolysis in skeletal muscle *in vivo* and murine myoblasts *in vitro* by a proteolysis-inducing factor (PIF), *Br. J. Cancer*, 85, 297, 2001.

33. Whitehouse A.S. and Tisdale M.J., Increased expression of the ubiquitin-proteasome pathway in murine myotubes by proteolysis-inducing factor (PIF) is associated with activation of the transcription factor NF-κB, *Br. J. Cancer*, 89, 1116, 2003.

34. Smith H.J., Wyke S.M., and Tisdale M.J., Role of protein kinase C and NF-kappa B in proteolysis-inducing factor-induced proteasome expression in C(2)C(12) myotubes, *Br. J. Cancer*, 90, 1850, 2004.
35. Watchorn T.M. et al., Proteolysis-inducing factor regulates hepatic gene expression via the transcription factors NF-κB and STAT-3, *FASEB J.*, 15, 562, 2001.
36. Bing C. et al., Expression of uncoupling proteins-1, -2 and -3 mRNA is induced by an adenocarcinoma-derived lipid mobilizing factor, *Br. J. Cancer*, 86, 612, 2002.
37. Russell S.T., Hirai K., and Tisdale M.J., Role of β_3-adrenergic receptors in the action of a tumour lipid mobilizing factor, *Br. J. Cancer*, 86, 424, 2002.
38. Bing C. et al., Increased gene expression of brown fat uncoupling protein (UCP) 1 and skeletal muscle UCP2 and UCP3 in MAC16-induced cancer cachexia, *Cancer Res.*, 60, 2405, 2000.
39. Clapham J.C. et al., Mice overexpressing human uncoupling protein-3 in skeletal muscle are hyperphagic and lean, *Nature*, 406, 415, 2000.
40. Busquets S. et al., Hyperlipaemia: a role in regulating UCP3 gene expression in skeletal muscle during cancer cachexia? *FEBS Lett.*, 505, 255, 2001.
41. Busquets S. et al., In the rat, tumor necrosis factor α administration results in an increase in both UCP2 and UCP3 mRNA in skeletal muscle: a possible mechanism for cytokine-induced thermogenesis? *FEBS Lett.*, 440, 348, 1998.
42. Dalgaard L.T. and Pedersen O., Uncoupling proteins: functional characteristics and role in the pathogenesis of obesity and type II diabetes, *Diabetologia*, 44, 946, 2001.
43. Collins P. et al., Muscle UCP-3 mRNA levels are elevated in weight loss associated with gastrointestinal adenocarcinoma in humans, *Br. J. Cancer*, 86, 372, 2002.
44. Wigmore S.J. et al., Cytokine regulation of constitutive production of interleukin-8 and -6 by human pancreatic cancer cell lines and serum cytokine concentrations in patients with pancreatic cancer, *Int. J. Oncol.*, 21, 881, 2002.
45. Wigmore S.J. et al., Endogenous production of IL-8 by human colorectal cancer cells and its regulation by cytokines, *Int. J. Oncol.*, 18, 467, 2001.
46. Sheen-Chen S.-M. et al., Serum concentration of tumour necrosis factor in patients with breast cancer, *Breast Cancer Res. Treat.*, 43, 211, 1997.
47. Bouffet E. et al., Phase I study of interleukin-6 in children with solid tumours in relapse, *Eur. J. Cancer*, 33, 1620, 1997.
48. Garcia-Martinez C., Lopez-Soriano F.J., and Argiles J.M., Interleukin-6 does not activate protein breakdown in rat skeletal muscle, *Cancer Lett.*, 76, 1, 1994.
49. Strassmann M. et al., Suramin interferes with interleukin-6 receptor binding *in vitro* and inhibits colon-26-mediated experimental cancer cachexia *in vivo*, *J. Clin. Invest.*, 92, 252, 1993.
50. Oliff A. et al., Tumours secreting human TNF/cachectin induce cachexia in mice, *Cell*, 50, 555, 1987.
51. Argiles J.M. et al., Catabolic mediators as targets for cancer cachexia, *D.D.T.*, 8, 838, 2003.
52. Matthys P. et al., Anti-interferon-γ antibody treatment, growth of Lewis lung tumours in mice and tumour-associated cachexia, *Eur. J. Cancer*, 27, 182, 1991.
53. Baumann H. and Gauldie J., The acute phase response, *Immunol. Today*, 15, 74, 1994.
54. Argiles J.M. and Lopez-Soriano F.J., The role of cytokines in cancer cachexia, *Med. Res. Rev.*, 19, 223, 1999.

55. Falconer J.S. et al., The acute phase response and survival duration in patients with pancreatic cancer, *Cancer*, 75, 2077, 1995.

56. Negrier S. et al., Prognostic factors of survival and rapid progression in 782 patients with metastatic renal carcinomas treated by cytokines: a report from the Groupe Francais d'Immunotherapie, *Ann. Oncol.*, 13, 1460, 2002.

57. Forrest L.M. et al., Comparison of an inflammation-based prognostic score (GPS) with performance status (ECOG) in patients receiving platinum-based chemotherapy for inoperable non-small cell lung cancer, *Br. J. Cancer*, 90, 704, 2004.

58. Deans C. et al., Adverse nutritional status is associated with systemic inflammation in gastro-oesophageal cancer and is an independent predictor of survival (Abstract), *Clin. Nutr.*, 23, 880, 2004.

59. Preston T., Slater C., McMillan D.C., Falconer J.S., Shenkin A., and Fearon, K.C.H. Fibrinogen synthesis is elevated in fasting cancer patients with an acute phase response, *J. Nutr.*, 128, 1355–1360, 1998.

60. O'Riordain M.G. et al., Peripheral blood cells from weight-losing cancer patients control the hepatic acute phase response by a primarily interleukin-6 dependent mechanism, *Int. J. Oncol.*, 15, 823, 1999.

61. Barber M.D., Fearon K.C.H., and Ross J.A., Relationship of serum levels of interleukin-6, soluble interleukin-6 receptor and tumour necrosis factor to the acute-phase response in advanced pancreatic cancer, *Clin. Sci. (Lond.)*, 96, 83, 1999.

62. Strassman G., Mechanisms of experimental cancer cachexia. Local involvement of IL-1 in colon-26 tumour, *J. Immunol.*, 150, 2341, 1993.

63. Deans C. et al., Tumour cytokine gene expression correlates with aspects of the acute phase response in patients with gastro-oesophageal cancer (Abstract), *Clin. Nutr.*, 880, 2004.

64. McKechnie A., Robins R.A., and Eremin O., Immunological aspects of head and neck cancer: biology, pathophysiology and therapeutic mechanisms, *J. R. Coll. Surg. Edin. Ire.*, 2, 187, 2004.

65. Duffield J.S., The inflammatory macrophage: a story of Jekyll and Hyde, *Clin. Sci.*, 104, 27, 2003.

66. Dalgliesh A.G. and O'Byrne K.J., Chronic immune activation and inflammation in the pathogenesis of AIDS and cancer, *Adv. Cancer Res.*, 84, 231, 2002.

67. O' Byrne K.J. and Dalgliesh A.G., Chronic immune activation and inflammation as the cause of malignancy, *Br. J. Cancer*, 85, 473, 2001.

68. Sharma R.A. et al., Angiogenesis and the immune response as targets for the prevention and treatment of colorectal cancer (Review), *Oncol. Rep.*, 10, 1625, 2003.

69. Wigmore S.J. et al., Effect of interleukin-2 on peripheral blood mononuclear cell cytokine production and the hepatic acute phase protein response, *Clin. Immunol.*, 104, 174, 2002.

70. Barber M.D. et al., A polymorphism of the interleukin-1β gene influences survival in pancreatic cancer, *Br. J. Cancer*, 83, 1483, 2000.

71. Barber M.D. et al., Two polymorphisms of the tumour necrosis factor gene do not influence survival in pancreatic cancer, *Clin. Exp. Immunol.*, 117, 425, 1999.

72. Hyltander A. et al., Elevated energy expenditure in cancer patients with solid tumours, *Eur. J. Cancer*, 27, 9, 1991.

73. Nixon D.W. et al., Resting energy expenditure in lung and colon cancer, *Metabolism*, 37, 1059, 1988.
74. Knox L.S. et al., Energy expenditure in malnourished cancer patients, *Ann. Surg.*, 197, 152, 1983.
75. Fearon K.C. et al., Influence of whole body protein turnover rate on resting energy expenditure in patients with cancer, *Cancer Res.*, 48, 2590, 1988.
76. Fredrix E.W. et al., Effect of different tumor types on resting energy expenditure, *Cancer Res.*, 51, 6138, 1991.
77. Falconer J.S. et al., Cytokines, the acute-phase response, and resting energy expenditure in cachectic patients with pancreatic cancer, *Ann. Surg.* 219, 325, 1994.
78. Fredrix E.W. et al., Resting energy expenditure in patients with newly detected gastric and colorectal cancers, *Am. J. Clin. Nutr.*, 53, 1318, 1991.
79. Bosaeus I. et al., Dietary intake and resting energy expenditure in relation to weight loss in unselected cancer patients, *Int. J. Cancer*, 93, 380, 2001.
80. DeWys W.D., Anorexia as a general effect of cancer, *Cancer*, 45, 2019, 1972.
81. Bing C. et al., Cachexia in MAC16 adenocarcinoma: suppression of hunger despite normal regulation of leptin, insulin and hypothalamic neuropeptide Y, *J. Neurochem.*, 79, 1004, 2001.
82. Laviano A., Meguid M.M., and Rossi-Fanelli F., Cancer anorexia: clinical implications, pathogenesis, and therapeutic strategies, *Lancet Oncol.*, 4, 686, 2003.
83. Cowley M.A., et al., Leptin activates anorexigenic POMC neurons through a neural network in the arcuate nucleus, *Nature*, 411, 480, 2001.
84. Jing E., et al., Deletion of the Nhlh2 transcription factor decreases the levels of the anorexigenic peptides alpha melanocyte-stimulating hormone and thyrotropin-releasing hormone and implicates prohormone convertases I and II in obesity, *Endocrinology* 145, 1503, 2004.
85. Wisse B.E. et al., Reversal of cancer anorexia by blockade of central melanocortin receptors in rats, *Endocrinology*, 142, 3292, 2001.
86. Plata-Salamn C.R., Central nervous system mechanisms contributing to the cachexia-anorexia syndrome, *Nutrition*, 16, 1009, 2000.
87. Chance W.T. et al., Hypothalamic concentration and release of neuropeptide Y into microdialysates is reduced in anorectic tumor-bearing rats, *Life Sci.*, 54, 1869, 1994.
88. King P.J. et al., Effect of cytokines on hypothalamic neuropeptide Y release *in vitro*, *Peptides*, 21, 143, 2000.
89. Lawrence C.B. and Rothwell, N.J., Anorexic but not pyrogenic actions of interleukin-1 are modulated by central melanocortin-3/4 receptors in the rat. *J. Neuroendocrinol.*, 13, 490, 2001.
90. Uehara Y. et al., The dipeptide Lys-Pro attenuates interleukin-1 beta-induced anorexia, *Peptide*, 14, 175, 1993.
91. Ciccocioppo R. et al., The bed nucleus is a neuroanatomical substrate for the anorectic effect of corticotropin-releasing factor and for its reversal by nociceptin/orphanin FQ, *J. Neurosci.*, 23, 9445, 2003.
92. Vergoni A.V. et al., Corticotropin-releasing factor (CRF) induced anorexia is not influenced by a melanocortin 4 receptor blockage, *Peptides*, 20, 509, 1999.
93. Oohara M. et al., Alpha-melanocyte stimulating hormone (MSH) antagonizes the anorexia by corticotropin releasing factor (CRF), *Life Sci.*, 53, 1473, 1993.

94. Laviano A. et al., Neurochemical mechanisms for cancer anorexia, *Nutrition*, 18, 1001, 2002.

95. Blundell J.E. et al., Cross-talk between physical activity and appetite control: does physical activity stimulate appetite? *Proc. Nutr. Soc.*, 62, 651, 2003.

96. Braunschweig C., Gomez S., and Sheean P.M., Impact of declines in nutritional status on outcomes in adult patients hospitalized for more than 7 days, *J. Am. Diet. Assoc.*, 100, 1316, 2000.

97. Ravera E. et al., Impact of hospitalization on the nutritional status of cancer patients, *Tumori*, 73, 375, 1987.

98. Donaldson S.S. and Lenon R.A., Alterations of nutritional status: impact of chemotherapy and radiation therapy, *Cancer*, 43, 2036, 1979.

99. Tohgo A. et al., Anticancer drugs that induce cancer-associated cachectic syndromes, *Expert Rev. Anticancer Ther.*, 2, 121, 2002.

100. Lawson D.H. et al., Enteral versus parenteral nutritional support in cancer patients, *Cancer Treat. Rep.*, 65, 101, 1981.

101. Samuels S.E. et al., Protein metabolism in the small intestine during cancer cachexia and chemotherapy in mice, *Cancer Res.*, 60, 4968, 2000.

102. Nelson K., Walsh D., and Sheehan F., Cancer and chemotherapy-related upper gastrointestinal symptoms: the role of gastric motor function and its evaluation in cancer patients, *Support. Care Cancer*, 10, 455, 2002.

103. Koch J., The role of body composition measurements in wasting syndromes, *Semin. Oncol.*, 25, 12, 1998.

104. Buccheri G. and Ferrigno D., Importance of weight loss definition in the prognostic evaluation of non-small-cell lung cancer, *Lung Cancer*, 34, 433, 2001.

105. Ottery F.D., Definition of standardized nutritional assessment and interventional pathways in oncology, *Nutrition*, 12, S15, 1996.

106. Bauer J. and Capra S., Comparison of a malnutrition screening tool with subjective global assessment in hospitalised patients with cancer-sensitivity and specificity, *Asia Pac. J. Clin. Nutr.*, 12, 257, 2003.

107. Slaviero K.A. et al., Baseline nutritional assessment in advanced cancer patients receiving palliative chemotherapy, *Nutr. Cancer*, 46, 148, 2003.

108. Ravasco P., Nutritional deterioration in cancer: the role of disease and diet, *Clin. Oncol. (R. Coll. Radiol.)*, 15, 443, 2003.

109. Nelson K.A. and Walsh D., The cancer anorexia-cachexia syndrome: a survey of the Prognostic Inflammatory and Nutritional Index (PINI) in advanced disease, *J. Pain Symp. Manage.*, 24, 424, 2002.

110. Walsh D., Mahmoud F., and Barna B., Assessment of nutritional status and prognosis in advanced cancer: interleukin-6, C-reactive protein, and the prognostic and inflammatory nutritional index, *Support. Care Cancer*, 11, 60, 2003.

111. Blagden S.P. et al., Performance status score: do patients and their oncologists agree? *Br. J. Cancer*, 89, 1022, 2003.

112. Buccheri G., Ferrigno D., and Tamburini M., Karnofsky and ECOG performance status scoring in lung cancer: a prospective, longitudinal study of 536 patients from a single institution, *Eur. J. Cancer*, 32A, 1135, 1996.

113. Roila F. et al., Intra- and interobserver variability in cancer patients' performance status assessed according to Karnofsky and ECOG scales, *Ann. Oncol.*, 2, 437, 1991.

114. Yates J.W., Chalmer B., and McKegney F.P., Evaluation of patients with advanced cancer using the Karnofsky performance status, *Cancer*, 45, 2220, 1980.

115. Batel-Copel L.M. et al., Do oncologists have an increasing interest in the quality of life of their patients? A literature review of the last 15 years, *Eur. J. Cancer*, 33, 29, 1997.
116. Schaafsma J. and Osoba D., The Karnofsky performance status scale re-examined: a cross validation with the EORTC-C30, *Qual. Life Res.*, 3, 413, 1994.
117. Foldvari M. et al., Association of muscle power with functional status in community-dwelling elderly women, *J. Gerontol. A. Biol. Sci. Med. Sci.*, 55, M192, 2000.
118. Chandler J.M. et al., Is lower extremity strength gain associated with improvement in physical performance and disability in frail, community dwelling-elders? *Arch. Phys. Med. Rehabil.*, 79, 24, 1998.
119. Bohannon R.W., Dynamometer measurements of hand-grip strength predict multiple outcomes, *Percept. Mot. Skills*, 93, 323, 2001.
120. Moses A.W. et al., Reduced total energy expenditure and physical activity in cachectic patients with pancreatic cancer can be modulated by an energy and protein dense oral supplement enriched with n-3 fatty acids, *Br. J. Cancer*, 8, 996, 2004.
121. Wilson K.A. et al., Perception of quality of life by patients, partners and treating physicians, *Qual. Life Res.*, 9, 1041, 2000.
122. Cohen S.R. and Leis A., What determines the quality of life of terminally ill cancer patients from their own perspective? *J. Palliat. Care*, 18, 48, 2002.
123. Cella D. et al., Advances in quality of life measurements in oncology patients. *Semin. Oncol.*, 29, 60, 2002.
124. Waldron D. et al., Quality-of-life measurement in advanced cancer: assessing the individual, *J. Clin. Oncol.*, 17, 3603, 1999.
125. Nordin K. et al., Alternative methods of interpreting quality of life data in advanced gastrointestinal cancer patients, *Br. J. Cancer*, 85, 1265, 2001.
126. Ballatori E., Unsolved problems in evaluating the quality of life of cancer patients, *Ann. Oncol.*, 12, S11, 2001.
127. Holzner B. et al., Quality of life measurement in oncology — a matter of the assessment instrument? *Eur. J. Cancer*, 37, 2349, 2001.
128. Conroy T., Bleiberg H., and Glimelius B., Quality of life in patients with advanced colorectal cancer: what has been learnt? *Eur. J. Cancer*, 39, 287, 2003.
129. Sarna L., Fluctuations in physical function: adults with non-small cell lung cancer, *J. Adv. Nurs.*, 18, 714, 1993.
130. O'Gorman P., McMillan D.C., and McArdle C.S., Impact of weight loss, appetite, and the inflammatory response on quality of life in gastrointestinal cancer patients, *Nutr. Cancer*, 32, 76, 1998.
131. O'Gorman P., McMillan D.C., and McArdle C.S., Longitudinal study of weight, appetite, performance status, and inflammation in advanced gastrointestinal cancer, *Nutr. Cancer*, 35, 127, 1999.
132. O'Gorman P., McMillan D.C., and McArdle C.S., Prognostic factors in advanced gastrointestinal cancer patients with weight loss, *Nutr. Cancer*, 37, 36, 2000.
133. Gibney E.R., Energy expenditure in disease: time to revisit? *Proc. Nutr. Soc.*, 59, 199, 2000.
134. Franssen F.M., Wouters E.F., and Schols A.M., The contribution of starvation, decondtitioning and ageing to the observed alterations in peripheral skeletal muscle in chronic organ diseases, *Clin. Nutr.*, 21, 1, 2002.

135. Mollinger L.A. et al., Daily energy expenditure and basal metabolic rates of patients with spinal cord injury, *Arch. Phys. Med. Rehabil.*, 66, 420, 1985.

136. Stallings V.A. et al., Energy expenditure of children and adolescents with severe disabilities: a cerebral palsy model, *Am. J. Clin. Nutr.*, 64, 627, 1996.

137. Jatoi A. et al., Daily activities: exploring their spectrum and prognostic impact in older, chemotherapy-treated lung cancer patients, *Support. Care Cancer*, 11, 460, 2003.

138. Rashid S.A. et al., Plasma protein profiles and prognosis in gastric cancer, *Br. J. Cancer*, 45, 390, 1982.

139. Ikeda M. et al., Significant host- and tumor-related factors for predicting prognosis in patients with esophageal carcinoma, *Ann. Surg.*, 238, 197, 2003.

140. McMillan D.C., Canna K., and McArdle C.S., Systemic inflammatory response predicts survival following curative resection of colorectal cancer, *Br. J. Surg.*, 90, 215, 2003.

141. Heys S.D. et al., Acute phase proteins in patients with large and locally advanced breast cancer treated with neo-adjuvant chemotherapy: response and survival, *Int. J. Oncol.*, 13, 589, 1998.

142. Bromwich E. et al., The systemic inflammatory response, performance status and survival in patients undergoing alpha-interferon treatment for advanced renal cancer, *Br. J. Cancer*, 91, 1236, 2004.

143. Daneryd P., Epoetin alfa for protection of metabolic and exercise capacity in cancer patients, *Semin. Oncol.*, 29, 69, 2002.

144. al-Majid S. and McCarthy D.O., Cancer-induced fatigue and skeletal muscle wasting: the role of exercise, *Biol. Res. Nurs.*, 2, 186, 2001.

145. al-Majid S. and McCarthy D.O., Resistance exercise training attenuates wasting of the extensor digitorum longus muscle in mice bearing the colon-26 adenocarcinoma, *Biol. Res. Nurs.*, 2, 155, 2001.

8 Cachexia Associated with AIDS

Kaspar Berneis and Katrin Utech

Contents

Abbreviations

AIDS	acquired immuno-deficiency syndrome
BCM	body cell mass
CB	cannabinoid
CC	cellular
CD	cluster of differentiation
CDC	Center for Disease Control
CTL	cytotoxic T lymphocyte
DHEA	dehydroepiandrosterone
FDA	Food and Drug Administration
FFM	free fat mass
FM	fat mass
GH	growth hormone
gp	glycoprotein
HAART	highly active antiretroviral therapy
HIV	human immunodeficiency virus
IFN	interferon
IGF-1	insulin-like growth factor 1
IL	interleukin
LBM	lean body mass
MHC	major histocompatibility complex
PI	protease inhibitor
PPARγ	peroxisome proliferator-activated receptor γ
SREBP1	sterol regulatory element-binding protein 1
TG	triglyceride
TNF	tumor necrosis factor
VLDL	very low density lipoprotein

SUMMARY

Cachexia is a common problem in HIV-infected patients and the wasting syndrome has been listed as an AIDS defining condition. Even in western societies where highly active antiretroviral therapies (HAART) have been established, wasting may still be present. Factors contributing to anorexia are decreased food intake, malabsorption, diarrhea, opportunistic infections, and hypermetabolism. Activation of the tumor necrosis factor system may contribute to wasting via inhibition of appetite. In addition, endocrinological abnormalities contributing to decreased lean body mass (LBM) include hypogonadism, decreased insulin-like growth factor 1 (IGF-1), and hypercortisolism. After the introduction of HAART, the mortality of HIV infection has decreased by 85%. However, approximately one third of patients may not control virus load and cachexia may persist or reoccur. There is no standardized approach to the management of AIDS-induced weight loss. However, it is commonly accepted that additional enteral nutritional supplements may help to reverse weight loss. In addition, various pharmacological agents have been shown to increase appetite and LBM. Cannabinoids increase appetite and decrease nausea, while megestrol acetate, testosterone, and its derivates such as nandrolone and oxandrolone result in increased body weight. Peptide hormones, such as growth hormone have strong anabolic properties and have been approved for the treatment of HIV-associated wasting.

It is concluded that careful nutritional assessment of patients suffering from HIV infection and corresponding treatment with nutritional counseling, increased nutritional support, and various pharmacological agents can lead to a substantial benefit in addition to the currently available HAART.

8.1 INTRODUCTION

A common manifestation of a severe disease such as the acquired immunodeficiency syndrome (AIDS) is the development of cachexia and malnutrition. Before its viral origin was recognized, AIDS was called "thin disease or slim disease" to describe the condition associated with profound wasting in individuals who were infected with the human immunodeficiency virus (HIV).[1] In 1987, the Center for Disease Control (CDC) recognized the "wasting syndrome" as an AIDS condition.[2] Wasting syndrome was defined by a loss of body weight of more than 10% with a lack of causes of wasting other than HIV. In addition, the "wasting syndrome" has been recognized as a prognostic factor in the advanced disease.[3, 4]

Malnutrition in patients with AIDS contributes to impaired host defense against infections,[5] to diminished physical mobility, impaired healing mechanisms, and to impaired quality of life.[6] The degree of cachexia is inversely correlated with survival time,[7, 8] even with adjustment for CD4 count and history of secondary events.[9]

In studies performed before the era of highly active artiretroviral therapy (HAART) estimates of the prevalence of wasting as first symptoms for AIDS defining diagnosis was up to 31%[10] and affected most of the patients by the time of death. With the use of HAART, the incidence of most of the complications associated with HIV infection has declined in western countries[11] and mortality decreased by about 85%,[12, 13] while lipodystrophy has been recognized as a new complication of the disease, which might be linked to HAART treatment.[14] However, the metabolic derangements associated with HIV infection have not been eliminated and wasting continues to occur even with HAART.[11] It has been suggested that in patients receiving HAART, even though weight loss is rare, loss of lean body mass (LBM) is common and is driven mainly by catabolic cytokines.[15]

8.2 OVERVIEW OF HUMAN IMMUNODEFICIENCY VIRUS INFECTION

8.2.1 Pathophysiology

The HIV is a member of the group lentiviruses (lentus, *lat.* = slow) of the retrovirus family. There are two types known: HIV-1 and HIV-2. The most common virus type in humans is HIV-1. The pathophysiology of the HIV-1 infection is explained by a progressive destruction of T helper (CD4) cells.[16] CD4 cells, cytotoxic T lymphocytes (CTL) and the humoral immune system, are affected by the virus infection resulting in a reduction of their function.[17] The immune response in HIV-1 infection is characterized by a weak or absent virus-specific CD4 function.[18-20] The reason is not yet completely understood. The pathogenetic process begins with the entry of the virus, most often through the genital mucosa, and contacts with a CD4 receptor carrying dendritic cell (CD4 cells, macrophages, and monocytes). There, the virus binds with an envelope glycoprotein (gp) 120. In addition to CD4 receptors, chemokine receptors (most are CC chemokine receptor 5 [CCR5] and fusin [CXCR4]) working like co-receptors are necessary.[21, 22] This results in a rapid spread of the virus in the lymphatic tissue. One theory to explain the lack of specific HIV-T helper cell activity proposes a selective clonal deletion during primary infection.[19, 23, 24] Normally, a massive expansion of virus-specific cells establishes a pool of memory cells for immunology recall during the next exposure to the virus. In primary HIV infection, viral replication occurs at an extremely rapid rate.[25] This leads to a first proliferation and activation of virus-specific CD4 T cells, resulting in subsequent infection and destruction of these cells by the virus, before virus-specific memory cells can be established.

8.2.2 Clinical Aspects and Symptoms of HIV Infection

The first recognition of AIDS cases in the 1980s was followed by a dramatic worldwide pandemic — a new dimension of a virus infection. For the purpose of surveillance, the American CDC published a case definition in 1992, which revised a first definition of AIDS from 1987 to include conditions indicative for severe immunodepression.[26] Since then, the CDC has expanded its definition[27-29] several times. The classification today has three ranges of CD4 cell counts and uses a matrix of nine categories. All patients with AIDS-indicator conditions or a CD4 cell count below 200 cells/mm^3 are reported as having AIDS. In the United States, AIDS can be defined solely on the CD4 criterion. AIDS-indicator conditions include candidiasis, cryptosporidiosis, cytomegalovirus infection, histoplasmosis, AIDS-associated wasting, dementia, Kaposi's sarcoma, lymphoma, mycobacterium infection, pneumocystis carinii infection, and salmonella infection. HIV infection is now recognized as a chronic and progressive illness with a broad spectrum of manifestations and complications from the acute, primary infection up to life-threatening opportunistic infections and malignancies. The primary infection may be present with nonspecific symptoms including fever (over 38–40°C), lymphadenopathy (most often axillary, cervical, and occipital nodes), mucocutaneous diseases (pharyngitis, generalized rash, and painful mucocutaneous ulcerations), myalgias, arthralgias, headache, aseptic meningitis, diarrhea, nausea, vomiting, hepatosplenomegaly, weight loss, and anorexia.

During the asymptomatic stage of the disease, usually following the primary HIV infection, there are no specific signs or symptoms. Diffuse lymphadenopathy, headache, and malaise may develop during this stage. The early manifestations of HIV disease are often constitutional symptoms such as headache, fatigue, fever, arthralgias, myalgias, night sweats, diarrhea, anorexia, malaise, and weight loss.[30] Late symptomatic HIV disease, also named AIDS, is characterized by declining CD4 cell count (<200 cells/mm^3), persistent or progressive constitutional symptoms, opportunistic infections, malignancies, wasting syndrome, and neurological syndromes. In very advanced HIV infection CD4 cell counts are below 50 cells/mm^3. Patients have often disseminated opportunistic infections, malignancies, and a severe wasting syndrome. Impaired cognitive function and changes in personality as signs of AIDS-dementia are commonly observed.[31]

8.3 MALNUTRITION AND CACHEXIA IN HIV-INFECTED PATIENTS: THE WASTING SYNDROME

8.3.1 Definition and Pathophysiology

The wasting syndrome has been added to the list of case definitions of AIDS since 1987 by the CDC. It has been defined as involuntary loss of more than 10% of the usual body weight plus either chronic diarrhea (at least two loose defecations daily during a period of more than 30 days) or chronic weakness, and chronic fever in the absence of a concurrent illness or conditions other than HIV infection that could explain the symptoms.[28] Wasting is a common situation in HIV-infected patients seen from the early to advanced infection with the potential to be a life-threatening factor. Even in western societies where the HAART has been established, wasting may still be present. Factors that have been demonstrated or hypothesized to contribute to wasting include opportunistic infections, diarrhea, malabsorption, metabolic alterations, increased cytokine production, and endocrinological abnormalities. Metabolic complications of HAART may also contribute (Figure 8.1).

Weight loss in HIV infection typically consists of both fat mass (FM) and LBM, that is, metabolically active cell mass (Figure 8.2). Whole body protein balance cycles in a corresponding fashion between periods of net protein anabolism and periods of net protein catabolism in response to nutritional state.[32, 33] Additional energy may be stored as fat when intake exceeds energy expenditure and fat may become the main energy source during caloric restriction and starvation to minimize the loss of body cell mass (BCM). In HIV infection, like in cancer and sepsis, there is often a prominent depletion of BCM,[34] which is mainly caused by accelerated muscle proteolysis. During inadequate caloric intake or increased energy requirements, the overall breakdown of cell proteins, particularly in muscle, is enhanced to provide the essential amino acids required

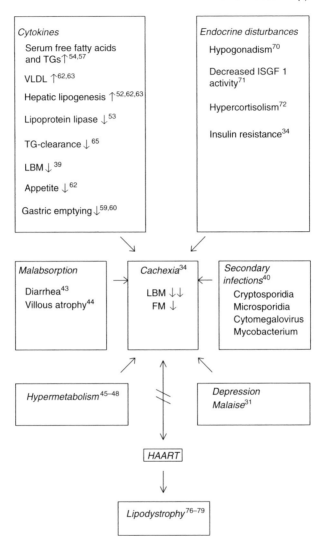

FIGURE 8.1 Mechanisms of HIV induced wasting. Several factors contribute to cachexia in HIV: Cytokines, particularly TNF-α, may suppress appetite and gastric emptying and lead to altered TG and lipoprotein metabolism and increased energy expenditure. Further, energy uptake may be reduced due to malabsorption and secondary infections may lead to increased energy requirements. Hypogonadism is associated with decreased activity of the anabolic hormone testosterone and hypercortisolism may contribute to increased protein turnover. In addition, alterations of mood and depression may enhance malnutrition, which finally results in cachexia. HAART, which may result in a complete control of viral replication, is able to prevent cachexia. However, HAART can result in other metabolic derangements, such as altered distribution of body fat. (IL-1 = interleukine 1, IL-6 = interleukine 6, IFN-α = interferone alpha, TNF-α = tumor necrosis factor-alpha, VLDL = very low density lipoprotein, TG = triglyceride, LBM = lean body mass, FM = fat mass, HAART = highly active antiretroviral therapy).

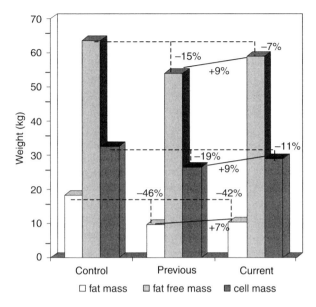

FIGURE 8.2 Body composition in healthy control, previous, and current HIV-infected men. Body composition is shown in age and sex matched healthy control subjects and HIV-infected men before (previous) and after 1996 (current) ($n = 77$ each group). The differences in body composition between previous and current HIV men may be due to antiretroviral therapy, which was received by 60 of current HIV-infected patients. 40 patients of the current group received protease inhibitors. In previous and current HIV-infected men there was a prominent depletion of the metabolic active cell mass, FM, and fat free mass (FFM) compared to healthy control men. However, current HIV-infected men demonstrated a less prominent reduction of body fat, FFM, and BCM compared to previous HIV-infected men. The differences in body composition between current and previous HIV men were only significant for FFM (Adapted from Kotler D.P., Rosenbaum K., Wang J., et al. (1999) *J. Acquir. Immune. Defic. Syndr. Hum. Retroviral.* 20: 228–37. With permission.).

for protein synthesis, energy metabolism, and function of immune system. This may result in negative nitrogen balance and in a catabolic state. Recent studies suggest that muscle wasting in catabolic states may be due to activation of the ubiquitin–proteasome pathway. Most proteins degraded using this pathway are first marked for degradation by linkage to the small protein cofactor ubiquitin.[35–37] The linkage reaction is repeated forming a chain of several ubiquitins linked to each other and to the protein substrate. After this modification, the substrate can be degraded by a large proteolytic complex, the proteasome. Proteasomes are the sites where most cell proteins are degraded and are the source of most peptides presented on MHC class I molecules.[36, 38] In addition, it has been reported that during conditions of altered metabolism, as observed in infection, cancer, sepsis, acidosis, and hormonal imbalances, cytokines may trigger muscle proteolysis.[39]

8.3.1.1 Secondary infections, diarrhea, and malabsorption in wasting syndrome of AIDS

There are different reasons why caloric intake can be decreased. As mentioned above, weight loss can be extremely severe in patients with AIDS and secondary infections. Patients who have marked anorexia and an average loss of 5% of their body weight in 28 days during episodes of opportunistic infections have been reported.[40] Other factors contributing to anorexia may be depression, fatigue, and malaise. In addition, malabsorption can decrease nutrient availability. Diarrhea, whether intermittent or chronic, is common in AIDS. The function of the gastrointestinal tract may be impaired by opportunistic infections (cryptosporidiosis, microsporidia, cytomegalovirus, and mycobacterium complex), malignancies, HIV itself, or pharmacological side effects. Malnutrition is often accompanied by lymphopenia (reductions in B and T lymphocytes) and reduced immunoglobulins. These immune system derangements may play a role in the increased risk of opportunistic infection and progression of HIV infection. In HIV-infected patients, malabsorption of D-xylose, pathological values for lactose breath tests, steatorrhea, and micronutrient deficiencies of zinc, iron, selenium, fat-soluble vitamins, vitamin B12, folate, and vitamin C and B are often present.[41–43] It has been reported that villous atrophy in AIDS patients is significantly associated with malabsorption.[44] But other studies suggested that malabsorption may also occur in the absence of wasting.[43]

8.3.1.2 Hypermetabolism in wasting syndrome of AIDS

Hypermetabolism characterized by increased resting energy expenditure is a common but not universal finding in HIV infection.[40, 45–48] There is no convincing causal relationship between wasting and hypermetabolism but decreased energy intake with metabolic alterations has been found to be a primary contributor to wasting.[40] Similar metabolic alterations were found in intensive care unit patients with sepsis. In such hypermetabolic states, it was shown that substantial losses of body protein may occur despite intensive nutritional support.[49] Hypermetabolism, particularly in the early stage of HIV infection was demonstrated as comparable to other wasting diseases such as cancer or sepsis. A variety of other metabolic alterations were also described in HIV-infected individuals, including increased[50] and decreased[51] rates of protein turnover, decreased rates of muscle protein synthesis,[52] and increased rates of novel hepatic lipogenesis,[53] lipid flux, and oxidative or nonoxidative lipid disposal.[54]

8.3.1.3 Wasting syndrome, cytokines, and hypertriglyceridemia

There are studies suggesting that cytokine disturbances are associated with metabolic disorders and wasting in HIV infection.[55] Cachectin has been identified as a responsible molecule for the wasting syndrome in many chronic diseases, such as

HIV.[56] This molecule has been shown to be identical to tumor necrosis factor (TNF) and increased levels of TNF were found in patients with AIDS.[53, 57, 58] By analyzing the role of TNF in wasting syndrome, a possible influence may be suggested by alterations in food intake and malabsorption; it has been described that TNF delays gastric emptying[59, 60] and also induces hemorrhagic lesions in the intestinal walls,[61] leading to malabsorption. Moreover, it has been suggested that an activation of the TNF-α system may contribute to inhibited appetite in AIDS.[62] In HIV infection, cytokine induced hypertriglyceridemia with increased hepatic production and decreased activity of lipoprotein lipase leading to decreased triglyceride clearance has been reported.[53, 63–65] IFN-α has been shown to decrease in parallel to triglycerides (TGs) after the beginning of antiretroviral therapy.[66] Therefore, it has been suggested that IFN-α may cause hypertriglyceridemia.[53, 67] However, there is no direct relationship between the prevalence of wasting syndrome and hypertriglyceridemia.

8.3.1.4 Endocrinological abnormalities in wasting syndrome of AIDS

It has been suggested that endocrinological abnormalities may facilitate the loss of LBM. One of these abnormalities is hypogonadism. It can be observed in up to 50% of men and women with advanced HIV infection.[68, 69] Moreover, hypogonadism also occurs in up to 20% of HIV-infected men receiving HAART.[70] A significant correlation between wasting and testosterone levels in men and women has been described.[68, 69] In addition, decreased levels of serum insulin-like growth factor 1 (IGF-1), which mediates many of the anabolic effects of growth harmone (GH) has been found.[71] In some HIV-infected patients, modestly increased serum cortisol and dehydroepiandrosterone (DHEA) levels were observed, suggesting a link between circulating concentrations of adrenal steroids and cachexia, which could lead to increased protein turnover. However, a correlation between cortisol levels and body weight has not been demonstrated.[72] Other abnormalities include insulin resistance, hyperlipidemia, and hypertension.[34] These findings are similar to those found in the metabolic syndrome. The pathogenesis of the metabolic syndrome is not fully understood although some studies suggested that there may be an underlying abnormality in the hypothalamic-pituitary-adrenal axis leading to chronic hypercortisolism and insulin resistance.[73] Particularly hyperlipidemia and insulin resistance have been reported in HIV-infected patients treated with protease inhibitors (PI).[14]

8.3.1.5 Wasting syndrome of AIDS and the role of modern antiretroviral therapy

Since the introduction of HAART, the morbidity and mortality of HIV infection have dramatically decreased.[74] But particularly PIs have side effects that are different from other antiretroviral substances.[14] They may cause significant gastrointestinal side effects such as nausea, vomiting, diarrhea, anorexia, hepatic stetosis, pancreatitis, and hyperglycemia due to insulin resistance.[34, 75] Furthermore, hypogonadism in men and abnormalities in lipid metabolism have been described.[70] In some cases, a combination of symptoms, referred to as "lipodystrophy" has been described with wasting of peripheral fat, central adiposity, hyperlipidemia, and insulin resistance. Lipodystrophy is characterized by subcutaneous lipoatrophy, most noticeable in the face, limbs, and buttocks, and accumulation of visceral fat.[76] It has been proposed that PIs inhibit proteins involved in lipid metabolism, especially by inhibiting sterol regulatory element-binding protein 1 (SREBP1)-mediated activation of the heterodimer consisting of adipocyte retinoid X receptor and peroxisome proliferator-activated receptor γ (PPARγ) or related transcription factors such as PPARγ coactivator 1.[77, 78] Nucleoside analogues may inhibit adipogenesis and adipocyte differentiation, promote lipolysis, and exert synergistic toxic effects when combined with PI.[79] Hyperglycemia may be caused by an effect on glucose disposal, while there is no significant compensation by insulin.[80] Murata et al.[81] have reported that PIs block the transport of glucose via a glucose transporter (Glut4). If lactate acidosis occurs during treatment with PIs, negative nitrogen balance and wasting may be accelerated.[82]

8.3.2 Prognosis of Wasting Syndrome in HIV-Infected Patients

Malnutrition, cachexia, and weight loss, particularly the wasting syndrome, have long been recognized as a common and severe complication of AIDS. Studies evaluating the nutritional status of hospitalized patients with AIDS have documented that 60% and more are underweight.[83–85] In developing countries, this number is probably much higher. Both retrospective and prospective studies have suggested a strong link between nutritional status and survival of patients. Kotler and colleagues have reported a critical level of BCM and weight at which death occurred (about 66% of ideal body weight).[86] Further observations have shown a strong association between reduced serum albumin concentrations and weight loss in AIDS patients, which persists even after the control of CD4 lymphocyte count.[87] In a study of 100 outpatients, BCM of above 30% body weight or a serum albumin concentration of above 3 mg/dL were associated with a longer survival, independent of CD4 cell count.[88] However, more data is needed about the effects of malnutrition on immune dysfunction and progression of HIV infection.

8.4 GENERAL TARGETS AND OBJECTIVES OF ANTICACHEXIA THERAPY IN AIDS

The most important measure to prevent cachexia is the complete control of viral replication. In most patients this will lead to weight gain and dramatic reduction of opportunistic diseases and constitutional symptoms.[12, 89] However, in up to 30 to 40% of patients, virus load may not be controlled in the long term and cachexia may either persist or reoccur.[90] Therefore, other measures are warranted (Table 8.1).

8.4.1   Enteral and Parenteral Nutrition

Significant advances have been made in the management of HIV infection. However, malnutrition that was treated not sufficiently from the first diagnosis of HIV infection remains a significant prognostic factor. A standardized approach to the management of AIDS-induced weight loss has not been established, but it is commonly accepted that early nutritional intervention is important in HIV patients to minimize the loss of LBM.

Nutritional counseling, enteral and parenteral nutritional support, and pharmacological agents have been demonstrated to reverse weight loss and increase LBM in HIV-infected patients. In a randomized controlled trial of nutritional counseling alone vs. nutrition counseling plus enteral supplementation for 6 weeks, nutritional counseling alone was able to substantially increase the energy intake in about 50% of malnourished HIV-infected patients suggesting that nutritional counseling has an important role in the management of malnourished HIV-infected patients.[91]

Enteral as well as parenteral nutrition have been demonstrated to be safe and effective in malnourished and immunodepressed patients with HIV disease.[92, 93] In general, additional enteral feeding should be preferred in those patients with intact gastrointestinal tract and additional parenteral feeding should be reserved for those with severe gastrointestinal disorders including malabsorption.[93, 94] We have demonstrated that nutritional intervention in HIV-infected subjects during 12 weeks with oral nutritional supplements combined with dietary counseling, decreased whole body protein catabolism as assessed using stable isotope measurements. Additional enteral nutrition increased LBM and decreased body FM as assessed by bioelectrical impedance analysis in the HIV patients with modest to moderate malnutrition but had no influence on lymphocyte subpopulations or cytokine receptor plasma concentrations.[95] Both standard and immune-enhancing nutritional formulas have been used to reverse AIDS-induced weight loss. There is no proof which of these formulas might be superior. In a randomized controlled multicenter study standard and immune-enhancing oral formulas consumed daily for 1 year had similar effects on nutrition and immune parameters.[96] A limitation of this study might be that it was performed in asymptomatic HIV patients.

8.4.2   Pharmacotherapy

Given the strong association between loss of lean tissue and survival, there has been a strong interest in pharmacological interventions targeted primarily at restoration of LBM. Placebo-controlled trials have demonstrated the efficacy of a variety of pharmacological agents promoting weight and lean tissue gain.

8.4.2.1   Appetite stimulants

8.4.2.1.1   Cannabinoids

Cannabinoids that are present in marijuana and derivatives, have a definitive effect on weight gain. Dronabinol, a synthetic form of delta-tetrahydrocannabinol has

TABLE 8.1
Pharmacological and Nutritional Effects on AIDS Associated Cachexia

(A) General targets of anticachexia therapy	Effects on cachexia	Side effects	FDA approval
Control of virus replication			
HAART	↓ Opportunistic diseases[12,89] ↓ Constitutional symptoms[12,89] ↑ Weight[12]	Diarrhea, vomiting, nausea, lipodystrophy, depression, lactate acidosis, polyneuropathy	Variable
Nutritional support			
Enteral and parenteral nutrition	↑ Weight ↑ Energy intake[92–94] ↑ LBM[92–94]	Diabetes, diarrhea, lactose intolerance	Variable
(B) Pharmacotherapy	**Effects on cachexia**	**Side effects**	**FDA approval**
Appetite stimulants			
Cannabinoids: dronabinol (Marinol®, Elevat®, Ronabin®)	↓ Nausea[97] ↑ Appetite[97]	Sedation, drowsiness, confusion, dizziness, depression	Yes
Megestrol acetate (Megace®, Megestrol®)	↑ FM[103] ↑ Weight[103]	Headache, fever, depression, skin rashes, hypertension	Yes

Protein anabolic agents			
Peptide hormones: GH (Serostim®, Humatrope®), IGF-1	↑ LBM alone[10] and especially in secondary infections[104–106] ↑ Anabolic effects[107]	Musculoskeletal discomfort, increased tissue turgor, diarrhea, nausea, fatigue, anemia	Yes
Anabolic steroids: Testosterone (Androderm®, Depo®-Testosterone, Testoderm®) Oxandrolone (Oxandrin®, Lonavar®)	↑ Weight[108, 109, 118] ↑ LBM[108, 109]	Acne, gynecomastia, benign prostata hyperplasia in men	No
Cytokine modulation			
Thalidomide (Thalidomid®, Thalix®)	↓ Diarrhea? ↑ Weight[117, 118]	Teratogenicity, sedation, neuropathy	No

The most important step to reduce HIV-induced wasting is the complete control of viral replication. In addition, nutritional supplementation and various pharmacological agents have demonstrated beneficial effects in HIV-associated cachexia.

(FDA = Food and Drug Administration, LBM = lean body mass, FM = fat mass, GH = growth hormone, IGF-1 = insulin like growth factor-1) been approved by the Food and Drug Administration (FDA) for treatment of HIV-associated anorexia. It has been shown in placebo-controlled trials to increase appetite and decrease nausea in patients with advanced HIV disease. These benefits were maintained at 12 months.[97] Different mechanisms by which cannabinoids exert their effects have been postulated: It has been suggested that they may act via endorphin receptors or by inhibiting prostaglandin synthesis.[98] Other reports suggest that they may act by inhibiting cytokine production or secretion.[99–101] Meanwhile two types of cannabinoid (CB) receptors were identified: CB(1) and CB(2). While CB(1) predominate on central and peripheral neurons, CB(2) receptors are present on immune cells. The mechanism following CB receptor activation is not completely understood, but a modulation of the release of cytokines may play a role.[102]

8.4.2.1.2 Megestrol acetate

Megestrol acetate is a progestational agent that has been tested for the prevention and treatment of HIV-associated wasting. In a randomized prospective study, Batterham and Garsia[103] compared three treatments, nandrolone decanoate, megestrol acetate, and dietary counseling, for HIV-associated weight loss for 12 weeks. The change in weight and percentage of body FM was significantly greater in those receiving megestrol acetate compared with the other two treatment arms. The increase in FM was significantly greater in both the nandrolone and megestrol arms than the dietary counseling arm.[103]

8.4.2.2 Protein anabolic agents

Peptide hormones that have been used are GH and IGF-1. Pharmacological use of GH has been shown to improve nitrogen balance and increase LBM in HIV-infected patients with wasting and to promote lean tissue maintenance in those with secondary infections.[104–106] The FDA accelerated approval for GH in HIV-associated wasting. However, the costs for this treatment limit its accessibility and maintenance-dosing regimens have not yet been identified. Most common side effects of GH therapy are arthralgia, myalgia, swelling, and fluid retention in the extremities and usually resolve with symptomatic treatment or dosage reduction. Anabolic effects of GH are mediated directly or indirectly via IGF-1[107] and it has been observed that during treatment with GH the increase in circulating levels of IGF-1 was less in patients with HIV-associated wasting compared to HIV negative controls.[54] However, treatment with IGF-1 alone has not been as successful as with GH. In addition, the increase in LBM achieved with a combination of IGF-1 and GH have not been superior compared to the increase in LBM achieved with GH alone[10] and dosing of IGF-1 is limited due to its hypoglycemic effects.

Other treatment options are anabolic steroids. Since anabolic steroids are derivatives or structural modifications of testosterone, they exhibit both anabolic and androgenic activity. Anabolic actions are an increase in protein synthesis and nitrogen retention, androgenic effects are the development and maintenance of sexual characteristics in males and females.

Replacement dosages of testosterone can increase LBM and muscle mass in men with wasting and low serum testosterone.[108, 109] Those situations can similarly improve lean tissue in eugonadal men with wasting.[110] In some cases, these changes have been coupled with improvements in self assessed quality of life and indices of depression.[108] Importantly, testosterone replacement has not been associated with significant side effects in these patients.

Nandrolone decanoate is a derivative of 10-nortesterone and has been shown to result in an increase in LBM in hypogonadal men with wasting[111] and eugonadal HIV positive men with no wasting.[112] Interestingly, in subjects who underwent resistance exercise during nandrolone treatment, it resulted in further increases in weight, LBM, and strength, suggesting that the protein anabolic effects of nandrolone can be augmented by concurrent resistance exercise.[112]

Oxandrolone has marked anabolic activity and few androgenic effects and in comparison with testosterone it has an anabolic/androgenic ratio of ten to one.[113] In studies with HIV/AIDS patients using the anabolic steroid oxandrolone, positive clinical outcomes regarding body composition, muscle function, and nutritional status have been reported. For example, in a study by Berger et al. [114] 15 mg/day of oxandrolone over 4 months resulted in significant improvements in body weight, appetite, and physical activity levels in men with HIV infection.

HIV-infected women with wasting also have reduced testosterone levels. However, there are only a few studies about the effects of anabolic/androgenic steroids in women with HIV-induced wasting. One of these studies demonstrated that replacement dosage of testosterone also resulted in weight gain, but mainly FM.[115]

8.4.2.3 Cytokine modulation

As described above, the excessive production of cytokines may contribute to wasting and suppression of cytokine production has been studied in HIV-associated wasting. Studies with thalidomide have shown small increases in weight[116, 117] and reductions of stool frequencies in patients with chronic diarrhea.[118] However, the future role of this medication in HIV-associated wasting is uncertain.

8.5 CONCLUSIONS

During the first two decades of HIV disease, the wasting syndrome was on the front page of the list of complications. The biggest influences were on life expectancy and quality of life. This major complication has decreased but not disappeared in western countries, particularly in patients with a long history of HIV infection, due to the availability of very potent and relatively well-tolerated antiretroviral therapies. It still remains a top issue in developing countries where access to antiviral therapies is limited. The nutritional assessment of patients suffering from HIV infection and corresponding treatment with nutritional counseling, increased nutritional support and various pharmacological agents can lead to a substantial benefit in addition to the currently available highly active antiretroviral therapies.

ACKNOWLEDGMENTS

We thank Prof. Manuel Battegay, Division of Infectious Diseases, and Prof. Ulrich Keller, Division of Endocrinology, Diabetes and Clinical Nutrition, University Hospital Basel, Switzerland, for their scientific advice during the preparation of the chapter.

REFERENCES

1. Serwadda D., Mugerwa R.D., Sewankambo N.K., et al. (1985) Slim disease: a new disease in Uganda and its association with HTLV-III infection. *Lancet* 2: 849–52.
2. Revision of the CDC surveillance case definition for acquired immunodeficiency syndrome. Council of State and Territorial Epidemiologists; AIDS Program, Center for Infectious Diseases. *MMWR Morb. Mortal. Wkly. Rep.* 1987; 36: 1S–15S.
3. Baum M.K. and Shor-Posner G. (1997) Nutritional status and survival in HIV-1 disease. *AIDS* 11: 689–90.
4. Wheeler D.A., Gibert C.L., Launer C.A., et al. (1998) Weight loss as a predictor of survival and disease progression in HIV infection. Terry Beirn Community Programs for Clinical Research on AIDS. *J. Acquir. Immune. Defic. Syndr. Hum. Retrovirol.* 18: 80–5.
5. Chandra R.K. (1983) Nutrition, immunity, and infection: present knowledge and future directions. *Lancet* 1: 688–91.
6. Wachtel T., Piette J., Mor V., et al. (1992) Quality of life in persons with human immunodeficiency virus infection: measurement by the medical outcomes study instrument. *Ann. Intern. Med.* 116: 129–37.
7. Harvey K.B., Bothe A., Jr., and Blackburn G.L. (1979) Nutritional assessment and patient outcome during oncological therapy. *Cancer* 43: 2065–9.
8. Nixon D.W., Heymsfield S.B., Cohen A.E., et al. (1980) Protein-calorie undernutrition in hospitalized cancer patients. *Am. J. Med.* 68: 683–90.
9. Melchior J.C., Niyongabo T., Henzel D., et al. (1999) Malnutrition and wasting, immunodepression, and chronic inflammation as independent predictors of survival in HIV-infected patients. *Nutrition* 15: 865–9.
10. Mulligan K. and Schambelan M. (2002) Anabolic treatment with GH, IGF-I, or anabolic steroids in patients with HIV-associated wasting. *Int. J. Cardiol.* 85: 151–9.
11. Moore R.D. and Chaisson R.E. (1999) Natural history of HIV infection in the era of combination antiretroviral therapy. *AIDS* 13: 1933–42.
12. Jaggy C., von Overbeck J., Ledergerber B., et al. (2003) Mortality in the Swiss HIV Cohort Study (SHCS) and the Swiss general population. *Lancet* 362: 877–8.
13. Palella F.J. Jr., Delaney K.M., Moorman A.C., Loveless M.O., Fuhrer J., Satten G.A., Aschman D.J., and Holmberg S.D. (1998) Declining morbidity and mortality among patients with advanced human immunodeficiency virus infection. HIV outpatient Study Investigators. *N. Engl. J. Med* 338: 853–60.
14. Carr A., Samaras K., Thorisdottir A., et al. (1999) Diagnosis, prediction, and natural course of HIV-1 protease-inhibitor-associated lipodystrophy, hyperlipidaemia, and diabetes mellitus: a cohort study. *Lancet* 353: 2093–9.
15. Roubenoff R., Grinspoon S., Skolnik P.R., et al. (2002) Role of cytokines and testosterone in regulating lean body mass and resting energy expenditure in HIV-infected men. *Am. J. Physiol. Endocrinol. Metab.* 283: E138–45.

16. Levy J.A. (1993) Pathogenesis of human immunodeficiency virus infection. *Microbiol. Rev.* 57: 183–289.
17. Stebbing J., Gazzard B., and Douek D.C. (2004) Where does HIV live? *N. Engl. J. Med.* 350: 1872–80.
18. Hay C.M., Ruhl D.J., Basgoz N.O., et al. (1999) Lack of viral escape and defective in vivo activation of human immunodeficiency virus type 1-specific cytotoxic T lymphocytes in rapidly progressive infection. *J. Virol.* 73: 5509–19.
19. Lane H.C. and Fauci A.S. (1985) Immunologic reconstitution in the acquired immunodeficiency syndrome. *Ann. Intern. Med.* 103: 714–8.
20. Rosenberg E.S., Billingsley J.M., Caliendo A.M., et al. (1997) Vigorous HIV-1-specific CD4+ T cell responses associated with control of viremia. *Science* 278: 1447–50.
21. Moore J.P. (1997) Coreceptors: implications for HIV pathogenesis and therapy. *Science* 276: 51–2.
22. Rucker J. and Doms R.W. (1998) Chemokine receptors as HIV coreceptors: implications and interactions. *AIDS Res. Hum. Retroviruses* 14: S241–6.
23. Miedema F., Petit A.J., Terpstra F.G., et al. (1988) Immunological abnormalities in human immunodeficiency virus (HIV)-infected asymptomatic homosexual men. HIV affects the immune system before CD4+ T helper cell depletion occurs. *J. Clin. Invest.* 82: 1908–14.
24. Wahren B., Morfeldt-Mansson L., Biberfeld G., et al. (1987) Characteristics of the specific cell-mediated immune response in human immunodeficiency virus infection. *J. Virol.* 61: 2017–23.
25. Kahn J.O. and Walker B.D. (1998) Acute human immunodeficiency virus type 1 infection. *N. Engl. J. Med.* 339: 33–9.
26. U.S. Congress, Office of Technology Assessment. The CDC's case definitions of AIDS: implications of the prognosed revisions-background paper, OTA-BP-H-89. Washington DC: US Government Printing Office, 1992.
27. Revision of the case definition of acquired immunodeficiency syndrome for national reporting — United States. Centers for Disease Control, Department of Health and Human Services. *Ann. Intern. Med.* 1985; 103: 402–3.
28. Centers of Disease Control. 1993 revised classification system for HIV infection and expanded surveillance case definition for AIDS. *MMWR* 1993; 41: 961–2.
29. Guidelines for national human immunodeficiency virus case surveillance, including monitoring for human immunodeficiency virus infection and acquired immunodeficiency syndrome. Centers for Disease Control and Prevention. *MMWR Recomm. Rep.* 1999; 48: 1–27, 29–31.
30. Hoover D.R., Saah A.J., Bacellar H., et al. (1993) Signs and symptoms of "asymptomatic" HIV-1 infection in homosexual men. Multicenter AIDS Cohort Study. *J. Acquir. Immune. Defic. Syndr.* 6: 66–71.
31. McArthur J.C. (2004) HIV dementia: an evolving disease. *J. Neuroimmunol.* 157: 3–10.
32. Braunwald E.F.A., Kasper D., Hauser S., Longo D., and Jameson J. (2001) Harrison's Principles of Internal Medicine. Mc Graw-Hill Incorporated, 15th ed.
33. McNurlan M.A. and Garlick P.J. (1989) Influence of nutrient intake on protein turnover. *Diabetes. Metab. Rev.* 5: 165–89.
34. Kotler D.P., Rosenbaum K., Wang J., et al. (1999) Studies of body composition and fat distribution in HIV-infected and control subjects. *J. Acquir. Immune. Defic. Syndr. Hum. Retrovirol.* 20: 228–37.

35. Goldberg A.L., Stein R., and Adams J. (1995) New insights into proteasome function: from archaebacteria to drug development. *Chem. Biol.* 2: 503–8.
36. Hershko A. and Ciechanover A. (1992) The ubiquitin system for protein degradation. *Annu. Rev. Biochem.* 61: 761–807.
37. Lowell B.B., Ruderman N.B., and Goodman M.N. (1986) Evidence that lysosomes are not involved in the degradation of myofibrillar proteins in rat skeletal muscle. *Biochem. J.* 234: 237–40.
38. Coux O., Tanaka K., and Goldberg A.L. (1996) Structure and functions of the 20S and 26S proteasomes. *Ann. Rev. Biochem.* 65: 801–47.
39. Mitch W.E. and Goldberg A.L. (1996) Mechanisms of muscle wasting. The role of the ubiquitin-proteasome pathway. *N. Engl. J. Med.* 335: 1897–905.
40. Grunfeld C., Pang M., Shimizu L., et al. (1992) Resting energy expenditure, caloric intake, and short-term weight change in human immunodeficiency virus infection and the acquired immunodeficiency syndrome. *Am. J. Clin. Nutr.* 55: 455–60.
41. Burkes R.L., Cohen H., Krailo M., et al. (1987) Low serum cobalamin levels occur frequently in the acquired immune deficiency syndrome and related disorders. *Eur J. Haematol.* 38: 141–7.
42. Harriman G.R., Smith P.D., Horne M.K., et al. (1989) Vitamin B12 malabsorption in patients with acquired immunodeficiency syndrome. *Arch. Intern. Med.* 149: 2039–41.
43. Miller T.L., Orav E.J., Martin S.R., et al. (1991) Malnutrition and carbohydrate malabsorption in children with vertically transmitted human immunodeficiency virus 1 infection. *Gastroenterology* 100: 1296–302.
44. Ullrich R., Zeitz M., Heise W., et al. (1989) Small intestinal structure and function in patients infected with human immunodeficiency virus (HIV): evidence for HIV-induced enteropathy. *Ann. Intern. Med.* 111: 15–21.
45. Kotler D.P., Tierney A.R., Brenner S.K., et al. (1990) Preservation of short-term energy balance in clinically stable patients with AIDS. *Am. J. Clin. Nutr.* 51: 7–13.
46. Macallan D.C., Noble C., Baldwin C., et al. (1993) Prospective analysis of patterns of weight change in stage IV human immunodeficiency virus infection. *Am. J. Clin. Nutr.* 58: 417–24.
47. Melchior J.C., Raguin G., Boulier A., et al. (1993) Resting energy expenditure in human immunodeficiency virus-infected patients: comparison between patients with and without secondary infections. *Am. J. Clin. Nutr.* 57: 614–9.
48. Sharpstone D.R., Ross H.M., and Gazzard B.G. (1996) The metabolic response to opportunistic infections in AIDS. *AIDS* 10: 1529–33.
49. Streat S.J., Beddoe A.H., and Hill G.L. (1987) Aggressive nutritional support does not prevent protein loss despite fat gain in septic intensive care patients. *J. Trauma* 27: 262–6.
50. Macallan D.C., McNurlan M.A., Milne E., et al. (1995) Whole-body protein turnover from leucine kinetics and the response to nutrition in human immunodeficiency virus infection. *Am. J. Clin. Nutr.* 61: 818–26.
51. Stein T.P., Nutinsky C., Condoluci D., et al. (1990) Protein and energy substrate metabolism in AIDS patients. *Metabolism* 39: 876–81.
52. Yarasheski K.E., Zachwieja J.J., Gischler J., et al. (1998) Increased plasma gln and Leu Ra and inappropriately low muscle protein synthesis rate in AIDS wasting. *Am. J. Physiol.* 275: E577–83.

53. Hellerstein M.K., Grunfeld C., Wu K., et al. (1993) Increased de novo hepatic lipo-genesis in human immunodeficiency virus infection. *J. Clin. Endocrinol. Metab.* 76: 559–65.
54. Mulligan K., Grunfeld C., Hellerstein M.K., et al. (1993) Anabolic effects of recom-binant human growth hormone in patients with wasting associated with human immunodeficiency virus infection. *J. Clin. Endocrinol. Metab.* 77: 956–62.
55. Grunfeld C. and Feingold K.R. (1991) The metabolic effects of tumor necrosis factor and other cytokines. *Biotherapy* 3: 143–58.
56. Kawakami M. and Cerami A. (1981) Studies of endotoxin-induced decrease in lipoprotein lipase activity. *J. Exp. Med.* 154: 631–9.
57. Beutler B. and Cerami A. (1986) Cachectin and tumour necrosis factor as two sides of the same biological coin. *Nature* 320: 584–8.
58. Grunfeld C., Kotler D.P., Shigenaga J.K., et al. (1991) Circulating interferon-alpha levels and hypertriglyceridemia in the acquired immunodeficiency syndrome. *Am. J. Med.* 90: 154–62.
59. Arbos J., Lopez-Soriano F.J., Carbo N., et al. (1992) Effects of tumour necrosis factor-alpha (cachectin) on glucose metabolism in the rat. Intestinal absorption and isolated enterocyte metabolism. *Mol. Cell. Biochem.* 112: 53–9.
60. Patton J.S., Peters P.M., McCabe J., et al. (1987) Development of partial tolerance to the gastrointestinal effects of high doses of recombinant tumor necrosis factor-alpha in rodents. *J. Clin. Invest.* 80: 1587–96.
61. Tracey K.J., Beutler B., Lowry S.F., et al. (1986) Shock and tissue injury induced by recombinant human cachectin. *Science* 234: 470–4.
62. Arnalich F., Martinez P., Hernanz A., et al. (1997) Altered concentrations of appetite regulators may contribute to the development and maintenance of HIV-associated wasting. *AIDS* 11: 1129–34.
63. Feingold K.R. and Grunfeld C (1987) Tumor necrosis factor-alpha stimulates hepatic lipogenesis in the rat *in vivo*. *J. Clin. Invest.* 80: 184–90.
64. Feingold K.R., Soued M., Serio M.K., et al. (1989) Multiple cytokines stimulate hepatic lipid synthesis *in vivo*. *Endocrinology* 125: 267–74.
65. Grunfeld C., Kotler D.P., Hamadeh R., et al. (1989) Hypertriglyceridemia in the acquired immunodeficiency syndrome. *Am. J. Med.* 86: 27–31.
66. Mildvan D., Machado S.G., Wilets I., et al. (1992) Endogenous interferon and triglyceride concentrations to assess response to zidovudine in AIDS and advanced AIDS-related complex. *Lancet* 339: 453–6.
67. Grunfeld C., Pang M., Doerrler W., et al. (1992) Lipids, lipoproteins, trigly-ceride clearance, and cytokines in human immunodeficiency virus infection and the acquired immunodeficiency syndrome. *J. Clin. Endocrinol. Metab.* 74: 1045–52.
68. Dobs A.S., Dempsey M.A., Ladenson P.W., et al. (1988) Endocrine disorders in men infected with human immunodeficiency virus. *Am. J. Med.* 84: 611–6.
69. Grinspoon S., Corcoran C., Miller K., et al. (1997) Body composition and endocrine function in women with acquired immunodeficiency syndrome wasting. *J. Clin. Endocrinol. Metab.* 82: 1332–7.
70. Rietschel P., Corcoran C., Stanley T., et al. (2000) Prevalence of hypogonadism among men with weight loss related to human immunodeficiency virus infection who were receiving highly active antiretroviral therapy. *Clin. Infect. Dis.* 31: 1240–4.

71. Frost R.A., Fuhrer J., Steigbigel R., et al. (1996) Wasting in the acquired immune deficiency syndrome is associated with multiple defects in the serum insulin-like growth factor system. *Clin. Endocrinol. (Oxf.)* 44: 501–14.

72. Christeff N., Gherbi N., Mammes O., et al. (1997) Serum cortisol and DHEA concentrations during HIV infection. *Psychoneuroendocrinology* 22: S11–8.

73. Bjorntorp P. (1988) Abdominal obesity and the development of noninsulin-dependent diabetes mellitus. *Diabetes. Metab. Rev.* 4: 615–22.

74. Cole S.R., Hernan M.A., Robins J.M., et al. (2003) Effect of highly active anti-retroviral therapy on time to acquired immunodeficiency syndrome or death using marginal structural models. *Am. J. Epidemiol.* 158: 687–94.

75. Merry C., McMahon C., Ryan M., et al. (1998) Successful use of protease inhibitors in HIV-infected haemophilia patients. *Br. J. Haematol.* 101: 475–9.

76. Carr A. and Cooper D.A. (1998) Images in clinical medicine. Lipodystrophy associated with an HIV-protease inhibitor. *N. Engl. J. Med.* 339: 1296.

77. Bastard J.P., Caron M., Vidal H., et al. (2002) Association between altered expression of adipogenic factor SREBP1 in lipoatrophic adipose tissue from HIV-1-infected patients and abnormal adipocyte differentiation and insulin res-istance. *Lancet* 359: 1026–31.

78. Caron M., Auclair M., Vigouroux C., et al. (2001) The HIV protease inhibitor ind-inavir impairs sterol regulatory element-binding protein-1 intranuclear localization, inhibits preadipocyte differentiation, and induces insulin resistance. *Diabetes* 50: 1378–88.

79. Roche R., Poizot-Martin I., Yazidi C.M., et al. (2002) Effects of antiretroviral drug combinations on the differentiation of adipocytes. *AIDS* 16: 13–20.

80. Grinspoon S. and Carr A. (2005) Cardiovascular risk and body-fat abnormalities in HIV-infected adults. *N. Engl. J. Med.* 352: 48–62.

81. Murata H., Hruz P.W., and Mueckler M. (2000) The mechanism of insulin resistance caused by HIV protease inhibitor therapy. *J. Biol. Chem.* 275: 20251–4.

82. Carr A. and Cooper D.A. (2000) Adverse effects of antiretroviral therapy. *Lancet* 356: 1423–30.

83. Chelluri L. and Jastremski M.S. (1989) Incidence of malnutrition in patients with acquired immunodeficiency syndrome. *Nutr. Clin. Pract.* 4: 16–8.

84. Kotler D.P., Wang J., and Pierson R.N. (1985) Body composition studies in patients with the acquired immunodeficiency syndrome. *Am. J. Clin. Nutr.* 42: 1255–65.

85. O'Sullivan P., Linke R.A., and Dalton S. (1985) Evaluation of body weight and nutritional status among AIDS patients. *J. Am. Diet. Assoc.* 85: 1483–4.

86. Kotler D.P., Tierney A.R., Wang J., et al. (1989) Magnitude of body-cell-mass depletion and the timing of death from wasting in AIDS. *Am. J. Clin. Nutr.* 50: 444–7.

87. Guenter P., Muurahainen N., Simons G., et al. (1993) Relationships among nutri-tional status, disease progression, and survival in HIV infection. *J. Acquir. Immune. Defic. Syndr.* 6: 1130–8.

88. Suttmann U., Ockenga J., Selberg O., et al. (1995) Incidence and prognostic value of malnutrition and wasting in human immunodeficiency virus-infected outpatients. *J. Acquir. Immune. Defic. Syndr. Hum. Retrovirol.* 8: 239–46.

89. Egger M., Hirschel B., Francioli P., et al. (1997) Impact of new antiretroviral com-bination therapies in HIV infected patients in Switzerland: prospective multicentre study. Swiss HIV Cohort Study. *BMJ* 315: 1194–9.

90. Ledergerber B., Egger M., Opravil M., et al. (1999) Clinical progression and virological failure on highly active antiretroviral therapy in HIV-1 patients: a prospective cohort study. Swiss HIV Cohort Study. *Lancet* 353: 863–8.

91. Rabeneck L., Palmer A., Knowles J.B., et al. (1998) A randomized controlled trial evaluating nutrition counseling with or without oral supplementation in malnourished HIV-infected patients. *J. Am. Diet. Assoc.* 98: 434–8.

92. Ockenga J., Suttmann U., Selberg O., et al. (1996) Percutaneous endoscopic gastrostomy in AIDS and control patients: risks and outcome. *Am. J. Gastroenterol.* 91: 1817–22.

93. Melchior J.C., Gelas P., Carbonnel F., et al. (1998) Improved survival by home total parenteral nutrition in AIDS patients: follow-up of a controlled randomized prospective trial. *AIDS* 12: 336–7.

94. Melchior J.C., Chastang C., Gelas P., et al. (1996) Efficacy of 2-month total parenteral nutrition in AIDS patients: a controlled randomized prospective trial. The French Multicenter Total Parenteral Nutrition Cooperative Group Study. *AIDS* 10: 379–84.

95. Berneis K., Battegay M., Bassetti S., et al. (2000) Nutritional supplements combined with dietary counselling diminish whole body protein catabolism in HIV-infected patients. *Eur. J. Clin. Invest.* 30: 87–94.

96. Keithley J.K., Swanson B., Zeller J.M., et al. (2002) Comparison of standard and immune-enhancing oral formulas in asymptomatic HIV-infected persons: a multicenter randomized controlled clinical trial. *JPEN. J. Parenter. Enteral. Nutr.* 26: 6–14.

97. Beal J.E., Olson R., Laubenstein L., et al. (1995) Dronabinol as a treatment for anorexia associated with weight loss in patients with AIDS. *J. Pain Symptom Manage.* 10: 89–97.

98. Mitchelson F (1992) Pharmacological agents affecting emesis. A review (Part I). *Drugs* 43: 295–315.

99. Watzl B., Scuderi P., and Watson R.R. (1991) Marijuana components stimulate human peripheral blood mononuclear cell secretion of interferon-gamma and suppress interleukin-1 alpha *in vitro*. *Int. J. Immunopharmacol.* 13: 1091–7.

100. Srivastava M.D., Srivastava B.I., Brouhard B. (1998) Delta9 tetrahydrocannabinol and cannabidiol alter cytokine production by human immune cells. *Immunopharmacology* 40: 179–85.

101. Facchinetti F., Del Giudice E., Furegato S., et al. (2003) Cannabinoids ablate release of TNFalpha in rat microglial cells stimulated with lypopolysaccharide. *Glia* 41: 161–8.

102. Pertwee R.G. and Ross R.A. (2002) Cannabinoid receptors and their ligands. *Prostaglandins Leukot. Essent. Fatty Acids* 66: 101–21.

103. Batterham M.J. and Garsia R. (2001) A comparison of megestrol acetate, nandrolone decanoate and dietary counselling for HIV associated weight loss. *Int. J. Androl.* 24: 232–40.

104. Daigle R.D. (1990) Anabolic steroids. *J. Psychoactive Drugs* 22: 77–80.

105. Paton N.I., Newton P.J., Sharpstone D.R., et al. (1999) Short-term growth hormone administration at the time of opportunistic infections in HIV-positive patients. *AIDS* 13: 1195–202.

106. Schambelan M., Mulligan K., Grunfeld C., et al. (1996) Recombinant human growth hormone in patients with HIV-associated wasting. A randomized, placebo-controlled trial. Serostim Study Group. *Ann. Intern. Med.* 125: 873–82.

107. Berneis K. and Keller U. (1996) Metabolic actions of growth hormone: direct and indirect. *Baillieres Clin. Endocrinol. Metab.* 10: 337–52.

108. Grinspoon S., Corcoran C., Stanley T., et al. (2000) Effects of hypogonadism and testosterone administration on depression indices in HIV-infected men. *J. Clin. Endocrinol. Metab.* 85: 60–5.

109. Bhasin S., Storer T.W., Berman N., et al. (1997) Testosterone replacement increases fat-free mass and muscle size in hypogonadal men. *J. Clin. Endocrinol. Metab.* 82: 407–13.

110. Grinspoon S., Corcoran C., Parlman K., et al. (2000) Effects of testosterone and progressive resistance training in eugonadal men with AIDS wasting. A randomized, controlled trial. *Ann. Intern. Med.* 133: 348–55.

111. Strawford A., Barbieri T., Neese R., et al. (1999) Effects of nandrolone decanoate therapy in borderline hypogonadal men with HIV-associated weight loss. *J. Acquir. Immune. Defic. Syndr. Hum. Retrovirol.* 20: 137–46.

112. Sattler F.R., Jaque S.V., Schroeder E.T., et al. (1999) Effects of pharmacological doses of nandrolone decanoate and progressive resistance training in immunodeficient patients infected with human immunodeficiency virus. *J. Clin. Endocrinol. Metab.* 84: 1268–76.

113. Kuhn C.M. (2002) Anabolic steroids. *Recent Prog. Horm. Res.* 57: 411–34.

114. Berger J.R., Pall L., Hall C.D., et al. (1996) Oxandrolone in AIDS-wasting myopathy. *AIDS* 10: 1657–62.

115. Miller K., Corcoran C., Armstrong C., et al. (1998) Transdermal testosterone administration in women with acquired immunodeficiency syndrome wasting: a pilot study. *J. Clin. Endocrinol. Metab.* 83: 2717–25.

116. Franks M.E., Macpherson G.R., and Figg W.D. (2004) Thalidomide. *Lancet* 363: 1802–11.

117. Sharpstone D., Rowbottom A., Nelson M., et al. (1995) The treatment of microsporidial diarrhoea with thalidomide. *AIDS* 9: 658–9.

118. Kaplan G., Thomas S., Fierer D.S., et al. (2000) Thalidomide for the treatment of AIDS-associated wasting. *AIDS Res. Hum. Retroviruses.* 16: 1345–55.

9 Rheumatoid Cachexia

Laura A. Coleman

Contents

Summary

Rheumatoid arthritis (RA) is a chronic, systemic, inflammatory disease that is characterized, in part, by increased production of the inflammatory cytokines, tumor necrosis factor (TNF)-α and interleukin (IL)-1β.[1] TNF-α seems to be a key

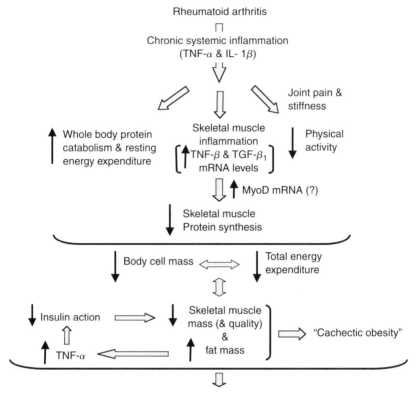

FIGURE 9.1   Summary of metabolic consequences of RA. RA is characterized by increased TNF and IL-1, leading to increased whole body protein catabolism and resting energy expenditure, skeletal muscle inflammation, and reduced physical activity. As a result, skeletal muscle protein synthesis is reduced, which contributes to decreased body cell mass and reduced total energy expenditure. The net effect is decreased skeletal muscle mass and quality, accompanied by increased fat mass, a condition that we have named "cachectic obesity." Ultimately, muscle weakness, physical disability, increased risk of infection and other diseases occur (Reprinted from Walsmith, J. and Roubenoff, R., *Int. J. Cardiol.*, 85, 89–99, 2002. With permission from Elsevier.).

mediator in the disease process, and IL-1β takes on a more permissive role, acting to shift whole body protein metabolism toward net catabolism, elevate resting energy expenditure, and increase joint pain and stiffness.[2] These relationships and the metabolic consequences of RA are summarized in Figure 9.1.

It is to be noted that while the metabolic abnormalities observed in RA (net whole body protein catabolism and hypermetabolism) may be exacerbated during disease flares, even when the disease is well-controlled and metabolic abnormalities are minimized, metabolic consequences — loss of body cell mass (BCM) and the functional sequelae — do occur and are not corrected without additional, direct intervention. While there remain many unanswered questions, we are able to conclude that increasing physical activity and maintaining a diet adequate in protein, energy, and micronutrients remain the cornerstones of good clinical management of the metabolic consequences of RA. Unfortunately, there is limited scientific evidence on interventions for the treatment of rheumatoid cachexia per se, and further research is needed in order to maintain optimal functional status for patients with RA.

9.1 INTRODUCTION

Cachexia, literally translated from Greek, means "bad condition," and while the terms "cachexia" and "wasting" may often be used synonymously, we now know that cachexia, more specifically, refers to a loss of body cell mass (BCM),[3] while wasting refers to loss of both BCM and body weight. A loss of BCM is important because in all situations in which it has been studied — starvation, critical illness, and normal aging — loss of greater than 40% of baseline BCM is associated with death[4–6]. Even a 5% decline in BCM is accompanied by a loss of muscle strength, altered energy metabolism, and increased susceptibility to infections.[7] We now know that cachexia is a common occurrence among patients with rheumatoid arthritis, and have termed this condition "rheumatoid cachexia."[8] The average loss of BCM for patients with RA is between 13 and 15%,[8,9] approximately one-third the maximum survivable loss of BCM, making rheumatoid cachexia an important contributor to increased morbidity and premature mortality in RA. Although patients do not die of RA per se, people with RA do have a 2 to 5-fold higher mortality rate, with the same causes of death as in the general population except for an increased risk of infections[10–12]. This chapter will focus on characterizing rheumatoid cachexia, reviewing potential contributing factors, and describing possible interventions.

9.2 RHEUMATOID ARTHRITIS

9.2.1 Epidemiology

Rheumatoid arthritis is a chronic, systemic, autoimmune, inflammatory disorder, and is the most common type of inflammatory arthritis, affecting approximately 1% of the population.[13] Women are affected two to three times more often than men, with the peak incidence between the fourth and sixth decades.[13] A higher prevalence rate of 5 to 6% has been found in some Native American populations, suggesting that there may be a set of high-risk genes for RA,[13] although this remains speculative. As a historical point of interest, Rothschild et al.[14] have argued that RA originated in the New World, prior to the arrival of Christopher Columbus; this may be consistent with the increased prevalence of RA observed among contemporary Native Americans if genetic factors are involved.[15]

9.2.2 Etiology

Although there seems to be some genetic basis for the development of RA, genetics alone does not fully account for the condition as evidenced by the finding that concordance rates in monozygotic twins are only 15 to 30%.[13] Additionally, environmental factors (i.e., diet, smoking), the aging process itself, and various infectious agents have all been explored as possible etiologic factors.[13] Clearly, cytokines, primarily IL-1β and TNF-α, play a central role in the development of inflammation in the synovium — a critical site for the onset of joint deterioration.

9.2.3 Clinical Features

Rheumatoid arthritis is most often characterized by morning stiffness, joint pain on motion or tenderness, and swelling, typically in a symmetric pattern.[16] Since RA is a systemic disease, extra-articular manifestations of the disease also occur. Most patients experience general malaise or fatigue; skin, ocular, respiratory, cardiac, neurologic, and hematologic manifestations may also occur.[16]

9.2.4 Treatment

Drugs are a mainstay of treatment for patients with RA, aimed at relieving pain, swelling, and fatigue; improving joint function; reducing joint damage; and preventing disability and disease-related morbidity.[17] However, while drug therapy may be a necessary part of treatment for RA, there are consequences in terms of nutritional status and body composition.

9.2.4.1 Nonsteroidal anti-inflammatory drugs (NSAIDs)

NSAIDs act by reducing one or both types of cyclooxygenase enzymes (COX 1 and COX 2), thereby reducing inflammation and pain. They do not prevent tissue injury or joint damage and are used primarily for symptom control. NSAID gastropathy is a common side-effect, and patients with RA have a 30% chance of hospitalization or death from gastrointestinal toxicity during the course of their disease.[17]

9.2.4.2 Corticosteroids

Corticosteroids are potent suppressors of inflammation but because of the well-known side-effects of maintaining a patient on steroids for an extended period of time, long-term use should be avoided.[17] Corticosteroids are known catabolic agents, and at high doses, such as those used to treat disease flares (1000 mg intravenous methylprednisolone for 3 days), cause clinically important nitrogen wasting during therapy.[18]

9.2.4.3 Disease-modifying antirheumatic drugs (DMARDs)

An effective DMARD should prevent joint damage and control active features of the disease, however these drugs do not provide a "cure" for RA.[17] Examples of these agents include hydroxychloroquine, sulfasalazine, methotrexate, aza-thioprine, cyclosporine, cyclophosphamide, and anti-TNF antibody agents.[17]

9.3 RHEUMATOID CACHEXIA

9.3.1 Definition

Body cell mass is the most metabolically active component of the human body, accounting for 95% of all metabolic activity. The amount of BCM determines energy expenditure, protein requirements, and the metabolic response to physiologic stress.[19] Body cell mass is made up primarily of muscle mass, with visceral mass (serum proteins, erythrocytes, granulocytes, lymphocytes, liver, kidneys, pancreas, heart) and immune cell mass contributing lesser amounts. Fat mass, extra-cellular water, connective tissue (cartilage, fibrous tissues, skeletal tissues) and bone account for the remaining components. It is to be noted that the various pools of protein are not readily exchangeable, so changes that occur in BCM during disease cannot be counteracted by mobilization of extra-cellular connective tissue.

There are numerous conditions that can lead to a loss of BCM. Wasting, involving a loss of both body weight (fat mass) and BCM, is typically due to inadequate dietary intake, as in starvation, for example. Cachexia refers to loss of BCM without loss of weight; in fact, loss of BCM is often accompanied by increased fat mass and stable body weight. For instance, patients with RA do not appear to be cachectic, and in fact, often appear to be overweight. A condition termed "rheumatoid cachectic obesity" has been used to describe the phenomenon of loss of lean mass, accompanied by increased fat mass; this condition is thought to affect nearly two-thirds of all patients with RA.[8, 9] One distinguishing feature of rheumatoid cachexia is the notable absence of certain impairments: there are no impairments in renal or hepatic function, there is no malabsorption, and there is relatively limited corticosteroid use among patients with RA. These features create an ideal situation in which to study hypercytokinemic cachexia, uncomplicated by these other confounding factors.

9.3.2 Loss of Lean Mass in Aging vs. Inflammation

Sarcopenia (literally translated from Greek means "lack of flesh") was first described as such approximately 15 years ago, referring to the loss of muscle mass that occurs with healthy, advancing age.[20] Although this happens to almost everyone to a certain extent, how rapidly muscle is lost and how debilitating the effects are vary depending on each individual's health status, physical activity, and perhaps diet.[21] Clearly, when disease or starvation, for example, occur in an elderly individual who has already experienced a loss of lean mass, they have less reserve for withstanding the additional insult, and the functional consequences will be more severe. Sarcopenia is thus distinct from cachexia caused by inflammatory disease, or from muscle wasting caused by starvation or advanced chronic disease.[22] However, sarcopenia has been referred to as the backdrop against which the drama of disease is played out: "normal" age-related protein loss creates a state whereby the body is less able to withstand the protein catabolism that occurs with illness.[23] Thus, in the case of rheumatoid cachexia, RA exacerbates a condition that is already present in healthy people with age. However, it remains unclear whether the mechanisms are the same in aging and inflammatory disease, but simply occurring more rapidly, and at a higher voltage, or whether there are unique pathways acting in RA, beyond mechanisms by which lean mass is lost in normal, healthy aging.[2] We have some evidence that there appear to be different factors at work: we have not found tumor necrosis factor (TNF)-α to be increased in the healthy elderly,[24] while it clearly is in RA, suggesting that RA exacerbates and accelerates sarcopenia, whereas sarcopenia per se may work more via interleukin (IL)-6 and IL-1β/IL-1 receptor antagonist.[24]

9.3.3   Mechanisms

9.3.3.1   Role of sarcoactive cytokines

The inflammatory cytokines TNF-α and IL-1β are thought to be centrally involved in the pathogenesis of RA. Both cytokines are produced primarily by monocytes and macrophages, but they are also produced by a variety of other immune and nonimmune system cells, including B-lymphocytes, T-lymphocytes, and skeletal muscle.[25–28] Levels of TNF-α and IL-1β are highest in patients with active RA[29, 30]; these cytokines act by stimulating the release of tissue-destroying matrix metalloproteinases, as well as by inhibiting the production of endogenous inhibitors of these metalloproteinases.[31] Ultimately, joint damage, characteristic of RA, results.

Not only are TNF-α and IL-1β centrally involved in contributing to joint damage in RA, but these cytokines also exert a powerful influence on whole body protein and energy metabolism. The so-called sarcoactive ("muscle active") cytokines include, in addition to TNF-α and IL-1β, IL-6, interferon-gamma (IFN-γ), and transforming growth factor-beta (TGF-β_1). In addition, the transcription factor MyoD is worth mentioning because of the integral role that MyoD has as a member of the TGF superfamily.

Over 20 years ago, researchers first demonstrated that circulating inflammatory cytokines such as TNF-α and IL-1β are released into the plasma by leukocytes and can stimulate protein degradation and whole body protein wasting.[32] The specific mechanism by which TNF-α and IL-1β exert their catabolic effect still remains to be elucidated, but we have shown that subjects with RA have higher rates of whole body protein breakdown as determined by ^{13}C-leucine infusion compared with young and elderly healthy subjects.[33] Furthermore, we have suggested that protein breakdown rates are directly associated with TNF-α production by peripheral blood mononuclear cells.[33] It is to be noted that after a 12-week strength training intervention, rates of protein breakdown were normalized in these subjects with RA such that there were no longer any differences between any of the subject groups (young, elderly, or RA).

Studies have also shown that TNF inhibits skeletal muscle differentiation in vitro by suppressing MyoD mRNA via nuclear factor kappa B and that the combination of TNF-α and IFN-γ induced reductions in MyoD.[34] These findings suggest that TNF-α alone or TNF-α and IFN-γ in combination may contribute to muscle loss via the inhibition of the skeletal muscle repair process and downregulation of MyoD. Work is currently underway in our laboratory to examine transcript (mRNA) levels of a panel of sarcoactive cytokines, including TNF-α, IL-1β, IL-6, IL-15, IFN-γ, and TGF-β_1 and MyoD.[35,36] Briefly, we found that the rate of skeletal muscle protein synthesis was 25% lower in patients with RA compared to healthy control subjects, and that mean TNF-α and TGF-β_1 were over three times higher in RA vs. healthy control subjects, while there was no difference in IL-1β.[36] Furthermore, mRNA levels of TNF-α and TGF-β_1 were positively correlated with one another, suggesting that the previously recognized protein-depleting effects of TNF-α may be at least partly mediated by TGF-β_1. Recent additional work by others has suggested that protein kinase B and Smad3 proteins may also play a role in mediating the action of these sarcoactive cytokines by affecting the protection and vulnerability of cells against TGF-β-induced apoptosis.[37]

9.3.3.2 Energy expenditure profile

Early studies from our research group suggested that after adjusting for BCM, resting energy expenditure (REE) is elevated in patients with RA.[8,9] Absolute REE, on the other hand, is no different between subjects with RA and healthy control subjects[9]. Furthermore, REE increased with RA severity, just as BCM fell. However, REE is only one component of total daily energy expenditure (TEE), the others being the energy expenditure of physical activity (EEPA) and the thermic effect of food (TEF), such that: TEE $=$ REE $+$ EEPA $+$ TEF. In general, the major component of TEE is REE (approximately 60–75%), with TEF being relatively constant at about 10–15% and EEPA representing the balance of energy expenditure.[38]

As a result of the multiple components of TEE, REE may be elevated in RA, but the net effect on TEE also depends on EEPA. We have recently examined this issue in women with RA and in age- and body mass index (BMI)-matched controls, and found that TEE is actually significantly lower in patients with RA than in control subjects because of lower EEPA.[39] In fact, the magnitude of the difference in EEPA was quite large in this study: 1034 kJ/d (approximately a 10% difference between RA and controls). Clearly, low EEPA plays an important role in determining TEE in patients with RA. Furthermore, in the 14 years since we first showed that cachexia is common in RA,[18] improvements in disease treatment have made overt hypermetabolism (elevated REE) less common. The implication of what is now known about energy metabolism in RA is that patients need to be cautioned to maintain a diet that is adequate — and not excessive — in terms of protein and calories in an attempt to maintain muscle but prevent fat gain. Optimally, incorporating a regular exercise regimen could help maintain energy balance, muscle mass, and ultimately, functional capacity, and thus should be considered an integral component of treatment for patients with RA.

9.3.3.3 Whole body protein turnover

We have consistently demonstrated that a loss of BCM is common in patients with RA.[8,9,18,39,40] By definition, catabolism (negative protein balance) must occur in order for cachexia to develop, and consistent with this theory, we have also found that adults with RA have increased whole body protein breakdown rates (measured by ^{13}C-leucine infusion).[33] Furthermore, we observed a direct relationship between peripheral blood mononuclear cell (PBMC) production of TNF-α and leucine flux — an indicator of whole body protein breakdown in the fasting state — where higher TNF-α production was associated with higher rates of whole body protein breakdown (catabolism).[33] Studies are currently underway in our laboratory to examine the skeletal muscle fractional synthetic rate in patients with RA.[35] For a discussion on this topic, refer to the previous section entitled Sarcoactive Cytokines (9.3.3.1).

9.3.3.4 Physical activity

We have demonstrated that patients with RA have low physical activity[9]; this is due to a variety of reasons, namely, joint pain and stiffness, metabolic changes leading to loss of muscle mass and strength, and simple disuse, perhaps related to a general cautiousness with regard to physical activity. This, however, appears to be changing as an escalation in physical activity is increasingly being recognized as an important, beneficial component of disease treatment for patients with RA.

9.3.3.5 Hormones

9.3.3.5.1 Growth hormone (GH) and insulin-like growth factor-I (IGF-I)

On the surface, rheumatoid cachexia appears to be similar to sarcopenia, and in some respects it is. In the case of sarcopenia (healthy aging), an association has been shown between a loss of lean mass and declining activity of the GH-IGF-I axis. This raised the question of whether GH is reduced in patients with RA, and whether a reduction in GH may contribute to some of the body composition changes observed in these patients. Evidence exists for abnormalities of various other hormones in RA.[41–43] However, a recent study measuring GH secretory kinetics by deconvolution analysis after 24-h blood sampling found no differences between patients with RA and healthy control subjects after adjusting for differences in fat mass between the two groups, as fat is known to suppress GH secretion.[44] Patients with RA did, however, have significantly reduced BCM. These findings suggest that persistent GH deficiency does not appear to be the cause of rheumatoid cachexia. However, it should be noted that a trend for lower serum IGF-I levels was found; these results have been supported by others who observed significant reductions in circulating IGF-I.[45] It is thus possible that reduced IGF-I could contribute to rheumatoid cachexia.

9.3.3.5.2 Insulin

Insulin is an important anabolic hormone, acting to inhibit muscle protein degradation. Several researchers have documented insulin resistance in inflammatory arthritis,[46,47] although its effect on protein metabolism remains unknown. We have hypothesized that the metabolic milieu created by a state of insulin resistance may be permissive to cytokine-driven muscle loss,[1] although this hypothesis remains to be investigated. The etiology of reduced peripheral insulin action in RA is not known, but TNF-α has been shown to interfere with insulin receptor signaling and may be a contributing factor.[48,49]

9.3.4 Interventions

While the mechanisms involved in contributing to rheumatoid cachexia remain unclear, what is already clear is that the functional consequences of this condition are to increase morbidity and mortality in patients with RA. Because of this, it is important to slow, if not reverse, the loss of BCM in RA and thus improve functional status. There are essentially three methods for doing so: exercise interventions, dietary interventions, and pharmacologic interventions.

9.3.4.1 Exercise

Historically, a mainstay in the treatment of inflammatory arthritis has been rest. With the recognition that loss of lean mass is a problem for people with RA, and the knowledge that exercise (particularly strength training) is an effective way to build muscle mass in healthy individuals, the practice of resting joints for patients with RA came under scrutiny. Two early studies demonstrated that high-intensity strength training is feasible and safe in selected patients with well-controlled RA.[40,50] We observed significant improvements in strength, pain, and fatigue without exacerbating disease activity or joint pain in these subjects who underwent 12 weeks of progressive resistance training, exercising at 80% of their 1-repetition maximum (the maximum amount of weight that can be lifted, once, with good form).[40] Furthermore, we observed no adverse effects on the immune system, as may sometimes occur after an acute bout of exercise.[51] Subsequently, additional studies have confirmed our early findings and have shown that regular progressive resistance strength training improves muscle strength and physical functioning in patients with RA,[52–55] even over the long-term.[56] Strength training is now considered to be an important part of the nonpharmacological treatment of RA and should be routinely prescribed and maintained in order to achieve maximal benefit.[57] It is important to note, however, that although strength training interventions do result in improvements in functional status, they have not been shown to reverse rheumatoid cachexia. In fact, we demonstrated an improvement in strength and functional status without a change in body composition.[40]

9.3.4.2 Diet

With regard to alteration or supplementation of various dietary components, many dietary modifications for RA have been suggested in the past, ranging from fasting or vegetarianism to supplementation with various fatty acids (primarily fish oil), protein, or vitamins/minerals.[58,59] A complete discussion of dietary modifications that have been studied for RA is beyond the scope of this chapter, however, the four major categories of dietary manipulation: energy and protein, supplementation with dietary fatty acids, vitamin/mineral supplementation, and fasting or vegetarianism, will be reviewed briefly.

9.3.4.2.1 Energy and protein

We have consistently found that energy and protein intakes among people with RA are not significantly different when compared with intakes from healthy individuals.[9,33,39,60] Typically, mean (±standard deviation) energy consumption ranges from approximately 25.5 ± 10^9 to 31 ± 3.8 kcal/kg/day,[33] and protein, from 0.7 ± 0.1 [60] to 1.3 ± 0.19 g/kg/d.[33] Inadequate dietary intake, therefore, does not seem to contribute significantly to rheumatoid cachexia, which is consistent with the definition of cachexia, distinguishing it from wasting. Supplementation with protein, amino acids, or additional calories, does not, therefore, appear to be warranted.

The current recommended protein intake, based upon the Dietary Reference Intakes (DRI), is 0.8 g/kg/day for adults, while the Estimated Energy Requirement is defined as the average dietary energy intake that is predicted to maintain energy balance in a healthy adult of a defined age, gender, weight, height and level of physical activity, consistent with good health.[61] Clearly, patients with RA are not healthy, but while the DRI may not apply directly in our studies, the concept of energy required for maintaining a healthy weight still applies. A combination of decreased EEPA and reduced BCM in patients with RA is a potent force favoring fat accretion, and in view of the overall obesity epidemic occurring worldwide, concern over the propensity for overweight among patients with RA is warranted. Clinicians should not recommend increased dietary energy intake to patients with RA despite their risk of elevated REE,[39] as this hypermetabolism is outweighed by a reduction in physical activity. The worst case scenario is the "fat frail" — increased weakness and a need to carry greater weight due to obesity.[62] Future studies are needed to determine optimal protein and energy intakes for overcoming the catabolic process in RA, particularly when combined with an exercise intervention. It is entirely possible that increased amounts of nitrogen are required in order to achieve maximal anabolic benefit.

9.3.4.2.2 Supplementation with dietary fatty acids

Most studies of supplementation with omega-3 fatty acids (eicosapentaenoic acid, or EPA and docosahexaenoic acid, or DHA) suggest a modest improvement in clinical symptoms associated with RA, which are to some extent dependent upon the dose and length of supplementation (reviewed in Reference 58). The goal of these interventions, in fact, has focused more on disease management (joint symptoms and pain) rather than on correcting metabolic abnormalities. On the other hand, increases in intake of EPA or DHA are associated with other health benefits, including reduced risk of cardiovascular disease. Thus, a prudent approach may be to promote fish consumption among patients with RA, particularly oily, marine fish.

9.3.4.2.3 Vitamin/mineral supplementation

Most studies of vitamin/mineral supplementation in RA have focused on either the antioxidant nutrients (vitamin E, vitamin C, beta carotene, selenium) or the B-vitamins (vitamin B6, in particular) (reviewed in Reference 58). In general, supplementation trials have failed to produce definitive results, and there remains little basis for recommending nutrient supplementation at a particular dosage. Dietary sources of antioxidant nutrients or vitamin B6 can certainly be recommended as part of a varied diet and may even provide additional benefits, related to the consumption of whole foods vs. isolated nutrients, that supplements do not offer.

9.3.4.2.4 Fasting or vegetarian diets

The concept that food sensitivity reactions contribute in any significant way to clinical symptoms of RA remains highly speculative. A small subgroup of patients may benefit from specific dietary modification involving the elimination of specific foods (animal products, for example) or food components, but any sort of elimination diet should be advised with extreme caution. Furthermore, it should be noted that a possible adverse effect of following a vegetarian diet in terms of further loss of lean mass has been observed,[63] so patients should be cautioned in this regard.

9.3.4.3 Cytokine antagonists

In order to prevent the inflammatory and destructive changes of RA that have been attributed primarily to TNF-α, a number of agents have been developed aimed at inhibiting TNF-α production. The U.S. Food and Drug Administration recently approved three TNF-α antagonists for the treatment of RA, one of which is a soluble receptor and the others, monoclonal antibodies.[64] While these agents have been shown to be effective in the treatment of signs and symptoms of RA,[64] concerns about their safety have been raised. Although there are some risks involved (i.e., reactivation of granulomatous diseases, increased incidence of non-Hodgkin's lymphoma, and exacerbation of advanced heart failure), with careful monitoring, it appears that these agents can be safely used in most instances. Although some have suggested that TNF inhibitors should be considered as part of standard therapy for RA,[65] others disagree[66] and view this class of medications as an alternative choice for patients who do not respond to other more traditional therapies.[67] Therefore, how widespread the use of anti-TNF therapies will become remains unclear. Furthermore, there have been no studies to date, to our knowledge, examining the effect of these agents on body composition in RA and future studies are warranted. One could speculate that by blocking the action of TNF-α — the primary cytokine that has been implicated in rheumatoid cachexia — BCM might be preserved in RA.

REFERENCES

1. Walsmith, J.M. and Roubenoff, R., Cachexia in rheumatoid arthritis, *Int. J. Cardiol.*, 85, 89–99, 2002.
2. Rall, L.C. and Roubenoff, R., Rheumatoid cachexia: the metabolic abnormalities, mechanisms, and interventions, *Rheumatology*, 43, 1219–1223, 2004.
3. Grunfeld, C. and Feingold, K.R., Metabolic disturbances and wasting in the acquired immunodeficiency syndrome, *N. Engl. J. Med.*, 327, 329–337, 1992.
4. Winick, M., Ed. *Hunger Disease*. 1979, John Wiley & Sons: New York. 261.
5. Kotler, D.P., Tierney, A.R., and Pierson, R.N., Magnitude of body cell mass depletion and the timing of death from wasting in AIDS, *Am. J. Clin. Nutr.*, 50, 444–447, 1989.
6. DeWys, W.D. et al., Prognostic effect of weight loss prior to chemotherapy in cancer patients. Eastern cooperative oncology group, *Am. J. Med.*, 69, 491–497, 1980.
7. Roubenoff, R. and Kehayias, J.J., The meaning and measurement of lean body mass, *Nutr. Rev.*, 46, 163–175, 1991.
8. Roubenoff, R. et al., Rheumatoid cachexia: depletion of lean body mass in rheumatoid arthritis. Possible association with tumor necrosis factor, *J. Rheumatol.*, 19, 1505–1510, 1992.
9. Roubenoff, R. et al., Rheumatoid cachexia: cytokine-driven hypermetabolism accompanying reduced body cell mass in chronic inflammation, *J. Clin. Invest.*, 93, 2379–2386, 1994.
10. Symmons, D.P.M., Mortality in rheumatoid arthritis, *Br. J. Rheumatol.*, 27, 44–54, 1988.
11. Wolfe, F. et al., The mortality of rheumatoid arthritis, *Arthritis Rheum.*, 37, 481–494, 1994.
12. Gabriel, S.B. et al., Survival in rheumatoid arthritis: a population-based analysis of trends over 40 years, *Arthritis Rheum.*, 48, 54–58, 2003.
13. Goronzy, J.J. and Weyand, C.M., Rheumatoid arthritis: epidemiology, pathology, and pathogenesis, in *Primer on the Rheumatic Diseases*, Klippel, J.H., Ed. 2001, Arthritis Foundation: Atlanta.
14. Rothschild, B.M. et al., Geographic distribution of rheumatoid arthritis in ancient North America: implications for pathogenesis, *Semin. Arthritis Rheum.*, 22, 181–187, 1992.
15. Del Puente, A. et al., High incidence and prevalence of rheumatoid arthritis in Pima Indians, *Am. J. Epidemiol.*, 129, 1170–1178, 1989.
16. Anderson, R.J., Rheumatoid arthritis: clinical and laboratory features, in *Primer on the Rheumatic Diseases*, Klippel, J.H., Ed. 2001, Arthritis Foundation: Atlanta.

17. Matteson, E.L., Rheumatoid arthritis: treatment, in *Primer on the Rheumatic Diseases*, Klippel, J.H., Ed. 2001, Arthritis Foundation: Atlanta.
18. Roubenoff, R. et al., Catabolic effects of high-dose corticosteroids persist despite therapeutic benefit in rheumatoid arthritis, *Am. J. Clin. Nutr.*, 52, 1113–1117, 1990.
19. Moore, F.D., Energy and the maintenance of body cell mass, *J. Parenter. Enteral. Nutr.*, 4, 228–260, 1980.
20. Rosenberg, I.H., Summary comments, *Am. J. Clin. Nutr.*, 50, 1231–1233, 1989.
21. Roubenoff, R., Sarcopenia: effects on body composition and function, *J. Gerontol. A Biol. Sci. Med. Sci.*, 58, 1012–1017, 2003.
22. Roubenoff, R. et al., Standardization of nomenclature of body composition in weight loss, *Am. J. Clin. Nutr.*, 66, 192–196, 1997.
23. Roubenoff, R. and Castaneda, C., Sarcopenia — understanding the dynamics of aging muscle, *JAMA*, 286, 1230–1231, 2001.
24. Roubenoff, R. et al., Monocyte cytokine production in an elderly population: effect of age and inflammation, *J. Gerontol. A Biol. Sci. Med. Sci.*, 53, M20–M26, 1998.
25. Choy, E.H.S. and Panayi, G.S., Cytokine pathways and joint inflammation in rheumatoid arthritis, *N. Engl. J. Med.*, 344, 907–916, 2001.
26. Koch, A.E., Kunkel, S.L., and Strieter, R.M., Cytokines in rheumatoid arthritis, *J. Invest. Med.*, 43, 28–38, 1995.
27. Greiwe, J.S. et al., Resistance exercise decreases skeletal muscle tumor necrosis factor-alpha in frail elderly humans, *FASEB J.*, 15, 475–482, 2001.
28. Cannon, J.G. et al., Increased interleukin-1 beta in human skeletal muscle after exercise, *Am. J. Physiol.*, 257, R451–R455, 1989.
29. Saxne, T. et al., Detection of tumor necrosis factor-alpha but not tumor necrosis factor beta in rheumatoid arthritis synovial fluid and serum, *Arthritis Rheum.*, 31, 1041–1045, 1988.
30. Chikanza, I.C., Kingsley, G., and Panayi, G.S., Peripheral blood and synovial fluid monocyte expression of interleukin-1 alpha and 1 beta during active rheumatoid arthritis, *J. Rheumatol.*, 22, 600–606, 1995.
31. Shingu, M. et al., The effects of cytokines on metalloproteinase inhibitors (TIMP) and collagenase production by human chondrocytes and TIMP production by synovial cells and endothelial cells, *Clin. Exp. Immunol.*, 94, 145–149, 1993.
32. Clowes, G.H. et al., Muscle proteolysis induced by a circulating peptide in patients with sepsis or trauma, *N. Engl. J. Med.*, 308, 545–552, 1983.
33. Rall, L.C. et al., Protein metabolism in rheumatoid arthritis and aging: effects of muscle strength training and tumor necrosis factor α, *Arthritis Rheum.*, 39, 1115–1124, 1996.
34. Guttridge, D. et al., NF-kappaB-induced loss of MyoD messenger RNA: possible role in muscle decay and cachexia, *Science*, 289, 2363–2366, 2000.
35. Walsmith, J.M. et al., Patients with rheumatoid arthritis (RA) have low protein synthesis rates and elevated tumor necrosis factor-alpha (TNF-alpha) and transforming growth factor-beta1 (TGF-beta) transcript levels in skeletal muscle, *Arthritis Rheum.*, 48, S104, 2003.
36. Walsmith, J.M. et al., A high ratio of interleukin-1beta (IL-1) to interleukin-1-receptor antagonist (IL-1RA) is associated with less sleep and higher physical activity in rheumatoid arthritis: a doubly labeled water (DLW) study, *FASEB J.*, 16, A26, 2002.

37. Remy, I., Montmarquette, A., and Michnick, S.W., PKB/Akt modulates TGF-beta signalling through a direct interaction with Smad3, *Natl. Cell. Biol.*, 6, 358–365, 2004.

38. Donahoo, W.T., Levine, J.A., and Melanson, E.L., Variability in energy expenditure and its components, *Curr. Opin. Clin. Nutr. Metab. Care*, 7, 599–605, 2004.

39. Roubenoff, R. et al., Low physical activity reduces total energy expenditure in women with rheumatoid arthritis: implications for dietary intake recommendations, *Am. J. Clin. Nutr.*, 76, 774–779, 2002.

40. Rall, L.C. et al., The effect of progressive resistance training in rheumatoid arthritis: increased strength without changes in energy balance or body composition, *Arthritis Rheum.*, 39, 415–426, 1996.

41. Cutolo, M. et al., Hypothalamic-pituitary-adrenocortical and gonadal functions in rheumatoid arthritis, *Ann. N.Y. Acad. Sci.*, 992, 107–117, 2003.

42. Tengstrand, B. et al., Abnormal levels of serum dehydroepiandrosterone, estrone, and estradiol in men with rheumatoid arthritis: high correlation between serum estradiol and current degree of inflammation, *J. Rheumatol.*, 30, 2338–2343, 2003.

43. Castagnetta, L.A. et al., Increased estrogen formation and estrogen to androgen ratio in the synovial fluid of patients with rheumatoid arthritis, *J. Rheumatol.*, 30, 2597–2605, 2003.

44. Rall, L.C. et al., Cachexia in rheumatoid arthritis is not explained by decreased growth hormone secretion, *Arthritis Rheum.*, 46, 2574–2577, 2002.

45. Lemmey, A. et al., Association between insulin-like growth factor status and physical activity levels in rheumatoid arthritis, *J. Rheumatol.*, 28, 29–34, 2001.

46. Dessein, P.H. et al., The acute phase response does not fully predict the presence of insulin resistance and dyslipidemia in inflammatory arthritis, *J. Rheumatol.*, 29, 462–466, 2002.

47. Svenson, K.L.G. et al., Impaired glucose handling in active rheumatoid arthritis: relationship to peripheral insulin resistance, *Metabolism*, 37, 125–130, 1988.

48. DeAlvaro, C. et al., Tumor necrosis factor alpha produces insulin resistance in skeletal muscle by activation of inhibitor kappa B kinase in a p38 MAPK-dependent manner, *J. Biol. Chem.*, 279, 17070–17078, 2004.

49. Yamaguchi, K. et al., The effect of tumor necrosis factor-alpha on tissue specificity and selectivity to insulin signaling, *Hypertens. Res.*, 26, 389–396, 2003.

50. Hakkinen, A., Hakkinen, K., and Hannonen, P., Effects of strength training on neuromuscular function and disease activity in patients with recent-onset inflammatory arthritis, *Scand. J. Rheumatol.*, 23, 237–242, 1994.

51. Rall, L.C. et al., Effects of progressive resistance training on immune response in aging and chronic inflammation, *Med. Sci. Sports Exerc.*, 28, 1356–1365, 1996.

52. Hakkinen, A. et al., Dynamic strength training in patients with early rheumatoid arthritis increases muscle strength but not bone mineral density, *J. Rheumatol.*, 26, 1257–1263, 1999.

53. Hakkinen, A. et al., A randomized two-year study of the effects of dynamic strength training on muscle strength, disease activity, functional capacity, and bone mineral density in early rheumatoid arthritis, *Arthritis Rheum.*, 44, 515–522, 2001.

54. Bearne, L., Scott, D., and Hurley, M., Exercise can reverse quadriceps sensorimotor dysfunction that is associated with rheumatoid arthritis without exacerbating disease activity, *Rheumatology*, 41, 157–166, 2002.

55. Hakkinen, A. et al., Effects of concurrent strength and endurance training in women with early or longstanding rheumatoid arthritis: comparison with healthy subjects, *Arthritis Rheum.*, 49, 789–797, 2003.

56. deJong, Z. et al., Is a long-term high-intensity exercise program effective and safe in patients with rheumatoid arthritis? Results of a randomized controlled trial, *Arthritis Rheum.*, 48, 2415–2424, 2003.

57. Hakkinen, A., Effectiveness and safety of strength training in rheumatoid arthritis, *Curr. Opin. Rheumatol.*, 16, 132–137, 2004.

58. Rall, L.C. and Roubenoff, R., Dietary aspects of the aeitology and nutritional management of inflammatory degenerative arthritis, in *Encyclopedia of Human Nutrition*, Sadler, M.J., Caballero, B., and Strain, J.J., Eds. 1999, Academic Press: London. pp. 138–143.

59. Rennie, K.L. et al., Nutritional management of rheumatoid arthritis: a review of the evidence, *J. Human Nutr. Diet.*, 16, 97–109, 2003.

60. Walsmith, J.M. et al., Tumor necrosis factor-alpha production is associated with less body cell mass in women with rheumatoid arthritis, *J. Rheumatol.*, 31, 23–29, 2004.

61. Panel on Macronutrients, Panel on the Definition of Dietary Fiber, Subcommittee on Upper Reference Levels of Nutrients, Subcommittee on Interpretation and Uses of Dietary Reference Intakes, and the Standing Committee on the Scientific Evaluation of Dietary Reference Intakes, *Dietary Reference Intakes for Energy, Carbohydrate, Fiber, Fat, Fatty Acids, Cholesterol, Protein, and Amino Acids*. 2002, Institute of Medicine: Washington, D.C.

62. Roubenoff, R., Sarcopenic obesity: the confluence of two epidemics, *Obes. Res.*, 12, 887–888, 2004.

63. Kjeldsen-Kragh, J., Rheumatoid arthritis treated with vegetarian diets, *Am. J. Clin. Nutr.*, 70, 594S–600S, 1999.

64. Khanna, D., McMahon, M., and Furst, D.E., Safety of tumour necrosis factor-alpha antagonists, *Drug Saf.*, 27, 307–324, 2004.

65. Messori, A., Santarlasci, B., and Vaiani, M., New drugs for rheumatoid arthritis, *N. Engl. J. Med.*, 351, 937–938, 2004.

66. Kalil, A.C., New drugs for rheumatoid arthritis, *N. Engl. J. Med.*, 351, 2660, 2004.

67. Roberts, L.J., New drugs for rheumatoid arthritis, *N. Engl. J. Med.*, 351, 2659, 2004.

10 Chronic Kidney Disease: Need for an Integrated Therapy Against Malnutrition and Wasting

Bengt Lindholm, Jonas Axelsson,
Olof Heimbürger, Abdul Rashid Qureshi,
Tommy Cederholm, and Peter Stenvinkel

CONTENTS

SUMMARY

Chronic kidney disease, and especially when it has progressed to end-stage renal
disease (ESRD), is associated with a variety of metabolic and nutritional abnormalities contributing to a high prevalence of protein-energy malnutrition (PEM)
and wasting, and high morbidity and mortality. These problems remain to a large
extent following the start of lifesaving renal replacement therapy by dialysis or
kidney transplantation. Factors contributing to PEM/wasting include reduced food
intake because of uremic anorexia, and increased net protein catabolism due to
hormonal derangements, accumulation of uremic toxins, fluid overload, acidosis
comorbidity, and a state of chronic systemic inflammation. Because the etiology
of PEM/wasting in ESRD is multifactorial, isolated interventions in the form of
energy and protein supplementation may not be sufficient in improving nutritional
status as long as other causes such as inflammation, acidosis, infections, and fluid
overload are present. In this brief review, we discuss the scope of the problem with
PEM/wasting in ESRD, its causes (in particular the role of chronic inflammation),

and propose an integrated therapeutic strategy consisting of a combination of traditional nutritional intervention and new forms of pharmacological treatments for the inflamed and wasted ESRD patients.

10.1 INTRODUCTION

The kidneys play a key role in maintaining fluid and electrolyte homeostatis, excretion of metabolic waste, and regulation of various hormonal and metabolic pathways. Therefore even a slight reduction in renal function may have metabolic and nutritional consequences. Patients with chronic kidney disease (CKD) display a variety of metabolic and nutritional abnormalities and a large proportion of the patients demonstrate signs of protein-energy malnutrition (PEM) and wasting,[1] and these problems become more severe when patients reach the stage of end-stage renal disease (ESRD)[2,3] a condition which by definition means the need to start lifesaving renal replacement therapy by dialysis or kidney transplantation. PEM and wasting are the consequence of multiple factors related to CKD and the development of a state of uremia with its subsequent disturbances in protein and energy metabolism, hormonal derangements, reduced food intake because of anorexia, nausea, and vomiting, caused by uremic toxicity, and numerous other causes linked to intercurrent or underlying comorbidity. As a result, ESRD patients starting dialysis are often suffering from PEM/wasting. Some patients develop cachexia but this late stage of PEM/wasting is associated with a very high mortality in ESRD, and therefore few patients with cachexia survive. It therefore follows from the above that prevention and treatment of PEM/wasting and cachexia is an important part of the care of ESRD patients.

The use of the word *malnutrition* has often not been used correctly in the renal literature.[4] While malnutrition is defined as a consequence of insufficient food intake and low serum protein levels, the loss of muscle mass (i.e., wasting) in the ESRD patients is usually a consequence of a number of catabolic mechanisms such as delayed gastric emptying, hormonal derangements, inadequate control of acidosis, comorbidity, and inflammation. Thus, the etiology of loss of lean body mass in ESRD as in other chronic diseases include numerous factors apart from poor food intake (i.e., "true malnutrition").[4] In this chapter, we will use the term PEM/wasting to underline that the nutritional problems in ESRD are usually due both to insufficient nutritional intake and accelerated breakdown of protein and energy stores due to increased net catabolism. We will focus on the problems in ESRD, rather than CKD, as the problems in ESRD patients are more severe and more is known about this patient group; however, the problems are often essentially of the same nature in the late and early stages of CKD. While we will discuss specific problems related to dialytic therapy we will not discuss the consequences of renal transplantation, which in general lead to improved, or even normalized nutritional status but also adverse effects of corticosteroid therapy.

During the period prior to the institution of dialysis therapy, ESRD patients usually spontaneously decrease their protein intake as a consequence of the decline in renal function[5] or because they are being treated with low protein diets. Also, some of the many drugs used by ESRD patients may worsen anorexia. In addition, underlying or intercurrent diseases, such as diabetes mellitus (DM), cardiovascular disease (CVD), infections, various complications of dialysis therapy as well as medications, in particular corticosteroid therapy, and not least, a state of chronic systemic inflammation, may all worsen PEM/wasting in these patients.[6]

Once dialysis therapy begins, accompanied by reduction of uremic symptoms and liberalization of the diet, some patients may show improved nutritional status.[7] However, many of the indicators of malnutrition that are present at the onset of therapy remain abnormal and some aspects of malnutrition may become even more severe. With dialysis therapy some of these factors, but far from all, can be partly or fully corrected. Also, additional metabolic and nutritional problems may be induced by the dialysis procedure, including losses of proteins, amino acids, water-soluble vitamins and other essential small molecular solutes into the dialysate as well as catabolism and suppression of appetite caused by inflammation due to bioincompatibility of the therapy or, in peritoneal dialysis (PD), from dialysate glucose absorption and abdominal discomfort induced by the treatment. Therefore, PEM/wasting, a common feature in ESRD patients before dialysis,[1] may become even more common after patients start on PD[3] or hemodialysis (HD).[2] Although the initiation of lifesaving dialysis therapy results in reduction of uremic symptoms and at least initially improved nutritional status, the mortality rate remains high, and increases in the dialysis dose in patients undergoing long-term dialysis treatment may result in only marginal, or not even significant, reductions in patient mortality, as recently shown in large studies in both HD[8] and PD[9] patients. Thus, factors other than dialysis dose, such as cardiovascular complications, inflammation, and wasting seem to be more important determinants of patient survival in this patient group. In fact, most premature deaths in ESRD patients are now thought to be associated with the presence of a malnutrition/wasting, inflammation and atherosclerosis (MIA) syndrome, which is an important part of the residual uremic syndrome although it is in large part also caused by coexisting comorbidity.[6, 10]

A state of systemic inflammation is present in most ESRD patients and may cause not only wasting but also the massive increase (10 to 100-fold compared to the general population) in the relative risk of CVD seen in ESRD patients. Low-grade chronic inflammation has been proposed to be one of the most important contributors to wasting in ESRD patients.[6, 11] Moreover, as inflammation plays a key role in the atherogenic process in nonrenal patients[12] it has been speculated that inflammation contributes to increased cardiovascular morbidity and mortality also in ESRD patients.[2, 6, 10, 11, 13] Thus, wasting is strongly associated with a persistent systemic inflammatory response, CVD, and impaired patient survival in ESRD (as well as in other chronic diseases). Evidence suggests that a facilitative interaction between inflammatory cytokines and other factors, such as poor appetite, comorbidity, acidosis, anemia, and hormonal derangements may be the most important cause of wasting in ESRD patients.[14]

As the etiology of PEM/wasting in ESRD is multifactorial, isolated interventions in the form of energy and protein supplementation may not be sufficient in improving nutritional status and outcome in ESRD patients as long as other

TABLE 10.1

Markers of Malnutrition and Cachexia in End-Stage Renal Disease (ESRD) Patients

- Body weight (% relative body weight, % preuremic body weight)
- Body mass index
- Visceral proteins (serum protein, albumin, prealbumin, transferrin and IGF-1)
- Subjective global assessment of nutritional status
- Handgrip strength
- Anthropometrics (skinfold thickness and midarm muscle circumference)
- Lean body mass (estimated by DEXA, bioimpedance or creatinine kinetics)
- Body fat mass (estimated by DEXA)
- Total body nitrogen and total body potassium
- Essential (low in ESRD) and nonessential (high in ESRD) plasma amino acids
- Protein intake estimated by nPNA[a]

[a] nPNA, protein equivalent of nitrogen appearance in urine and dialysate, normalized to actual or desirable body weight.

causative factors such as inflammation are not eliminated. Instead, to be successful, the treatment of PEM/wasting should be multifactorial, including not only one, but a number of concomitant measures against this devastating complication. New treatment strategies to improve nutritional status and the outcome of ESRD patients may include appetite stimulants, "anti-inflammatory diets," anti-inflammatory pharmacological agents, in combination with more traditional forms of nutritional support.[14]

In the present review, we will briefly discuss the scope of the problem of PEM/wasting in ESRD patients, its causes (in particular the role of chronic inflammation in the wasting syndrome of ESRD), and possible therapies including new emerging dietary and pharmacological treatments for the inflamed and wasted ESRD patient. We also propose an integrated therapeutic strategy against PEM/wasting in ESRD patients, consisting of combinations of different dietary and pharmacological treatments.

10.2 PROTEIN-ENERGY MALNUTRITION/WASTING IN ESRD

10.2.1 Signs of Protein-Energy Malnutritions

Assessment and monitoring of nutritional status are essential to prevent, diagnose, and treat uremic malnutrition.[15] Common signs of malnutrition in ESRD patients (Table 10.1) are reduced muscle mass as assessed by anthropometric methods, low serum concentrations of albumin, transferrin, prealbumin and other liver-derived proteins. Serum albumin and the method of subjective global assessment of nutritional status (SGA)[1–3, 16] are the most commonly used methods to identify malnutrition in these patients.

TABLE 10.2
Catabolic Factors in End-Stage Renal Disease Patients

Factors due to uremia per se
- Anorexia caused by increased circulating levels of anorexins like cholecystokinin, leptin, and pro-inflammatory cytokines.
- Abnormal protein and amino acid metabolism.
- Metabolic acidosis.
- Decreased biologic activity of anabolic hormones, such as insulin and IGF-1.
- Increased circulating levels of pro-inflammatory cytokines and catabolic hormones, such as parathyroid hormone.
- Abnormal energy metabolism, carbohydrate intolerance, and impaired lipid metabolism.

Other catabolic factors
- Fluid overload and congestive heart failure.
- Comorbidity such as cardiovascular disease, diabetes, chronic obstructive pulmonary disease, and infections.
- Gastrointestinal problems.
- Physical inactivity.

Factors related to peritoneal dialysis
- Loss of proteins, amino acids, and other essential nutrients in dialysate.
- Suppression of appetite by glucose absorption from dialysate and abdominal discomfort.
- Infectious complications, such as peritonitis and exit site infection.

Factors related to hemodialysis
- Blood access infections and sepsis.
- Bioincompatible dialyzer membranes.
- Exposure to endotoxins in dialysate.
- Loss of nutrients such as amino acids in dialysate.
- The unphysiological nature of intermittent dialysis with its marked fluctuations in fluid status, acid base balance, electrolytes, and other dialyzable substances.

10.2.1.1 Subjective global assessment of nutritional status

SGA has been widely used in both dialysis patients[2, 3, 16] as well as in patients with ESRD starting dialysis therapy.[1] SGA has been demonstrated to have good correlations to other nutritional markers in dialysis patients[3, 16] as well as to have a high predictive value for mortality in patients starting dialysis therapy.[17] However, one potential problem with SGA is its subjective nature, which may reduce its reproducibility, therefore small differences in SGA score must be interpreted with great caution. In general, SGA may differentiate severely malnourished patients from those with normal nutrition, but it is not a reliable predictor of the degree of malnutrition.[18] Furthermore, the definition of SGA may vary between different studies in CKD patients and there is a need to establish the validity and reliability of different SGA versions among these patients.[19]

10.2.1.2 Serum albumin, other plasma proteins, and amino acids in ESRD

Serum albumin has by far been the most commonly used marker of nutritional status in ESRD patients.[20] However, serum albumin levels in ESRD patients are influenced by several other factors (Table 10.2)[20–22]; a low serum albumin level mainly represents the acute phase response and albumin losses in dialysate and urine, and only to a lesser extent reflects a poor nutritional status.[20, 22, 23] The same is true also for other plasma proteins such as pre-albumin, transferrin, and retinol binding protein[24] as for serum albumin there is considerable overlap between serum levels of these proteins between malnourished and well-nourished patients.[3] They predict a clinical outcome mainly because they are also negative acute phase proteins.[1] Plasma amino acid levels are also abnormal in ESRD patients, the levels of essential amino acids are low while nonessential amino acids are often high, and dialysis does not reverse these abnormalities.[2, 25]

10.2.1.3 Other methods for assessment of nutritional status

Other common methods are anthropometry (such as skinfold thickness and calculation of midarm muscle circumference, MAMC), handgrip strength, bioimpedance (BIA) and dual-energy x-ray absorptiometry (DEXA) (Table 10.1). Measurements of actual body weight, height, and frame size, are easy to obtain and provide estimates of ideal body weight, body mass index, and other height-to-weight indices. In ESRD patients the preuremic weight for a patient may represent a target weight. Weight gains or losses in ESRD patients are influenced by the fluctuating state of hydration in these patients which is a consequence of dialytic therapy, use of diuretics, and fluctuating fluid intake. *Handgrip strength* is not only a marker of lean body mass but also provides important prognostic information,[1, 2, 26] at least in males.[27] *Bioimpedance* is affected by fluid status and may therefore be a less reliable tool in dialysis patients. *DEXA* is increasingly used and is now considered as a practical tool to estimate lean body mass and fat mass[1] although skinfold thickness may be a better method to assess subcutaneous fat.[28] However, estimation of truncal fat mass by waist circumference[29] or DEXA[30] is of particular interest because of its link with adipokine-related inflammation and the metabolic syndrome.

10.2.2 Prevalence of PEM/Wasting in ESRD

Several reports have shown a high prevalence (about 20 to 60%) of PEM/wasting in ESRD patients, before start of dialysis and following dialysis.[2, 3, 7, 31] Whereas the prevalence and severity of PEM/wasting is clearly related to residual renal function,[32, 33] the choice of dialysis modality does not seem to play a major role. A few studies have compared the prevalence of malnutrition in PD and HD patients, and in most of these studies, no major differences were found. For example, Nelson et al.[34] noted no difference in nutritional status between the two groups whereas in another study there was a greater prevalence of malnutrition among PD patients

(42%) compared to HD patients (31%).[35] The serum albumin levels among PD patients are often found to be lower than in HD patients, possibly due to the protein losses in dialysate; however, this is not a consistent finding.[7] Initiation of HD therapy is associated with improvements in most nutritional markers.[36] A common finding is that there is a larger gain in body fat in PD compared to HD patients,[7] most likely due to the impact of glucose absorption from the dialysate, and the absence of the intermittent catabolic stimulus associated with the HD procedure. Thus, after initiation of PD, there is improvement of nutritional status with weight gain,[37] improved anthropometric parameters[7, 38] and a rise in plasma proteins[7, 37, 38] indicating an increased net anabolism. Detailed prospective studies using anthropometrics,[7, 39] DEXA[40] and the combination of total body potassium and tritium dilution[41] have all shown that whereas lean body mass seems to be stable, body fat mass increases considerably, at least during the first years of treatment with PD. In addition, part of the reported weight gain may be due to increased body water.

10.2.3 PEM/Wasting as Predictor of Mortality in ESRD Patients

PEM/wasting is an important risk factor for morbidity and mortality in ESRD patients, both those treated with PD[7, 17, 39, 42, 43] and HD.[2, 43–47] Cachexia has been reported to be a cause of death in both PD patients[48, 49] as well as in HD patients.[49] However, PEM/wasting usually does not appear as a documented direct cause of mortality. Instead, CVD is by far the most common documented cause of mortality in the dialysis population, while malnutrition is a reported direct cause of less than 5% of deaths.[50] Although inflammation is often intertwined with PEM/wasting (vide infra), uremic malnutrition may also predict death independent of inflammatory status.[51]

10.3 Causes of PEM/Wasting in ESRD Patients

A variety of causes contribute to impaired nutritional status in patients with ESRD. When dialysis therapy begins, uremic symptoms are reduced, the diet is less restricted and most patients may thus show an improved nutritional status. However, anorexia may persist and many of the catabolic factors found at onset of therapy remain abnormal (Table 10.2). Furthermore, the dialytic procedure may induce additional catabolic factors (Table 10.2), in particular the blood membrane contact in HD and the dialytic loss of proteins (5 to 15 g/24 h) in PD, and these factors may increase the protein requirements in dialysis patients more than that of nondialyzed uremic patients. However, despite the increased nutritional requirements, ESRD patients often maintain a low nutritional intake because of anorexia, nausea and vomiting, caused by uremic toxicity (due to underdialysis), medications and unpalatable diets, infections, and other complicating illnesses. In addition, HD patients may have difficulties in eating during and after the HD session, and PD patients may experience aggravated anorexia due to absorption of glucose or amino acids from the dialysate as well as abdominal discomfort induced by the dialysate (Table 10.2).

The blood-membrane contact during the HD procedure may give rise to an inflammatory reaction, the intensity of which depends on the membrane material that is used.[52–54] The blood membrane interaction in a dialyzer without dialysate during sham-HD has been shown to stimulate net protein catabolism in normal individuals, especially when membranes of low biocompatibility were used, demonstrating that the effect is due to the interaction between blood and the membrane per se.[55, 56]

Compared to the HD procedure, PD appears to be a less strong catabolic stimulus, provided that the patient is free from peritonitis. However, there is a possibility that in PD the dialytic procedure per se elicits a low-grade inflammatory response stimulating protein catabolism through substances other than live bacteria present in the sterile dialysate. These substances could be microbial products (endotoxins), plastics, silicon, glucose break-down products[57] or other as yet unknown products which elute into the peritoneal cavity. Treatment with PD also involves other catabolic factors, such as loss of appetite, insufficient removal of small solutes, dialysate loss of proteins and amino acids, and recurrent peritonitis.[58, 59] In addition, the continuous supply of glucose (100 to 200 g/24 h) and lactate from the dialysate represents a sizeable and perhaps undesirable energy load that may induce or accentuate hyperglycemia, hyperinsulinemia, hypertriglyceridemia, and other metabolic abnormalities.[60]

10.3.1 Inflammation as Cause of PEM/Wasting in ESRD

Increased levels of C-reactive protein (CRP),[13] or pro-inflammatory cytokines[61, 62] are strong predictors of wasting and mortality in ESRD patients. Moreover, wasting and inflammation are strongly associated with CVD. Thus, it has been suggested that malnutrition, in part, is the consequence of chronic heart failure (CHF), or infection/inflammation, which triggers the development of not only PEM/wasting, but also atherosclerotic CVD, resulting in higher mortality rates.[10] Indeed, wasting (evaluated by SGA) is more common in patients with inflammation and CVD,[11] and Beddhu et al.[63] have demonstrated a significantly greater prevalence of coronary artery disease in dialysis patients with low serum albumin. However, in another study by Beddhu et al.[64] there was no association between malnutrition (defined as BMI <18.5 kg/m^2 and by a low urinary creatinine excretion) and coronary artery disease events in dialysis patients. On the other hand, uremic malnutrition as such is reported to be a predictor of death independent of inflammatory status.[51]

The strong impact of inflammation on nutritional status is of importance for understanding how to best treat or prevent PEM/wasting in ESRD patients. We have proposed that there may be at least two types of PEM/wasting in ESRD.[6] In type-1 PEM/wasting (i.e., true PEM), in which inadequate nutritional intake is the predominant cause, one would expect that nutritional supplementation alone would be effective in restoring nutritional deficits. However, as oral nutritional supplements or intradialytic supplementation have proved either only partially effective[65] or totally ineffectual[66] in the repair of nutritional status in the majority of patients it is suggestive that other factors may be responsible for inadequate nutritional status in this group of patients. As inflammation may not only decrease protein synthesis but also raise resting energy expenditure and protein catabolic rate, another type of malnutrition (i.e., wasting) has been proposed, in which a consistent inflammatory response and advanced comorbidity are the predominant causes. This inflammation-driven type-2 PEM/wasting with hypercatabolism and failure to deposit nutrients caused by inflammation could be the main reason for a poor nutritional status, and this condition is therefore much more difficult to treat by pure nutritional means, unless one also treats inflammation and comorbidities.[6] Interestingly, in most studies on nutritional supplementation in ESRD patients, a low serum albumin level, a hallmark of inflammation, was used as an inclusion criterion. Accordingly, based on the discussion above, it is not surprising that nutritional intervention may not always result in beneficial effects on nutritional status. As cytokines have catabolic effects on protein metabolism as well as central nervous effects resulting in anorexia, an upregulated cytokine activity may be a cause of inadequate food intake in type-2 PEM/wasting. However, as proposed by Dinarello and Roubenoff[67] cytokines may perhaps play a decreasing role in the pathogenesis of loss of muscle mass as muscle wasting increases and curtails the synthesis of cytokines, which can be viewed as a protective mechanism. Nevertheless, in the clinical setting, type-1 and type-2 PEM/wasting may be combined and difficult to disentangle and it is likely that inflammation in most cases may act in concert with undernutrition to cause wasting.

In patients who have inflammation, anorexia, loss of body weight, malaise, fatigue, and depression (i.e., sickness behaviors) are commonly observed. As weight loss is such a common feature of many inflammatory disorders, it has been suggested that subclinical inflammation is a principal component of the pathophysiology of muscle wasting by causing both low nutritional intake as well as increased muscle catabolism.[68] A recent study by Eustace et al.[69] demonstrated that both elevated CRP (a precise and objective index of the inflammatory activity reflecting the generation of pro-inflammatory cytokines) and low serum bicarbonate levels were independently associated with wasting. Indeed, inflammatory biomarkers, such as CRP,[70] tumor necrosis factor-α (TNF-α),[71] and interleukin-6 (IL-6)[61] are all primary predictors of S-albumin levels in ESRD patients.

It is therefore not surprising that pro-inflammatory cytokines have been shown to have significant effects on wasting and sickness behaviors in dialysis patients[72] and may play a central role in the development and maintenance of the MIA syndrome.[73, 74] TNF-α has been recognized as the prototype of an anorectic cytokine.[72] Indeed, cachexia is usually associated with elevated concentrations of TNF-α in older patients without renal disease.[75] In dialysis patients, hypoalbuminemia[71] has also been associated with elevated circulating levels of TNF-α. Another important mechanism for the proposed TNF-α-induced malnutrition is the effect of this cytokine on appetite and eating behavior. Elevated TNF-α levels has been reported to be associated with anorexia in dialysis patients.[76] Furthermore, Grunfeld et al.[77] found that administration of TNF-α resulted in an increase in leptin mRNA levels in hamsters and noted a strong inverse correlation between leptin mRNA level and subsequent food intake.

IL-6 is another pro-inflammatory cytokine proposed to play an important role in the muscle wasting which occurs with normal aging.[78] It has been suggested that the process of age-related muscle loss (sarcopenia) is related to loss of sex hormones (particularly testosterone) that act to suppress IL-6 expression.[79] Also genetic factors may contribute to the muscle wasting seen with older age as individuals who are genetically predisposed to produce high IL-6 levels during aging are disadvantaged against longevity.[79] Indeed, IL-6 has been shown to stimulate the breakdown of muscle protein[80] and promote cancer-related wasting[81] while the administration of IL-6 receptor antibody has been shown to inhibit muscle atrophy that occurs in IL-6 transgenic mice.[82] Since IL-6 also inhibits the secretion of insulin-like growth factor (IGF-1), reduced IGF-1 signaling may be yet another reason for IL-6-associated muscle atrophy.[83] The close association between IL-6 and both hypoalbuminemia[84] and muscle wasting (evaluated by thigh muscle area) in HD patients[85] provides compelling evidence that this cytokine may also be catabolic in this patient group. Finally, since intracerebroventricular injection of IL-6 increases energy expenditure and decreases body fat in rodents, IL-6 may play an important role in appetite and body weight control mediated by the central nervous system.[86]

Among the large number of other cytokines that may have nutritional impact, IL-15 has recently attracted much interest as it exerts important effects in skeletal muscle where it behaves as an anabolic cytokine both *in vitro*[87] and *in vivo*[88] and favors muscle fiber hypertrophy. This cytokine exerts its effects in muscle cells by both decreasing the rate of protein degradation and increasing the rate of proliferation. This distinct mode of IL-15 action suggests that it may be of potential usefulness in the treatment of wasting disorders. Therefore, studies on the effects of IL-15 administration are warranted in experimental renal failure models.

10.4 PREVENTION AND TREATMENT OF MALNUTRITION IN ESRD PATIENTS

In order to prevent and treat PEM/wasting among ESRD patients it is obvious that the patients should obtain a sufficient intake of energy and protein (Table 10.3). Although this is necessary and thus must be considered to be the cornerstone in all nutritional intervention strategies, improved nutrition alone is not sufficient because of the multifactorial nature of PEM/wasting in ESRD. Instead a whole battery of measures other than just nutritional support must be instituted as inadequate therapy of just one of the causative factors may be enough to inhibit any improvement (Figure 10.1).

10.4.1 General Aspects

In ESRD, there are a number of causative factors which need to be addressed such as adequate dialysis, acidosis, catabolic factors, and the like. Even slight acidosis should be corrected by oral supplementation with sodium bicarbonate (or altered dialysate buffer concentration). Physical exercise should be encouraged and intercurrent diseases should be actively treated. The prevention and treatment of

TABLE 10.3
Recommended Nutritional Intakes Per Day for End-Stage Renal Disease
(ESRD) Patients

Protein	$\geq$1.2 g/kg body weight ($\geq$50% of high biological value)
Energy	$\geq$35 kcal/kg body weight (including glucose absorption from the dialysate)
Fat	30% of total energy supply (high content of unsaturated lipids)
Water and sodium	As tolerated by fluid balance
Potassium	40–80 mmol
Calcium	Individualized, 800–1000 mg
Phosphorous	8–17 mg/kg body weight (phosphate binders are often needed)
Magnesium	200–300 mg
Iron	10–15 mg (supplements are often needed)
Zinc	15 mg (supplements may be required)

Vitamins (recommended supplementation)

Pyridoxine (B_6)	10 mg
Ascorbic acid	60–100 mg
Thiamine	1–5 mg
Folic acid	1 mg
Vitamin A, E, and K	Not routinely
Vitamin D	Individual supplementation

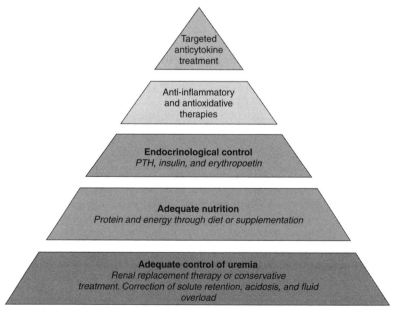

FIGURE 10.1 Integrated therapy of wasting consists of a combination of several treatment components, each of which is necessary, but not in itself sufficient to prevent and treat uremic wasting and malnutrition.

infectious complications is also crucial for the maintenance of adequate nutrition in ESRD patients, but lies outside the scope of the present review. Optimal treatment of comorbidity is an obvious goal and, in particular, if there are signs of inflammation, like increased CRP levels, it is important to elucidate and, if possible, eliminate the etiology. An increased dialysis dose should always be considered in ESRD patients with very low nutritional intakes. Indeed, there are several cross-sectional studies that report a correlation between the dose of dialysis and Protein Equivalent of Nitrogen Appearance (PNA), although there is no consensus on whether this is mainly due to a physiological relationship or to a mathematical artifact.[89–91]

10.4.1.1 Treatment of metabolic acidosis

Metabolic acidosis, a common condition in ESRD, is an important stimulus for net protein catabolism and a common cause of PEM/wasting in ESRD patients.[92] A low serum bicarbonate level is independently associated with hypoalbuminemia in chronic kidney disease patients.[93] A study of leucine kinetics in normal subjects during acute acidosis and alkalosis showed that total body protein breakdown as well as apparent leucine oxidation increases more during acidosis than during alkalosis.[94] In nondialysed chronic uremic patients the correction of metabolic acidosis improves nitrogen balance.[95] Acidosis, rather than uremia per se, appears to enhance protein catabolism in rats with CKD,[96] an effect that seems to be mediated by the stimulation of skeletal muscle branched-chain keto acid decarboxylation, resulting in increased catabolism of the branched-chain amino acids (valine, leucine, and isoleucine) which are mainly metabolized in muscle tissue.[97, 98] Mitch et al.[98] have shown that metabolic acidosis elicits the transcription of genes for proteolytic enzymes in muscles. Our group has previously reported that low intracellular valine concentration in muscle correlated with the predialysis blood standard bicarbonate level (which varied between 18 and 24 mmol/l) in a group of HD patients.[99] This finding indicates that even slight, and intermittent, acidosis may have stimulated the catabolism of valine in muscle, resulting in valine depletion that may be a limiting factor for protein synthesis.

Note that acidosis is currently the only identified uremic "toxic" factor which induces catabolism and impairs nitrogen utilization. Correction of acidosis in seven CKD patients studied with L-$(1-^{13}C)$ leucine kinetics resulted in decreased protein degradation whereas protein synthesis capacity was preserved and the patients adapted successfully to lower dietary protein intake.[100] Oral treatment with sodium bicarbonate in HD patients with acidosis resulted in increased plasma levels of branched-chain amino acids.[101] Oral sodium bicarbonate treatment of metabolic acidosis in PD patients resulted in improved nutritional status and shorter duration of hospitalization.[102] Thus, full correction of acidosis is an obvious goal for treatment in ESRD patients and oral bicarbonate should be prescribed even when the blood standard bicarbonate level is only marginally decreased.[103] It is recommended that the serum bicarbonate level in ESRD patients should be at least 22 mmol/l.[104]

10.4.1.2 Infections

Infections are the second most common cause of death in ESRD patients and an important cause of PEM/wasting. Uremia leads to disturbances in the immune response, with cutaneous anergy and impaired granulocyte function, thus increasing the susceptibility to infections. Although marked improvement of cell-mediated immunity (as assessed by hypersensitivity skin testing) has been reported during the first year of dialysis treatment,[105] most ESRD patients still show relative anergy.[105]

10.4.1.3 Fluid overload

Several recent studies have demonstrated the importance of fluid balance as a predictor of outcome in dialysis patients.[106] Fluid overload may contribute to inadequate dietary protein and energy intakes in these patients.[107] Thus, given the documented link between volume overload and inflammation in nonrenal[108] and renal[109] patients, various treatment strategies that improve fluid balance may be of potential benefit in reducing inflammation in ESRD patients. These strategies include not only strict fluid and salt compliance but also the use of diuretics and the PD-solution icodextrin to reduce extracellular volume.[110] Although studies in nonrenal patients have demonstrated that improvement of volume status may be associated with lower circulating levels of inflammatory biomarkers[108, 111] no such studies have, to the best of our knowledge, yet been reported in ESRD patients. As selective aldosterone blockade prevents angiotensin II/salt-induced vascular inflammation in the rat heart[112] it is also possible that certain diuretics may have significant anti-inflammatory effects per se.

10.4.2 Nutritional Requirements in ESRD Patients

If a macronutrient (protein, energy) or an essential nutrient (e.g., essential amino and fatty acids, vitamins, and trace minerals) are not provided in sufficient amounts, the individual will sooner or later suffer adverse reactions. Although, it may take quite some time before a nutritional deficiency is reflected in vital statistics, these subclinical deficiencies may indirectly have a negative effect by sensitizing the individual to other morbid factors. For example, protein malnutrition may impair the immune response, resulting in an increased risk of severe infections. Thus, it is necessary (but not always sufficient) that nutritional requirements of macronutrients and essential nutrients are met to prevent the development of PEM/wasting in ESRD patients. In the following sections, we highlight some of the specific nutritional requirements in ESRD.

10.4.2.1 Protein requirements in ESRD patients

The variation in protein requirements is much larger among ESRD patients than in healthy subjects, due to additional sources of variability, such as endocrine and biochemical abnormalities, anemia, drugs, physical inactivity, and comorbid conditions, for example, CVD, DM and infections. In addition, specific effects of the dialytic process may increase the protein requirements, especially in HD patients.[59]

It is currently recommended that ESRD patients undergoing dialysis should have a protein intake of at least 1.2 g/kg body weight/day, a large part of which should be of high biological value (Table 10.3). Whereas many patients may require less than this to maintain nitrogen equilibrium,[113, 114] patients who develop signs of PEM/wasting may require even higher amounts of protein (and energy) for repleting the protein and energy stores. Some patients may benefit by eating as much as 1.4 to 2.1 g protein/kg body weight/day, especially during the initial months of dialysis treatment.[113] Other patients may be in neutral or positive nitrogen balance with a protein intake as low as 0.7 g/kg body weight/day.[113]

10.4.2.2   Energy intake in ESRD patients

There is as yet no evidence that the energy requirements of ESRD patients are systematically different from those of normal subjects. Decreased resting energy expenditure has been reported in nondialyzed CKD patients.[115] Resting energy expenditure however does not seem to be influenced by the degree of renal function as such but may be elevated due to inflammation,[116] and may therefore predict mortality.[117] Monteon et al.[118] measured energy expenditure in normal subjects, nondialysed CKD patients, and HD patients and found no differences between the three groups with the subjects sitting, exercising, or in the postprandial state. This suggests that the energy expenditure of chronic HD patients does not differ from that of normal subjects during a given physical activity. Nor is there any data indicating that the energy requirements of PD patients differ from normal subjects. Thus, ESRD patients are recommended an energy intake of at least 35 kcal/kg body weight (147 kJ/kg body weight) (1) (Table 10.3). However, most ESRD patients have much lower energy intake. It should be noted that in addition to the oral energy intake, PD patients may absorb about 60% of the daily dialysate glucose load, resulting in a glucose absorption of about 100 to 200 g glucose/24 h.[60] Inadequate energy intake is a major cause of PEM in ESRD patients but energy intake is often underestimated in these patients.[119]

10.4.2.3   Monitoring and maintenance of adequate protein and energy intakes

It is strongly advisable to monitor the estimated protein intake as assessed by urea kinetics on a regular basis for all ESRD patients, in order to identify patients with a suboptimal protein intake. Repeated values below 1.0 g/kg body weight/day should raise the suspicion that patient protein intake is too low, and the patient should be advised to increase dietary protein. In ESRD patients, PNA is calculated from urea appearance rate in urine (and in the dialysate of dialysis patients).[113] Note that PNA previously was often denoted protein catabolic rate (PCR); however, PNA has been suggested as a more accurate term because true protein catabolic rate is about six times higher than PNA as estimated from urea appearance rate.[113, 120]

The recommended as well as the estimated nutritional intakes of various nutrients are usually normalized for body size. Several different ways of normalization have been applied; although there is no general agreement on which way of normalization is the best. When PNA has been calculated, it is usually normalized to g/kg body weight/day (*normalized* PNA, *nPNA*). However, this normalization has been performed using several different weight standards.[89, 121, 122] The easiest, and most common, way to normalize PNA is simply to divide it by the patient's *actual* dry body weight. This method is often applied although it does not take into account that the protein requirements per kg body weight possibly are lower in obese subjects and, furthermore, this method will yield high values of nPNA in malnourished patients with a low body weight.[89, 121] Other possibilities are to normalize PNA to normal body weight (from the National Health and Nutrition Examination Survey, NHANES),[123] desirable body weight (from the Metropolitan Life Insurance Company data)[124] or other weight standards,[122] alternatively to the patients preuremic weight (if it is known).

10.4.2.4 Vitamin and trace mineral requirements of ESRD patients

The vitamin and trace mineral requirements of ESRD patients are noted in Table 10.3. Although PEM occurs commonly in ESRD patients, it is not the only form of malnutrition that may exist in these patients. They may also suffer from deficiencies of micronutrients, particularly vitamins and trace elements.[125] Inadequate dietary intake, altered metabolism in uremia and vitamin loss into dialysate may lead to vitamin deficiencies, in particular deficiencies of water-soluble vitamins, in patients treated with dialysis.[126, 127] Thus, the serum levels of ascorbic acid, thiamine (B_1), pyridoxine (B_6), and folic acid have been reported to be low in dialysis patients.[126] Thiamin (vitamin B_1) deficiency with encephalopathy has been described in dialysis patients[128] and may be confounded with other neurologic diseases. A common dietary intake of 0.5–1.5 mg per day can be supplemented with a daily dose of 1–5 mg of thiamin hydrochloride. Vitamin B_6 coenzymes play a vital role in several aspects of amino acid utilization and the need for vitamin B_6 is particularly critical if protein and amino acid intake is limited.[129] Indeed, changes in fasting plasma amino acid and serum high-density lipoprotein levels after correction of the vitamin B_6 deficiency in dialysis patients indicates its role in the pathogenesis of the abnormal amino acid and lipid metabolism.[130] The daily requirement of pyridoxine may be higher in dialysis patients than in normal subjects and dialysis patients should be supplemented with a minimum of 10 mg of vitamin B_6 per day.[129] Folate is lost in dialysate and as folate levels have been reported to be reduced in serum, a daily dose of 1 mg of folic acid is recommended. High doses of folic acid (5–10 mg/day) have been shown to reduce the markedly elevated plasma homocysteine levels in ESRD patients by about two-thirds, which is still not fully within normal limits.[131] The question whether these high doses of folate should be prescribed in order to lower plasma homocysteine levels and whether this improves cardiovascular morbidity and mortality is still open, and prospective studies are needed in this area. Furthermore, supplementation with vitamin C has also been recommended.[127] However, high intake of vitamin C may aggravate hyperoxalemia in dialysis patients.[132]

Supplementation of the fat-soluble vitamins A, D, E, and K is not recommended on a routine basis.[126] Vitamin A tends to accumulate in CKD patients and supplementation should be avoided as vitamin A may have potentially harmful effects. Vitamin D and its active forms should be given on the basis of an evaluation of the bone metabolic status and taking the risks of hyperphosphatemia and hypercalcemia into consideration. Blood levels of vitamin E have been found to be normal or high in most studies in uremic patients, and vitamin K deficiency has not been reported in renal failure patients.[133] However, as vitamin E (tocopherol) is a strong antioxidant compound it may be worth while to consider treating ESRD patients with vitamin E. A randomized controlled high dose vitamin E (800 IU alpha-tocopherol/day) supplementation in high cardiovascular risk HD patients showed a significant 50% decrease in a cardiovascular composite index as compared to the placebo group.[134] However, studies in nonrenal patient groups have not shown similar results, possibly partly because different doses and forms of vitamin E were used.

10.4.2.5 Trace elements

Dietary requirements for trace elements are not well defined in ESRD patients.[125] Trace element metabolism is frequently altered in these patients[126] and high levels of trace elements have been attributed to impaired renal elimination or contamination of dialysis fluids, while low levels of trace elements may occur due to inadequate dietary intake or loss of protein bound trace elements into the dialysate.[126] Zinc deficiency has been reported to be associated with anorexia, hypogeusia, hyperprolactinemia, and impotence which have been alleviated by zinc administration. However, these results have not been generally confirmed and the role of zinc deficiency and requirements for extra supply of zinc in the diet of ESRD patients remains at present controversial,[126] although supplementation with zinc has been suggested in patients with hypogeusia, anorexia, and muscle weakness.

10.4.2.6 Carnitine depletion

A deficiency of L-carnitine may lead to impaired oxidation of long-chain fatty acids, inefficient energy production, and derangements of intermediary metabolism. Low as well as normal plasma and muscle concentrations of carnitine and carnitine esters have been reported in ESRD patients,[135] although only a minor part of HD patients seem to exhibit severe carnitine deficiency.[136] Also, several positive clinical effects of L-carnitine administration to HD patients have been reported, including improved nutritional status and muscle strength, increased well-being, and reduction of cardiac arrhythmias and angina, suggesting that carnitine depletion may be a pathogenic factor in skeletal muscle weakness and uremic heart disease.[136] However, a meta-analysis of all randomized controlled clinical trials of L-carnitine supplementation did not show any effect of L-carnitine supplementation on cholesterol and triglyceride profiles in hemodialysis patients.[137]

10.4.3 Nutritional Supplementation in ESRD

The addition of oral essential amino acids in tablets or to the diet may increase the biological value and the total intake of ingested protein and may improve the nutritional status.[138] Special amino acid formulas with a modified amino acid composition (high valine, addition of tyrosine and serine) have been designed to compensate for amino acid deficiencies present in uremia.[139, 140] Treatment with oral amino acid supplements in the form of tablets has been reported to result in significant improvement in the serum albumin concentration in HD patients.[141]

For a more complete supplementation of both protein and energy, several liquid formulas containing large amounts of protein of high biological value, as well as lipids and carbohydrates (but low amounts of phosphate, potassium, and sodium which must be restricted in most ESRD patients) in a small amount of fluid are available, suitable for the supplementary nutrition of dialysis patients.[140]

If severe malnutrition develops in ESRD patients despite adequate dialysis, measures to eliminate anorectic and catabolic factors, and food supplements, it may be necessary to give enteral or parenteral nutritional supplementation with energy and amino acids. Severely malnourished patients may have to be hospitalized temporarily for such treatment. Enteral nutrition through a thin nasogastric tube is preferable whenever possible as it is less expensive than parenteral nutrition and does not carry the risk of catheter sepsis. In patients who need parenteral nutrition with amino acids, a mixture of essential and nonessential amino acids seems better than a solution with only essential amino acids.[58] Energy should be provided simultaneously as hypertonic glucose or a mixture of glucose and lipid emulsion. Parenteral nutrition may be needed especially during peritonitis or sepsis.[142]

10.4.3.1 Clinical cause of amino acid-based peritoneal dialysis fluids

PD dialysis solutions containing amino acids may supplement in excess the daily loss of amino acids during dialysis with glucose-based solutions.[143] The amino acid solutions produce ultrafiltration and solute transport patterns that are similar to those with the standard glucose solutions although the period of effective ultrafiltration, for the same concentration of the osmotic agent, is slightly shorter.[144] The treatment with intraperitoneal amino acid solution may result in a markedly positive nitrogen balance, a significant increase in net protein anabolism, a more normal fasting plasma amino acid pattern and significant increases in serum total protein and transferrin.[114] It has also been demonstrated that the absorption of amino acids from the dialysate during one exchange of amino acid solution resulted in amino acid absorption (on average 17.6 g/day) that was twice as large as the dialysate losses of amino acids and protein (on average 9.2 g/day).[143] In a large randomized study, 134 malnourished PD patients were randomized to either use one or two exchanges of this amino acid solution per day ($n = 71$) or to continue with their usual glucose-based dialysis solution for three months.[145] At one month of study, there were (by analysis of covariance) significant increases in albumin, prealbumin, transferrin, and total proteins compared to baseline values in the amino acid group. Midarm muscular circumference also increased significantly in the amino acid group. At three months of the study, 70% of the patients in the amino acid group had improved in two or more nutritional variables vs. only 45% among the patients that were randomized to the control group, and, furthermore, there was a significant difference between the two groups in insulin-like growth factor-1 (IGF-1) compared to baseline values.[145] Another randomized study showed similar positive results.[146]

10.4.3.2 Intradialytic parenteral nutrition during HD

Intradialytic parenteral nutrition (IDPN) is a convenient and safe therapeutic intervention to provide nutrition during the HD procedure. IDPN may have a significant positive effect on protein and energy metabolism in stable HD patients. Pupim et al. observed improvements following IDPN in both net whole-body and skeletal muscle protein balance, reversal of a catabolic state, and resulting in net protein accretion, based on stable isotope infusion techniques.[147] IDPN was reported to result in concomitant increases in whole-body protein synthesis and the fractional synthetic rate of albumin.[148] Interestingly, exercise seems to augment the acute anabolic effects of IDPN.[149] However, not all studies have shown clear-cut benefits for IDPN, and the lack of larger prospective randomized studies has hampered definitive conclusions.

10.4.4 Is There a Role for Appetite Stimulants in ESRD?

Anorexia is a common phenomenon in ESRD patients that is associated with higher levels of pro-inflammatory cytokines, greater hospitalization rates, and poor clinical outcomes.[150] Thus, a simple inquiry about appetite may yield important information about the future risk of poor outcome in the ESRD patient.[150] There are several different mechanisms contributing to anorexia in dialysis patients, including inadequate dialysis, delayed gastric emptying, elevated levels of anorectic substances, like leptin and pro-inflammatory cytokines, as well as effects of the hemodialysis procedure, and the intraperitoneal infusion of dialysis fluid in PD patients.[151] Although some of these causes of anorexia can be treated through measures such as an increased dialysis dose in underdialysed patients and decreased intraperitoneal volume in PD patients with local symptoms, it is evident that these measures are not always effective.

The use of appetite stimulants may be a tempting part of an *integrated therapy* against malnutrition in ESRD patients. Unfortunately, most appetite stimulatory drugs available are relatively ineffective and they have not been systematically tested in dialysis patients. *Megesterol acetate* is the most extensively studied appetite stimulant and is widely used in cancer and AIDS patients.[152] However, megesterol acetate is associated with several side effects including hypogonadism, impotence, and increased risk of thromboembolism. Megestrol acetate has been shown to stimulate appetite in ESRD patients; however, the treatment may be risky and must be monitored closely.[153] *Cannabinoids*, such as dronabinol (∂-9 tetrahydrocannabinol), have been reported to increase appetite and weight gain and this drug class has been evaluated in cancer, HIV, and Alzheimer's disease patients. A comparison of dronabinol against megesterol acetate in a group of patients with advanced cancer found that megesterol acetate had a significantly better effect on appetite and weight gain.[154] To the best of our knowledge, no studies have been performed to study the putative appetite stimulating effects of cannaboids in ESRD patients. Several studies have demonstrated that *corticosteroids* increase appetite and well-being in cancer patients although a long lasting weight gain is often not observed.[155] However, the side effects (including the stimulation of muscle proteolysis through the ubiquitin-proteasome pathway) may be serious, and therefore corticosteroids have no primary role in the treatment of malnutrition in ESRD patients. *Cyproheptadine*, an antihistamine with serotonin antagonist properties has mainly been used for the treatment of cancer-induced weight loss, and in anorexia nervosa, but the effects have been questionable.[156] Furthermore, cyproheptadine has anticholinergic side effects[156]; it is therefore probably not ideal for ESRD patients. Other possible anticachectic drugs, such as thalidomide (because of its anti-inflammatory effect) and melatonin (because of its effect on muscle metabolism), are presently reaching the clinical trial stage.[157]

10.4.5 Hormonal Treatment of Malnutrition in Peritoneal Dialysis Patients

The dramatic evolution of molecular biology and new biotechnological tools has resulted in the possibility of producing recombinant human hormones which may be utilized to treat the disease. In nephrology, the use of recombinant human erythropoietin (rh-EPO) has dramatically changed the ability to treat renal anemia, and, furthermore, the possibility to treat malnutrition among dialysis patients with recombinant human growth stimulating hormones seems to be a promising perspective in the future.

10.4.5.1 Correction of anemia and erythropoietin treatment

Anemia is present in most ESRD patients and may be severe, especially in anephric patients and in patients who are inadequately dialyzed. Anemia leads to fatigue, diminishing exercise capacity, and physical inactivity, which may contribute to muscle wasting and malnutrition. Correction of anemia with rh-EPO has been reported to improve nutritional status to a moderate degree in HD patients,[158] presumably a secondary effect of anemia correction on physical work capacity, general well-being and appetite, rather than a specific anabolic effect of rh-EPO. The use of rh-EPO for correction of renal anemia, should be accompanied by an increased supply of iron to non-overloaded patients as the hemoglobin mass increases. Therefore the use of rh-EPO in ESRD patients requires assessment of iron stores because iron depletion will impair the response to rh-EPO and rh-EPO can cause iron deficiency. The increased iron requirements should be met if possible by oral substitution with iron[159] although parenteral iron supplementation may be needed in many cases.

10.4.5.2 Anabolic steroids

Anabolic steroids may exert a beneficial effect on the malnutrition of renal failure,[160] but larger prospective studies are needed to clarify the role of treatment with anabolic steroids among these patients as well as the severity of possible adverse effects.

10.4.5.3 Growth hormone and insulin-like growth factor-1

Recombinant human growth hormone (rhGH) administration enhances the growth velocity of children undergoing dialysis[161] and may reduce urea generation and improve the efficiency of dietary protein utilization in stable adult HD patients.[162] Furthermore, the combination of intradialytic parenteral nutrition and rhGH treatment in malnourished HD patients resulted in improved nutritional parameters (increased serum albumin, transferrin, and IGF-1) as well as in a decreased intradialytic urea appearance indicating that the treatment promoted net anabolism.[163] Short-term rhGH treatment in 10 PD patients resulted in signs of anabolism including a marked increase in IGF-1 levels.[164] Furthermore, administration of rhGH resulted in a decline of plasma essential amino acids suggesting an increased utilization of essential amino acids for protein anabolism.[165]

Many of the effects of growth hormone are mediated via IGFs (previously denoted somatomedins).[166] Indeed, some individuals may become resistant to the anabolic affects of rhGH due to malnutrition or uremia,[167–169] and this resistance seems to be partly associated with a low IGF-1 response to GH as well as a reduced bioactivity of IGF-1 in uremic serum.[170] PD patients with documented rhGH resistance, showed signs of anabolism within hours of commencing rhIGF-1 treatment, with a strongly positive nitrogen balance ($+2.0\,g/day$) that was sustained over the 20 days of the study without any sign of attenuation.[171]

Growth hormone and IGF-1 may turn out to be a useful adjunctive therapy to diminish body protein catabolism in ESRD patients, perhaps in combination of rhGH or IGF-1 treatment with supplements containing amino acids. However, larger long-term studies are clearly needed to clarify if the positive effects rhGH and IGF-1 treatment can be sustained for longer time periods, for patients to benefit from the different treatments, and as well for the assessment of the potential side-effects of these hormones.

10.5 POTENTIAL NEW TREATMENT STRATEGIES USING INFLAMMATION AS A TARGET

Given the hitherto rather poor results of energy and protein supplementation and the fact that the cachexia of ESRD seems to be associated with an upregulated proinflammatory cytokine system activity, new treatment strategies are needed. Below is a brief outline of the novel therapies and pathways currently being investigated in the ongoing quest to improve morbidity and mortality in the ESRD population.

10.5.1 Anti-Inflammatory Nutritional Treatment Strategies

Based on epidemiological studies in both renal and nonrenal populations, it is obvious that important differences regarding the prevalence and outcome of cachexia, inflammation, and atherosclerosis exist in different parts of the world.[172] In general, the population in Southeast Asia, China, and Japan consumes a substantial amount of fish and soy, resulting in a lower fat content and a higher fiber diet than the typical Western diet. Soybeans are a unique source of the phytoestrogens genistein and diadzein (estrogen-like substances) and in the Japanese population the phytoestrogen serum concentration is markedly higher compared to other populations.[173] As the phytoestrogen genistein is effective in blocking inflammatory gene expression,[174] dietary phytoestrogens may not only have a possible role in renal disease protection,[175] but could also provide significant anti-inflammatory properties, which could be of value for ESRD patients. Based on these findings, prospective studies on the impact of a high-soy diet on both the prevalence of MIA and outcome in ESRD patients are warranted.

The importance of *dietary fibers* is underscored by a recent evaluation study demonstrating that nonrenal subjects with a high fiber consumption had a lower risk of elevated CRP.[176] The anti-inflammatory and cardioprotective effects of the *omega-3 fatty acids* of fish oil, mainly eicosapentaenoic acid, are well recognized. For example, dietary fish oil decreases CRP and IL-6 levels in nonrenal subjects.[177] In addition, Kutner et al.[178] found that dialysis patients who reported fish consumption were 50% less likely to die during the observation period. Therefore, the potential beneficial effects of a diet with high content of dietary fibers and omega-3 fatty acids in ESRD patients certainly merits further investigations.

Advanced glycation end-products (AGEs), the result of the nonenzymatic reaction of reducing sugars with proteins, lipids, and nucleic acids, are usually markedly elevated in ESRD patients. It has been proposed that AGEs promote atherosclerosis through interaction with endothelial receptors.[179] Although reduced renal clearance and increased oxidative stress may be the most important causes of elevated AGEs in ESRD patients, diet may be an important source of highly reactive AGEs. As correlations have been found between one form of AGE, pentosidine, and CRP in both renal[180] and nonrenal[181] patients, it has been suggested that AGEs can trigger an inflammatory response.[182] Uribarri et al.[183] have shown that dietary glycotoxins contribute to significantly elevated AGE levels in ESRD patients.

Further studies are thus warranted to elucidate if dietary restrictions of the intake of AGE may reduce both excess toxic AGE and inflammation in this patient group. Importantly, a reduction in dietary AGE content can be obtained safely without compromising the content of vital nutrients, such as dietary protein, fat, and carbohydrates.[184] It may also be possible to use drug therapy to induce the breakdown of preexisting AGEs. One example of such a drug, which is presently being evaluated in clinical trials, is ALT-711. This cross-linked breaker also has beneficial effects on putative mediators of renal injury, such as prosclerotic cytokines and oxidative stress,[185] that might be beneficial for ESRD patients.

10.5.2 Anti-Inflammatory Pharmacological Treatment Strategies

As there are strong associations between pro-inflammatory cytokines and PEM/wasting in ESRD patients, various pharmacological anticytokine treatment strategies have been proposed for these patients (Table 10.4).

10.5.2.1 Anti-inflammatory effects of regularly used drugs

Several commonly used drugs in ESRD patients, such as statins and angiotensin-converting enzyme inhibitors (ACEI), have significant anti-inflammatory effects.[186] Indeed, two studies have demonstrated that statins, in addition to their lipid-lowering effect, also had an anti-inflammatory effect in HD patients.[187, 188]

Paradoxically, since lipoproteins isolated from normal human plasma can bind and neutralize bacterial lipopolysaccharide[189] the cholesterol-lowering effect of statins may represent a disadvantage of this treatment in wasted ESRD patients. In fact, it has been speculated that "non-lipid lowering statins" may be as effective and even more beneficial than "lipid-lowering statins" in wasted and inflamed patients.[190] This hypothesis has been challenged by the observation by Liu et al.[191] that the inverse association between total cholesterol with mortality in dialysis patients is related to the cholesterol-lowering effect of systemic inflammation and wasting. Therefore, at the present time, these novel findings support the treatment of hypercholesterolemia in this patient population.

Also, the renin–angiotensin system may contribute to inflammatory processes within the vascular system. Brull et al.[192] noted that ACEI treatment was associated with a reduction in IL-6 in response to coronary artery graft surgery, and we[193] have found lower plasma levels of TNF-α in ESRD patients treated with ACEI. A recent study demonstrated that treatment with an ACEI reduced the risk of weight loss,[194] supporting the hypothesis of strong relationships among wasting, the renin–angiotensin system, and inflammation. In addition, since the use of ACEI is an independent factor associated with a decreased rate of decline of residual renal function in dialysis patients[195] it is not surprising that ACEI improves the prognosis of renal patients.[196]

A variety of other agents that are not commonly used in ESRD patients may also have a role in the treatment of inflammatory-associated wasting. There have

TABLE 10.4

Examples of Pharmacological Interventions with Proposed Direct or Indirect Anticytokine Effects

Drugs that may have secondary anticytokine effects	Current indications	Use in CKD
Diuretics	Overhydration	Yes
Tocopherol	Vitamin supplementation	Yes
N-acetylcysteine	Bronchits	Yes
HMG-CoA reductase inhibitors	Dyslipidemia	Yes
Angiotensin-converting enzyme inhibitors	Hypertension, heart failure	Yes
PPAR-γ-activators (glitazones)	Diabetes mellitus	Restricted
Macrolides	Bacterial infections	As appropriate
Non-steroidal anti-inflammatory drugs	Pain	Restricted
COX-2 inhibitors	Pain	Restricted
Testosterone	Hypogonadism	No
AGEs inhibitors	—	No
Targeted anticytokine treatment strategies		
Thalidomide	Myeloma, leprosy	No
Etanercept	RA, Crohn's disease	No
Infliximab	RA, Crohn's disease	No
Anakinra	RA	No

Current indications and use in chronic kidney disease (CKD) patients.
Abbreviations: *HMG-CoA*, 3-hydroxy-3-methylglutaryl coenzyme A; *PPAR-γ*, peroxisome proliferator-activated receptor γ; *COX-2*, cyclooxygenase-2; *AGEs*, advanced glycated end-products; *RA*, rheumatoid arthritis.

been numerous reports suggesting that inhibition of prostaglandin production by nonsteroidal anti-inflammatory drugs may impact tumor-mediated wasting.[197] However, as these drugs may have serious side effects in ESRD patients, such as bleeding from the gastrointestinal tract, hematopoietic toxicities, and reduced residual renal function, nephrologists have been reluctant to use them widely. As COX-2 inhibitors also have been reported to reverse tumor-mediated wasting and associated humoral factors, such as IL-6 in different experimental models of cachexia,[198] the potential of this new class of drug in reversing uremic wasting needs to be addressed. Again, adverse effects and increased CVD risk may be an issue. Anabolic androgenic steroids, such as testosterone, nandrolone, oxandrolone, and stanozol have been studied in a number of settings of catabolic weight loss, such as AIDS, cancer, cardiac wasting, and ESRD. Sex hormones may be important for atheroprotection as men with low testosterone levels as well as women with low estrogen levels are at increased risk of CVD. The reported ability of sex hormones to interfere with cytokine production by expression of IL-6 mRNA may contribute to these protective effects.[199] As discussed above in the section on hormonal treatment of malnutrition (Section 10.4.5), recombinant rhIGF-1 may induce an anabolic response in ESRD patients in whom the primary cause of malnutrition is a low protein intake (see above). Indeed, injection of rhIGF-1 has been shown to induce a strong and sustained anabolic effect in PD patients.[200] However, the effectiveness of anabolic hormones, such as rhGH and rhIGF-1, has proved to be blunted if inflammation is present.[201] This again underlines the need to provide an integrated therapy for wasting in ESRD. Clearly, larger studies are needed to establish the respective role of anabolic hormones in inflamed and noninflamed ESRD patients with wasting.

Peroxisome proliferator-activated receptor γ (PPAR-γ activators; e.g., glitazones) have also been shown to inhibit the activation of inflammatory response genes, and promote a deviation of the immune system away from Th1 toward Th2 cytokine production.[202, 203] As DM is the most common cause of ESRD and insulin resistance is present in most ESRD patients, this class of drugs may have beneficial effects in these patients. Clarithromycin, which belongs to the macrolide group of antibiotics, has been found to reduce IL-6 levels and increase body weight in a small group of cachectic lung cancer patients.[204, 205] Thus, further studies are warranted to investigate the putative anticachectic and anti-inflammatory effects of both of these classes of agents in ESRD patients.

10.5.2.2 Antioxidants in ESRD

As wasting may be associated with increased oxidative stress[206] and poor nutritional intake may lead to a low nutritional intake of antioxidants, it could be speculated that antioxidative treatment strategies may be beneficial in wasted and inflamed ESRD patients.[207] It is of interest that vitamin E and N-acetylcysteine, have not only an anti-inflammatory potential,[208, 209] and improve endothelial dysfunction,[210, 211] but have also been shown to reduce cardiovascular events in dialysis patients.[134, 212] Indeed, Boaz et al.[134] demonstrated that supplementation of vitamin E (800 IU/day) reduced composite CVD endpoints and myocardial infarction in HD-patients followed for 519 days and Tepel et al.[212] showed that also N-acetylcysteine (600 mg BID) reduced composite cardiovascular endpoints in a group of 134 HD-patients followed for 14.5 months. Clearly, as both studies were small, larger randomized studies are needed to confirm these intriguing findings.

10.5.2.3 Targeted anticytokine treatment strategies

Targeted anticytokine treatment strategies, which have been tested in patients with other inflammatory states, may also be of interest in ESRD patients. However, so far no studies exist investigating the efficacy and safety of drugs such as thalidomide, IL-1 receptor antagonist (IL-1Ra; anakinra), TNF-receptor blockade (etanercept) or antibody treatment against TNF (infliximab). Indeed, central administration of specific IL-1Ra attenuates some, but not all, of the metabolic responses secondary to systemic infection and endotoxins and prevents sepsis-induced inhibition of protein synthesis in rats[213] and an interesting link between stimulus-dependent variability in IL-1Ra synthesis by mononuclear cells and clinical outcome of HD patients has been reported.[214] However, it should be noted that although the use of targeted anti-inflammatory treatment would seem to be logical in inflamed and wasted ESRD patients with hypercytokinemia, available toxicity data should caution clinicians against liberal use of these agents until large randomized trials have been conducted to prove their efficacy in this patient group. Indeed, there is a growing body of evidence suggesting that neutralization of TNF-α[215] but not of IL-1[216] may be associated with an increased risk of opportunistic infections, such as with *Mycobacterium tuberculosis*.

10.6 Summary and Conclusions

A large proportion of ESRD patients demonstrate signs of protein-energy malnutrition and wasting due to various factors including disturbances in protein and energy metabolism, hormonal derangements, infections and other superimposed illnesses, and poor food intake because of anorexia and nausea. Importantly, signs of PEM/wasting have been found to correlate to the malnutrition, inflammation, and atherosclerosis (MIA)-syndrome which may play a role in most premature deaths in ESRD patients. As conventional measures to improve nutritional status, as well as increased dialysis dose, have failed to improve the excessive mortality in ESRD patients, there is new hope that an integrated therapy against PEM/wasting may improve the abysmal clinical outcome in ESRD patients.

The protein requirement in ESRD patients appears to be increased to about 1.2 g protein/kg/day, which is twice that of normal individuals, although some patients are in neutral balance with a protein intake as low as 0.7 g/kg/day. The nitrogen balance is strongly dependent on the energy intake which often is lower than the recommended 35 kcal/kg/day in these patients. Efforts to provide adequate protein and energy intakes are fundamental, and provide a necessary platform on which other interventional measures should be added.

It is important to monitor protein intake (preferably by urea kinetics) and nutritional status (preferably by subjective global assessment, body weight, serum albumin, and an estimation of lean body mass) in all patients. Note however that a low serum albumin level does not always reflect a poor nutritional status of the patient and that, in the medical decision process, it is important to consider the multiple factors, other than poor nutrition, that may contribute to low serum albumin concentrations in ESRD patients, in particular inflammation and losses of albumin in the dialysate. If the patients have signs of inflammation, such as increased CRP levels, it is important to seek for and, if possible, treat the cause.

Anorexia with low protein and energy intake results from a variety of factors of which underdialysis with insufficient control of uremic toxicity seems to be a major one. Malnourished ESRD patients should be recommended to increase protein intake, as well as energy intake, while acidosis should be corrected and the dialysis dose should be increased if possible. Amino acid based dialysis fluids may provide new opportunities to improve the nutritional status in malnourished PD patients, and treatment with hormones promoting net anabolism may turn out to be a useful adjunctive therapy to diminish body protein catabolism in ESRD patients in the future.

PEM/wasting in ESRD patients is multifactorial and, as such, single therapeutic strategies are not likely to be successful. Given the hitherto rather poor results of energy and protein supplementation, new treatment strategies are needed. We believe that much could be learned from other wasted and inflamed patients groups, such as HIV, CHF, and cancer-cachexia patients, in which various anti-inflammatory treatment strategies in combination with efforts to improve nutritional intake seems to improve nutritional status and outcome. Thus, since there is now the possibility of addressing the uremic wasting syndrome on different pathophysiologic levels, we should grasp the opportunity to utilize new therapeutic modalities to try to improve the quality of life and outcome of wasted and inflamed ESRD patients. We envisage that based on such studies, it will be possible to establish an integrated therapy against wasting in ESRD patients consisting of both traditional (such as increased nutrient supply and correction of acidosis and anemia) and nontraditional (such as various anti-inflammatory treatment strategies) methods.

REFERENCES

1. Heimburger, O., Qureshi, A.R., Blaner, W.S., Berglund, L., and Stenvinkel, P., Hand-grip muscle strength, lean body mass, and plasma proteins as markers of nutritional status in patients with chronic renal failure close to start of dialysis therapy, *Am. J. Kidney Dis.*, 36, 1213, 2000.
2. Qureshi, A.R., Alvestrand, A., Danielsson, A., Divino-Filho, J.C., Gutierrez, A., Lindholm, B. et al., Factors predicting malnutrition in hemodialysis patients: A cross-sectional study, *Kidney Int.*, 53, 773, 1998.
3. Young, G.A., Kopple, J.D., Lindholm, B., Vonesh, E.F., DeVecchi, A., Scalamogna, A. et al., Nutritional assessment of continuous ambulatory peritoneal dialysis patients: An international study, *Am. J. Kidney Dis.*, 17, 462, 1991.
4. Mitch, W.E., Malnutrition: A frequent misdiagnosis for hemodialysis patients, *J. Clin. Invest.*, 110, 437, 2002.
5. Ikizler, T.A., Greene, J.H., Wingard, R.L., Parker, R.A., and Hakim, R.M., Spontaneous dietary intake during progression of chronic renal failure, *Kidney Int.*, 6, 1386, 1995.
6. Stenvinkel, P., Heimburger, O., Lindholm, B., Kaysen, G.A., and Bergstrom, J., Are there two types of malnutrition in chronic renal failure? Evidence for relationships between malnutrition, inflammation and atherosclerosis (MIA syndrome), *Nephrol. Dial. Transplant.*, 15, 953, 2000.
7. Jager, K.J., Merkus, M.P., Huisman, R.M., Boeschoten, E.W., Dekker, F.W., Korevaar, J.C. et al., Nutritional status over time in haemodialysis and peritoneal dialysis, *J. Am. Soc. Nephrol.*, 12, 1272, 2001.

8. Eknoyan, G., Beck, G.J., Cheung, A.K., Daugirdas, J.T., Greene, T., Kusek, J.W. et al., Effect of dialysis dose and membrane flux in maintenance hemodialysis, *N. Engl. J. Med.*, 347, 2010, 2002.

9. Paniagua, R., Amato, D., Vonesh, E., Correa-Rotter, R., Ramos, A., Moran, J. et al., Effects of increased peritoneal clearances on mortality rates in peritoneal dialysis: ADEMEX, a prospective, randomized, controlled trial, *J. Am. Soc. Nephrol.*, 13, 1307, 2002.

10. Bergström, J. and Lindholm, B., Malnutrition, cardiac disease and mortality — An integrated point of view, *Am. J. Kidney Dis.*, 32, 834, 1998.

11. Stenvinkel, P., Heimbürger, O., Paultre, F., Diczfalusy, U., Wang, T., Berglund, L. et al., Strong association between malnutrition, inflammation, and atherosclerosis in chronic renal failure, *Kidney Int.*, 55, 1899, 1999.

12. Ross, R., Atherosclerosis: An inflammatory disease, *N. Engl. J. Med.*, 340, 115, 1999.

13. Zimmermann, J., Herrlinger, S., Pruy, A., Metzger, T., and Wanner, C., Inflammation enhances cardiovascular risk and mortality in hemodialysis patients, *Kidney Int.*, 55, 648, 1999.

14. Stenvinkel, P., Lindholm, B., and Heimburger, O., Novel approaches in an integrated therapy of inflammatory-associated wasting in end-stage renal disease, *Semin. Dial.*, 17, 505, 2004.

15. Pupim, L.B. and Ikizler, T.A., Assessment and monitoring of uremic malnutrition, *J. Ren. Nutr.*, 14, 6, 2004.

16. Enia, G., Sicuso, C., Alati, G., and Zocalli, C., Subjective global assessment of nutrition in dialysis patients, *Nephrol. Dial. Transplant.*, 8, 1094, 1993.

17. Adequacy of dialysis and nutrition in continuous peritoneal dialysis: Association with clinical outcomes. Canada-USA (CANUSA) Peritoneal Dialysis Study Group, *J. Am. Soc. Nephrol.*, 7, 198, 1996.

18. Cooper, B.A., Bartlett, L.H., Aslani, A., Allen, B.J., Ibels, L.S., and Pollock, C.A., Validity of subjective global assessment as a nutritional marker in end-stage renal disease, *Am. J. Kidney Dis.*, 40, 126, 2002.

19. Steiber, A.L., Kalantar-Zadeh, K., Secker, D., McCarthy, M., Sehgal, A., and McCann, L., Subjective Global Assessment in chronic kidney disease: A review, *J. Ren. Nutr.*, 14, 191, 2004.

20. Heimbürger, O., Bergström, J., and Lindholm, B., Is serum albumin an indication of nutritional status in CAPD patients?, *Perit. Dial. Int.*, 14, 108, 1994.

21. Kaysen, G.A., Rathore, V., Shearer, G.C., and Depner, T.A., Mechanisms of hypoalbuminemia in hemodialysis patients, *Kidney Int.*, 48, 510, 1995.

22. Schoenfeld, P.Y., Albumin is an unreliable marker of nutritional status, *Semin. Dial.*, 5, 218, 1992.

23. Yeun, J.Y. and Kaysen, G.A., Acute phase proteins and peritoneal dialysate albumin loss are the main determinants of serum albumin in peritoneal dialysis patients, *Am. J. Kidney Dis.*, 30, 923, 1997.

24. Lowrie, E.G. and Lew, N.L., Commonly measured laboratory variables in hemodialysis patients: Relationships among them and to death risk, *Semin. Nephrol.*, 12, 276, 1992.

25. Lindholm, B., Alvestrand, A., Furst, P. and Bergstrom, J., Plasma and muscle free amino acids during continuous ambulatory peritoneal dialysis, *Kidney Int.*, 35, 1219, 1989.

26. Wang, A.Y., Sea, M.M., Ho, Z.S., Lui, S.F., Li, P.K., and Woo, J., Evaluation of handgrip strength as a nutritional marker and prognostic indicator in peritoneal dialysis patients, *Am. J. Clin. Nutr.*, 81, 79, 2005.

27. Stenvinkel, P., Barany, P., Chung, S.H., Lindholm, B., and Heimburger, O., A comparative analysis of nutritional parameters as predictors of outcome in male and female ESRD patients, *Nephrol. Dial. Transplant.*, 17, 1266, 2002.

28. Avesani, C.M., Draibe, S.A., Kamimura, M.A., Cendoroglo, M., Pedrosa, A., Castro, M.L. et al., Assessment of body composition by dual energy X-ray absorptiometry, skinfold thickness and creatinine kinetics in chronic kidney disease patients, *Nephrol. Dial. Transplant.*, 19, 2289, 2004.

29. Wisse, B.E., The inflammatory syndrome: The role of adipose tissue cytokines in metabolic disorders linked to obesity, *J. Am. Soc. Nephrol.*, 15, 2792, 2004.

30. Axelsson, J., Rashid Qureshi, A., Suliman, M.E., Honda, H., Pecoits-Filho, R., and Heimburger, O. et al., Truncal fat mass as a contributor to inflammation in end-stage renal disease, *Am. J. Clin. Nutr.*, 80, 1222, 2004.

31. Jansen, M.A.M., Korevaar, J.C., Dekker, F.W., Jager, K.J., Boeschoten, E.W., and Krediet, R.T., Renal function and nutritional status at the start of chronic dialysis treatment., *J. Am. Soc. Nephrol.*, 12, 157, 2001.

32. Cooper, B.A., Aslani, A., Ryan, M., Ibels, L.S., and Pollock, C.A., Nutritional state correlates with renal function at the start of dialysis, *Perit. Dial. Int.*, 23, 291, 2003.

33. Duenhas, M.R., Draibe, S.A., Avesani, C.M., Sesso, R., and Cuppari, L., Influence of renal function on spontaneous dietary intake and on nutritional status of chronic renal insufficiency patients, *Eur. J. Clin. Nutr.*, 57, 1473, 2003.

34. Nelson, E.N., Hong, C.D., Pesce, A.L., Peterson, D.W., Singh, S., and Pollak, V.E., Anthropometric norms for the dialysis population, *Am. J. Kidney Dis.*, 16, 32, 1990.

35. Cianciaruso, B., Brunori, G., Kopple, J.D. et al., Cross-sectional comparison of malnutrition in continuous ambulatory dialysis and hemodialysis patients, *Am. J. Kidney Dis.*, 26, 475, 1995.

36. Pupim, L.B., Kent, P., Caglar, K., Shyr, Y., Hakim, R.M., and Ikizler, T.A., Improvement in nutritional parameters after initiation of chronic hemodialysis, *Am. J. Kidney Dis.*, 40, 143, 2002.

37. Tranaeus, A., Heimbürger, O., Lindholm, B., and Bergström, J., Six years' experience of CAPD at one centre: A survey of major findings, *Perit. Dial. Int.*, 8, 31, 1988.

38. Pollock, C.A., Ibels, L.S., Caterson, R.J., Mahony, J.F., Waugh, D.A., and Cocksedge, B., Continuous ambulatory peritoneal dialysis, eight years of experience at a single center, *Medicine*, 68, 293, 1989.

39. Pollock, C.A., Allen, B.J., Warden, R.A., Caterson, R.J., Blagojevic, N., Cocksedge, B. et al., Total body nitrogen by neutron activation in maintenance dialysis, *Am. J. Kidney Dis.*, 16, 38, 1990.

40. Stenvinkel, P., Lindholm, B., Lönnqvist, F., Katzarski, K., and Heimbürger, O., Increases in serum leptin during peritoneal dialysis are associated with inflammation and a decrease in lean body mass, *J. Am. Soc. Nephrol.*, 11, 1303, 2000.

41. Johansson, A., Samuelsson, O., Haraldsson, B., Bosaeus, I., and Attman, P.-O., Body composition in patients treated with peritoneal dialysis, *Nephrol. Dial. Transplant.*, 13, 1511, 1998.

42. Chung, S.H., Lindholm, B., and Lee, H.B., Influence of initial nutritional status on continuous ambulatory peritoneal dialysis patient survival, *Perit. Dial. Int.*, 20, 19, 2000.

43. Marckmann, P., Nutritional status of patients on hemodialysis and peritoneal dialysis, *Clin. Nephrol.*, 29, 75, 1988.

44. Lowrie, E.G. and Lew, N.L., Death risk in hemodialysis patients: The predictive value of commonly measured variables and an evaluation of death rate differences between facilities, *Am. J. Kidney Dis.*, 15, 458, 1990.

45. Oksa, H., Ahonen, K., Pasternack, A., and Marnela, K.M., Malnutrition in hemodialysis patients, *Scand. J. Urol. Nephrol.*, 25, 157, 1991.

46. Degoulet, P., Legrain, M., Réach, I., et al., Mortality risk factors in patients treated by chronic hemodialysis, *Nephron*, 31, 103, 1982.

47. Bilbrey, G.L. and Cohen, T.L., Identification and treatment of protein calorie malnutrition in chronic hemodialysis patients, *Dial. Transplant.*, 18, 669, 1989.

48. Maiorca, R., Vonesh, E., Cancarini, G.C., Cantaluppi, A., Manili, L., Brunori, G. et al., A six years comparison of patient and technique survivals in CAPD and HD, *Kidney Int.*, 34, 518, 1988.

49. Maiorca, R., Vonesh, E.F., Cavalli, P.L., De Vecchi, A., Giangrande, A., La Greca, G. et al., A multicenter, selection-adjusted comparison of patient and technique survivals on CAPD and hemodialysis, *Perit. Dial. Int.*, 114, 118, 1991.

50. Foley, R.N., Parfrey, P.S., and Sarnak, M.J., Clinical epidemiology of cardiovascular disease in chronic renal failure., *Am. J. Kidney Dis.*, 32 (Suppl. 5), S112, 1998.

51. Pupim, L.B., Caglar, K., Hakim, R.M., Shyr, Y., and Ikizler, T.A., Uremic malnutrition is a predictor of death independent of inflammatory status, *Kidney Int.*, 66, 2054, 2004.

52. Betz, M., Haenisch, G.M., Rauterberg, E.W., Bommer, J., and Ritz, E., Cuprammonium membranes stimulates interleukin-1 release and archidonic acid metabolism in monocytes in the absence of complement, *Kidney Int.*, 34, 67, 1988.

53. Cheung, A.K., Biocompatibility of hemodialysis membranes, *J. Am. Soc. Nephrol.*, 1, 150, 1990.

54. Lonnemann, G., Bingel, M., Floege, J., Koch, K.M., Shaldon, S., and Dinarello, C.A., Detection of endotoxin-like interleukin-1-inducing activity during in vitro dialysis, *Kidney Int.*, 33, 29, 1988.

55. Gutierrez, A., Alvestrand, A., Wahren, J., and Bergström, J., Effect of in vivo contact between blood and dialysis membranes on protein catabolism in humans, *Kidney Int.*, 38, 487, 1990.

56. Gutierrez, A., Bergström, J., and Alvestrand, A., Protein catabolism in sham-hemodialysis: The effect of different membranes, *Clin. Nephrol.*, 38, 20, 1992.

57. Nilsson-Thorell, C.B., Muscalu, N., Andren, A.H., Kjellstrand, P.T., and Wieslander, A.P., Heat sterilization of fluids for peritoneal dialysis gives rise to aldehydes, *Perit. Dial. Int.*, 13, 208, 1993.

58. Lindholm, B. and Bergström, J. Nutritional requirements of peritoneal dialysis. In: Gokal, R., and Nolph, K.D., eds. *Textbook of Peritoneal Dialysis*. Dordrecht: Kluwer Academic Publishers, 1994; p. 443.

59. Bergström, J. and Lindholm, B., Nutrition and adequacy of dialysis. How do hemodialysis and CAPD compare?, *Kidney Int.*, 43 (Suppl. 40), S39, 1993.

60. Heimbürger, O., Waniewski, J., Werynski, A., and Lindholm, B., A quantitative description of solute and fluid transport during peritoneal dialysis, *Kidney Int.*, 41, 1320, 1992.

61. Bologa, R.M., Levine, D.M., Parker, T.S., Cheigh, J.S., Seur, D., Stenzel, K.H. et al., Interleukin-6 predicts hypoalbuminemia, hypocholesterolemia, and mortality in hemodialysis patients, *Am. J. Kidney Dis.*, 32, 107, 1998.

62. Kimmel, P.L., Phillips, T.M., Simmens, S.J., Peterson, R.A., Weihs, K.L., Alleyne, S. et al., Immunologic function and survival in hemodialysis patients, *Kidney Int.*, 54, 236, 1998.

63. Beddhu, S., Kaysen, G.A., Yan, G., Sarnak, M., Agodoa, L., Ornt, D. et al., Association of serum albumin and atherosclerosis in chronic hemodialysis patients, *Am. J. Kidney Dis.*, 40, 721, 2002.

64. Beddhu, S., Pappas, L.M., Ramkumar, N., and Samore, M.H., Malnutrition and atherosclerosis in dialysis patients, *J. Am. Soc. Nephrol*, 15, 733, 2004.

65. Capelli, J.P., Kushner, H., Canmiscioli, T.C., Chen, S.-M., and Tores, M.A., Effect of intradialytic pareneteral nutrition on mortality rates in end-stage renal disease care, *Am. J. Kidney Dis*, 23, 808, 1994.

66. Wolfson, M., Use of nutritional supplements in dialysis patients, *Semin. Dial.*, 5, 285, 1992.

67. Dinarello, C.A. and Roubenoff, R.A., Mechanisms of loss of lean body mass in patients on chronic dialysis, *Blood Purif.*, 14, 388, 1996.

68. Roubenoff, R., Catabolism of aging: Is it an inflammatory process? *Curr. Opin. Clin. Nutr. Metab. Care*, 6, 295, 2003.

69. Eustace, J.A., Astor, B., Muntner, P.M., Ikizler, T.A., and Coresh, J., Prevalence of acidosis and inflammation and their association with low serum albumin in chronic kidney disease, *Kidney Int.*, 65, 1031, 2004.

70. Kaysen, G.A., Dublin, J.A., Müller, H.G., Rosales, L.M., and Levin, N.W., The acute-phase response varies with time and predicts serum albumin levels in hemodialysis patients, *Kidney Int.*, 58, 346, 2000.

71. Odamaki, M., Kato, A., Takita, T., Furuhashi, M., Maruyama, Y., Yonemura, K. et al., Role of soluble receptors for tumor necrosis factor alpha in the development of hypoalbuminemia in hemodialysis patients, *Am. J. Nephrol.*, 22, 73, 2002.

72. Stenvinkel, P., Heimburger, O., and Lindholm, B., Wasting, but not malnutrition, predicts cardiovascular mortality in end-stage renal disease, *Nephrol. Dial. Transplant.*, 19, 2181, 2004.

73. Stenvinkel, P., Heimbürger, O., and Jogestrand, T., Elevated interleukin-6 predicts progressive carotid atherosclerosis in dialysis patients: Association to chlamydia pneumoniae seropositivity, *Am. J. Kidney Dis.*, 39, 274, 2002.

74. Stenvinkel, P., Barany, P., Heimburger, O., Pecoits-Filho, R., and Lindholm, B., Mortality, malnutrition and atherosclerosis in end-stage renal disease: What is the role of interleukin-6? *Kidney Int.*, 61, S103, 2002.

75. Yeh, S.S. and Schuster, M.W., Geriatric cachexia: The role of cytokines, *Am. J. Clin. Nutr.*, 70, 183, 1999.

76. Aguilera, A., Codoceo, R., Selgas, R., Garcia, P., Picornell, M., Diaz, C. et al., Anorexigen (TNF-alpha, cholecystokinin) and orexigen (neuropeptide Y) plasma levels in peritoneal dialysis (PD) patients. Their relationship with nutritional parameters, *Nephrol. Dial. Transplant.*, 13, 1476, 1998.

77. Grunfeld, C., Zhao, C., Fuller, J., Pollock, A., Moser, A., Friedman, J. et al., Endotoxin and cytokines induce expression of leptin, the ob gene product in hamsters. A role for leptin in the anorexia of infection, *J. Clin. Invest.*, 97, 2152, 1996.

78. Payette, H., Roubenoff, R., Jacques, P.F., Dinarello, C.A., Wilson, P.W., Abad, L.W. et al., Insulin-like growth factor-1 and interleukin 6 predict sarcopenia in very old community-living men and women: The Framingham Heart Study, *J. Am. Geriatr. Soc.*, 51, 1237, 2003.

79. Bonafe, M., Olivieri, F., Cavallone, L., Giovagnetti, S., Mayegiani, F., Cardelli, M. et al., A gender-dependent genetic predisposition to produce high levels of IL-6 is detrimental for longevity, *Eur. J. Immunol.*, 31, 2357, 2001.

80. Goodman, M.N., Interleukin-6 induces skeletal muscle protein breakdown in rats, *Proc. Soc. Exp. Biol. Med.*, 205, 182, 1994.

81. Strassman, G., Fong, M., Kenney, J.S., and Jacob, C.O., Evidence for the involvement of interleukin 6 in experimental cancer cachexia, *J. Clin. Invest.*, 89, 1681, 1992.

82. Tsujinaka, T., Fujita, J., Ebisuri, C., Yano, M., Kominami, E., Suzuki, K. et al., Interleukin-6 receptor antibody inhibits muscle atrophy and modulates proteolytic systems in interleukin-6 transgenic mice, *J. Clin. Invest.*, 97, 244, 1996.

83. Barbieri, M., Ferruci, L., Ragno, E., Corsi, A., Bandinelli, S., Bonafe, M. et al., Chronic inflammation and the effect of IGF-1 on muscle strength and power in older persons, *Am. J. Physiol. Endocrinol. Metab.*, 284, E481, 2003.

84. Kaizu, Y., Kimura, M., Yoneyama, T., Miyaji, K., Hibi, I., and Kumagai, H., Interleukin-6 may mediate malnutrition in chronic hemodialysis patients, *Am. J. Kidney Dis.*, 31, 93, 1998.

85. Kaizu, Y., Ohkawa, S., Odamaki, M., Ikegaya, N., Hibi, I., Miyaji, K. et al., Association between inflammatory mediators and muscle mass in long-term hemodialysis patients, *Am. J. Kidney Dis.*, 42, 295, 2003.

86. Stenlof, K., Wernstedt, I., Fjallman, T., Wallenius, V., Wallenius, K., and Jansson, J.O., Interleukin-6 levels in the central nervous system are negatively correlated with fat mass in overweight/obese subjects, *J. Clin. Endocrinol. Metab.*, 88, 4379, 2003.

87. Quinn, L.S., Anderson, B.G., Drivdahl, R.H., Alvarez, B., and Argiles, J.M., Overexpression of interleukin-15 induces skeletal muscle hypertrophy in vitro: Implications for treatment of muscle wasting disorders, *Exp. Cell. Res*, 280, 55, 2002.

88. Carbo, N., Lopez-Soriano, J., Costelli, P., Busquets, S., Alvarez, B., Baccino, F.M. et al., Interleukin-15 antagonizes muscle protein waste in tumour-bearing rats, *Br. J. Cancer*, 83, 526, 2000.

89. Harty, J.C., Boulton, H., Curwell, J., Heelis, N., Uttley, L., Venning, M.C. et al., The normalized protein catabolic rate is a flawed marker of nutrition in CAPD patients, *Kidney Int.*, 45, 103, 1994.

90. Stein, A. and Walls, J., The correlation between Kt/V and protein catabolic rate — A self-fulfilling prophecy, *Nephrol. Dial. Transplant.*, 9, 743, 1994.

91. Harty, J., Venning, M., and Gokal, R., Does CAPD guarantee adequate dialysis delivery and nutrition, *Nephrol. Dial. Transplant.*, 9, 1721, 1994.

92. Szeto, C.C. and Chow, K.M., Metabolic acidosis and malnutrition in dialysis patients, *Semin. Dial.*, 17, 371, 2004.

93. Eustace, J.A., Astor, B., Muntner, P.M., Ikizler, T.A., and Coresh, J., Prevalence of acidosis and inflammation and their association with low serum albumin in chronic kidney disease, *Kidney Int.*, 65, 1031, 2004.

94. Straumann, E., Keller, U., Küry, D., Bloesch, D., Thélin, A., Arnaud, M.J. et al., Effect of acute acidosis and alkalosis on leucine kinetics in man, *Clin. Physiol.*, 12, 39, 1992.
95. Papadoyannakis, N.J., Stefanidis, C.J., and McGeown, M., The effect of the correction of metabolic acidosis on nitrogen and potassium balance of patients with chronic renal failure, *Am. J. Clin. Nutr.*, 40, 623, 1984.
96. Hara, Y., May, R.C., Kelly, R.C., and Mitch, W.E., Acidosis, not azotemia, stimulates branched-chain, amino acid catabolism in uremic rats, *Kidney Int.*, 32, 808, 1987.
97. May, R.C., Hara, Y., Kelly, R.A., Block, K.P., Buse, M., and Mitch, W.E., Branched-chain amino acid metabolism in rat muscle: Abnormal regulation in acidosis, *Am. J. Physiol.*, 252, E712, 1987.
98. Mitch, W.E., Jurkovic, C., and England, B.K., Mechanisms that cause protein and amino acid catabolism in uremia, *Am. J. Kidney Dis.*, 21, 91, 1993.
99. Bergström, J., Alvestrand, A., and Fürst, P., Plasma and muscle free amino acids in maintenance hemodialysis patients without protein malnutrition, *Kidney Int.*, 38, 108, 1990.
100. Lim, V.S., Yarasheski, K.E., and Flanigan, M.J., The effect of uraemia, acidosis, and dialysis treatment on protein metabolism: A longitudinal leucine kinetic study, *Nephrol. Dial. Transplant.*, 13, 1723, 1998.
101. Kooman, J.P., Deutz, N.E., Zijlmans, P., van den Wall Bake, A., Gerlag, P.G., van Hooff, J.P. et al., The influence of bicarbonate supplementation on plasma levels of branched-chain amino acids in haemodialysis patients with metabolic acidosis, *Nephrol. Dial. Transplant.*, 12, 2397, 1997.
102. Szeto, C.C., Wong, T.Y., Chow, K.M., Leung, C.B., and Li, P.K., Oral sodium bicarbonate for the treatment of metabolic acidosis in peritoneal dialysis patients: A randomized placebo-control trial, *J. Am. Soc. Nephrol.*, 14, 2119, 2003.
103. Stein, A., Baker, F., Larratt, C., Bennett, S., Harris, K., Feehally, J. et al., Correction of metabolic acidosis and the protein catabolic rate in PD patients, *Perit. Dial. Int.*, 14, 187, 1994.
104. Kalantar-Zadeh, K., Mehrotra, R., Fouque, D., and Kopple, J.D., Metabolic acidosis and malnutrition-inflammation complex syndrome in chronic renal failure, *Semin. Dial.*, 17, 455, 2004.
105. Young, G.A., Young, J.B., Young, S.M., Hobson, S.M., Hildreth, B., Brownjohn, A.M. et al., Nutrition and delayed hypersensitivity during continuous ambulatory peritoneal dialysis in relation to peritonitis, *Nephron*, 43, 177, 1986.
106. Ates, K., Nergizoglu, G., Keven, K., Sen, A., Kutlay, S., Erturk, S. et al., Effect of fluid and sodium removal on mortality in peritoneal dialysis patients, *Kidney Int.*, 60, 767, 2001.
107. Wang, A.Y., Sanderson, J., Sea, M.M., Wang, M., Lam, C.W., Li, P.K. et al., Important factors other than dialysis adequacy associated with inadequate dietary protein and energy intakes in patients receiving maintenance peritoneal dialysis, *Am. J. Clin. Nutr.*, 77, 834, 2003.
108. Niebauer, J., Volk, H.-D., Kemp, M., Dominguez, M., Schumann, R.R., Rauchhaus, M. et al., Endotoxin and immune activation in chronic heart failure: A prospective cohort study, *Lancet*, 353, 1838, 1999.
109. Konings, C.J., Kooman, J.P., Schonck, M., Struijk, D.G., Gladziwa, U., Hoorntje, S.J. et al., Fluid status in CAPD patients is related to peritoneal transport

and residual renal function: Evidence from a longitudinal study, *Nephrol. Dial. Transplant.*, 18, 797, 2003.

110. Davies, S.J., Woodrow, G., Donovan, K., Plum, J., Williams, P., Johansson, A.C. et al., Icodextrin improves the fluid status of peritoneal dialysis patients: Results of a double-blind randomized controlled trial, *J. Am. Soc. Nephrol.*, 14, 2338, 2003.

111. Sato, Y., Takatsu, Y., Kataoka, K., Yamada, T., Taniguchi, R., Sasayama, S. et al., Serial circulating concentrations of C-reactive protein, interleukin (IL)-4, and IL-6 in patients with acute left heart decompensation, *Clin. Cardiol.*, 22, 811, 1999.

112. Rocha, R., Martin-Berger, C.L., Yang, P., Scherrer, R., Delyani, J., and McMahon, E., Selective aldosterone blockade prevents angiotensin II/salt-induced vascular inflammation in the rat heart, *Endocrinology*, 143, 4828, 2002.

113. Bergström, J., Fürst, P., Alvestrand, A., and Lindholm, B., Protein and energy intake, nitrogen balance and nitrogen losses in patients treated with continuous ambulatory peritoneal dialysis, *Kidney Int.*, 44, 1048, 1993.

114. Kopple, J., Bernard, D., Messana, J., Swartz, R., Bergström, J., Lindholm, B. et al., Treatment of malnourished CAPD patients with an amino acid based dialysate, *Kidney Int.*, 47, 1148, 1995.

115. Avesani, C.M., Draibe, S.A., Kamimura, M.A., Dalboni, M.A., Colugnati, F.A., and Cuppari, L., Decreased resting energy expenditure in non-dialysed chronic kidney disease patients, *Nephrol. Dial. Transplant.*, 19, 3091, 2004.

116. Avesani, C.M., Draibe, S.A., Kamimura, M.A., Colugnati, F.A., and Cuppari, L., Resting energy expenditure of chronic kidney disease patients: Influence of renal function and subclinical inflammation, *Am. J. Kidney Dis.*, 44, 1008, 2004.

117. Wang, A.Y., Sea, M.M., Tang, N., Sanderson, J.E., Lui, S.F., Li, P.K. et al., Resting energy expenditure and subsequent mortality risk in peritoneal dialysis patients, *J. Am. Soc. Nephrol.*, 15, 3134, 2004.

118. Monteon, F.J., Laidlaw, S.A., Shaib, J.K., and Kopple, J.D., Energy expenditure in patients with chronic renal failure, *Kidney Int.*, 30, 741, 1986.

119. Avesani, C.M., Kamimura, M.A., Draibe, S.A., and Cuppari, L., Is energy intake underestimated in nondialyzed chronic kidney disease patients? *J. Ren. Nutr.*, 15, 159, 2005.

120. Kopple, J.D., Jones, M.R., Eshaviah, K.P.R., Bergström, J., Lindsay, R.M., Moran, J. et al., A proposed glossary for dialysis kinetics, *Am. J. Kidney Dis.*, 26, 963, 1995.

121. Jones, M.R., Etiology of severe malnutrition: Results of an international cross-sectional study in continuous ambulatory peritoneal dialysis patients, *Am. J. Kidney Dis.*, 23, 412, 1994.

122. Dwyer, J. and Kenler, S.R., Assessment of nutritional status in renal disease. In: Mitch, W.E. and Klahr, S., eds. *Nutrition and the Kidney*. Boston: Little, Brown and Co., 1993; p. 61.

123. Frisancho, A.R., New standards of weight and body composition by frame size and height for assessment of nutritional state of adults and the elderly, *Am. J. Clin. Nutr.*, 40, 808, 1984.

124. Metropolitan height and weight tables, Stat bull 64. January–June, 1983: Metropolitan Life Insurance Company, 1983.

125. Kalantar-Zadeh, K., and Kopple, J.D., Trace elements and vitamins in maintenance dialysis patients, *Adv. Ren. Replace. Ther.*, 10, 170, 2003.

126. Gilmour, E.R., Hartley, G.H., and Goodship, T.H.J. Trace elements and vitamins in renal disease. In: Mitch, W.E. and Klahr, S., eds. *Nutrition and the Kidney*. Boston: Little, Brown and Co, 1993; p. 114.

127. Boeschoten, E.W., Schrijver, J., Krediet, R.T., Schreurs, W.H., and Arisz, L., Deficiencies of vitamins in CAPD patients: The effect of supplementation, *Nephrol. Dial. Transplant.*, 3, 187, 1988.

128. Hung, S.C., Hung, S.H., Tarng, D.C., Yang, W.C., Chen, T.W., and Huang, T.P., Thiamine deficiency and unexplained encephalopathy in hemodialysis and peritoneal dialysis patients, *Am. J. Kidney Dis.*, 38, 941, 2001.

129. Kopple, J.D., Mercurio, K., Blumenkrantz, M.J., Jones, M.R., Tallos, J., Roberts, C. et al., Daily requirement for pyridoxine supplements in chronic renal failure, *Kidney Int.*, 19, 694, 1981.

130. Kleiner, M.J., Tate, S.S., Sullivan, J.F., and Charmi, J., Vitamin B6 deficiency in maintenance dialysis patients: Metabolic effects of repletion, *Am. J. Clin. Nutr.*, 33, 1612, 1980.

131. Arnadottir, M., Brattstrom, L., Simonsen, O., Thysell, H., Hultberg, B., Andersson, A. et al., The effect of high-dose pyridoxine and folic acid supplementation on serum lipid and plasma homocysteine concentrations in dialysis patients, *Clin. Nephrol.*, 40, 236, 1993.

132. Canavese, C., Petrarulo, M., Massarenti, P., Berutti, S., Fenoglio, R., Pauletto, D. et al., Long-term, low-dose, intravenous vitamin C leads to plasma calcium oxalate supersaturation in hemodialysis patients, *Am. J. Kidney Dis.*, 45, 540, 2005.

133. Harty, J., Conway, L., Keegan, M., Curwell, J., Venning, M., Campbell, I. et al., Energy metabolism during CAPD; a controlled study, *Adv. Perit. Dial.*, 11, 229, 1995.

134. Boaz, M., Smetana, S., Weinstein, T., Matas, Z., Gafter, U., Iaina, A. et al., Secondary prevention with antioxidants of cardiovascular disease in end stage renal disease (SPACE): Randomised placebo-controlled trial, *Lancet*, 356, 1213, 2000.

135. Moorthy, A.V., Rosenbaum, M., Rajaram, R., and Shug, A.L., A comparison of plasma and muscle carnitine levels in patients on peritoneal and hemodialysis for chronic renal failure, *Am. J. Nephrol.*, 3, 205, 1983.

136. Wanner, C. and Hörl, W.H., Carnitin abnormalities in patients with renal insufficiency, *Nephron*, 50, 89, 1988.

137. Hurot, J.M., Cucherat, M., Haugh, M. and Fouque, D., Effects of L-carnitine supplementation in maintenance hemodialysis patients: A systematic review, *J. Am. Soc. Nephrol.*, 13, 708, 2002.

138. Ikizler, T.A., Protein and energy: Recommended intake and nutrient supplementation in chronic dialysis patients, *Semin. Dial.*, 17, 471, 2004.

139. Alvestrand, A., Ahlberg, M., Furst, P., and Bergstrom, J., Clinical results of long-term treatment with a low protein diet and a new amino acid preparation in patients with chronic uremia, *Clin. Nephrol.*, 19, 67, 1983.

140. Garibotto, G., Defarrari, G., Robaudo, C., Saffioti, S., Sala, M.R., Paoletti, E. et al., Effects of a new amino acid supplement on blood AA pools in patients with chronic renal failure, *Amino Acids*, 1, 319, 1991.

141. Eustace, J.A., Coresh, J., Kutchey, C., Te, P.L., Gimenez, L.F., Scheel, P.J. et al., Randomized double-blind trial of oral essential amino acids for dialysis-associated hypoalbuminemia, *Kidney Int.*, 57, 2527, 2000.

142. Gahl, G. and Hain, H. Nutrition and metabolism in continuous ambulatory peritoneal dialysis. In: Scarpione, L.L. and Ballocchi, S., eds. *Evolution and Trends in Peritoneal Dialysis*, vol. 84. Basel: Karger, 1990; p.36.

143. Jones, M.R., Gehr, T.W., Burkart, J.M., Hamburger, R.J., Kraus, A.P., Jr., Piraino, B.M. et al., Replacement of amino acid and protein losses with 1.1% amino acid peritoneal dialysis solution, *Perit. Dial. Int.*, 18, 210, 1998.

144. Park, M.S., Heimbürger, O., Bergström, J., Waniewski, J., Werynski, A., and Lindholm, B., Peritoneal transport during dialysis with amino acid-based solutions, *Perit. Dial. Int.*, 13, 280, 1993.

145. Jones, M., Hagen, T., Boyle, C.A., Vonesh, E., Hamburger, R., Charytan, C. et al., Treatment of malnutrition with 1.1% amino acid peritoneal dialysis solution: Results of a multicenter outpatient study, *Am. J. Kidney Dis.*, 32, 761, 1998.

146. Li, F.K., Chan, L.Y., Woo, J.C., Ho, S.K., Lo, W.K., Lai, K.N. et al., A 3-year, prospective, randomized, controlled study on amino acid dialysate in patients on CAPD, *Am. J. Kidney Dis.*, 42, 173, 2003.

147. Pupim, L.B., Flakoll, P.J., Brouillette, J.R., Levenhagen, D.K., Hakim, R.M., and Ikizler, T.A., Intradialytic parenteral nutrition improves protein and energy homeostasis in chronic hemodialysis patients, *J. Clin. Invest.*, 110, 483, 2002.

148. Pupim, L.B., Flakoll, P.J., and Ikizler, T.A., Nutritional supplementation acutely increases albumin fractional synthetic rate in chronic hemodialysis patients, *J. Am. Soc. Nephrol.*, 15, 1920, 2004.

149. Pupim, L.B., Flakoll, P.J., Levenhagen, D.K., and Ikizler, T.A., Exercise augments the acute anabolic effects of intradialytic parenteral nutrition in chronic hemodialysis patients, *Am. J. Physiol. Endocrinol. Metab.*, 286, E589, 2004.

150. Kalantar-Zadeh, K., Block, G., McAllister, C.J., Humphreys, M.H., and Kopple, J.D., Appetite and inflammation, nutrition, anemia and clinical outcome in hemodialysis patients, *Am. J. Clin. Nutr.*, 80, 299, 2004.

151. Mehrotra, R. and Kopple, J.D. Causes of protein-energy malnutrition in chronic renal failure. In: Kopple, J.D. and Massry, S.G., eds. *Nutritional Management of Renal Disease*, 2nd ed, Vol. 167–182. Philadelphia, PA: Lippincott Williams & Wilkins, 2004.

152. Argiles, J.M., Meijsing, S.H., Pallares-Trujillo, J., Guirao, X., and Lopez-Soriano, F.J., Cancer cachexia: A therapeutic approach, *Med. Res. Rev.*, 21, 83, 2001.

153. Boccanfuso, J.A., Hutton, M., and McAllister, B., The effects of megestrol acetate on nutritional parameters in a dialysis population, *J. Ren. Nutr.*, 10, 36, 2000.

154. Jatoi, A., Windschitl, H.E., Loprinzi, C.L., Sloan, J.A., Dakhil, S.R., Mailliard, J.A. et al., Dronabinol versus megestrol acetate versus combination therapy for cancer-associated anorexia: A North Central Cancer Treatment Group study, *J. Clin. Oncol.*, 15, 567, 2020.

155. Nelson, K.A., Modern management of the cancer anorexia-cachexia syndrome, *Curr. Pain Headache Rep.*, 5, 250, 2001.

156. Golden, A.C., Daiello, L.A., Silverman, M.A., Llorenete, M., and Preston, R.A., University of Miami Division of clinical Pharmacology therapeutic rounds: Medications used to treat anorexia in the frail elderly, *Am. J. Ther.*, 10, 292, 2003.

157. Gagnon, B. and Bruera, E., A review of the drug treatment of cachexia associated with cancer, *Drugs*, 55, 675, 1998.

158. Barany, P., Pettersson, E., Ahlberg, M., Hultman, E., and Bergstrom, J., Nutritional assessment in anemic hemodialysis patients treated with recombinant human erythropoietin, *Clin. Nephrol.*, 35, 270, 1991.

159. Barany, P., Eriksson, L.C., Hultcrantz, R., Pettersson, E., and Bergstrom, J., Serum ferritin and tissue iron in anemic dialysis patients, *Miner. Electrolyte Metab.*, 23, 273, 1997.

160. Soliman, G. and Oreopoulos, D.G., Anabolic steroids and malnutrition in chronic renal failure, *Perit. Dial. Int.*, 14, 362, 1994.

161. Fine, R.N., Growth in children undergoing CAPD/CCPD/APD, *Perit. Dial. Int.*, 13 (Suppl. 2), S247, 1993.

162. Ziegler, T.R., Lazarus, J.M., Young, L.S., Hakim, R., and Wilmore, D.W., Effects of recombinant human growth hormone in adults receiving maintenance hemodialysis, *J. Am. Soc. Nephrol.*, 2, 1130, 1991.

163. Schulman, G., Wingard, R.L., Hutchinson, R.L., Lawrence, P., and Hakim, R., The effects of recombinant human growth hormone and intradialytic parenteral nutrition in malnourished hemodialysis patients, *Am. J. Kidney Dis.*, 21, 527, 1993.

164. Ikizler, T.A., Wingard, R.L., Breyer, J.A., Schulman, G., Parker, R.A., and Hakim, R.M., Short-term effects of recombinant human growth hormone in CAPD patients, *Kidney Int.*, 46, 1178, 1994.

165. Ikizler, T.A., Wingard, R.L., Flakoll, P.J., Schulman, G., Parker, R.A., and Hakim, R.M., Effects of recombinant human growth hormone on plasma and dialysate amino acid profiles in CAPD patients, *Kidney Int.*, 50, 229, 1996.

166. Hammerman, M.R., The growth hormone-insulin-like growth factor axis in kidney, *Am. J. Physiol.*, 26, F503, 1989.

167. Kopple, J.D., The rationale for the use of growth hormone or insulin-like growth factor 1 in adult patients with renal failure, *Miner. Electrolyte Metab.*, 18, 269, 1992.

168. Tönshoff, B., Edén, S., Weiser, E., Carlsson, B., Robinson, I.C.A.F., Blum, W.F. et al., Reduced hepatic growth hormone (GH) receptor gene expression and increased plasma GH binding protein in experimental uremia, *Kidney Int.*, 45, 1085, 1994.

169. Tönshoff, B., Schäfer, F., and Mehls, O., Disturbance of growth hormone-insulin-like growth factor axis in uremia, *Pediatr. Nephrol*, 4, 654, 1990.

170. Blum, W.F., Insulin-like growth factors (IGFs) and IGF binding proteins in chronic renal failure: Evidence for reduced secretion of IGFs, *Acta Paediatr. Scand.*, 379 (Suppl.), 24, 1991.

171. Fouque, D., Peng, S.C., Shamir, E., and Kopple, J.D., Recombinant human insulin-like growth factor-1 induces an anabolic response in malnourished CAPD patients, *Kidney Int.*, 57, 646, 2000.

172. Nascimento, M.M., Pecoits-Filho, R., Lindholm, B., Riella, M.C., and Stenvinkel, P., Inflammation, malnutrition and atherosclerosis in end-stage renal disease: A global perspective, *Blood Purif.*, 20, 454, 2002.

173. Morton, M.S., Arisaka, O., Miyake, N., Morgan, L.D., and Evans, B.A., Phytoestrogen concentrations in serum from Japanese men and women over forty years of age, *J. Nutr.*, 132, 168, 2002.

174. Evans, M.J., Eckert, A., Lai, K., Adelman, S.J., and Harnish, D.C., Reciprocal antagonism between estrogen receptor and NF-kappaB activity in vivo, *Circ. Res.*, 89, 823, 2001.

175. Velasquez, M.T. and Bhathena, S.J., Dietary phytoestrogens: A possible role in renal disease protection, *Am. J. Kidney Dis.*, 37, 1056, 2001.

176. King, D.E., Egan, B.M., and Geesey, M.E., Relation of dietary fat and fiber to elevation of C-reactive protein, *Am. J. Cardiol.*, 92, 1335, 2003.

177. Ciubotaru, I., Lee, Y.S., and Wander, R.C., Dietary fish oil decreases C-reactive protein, interleukin-6 and tricylglycerol to HDL-cholesterol ratio in postmenopausal women on HRT, *J. Nutr. Biochem.*, 14, 513, 2003.

178. Kutner, N.G., Clow, P.W., Zhang, R., and Aviles, X., Association of fish intake and survival in a cohort of incident dialysis patients, *Am. J. Kidney Dis.*, 39, 1018, 2002.

179. Miyata, T., Ishikawa, S., Asahi, K., Inagi, R., Suzuki, D., Horie, K. et al., 2-isopropylidenehydrazono-4-oxo-thiazolidin-5-ylacetanilide (OPB-9195) treatment inhibits the development of intimal thickening after ballon injury of rat carotid artery: Role of glycoxidation and lipooxidation reactions in vascular tissue damage, *FEBS Lett.*, 445, 202, 1999.

180. Suliman, M., Heimburger, O., Barany, P., Anderstam, B., Pecoits-Filho, R., Ayala, E.R. et al., Plasma pentosidine is associated with inflammation and malnutrition in end-stage renal disease patients starting on dialysis therapy, *J. Am. Soc. Nephrol.*, 14, 1614, 2003.

181. Miyata, T., Ishiguro, N., Yasuda, Y., Ito, T., Nangaku, M., Iwata, H. et al., Increased pentosidine, an advanced glycation end product, in plasma and synovial fluid from patients with rheumatoid arthritis and its relation to inflammatory markers, *Biochem. Biophys. Res. Commun.*, 244, 45, 1998.

182. Schwedler, S., Schinzel, R., Vaith, P., and Wanner, C., Inflammation and advanced glycation end products in uremia: Simple coexistence, potentiation or causal relationship? *Kidney Int.*, 59 (Suppl. 78), S32, 2001.

183. Uribarri, J., Peppa, M., Cai, W., Goldberg, T., Lu, M., He, C. et al., Restriction of dietary glycotoxins reduces excessive advanced glycation end products in renal failure patients, *J. Am. Soc. Nephrol.*, 14, 728, 2003.

184. Uribarri, J., Peppa, M., Cai, W., Goldberg, T., Lu, M., Baliga, S. et al., Dietary glycotoxins correlate with circulating advanced glycation end product levels in renal failure patients, *Am. J. Kidney Dis.*, 42, 532, 2003.

185. Forbes, J.M., Thallas, V., Thomas, M.C., Founds, H.W., Burns, W.C., Jerums, G. et al., The breakdown of preexisting advanced glycation end products is associated with reduced renal fibrosis in experimental diabetes, *FASEB J.*, 17, 1762, 2003.

186. Libby, P., Inflammation in atherosclerosis, *Nature*, 420, 868, 2002.

187. Chang, J.W., Yang, W.S., Min, W.K., Lee, S.K., Park, J.S., and Kim, S.B., Effects of simvastatin on high-sensitivity C-reactive protein and serum albumin in hemodialysis patients, *Am. J. Kidney Dis.*, 39, 1213, 2002.

188. Vernaglione, L., Cristofano, C., Muscogiuri, P., and Chimienti, S., Does atorvastatin influence serum C-reactive protein levels in patients on long-term hemodialysis?, *Am. J. Kidney Dis.*, 43, 471, 2004.

189. Wurfel, M.M., Kunitake, S.T., Lichenstein, H., Kane, J.P., and Wright, S.D., Lipopolysaccharide (LPS)-binding protein is carried on lipoproteins and acts as a cofactor in the neutralization of LPS, *J. Exp. Med.*, 180, 1025, 1994.

190. Rauchhaus, M., Coats, A.J., and Anker, S.D., The endotoxin-lipoprotein hypothesis, *Lancet*, 356, 930, 2000.

191. Liu, Y., Coresh, J., Eustace, J.A., Longnecker, J.C., Jaar, B., Fink, N.E. et al., Association between cholesterol level and mortality in dialysis patients. Role of inflammation and malnutrition, *JAMA*, 291, 451, 2004.

192. Brull, D.J., Sanders, J., Rumley, A., Lowe, G.D., Humphries, S.E., and Montomery, H.E., Impact of angiotensin converting enzyme inhibition on post-coronary artery bypass interleukin 6 release, *Heart*, 87, 252, 2002.

193. Stenvinkel, P., Andersson, A., Wang, T., Lindholm, B., Bergström, J., Palmblad, J. et al., Do ACE-inhibitors suppress tumor necrosis factor-α production in advanced chronic renal failure?, *J. Intern. Med.*, 246, 503, 1999.

194. Anker, S.D., Negassa, A., Coats, A.J., Afzal, R., Poole-Wilson, P.A., Cohn, J.N. et al., Prognostic importance of weight loss in chronic heart failure and the effect of treatment with angiotensin-converting-enzyme inhibitors: An observational study, *Lancet*, 361, 1077, 2003.

195. Moist, L.M., Port, F.K., Orzol, S.M., Young, E.W., Ostbye, T., Wolfe, R.A. et al., Predictors of loss of residual renal function among new dialysis patients, *J. Am. Soc. Nephrol.*, 11, 556, 2000.

196. Mann, J.F., Gerstein, H.C., Pogue, J., Bosch, J., and Yusuf, S., Renal insufficiency as a predictor of cardiovascular outcomes and the impact of ramipril: The HOPE randomized trial, *Ann. Intern. Med.*, 134, 629, 2001.

197. Okamoto, T., NSAID zaltoprofen improves the decrease in body weight in rodent sickness behavior models: Proposed new applications of NSAIDs, *Int. J. Mol. Med.*, 9, 369, 2002.

198. Davis, T.W., Zweifel, B.S., O'Neal, J.M., Heuvelman, D.M., Abegg, A.L., Hendrich, T.O. et al., Inhibition of COX-2 by celecoxib reverses tumor induced wasting, *J. Pharmacol. Exp. Ther.*, 308, 929, 2004.

199. Ershler, W.B. and Keller, E.T., Age-associated increased interleukin-6 gene expression, late-life diseases, and frailty, *Ann. Rev. Med.*, 51, 245, 2000.

200. Fouque, D., Peng, S.C., Shamir, E., and Kopple, J.D., Recombinant human insulin-like growth factor-1 induces an anabolic response in malnourished CAPD patients, *Kidney Int.*, 57, 646, 2000.

201. Ericsson, F., Filho, J.C., and Lindgren, B.F., Growth hormone treatment in hemo-dialysis patients — A randomized, double-blind, placebo-controlled study, *Scand. J. Urol. Nephrol.*, 38, 340, 2004.

202. Saubermann, L.J., Nakajima, A., Wada, K., Zhao, S., Terauchi, Y., Kadowaki, T. et al., Peroxisome proliferator-activated receptor gamma agonist ligands stim-ulate a Th2 cytokine response and prevent acute colitis, *Inflamm. Bowel Dis.*, 8, 330, 2002.

203. Chinetti, G., Fruchart, J.C., and Staels, B., Peroxisome proliferator-activated receptors (PPARs): Nuclear receptors at the crossroads between lipid metabolism and inflammation, *Inflamm. Res.*, 49, 497, 2000.

204. Sakamoto, M., Mikasa, K., Majima, T., Hamada, K., Konishi, M., Maeda, K. et al., Anti-cachectic effect of clarithromycin for patients with unresectable non-small cell lung cancer, *Chemotherapy*, 47, 444, 2001.

205. Joniau, S. and Jorissen, M., Macrolides: More than just antibiotics?, *Acta Otorhinolaryngol. Belg.*, 57, 209, 2003.

206. Stenvinkel, P., Holmberg, I., Heimbürger, O., and Diczfalusy, U., A study of plasmalogen as an index of oxidative stress in patients with chronic renal failure. Evidence of increased oxidative stress in malnourished patients, *Nephrol. Dial. Transplant.*, 13, 2594, 1998.

207. Stenvinkel, P., Pecoits-Filho, R., and Lindholm, B., Coronary artery disease in end-stage renal disease — no longer a simple plumbing problem, *J. Am. Soc. Nephrol.*, 14, 1927, 2003.

208. Jiang, Q., Elson-Schwab, I., Courtemanche, C., and Ames, B.N., Gamma-tocopherol and its major metabolite, in contrast to alpha tocopherol, inhibit cycloxygenase activity in macrophages and epithelial cells, *Proc. Natl Acad. Sci.*, 97, 11494, 2000.

209. Lappas, M., Permezel, M., and Rice, G.E., N-acetyl-cysteine inhibits phospholipid metabolism, proinflammatory cytokine release, protease activity, and nuclear factor-kappaB deoxyribonucleic acid-binding activity in human fetal membranes in vitro, *J. Clin. Endocrinol. Metab.*, 88, 1723, 2003.

210. Kinlay, S., Fang, J.C., Hikita, H., Ho, I., Delagrange, D.M., Frei, B. et al., Plasma alpha-tocopherol and coronary endothelium-dependent vasodilator function, *Circulation*, 100, 219, 1999.

211. Scholze, A., Rinder, C., Beige, J., Riezler, R., Zidek, W., and Tepel, M., Acetylcysteine reduces plasma homocysteine concentration and improves pulse pressure and endothelial function in patients with end-stage renal failure, *Circulation*, 109, 369, 2004.

212. Tepel, M., van der Giet, M., Statz, M., Jankowski, J., and Zidek, W., The antioxidant acetylcysteine reduces cardiovascular events in patients with end-stage renal failure, *Circulation*, 107, 992, 2003.

213. Lloyd, C.E., Palopoli, M., and Vary, T.C., Effect of central administration of interleukin-1 receptor antagonist on protein synthesis is skeletal muscle, kidney, and liver during sepsis, *Metabolism*, 52, 1218, 2003.

214. Balakrishnan, V.S., Schmid, C.H., Jaber, B.L., Natov, S.N., King, A.J., and Pereira, B.J., Interleukin-1 receptor antagonists synthesis by peripheral blood mononuclear cells: A novel predictor of morbidity among hemodialysis patients, *J. Am. Soc. Nephrol.*, 11, 2114, 2000.

215. Dinarello, C.A., Anti-cytokine therapy and infections, *Vaccine*, 21, 24, 2003.

216. Fleischmann, R.M., Addressing the safety of anakinra in patients with rheumatoid arthritis, *Rheumatology*, 42 (Suppl. 2), 29, 2003.

11 Cardiac Failure

Philip A. Poole-Wilson

Contents

Summary

Heart failure is a condition where the key body system for survival, namely the circulation, is threatened. The response of the body to heart failure is the paradigm for the mechanisms and responses of the body to maintain survival. These are ancient reflexes built into the body during evolution. Thus maintenance of the blood pressure, the elimination of external microorganisms and the prevention of hemorrhage are the three key requirements for survival and it is precisely these systems that are activated when the body is challenged by diseases such as coronary heart disease which occur way beyond the reproductive age. Consequently the response of the body to many chronic diseases is identical to that observed in terminal heart failure when the entity referred to as cardiac cachexia is present.

11.1 Introduction

There have been many attempts over the years to define the clinical syndrome which physicians recognize as congestive heart failure. Many of these definitions have focused on physiological or metabolic entities such as cardiovascular haemodynamics or oxygen consumption.[1] Such definitions have not proved useful either in epidemiology or made any significant difference to the patient. A recent definition which is typical of the more modern approach is that "heart failure is a complex syndrome that can result from any structural or functional cardiac disorder that impairs the ability of the heart to function as a pump to support a physiological circulation. The syndrome of heart failure is characterized by symptoms such as breathlessness and fatigue, and signs such as fluid retention."[2] The key features of this definition are that there should be an abnormality of the heart and the presence of symptoms as a consequence. That definition is similar to one proposed previously by the European Society of Cardiology[3,4] and allows application in epidemiological studies and health surveys. Furthermore, the diagnosis can be made by straightforward clinical investigation.

Heart failure is common, becoming more common, is a disease of the elderly, can be detected in the community, carries a poor prognosis and is treatable. The prevalence rate is about 3% in the developed countries and the incidence is about 1% per annum. In younger age groups the disease predominates in men but in the elderly it is equally common in both sexes.[5,6] Because women live longer than men the absolute number of the elderly with heart failure may be greater in women.

Coronary heart disease is the commonest cause (60 to 75%) of heart failure in developed countries.[5,7,8] Cardiomyopathies (approximately 10%), valve disorders, and other conditions make up the remaining cases. In some countries, particularly lower and middle income countries, rheumatic fever is still a common cause of heart failure and in South America, Chagas disease accounts for a high proportion of cases.

The mean age of patients with heart failure in the community is 75 years. Most clinical trials have focused on patients in their 60s, with about 80% being female, largely because patients who have been placed in these trials after heart failure have been detected in hospital. Heart failure is more common in men in this age group. Although the mean age of 75 years is high, because heart failure is so common, there are large numbers of patients in their 60s with this condition.[6] The prevalence has increased because of the success in reducing the mortality from myocardial infarction and because in many developed countries there have been demographic changes increasing the proportion of elderly patients in the population.

Heart failure is a progressive disease. There is a high early rate of attrition after diagnosis and this is followed by a mortality of approximately 10% per annum.[6,9] Thus overall death occurs at 3 years for 50% of those with the diagnosis. The prognosis varies greatly with the severity of the disease at presentation, the ejection fraction (size of the heart), the exercise capacity (maximum oxygen consumption and function of the heart) and the extent of neurohormonal activation.

The nature of the progression of heart failure is still not fully understood. Usually patients respond to treatment and remain well for a period of time (Figure 11.1). It is during this period that they may die from arrhythmias (sudden death) or heart failure, progress due to further ischaemic episodes or to repetition of the causative mechanism such as excessive alcohol consumption. Subsequently patients deteriorate with increasing rapidity. The reason for this deterioration is the subject of considerable research interest. It is during this latter phase of the natural history that the entity cardiac cachexia is observed.

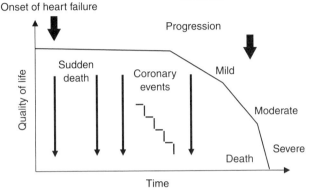

| Loss of myocardium
Fall of blood pressure stimulates
baroreceptors. Ergoreflexes and
chemoreflexes activated.
Maintains hormone activation | Bacterial invasion
Immune and inflammatory
response
Onset of cachexia
hastens demise |

FIGURE 11.1   Progression of heart failure. Early activation of body responses is later followed by more rapid deterioration.

11.2 DIAGNOSIS OF CARDIAC CACHEXIA

Cardiac cachexia is not a new clinical entity. The earliest reports come from the School of Medicine on the island of Kos. Hippocrates (459 B.C. – 357 B.C.) wrote "the flesh is consumed and becomes water, . . . the abdomen fills with water, the feet and legs swell, the shoulders, clavicles, chest and thighs melt away . . . . This illness is fatal." No doubt physicians both before and after were very familiar with this syndrome.[10, 11] In the modern world it is necessary to define cachexia more carefully. There have been many attempts to do so. One reasonable definition is that the cachexic state exists in patients with heart failure when there is no evidence of other diseases, for example, cancer, liver disease, thyroid disease, malabsorption and the like, and there is a loss of more than 6% of the previous normal body weight over a period of more than six months (Figure 11.2).[12] The loss of weight must not be due to the removal of oedema by the use of diuretics. In middle-aged patients, the diagnosis is rather straightforward. In the elderly, sarcopenia is common and the diagnosis more difficult. A further problem in the elderly is that the diagnosis of heart failure itself is more complex because of the increasing occurrence (up to 50%) of the entity called diastolic heart failure. Diastolic heart failure is currently diagnosed when patients have the symptoms of heart failure, an abnormality of the heart but an ejection fraction which is greater than about 40% and therefore nearly within the normal range. A near normal ejection fraction means that the end diastolic volume is near normal, because the stroke volume in unchanged, which in turn means that the heart size is small. This entity is common in the elderly and is related to the presence of myocardial ischaemia, hypertension, and fibrosis. The reasons for the heart, in some patients with heart failure, to progressively enlarge

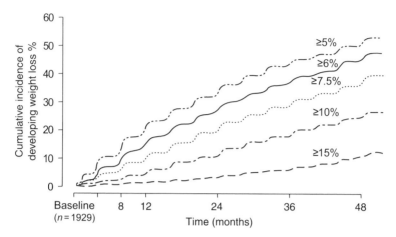

FIGURE 11.2 Incidence of weight loss in the SOLVD trial (From Anker, S.D., Negass, A., Coats, A.J., Afzal, R., Poole-Wilson, P.A., Cohn, J.N. et al. *Lancet* 2003;361:1077–1083. With permission.).

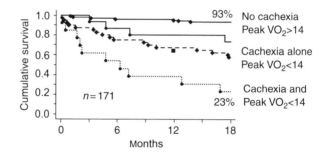

FIGURE 11.3 Survival from cardiac cachexia over 1.5 years (From Anker, S.D., Ponikowski, P., Varney, S., Chua, T.P., Clark, A.L., Webb-Peploe, K.M. et al. *Lancet* 1997;349:1050–1053. With permission.).

(systolic heart failure) and in others remain a near normal size (diastolic heart failure) is still unknown.

Cardiac cachexia is not uncommon and indeed may be a feature of almost all patients with heart failure who do not die suddenly and early in the natural history of the disease. In one early study 16% of patients were found to be cachectic and the eighteen month mortality was high (50%) (Figure 11.3).[12, 13] That outcome is worse than almost all types of cancer except lung cancer. A reanalysis of the data from the SOLVD study confirmed these early findings (Figure 11.2).[12] Weight loss greater than 6% occurred in more than 35% of patients over a period of three years. Evaluation of patients in other large trials have provided identical information (Figure 11.4)

	HR (95%CI)	χ^2	p
NYHA class	1.914 (1.655–2.213)	76.9	<0.0001
BMI (kg/m^2)	0.946 (0.926–0.967)	24.9	<0.0001
UA (100 μmol/l)	1.167 (1.098–1.237)	22.1	<0.0001
Age (y)	1.023 (1.010–1.036)	11.8	=0.0006
LVEF (%)	0.989 (0.977–1.001)	3.00	=0.08
Sex (f vs. m)	0.850 (0.696–1.038)	2.54	=0.11
Therapy (L vs. C)	1.063 (0.897–1.274)	0.55	=0.46

HR = Hazard ratio, NYHA = New York Heart Association, BMI = Body mass index, UA = Uric acid, LVEF = Left ventricular ejection fraction, L = Losartan, C = Captopril

FIGURE 11.4 Survival by multivariate analysis in ELITE II trial. Body mass index is an important predictor of survival.

Morphology	Quantity	Loss of muscle mass (or bulk)
	Site	Localized to legs or general abnormality
		Orientation and fiber position
	Quality	Atrophy, damage, and/or necrosis
		Change of fiber type
Function		Weakness and/or increased fatigue
Blood flow		Reduced when expressed as ml/min
		Variable when expressed as ml/min/100 ml
Metabolism		An inevitable consequence of atrophy
		and damage, or a specific change

FIGURE 11.5 The changes in skeletal muscle in heart failure (From Clark, A.L., Poole-Wilson, P.A., and Coats, A.J.J. *Am. Coll. Cardiol.* 1996;28:1092–1102; Poole-Wilson, P.A. and Ferrari R. *J. Mol. Cell. Cardiol.* 1996;28:2275–2285; Harrington, D., Anker, S.D., Chua, T.P., Webb-Peploe, K.M., Ponikowski, P.P., Poole-Wilson, P.A. et al. *J. Am. Coll. Cardiol.* 1997;30:1758–1764. With permission.).

11.3 COMPOSITION OF BODY TISSUES

Patients with cardiac cachexia are often weak and the physician is able to detect atrophy of skeletal muscle (Figure 11.5).[14–17] This atrophy is probably a major cause of the symptoms of shortness of breath and fatigue in these patients and may also be one of the sources of signals to the brain which perpetuate the activation of the neurohormonal, immunological, and inflammatory systems in congestive heart failure. However the cachexia is not solely due to atrophy of skeletal muscle. Recent evidence has shown that fat tissue is diminished[18] and bone mineral density is decreased.[19]

Of particular importance is that the heart itself is involved in the cachectic process. Reduction of left ventricular muscle mass has been shown by both magnetic resonance imaging[20] and echocardiography[21] in patients with cachexia. It would seem that the perpetuation of the syndrome of heart failure establishes a vicious circle impacting on the heart itself.

11.4 CAUSES OF CARDIAC CACHEXIA

The underlying causes of cardiac cachexia are not different from those in many other diseases. Fatigue, tiredness and loss of energy may lead to a reduction in the intake of calories.[22] Another feature of the diet may be a lack of intake of selected nutrients. The gut has not been widely studied in heart failure but malabsorption is another potential contributor to cachexia in these patients. Furthermore loss of nutrients in the urine or gut may occur. Overall in heart failure the metabolic rate is increased[23,24] and increases further with the severity of the disease.[25] Almost all these abnormalities relate to the severity of heart failure.

11.5 UNDERLYING CAUSES OF CARDIAC CACHEXIA

Many studies in the literature have defined the characteristics of body systems in heart failure and worsening heart failure. The focus in recent years has been on neural systems (sympathetic and parasympathetic systems), the endocrine system, and more recently immune and inflammatory systems. Key papers demonstrated that catecholomines[26] and TNF alpha[27] were linked not just to the severity of heart failure but to mortality. More recent work has demonstrated activation of the immune and inflammatory systems.[28-30] Again activation of these systems relates to severity of the disease and to the outcome. Elevation of plasma cytokine concentration is a harbinger of death.[31]

One key feature is that the severity of the abnormalities of the neurohormonal system are associated with cachexia (Figure 11.6). Indeed early in heart failure, most of the hormonal systems (Figure 11.7) and particularly in the untreated state, do not show any activation at all.[28] It is easy then to come to the conclusion that cachexia is a consequence of activation of these systems.[28-30] But that is a naive argument which essentially is stating that in patients with a chronic condition, illness predicts a worsening illness which in turn foresees death. The critical question is what are the fundamental causes and what are the causal relationships.

There are currently a number of hypotheses as to the underlying causes (Figure 11.8). The first is that the clinical picture of cachexia is a consequence of long-term activation of the sympathetic system.[32] The second is that the heart itself is the cause of activation of these symptoms through the release of cytokines such as TNF alpha.[33] The evidence for that theory is weak. The third hypothesis is that endotoxin enters the body either from the gut or lung and stimulates macrophages to release cytokines (Figure 10.9).[34,35] The endotoxin promotes release of cytokines from the macrophage although complex interactions occur with lipids in the blood. The fourth is that cytokine release is a consequence of tissue hypoxia.[32,36] In heart failure the arterial oxygen saturation is rarely reduced but the flux of oxygen into tissues may be substantially diminished due to a reduction in blood flow. The fifth hypothesis is that damage to the myocardium activates neurological pathways which activate the midbrain to pass neurological signals to the immune and inflammatory systems.

	Controls	All HF	Non-cachectic	Cachectic
Sodium mmol/l	139.9	137.7*	138.6	135.6*
Potassium mmol/l	3.9	4.0	4.0	3.9
Creatinine μmol/l	92	125*	121	134
Aldosterone pmol/l	279	699**	552	1039*
PRA ng/ml/h	1.3	11.7*	9.2	17**
TSH μiu/l	1.5	2.1	2.1	2.2
Reverse T_3 nmol/l	0.31	0.52*	0.48	0.62
Norepinephrine nmol/l	1.94	3.41**	2.58	5.31*
Epinephrine nmol/l	0.51	1.29	0.68	2.7*
TNF-α pg/ml	7.0	9.5	6.9	15.3*
Cortisol nmol/l	372	415	379	498*
Testostenone nmol/l	9.9	11.2	11.5	10.5
Estradiol pmol/l	45.4	69.2	62.6	84.6
DHEA nmol/l	15.3	10.4**	11.3	8.3
Basal insulin pmol/l	40.6	77.8*	81.7	68
hGH ng/ml	1.1	1.8	0.9	3.8*
IGF-1 ng/ml	148	146	150	137

PRA = plasma renin activity, TSH = thyroid stimulating hormone, DHEA = Dihydroepiandrosterone, hGH = human growth hormone, IGF = Insulin like growth factor

FIGURE 11.6 Hormonal changes in heart failure and cachexia. *$P < .01$, **$P < .001$ (From Anker, S.D., Chua, T.P., Ponikowski, P., Harrington, D., Swan, J.W., Kox, W.J. et al. *Circulation* 1997;96:526–534. With permission.).

Constrictors	Dilators	Growth factors
Noradrenaline	ANP	Insulin
Angiotensin II	Prostaglandin E2	TNF alpha
Endothelin	metabolites	Growth hormone
Vasopressin	EDRF	Angiotensin II
NPY	Dopamine	Catecholamines
	CGRP	NO
		Cytokines
		Oxygen radicals

NPY = Neuropeptide Y, ANP = Atrial natriuretic peptide, EDRF = Endothelial derived relaxing factor, CGRP = Calcitonin gene related peptide, TNF = Tumor necrosis factor, NO = Nitric oxide

FIGURE 11.7 Some hormonal mediators in heart failure.

11.6 CLINICAL RELEVANCE

The identification and the existence of cardiac cachexia in a patient with congestive heart failure indicates serious disease and predicts a poor outcome. At present our understanding of the condition is limited and attempts to use drugs or interventions[37] which are aimed at preventing cardiac cachexia in a specific

1. Persistent activation of early markers such as the sympathetic
   nervous system or renin/angiotensin systems
   Maintenance of activation by signals emanating
   from skeletal muscle
2. A manifestation of inflammation within an organ
3. A consequence of increased propensity for invasion
   by bacteria or bacterial products from the gut or lung
4. A consequence of tissue hypoxia
5. Activation of the autonomic system centrally

FIGURE 11.8   Signal for the activation of the hormonal and immune systems in cachexia.

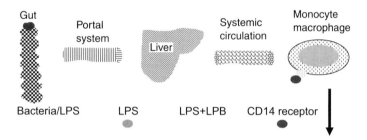

Release of: IL-1, IL-6, IL-8, IL-10
IFN-γ TGF-β, chemokines,
adhesion molecules

LPS = lipopolysaccharides (endotoxin), LPB = lipopolysaccharide binding protein,
IL = interleukin, IFN = interferon, TGF = transforming growth factor

FIGURE 11.9   The endotoxin hypothesis (From Anker, S.D., Egerer, K.R., Volk, H.D.,
Kox, W.J., Poole-Wilson, P.A., and Coats, A.J. *Am. J. Cardiol.* 1997;79:1426–1430. With
permission.).

- Growth factors	Growth hormone, Ghrelin
	IGF-I (insulin like growth factor)
- Anabolic steroids	
- Other hormones	Erythropoietin
- Immune modulators	Many possibilities
- Anti-muscle wasting	Myostatin-antibodies etc.
- Appetite, nutrition	Megace, calories
- Alternatives	Insulin sensitizers
	Xanthin oxidase inhibitors,
	Statins

FIGURE 11.10   Potential treatments for cardiac cachexia.

manner have in general been ineffective (Figure 11.10). In particular inhibitors of
cytokines have not shown benefit. On the other hand trials which focus not on a
single pathway but impact on the entire immunological system have provided at
least a glimmer of hope.[38]

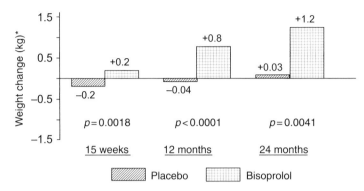

FIGURE 11.11 Weight change during follow-up in CIBIS II. *Excluding any weight measurements when oedema was present.

The problem with most studies into treatments for cachexia is that the treatments improve heart failure itself. Any treatment or intervention which impacts effectively on the underlying illness will inevitably reverse cachexia eventually. The best example of this is, of course, the consequences of transplantation or the use of left ventricular assist devices. The clinical problem then is whether any effect of drugs such as beta blockers (Figure 11.11) and ACE inhibitors[12] on the function of the cardiovascular system can be separated from mechanisms which might independently, from a direct impact on the cardiovascular system, contribute to the progression of heart failure.

REFERENCES

1. Poole-Wilson, P.A. Chronic heart failure: Definition, epidemiology, pathophysiology, clinical manifestations and investigations. In: Julian, D.G., Camm, A.J., Fox, K.M., Hall, R.J.C., Poole-Wilson, P.A., Ed., *Diseases of the Heart*. 2 ed. London: W.B. Saunders Co. Ltd; 1996. p. 467–481.
2. NICE Guidelines. Developed by The National Collaborating Centre for Chronic Conditions. Chronic Heart Failure. National clinical guidelines for diagnosis and management in primary and secondary care NICE guideline No. 5. 1-163. 2005. London, Royal College of Physicians.
3. The Task Force on Heart Failure of the European Society of Cardiology. Guidelines for the diagnosis of heart failure. *Eur. Heart J.* 1995;16:741–751.
4. Remme, W.J. and Swedberg, K. Guidelines for the diagnosis and treatment of chronic heart failure. *Eur. Heart J.* 2001;22:1527–1560.
5. Cowie, M.R., Mosterd, A., Wood, D.A., Deckers, J.W., Poole-Wilson, P.A., Sutton, G.C. et al. The epidemiology of heart failure [see comments]. *Eur. Heart J.* 1997;18:208–225.
6. De Giuli, F., Khaw, K.T., Cowie, M.R., Sutton, G.C., Ferrari, R., Poole-Wilson, P.A. Incidence and outcome of persons with a clinical diagnosis of heart failure in a general practice population of 696,884 in the United Kingdom. *Eur. J. Heart Fail.* 2005;7:295–302.

7. Fox, K.F., Cowie, M.R., Wood, D.A., Coats, A.J., Gibbs, J.S., Underwood, S.R. et al. Coronary artery disease as the cause of incident heart failure in the population. *Eur. Heart J.* 2001;22:228–236.

8. Cowie, M.R., Wood, D.A., Coats, A.J.S., Thompson, S.G., Poole-Wilson, P.A., Suresh. V. et al. Incidence and aetiology of heart failure. A population-based study. *Eur. Heart J.* 1999;20:421–428.

9. Cowie, M.R., Wood, D.A., Coats, A.J., Thompson, S.G., Suresh, V., Poole-Wilson, P.A. et al. Survival of patients with a new diagnosis of heart failure: A population based study. *Heart* 2000;83:505–510.

10. Pittman, J. and Cohen, P. The pathogenesis of cardiac cachexia. *N. Engl. J. Med.* 1964;271:403–409.

11. Doehner, W. and Anker, S.D. Cardiac cachexia in early literature: A review of research prior to Medline. *Int. J. Cardiol.* 2002;85:7–14.

12. Anker, S.D., Negassa, A., Coats, A.J., Afzal, R., Poole-Wilson, P.A., Cohn, J.N. et al. Prognostic importance of weight loss in chronic heart failure and the effect of treatment with angiotensin-converting-enzyme inhibitors: An observational study. *Lancet* 2003;361:1077–1083.

13. Anker, S.D., Ponikowski, P., Varney, S., Chua, T.P., Clark, A.L., Webb-Peploe, K.M. et al. Wasting as independent risk factor for mortality in chronic heart failure. *Lancet* 1997;349:1050–1053.

14. Clark, A.L., Poole-Wilson, P.A., and Coats, A.J. Exercise limitation in chronic heart failure: Central role of the periphery. *J. Am. Coll. Cardiol.* 1996;28:1092–1102.

15. Poole-Wilson, P.A. and Ferrari, R. Role of skeletal muscle in the syndrome of chronic heart failure. *J. Mol. Cell. Cardiol.* 1996;28:2275–2285.

16. Harrington, D. and Coats, A.J. Skeletal muscle abnormalities and evidence for their role in symptom generation in chronic heart failure. *Eur. Heart J.* 1997;18:1865–1872.

17. Libera, L.D. and Vescovo, G. Muscle wastage in chronic heart failure, between apoptosis, catabolism and altered anabolism: A chimaeric view of inflammation? *Curr. Opin. Clin. Nutr. Metab. Care* 2004;7:435–441.

18. Anker, S.D., Ponikowski, P.P., Clark, A.L., Leyva, F., Rauchhaus, M., Kemp, M. et al. Cytokines and neurohormones relating to body composition alterations in the wasting syndrome of chronic heart failure. *Eur. Heart J.* 1999;20:683–693.

19. Anker, S.D., Clark, A.L., Teixeira, M.M., Hellewell, P.G., and Coats, A.J. Loss of bone mineral in patients with cachexia due to chronic heart failure. *Am. J. Cardiol.* 1999;83:612–615, A10.

20. Florea, V.G., Moon, J., Pennell, D.J., Doehner, W., Coats, A.J., and Anker, S.D. Wasting of the left ventricle in patients with cardiac cachexia: A cardiovascular magnetic resonance study. *Int. J. Cardiol.* 2004;97:15–20.

21. Florea, V.G., Henein, M.Y., Rauchhaus, M., Koloczek, V., Sharma, R., Doehner, W. et al. The cardiac component of cardiac cachexia. *Am. Heart J.* 2002;144:45–50.

22. Mustafa, I. and Leverve, X. Metabolic and nutritional disorders in cardiac cachexia. *Nutrition* 2001;17:756–760.

23. Poehlman, E.T., Scheffers, J., Gottlieb, S.S., Fisher, M.L., and Vaitekevicius, P. Increased resting metabolic rate in patients with congestive heart failure. *Ann. Intern. Med.* 1994;121:860–862.

24. Riley, M., Elborn, J.S., McKane, W.R., Bell, N., Stanford, C.F., and Nicholls, D.P. Resting energy expenditure in chronic cardiac failure. *Clin. Sci. (London)* 1991;80:633–639.
25. Obisesan, T.O., Toth, M.J., Donaldson, K., Gottlieb, S.S., Fisher, M.L., Vaitekevicius, P. et al. Energy expenditure and symptom severity in men with heart failure. *Am. J. Cardiol.* 1996;77:1250–1252.
26. Cohn, J.N., Levine, T.B., Olivari, M.T., Garberg, V., Lura, D., Francis, G.S. et al. Plasma norepinephrine as a guide to prognosis in patients with chronic congestive heart failure. *N. Engl. J. Med.* 1984;311:819–823.
27. Levine, B., Kalman, J., Mayer, L., Fillit, H.M., and Packer, M. Elevated circulating levels of tumour necrosis factor in severe chronic heart failure. *N. Engl. J. Med.* 1990;323:236–241.
28. Anker, S.D., Chua, T.P., Ponikowski, P., Harrington, D., Swan, J.W., Kox, W.J. et al. Hormonal changes and catabolic/anabolic imbalance in chronic heart failure and their importance for cardiac cachexia. *Circulation* 1997;96:526–534.
29. Anker, S.D., Ponikowski, P.P., Clark, A.L., Leyva, F., Rauchhaus, M., Kemp, M. et al. Cytokines and neurohormones relating to body composition alterations in the wasting syndrome of chronic heart failure. *Eur. Heart J.* 1999;20:683–693.
30. Anker, S.D., Chua, T.P., Ponikowski, P., Harrington, D., Swan, J.W., Kox, W.J. et al. Hormonal changes and catabolic/anabolic imbalance in chronic heart failure and their importance for cardiac cachexia. *Circulation* 1997;96:526–534.
31. Rauchhaus, M., Doehner, W., Francis, D.P., Davos, C., Kemp, M., Liebenthal, C. et al. Plasma cytokine parameters and mortality in patients with chronic heart failure. *Circulation* 2000;102:3060–3067.
32. Bachetti, T. and Ferrari, R. The dynamic balance between heart function and immune activation. *Eur. Heart J.* 1998;19:681–682.
33. Torre-Amione, G., Kapadia, S., Lee, J., Durand, J.B., Bies, R.D., Young, J.B. et al. Tumor necrosis factor-alpha and tumor necrosis factor receptors in the failing human heart. *Circulation* 1996;93:704–711.
34. Anker, S.D., Egerer, K.R., Volk, H.D., Kox, W.J., Poole-Wilson, P.A., and Coats, A.J. Elevated soluble CD14 receptors and altered cytokines in chronic heart failure. *Am. J. Cardiol.* 1997;79:1426–1430.
35. Rauchhaus, M., Coats, A.J., and Anker, S.D. The endotoxin-lipoprotein hypothesis. *Lancet* 2000;356:930–933.
36. Hasper, D., Hummel, M., Kleber, F.X., Reindl, I., and Volk, H.D. Systemic inflammation in patients with heart failure. *Eur. Heart J.* 1998;19:761–765.
37. Argiles, J.M., Almendro, V., Busquets, S., and Lopez-Soriano, F.J. The pharmacological treatment of cachexia. *Curr. Drug Targets* 2004;5:265–277.
38. Torre-Amione, G., Sestier, F., Radovancevic, B., and Young J. Effects of a novel immune modulation therapy in patients with advanced chronic heart failure: Results of a randomized, controlled, phase II trial. *J. Am. Coll. Cardiol.* 2004;44:1181–1186.
39. Harrington, D., Anker, S.D., Chua, T.P., Webb-Peploe, K.M., Ponikowski, P.P., Poole-Wilson, P.A. et al. Skeletal muscle function and its relation to exercise tolerance in chronic heart failure. *J. Am. Coll. Cardiol.* 1997;30:1758–1764.
40. Anker, S.D., Egerer, K.R., Volk, H.D., Kox, W.J., Poole-Wilson, P.A., and Coats, A.J. Elevated soluble CD14 receptors and altered cytokines in chronic heart failure. *Am. J. Cardiol.* 1997;79:1426–1430.

12 Chronic Obstructive Pulmonary Disease

Annemie M. W. J. Schols

Contents

Summary

Skeletal muscle wasting and dysfunction are strong independent predictors of mortality and disability in patients with Chronic Obstructive Pulmonary Disease (COPD). Current insight into the pathogenesis of weight loss, muscle wasting and intrinsic muscle abnormalities has led to nutritional and pharmacological therapeutic strategies that may reverse skeletal muscle dysfunction, particularly when incorporated as integrated part of pulmonary rehabilitation.

12.1 Introduction

Chronic obstructive pulmonary disease (COPD) is a progressive disorder leading to significant debilitation. While traditionally it has been considered as an irreversible lung disease, there is growing evidence that COPD is a multiorgan systemic disease. Parallel to this awareness, the interest in weight loss and muscle wasting in the management of COPD has changed remarkably during the past two decades. Involuntary weight loss is a well-recognized clinical finding and a substantial number of patients suffering from COPD, particularly emphysema, become emaciated during the course of the disease. Interestingly attempts to classify COPD patients found that body weight might be a discriminating factor. This led to the classical description of the pink puffer (emphysematous type) and the blue bloater (bronchitic type). Since weight loss was thought to be an epiphenomenon of severe disease it was not considered as a therapeutic target until the 1980s. Recent studies however have yet convincingly challenged this viewpoint and consistently shown that a low body mass index is associated with increased mortality independent of disease severity.[1,2] Furthermore two studies have even shown that weight gain is associated with decreased mortality risk[3,4] in COPD patients with a body mass index below 25 kg/m^2.

12.2 CONSEQUENCES OF WEIGHT LOSS AND MUSCLE WASTING IN COPD

Prominent symptoms of COPD are dyspnoea and exercise intolerance. Independently of pulmonary impairment, skeletal muscle weakness is an important determinant of these symptoms. Body composition studies have shown that skeletal muscle strength is largely determined by skeletal muscle mass in COPD.[5–7] Besides effects on muscle strength, muscle wasting is also a significant determinant of decreased exercise capacity and altered metabolic and ventilatory response.[8,9] These associations imply that functional consequences of weight loss are not only related to muscle wasting per se, but also to intrinsic alterations in muscle morphology and energy metabolism as will be discussed later. Muscle wasting in COPD is fiber type specific affecting predominantly the "fast-twitch" type IIx fibers.[10]

The relationship between weight loss and mortality in COPD has been subject of investigation since the 1960s. In the early years a significant association was reported between weight loss and survival.[11] In line, several retrospective studies using different COPD populations from the USA,[1] Canada,[12] Denmark,[2] and the Netherlands[3] reported more recently a relationship between low body mass index and mortality independent of disease severity. Remarkably in all these studies a decreased mortality risk was observed in overweight patients not only compared to underweight patients but also to normal weight subjects. This observation can be explained by the fact that COPD is not only linked to weight loss but also to a shift in body composition. The normal adaptive response to weight loss is a preferential loss of fat mass to spare loss of the metabolically and functionally active fat-free mass (FFM). In contrast weight loss in COPD is accompanied by significant loss of FFM, out of proportion to the loss of fat mass.[13] It is specifically the loss of fat-free mass as marker of muscle mass that causes impaired skeletal muscle strength and exercise capacity. These studies have furthermore shown that muscle wasting may also occur in normal weight stable subjects.[13] In stable patients with moderate to severe COPD, a low FFM-index defined as FFM/height^2 <16 kg/m^2 (males) and <15 kg/m^2 (females) has been reported in 20% of COPD out-patients,[14] in 35% of those eligible for pulmonary rehabilitation[13] and even in 45% of lung transplant candidates.[15] Limited data are available regarding the prevalence of FFM depletion in representative groups of mild COPD, and on the other side of the spectrum, in patients suffering from acute respiratory failure. Two recent studies in patients with moderate to severe COPD have shown that muscle wasting is not only a significant determinant of functional impairment, but also of decreased health related quality of life.[16, 17] Furthermore muscle mass has been identified as a better predictor of survival than body weight.[18]

Weight loss and underweight are more frequently observed in emphysematous patients compared to those with chronic bronchitis, but differences in body weight between the two COPD subtypes, if present, merely reflect a difference in fat mass.[19] Depletion of FFM even despite a relative preservation of fat mass also occurs in chronic bronchitis.[19]

The effect of weight loss in relation to lung function has predominantly focused on ventilatory pump function and has shown that respiratory muscle strength is affected to a comparable degree as limb muscle strength. Potential effects of weight loss on the lung parenchyma in humans are difficult to explore. It is possible to obtain lung tissue resected during surgery or whole lungs at autopsy in order to study the influence of weight loss on lung structure and function. The presence of coexisting pathological processes, which would have led to lung resection or death, however, could hinder investigation of the effects of nutritional deficiency on lung parenchyma. Developments in high resolution CT scanning allow noninvasive measurement of lung mass and density. Using this technique emphysematous changes were recently detected in patients with severe anorexia nervosa.[20] These observations are in line with dramatic observations during the second World War in a Polish Ghetto. In a total of 370 autopsies of Jewish people that were starved to death, 50 cases of emphysema (13.5%) were observed. Of the 50 cases, 14 were reported in individuals under the age of 30 years, and 20 in persons under the age of 40 years. The pathological features seen in these mostly young adults were similar to those seen in senile emphysema.[21]

12.3 WEIGHT LOSS AND ENERGY BALANCE

12.3.1 Energy Expenditure

It is not fully understood why COPD patients become underweight but weight loss and specifically loss of fat mass, is generally the result of a negative energy balance and appears to be particularly prevalent in patients with emphysema.[19] In contrast to an adaptive decreased energy metabolism during (semi) starvation, increased resting energy requirements have been observed in some of the COPD patients, linked to low-grade systemic inflammation.[22] Studies in other chronic wasting diseases characterized by hypermetabolism and systemic inflammation (e.g., cancer, chronic heart failure, acquired immunodeficiency disease syndrome (AIDS)) have shown an adaptive decrease in activity-induced energy expenditure so that daily energy expenditure is normal. In contrast, elevated activity induced and total energy expenditure has been measured in free-living ambulatory COPD patients.[23] The cause of this disease-specific increase in energy metabolism is not yet clear. A decreased mechanical efficiency in leg exercise has been described that could result from a decreased efficiency of skeletal muscle energy metabolism. Furthermore some studies reported an increased oxygen cost of respiratory muscle activity due to lung hyperinflation. An obvious choice to improve energy balance might thus be to decrease energy expenditure. However, according to the recent Global Initiative for Obstructive Lung Disease (GOLD) guidelines, (www.goldcopd.com) pulmonary rehabilitation is an evidence-based intervention and exercise training a key component to improve limited functional abilities and maintain an active lifestyle. Since COPD patients may have an elevated energy metabolism and should at the same time be advised to increase exercise, restricting energy output will be hard to realize and may not be desirable. This implies that COPD patients who suffer from weight loss, and even some weight stable patients, should be encouraged to increase their apparently normal energy intake. This could avoid weight loss, specific loss of muscle mass, and a related decrease in functional ability, or could even help them regain weight. Besides optimizing the treatment of patients who are already underweight, it is therefore important to detect and reverse involuntary weight loss in order to avoid functional decline. This may be achieved by increasing dietary intake per se or by altering dietary habits to include different (energy-dense) foods and optimum timing of meals/snacks in relation to symptoms and activity patterns.

12.3.2 Dietary Intake

Several factors may limit dietary intake in COPD. Dyspnea and fatigue are prominent symptoms that may affect appetite, particularly during acute exacerbations of the disease.[24] Furthermore arterial hypoxemia is associated with weight loss and decreased dietary intake.[25] There is increasing evidence that, besides local upregulation of inflammatory processes in the lung, COPD is characterized by an elevated systemic inflammatory response as reflected by elevated concentrations of proinflammatory cytokines and acute phase proteins in peripheral blood.[26-28] Like in other chronic inflammatory diseases, weight loss has specifically been associated with increased markers of the TNFα and soluble TNF receptors.[29-30] Inflammatory cytokines have been shown to increase circulating leptin. Leptin, a 167 amino-acid protein synthesized and secreted by white adipose tissue, is a component of a lipostatic signaling pathway that alters energy balance by central and peripheral mechanisms. Administration of leptin in animals results in a reduction in food intake and an increase in energy expenditure. These effects seem to be mediated by a leptin-induced decrease in the hypothalamic biosynthesis and release of neuropeptide Y, a hormone that potently stimulates appetite and food intake and reduces energy expenditure. In patients with emphysema a positive correlation between leptin and the soluble TNF receptor 55 has been shown that theoretically could in turn affect dietary intake.[31]

Substrate oxidation and ventilation are intrinsically related and theoretically meal-related dyspnoea and impaired ventilatory reserves might restrict the caloric amount and specifically the carbohydrate content of nutritional support in respiratory disease. Earlier studies indeed showed adverse effects of a carbohydrate-rich energy overload on carbon dioxide production and exercise capacity[32] but these results were not confirmed when using a normal energy load.[33] In fact even positive effects of a carbohydrate-rich supplement relative to a fat-rich supplement on lung function and dyspnoea sensation were reported.[34] Other factors that have been shown to affect the efficacy of nutritional support to increase dietary intake are the timing of supplements in relation to daily activities and the portion size of the supplements.[35,36]

12.4 STRATEGY AND EFFICACY OF NUTRITIONAL INTERVENTION

Nutritional interventions for COPD patients have focused mainly on therapeutic caloric support, but few well-controlled studies (i.e., randomized controlled trials RCTs) have so far been conducted. Ferreira et al.[37] have recently reviewed the available studies on therapeutic dietary supplementation in a meta-analysis. They managed to select only six RCTs that were considered to be of sufficient quality, of which two were double-blinded. The pooled effects, based on the analysis of a total of 277 subjects, as well as the results of the individual studies, showed that the effect of nutritional support on anthropometrics was minor at best and generally did not achieve clinical importance or statistical significance. Five of these studies used oral supplementation and four were conducted among outpatients. These results contradict to some extent the results reported by Baldwin et al.,[38] who reviewed the literature on dietary advice and supplementation interventions for patients with disease-related malnutrition in general (including COPD). Their conclusion was that dietary supplementation resulted in better effects on body weight than dietary advice. The review by Ferreira et al.[37] made no distinction between what may be called "failure to intervene" on the one hand and "failure of the intervention" on the other. In some of the papers on which the meta-analysis was based, patients took the prescribed dietary supplements to replace regular meals instead of as additional calorie input. In such cases the intervention did not result in a relevant increase in energy intake and therefore no weight gain could be expected. In the studies that did manage to increase energy intake, functional improvements were also observed. Further, studies investigating the effect of dietary supplementation were often conducted among severe COPD cases, in whom besides a negative energy balance, also a specific negative protein balance is often observed (as discussed below). Nevertheless, the meta-analysis and related studies do show that increasing energy intake among severe COPD cases is difficult to accomplish, and if energy intake is not increased, weight and functionality will certainly not improve.

Interventions should also be extended to prevention and early treatment of weight loss, that is, before patients are extremely wasted. This means expanding the target group to include COPD outpatients and primary care patients before they have become underweight, and putting more emphasis on dietary change than on medically prescribed supplementation. Few studies have been published on the possibilities and effects of voluntary dietary change among outpatients. Diet is often part of the focus in self-help or self-management programs for COPD patients and some of these programs have been evaluated. However there have not yet been any well-controlled studies on the prevention of weight loss in COPD.

12.5 MUSCLE WASTING AND PROTEIN BALANCE

Wasting of muscle mass is due to an impaired balance between protein synthesis (anabolism) and protein breakdown (catabolism). Besides nutritional abnormalities and physical inactivity, altered neuro-endocrine responses and the presence of a systemic inflammatory response may contribute to a negative protein balance in chronic diseases. From a therapeutic perspective it is important to know the relative contribution of these factors to altered protein synthesis and protein breakdown respectively. While increased dietary intake can compensate for elevated energy requirements and vice versa, uncontrolled protein breakdown cannot be overcome by only increasing protein synthesis.

12.6 PROTEIN METABOLISM

Several studies have investigated in COPD and other chronic wasting disorders the effects of pharmacological anabolic stimuli to promote protein synthesis including anabolic steroids, growth hormone, and insulin-like growth factor. Most studies in COPD were able to document a modest but significant gain in muscle mass after intervention[39, 40] illustrating that in some of the patients, stimulation of protein synthesis is an effective therapeutic strategy. No studies have yet specifically investigated the ability to induce or enhance muscle weight gain by nutritional modulation of protein synthesis or protein breakdown rates. Optimizing protein intake and essential amino acid intake may not only stimulate protein synthesis per se, but also enhance the efficacy of anabolic drugs[41] as well as physiological stimuli such as resistance exercise.[42] Protein requirements in COPD patients and many other chronic diseases are not well established. In normal weight COPD patients, whole body protein synthesis and breakdown rates were elevated compared to a well-matched healthy control group.[43] A potential explanation for this elevated protein turnover is thought to be enhanced acute phase protein synthesis, associated with low-grade inflammation. This is balanced by increased amino acid release from the skeletal muscle compartment (ultimately) resulting in net muscle protein breakdown. Indirect support for this hypothesis is given by two studies that demonstrated a specific association between muscle wasting and markers of systemic inflammation in COPD,[44] as well as by another study linking hypermetabolism, increased levels of acute phase proteins, and decreased plasma amino acid status.[45] However, these studies were merely descriptive and limited by a cross-sectional study design. Investigating whole body as well as acute phase protein turnover simultaneously[46] may provide more insight into the proposed link between protein metabolism and systemic inflammation in COPD. Skeletal muscle protein turnover has so far been investigated in only one study. Muscle protein synthesis rate was decreased in a group of under-weight clinically stable patients with emphysema while protein breakdown rate was normal.[47]

12.7 AMINO ACID METABOLISM

Amino acids are the building blocks of protein and several studies have to date reported an abnormal plasma amino acid pattern in COPD. Of interest are the consistently reduced plasma levels of branched chain amino acids (BCAAs) in underweight COPD patients and in those with low muscle mass.[48,49] There are some indications that low plasma BCAAs in COPD patients are due to specific alterations in leucine metabolism possibly mediated by altered insulin regulation and increased leucine oxidation in skeletal muscle to a noncarbohydrate energy substrate.[49] Leucine is an interesting nutritional substrate since it not only serves as a precursor, but also activates signaling pathways that enhance activity and synthesis of proteins involved in messenger ribonucleic acid (RNA) translocation to upregulate protein synthesis in skeletal muscle.[50]

In addition to fostering a higher rate of postprandial protein synthesis, increased availability of amino acids also enhances the stimulation of protein synthesis that occurs in response to exercise.[51] The magnitude of stimulation however depends on the timing of amino acids administration relative to the period of exercise.[51] BCAAs are also important precursors for glutamate (GLU), which is one of the most important nonessential amino acids in muscle. BCAAs derived from net protein breakdown and by uptake into the muscle pool, undergo transamination to yield branched-chain keto acids and GLU. Intracellular GLU is involved in numerous metabolic processes including substrate phosphorylation and replenishment of tricarboxylic acid intermediates to preserve high-energy phosphates at rest and during exercise. Moreover intracellular GLU is known as an important precursor for antioxidant glutathione (GSH) and glutamine synthesis in muscle.[52] Recently, a consistently reduced muscle GLU status of severe COPD patients was reported,[45,52] that further decreased during a submaximal exercise bout.[53] While muscle redox potential (glutathione disulphide/GSH) increases after endurance exercise training in healthy subjects, patients with COPD showed a reduced ability to adapt in this way as reflected by a lower capacity to synthesize GSH.[54] These observations may provide perspective for amino acid supplementation to modulate exercise-induced protein synthesis as well as exercise-induced oxidative stress.

12.8 INFLAMMATORY MODULATION

Despite anabolic nutritional and/or pharmacological stimulation, (muscle) weight gain is limited in some COPD patients. Like in other chronic inflammatory disorders, a poor therapeutic response is related to presence of systemic inflammation.[55] The disproportionate muscle wasting linked to systemic inflammation and unresponsive to nutritional supplementation is commonly referred to as the cachexia syndrome.[56] Current insight into the molecular mechanisms of cachexia indicates a complex interaction between inflammatory mediators, oxidative stress and growth factors, not only involved in an imbalance between muscle protein synthesis and breakdown, but also in processes that govern the maintenance of skeletal muscle and muscle plasticity such as skeletal muscle fiber degeneration, apoptosis, and regeneration.[56]

The fatty acid composition of inflammatory and immune cells is sensitive to change according to the fatty acid composition of the diet. The n-3 polyunsaturated fatty acids (PUFA) eicosapentaenoic acid (EPA) and docosahexaenoic acid are found in high proportions in oily fish and fish oils. The n-3 PUFA are structurally and functionally distinct from the n-6 PUFA. Typically human inflammatory cells contain high proportions of the n-6 PUFA, arachidonic acid and low proportions of n-3 PUFA. The significance of this difference is that arachidonic acid is the precursor of 2-series prostaglandins and 4-series leukotrienes, which are highly active mediators of inflammation. Feeding fish oil results in partial replacement of arachidonic acid in inflammatory cell membranes by EPA. This change leads to decreased production of arachidonic acid-derived mediators. This response alone is a potentially beneficial anti-inflammatory effect of PUFA. Supplementation of the diet of healthy volunteers with fish oil-derived n-3 PUFA resulted in decreased monocyte and neutrophil chemotaxis and decreased production of pro-inflammatory cytokines.[57] Clinical studies have reported that fish oil supplementation has beneficial effects on the systemic inflammatory response and disease activity in conditions such as rheumatoid arthritis and inflammatory bowel disease.[57] In patients with cancer cachexia encouraging effects of PUFA supplementation were shown in terms of a decreased systemic inflammatory response and body weight gain.[58] The latter may be related to other effects of n-3 PUFA, which occur downstream of altered eicosanoid production or might be independent of this activity. Nuclear factor kappa B (NF-κB) is a critical mediator of the intracellular signaling events triggered by tumor necrosis factor (TNF)-α and other inflammatory cytokines including skeletal muscle specific gene expression.[56] NF-κB activation has indeed been demonstrated in skeletal muscle of severely underweight patients with COPD.[59] Recent studies have shown that n-3 PUFA can down-regulate the activity of NF-κB. The interaction of n-3 PUFA and cytokine biology is however complex. In healthy volunteers the sensitivity of a person to the suppressive effects of n-3 PUFA on TNF-α production is linked to the inherent level of production of the cytokines by cells from the person before supplementation and genetic variation encoded by, or associated with the TNF-α $-$ 308 and lymphotoxin a $+$252 single nucleotide polymorphisms.[60] Some studies suggest that TNF-α levels are particularly increased in COPD patients with weight loss and/or muscle wasting.[44] TNF-α − 308 polymorphism however has also been associated with the presence of COPD[61] and even specifically with the extent of emphysematous changes in these patients.[62]

12.9 INTRINSIC ABNORMALITIES IN SKELETAL MUSCLE

Besides muscle wasting, intrinsic abnormalities in peripheral skeletal muscle morphology and metabolism have been described in COPD patients, pointing toward a decreased oxidative capacity. These abnormalities include muscle fiber type shifts from the oxidative type I fibers toward the glycolytic type IIx fibers,[63] accompanied by a decrease in oxidative enzymes involved in carbohydrate and fatty acid oxidation.[64] Detailed information on the effects in substrate metabolism at the whole-body and skeletal muscle level is lacking. Nevertheless, the metabolic adaptations have clinical consequences as illustrated, for example, by a decreased mechanical efficiency[65] and enhanced lactic acid production during exercise[66] relative to healthy control subjects. In addition, nuclear magnetic resonance studies using single limb exercise models showed a rapid decline and impaired recovery of phosphocreatine stores.[67] There are indications that the decreased oxidative capacity is more pronounced in patients with emphysema than in chronic bronchitis, possibly related to altered oxygen availability.[63, 68, 69]

Positive effects of pulmonary rehabilitation, in particular of endurance exercise training, illustrate that decreased muscle oxidative capacity in COPD is at least partly reversible.[67] While overall effects of endurance type exercise are positive, the available studies also clearly show that it is difficult to enhance this response by modulating exercise type and intensity only.[70, 71] It is therefore tempting to explore the potential of nutritional modulation on muscle substrate metabolism.

The muscle fiber type shift from type I to type IIx together with the enhanced lactic acid production during exercise points toward a decreased oxidative capacity specifically for fatty acids. This suggestion is consistent with the finding that 3-hydroxyacyl-coenzyme A dehydrogenase (HADH) (an enzyme involved in the β-oxidation of fatty acids) was shown to be decreased in COPD.[64] Polyunsaturated fatty acids, particularly those of the (n-3) family and to a lesser extent of the (n-6) family, have been shown to specifically alter expression of genes involved in substrate metabolism. They may upregulate the expression of genes encoding proteins involved in fatty acid oxidation while simultaneously down-regulating genes encoding proteins of lipid synthesis (72). Therefore polyunsaturated fatty acids can be of potential interest in improving oxidative capacity in COPD. A recent randomized prospective clinical trial indeed showed beneficial effects of PUFA on exercise capacity.[73] A potential mechanism could be related to the fact that polyunsaturated fatty acids govern oxidative gene expression by activation of the peroxisome proliferators-activated receptors.[74]

Creatine monohydrate is another commonly available nutritional supplement that could aid exercise performance. In skeletal muscle creatine undergoes rapid and reversible phosphorylation catalyzed by the enzyme creatine kinase to form phosphocreatine (PCr). PCr provides an immediate source of high energy phosphate that is crucial for maintaining the rate of ATP resynthesis during the initial stages of exercise, especially when metabolic demands are high. Uptake and incorporation of exogeneous creatine into skeletal muscle is evident from increased intramuscular PCr levels. In a placebo controlled clinical trial, 2 weeks of oral loading creatine supplementation (15 g/day) increased FFM and skeletal muscle function in patients with moderate to severe COPD. Maintenance creatine supplementation (5 g/day) potentiated the effects of pulmonary rehabilitation with additional improvements in FFM and muscle function but produced no improvement in whole body exercise performance.[75] These two clinical trials nicely illustrate the need for a targeted multicomponent therapeutic approach by physiological and pharmacological means to reverse systemic manifestations of the disease and ultimately improve health status and mortality.

REFERENCES

1. Wilson, D.O., Rogers, R.M., Wright, E.C., and Anthonisen, N.R. Body weight in chronic obstructive pulmonary disease. The National Institutes of Health Intermittent Positive-Pressure Breathing Trial. *Am. Rev. Respir. Dis.* 1989;139:1435–1438.
2. Landbo, C., Prescott, E., Lange, P., Vestbo, J., and Almdal, T.P. Prognostic value of nutritional status in chronic obstructive pulmonary disease. *Am. J. Respir. Crit. Care Med.* 1999;160:1856–1861.
3. Schols, A.M., Slangen, J., Volovics, L., and Wouters, E.F. Weight loss is a reversible factor in the prognosis of chronic obstructive pulmonary disease. *Am. J. Respir. Crit. Care Med.* 1998;157:1791–1797.
4. Prescott, E., Almdal, T., Mikkelsen, K.L., Tofteng, C.L., Vestbo, J., Lange, P. Prognostic value of weight change in chronic obstructive pulmonary disease: results from the Copenhagen City Heart Study. *Eur. Respir. J.* 2002;20:539–544.
5. Engelen, M.P., Schols, A.M., Does, J.D., and Wouters, E.F. Skeletal muscle weakness is associated with wasting of extremity fat-free mass but not with airflow obstruction in patients with chronic obstructive pulmonary disease. *Am. J. Clin. Nutr.* 2000;71:733–738.
6. Bernard, S., LeBlanc, P., Whittom, F., Carrier, G., Jobin, J., Belleau, R. et al. Peripheral muscle weakness in patients with chronic obstructive pulmonary disease. *Am. J. Respir. Crit. Care Med.* 1998; 158: 629–634.
7. Murciano, D., Riagaud, D., Pingleton, S., Armengaud, M.H., Melchior, J.C., and Aubier, M. Effects of renutrition. *Am. J. Respir. Crit. Care Med.* 1994:1569–1574.
8. Baarends, E.M., Schols, A.M., Mostert, R., and Wouters, E.F. Peak exercise response in relation to tissue depletion in patients with chronic obstructive pulmonary disease. *Eur. Respir. J.* 1997; 10: 2807–2813.
9. Palange, P., Forte, S., Onorati, P., Paravati, V., Manfredi, F., Serra, P., and Carlone, S. Effect of reduced body weight on muscle aerobic capacity in patients with COPD. *Chest* 1998; 114:12–18.
10. Gosker, H.R., Engelen, M.P., van Mameren, H., van Dijk, P.J., van der Vusse G.J., Wouters, E.F., and Schols, A.M. Muscle fiber type IIX atrophy is involved in the

loss of fat-free mass in chronic obstructive pulmonary disease. *Am. J. Clin. Nutr.* 2002: 76; 113–119.

11. Vandenbergh, E., van de Woestijne, K.P., Gyselen, A. Weight changes in the terminal stages of chronic obstructive pulmonary disease. *Am. Rev. Respir. Dis.* 1967; 95:556–566.

12. Gray Donald, K., Gibbons, L., Shapiro, S.H., Macklem, P.T., and Martin, J.G. Nutritional status and mortality in chronic obstructive pulmonary disease. *Am. J. Respir. Crit. Care Med.* 1996; 153:961–966.

13. Schols, A.M., Soeters, P.B., Dingemans, A.M., Mostert, R., Frantzen, P.J., and Wouters, E.F. Prevalence and characteristics of nutritional depletion in patients with stable COPD eligible for pulmonary rehabilitation. *Am. Rev. Respir. Dis.* 1993;147:1151–1156.

14. Engelen, M.P., Schols, A.M., Baken, W.C., Wesseling, G.J., and Wouters, E.F. Nutritional depletion in relation to respiratory and peripheral skeletal muscle function in out-patients with COPD. *Eur. Respir. J.* 1994;7:1793–1797.

15. Schwebel, C., Pin, I., Barnoud, D., Devouassoux, G., Brichon, P.Y., Chaffanjon, P., Chavanon, O., Sessa, C., Blin, D., Guignier, M., Leverve, X., and Pison, C. Prevalence and consequences of nutritional depletion in lung transplant candidates. *Eur. Respir. J.* 2000 Dec;16:1050–1055.

16. Shoup, R., Dalsky, G., Warner, S., Davies, M., Connors, M., Khan, M., Khan, F., and ZuWallack, R. Body composition and health-related quality of life in patients with obstructive airways disease. *Eur. Respir. J.* 1997; 10:1576–1580.

17. Mostert, R., Goris, A., Weling-Scheepers, C., Wouters, E.F.M., and Schols, A.M.W.J. Tissue depletion and health related quality of life in patients with chronic obstructive pulmonary disease. *Respir. Med.* 2000;9:859–867.

18. Marquis, K., Debigare, R., Lacasse, Y. et al. Midthigh muscle cross-sectional area is a better predictor of mortality than body mass index in patients with chronic obstructive pulmonary disease. *Am. J. Respir. Crit. Care Med.* 2002;166:809–813.

19. Engelen, M.P., Schols, A.M., Lamers, R.J., and Wouters, E.F. Different patterns of chronic tissue wasting among patients with chronic obstructive pulmonary disease. *Clin. Nutr.* 1999;18:275–280.

20. Coxson, H.O., Chan, I.H., Mayo, J.R., Hlynsky, J., Nakano, Y., and Birmingham, C.L. Early emphysema in patients with anorexia nervosa. *Am. J. Respir. Crit. Care Med..* 2004;170:748–752.

21. American Joint Distribution Committee. Maladie de famine. Recherches cliniques sur la famine executes dans le ghetto de Varsovie en 1942. *Varsovie*, 1946.

22. Nguyen, L.T., Bedu, M., Caillaud, D. et al. Increased resting energy expenditure is related to plasma TNF-alpha concentration in stable COPD patients. *Clin. Nutr.* 1999;18:269–274.

23. Baarends, E.M., Schols, A.M., Pannemans, D.L., Westerterp, K.R., and Wouters, E.F. Total free living energy expenditure in patients with severe chronic obstructive pulmonary disease. *Am. J. Respir. Crit. Care Med.* 1997;155:549–554.

24. Vermeeren, M.A., Schols, A.M., and Wouters, E.F. Effects of an acute exacerbation on nutritional and metabolic profile of patients with COPD. *Eur. Respir. J.* 1997;10:2264–2269.

25. Vermeeren, M.A.P., Wouters, E.F.M., Geraerts-Keeris A.J.W., and Schols A.M.W.J. Nutritional support in patients with chronic obstructive pulmonary disease during hospitalization for an acute exacerbation; a randomized controlled feasibility trial. *Clin. Nutr.* 2004;23:1184–1192.

26. Schols, A.M., Buurman, W.A., Staal van den Brekel, A.J., Dentener, M.A., and Wouters, E.F. Evidence for a relation between metabolic derangements and increased levels of inflammatory mediators in a subgroup of patients with chronic obstructive pulmonary disease. *Thorax* 1996;51:819–824.

27. Takabatake, N., Nakamura, H., Abe, S., Hino, T., Saito, H., Yuki, H. et al. Circulating leptin in patients with chronic obstructive pulmonary disease. *Am. J. Respir. Crit. Care Med.* 1999;159:1215–19.

28. Dentener, M.A., Creutzberg, E.C., Schols, A.M., Mantovani, A., van't Veer, C., Buurman, W.A., and Wouters, E.F. Systemic anti-inflammatory mediators in COPD: increase in soluble interleukin 1 receptor II during treatment of exacerbations. *Thorax* 2001;56:721–726.

29. Di Francia, M., Barbier, D., Mege, J.L., and Orehek, J. Tumor necrosis factor-alpha levels and weight loss in chronic obstructive pulmonary disease. *Am. J. Respir. Crit. Care Med.* 1994;150:1453–1455.

30. De Godoy, I., Donahoe, M., Calhoun, W.J., Mancino, J., and Rogers, R.M. Elevated TNF-alpha production by peripheral blood monocytes of weight-losing COPD patients. *Am. J. Respir. Crit. Care Med.* 1996;153:633–637.

31. Schols, A.M., Creutzberg, E.C., Buurman, W.A., Campfield, L.A., Saris, W.H., and Wouters, E.F. Plasma leptin is related to proinflammatory status and dietary intake in patients with chronic obstructive pulmonary disease. *Am. J. Respir. Crit. Care Med.* 1999;160:1220–6.

32. Ferreira, I., Brooks, D., Lacasse, Y., and Goldstein, R. Nutritional intervention in COPD: A systematic overview. *Chest* 2001;119:353–363.

33. Akrabawi, S.S., Mobarhan, S., Stoltz, R.R., and Ferguson, P.W. Gastric emptying, pulmonary function, gas exchange, and respiratory quotient after feeding a moderate versus high fat enteral formula meal in chronic obstructive pulmonary disease patients. *Nutrition* 1996;12:260–265.

34. Vermeeren, M.A., Wouters, E.F., Nelissen, L.H., van Lier, A., Hofman, Z., and Schols, A.M. Acute effects of different nutritional supplements on symptoms and functional capacity in patients with chronic obstructive pulmonary disease. *Am. J. Clin. Nutr.* 2001;73:295–301.

35. Goris, A.H.C., Vermeeren, M.A.P., Wouters, E.F.M., Schols, A.M.W.J., and Westerterp, K.R. Energy balance in depleted ambulatory patients with chronic obstructive pulmonary disease: The effect of physical activity and oral nutritional supplementation. *Br. J. Nutr.* 2003;89:725–729.

36. Broekhuizen, R., Creutzberg, E.C., Weling-Scheepers, C.A., Wouters, E.F., and Schols, A.M. Optimizing oral nutritional drink supplementation in patients with chronic obstructive pulmonary disease. *Br. J. Nutr.* 2005; 93: 965–971.

37. Ferreira, I.M., Brooks, D., Lacasse, Y., and Goldstein, R.S. Nutritional support for individuals with COPD: A meta-analysis. *Chest* 2000;117:672–678.

38. Baldwin, C., Parsons, T., and Logan, S. Dietary advice for illness-related malnutrition in adults. *Cochrane. Database. Syst. Rev.* 2001; CD002008.

39. Burdet, L., de Muralt, B., Schutz, Y., Pichard, C., and Fitting, J.W. Administration of growth hormone to underweight patients with chronic obstructive pulmonary disease. A prospective, randomized, controlled study. *Am. J. Respir. Crit. Care Med.* 1997;156:1800–1806.

40. Schols, A.M., Soeters, P.B., Mostert, R., Pluymers, R.J., and Wouters, E.F. Physiologic effects of nutritional support and anabolic steroids in patients with

chronic obstructive pulmonary disease. A placebo-controlled randomized trial. *Am. J. Respir. Crit. Care Med.* 1995;152:1268–1274.

41. Basaria, S., Wahlstrom, J.T., and Dobs, A.S. Clinical review 138: Anabolic-androgenic steroid therapy in the treatment of chronic diseases. *J. Clin. Endocrinol. Metab.* 2001;86:5108–5117.

42. Borsheim, E., Tipton, K.D., Wolf, S.E., and Wolfe, R.R. Essential amino acids and muscle protein recovery from resistance exercise. *Am. J. Physiol. Endocrinol. Metab.* 2002;283:E648–E657.

43. Engelen, M.P., Deutz, N.E., Wouters, E.F., and Schols, A.M. Enhanced levels of whole-body protein turnover in patients with chronic obstructive pulmonary disease. *Am. J. Respir. Crit. Care Med.* 2000;162:1488–1492.

44. Eid, A.A., Ionescu, A.A., Nixon, L.S. et al. Inflammatory response and body composition in chronic obstructive pulmonary disease. *Am. J. Respir. Crit. Care Med.* 2001;164:1414–1418.

45. Pouw, E.M., Schols, A.M., Deutz, N.E., and Wouters, E.F. Plasma and muscle amino acid levels in relation to resting energy expenditure and inflammation in stable chronic obstructive pulmonary disease. *Am. J. Respir. Crit. Care Med.* 1998;158:797–801.

46. Mansoor, O., Cayol, M., Gachon, P. et al. Albumin and fibrinogen syntheses increase while muscle protein synthesis decreases in head-injured patients. *Am. J. Physiol.* 1997;273:E898–E902.

47. Morrison, W.L., Gibson, J.N., Scrimgeour, C., and Rennie, M.J. Muscle wasting in emphysema. *Clin. Sci.* 1988;75:415–420.

48. Yoneda, T., Yoshikawa, M., Fu, A., Tsukaguchi, K., Okamoto, Y., and Takenaka, H. Plasma levels of amino acids and hypermetabolism in patients with chronic obstructive pulmonary disease. *Nutrition* 2001;17:95–99.

49. Engelen, M.P., Wouters, E.F., Deutz, N.E., Menheere, P.P., and Schols A.M. Factors contributing to alterations in skeletal muscle and plasma amino acid profiles in patients with chronic obstructive pulmonary disease. *Am. J. Clin. Nutr.* 2000;72:1480–1487.

50. Anthony, J.C., Anthony, T.G., Kimball, S.R., and Jefferson, L.S. Signaling pathways involved in translational control of protein synthesis in skeletal muscle by leucine. *J. Nutr.* 2001;131:856S–860S.

51. Levenhagen, D.K., Gresham, J.D., Carlson, M.G., Maron, D.J., Borel M.J., and Flakoll P.J. Postexercise nutrient intake timing in humans is critical to recovery of leg glucose and protein homeostasis. *Am. J. Physiol. Endocrinol. Metab.* 2001;280:E982–E993.

52. Engelen, M.P., Schols, A.M., Does, J.D., Deutz, N.E., and Wouters, E.F. Altered glutamate metabolism is associated with reduced muscle glutathione levels in patients with emphysema. *Am. J. Respir. Crit. Care Med.* 2000;161:98–103.

53. Engelen, M.P., Wouters, E.F., Deutz, N.E., Does, J.D., and Schols, A.M. Effects of exercise on amino acid metabolism in patients with chronic obstructive pulmonary disease. *Am. J. Respir. Crit. Care Med.* 2001;163:859–864.

54. Rabinovich, R.A., Ardite, E., Troosters, T. et al. Reduced muscle redox capacity after endurance training in patients with chronic obstructive pulmonary disease. *Am. J. Respir. Crit. Care Med.* 2001;164:1114–1118.

55. Creutzberg, E.C., Schols, A.M., Weling-Scheepers, C.A., Buurman, W.A., and Wouters, E.F. Characterization of nonresponse to high caloric oral nutritional

therapy in depleted patients with chronic obstructive pulmonary disease. *Am. J. Respir. Crit. Care Med.* 2000;161:745–752.

56. Debigare, R., Cote, C.H., and Maltais, F. Peripheral muscle wasting in chronic obstructive pulmonary disease. Clinical relevance and mechanisms. *Am. J. Respir. Crit. Care Med.* 2001;164:1712–1717.

57. Barber, M.D., Ross, J.A., Voss, A.C., Tisdale, M.J., and Fearon, K.C. The effect of an oral nutritional supplement enriched with fish oil on weight-loss in patients with pancreatic cancer. *Br. J. Cancer.* 1999;81:80–86.

58. Fearon, K.V., Von Meyenfeldt, M.F., Moses, A.G., Van Geenen, R., Roy, A., Gouma D.J., Giacosa, A., van Gosum, A., Bauer, J., Barber, M.D., Aarenson, N.K., Voss A.C., and Tisdale, M.J. Effect of a protein and energy dense N-3 fatty acid enriched oral supplement on loss of weight and lean tissue in cancer cachexia: A randomised double blind trial. *Gut* 2003; 52:1479–86.

59. Agusti, A., Morla, M., Sauleda, J., Sasua, C., and Busquets, X. NF-kappaB activation and iNOS upregulation in skeletal muscle of patients with COPD and low body weight. *Thorax* 2004;59:483–7.

60. Grimble, R.F., Howell, W.M., O'Reilly, G. et al. The ability of fish oil to suppress tumor necrosis factor alpha production by peripheral blood mononuclear cells in healthy men is associated with polymorphisms in genes that influence tumor necrosis factor alpha production. *Am. J. Clin. Nutr.* 2002;76:454–459.

61. Sakao, S., Tatsumi, K., Igari, H., Shino, Y., Shirasawa, H., and Kuriyama, T. Association of tumor necrosis factor alpha gene promoter polymorphism with the presence of chronic obstructive pulmonary disease. *Am. J. Respir. Crit. Care Med.* 2001;163:420–422.

62. Sakao, S., Tatsumi, K., Igari, H. et al. Association of tumor necrosis factor-alpha gene promoter polymorphism with low attenuation areas on high-resolution CT in patients with COPD. *Chest* 2002;122:416–420.

63. Gosker, H.R., van Mameren, H., van Dijk, P.J. et al. Skeletal muscle fibre-type shifting and metabolic profile in patients with chronic obstructive pulmonary disease. *Eur. Respir. J.* 2002;19:617–625.

64. Maltais, F., LeBlanc, P., Whittom, F. et al. Oxidative enzyme activities of the vastus lateralis muscle and the functional status in patients with COPD. *Thorax* 2000;55:848–853.

65. Baarends, E.M., Schols, A.M., Akkermans, M.A., and Wouters, E.F. Decreased mechanical efficiency in clinically stable patients with COPD. *Thorax* 1997;52:981–986.

66. Maltais, F., Simard, A.A., Simard, C., Jobin, J., Desgagnes, P., and LeBlanc, P. Oxidative capacity of the skeletal muscle and lactic acid kinetics during exercise in normal subjects and in patients with COPD. *Am. J. Respir. Crit. Care Med.* 1996;153:288–293.

67. Sala, E., Roca, J., Marrades, R.M. et al. Effects of endurance training on skeletal muscle bioenergetics in chronic obstructive pulmonary disease. *Am. J. Respir. Crit. Care Med.* 1999;159:1726–1734.

68. Pouw, E.M., Schols, A.M.W.J., Vusse van der, G.J., and Wouters, E.F.M. Elevated inosine monophosphate levels in resting muscle of patients with stable COPD. *Am. J. Respir. Crit. Care Med.* 1998;157:453–457.

69. Satta, A., Migliori, G.B., Spanevello, A. et al. Fibre types in skeletal muscles of chronic obstructive pulmonary disease patients related to respiratory function and exercise tolerance. *Eur. Respir. J.* 1997;10:2853–2860.

70. Coppoolse, R., Schols, A.M., Baarends, E.M. et al. Interval versus continuous training in patients with severe COPD: A randomized clinical trial. *Eur. Respir. J.* 1999;14:258–263.

71. Ortega, F., Toral, J., Cejudo, P. et al. Comparison of effects of strength and endurance training in patients with chronic obstructive pulmonary disease. *Am. J. Respir. Crit. Care Med.* 2002;166:669–674.

72. Clarke, S.D. Polyunsaturated fatty acid regulation of gene transcription: A mechanism to improve energy balance and insulin resistance. *Br. J. Nutr.* 2000;83: Suppl. 1,S59–S66.

73. Broekhuizen, R., Wouters, E.F.M., Creutzberg, E.C., Weling-Scheepers, C.A.P.M., and Schols, A.M.W.J. Polyunsaturated fatty acids improve exercise capacity in chronic obstructive pulmonary disease. *Thorax* 2005;60:376–382.

74. Schoonjans, K., Staels, B., and Auwerx, J. The peroxisome proliferator activated receptors (PPARS) and their effects on lipid metabolism and adipocyte differentiation. *Biochim. Biophys. Acta.* 1996;1302:93–109.

75. Fuld, J.P., Kilduff, L.P., Neder, J.A., Pitsiladis, Y., Lean, M.E.J., Ward S.A., and Cotton, M.M. Creatine supplementation during pulmonary rehabilitation in chronic obstructive pulmonary disease, *Thorax* 2005; 60: 531–537.

13 Cachexia in Older Persons

John E. Morley

CONTENTS

SUMMARY

Weight loss in older persons causes hip fracture, institutionalization, and death. There are four causes of weight loss in older persons, that is, sarcopenia, anorexia, cachexia, and dehydration. Sarcopenia is pure muscle wasting that is associated with frailty and disability. The causes of sarcopenia are multifactorial including a decline in anabolic hormones, increased cytokines, and a decrease in food intake and physical activity. There is a physiological anorexia of aging due to decreased taste and smell, altered gastric emptying and an increase in cholecystokinin. Depression and medications are the major causes of weight loss in older persons. The orexigenic drugs most commonly used in older persons are megestrol acetate and dronabinol. Many of the causes of weight loss in older persons are reversible.

Weight loss in older persons is associated with a marked increase in mortality whether the person is living in the community,[1-4] hospitalized,[5] or in a nursing home.[6] Weight loss is also associated with increase in hip fracture[7] and increased institutionalization.[8] Protein energy malnutrition leads to pressure ulcers, fatigue, anemia, cognitive dysfunction, immune dysfunction, increased infections, falls and death. In particular, protein energy undernutrition is associated with a marked decline in CD_4+ T cells.[9]

There are 4 major causes of weight loss in older persons:

1. Dehydration
2. Anorexia
3. Sarcopenia
4. Cachexia

This chapter describes the causes of each of these conditions and the approaches to their management.

13.1 DEHYDRATION

Dehydration accounts for 10% of the weight loss seen in older persons. It is a major cause of orthostatic hypotension and is associated with functional decline. Severe dehydration leads to delirium.

It is now well recognized that older persons have a decrease in thirst perception. This decline in thirst perception is related both to a decrease in the angiotensin II and the mu opioid thirst drives.[10, 11] The diagnosis is based on an elevated plasma osmolality. Elevated serum sodium levels and an increased in blood urea nitrogen (BUN)/creatinine ratio greater than 20:1 may suggest dehydration. However, an elevated BUN/creatinine ratio also occurs with gastrointestinal bleeding, congestive heart failure, and chronic renal failure. Many older persons have some degree of renal failure and this occurs in the face of creatinine of less than 1.2 mg/dl due to the decline in muscle mass with aging.

Older persons require between 800 and 1500 ml water per day to maintain adequate hydration, that is, 4 to 8 glasses of fluid daily. It is essential to continually remind older persons to ingest fluid. This is particularly true when the older persons' food intake falls as most fluid is ingested with meals. Dehydration develops very rapidly when older persons increase insensible fluid loss, such as when they develop fever or hyperpnea associated with a respiratory tract infection. There is a tendency for physicians to overutilize diuretics in older persons without recognizing their lack of ability to respond to thirst. Persons being tube fed often are given insufficient free water. Some older persons need as much as 250 to 300 ml of water every 4 h to maintain their hydration status. Alterations in responsiveness to arginine vasopressin, salt deficiency, and/or medications (e.g., selective serotonin inhibitors) can lead to profound hyponatremia in older persons. This requires a delicate balance in managing fluids in older persons making sure to produce neither over- nor under-hydration.

At present, no medications are used to enhance thirst perception in older persons.

13.2 ANOREXIA

There is a substantial decline in food intake over the lifetime.[12] This decline in food intake is greater in men than women. The causes of this decline in food intake are multifactorial. A number of physiological changes that occur with aging result in a reduction of the desire to eat. There is a marked decline in the ability to smell with aging. Persons with Alzheimer's disease often have anosmia. Taste thresholds increase slightly with aging. This predominantly results in older persons choosing less variety with their meals. Medications can lead to a further decrease in taste. Zinc deficiency, associated with diuretic use and diabetes mellitus, markedly decreases the ability to taste. The use of taste enhancers, such as monosodium glutamate, results in weight gain in older persons.[13]

There is a decline in adaptive relaxation of the fundus of the stomach with aging.[14] This is due to a reduction in the release of nitric oxide from the wall of the fundus. Nitric oxide produces local smooth muscle relaxation, thus increasing the fundal volume available to accept food. Smaller fundal volume leads to food more rapidly passing into the antrum. This produces antral stretch and early satiation.

There is an increase in both basal and fat stimulated cholecystokinin (CCK) plasma levels with aging.[15] This is due to decreased clearance of CCK in older persons. In addition, CCK is a more potent satiating agent in older compared to younger persons.[16]

Amounts of food under 500 calories and liquid food is less satiating in older persons than are larger volumes. More than 500 calories at a single meal results in a delay in gastric emptying.[17]

The decline in testosterone in older males leads to an increase in the satiating hormone, leptin.[18] Testosterone replacement lowers leptin levels.[19] The increase of leptin in older males appears to account for the greater decline in food intake that occurs in males with aging.

Ghrelin is a peptide hormone released from the fundus of the stomach. It increases food intake, produces growth hormone release, and increases muscle mass. Its role in the anorexia of aging is yet to be determined.

With the onset of minor diseases and/or obesity in middle to early old age, there is an increase in pro-inflammatory cytokines. Tumor necrosis factor α and interleukin-1 are potent anorectic agents.[20] The role of circulating cytokines in the anorexia of aging has not been well established.

Changes in central neurotransmitters such as neuropeptide Y and kappa opioids and their effectiveness have been demonstrated to occur in aging animals.[21] Whether these same changes occur in humans is uncertain.

Finally, it needs to be recognized that numerous reversible diseases result in anorexia (Table 13.1). Depression accounts for 30% of weight loss in older persons while cancer accounts for less than 10%.[22,23]

13.2.1 Pharmacotherapy

The different approaches to pharmacotherapy of the anorexia of aging are outlined in Table 13.2.

Nitroglycerine tablets can be used to reduce the decreased fundal compliance that occurs with aging. The use of drugs that increase the rate of stomach emptying

TABLE 13.1
Reversible Diseases that Result in Anorexia — The MEALS-ON-WHEELS Mnemonic

*M*edications (e.g., digoxin, theophylline, cimetidine)
*E*motional (e.g., depression)
*A*lcoholism, elder abuse, anorexia tardive
*L*ate life paranoia
*S*wallowing problems

*O*ral factors
*N*osocomial infections (e.g., tuberculosis)

*W*andering and other dementia-related factors
*H*yperthyroidism, Hypercalcemia, Hypoadrenalism
*E*nteral problems (e.g., gluten enteropathy)
*E*ating problems
*L*ow salt, low cholesterol, and other therapeutic diets
*S*tones (cholecystitis)

may also allow ingestion of larger meals. CCK antagonists have the potential to decrease anorexia associated with high fat meals. Ghrelin, or ghrelin agonists, such as MK-771, may increase food intake.

Testosterone may be used to reduce leptin levels in males. However, while it increases muscle mass there is limited evidence that it will increase food intake.[24] Another anabolic hormone, growth hormone, has been shown to produce weight gain, enhance nitrogen retention, and improve functional performance.[25, 26] However, in acutely ill malnourished persons, growth hormone resulted in increased mortality.[27]

Numerous cytokine antagonists exist.[28] Of these, megestrol acetate, has been most used in older persons to increase weight.[28, 29] Megestrol acetate is a glucocorticoid-progestational agent that has equivalent glucocorticoid activity to prednisone.[30] Its progestational component leads to an increase in central neuropeptide Y effects. Weight gain continues after the drug is discontinued.[29] The effect on weight gain is associated with elevated cytokine levels.[31] It does not increase mortality.[32] It increases albumin and hematocrit.[29] Its side effects include a small increase in deep vein thrombosis, though this was not present in a systematic analysis of controlled trials.[33] The increase is less than that seen with antipsychotics. Megestrol produces a decline in testosterone levels[34] but has its own androgenic activity.[35] The decline in cortisol response to adrenocorticotrophic hormone (ACTH), is the expected response due to its glucocorticoid activity. Addisonian crisis is extremely rare after short-term use.

Dronabinol is a synthetic tetrahydrocannabinol. In a controlled study in older persons, it produced a small, nonsignificant weight gain.[36] It may reduce agitation

TABLE 13.2
Approaches to the Management of the Anorexia of Aging

Drug	Effect
Zinc[a]	Improves taste sensation and reduces anorexia in zinc-deficient persons
Monosodium glutamate	Taste enhancer leading to weight gain
Nitroglycerine	Enhances fundal compliance allowing ingestion of larger meals
Promotility agents[a] (e.g., metoclopramide, cisapride, tegaserod)	Increase rate of gastric emptying thus decreasing antral stretch
Cholecystokinin antagonists	Decrease anorexia associated with high fat meals
Ghrelin and its receptor agonists (MK-771)	Increase food intake, muscle mass, memory, and growth hormone release
Testosterone[a]	Decreases leptin levels and increases muscle mass and strength
Growth hormone[b]	Increases weight, nitrogen retention and function
Leptin antagonists	Unknown
Megestrol acetate[a]	Increases appetite, weight gain, and quality of life
Dronabinol[a]	Increases appetite with minimal weight gain
Other anticytokines[b] (e.g., thalidomide)	Weight gain but not studied in older persons
Cyproheptadine acetate[b]	Minimal to no weight gain
Antidepressants[a] (e.g., Monoamine Oxidase Inhibitors or mirtazapine)	Weight gain in depressed persons
Olanzapine[b]	Weight gain associated with diabetes mellitus
Proteasome inhibitors (e.g., Bortezomil)	Unknown effects

[a] Recommended drugs that are clinically available.
[b] Not recommended but available.

in some demented persons. It decreases pain and nausea and produces a feeling of well-being. In high doses it can produce delirium.

Cyproheptadine acetate, an antiserotonergic agent has small appetite stimulant effects. It produces delirium that limits its usefulness in older person. No drugs that modulate central nervous system orexigenic neurotransmitters are available.

Depression, when present needs to be treated. Either monoamine oxidase inhibitors or mirtazapine appear to be the drugs of choice to treat weight loss associated with depression. Olanzapine, an antipsychotic, produces weight gain associated with diabetes mellitus.

No ideal drug for treating anorexia in older persons exists. At present, megestrol acetate in doses of 400 to 800 mg appears to be the best treatment. Megestrol acetate is a progestational glucocorticoid that inhibits cytokines and activates neuropeptide Y within the Hypothalamus. Megestrol acetate is poorly absorbed in the fasting state. A new formulation of megestrol acetate using a nanocrystal formulation that is absorbed in the fasting state, should produce better weight gain at a lower dose.

13.3 SARCOPENIA

Sarcopenia is defined as an appendicular muscle mass two standard deviations below that of normal young persons.[37] Sarcopenia is associated with functional decline and death.[18] As with the causes of anorexia, the causes of sarcopenia are multifactorial.

Age, lack of physical activity, and decreased food intake, especially intake of the amino acid creatine, are all causes of sarcopenia. Peripheral vascular disease leads to decreased blood flow to the lower limbs and loss of muscle mass. Decreased neuronal input to muscles further results in loss of muscle mass.

The longitudinal decline of testosterone is the major factor in males associated with the development of sarcopenia.[38] Testosterone increases muscle protein synthesis and stimulates skeletal muscle satellite precursors and inhibits preadipocytes. Testosterone increases muscle mass and strength in older men.[39] Low testosterone levels are associated with poor function.[40] Testosterone replacement enhances functional status following hospitalization for general medical problems[41] or knee surgery.[42]

Insulin-like growth factor-1 (IGF-1) plasma levels decline with aging. Testosterone increases IGF-1 levels. IGF-1 stimulates muscle protein synthesis and satellite cell formation. Transgenic IGF-1 expression inhibits the loss of muscle in older mice.[43] There are two isoforms of IGF-1, and the muscle isoform appears to play the major role in preventing sarcopenia.

Cytokines, especially TNFα, interleukin-1, and interleukin-6, cause muscle loss and functional deterioration.[44] These effects seem to utilize similar mechanisms as those that cause cachexia (*vide infra*).

Myostatin inhibits muscle growth. Myostatin transgene leads to cachexia[45] and myostatin deletion leads to muscle hypertrophy in mice, cows, and humans.[46]

At present, only testosterone is available for the pharmacotherapy of sarcopenia. Myostatin antibodies and myostatin inhibitors are potential therapeutic agents of the future.

13.4 CACHEXIA

Cachexia is severe muscle and fat loss associated with a marked increase in proinflammatory cytokines. In addition, cytokines slow intestinal mobility, produce anorexia, cognitive impairment, and sickness behavior. They result in an increase in epinephrine and cortisol plasma levels.

Cytokines cause muscle wasting by activating the ubiquitin-proteasome system, inhibiting NFkappaB and thus protein synthesis and producing insulin resistance, thus diverting glucose from muscle to the liver.

Pharmacotherapy of cachexia includes the use of specific and nonspecific cytokine inhibitors. At present, megestrol acetate is the only drug with Class A evidence for the treatment of cachexia.[33] A future possibility represents the use of proteasome inhibitors or anti-inflammatory cytokines.

13.5 CONCLUSION

The development of pharmacotherapy to treat weight loss in older person is in its infancy. The majority of persons who develop cachexia are older. The understanding of the interaction of the anorexia of aging and cachexia is key to the eventual appropriate therapy of weight loss. The effects of aging on the drugs developed to treat cachexia must be better understood.

REFERENCES

1. Morley, J.E. International aging: Why does the United States do so poorly? *J. Am. Geriatr. Soc.* 1991;39:836–838.
2. Pamuk, E.R., Williamson, D.F., Madans, J., Serdula, M.K., Kleinmon, J.C., and Byers, T. Weight loss and mortality in a national cohort of adults. *Am. J. Epidemiol.* 1992;136:686–697.
3. de Groot, C.P., Enzi, G., Matthys, C., Moreiras, O., Roszkowski, W., and Schroll, M. Ten-year changes in anthropometric characteristics of elderly Europeans. *J. Nutr. Health Aging* 2002;6:4–8.
4. Losonczy, K.G., Harris, T.B., Cornoni-Huntley, J., Simonsick, E.M., Wallace, R.B., Cook, N.R., Ostfeld, A.M., and Blazer, D.G. Does weight loss from middle age to old age explain the inverse weight mortality relation in old age? *Am. J. Epidemiol.* 1995;141:312–321.
5. Liu, L., Bopp, M.M., Roberson, P.K., and Sullivan, D.H. Undernutrition and risk of mortality in elderly patients within 1 year of hospital discharge. *J. Gerontol. Med. Sci.* 2002;57A:M741–M746.
6. Yamashita, B.D., Sullivan, D.H., Morley, J.E., Johnson, L.E., Barber, A., Olson, J.S., Stevens, M.R., Reinhart, S.P., Trotter, J.P., and Olave, X.E. The GAIN (Geriatric Anorexia Nutrition) Registry: The impact of appetite and weight on mortality in a long-term care population. *J. Nutr. Health Aging* 2002;6:275–281.
7. Ensrud, K.E., Ewing, S.K., Stone, K.L., Cauley, J.A., Bowman, P.J., and Cummings, S.R. Study of Osteoporotic Fractures Research Group. Intentional and unintentional weight loss increase bone loss and hip fracture risk in older women. *J. Am. Geriatr. Soc.* 2003;51:1740–1747.
8. Payette, H., Coulombe, C., Boutier, V., and Gray-Donald, K. Weight loss and mortality among free-living frail elders: A prospective study. *J. Gerontol. Med. Sci.* 1999;54A:M440–M445.
9. Kaiser, F.E. and Morley, J.E. Idiopathic CD_4+ T lymphopenia in older persons. *J. Am. Geriatr. Soc.* 1994;42:1291–1294.
10. Silver, A.J. and Morley, J.E. Role of the opioid system in the hypodipsia associated with aging. *J. Am. Geriatr. Soc.* 1992;40:556–560.
11. Silver, A.J., Morley, J.E., Ishimaru-Tseng, T.V., and Morley, P.M. Angiotensin II and fluid ingestion in old rats. *Neurobiol. Aging* 1993;14:519–522.
12. Morley, J.E. and Silver, A.J. Anorexia in the elderly. *Neurobiol. Aging* 1988;9:9–16.
13. Mathey, M.F., Siebelink, E., de Graaf, C., and van Staveren, W.A. Flavor enhancement of food improves dietary intake and nutritional status of elderly nursing home residents. *J. Gerontol. Med. Sci.* 2001;56A:M200–M205.
14. Rayner, C.K., MacIntosh, C.G., Chapman, I.M., Morley, J.E., and Horowitz, M. Effects of age on proximal gastric motor and sensory function. *Scand. J. Gastroenterol.* 2000;35:1041–1047.

15. MacIntosh, C.G., Horowitz, M., Verhagen, M.A., Smouth, A.J., Wishart, J., Morris, H., Goble, E., Morley, J.E., and Chapman, I.M. Effect of small intestinal nutrient infusion on appetite, gastrointestinal hormone release, and gastric myoelectrical activity in young and older men. *Am. J. Gastroenterol.* 2001;96:997–1007.

16. MacIntosh, C.G., Morley, J.E., Wishart, J., Morris, H., Jansen, J.B., Horowitz, M., and Chapman, I.M. Effect of exogenous cholecystokinin (CCK)-8 on food intake and plasma CCK, leptin, and insulin concentrations in older and young adults: Evidence for increased CCK activity as a cause of the anorexia of aging. *J. Clin. Endocrinol. Metab.* 2001;86:5830–5837.

17. Chapman, I.M., MacIntosh, C.G., Morley, J.E., and Horowitz, M. The anorexia of aging. *Biogerontology* 2002;3:67–71.

18. Baumgartner, R.N., Ross, R.R., Waters, D.L., Brooks, W.M., Morley, J.E., Montoya, G.D., and Garry, P.J. Serum leptin in elderly people: Associations with sex hormones, insulin, and adipose tissue volumes. *Obesity Res.* 1999;7:141–149.

19. Sih, R., Morley, J.E., Kaiser, F.E., Perry, H.M., III, Patrick, P., and Ross, C. Testosterone replacement in older hypogonadal men: A 12-month randomized controlled trial. *J. Clin. Endocrinol. Metab.* 1997;82:1661–1667.

20. Baez-Franceschi, D. and Morley, J.E. Pathophysiology of catabolism in undernourished elderly patients. *Zeitschrift fur Gerontologie und Geriatrie* 1999;32(Suppl 1):I12–I19.

21. Wilson, M.M. and Morley, J.E. Invited review: Aging and energy balance. *J. Appl. Physiol.* 2003;95:1728–1736.

22. Morley, J.E. and Kraenzle, D. Causes of weight loss in a community nursing home *J. Am. Geriatr. Soc.* 1994;42:583–585.

23. Wilson, M.M., Vaswani, S., Liu, D., Morley, J.E., and Miller, D.K. Prevalence and causes of under nutrition in medical outpatients. *Am. J. Med.* 1998;104:56–63.

24. Morley, J.E. and Perry, H.M., III. Androgen treatment of male hypogonadism in older males *J. Steroid Biochem. Mol. Biol.* 2003;85:367–373.

25. Chu, L.W., Lam, K.S., Tam, S.C., Hu, W.J., Hui, S.L., Chiu, A., Chiu, K.C., and Ng, P. A randomized controlled trial of low-dose recombinant human growth hormone in the treatment of malnourished elderly medical patients. *J. Clin. Endocrinol. Metab.* 2001;86:1913–1920.

26. Kaiser, F.E., Silver, A.J., and Morley, J.E. The effect of recombinant human growth hormone on malnourished older individuals. *J. Am. Geriatr. Soc.* 1991;39:235–240.

27. Takala, J., Ruokonen, E., Webster, Nr., Nielsen, M.S., Zandstra, D.F., Vundelinckx, G., and Hinds, C.J. Increased mortality associated with growth hormone treatment in critically ill adults. *N. Engl. J. Med.* 1999;341:785–792.

28. Karcic, E., Philpot, C., and Morley, J.E. Treating malnutrition with megestrol acetate: Literature review and review of our experience. *J. Nutr. Health Aging* 2002;6:191–200.

29. Yeh, S.S., Wu, S.Y., Le, T.P., Olson, J.S., Stevens, M.R., Dixon, T., Porcelli, R.J., and Schuster, M.W. Improvement in quality-of-life measures and stimulation of weight gain after treatment with megestrol acetate oral suspension in geriatric cachexia: Results of a double-blind, placebo-controlled study. *J. Am. Geriatr. Soc.* 2000;48:485–492.

30. Kontula, K., Paavonen, T., Luukkainen, T., and Andersson, L.C. Binding of progestins to the glucocorticoid receptor. Correlation to their glucocorticoid-like

effects on *in vitro* functions of human mononuclear leukocytes. *Biochem. Pharmacol.* 1983;32:1511–1518.

31. Yeh, S.S., Wu, S.Y., Levine, D.M., Parker, T.S., Olson, J.S., Stevens, M.R., and Schuster, M.W. The correlation of cytokine levels with body weight after megestrol acetate treatment in geriatric patients *J. Gerontol. Med. Sci.* 2001;56A:M48–M54.

32. Yeh, S.S., Hafner, A., Chang, C.K., Levine, D.M., Parker, T.S., and Schuster, M.W. Risk factors relating blood markers of inflammation and nutritional status to survival in cachectic geriatric patients in a randomized clinical trial. *J. Am. Geriatr. Soc.* 2004;52:1708–1712.

33. Pascual Lopez, A., Roque, I., Figuls, M., Urrutia Cuchi, G., Berenstein, E.G., Almenar Pasies, B., Balcells Alegre, M., and Herdman, M. Systematic review of megestrol acetate in the treatment of anorexia-cachexia syndrome. *J. Pain Symp. Manage.* 2004;27:360–369.

34. Lambert, C.P., Flynn, M.G., Sullivan, D.H., and Evans, W.J. Effects of megestrol acetate on circulating interleukin-15 and interleukin-18 concentrations in healthy elderly men. *J. Gerontol. Med. Sci.* 2004;59A:855–858.

35. Labrie, C., Cusan, L., Plante, M., Lapointe, S., and Labrie, F. Analysis of the androgenic activity of synthetic "progestins" currently used for the treatment of prostate cancer. *J. Steroid Biochem.* 1987;28:379–384.

36. Volicer, L., Stelly, M., Morris, J., McLaughlin, J., and Volicer, B.J. Effects of dronabinol on anorexia and disturbed behavior in patients with Alzheimer's disease. *Int. J. Geriatr. Psychiatry* 1997;12:913–919.

37. Baumgartner, R.N., Koehler, K.M., Gallagher, D., Romero, L., Heymsfield, S.B., Ross, R.R., Garry, P.J., and Lindeman, R.D. Epidemiology of sarcopenia among the elderly in New Mexico. *Am. J. Epidemiol.* 1998;147:755–763.

38. Morley, J.E. and Perry, H.M., III. Androgen treatment of male hypogonadism in older males. *J. Steroid Biochem. Mol. Biol.* 2003;85:367–373.

39. Bhasin, S., Taylor, W.E., Singh, R., Artaza, J., Sinha-Hikim, I., Jasuja, R., Choi, H., and Gonzalex-Cadavid, N.F. The mechanisms of androgen effects on body composition: Mesenchymal pluripotent cell as the target of androgen action. *J. Gerontol. Med. Sci.* 2003;58A:M1101–M1110.

40. Perry, H.M., III, Miller, D.K., Patrick, P., and Morley, J.E. Testosterone and leptin in older African-American men: Relationship to age, strength, function, and season. *Metab. Clin. Exp.* 2000;49:1085–1091.

41. Bakhshi, V., Elliott, M., Gentili, A., Godschalk, M., and Mulligan, T. Testosterone improves rehabilitation outcomes in ill older men. *J. Am. Geriatr. Soc.* 2000;48:550–553.

42. Amory, J.K., Chansky, H.A., Chansky, K.L., Camuso, M.R., Hoey, C.T., Anawalt, B.D., Matsumoto, A.M., and Bremner, W.J. Preoperative supraphysiological testosterone in older men undergoing knee replacement surgery. *J. Am. Geriatr. Soc.* 2002;50:1698–1701.

43. Musaro, A., McCullagh, K., Paul, A., Houghton, L., Dobrowolny, G., Molinaro, M., Barton, E.R., Sweeney, H.L., and Rosenthal, N. Localized Igf-1 transgene expression sustains hypertrophy and regeneration in senescent skeletal muscle. *Nat. Genet.* 2001;27:195–200.

44. Morley, J.E. and Baumgartner, R.N. Cytokine-related aging process. *J. Gerontol. Med. Sci.* 2004;59A:924–929.

45. Zimmers, T.A., Davies, M.V., Koniaris, L.G., Haynes, P., Esquela, A.F., Tomkinson, K.N., McPherron, A.C., Wolfman, N.M., and Lee, S.J. Induction of cachexia in mice by systemically administered myostatin. *Science* 2002;296:1486–1488.

46. McNally, E.M. Powerful genes — myostatin regulation of human muscle mass. *N. Engl. J. Med.* 2004;350:2642–2644.

III

General Therapeutic Aspects

14 Treatment Goals

Andrew J. S. Coats

CONTENTS

SUMMARY

Cachexia is a common and complex disorder that complicates multiple disease processes. Few if any treatments of modern standards have proven to be effective in cachexia. To prove that a new treatment is effective and worthwhile would require either unequivocal demonstration of improved symptoms and quality of life or protection from mortality and major morbidity. Despite this, multiple attractive modes of action exist that could be tested to establish effective cachexia cures. Our patients require us to continue this effort.

14.1 INTRODUCTION

Cachexia is a common complication of many separate disease processes.[1] We have learnt much of the observational pathophysiology of this condition, and some about its epidemiology, but very little about scientifically based and proven therapies. We would like to have documented treatments, effective both in prevention and in cure, and useful for amelioration as well as in improving the prognosis of affected patients. Unfortunately these targets remain distant. In assessing and developing new treatments for cachexia we need to define our treatment objectives as well as assess how we will set about proving their effectiveness and cost/benefit ratio. None of these tasks will be easy. This chapter reviews the aims of cachexia therapy and assesses the appropriate methodologies for proving these effects.

14.2 THE ROLE OF PHARMACOTHERAPY IN CACHEXIA

The management of a cachectic patient should include optimal management of intercurrent or underlying illnesses, psychological and physical support, education, exercise, and nutritional advice and a multidisciplinary approach to the patient's condition, living circumstances and expectations. Drugs should never be advised in ignorance of the potential of these lifestyle and nonpharmaceutical factors in improving a patient's condition.[2] In devising treatments we need to factor in the most appropriate background treatment of the patients so that trials will inevitably include many background therapies in addition to the test treatment under study.

14.3 CHOICE OF A TREATMENT EFFECT

Treatments almost always carry risk as well as cost and inconvenience. Too infrequently do we ask, "Why should we recommend this treatment?" Patients in the Internet era are now much more likely to ask for justification for taking courses of therapy, particularly open-ended therapies. They are also more likely to have researched their condition and to have received patchy information about their disease and treatments that have been proposed. It is essential, therefore, to have a checklist of justification for treatment as a background to recommending treatments (see Table 14.1).

Cachexia is a progressive complication of many separate disease processes. These include cancer, acquired immune deficiency syndrome (AIDS), chronic obstructive pulmonary disease (COPD), chronic heart failure (CHF), and chronic renal failure. As a result treatments cannot in a true sense cure the disease, as cachexia is not a disease with a single aetiology. It is a complication comprising a series of pathophysiological processes that arise secondary to another underlying disease process. In most cases the underlying disease will be undergoing treatment interventions proven in their own right to be effective. It is a moot point whether these treatments should be continued, discontinued, or accelerated when cachexia intervenes, for in most conditions we know little about whether

---

TABLE 14.1
Desired Effects of a Treatment

Must do at least one of:

1. Prolonging life
2. Improving or controlling symptoms
3. Delaying disease progression
4. Avoiding complications

preferably with few side-effects and good affordability

---

the development of cachexia modifies the effectiveness of treatments effective for the underlying disease. In most cases such as cancer, AIDS, COPD, or CHF, the disease expert would continue effective treatment even when cachexia develops, for the simple reason that there is no evidence to suggest the treatment has ceased to be effective. In other words treatment efficacy is blind to the development of cachexia. In heart failure, for example, we know that several treatments are effective. These include angiotensin converting enzyme (ACE) inhibitors, beta-blockers, angiotensin receptor antagonists, and anti-aldosterone agents for prognosis and diuretics for relief of the symptom of oedema. The trials and studies that have proved this have not specifically excluded patients with cachexia at trial entry and mostly have included patients who developed cachexia during the trial. We would therefore err on the side of continuing effective treatments. We have conducted retrospective surveys of the experience of weight loss during both beta-blocker and ACE inhibitor trials[3] and concluded that these treatments appear to partially prevent the development of cachexia. For this reason we would therefore recommend that the treatments be continued even in patients with heart failure who develop cachexia. We would like to have specific trials of cardiac cachexia (heart failure with a complication of cachexia) but it would be unethical to randomize patients in this situation not to receive these proven therapies. For cardiac transplantation we also have evidence that cachexia per se does not appear to prevent a beneficial outcome, if anything there is an apparent added improvement in prognosis (derived by the finding that low body weight heart failure patients have a worse survival than normal body weight patients, whereas post transplant their survival experience is similar) suggesting that cachectic heart failure patients may actually gain more from transplantation (unpublished observations). By the same reasoning, unless we have convincing evidence that cachexia changes the response of patients to their underlying disease then the treatment for the underlying disease should continue.

A second issue is what is required of treatments that are given on the basis of cachexia itself. These may be either treatments that appear to work for all cachexias or treatments that have been shown to be effective in cachexia complicating one or more specific diseases. The important point is that they are not given on the basis of the underlying disease but because of the cachexia. This raises the question of why we are treating the cachexia. Referring to Table 14.1 and noting that cachexia is a condition with a very high mortality rate and multiple disabling symptoms, we should conclude that either a reduction in mortality or an improvement in cachexia-related symptoms (e.g., weakness, fatigue, dyspnoea) would be valid treatment targets. Developing treatments that modify the disease process is much more problematic as the pathophysiology of cachexia is so complex and so poorly understood that we do not know which aspect of the pathophysiology should be targeted. Cachexia is associated with weight loss, immune activation, growth hormone resistance, insulin resistance, cytokine activation, neurohormonal activation, reflex alterations, muscle apoptosis, and gross changes in many body systems and organs. We do not know which of these alterations are most important to the progression of cachexia, and we cannot recommend treatments purely on the basis of a change in a laboratory parameter that may not be critical to the disease process. As a result we are left looking for improvements in survival or in symptoms that are of significant importance to the patients. The first requires very large trials. Some reduction of trial size can be achieved with the use of a composite trial end-point that includes elements of major importance and which may be more frequent than death. This is not a clear-cut choice, for the other end-point must be clearly disease related, must be of major importance to the patient, and must be close to death in terms of its importance to the patient and the health care system. For many heart failure trials the composite of death or hospitalization is used. Hospitalization is expensive, related to worsening disease, and of major importance to the patient so it has become acceptable as a composite with mortality. A more minor end-point such as stroke or myocardial infarction (which may only be demonstrated by a change in a laboratory test such as MRI scan or ECG) is not accepted as being of grave enough importance to be combined with death in a composite end-point and still have the result considered a major mortality and morbidity (M + M) outcome. There is as yet no consensus as to which if any other end-points could be added to mortality as an acceptable composite end-point in an M + M trial. For symptoms the choice will be dictated by which symptoms matter to the patient and are sufficiently robust to be reproducible in measurement, accurate in elucidation, and modifiable in response to treatment. Mention should specifically be made concerning weight loss, or similar measures such as body weight, body mass index, or even component specific measures such as skeletal muscle mass. These of themselves are not markers of the disease process, nor disabling symptoms in their own right. We should accept therefore that treatments proven only to increase weight should not necessarily be recommended to cachectic patients for they may be of limited true value to our patients unless the weight gain necessarily or reliably is associated with improved symptoms or survival, a level of proof that is, probably in our present state of knowledge, unavailable.

14.4 Types of Indications

As reviewed in the preceding section a treatment for cachexia needs to improve a meaningful end-point. It would only be approved for that indication, that is, for treatment of a defined set of patients with the express purpose of producing the effect demonstrated in the trial(s). Meaningful end-points might not include weight gain if the patients studied did not necessarily need weight gain. An irrelevantly small weight gain, that is, 2 kg or less or a gain in a patient population who would perceive no self-evident benefit from a weight gain would not be sufficient. If the patient group was demonstrably complaining of muscle weakness or inability to exercise or perform certain tasks then an indication might be given specifically to improve the patients' ability to perform these tasks or to ameliorate the symptoms of which they complain. There may therefore be a variety of functional or symptom based end-points that could form the basis of an approvable trial end-point. These include certain exercise or functional end-points (see Table 14.2) and a variety of

TABLE 14.2
Functional End-Points of Potential Relevance
to Cachexia

1. Maximal exercise testing (time, peak oxygen uptake etc.)
2. Sub-maximal exercise (corridor walk tests, endurance etc.)
3. Muscle-specific tests (peak strength, fatigability etc.)

symptom assessment scores that concentrate on symptoms of significance to these patients. Measures of cachexia development or regression such as body weight, muscle bulk, muscle mass, fat mass, protein or fat turnover, total nitrogen etc. may be useful in determining proof of principle that an agent or intervention might be effective in cachexia, but they alone will not produce in general an approvable effect, as changes in these parameters are not in themselves of use to patients. These changes need to lead to changes in symptoms, quality of life, or prognosis.

14.5 Design of a Trial

A treatment for cachexia will not be established with a single trial. Any new chemical entity will require preclinical toxicology, animal models of efficacy, phase 1 dosage, and pharmacodynamic and kinetic studies even before treatment is given to a cachectic patient. For established agents with a licensed indication for another disease, trials can start immediately with cachectic patients but some phase 2 studies will be needed to establish safe and effective dosages and to establish proof of principle that the treatment can benefit at least an aspect of the cachectic process that might be beneficial. These studies could be small and single centered as long as some effect is demonstrable and safety and acceptable tolerance can also be demonstrated. End-points at this stage could include weight gain for we need an end-point that is relatively easy to achieve with acceptable patient numbers. Once the possibility of useful effect can be established definitive phase 3 trials can be planned. In general for approval two independent trials each with a clinical useful and important end-point (see above) need to be achieved. In rare circumstances (overwhelming statistical certainty) or for certain "orphan" indications where the number of sufferers is very limited and further funded trials very unlikely, approval might be obtained with one outcome trial.

14.6 Extrapolation to all Cachectic Patients

Cachexia is a complication of many separate diseases. Much of the study of the pathophysiology has suggested that very similar processes complicate chronic disorders and lead to cachexia through substantially similar mechanisms. It is not unreasonable to suggest, therefore, that an effective treatment of AIDS-related cachexia has a good chance of being similarly effective in cancer-related cachexia. It is not certain if the safety will be similar for there may be particular safety issues relating to the underlying diseases such as fluid retaining effects of anabolic steroids in heart failure-related cachexia or cell growth promoting effects of growth factors or hormones in cancer-related cachexia.[4] Does this mean that an effective treatment for cachexia needs to be established separately for each cachectic condition? This would be an extremely onerous undertaking. A logical compromise would be to power a trial program to establish efficacy in a range of the major cachexia-related conditions and to establish reasonable safety data in each. Any major area of apparently discordant effect or specific safety concerns would then need to be subjected to specific investigation. The issue of age-related cachexia or idiopathic cachexia would probably justify specific trials in these conditions alone as their pathogenesis may be unique to these settings.

14.7 THE FUTURE

Whilst at present we have few proven therapies for any cachexia, in the future the development of gene, growth factor, and cell therapy raises enormous potential to inhibit and even reverse cachexia.[5, 6] This may have major implications for the mortality rate of the common antecedents of cachexia and for the health and vitality of the population in an ageing society.

REFERENCES

1. Anker, S.D. Cachexia: Time to receive more attention. *Int. J. Cardiol.* 2002; 85: 5–6.
2. McCarthy, D.O. Rethinking nutritional support for persons with cancer cachexia. *Biol. Res. Nurs.* 2003; 5: 3–17.
3. Anker, S.D., Negassa, A., Coats, A.J., Afzal, R., Poole-Wilson, P.A., Cohn, J.N., and Yusuf, S. Prognostic importance of weight loss in chronic heart failure and the effect of treatment with angiotensin-converting-enzyme inhibitors: An observational study. *Lancet* 2003; 361: 1077–1083.
4. Tassinari, D., Fochessati, F., Panzini, I., Poggi B., Sartori, S., and Ravaioli, A. Rapid progression of advanced "hormone-resistant" prostate cancer during palliative treatment with progestins for cancer cachexia. *J. Pain Symptom Manage.* 2003; 25: 481–484.
5. Rosenthal, N, and Musaro, A. Gene therapy for cardiac cachexia? *Int. J. Cardiol.* 2002; 85: 185–191.
6. Sturlan, S., Beinhauer, B.G., Oberhuber, G., Huang, L., Aasen, A.O., and Rogy, M.A. *In vivo* gene transfer of murine interleukin-4 inhibits colon-26-mediated cancer cachexia in mice. *Anticancer Res.* 2002; 22: 2547–2554.

15 Nutritional Treatment of Cachexia

Claudine Falconnier and Ulrich Keller

Contents

Summary

Nutritional treatment of malnourished and cachectic patients is of major importance to improve their well-being and to diminish somatic and psychological complications. Nevertheless, malnutrition is often diagnosed late during an illness. The essential basis to treat malnutrition adequately is the understanding of associated physiological and pathophysiological abnormalities.

Starvation and disease-related malnutrition share similarities but also contrasting features. During starvation, physiological adaptations, for example, protein sparing mechanisms assure a prolonged survival, whereas stress metabolism in severe illnesses results in mobilization of metabolic substrates, accelerated catabolism, and finally cachexia.

Assessment of the nutritional risk using a nutrition risk score system therefore includes two parameters — one related to the degree of nutritional depletion and rated according to the amount of weight loss per month, and the second parameter represents a score of the severity of the illness.

Other measurements aim to assess changes in body composition, like triceps skinfold thickness, arm muscle circumference, and bioelectric impedance analysis. These latter parameters are necessary for a systematic assessment but not necessarily required for the clinical management of the illness. Laboratory parameters are used for the evaluation of specific deficits, for example, micronutrient deficiencies or to determine immune competence.

An algorithm for the selection of appropriate nutritional treatments is presented, using a step-by-step increase in interventions according to the individual nutritional problems, starting with oral supplements, followed by tube feeding and finally parenteral nutrition, all levels having their own pros and cons. Enteral nutrition is always preferable to parenteral nutrition when the gut can be used since it is more physiological, less expensive, and associated with less complications than the latter. Special attention has to be paid to the refeeding syndrome upon resumption of feeding in severly malnourished patients. This potentially lethal condition is characterized by a severe electrolyte and fluid shift upon too aggressive nutrition after severe and long-lasting undernutrition.

Specific nutritional treatments in certain types of malnutrition (e.g., fish oil with polyunsaturated long-chain omega-3 fatty acids in cancer cachexia), and branched-chain and other amino acids, for example, in severely ill patients with sepsis, exert beneficial effects beyond that of providing energy and metabolic substrates. Few data of controlled trials with combinations of drugs and nutritional substrates in malnourished patients have been reported previously.

15.1 ROLE OF UNDERNUTRITION VS. UNDERLYING DISEASE IN THE PATHOGENESIS AND TREATMENT OF CACHEXIA

Nutritional deprivation due to starvation and disease-related malnutrition share many similarities but are characterized by different metabolic alterations. Adaptive processes during starvation of healthy subjects are physiological mechanisms which result in prolonged survival. In contrast, stress-induced metabolism represents the attempt of the individual to fight the consequences of acute trauma or severe disease and results in, for example, the mobilization of metabolic substrates. Long-term stress ends up in uncontrolled catabolism, malnutrition, and finally in cachexia.

Acute stress events such as major surgery, trauma, or sepsis in previously healthy people decrease transiently the mobilization of endogenous energy substrates, and energy production is low for about 1 to 2 days (Ebb phase). Later on, circulating stress hormones and cytokines (particularly during severe infection) induce a hypermetabolic state with increased gluconeogenesis, proteolysis, lipolysis, and decreased ketogenesis (Flow phase). These alterations are associated with anorexia and often inadequate nutritional intake. However, even generous provision of nutrition can only partially reduce this catabolic state. In contrast to caloric deprivation during starvation, plasma glucose concentrations are slightly elevated, and there is a state of insulin resistance. The increased plasma insulin levels diminish lipolysis and ketogenesis, because these processes are relatively insulin-sensitive; these effects explain, at least in part, that protein saving mechanisms are impaired. This is a cardinal difference between metabolism in health during starvation on one hand, and stress metabolism in severe acute and chronic diseases on the other.

The term "cachexia" is derived from Greek ("kakos hexis") and means: "wasting," "state of weakness," that is, an advanced state of malnutrition. The term is not clearly defined in the literature although it is generally understood that it represents a condition of disease-related severe loss of body mass, beyond the condition of "malnutrition" for which there exist established definitions, and where loss of body mass may not readily be apparent.

Cachexia is a frequent complication in patients suffering from severe illnesses such as advanced cancer, chronic heart failure, respiratory failure, gastrointestinal disease, or infections. It is characterized by anorexia, wasting, and other signs of malnutrition, which combined with psychological and social sequelae leads to deterioration of a patient's quality of life. Various factors play a role in the development of cachexia; they include the release of cytokines and stress hormones (Cederholm et al., 1997), which result in whole body catabolism, anorexia, and malnutrition-related deficits of micro- or macronutrients.

Prevention and treatment is primarily oriented toward the underlying disease, if possible. The effectiveness of this approach is illustrated by the observation that patients with AIDS in the first years of the epidemic often suffered from severe cachexia; nowadays, due to the availability of effective antiretroviral therapy, these patients are rarely cachectic when these treatments are prescribed.

Nonpharmacological treatment of malnutrition consists of nutritional support using adapted oral nutrition, sip feedings, enteral (tube) or parenteral nutrition. Other important aspects are psychological support, treatment of associated symptoms such as pain, and prevention of complications.

It is surprising to note that despite its importance, relatively few data in the literature exist about the pathophysiology and the clinical management of cachexia.

Specifically, there is a substantial shortage of randomized, controlled trials investigating the effect of nutritional support and treatment in ameliorating morbidity or mortality. Ethical concerns make it difficult to randomize malnourished patients to control groups receiving no adequate nutrition. And what should be the endpoint of such trials? Is it mortality and morbidity, body weight, immune defense, or quality of life? Furthermore, it is difficult to perform controlled clinical trials with nutritional interventions, and there is a modest interest by pharmaceutical companies to support such expensive trials. Nevertheless, adequate nutrition is a fundamental requirement of human well-being (Barac-Nieto et al. 1979). Insufficient supply of nutrients has a detrimental impact on development and health. Therefore, it is obvious that inadequate nutrition has negative effects on severely ill patients, even if they suffer from a severe illness with limited life expectancy (Lligmore et al. 1997).

15.2 UNDERNUTRITION IN HOSPITALS: THE "FOOD AND NUTRITIONAL CARE IN HOSPITALS" REPORT OF THE COUNCIL OF EUROPE

In 1999, an expert group of the Council of Europe collected information regarding nutrition programs in hospitals. The background of this initiative was the observation that disease-related undernutrition has been frequently observed in European hospitals but often neglected and not treated. The aim was to review the current hospital food provision systems in European hospitals and the awareness and activities to combat the problem of malnutrition, to uncover deficiencies, and to issue recommendations. The collected data of surveys showed that the use of nutritional risk screening and nutritional support was usually sparse. A major problem was the frequent lack of assignment of responsibilities for these tasks. The different groups of health care professionals did not collaborate to achieve these tasks. Since all patients have the right to expect that their nutritional needs will be fulfilled during a hospitalization, the duties of different staff categories and the hospital management for nutritional care and support should be clearly defined.

Nutritional support for patients at nutritional risk was described as often insufficient, and it was provided only in severely undernourished patients.

Hospital physicians were reported to be often poorly trained in clinical nutrition, medical schools provided only few lectures on nutrition-related topics. Similarly, nurses and particularly hospital management staff were poorly trained in clinical nutrition.

Dieticians seem to receive the most up-to-date training: however, their responsibilities are in practice very varied, and many hospitals have only a few dieticians available.

The report concluded that a general improvement in the educational level of all staff groups was needed.

15.2.1 Assessment of the Nutritional State Using a Nutrition Risk Score

The deterioration of the nutritional state is linked to several factors involved with food supply and intake, apart from the disease process itself (Allison, 1992; Kondrup, 2001).

The lack of a proper screening tool is seen as one of the major reasons for not initiating nutritional support (Rasmussen et al., 1999). In 2003, a useful screening tool was introduced and validated as a means to score the risk for malnutrition (Kondrup et al., 2003) (Table 15.1).

This risk score was recommended by the ad hoc *Group of the Council of Europe on Undernutrition in Hospitals*. It was developed with the assumption that the risk for malnutrition and thus the indication for nutritional support was expressed by two main elements, the degree of undernutrition and the increase in nutritional requirements due to the disease. Thus, severe undernutrition or severe disease by themselves or in varying combinations may indicate a need for nutritional support. The advantage of the score is that it has been validated against other parameters of malnutrition and that it is rapidly and easily employed at the bedside.

However, being a score for risk of malnutrition it is not equivalent to a specific assessment of malnutrition. The latter may include the patient's history, physical examination, anthropometric measurements, laboratory data, and assessment of immunocompetence. Anthropometric assessment like triceps skinfold thickness and arm muscle circumference measurements have frequently been used to assess muscle and fat mass. However, these measurements are not very accurate and are rarely performed in the clinical setting.

Bioelectric impedance analysis (BIA) is a simple, noninvasive bedside method for the assessment of body fat, fat-free mass, and total body water. BIA is performed by measuring the electric conductivity of a weak current between electrodes placed on the dorsal surfaces of hands and feet. The measurement reflects differences in the impedance to an electric current, which is greatest through fat and least through water. Lean body mass can be calculated by subtracting fat mass from body weight or by dividing total body water by 0.73 (Garrow, 1982; Wanke et al., 2002).

BIA is most useful in assessing body fat and fat-free mass in stable patients and in those who suffer from conditions leading to relative starvation. Loss of body cell mass measured by BIA has been closely correlated with mortality in patients with AIDS. However, BIA measurements may be confounded in patients receiving protease inhibitor therapy if they suffer from lipodystrophy with associated redistribution of interscapular, abdominal, and breast fat (Currier et al., 2002; Wanke et al., 2002).

Lowered serum albumin concentrations are a hallmark of protein calorie malnutrition, however, multiple factors beside malnutrition influence the concentrations of serum proteins; therefore, serum albumin has a low specificity but a high sensitivity for protein malnutrition (Puchstein et al., 1989). The responsiveness to treatment is slow due to a half-life of 2 to 3 weeks of serum albumin.

A normal serum albumin level in a well-hydrated patient is inconsistent with protein malnutrition. In contrast, a low serum albumin level could reflect protein malnutrition, but it may also occur when the plasma volume is increased in an overhydrated patient, or when there is chronic liver, renal, or cardiopulmonary failure. The serum albumin level decreases during acute stress of surgery, sepsis, or during any other acute inflammatory illness due to increased circulating extracellular volume.

Most hospitals use laboratory parameters such as serum albumin, which also represents an indirect measurement of the severity of illness (Allison et al., 2001).

TABLE 15.1
Screening of Nutritional Risk at the Bedside

Impaired nutritional state			Severity of disease (stress metabolism)		
Absent	Score 0	Normal nutritional state	Absent	Score 0	Normal nutritional requirements
Mild	Score 1	Weight loss >5% in 3 months Or Food intake below 50–75% of normal requirement in preceding week	Mild	Score 1	Hip fracture Chronic patients, particularly with acute complications Cirrhosis (Cabre et al., 1990) COPD (Schols et al., 1995) Chronic hemodialysis, diabetes, Oncology
Moderate	Score 2	Weight loss >5% in 2 months Or BMI 18.5–20.5 + impaired general condition Or Food intake 25–50% of normal requirement in preceding week	Moderate	Score 2	Major abdominal surgery (Beier-Holgersen and Boesby, 1996; Keele et al., 1997) Stroke (Gariballa et al., 1998) Severe pneumonia, hematologic malignancy
Severe	Score 3	Weight loss >5% in 1 month Or BMI <18.5 + impaired general condition (Keys et al., 1950) Or Food intake 0–25% of normal requirement in preceding week	Severe	Score 3	Head injury (Grahm et al., 1989; Rapp et al., 1983) Bone marrow transplantation (Weisdorf et al., 1987) Intensive care patients (APACHE 10)

To calculate the total score:
1. Find score (0–3) for Impaired nutritional state (only one: choose the variable with highest score) and Severity of disease (stress metabolism, i.e., increased nutritional requirements).
2. Add the two scores (≥total score).
3. If age ≥70 years: add 1 to the total score to correct for frailty of elderly.
4. If age-corrected total score ≥3: start nutritional support.
Source: Kondrup, J., Allison, S.P., Elia, M., Vellas, B., and Plauth, M. (2003). *Clin. Nutr. 22*, 415–421. With permission.

It helps the clinician to identify the patients with the greatest risk of complications of malnutrition. These patients usually require aggressive and closely monitored nutritional interventions (Fuhrman et al., 2004).

Protein malnutrition is often associated with low serum levels of vitamin A, zinc, and magnesium. Abnormal digestion and absorption of dietary fats are associated with deficiencies of the fat soluble vitamins A, D, E, and K, whereas intestinal mucosal malabsorption is commonly associated with additional deficiencies of iron and folate.

15.3 PRINCIPLES OF NUTRITIONAL THERAPY IN MALNOURISHED PATIENTS

The presence of malnutrition is associated with increased morbidity (Chandra, 1983; Windsor and Hill, 1988) and mortality (Herrmann et al., 1992). The difficult question to answer is whether nutritional support improves the well-being or the prognosis of the individual patient. While it is obvious that all human beings in the very terminal phase of severe illnesses undergo a certain period of malnutrition, it is also clear that insufficient provision of vital nutrients leads to adverse effects in patients, which negatively affect the patient's quality of life, host defense, and overall prognosis. Therefore, a clinical judgment is always necessary to assess the possible advantages of improved nutrition. On the other hand the general condition of the patient, particularly the prognosis and the patient's will, have to be considered.

Nutritional support is particularly efficient in patients with inadequate intake or absorption due to anorexia or to gastrointestinal illnesses. But it is also a step-by-step escalation to cover the nutritional needs, as outlined in Figure 15.1. It begins usually with adapted oral nutrition, supported by specific nutritional supplements. When this approach proves to be insufficient, enteral and parenteral formulations are able to deliver all essential nutrients, and many patients who cannot eat normally, live sustainable lives over years while being nourished by one of these two routes.

15.3.1 Fluid Requirements

Fluid requirements can be determined by adding the normal requirement (~35 ml/kg of body weight) to any abnormal loss. Since abnormal loss of enteric fluid implies significant mineral losses, extra amounts of these nutrients, as well as fluid, must be added.

15.3.2 Energy Requirements and Energy Substrates

Malnourished patients may be normo- or hypermetabolic. In the absence of metabolic stress, they may expend approximately 85 kJ/kg (20 kcal/kg) per day, but

30–50% more calories are needed both for tissue repletion and because the metabolic rate increases with refeeding. In contrast, a highly stressed patient with sepsis or trauma may expend up to 165 kJ/kg (35–40 kcal/kg) per day, with a significant proportion of the calories being derived from protein breakdown gluconeogenesis and from catecholamine-induced lipolysis.

Carbohydrates usually provide 45 to 60% of all calories. They are available in the form of disaccharides (sucrose) and oligosaccharides in enteral formulas for tube feeding to maximize absorption while limiting the intraluminal osmolar load, and fructooligosaccharides have been shown to promote normal gut flora. Vegetable oils which are relatively rich in polyunsaturated fats are used in most enteral formulas for tube feeding because of their fluidity. Median chain triglycerides may be added because of their more efficient absorption. Supplying omega-3 long-chain polyunsaturated fats as contained in fish oil, in addition to polyunsaturated vegetable fats may blunt the catabolic response to burn, injury, trauma, and radiation by reducing the synthesis of prostaglandins which enhance the inflammatory response (see also Section 15.4.1).

15.3.3 Protein or Amino Acid Requirements

The recommended dietary protein allowance of 0.8 g/kg per day is adequate for nonstressed patients. Catabolic patients may require up to 1.5 g/kg per day of protein to diminish negative nitrogen balance. The adequacy of protein support can be assessed in a stable patient by analyzing the protein balance:

$$\text{Protein balance} = \text{protein intake} - \text{protein loss}$$

where protein loss (in g) $= [(24\text{-h urinary urea nitrogen (g)} + 4^*) \times 6.25]$.
*This number estimates nitrogen losses in urine other than urea and in stool.

Due to the high but slow turnover of the urea pool this measurement is unsuitable in patients with rapid changes of protein turnover, for example, during acute protein losses.

15.3.4 Mineral and Vitamin Requirements

Adjustments of usual electrolyte needs (recommended daily allowances, RDA) are necessary when the patient has significant gastrointestinal losses or renal failure. Requirements of some minerals and vitamins are higher when administered parenterally rather than enterally for several reasons: (1) many micronutrients delivered into the systemic rather than the portal circulation are not processed by the liver and are excreted by the kidney; (2) patients with bowel disease often have increased enteric losses of sodium, potassium, chloride, and bicarbonate; malabsorption of divalent cations, fat-soluble vitamins, and vitamin B_{12}.

15.3.5 Regular Food

To prevent or at least to limit the development of malnutrition of patients, they should receive appropriate nutritional advice. Specific counseling may ameliorate problems such as loss of appetite for specific foods, nausea, gustatory disturbances or dry mouth. A study by Dupertuis et al. (2003) at the Geneva University Hospital demonstrated in 1707 patients of medical and surgical wards, that 57% did not eat enough calories or protein; on the other hand, daily meals provided on average an excess of 41% of calories and of 15% of proteins. These data indicated that paradoxically, both food oversupply and insufficient intake were extremely common, calling for a more patient-adapted nutritional supply in hospitals.

Enteral liquid diets and commercially available balanced nutritional supplements offer a good alternative to regular diet, or may be used as protein-rich supplements.

Patients at risk of malnutrition have to be encouraged to ingest as much energy as possible and they have to be counseled on the ways to integrate the necessary components into their daily diet.

When malnutrition is considered likely, the first step is to define the problem:

1. Dietary advice → dietician
2. Dysphagia → logopedic evaluation
3. Dental problems → dentist
4. Eating aids → occupational therapy
5. Social problems → include relatives, neighbors, or friends
6. Complex nutritional problem → evaluation of supplementation and/or enteral and parenteral feeding by a nutrition specialist

15.3.6 Sip Feedings (Balanced Nutritional Supplements)

Commercially produced sip feedings are a valuable addition to or substitute of the ordinary hospital food provided by the kitchen. The products are defined in their composition, they are nutritionally complete, include macro- and micronutrients and are available in a wide range of flavors (sweet, fruity, salty, spicy). Most products are easily digested and are free of lactose and gluten; only few contain fiber. They are especially useful to meet the increased requirements of proteins; they should be given between main meals, to avoid appetite suppression. According to a Cochrane analysis, the use of liquid nutritional supplements resulted in short-term improvement of several markers of malnutrition (Baldwin et al., 2001).

15.3.7 Enteral Nutrition (Tube Feeding)

Nutritional support by enteral or parenteral nutrition is appropriate for patients who cannot take adequate food by mouth and who consequently are at risk of malnutrition. This may lead to complications, such as susceptibility to infections, weakness, and immobility, which result in delayed recovery from illnesses and increased mortality. Enteral nutrition is always preferable to parenteral if possible because it has fewer complications, is more physiological, and less expensive than the latter. Patients with intestinal disorders such as short bowel syndrome or chronic intestinal obstruction (Messing et al., 1995), neurological impairment, or oropharyngeal dysfunction, and premature infants (Driscoll et al., 1972) need long-term nutritional support to prevent death from starvation.

Enteral nutrition is indicated when the gastrointestinal tract is functioning but swallowing or mastication is compromised by disease, or if it is necessary to pass an obstructed area in the upper gastrointestinal tract. Like oral feeding, it relies on the absorption of nutrients by the intestinal mucosa. Specific solutions are used to provide complete nutrition or single nutrients that are not easily tolerated orally.

The types of tube feeding have different indications, advantages, and complications. The solution can be delivered at a slow, continuous or intermittent rate to optimize the intestinal absorptive capacity when it is compromised by chemo- and radiotherapy, short bowel, or partial obstruction (Mercadante, 1998; Shike, 1996). A nasogastric tube is easy to insert but is intended for short-term use (less than 3 weeks) because it may cause pharyngeal ulceration, dysphagia, sinusitis, and problems with reinsertion. It requires frequent flushing and is often considered as an incommoding foreign body. The risk of aspiration pneumonia is relatively high in patients with delayed gastric emptying. Frequent tube aspiration is necessary to prevent this complication (Mercadante, 1996). With prolonged use it becomes inconvenient and uncomfortable.

Gastrostomy and jejunostomy have been proven to be efficient methods of providing long-term enteral nutrition. Gastric and intrajejunal tubes are better tolerated than nasogastric ones and improve well-being. Percutaneous endoscopic gastrostomy (PEG) has the advantage of being hidden under the patient's clothes and is intended for long-term use; it may be unsuitable in patients with gastric disorders.

A jejunostomy tube can overcome this limitation, but surgical placement is necessary. Diarrhea is sometimes associated with enteral feeding. This complication is attributed to the rapid initiation of feeding, to the high osmolality of the mixture, and to the concomitant antibiotic therapy (Keohane et al., 1984).

Patients with impaired intestinal absorptive capacity may benefit from tube feeding solutions which require minimal digestion; they contain simple sugars, amino acids or oligopeptides, and medium-chain triglycerides. However, such solutions are hypertonic, and they may cause diarrhea; they may be diluted to diminish the osmolar load (Mercandate, 1997).

Table 15.2 gives an overview of standard and modified enteral formulas.

15.3.8 Parenteral Nutrition

Total parenteral nutrition consists of intravenous administration of all necessary substrates (carbohydrates, lipids, and amino acids as well as vitamins and minerals), usually mixed in a container and administered continuously over 24 h by a pump. The timing of the infusion can be flexible and adjusted to any activities

TABLE 15.2
Standard and Modified Formulas for Enteral Nutrition

Composition characteristics	Clinical indications
Standard enteral formula[a]	
1. Caloric density approximately 1 kcal/ml(+)	Suitable for most patients requiring tube feeding
2. Protein 10–20% of calories, caseinates, soy, lactalbumin	
3. Fat 20–30% of calories, corn, soy, safflower oils	
4. CHO 50–60% calories, hydrolyzed corn starch, maltodextrin, sucrose	
5. Recommended daily intake of all minerals and vitamins in a flask containing >1500 kcal/d	
6. Osmolality (mosmol/kg): 300	
Modified enteral formulas[a]	
1. Caloric density 1.5–2 kcal/ml (+)	1. Fluid-restricted and malnourished patients
2. a. High protein (20 cal-percent) (+)	2. a. Protein malnutrition and wound healing
b. Hydrolyzed protein to small peptides (++)	b. Protein digestion/absorption or allergy
c. Enriched with glutamine, arginine, *S*-containing amino acids, nucleotides (+++)	c. Severely immunocompromised patients
d. Enriched with branched chain amino acids, aromatic amino acids (+++)	d. Liver failure patients intolerant of 0.8 g/kg per day of regular protein
e. Low protein of high biological value	e. Renal failure patients
3. Fats with ω3 (fish oil) and ω6 fatty acids (+++)	3. Immunocompromised patients and autoimmune disorders

[a] Cost: + inexpensive; ++ moderately expensive; +++ very expensive.
Note: CHO, carbohydrate; ω3 or ω6, polyunsaturated fat with first double bond at carbon 3 (fish oils) or carbon 6 (vegetable oils).
Source: Adapted from Harrison's Online, 2004, The McGraw-Hill Companies.

planned and to the patient's lifestyle. Due to the hypertonicity of the solution it has to be infused into a central vein. Parenteral nutrition has been shown to increase body weight, to improve nitrogen balance and to reverse several markers of malnutrition (Jeejeebhoy, 2001). Parenteral nutrition should be reserved for patients with impaired gastrointestinal tract function due to for example, obstruction, surgery, high output fistula, or gastrointestinal diseases.

Complications due to parenteral nutrition are due to the central vein catheter (mechanical and infectious complications), or to metabolic derangements. Fluid overload can cause congestive heart failure, particularly in elderly and debilitated patients. Glucose overload may produce hyperglycemia and osmotic diuresis or stimulate insulin secretion, which in turn promotes shifts of potassium and phosphate from the extracellular to the intracellular space. Such shifts are most dangerous in malnourished patients with depleted potassium and phosphate stores;

they may cause arrhythmias, cardiopulmonary dysfunction and neurologic symptoms. To avoid these problems, parenteral nutrition should be started slowly in cachectic patients, and monitoring of serum glucose and electrolytes is essential. Glucose supply is increased gradually as the patient develops tolerance to the high glucose load. Late metabolic complications of parenteral nutrition include cholestatic liver disease, with bile sludging and gallstone formation. The exact cause of the liver disease is not understood, but lack of enteral stimulation of bile flow and defective sulfur amino acid metabolism and cholesterol solubilization appear to play a role. Cholestasis is less likely to occur if at least some enteral feeding is maintained. Parenteral nutrition induces hypercalciuria and frequently results in negative calcium balance and osteopenia. Once patients on long-term parenteral nutrition change to sustained anabolism, deficiencies of micronutrients such as essential fatty acids, trace minerals, and vitamins may develop unless they are supplied in adequate amounts.

The decision tree demonstrates the indications for oral supplements, enteral, and parenteral nutrition (Figure 15. 1).

15.3.9 Refeeding Syndrome

Patients with cachexia undergoing aggressive refeeding are at risk for the refeeding syndrome during the first two to three weeks (Crook et al., 2001; Solomon and Kirby, 1990). This potentially lethal condition can be defined as severe electrolyte and fluid shifts associated with metabolic abnormalities. It bears a significant risk for morbidity and mortality. The clinical features are fluid-balance disturbances, abnormal glucose metabolism, hypophosphatemia, hypomagnesiemia and hypokalemia. In addition, thiamine deficiency may occur.

Figure 15.2 illustrates the main consequences of severe malnutrition and of rapid refeeding. During refeeding there is a shift from fat to carbohydrate oxidation. Administration of glucose stimulates the release of insulin, causing increased cellular uptake of glucose, phosphate, potassium, magnesium and water, and protein synthesis (Figure 15.2).

Fluid retention: Carbohydrate refeeding can reduce water and sodium excretion, resulting in an expansion of the extracellular fluid compartment and weight gain. Patients are particularly at risk if sodium intake is increased. Fluid intolerance may produce fluid overload, edema, and heart failure. This may particularly occur in patients with compromised heart function or myocardial atrophy due to prolonged and severe malnutrition.

Abnormal glucose and lipid metabolism: Large amounts of intravenous glucose administered to cachectic patients which are relatively insulin resistant due to prolonged starvation may cause severe and prolonged hyperglycemia, resulting in hyperosmolar nonketotic coma, ketoacidosis and metabolic acidosis, osmotic diuresis, and dehydration (Crook et al., 2001). Large amounts of glucose increase carbon dioxide production and may provoke hypercapnia and respiratory failure. Glucose can be

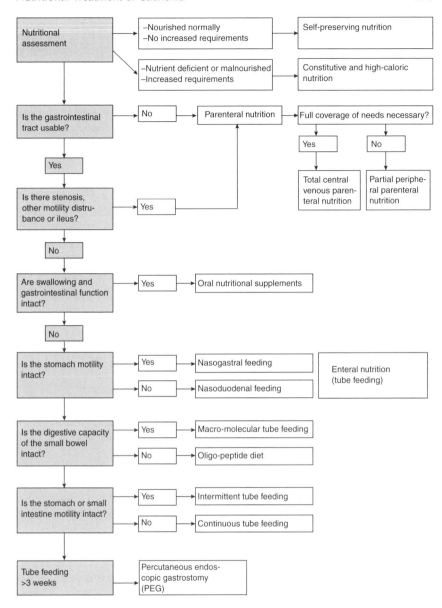

Figure 15.1 Clinical decision tree for various modes of oral, enteral, and parenteral nutrition.

converted to fat via lipogenesis, which can evoke hypertriglyceridemia, fatty liver, and abnormal liver function tests.

Hypophosphatemia: One of the dominant features of the refeeding syndrome is a low serum phosphate level. In catabolic states there is a loss

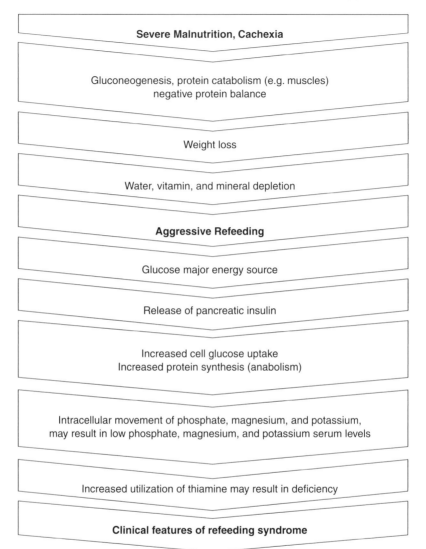

FIGURE 15.2   Metabolic sequelae of severe malnutrition and refeeding.

of phosphate from the intracellular compartment. Subsequent to high intakes of calories and carbohydrates there is an insulin-mediated shift of glucose and phosphate into cells. Low phosphate levels lead to impaired neuromuscular function (paraesthesia, seizures, and perturbed mental state). Involvement of ventilatory muscles may cause hypoventilation and eventually respiratory failure. Severe hypophosphatemia has been described to produce rhabdomyolysis, thrombocytopenia, impaired blood clotting and deficient leukocyte function as well as mental disturbances, and eventually coma.

Hypomagnesiemia and hypokalemia: The refeeding syndrome is associated with hypomagnesiemia and hypokalemia (for the same reasons as hypophosphatemia). This may lead to cardiac arrhythmias and even arrest, to general muscle weakness and to paralysis, confusion, rhabdomyolysis, and respiratory depression.

Thiamine (vitamin B_1) deficiency: This is the principal vitamin deficiency in the refeeding syndrome owing to its rapid consumption during glycolysis. Thiamine deficiency impairs glucose metabolism (pyruvate dehydrogenase reaction) with subsequent lactic acidosis. Thiamine deficiency can result in Wernicke's encephalopathy (ocular disturbances, confusion, ataxia, and coma) or Korsakov's syndrome (loss of short term memory and confabulation) (Crook et al., 2001). Provision of thiamine with refeeding reduces symptoms of postfeeding thiamine deficiency (Solomon and Kirby, 1990). Administration of 50 to 250 mg of thiamine at least 30 min before the start of refeeding is recommended (Crook et al., 2001).

Before refeeding, electrolyte disorders should be corrected, and the circulatory volume should be carefully restored. This may delay the administration of complete nutrition but is usually accomplished within 12 to 24 h.

Caloric repletion should be slow at a rate of approximately 20 kcal/kg per day or 1000 kcal per day initially. However this rate may not meet the patients' fluid, sodium, potassium, or vitamin requirements unless these are specifically covered. The usual protein requirement is about 1.2 to 1.5 g/kg per day. Gradual introduction of calories, particularly over the first week of refeeding, may be prudent until the patient is metabolically stable (Solomon and Kirby, 1990).

Hypophosphatemia has to be treated if the serum level is less than 0.30 mmol/l or the patient is symptomatic. Supplementation should be given intravenously at 40 to 80 mmol per day together with magnesium (8 to 16 mmol per day) and potassium (80 to 120 mmol per day). These dosages should be adjusted according to monitored serum levels (Sobotka, 2004).

15.3.10   Costs and Economical Aspects

The costs of identifying patients at increased nutritional risk are low. Early and adequate nutritional support is associated with a beneficial cost-benefit relationship (Green, 1999). Regarding costs of parenteral nutrition, catheter-related complications should be considered. Up to 25% of patients treated in acute care facilities have central vein catheters placed, and 20 to 30% of these catheters are used for parenteral nutrition. The incidence of catheter-related infection varies from 2 to 30 per thousand catheter days, depending on the type of patients involved. If the handling of these lines is performed by specially trained teams, the complications can be reduced by almost 80%, impacting significantly on outcome and costs. The current trend from parenteral to enteral nutrition may also be cost-saving, but the true cost of complex enteral feeding is unknown. Home parenteral nutrition costs approximately half as much as similar treatment in the hospital, and home enteral nutrition costs significantly less (Cowl et al., 2000; Polderman and Girbes, 2002) than parenteral nutrition.

15.4 SPECIFIC NUTRITIONAL TREATMENTS OF CACHEXIA

15.4.1 Long-Chain Omega-3 Polyunsaturated Fatty Acids (Fish Oil)

Cachexia has been suggested to be mediated by cytokines, such as tumor necrosis factor-α, interleukins, and interferon. Such cytokines are produced by host tissues or by certain tumor cells. Catabolic mediators of lipolysis and proteolysis, such as lipid mobilizing factor (LMF) (Todorov et al., 1998) and proteolysis-inducing factor (PIF) were originally thought to be produced by the tumor only, although LMF has recently been shown to be produced by both white and brown adipose tissue, in addition to the tumor (Bing et al., 2004).

The polyunsaturated fatty acid eicosapentaenoic acid (EPA) has been shown to antagonize the action of the proteolysis-inducing factor (Tisdale, 2004; Todorov et al., 1996). It attenuates protein degradation by diminishing the ubiquitin–proteasome proteolytic pathway which is enhanced by the proteolysis-inducing factor. EPA has been investigated as an anticachectic agent, both as a triglyceride (Wigmore et al., 1996) and as the free acid (Wigmore et al., 2000) in patients with pancreatic cancer. It was found to slow down weight loss from an initial loss of 2 to 2.9 kg per month. When a stimulus for protein synthesis was provided in the form of an energy-rich nutritional supplement, EPA produced significant weight gain at 3 weeks (1 kg) and 7 weeks (2 kg) (Barber et al., 1999). Both performance status and appetite were significantly improved, and there was a fall in resting energy expenditure. Weight gain was attributed solely to an increase in lean body mass. After patients had consumed the fish-oil enriched supplement for 3 weeks, there was a significant fall in the production of IL-6, a rise in serum insulin and a fall in the cortisol-to-insulin ratio (Barber et al., 2001).

A randomized double-blind, placebo-controlled trial in weight losing cancer patients confirmed that if the supplement was taken in sufficient quantity there was a net gain of body weight and lean tissue, and quality of life improved (Fearon et al., 2003). However, direct comparison with protein-rich and energy-dense oral supplements demonstrated no clear advantage of n-3 fatty acids consumed at a mean dose of 1.4 cans per day. At baseline, patients had lost on average 17% of their pre-illness weight and were progressively losing weight at a rate of approximately 3.3 kg per month. When protein and energy dense oral supplements (whether or not enriched with n-3 fatty acids and antioxidants) were administered there was a stabilization of body weight. This suggests a net benefit from the use of such supplements in this particular group of patients. However, due to the advanced and malignant nature of their disease, only 55% of them reached the eight-week end point. The experimental supplement used was designed to deliver a high dose of n-3 fatty acids within a nutrient matrix to provide extra protein and energy and to allow net deposition of lean tissue. However, by incorporating EPA in a nutritional supplement, the dose of EPA was then linked to overall supplement intake, and a dose of 2.1 cans per day compared to 1.9 cans per day was associated with net gain of lean body mass (Barber et al., 1999). Although spontaneous meal intake was partially reduced, the net gain in total energy and protein intake appeared sufficient to allow the patients in both groups to achieve weight stability.

Previous trials of dietary advice and conventional (1 kcal/ml) oral supplements in weight stable cancer patients undergoing outpatient chemotherapy suggested no advantage compared to no dietary intervention (Evans et al., 1987; Ovesen et al., 1993).

15.4.2 Branched-Chain and Other Amino Acids

Accelerated peripheral muscle proteolysis as observed in cancer cachexia provides amino acids required for the synthesis of liver and tumor proteins. Administration of increased amounts of amino acids may theoretically serve as a protein-sparing metabolic fuel by providing substrates for both muscle proteins and gluconeogenesis (Inui, 2002; Argiles et al., 2001). However, this concept has not been proven in large intervention trials.

Branched-chain amino acids (leucine, isoleucine, and valine) have been used with the aim of improving nitrogen balance, particularly muscle protein metabolism. Earlier studies demonstrated that administration of branched-chain amino acid enriched total parenteral nutrition to cachectic cancer patients results in improved protein accretion and albumin synthesis (Tayek et al., 1986).

Branched-chain amino acids may also serve to counteract anorexia and cachexia by competing for tryptophane, the precursor of brain serotonin, at the blood-brain barrier, thereby preventing increased hypothalamic activity of serotonin.

It is known that increased plasma levels of tryptophane can lead to increased cerebrospinal fluid tryptophane concentrations and increased serotonin synthesis in patients with cancer (Laviano et al., 1996). Oral supplementation of branched-chain amino acids has been reported to decrease the severity of anorexia in cancer patients (Cangiano et al., 1996). The quality and the size of these studies appear to be too limited to alter nutritional therapy in cancer patients.

15.5 PERSPECTIVES

Balanced nutritional support improves body composition in patients with malnutrition and cachexia. A main goal of nutritional support is the restoration of lean body mass. Importantly, improved clinical function and well-being is not necessarily related to increased weight or size of body compartments because nutritional treatment can affect organ function faster than its size and mass (Lopes et al., 1982).

Therefore, in many patients with advanced stages of disease associated with malnutrition and cachexia, the effect of nutritional support on clinical symptoms and well-being is more important than for example, its effect on body composition or on mortality.

REFERENCES

Allison, S.P. (1992). The uses and limitations of nutritional support. The Arvid Wretlind Lecture given at the 14th ESPEN Congress in Vienna, 1992. *Clin. Nutr. 11*, 319–330.

Allison, S.P., Lobo, D.N., and Stanga, Z. (2001). The treatment of hypoalbuminaemia. *Clin. Nutr. 20*, 275–279.

Argiles, J.M., Meijsing, S.H., Pallares-Trujillo, J., Guirao, X., and Lopez-Soriano, F.J. (2001). Cancer cachexia: A therapeutic approach. *Med. Res. Rev. 21*, 83–101.

Baldwin, C., Parsons, T., and Logan, S. (2001). Dietary advice for illness-related malnutrition in adults. *Cochrane Database Syst. Rev.*, 2.

Barac-Nieto, M., Spurr, G.B., Lotero, H., Maksud, M.G., and Dahners, H.W. (1979). Body composition during nutritional repletion of severely undernourished men. *Am. J. Clin. Nutr. 32*, 981–991.

Barber, M.D., Fearon, K.C., Tisdale, M.J., McMillan, D.C., and Ross, J.A. (2001). Effect of a fish oil-enriched nutritional supplement on metabolic mediators in patients with pancreatic cancer cachexia. *Nutr. Cancer 40*, 118–124.

Barber, M.D., Ross, J.A., Voss, A.C., Tisdale, M.J., and Fearon, K.C. (1999). The effect of an oral nutritional supplement enriched with fish oil on weight-loss in patients with pancreatic cancer. *Br. J. Cancer 81*, 80–86.

Beier-Holgersen, R. and Boesby, S. (1996). Influence of postoperative enteral nutrition on postsurgical infections. *Gut 39*, 833–835.

Bing, C., Bao, Y., Jenkins, J., Sanders, P., Manieri, M., Cinti, S., Tisdale, M.J., and Trayhurn, P. (2004). Zinc-alpha2-glycoprotein, a lipid mobilizing factor, is expressed in adipocytes and is up-regulated in mice with cancer cachexia. *Proc. Natl. Acad. Sci. USA 101*, 2500–2505.

Cabre, E., Gonzalez-Huix, F., Abad-Lacruz, A., Esteve, M., Acero, D., Fernandez-Banares, F., Xiol, X., and Gassull, M.A. (1990). Effect of total enteral nutrition on the short-term outcome of severely malnourished cirrhotics. A randomized controlled trial. *Gastroenterology 98*, 715–720.

Cangiano, C., Laviano, A., Meguid, M.M., Mulieri, M., Conversano, L., Preziosa, I., and Rossi-Fanelli, F. (1996). Effects of administration of oral branched-chain amino acids on anorexia and caloric intake in cancer patients. *J. Natl. Cancer. Inst. 88*, 550–552.

Cederholm, T., Wretlind, B., Hellstrom, K., Andersson, B., Engstrom, L., Brismar, K., Scheynius, A., Forslid, J., and Palmblad, J. (1997). Enhanced generation of interleukins 1 beta and 6 may contribute to the cachexia of chronic disease. *Am. J. Clin. Nutr. 65*, 876–882.

Chandra, R.K. (1983). Nutrition, immunity, and infection: Present knowledge and future directions. *Lancet 1*, 688–691.

Cowl, C.T., Weinstock, J.V., Al-Jurf, A., Ephgrave, K., Murray, J.A., and Dillon, K. (2000). Complications and cost associated with parenteral nutrition delivered to hospitalized patients through either subclavian or peripherally-inserted central catheters. *Clin. Nutr. 19*, 237–243.

Crook, M.A., Hally, V., and Panteli, J.V. (2001). The importance of the refeeding syndrome. *Nutrition 17*, 632–637.

Currier, J., Carpenter, C., Daar, E., Kotler, D., and Wanke, C. (2002). Identifying and managing morphologic complications of HIV and HAART. *AIDS Read 12*, 114–119, 124–115.

Driscoll, J.M., Jr., Heird, W.C., Schullinger, J.N., Gongaware, R.D., and Winters, R.W. (1972). Total intravenous alimentation in low-birth-weight infants: A preliminary report. *J. Pediatr 81*, 145–153.

Dupertuis, Y.M., Kossovsky, M.P., Kyle, U.G., Raguso, C.A., Genton, L., and Pichard, C. (2003). Food intake in 1707 hospitalised patients: A prospective comprehensive hospital survey. *Clin. Nutr.* 22, 115–23.

Evans, W.K., Nixon, D.W., Daly, J.M., Ellenberg, S.S., Gardner, L., Wolfe, E., Shepherd, F.A., Feld, R., Gralla, R., Fine, S., et al. (1987). A randomized study of oral nutritional support versus ad lib nutritional intake during chemotherapy for advanced colorectal and non-small-cell lung cancer. *J. Clin. Oncol. 5*, 113–124.

Fearon, K.C., Von Meyenfeldt, M.F., Moses, A.G., Van Geenen, R., Roy, A., Gouma, D.J., Giacosa, A., Van Gossum, A., Bauer, J., Barber, M.D., et al. (2003). Effect of a protein and energy dense N-3 fatty acid enriched oral supplement on loss of weight and lean tissue in cancer cachexia: A randomised double blind trial. *Gut 52*, 1479–1486.

Fuhrman, M.P., Charney, P., and Mueller, C.M. (2004). Hepatic proteins and nutrition assessment. *J. Am. Diet. Assoc. 104*, 1258–1264.

Gariballa, S.E., Parker, S.G., Taub, N., and Castleden, C.M. (1998). A randomized, controlled, a single-blind trial of nutritional supplementation after acute stroke. *JPEN J. Parenter. Enteral. Nutr. 22*, 315–319.

Garrow, J.S. (1982). New approaches to body composition. *Am. J. Clin. Nutr. 35*, 1152–1158.

Grahm, T.W., Zadrozny, D.B., and Harrington, T. (1989). The benefits of early jejunal hyperalimentation in the head-injured patient. *Neurosurgery 25*, 729–735.

Green, C.J. (1999). Existence, cause and consequences of disease-related malnutrition in hospitals and the community, and clinical and financial benefits of nutritional intervention. *Clin. Nutr. 18*, 3–28.

Harrison's online (2004–2005). The McGraw Hill Company Part 4. Nutrition, Chapter 63. Enteral and Parenteral Nutrition Therapy, Table 63-10. Enteral Formulas.

Herrmann, F.R., Safran, C., Levkoff, S.E., and Minaker, K.L. (1992). Serum albumin level on admission as a predictor of death, length of stay, and readmission. *Arch. Intern. Med. 152*, 125–130.

Inui, A. (2002). Cancer anorexia-cachexia syndrome: Current issues in research and management. *CA Cancer J. Clin. 52*, 72–91.

Jeejeebhoy, K.N. (2001). Total parenteral nutrition: Potion or poison? *Am. J. Clin. Nutr. 74*, 160–163.

Keele, A.M., Bray, M.J., Emery, P.W., Duncan, H.D., and Silk, D.B. (1997). Two phase randomised controlled clinical trial of postoperative oral dietary supplements in surgical patients. *Gut 40*, 393–399.

Keohane, P.P., Attrill, H., Love, M., Frost, P., and Silk, D.B. (1984). Relation between osmolality of diet and gastrointestinal side effects in enteral nutrition. *Br. Med. J. (Clin. Res. Ed.) 288*, 678–680.

Keys, A., Brozek, J., Henschel, A., et al. (1950). *The Biology of Human Starvation* (Minneapolis, MN, University of Minnesota Press).

Kondrup, J. (2001). Can food intake in hospitals be improved. *Clin. Nutr. 20*, 153–160.

Kondrup, J., Allison, S.P., Elia, M., Vellas, B., and Plauth, M. (2003). ESPEN guidelines for nutrition screening 2002. *Clin. Nutr. 22*, 415–421.

Laviano, A., Meguid, M.M., Yang, Z.J., Gleason, J.R., Cangiano, C., and Rossi Fanelli, F. (1996). Cracking the riddle of cancer anorexia. *Nutrition 12*, 706–710.

Lopes, J., Russell, D.M., Whitwell, J., and Jeejeebhoy, K.N. (1982). Skeletal muscle function in malnutrition. *Am. J. Clin. Nutr. 36*, 602–610.

Mercadante, S. (1996). Nutrition in cancer patients. *Support Care Cancer 4*, 10–20.

Mercadante, S. (1998). Parenteral versus enteral nutrition in cancer patients: Indications and practice. *Support Care Cancer 6*, 85–93.

Mercadante, S. (1998). Diarrhea, absorption, constipation. In *Principles and Practice of Supportive Oncology*, A. Berger, M.H. Levy, R.K. Protenoy, and D.E. Wessmann, eds. Lippencott-Raven, Philadelphia, 191–206.

Messing, B., Lemann, M., Landais, P., Gouttebel, M.C., Gerard-Boncompain, M., Saudin, F., Vangossum, A., Beau, P., Guedon, C., Barnoud, D., et al. (1995). Prognosis of patients with nonmalignant chronic intestinal failure receiving long-term home parenteral nutrition. *Gastroenterology 108*, 1005–1010.

Ovesen, L., Allingstrup, L., Hannibal, J., Mortensen, E.L., and Hansen, O.P. (1993). Effect of dietary counseling on food intake, body weight, response rate, survival, and quality of life in cancer patients undergoing chemotherapy: A prospective, randomized study. *J. Clin. Oncol. 11*, 2043–2049.

Polderman, K.H. and Girbes, A.R. (2002). Central venous catheter use. Part 2: Infectious complications. *Intensive Care Med. 28*, 18–28.

Puchstein, C., Mertes, N., and Nolte, G. (1989). [Assessing nutritional status]. *Infusionstherapie 16*, 222–228.

Rapp, R.P., Young, B., Twyman, D., Bivins, B.A., Haack, D., Tibbs, P.A., and Bean, J.R. (1983). The favorable effect of early parenteral feeding on survival in head-injured patients. *J. Neurosurg. 58*, 906–912.

Rasmussen, H.H., Kondrup, J., Ladefoged, K., and Staun, M. (1999). Clinical nutrition in Danish hospitals: A questionnaire-based investigation among doctors and nurses. *Clin. Nutr. 18*, 153–158.

Schols, A.M., Soeters, P.B., Mostert, R., Pluymers, R.J., and Wouters, E.F. (1995). Physiologic effects of nutritional support and anabolic steroids in patients with chronic obstructive pulmonary disease. A placebo-controlled randomized trial. *Am. J. Respir. Crit. Care Med. 152*, 1268–1274.

Shike, M. (1996). Nutrition therapy for the cancer patient. *Hematol. Oncol. Clin. North. Am. 10*, 221–234.

Sobotka, L. (2004). Refeeding syndrome. In *Basics in Clincal Nutrition*, L.E.-i.-C. Sobotka, S.P. Allison, P. Fürst, R. Meier, M. Perpkiewicz, and P. Soeters, eds. (Prague, Galén), pp. 288–291.

Solomon, S.M. and Kirby, D.F. (1990). The refeeding syndrome: A review. *JPEN J. Parenter. Enteral. Nutr. 14*, 90–97.

Tayek, J.A., Bistrian, B.R., Hehir, D.J., Martin, R., Moldawer, L.L., and Blackburn, G.L. (1986). Improved protein kinetics and albumin synthesis by branched chain amino acid-enriched total parenteral nutrition in cancer cachexia. A prospective randomized crossover trial. *Cancer 58*, 147–157.

Tisdale, M.J. (2004). Cancer cachexia. *Langenbecks Arch. Surg. 389*, 299–305.

Todorov, P., Cariuk, P., McDevitt, T., Coles, B., Fearon, K., and Tisdale, M. (1996). Characterization of a cancer cachectic factor. *Nature 379*, 739–742.

Todorov, P.T., McDevitt, T.M., Meyer, D.J., Ueyama, H., Ohkubo, I., and Tisdale, M.J. (1998). Purification and characterization of a tumor lipid-mobilizing factor. *Cancer Res. 58*, 2353–2358.

Wanke, C., Polsky, B., and Kotler, D. (2002). Guidelines for using body composition measurement in patients with human immunodeficiency virus infection. *AIDS Patient Care STDS 16*, 375–388.

Weisdorf, S.A., Lysne, J., Wind, D., Haake, R.J., Sharp, H.L., Goldman, A., Schissel, K., McGlave, P.B., Ramsay, N.K., and Kersey, J.H. (1987). Positive effect of prophylactic total parenteral nutrition on long-term outcome of bone marrow transplantation. *Transplantation 43*, 833–838.

Wigmore, S.J., Barber, M.D., Ross, J.A., Tisdale, M.J., and Fearon, K.C. (2000). Effect of oral eicosapentaenoic acid on weight loss in patients with pancreatic cancer. *Nutr. Cancer 36*, 177–184.

Wigmore, S.J., Plester, C.E., Richardson, R.A., and Fearon, K.C. (1997). Changes in nutritional status associated with unresectable pancreatic cancer. *Br. J. Cancer 75*, 106–109.

Wigmore, S.J., Ross, J.A., Falconer, J.S., Plester, C.E., Tisdale, M.J., Carter, D.C., and Fearon, K.C. (1996). The effect of polyunsaturated fatty acids on the progress of cachexia in patients with pancreatic cancer. *Nutrition 12*, S27–S30.

Windsor, J.A. and Hill, G.L. (1988). Risk factors for postoperative pneumonia. The importance of protein depletion. *Ann. Surg. 208*, 209–214.

IV

Drugs in Clinical Use

16 Appetite Stimulants

Matthias John

CONTENTS

SUMMARY

The article focuses on the clinical use of progestational drugs, cannabinoids, and cyproheptadine as appetite stimulants in the therapy of the anorexia and cachexia syndrome. These drugs have been shown to be partially effective in improving the symptom of body weight loss in patients with chronic illnesses.

    The following article gives an up-to-date overview about progestational drugs, cannabinoids, and cyproheptadine for the therapy of the anorexia and cachexia syndrome.

    Megestrol acetate (MA) and medroxyprogesterone acetate (MPA) are synthetic orally active progestational agents. These drugs have been shown to stimulate appetite and weight gain in patients with nonhormone responsive tumors and anorexia/cachexia syndrome. Cannabinoids have many potential therapeutic effects. Especially the effects on appetite stimulation are used in the clinical routine to treat cachexia and malnutrition in patients with chronic illnesses. Cyproheptadine is an antiserotoninergic drug, which appears to cause slight appetite stimulation. Corticosteroids are frequently used in clinical practice for appetite stimulation in patients with advanced malignancies since randomized clinical trials showed that corticosteroid medications could stimulate the appetites of advanced cancer patients.

16.1 INTRODUCTION

Cachexia is a debilitating and life-threatening syndrome of tissue wasting characterized by anorexia, involuntary weight loss, exercise limitation, and a poor clinical prognosis.

New pharmacologic therapies for primary anorexia and cachexia syndrome are expected to enter clinical practice soon; however, until then, treatment with conventional appetite stimulants such as progestational drugs, cannabinoids, and cyproheptadine is the only evidence-based therapy for this syndrome.

There are now effective drugs that have been shown to improve appetite, food intake, and sensation of well being, and which elicit body weight gain. These drugs make it possible to address the clinical syndrome of anorexia and cachexia at a different level and also provide an opportunity for a combined therapy in order to improve the quality of life of these patients.[1-7]

This chapter gives an up-to-date overview about progestational drugs, cannabinoids, and cyproheptadine for the therapy of the anorexia and cachexia syndrome. These drugs have been shown to be partially effective in improving the symptoms of the patients with chronic illnesses. The promotion of appetite stimulation and weight gain is a necessary precondition to regain quality of life and physical activity of patients with chronic diseases. Therefore, appetite stimulants for the treatment of malnutrition and weight loss can help to ameliorate many different medical conditions.

16.2 MEGESTROL ACETATE AND MEDROXYPROGESTERONE ACETATE

Megestrol acetate (MA) and medroxyprogesterone acetate (MPA) are synthetic orally active progestational agents. In several controlled randomized studies it has been reported that these drugs stimulate appetite and weight gain in patients with nonhormone responsive tumors and anorexia/cachexia syndrome.[8-10] Megestrol has demonstrated a dose-related beneficial effect in a dose range from 160 to 1600 mg/day on appetite, caloric intake, body weight gain (mainly fat) with an optimal dosage of 800 mg daily.[11] The patient's sense of well being was improved as well. Doses above 800 mg megestrol per day do not lead to a further enhancement of the initial response to the dose range from 160 to 800 mg/day. It is recommended that a patient be started on the lowest dosage [160 mg/day] and the dose be titrated upward according to the clinical response.[12, 13] Quality of life measures such as the Karnofsky index (scores the activity of patients in oncology) may or may not be influenced by progesterone agents.[13-15] Medroxyprogesterone has similarly been shown to increase appetite and food intake with stabilization of body weight at a dose of 1000 mg (500 mg twice daily).[13] Although the drug may be used at 500 to 4000 mg daily, side effects increase above oral doses of 1000 mg daily.[16]

There is an increasing dose/response curve for appetite stimulation from 160 to 800 mg/day; doses above 800 mg/day have no additional benefit. Different strategies include beginning at 400 mg/day and titrating for effect to 800 mg/day. Alternatively, dosing can begin at 800 mg/day (Table 16.1).

TABLE 16.1
Dose Regimen for Appetite-Stimulating Substances

Substance	Optimal daily dosage (mg)	Reference
MA	800	Loprinzi 1993, Gagnon 1998
Medroxyprogesterone	1000	Mantovani 2001, Nelson 2000
Cyproheptadine	4–8	Daviss 2004, Kardinal 1990
THC	5	Walsh 2003, Nelson 1994

Medroxyprogesterone can also be given in a depot formulation. There is a trend of an increased prescription of megestrol or medroxyprogesterone in oral suspensions rather than tablets because of an improved patient compliance and a better cost efficiency.[3, 17] There is, at present, considerable evidence for the effect of synthetic progestins on appetite and body weight in patients with cancer anorexia–cachexia syndrome.[18] However, further issues regarding the optimal treatment duration, the best time to start treatment during the natural history of the disease, and the eventual impact on the overall quality of life need to be clarified.[18] Moreover, optimal dose regimens for MA in different indications, such as appetite improvement, patients' sense of well-being, weight gain, are still to be identified.

A study by De Conno et al.[19] investigated in a double-blind controlled clinical trial the effect of MA in a low dose range on anorexia in patients with far-advanced cancer. The aim of this study was to evaluate the safety and symptomatic effectiveness of a 320 mg/day regimen of MA on anorexia defined as loss of appetite. Outpatients with far-advanced nonhormone responsive tumors and loss of appetite were randomized in this placebo-controlled phase III trial. Appetite, food intake, body weight, performance status, mood, and quality of life were evaluated with standardized measures; patient's global judgment about treatment efficacy was also requested. Of 42 patients entering the study, 33 (17 on MA and 16 on placebo) were evaluable for efficacy. The appetite score (a scale from 1 to 10 that reflects the subjective desire to eat) improved significantly with MA after 7 days ($P = .0023$), and this effect was still significant after 14 days ($P = .0064$). Patients judged the treatment with MA effective in 88.2% of cases (14th day), while placebo was considered effective in 25% ($P = .0003$). None of the other measures showed significant changes during treatment. The beneficial effect on appetite evident after 7 days, without serious side effects, shows that MA can produce significant subjective effects even in patients with far-advanced cancer. Both are considered as relatively non-toxic drugs with low incidence of adverse effects when taken in the usual recommended dose. The following adverse events have been reported: thromboembolic phenomena, breakthrough uterine bleeding, peripheral edema, hyperglycemia, hypertension, adrenal suppression, and adrenal insufficiency if the drug is abruptly discontinued.[1, 10, 12, 14, 16, 20–23] There is evidence that patients rarely need to stop MA or MPA medication because of adverse effects. As a precaution, these drugs should not be prescribed in cases of thromboembolic/thrombotic disease, heart failure, or for patients at risk for serious fluid retention.[16] As shown recently, the risk of deep vein thrombosis must be considered when prescribing MA, and its use should be limited in treating nutritionally at-risk nursing home residents.[24] Although prolonged use of the drug appears to be safe, there are reports that long-term administration of MA may have induced secondary adrenal suppression and adrenal insufficiency.[25] Short-term administration (12 weeks) of MA in a pilot study of geriatric nursing home residents with cachexia ($n = 69$) revealed the drug to be safe without apparent evidence of adrenal suppression. Rare other side effects can be alteration of menstrual pattern with unpredictable bleeding, pain in chest, visual disturbances, headache, insomnia, pain in abdomen, groin, calf, or leg, loss of coordination, slurred speech, weakness or numbness in extremities, yellow eyes or skin, depression, skin rashes, swelling in hands and feet, brown spots on skin, acne, increased body hair, increased breast tenderness, and loss of scalp hair. MA induced reduction of testosterone in males. Apart from its use in appetite stimulation, MA has been used for the therapy of endometrial or some types of breast cancer.

Although the mechanism of weight gain of progestational drugs is presently uncertain, it might be related to glucocorticoid activity.[12] McCarthy et al.[26] found that MA stimulated food and water intake in rats and that the effect may involve neuropeptide Y. Furthermore, modulation of calcium channels by MA or MPA in the ventromedial hypothalamus (VMH) — a well known satiety-center may lead to a reduction in the firing rate of VMH neurons.[27-36] This was confirmed by Costa et al.[37] showing that MA stimulation of appetite involves calcium channels in the VMH. A further pharmacological mechanism of MA is the inhibition of the activity of proinflammatory cytokines such as IL-1, IL-6, and tumor necrosis factor-α(TNF-α).[13,38-41] Serum levels of such cytokines were reported to be decreased in cancer patients after MA or MPA treatment.[13] The involvement of cytokines was demonstrated in a case report by Castle et al.[42] They reported results from four patients, in which an oral suspension of MA was used to treat geriatric cachexia and anorexia. Although the study size was small, the two patients with the highest IL-6 concentrations showed the greatest weight gain in response to the treatment. This is similar to cachectic HIV-infected patients[43] and supports the hypothesis of an inflammatory cytokine driven weight loss in patients with chronic diseases. Mantovani et al.[44] reported that MA reduces the *in vitro* production of certain cytokines (IL-1, IL-6, TNF-α, and serotonin) by peripheral mononuclear cells of cancer patients. This group[45] also found in a small study of nine patients that appetite, body weight, and sense of well being improved with MA in cancer patients. Mantovani also found that these patients' cytokine concentrations decreased with MA treatment but not with chemotherapy alone. A combined therapy with MA and ibuprofen led to weight gain and reduction of the inflammatory response as has been described by McMillan et al.[46] in cachectic patients with advanced gastrointestinal cancer after a 6 week therapy with the above mentioned combination. There is also a growing evidence that progesterone plays an important role in the regulation of nutritional status. Lapp et al.[47] found that higher progesterone concentrations are common in late pregnancy, resulting in decreased IL-6 concentrations that were 40 to 50% of those in control subjects, suggesting that progestational drugs may also have anti-inflammatory properties.

More studies are needed to finally clarify the pharmacologic effects of MA and MPA drugs and to confirm the antiinflammatory effects.

16.3 CANNABINOIDS

The cannabinoids are highly lipid-soluble substances that work synergistically, additively, or even antagonistically when ingested together by smoking marijuana. This has significant implications for the clinical usefulness of marijuana or its individual compounds. The most common method of cannabis ingestion in the United States and Europe is by smoking marijuana that is derived from any part of the dried plant. Other regional preparations are hashish (the Middle East, North Africa), charas (India), and ganga (Jamaica) that are derived from the resin of the cannabis plant.[48]

The active ingredient of cannabis is delta-9-tetrahydrocannabinol (THC). There are now three other synthetic cannabinoids that have been evaluated in clinical trials: dronabinol, nabilone, and levonantradol. Dronabinol is the synthetic form of THC that is commercially available today. THC has been proved to be useful for medical purposes. At present, there are only two approved indications: cytotoxic chemotherapy-induced nausea and vomiting and AIDS-associated anorexia and wasting. In tumor patients, analgesia, muscle relaxation, and mood elevation are additional beneficial effects.[49,50]

The effects of cannabinoids are mediated via specific receptors. Two types of cannabinoid receptors, CB1 and CB2 have been detected.

Cannabinoids can be administered via inhalation, intravenously, or orally. The onset of their effects depends on the route of administration. The peak concentration is reached within minutes following inhalation or intravenous application and about 2 h following oral ingestion. The peak effects appear between 15 and 30 min after intake.

The effects of the inhalative application of a purified, single, cannabinoid via a mechanical device for medical purposes are better predictable than smoking marijuana. When marijuana is smoked, a variable fraction of THC gets lost into the air or the respiratory dead space and a small amount disappears by pyrolysis. Other subjective factors such as smoking habits or inhalation technique may influence the pharmacokinetics of THC.[48–51]

Acute effects of cannabinoids occur in the cardiovascular and central nervous system. Tachycardia, hypotension, and decreased cardiac function are the first effects of cannabinoids. Psychomimetic effects occur in a biphasic manner. An initial euphoria is followed by drowsiness and relaxation. The onset and duration of these effects correlate with plasma THC concentration. Use of cannabis moderately increases the risk of psychotic symptoms in young people but has a much stronger effect in those with a predisposition for psychosis.[52] There is also substantial evidence that cannabis sativa derivates act on the brain system in a way similar to other drugs of abuse. The endogenous opioids are responsible for the modulation of the addictive properties of cannabinoids. Recent investigations revealed an interaction between the endogenous cannabinoid and opioid systems.[53] Abstinent chronic cannabis smokers produce reduced activation in motor cortical areas in response to finger sequencing compared to controls indicating that chronic cannabis abuse may lead to chronic neurological impairments.[54]

Furthermore, pulmonary (bronchodilation), gastrointestinal (decreased bowel motility), and musculoskeletal (muscle relaxation) effects of cannabinoids have been described.

The chronic effects of cannabinoids have been investigated in several epidemiological studies in Jamaica, Costa Rica, and other countries where cannabis use is prevalent. Taken together, no or only slight differences were found between user and nonuser groups in body weight, blood pressure, heart rate, and hematologic and biochemical blood tests. Furthermore, no differences between the study group and controls regarding bronchitis, emphysema, tooth decay, central/peripheral nervous system abnormality, electroencephalogram, and third ventricle size were observed. There were obvious differences in the psychiatric testing, indicating that the user groups had a significantly higher incidence of psychopathological impairments. In most cases, the psychopathology took the form of personality disorders. The psychological assessment was also different. Controls scored better in several subtests, but there were no differences in total I.Q. measurement between the two groups.[54]

The medical use of cannabinoids is under consideration for some clinical indications. The most promising application of any form of THC is to counteract the nausea associated with cancer chemotherapy and to stimulate appetite. The appetite-stimulating effect of THC may be beneficial for patients with wasting related to the acquired immuno deficiency syndrome (AIDS) and those with severe cancer-related anorexia. A few studies on the use of either pure THC or crude marijuana for appetite stimulation have been reported in the literature. One prospective, unblinded, uncontrolled study demonstrated that appetite improved in patients with terminal cancer who received low-dose oral THC (2.5 mg twice daily, 1 h after meals).[55] Twenty two percent of patients withdrew from the trial because of typical cannabinoid toxicity. Furthermore, the study revealed that only low doses of oral THC were necessary to induce appetite.

In a double-blind, placebo-controlled, parallel group study[56] 2.5 mg of oral THC twice daily effectively stimulated appetite in patients with AIDS. The investigators did not evaluate muscle mass or total body fat but did find that in patients who received oral THC, weight was maintained or increased slightly.

Mattes et al.[49] compared the effects of oral and rectal suppository preparations of THC on appetite stimulation and calorie intake with those of smoked marijuana in healthy persons. The design of this study was double-blind, placebo-controlled. Smoked marijuana was no more effective than suppository THC in stimulating appetite, as measured by calorie intake. Rectal suppositories and oral THC were given at a dosage of 2.5 mg twice daily. The plasma THC levels of the group assigned to smoked marijuana peaked more quickly with the inhaled THC but also decreased more quickly; in contrast, the levels achieved with suppository THC were more sustained.

The appetite-stimulating properties of THC have also been observed in the AIDS population in a double-blind, randomized, placebo-controlled crossover trial with two 5-week treatment periods.[57] The patients had a mean age of 37 years, and more than 50% had previous experience of smoking marijuana. With 2.5 mg once daily (10% of the study population) or twice daily (90% of the population) they reported an appetite increase over baseline for 12 months in the 24% of patients who were evaluable. Despite low dose and young age, 38% experienced typical psychomimetic side effects, and half withdrew from the study because of side effects or declining performance status (Table 16.1).

Appetite stimulation was not dependent upon the development of euphoria associated with the drug, as noted in some antiemetic studies. The low dose and timing were thought to be important in preventing the significant side effects seen in the other studies.[58]

Cannabinoids have also antiemetic effects that were reported in several studies on chemotherapy induced nausea and vomiting. These studies have documented the superior efficacy of THC over placebo and prochlorperazine.[59,60] All these studies used considerably higher doses than the subsequent appetite studies and were done in younger populations. THC was effective against mild and moderately emetogenic chemotherapy, but not against high-dose cisplatin chemotherapy.[56] THC was superior to low doses of metoclopramide in patients receiving moderately emetogenic chemotherapy, but inferior to high-dose metoclopramide in cisplatin-based chemotherapy.[56,61,62] Cannabinoids have many potential therapeutic effects. In particular, the benefit of appetite stimulation will be the focus in future studies to introduce cannabinoids in the clinical routine to treat cachexia and malnutrition in patients with chronic illnesses.

16.4 OTHER APPETITE STIMULANTS — CYPROHEPTADINE, CORTICOSTEROIDS, AND HYDRAZINE SULFATE

Several other drugs have been evaluated as agents to ameliorate cancer-associated anorexia/cachexia. Cyproheptadine is an antiserotoninergic drug, which appears to cause slight appetite stimulation in patients. A randomized clinical trial, however, was unable to demonstrate any weight gain from this agent. A retrospective chart review was conducted by Daviss and colleagues on 28 consecutive pediatric psychiatry outpatients who were prescribed cyproheptadine for weight loss or insomnia while on stimulants due to attention deficit hyperreactivity disorder.[63] Of these, four patients never took cyproheptadine consistently, and three discontinued it within the first 7 days due to intolerable side effects. Data were analyzed for 21 other patients who continued with 4 to 8 mg of cyproheptadine daily at night for at least 14 days (mean duration = 104.7 days). All 21 patients gained weight taking concomitant cyproheptadine, with a mean gain of 2.2 kg and a mean weight velocity of 32.3 g/day. The results led to the conclusion that concomitant cyproheptadine may be useful in youths with attention deficit hyperactivity disorder for stimulant-induced weight loss.[63]

The results of a study from Genazzani et al.[64] demonstrated a specific central effect of cyproheptadine on the serotoninergic pathway controlling food intake at the hypothalamic level. This study aimed to evaluate the possible interactions and modulations of the serotoninergic system on hormonal parameters and the reproductive axis in amenorrheic subjects. Hypogonadotropic, underweight, amenorrheic patients ($n = 8$) were studied before and during cyproheptadine chlorhydrate administration (4 mg/day for 3 months). Serotoninergic receptor blockade affected neither the naloxone-inducted LH response nor the gonadotropin pulsatile parameters. Body mass index (BMI) did not vary; conversely, integrated mean gonadotropins, growth hormone, and free trijodthyronine (fT3) concentrations increased during the treatment. The results led to the conclusion that cyproheptadine administration affects some of the abnormal endocrine parameters of underweight amenorrheic subjects with no modulation of the opioidergic system.

A randomized, placebo-controlled, double-blinded clinical trial using cyproheptadine, 8 mg orally three times a day in 295 patients with advanced malignant disease was performed by Kardinal et al.[65] Patients' appetite, measured by serial patient-completed questionnaires, appeared to be mildly enhanced by cyproheptadine. However, a significant weight gain was not observed.[65]

Patients assigned to cyproheptadine had less nausea ($p = .02$), less emesis ($p = .11$), more sedation ($p = .07$), and more dizziness ($p = .01$) than placebo patients.

Taken together, the studies on the effects of cyproheptadine in progressive weight loss in patients with cancer or other causes of cachexia suggested that this substance has a beneficial effect on appetite stimulation but only slight effects on weight gain. In addition, basic research suggests that cyproheptadine may be helpful in patients with cancer anorexia/cachexia. However, larger studies need to be performed to prove the effect of cyproheptadine in larger patient populations (Table 16.1).

Corticosteroids are frequently used in clinical practice for appetite stimulation in patients with advanced malignancies since randomized clinical trials showed that corticosteroid medications could stimulate the appetites of advanced cancer patients. However, these studies were not able to show any substantial nonfluid weight gain in treated patients. Efforts are also ongoing to evaluate both anabolic steroids and hydrazine sulfate as drugs for the treatment of patients with cancer anorexia/cachexia. Hydrazine is a substance that inhibits the enzyme phosphoenol pyruvate carboxykinase (PEP-CK) and interferes with gluconeogenesis. However, hydrazine and hydrazine sulfate might be a human carcinogen based on evidence for carcinogenicity in animal studies.[66] The preliminary nature of these investigations, however, precludes recommendations for the use of these drugs in routine clinical practice.

References

1. Argiles J.M., Meijsing S.H., and Pallares-Trujillo J. Cancer cachexia: a therapeutic approach. *Med. Res. Rev.* 2001; 21: 83–101.
2. Woods S.C., Seeley R.J., and Porte D. Signals that regulate food intake and energy homeostasis. *Science* 1998; 280:1378–1383.
3. Chlebowski R.T., Palomares M.R., Lillington L., and Grosvenor M. Recent implications of weight loss in lung cancer management. *Nutrition* 1996; 12: S43–S47.
4. Inui A. Cancer anorexia-cachexia syndrome: current issues in research and management. *CA Cancer J. Clin.* 2002; 52: 72–91.
5. Yeh S.S. and Schuster M.W. Geriatric cachexia: the role of cytokines. *Am. J. Clin. Nutr.* 1999; 70: 183–197.
6. Argiles J.M., Almendro V., Busquets S., and Lopez-Soriano F.J. The pharmacological treatment of cachexia. *Curr. Drug Targets* 2004; 5: 265–277.
7. Desport J.C., Bachmann P., and Senesse P. Standards, options and recommendations for the use of appetite stimulants in oncology. *BMJ* 2000; 89: S98–S100.
8. Bruera E., Macmillan K., Kuehn N., Hanson J., and MacDonald R.N. A controlled trial of megestrol acetate on appetite caloric intake, nutritional status, and other symptoms in patients with advanced cancer. *Cancer* 1990; 66: 1279–1282.
9. Loprinzi C.L., Ellison N.M., and Schaid D.J. Controlled trial of megestrol acetate for the treatment of cancer anorexia and cachexia. *J. Natl Cancer Inst.* 1990; 82: 1127–1132.
10. Tchekmedyian N.S., Hickman H., Siau J., Greco A., and Aisner J. Treatment of cancer anorexia with megestrol acetate: impact on quality of life. *Oncology* 1990; 4: 185–192.
11. Loprinzi C.L., Michalak J.C., and Schaid D.J. Phase III evaluation of four doses of megestrol acetate as therapy for patients with cancer anorexia and/or cachexia. *J. Clin. Oncol.* 1993; 11: 762–767.
12. Gagnon B. and Bruera E. A review of the drug treatment of cachexia associated with cancer. *Drugs* 1998; 55: 675–688.
13. Mantovani G., Maccio A., Massa E., et al. Managing cancer-related anorexia/cachexia. *Drugs* 2001; 61: 499–514.
14. Bruera E. Pharmacological treatment of cachexia: any progress? *Support Care Cancer* 1998; 6: 109–113.
15. Rowland K.M. Jr, Loprinzi C.L., and Shaw E.G. Randomized double-blind placebo-controlled trial of cisplatin and etoposide plus megestrol acetate/placebo in extensive-stage small-cell lung cancer: a North Central Cancer Treatment Group study. *J. Clin. Oncol.* 1996; 14: 135–141.
16. Nelson K.A. The cancer anorexia-cachexia syndrome. *Semin. Oncol.* 2000; 27: 64–68.
17. Ottery F.D., Walsh D., and Strawford A. Pharmacologic management of anorexia/cachexia. *Semin. Oncol.* 1998; 25: 35–44.
18. Maltoni M., Nanni O., and Scarpi E. High-dose progestins for the treatment of cancer anorexia–cachexia syndrome: a systemic review of randomized clinical trials. *Ann. Oncol.* 2001; 12: 289–300.
19. De Conno F., Martini C., Zecca E., Balzarini A., Venturino P., Groff L., and Caraceni A. Megestrol acetate for anorexia in patients with far-advanced cancer: a double-blind controlled clinical trial. *Eur. J. Cancer* 1998; 34: 1705–1709.

20. Lapp C.A., Thomas M.E., and Lewis J.B. Modulation by progesterone of IL-6 production by gingival fibroblasts. *J. Periodontol.* 1995; 66: 279–84.

21. Karcic E., Philpot C., and Morley J.E. Treating malnutrition with megestrol acetate: literature review of our experience. *J. Nutr. Health Aging* 2002; 6: 191–200.

22. Feliu J., Gonzalez-Baron M., and Berrocal A. Treatment of cancer anorexia with megestrol acetate: which is the optimal dose? *J. Natl Cancer Inst.* 1991; 83: 449–450.

23. Tchekmedyian N.S., Hickman M., and Siau J. Megestrol acetate in cancer anorexia and weight loss. *Cancer* 1992; 69: 1268–1274.

24. Bolen J.C., Andersen R.E., and Bennett R.G. Deep vein thrombosis as a complication of megestrol acetate therapy among nursing home residents. *J. Am. Med. Dir. Assoc.* 2000; 1: 248–252.

25. Leinung M.C., Liporace R., and Miller C.H. Induction of adrenal suppression by megestrol acetate in patients with AIDS. *Ann. Intern. Med.* 1995; 122: 843–845.

26. McCarthy H.D., Crowder R.E., and Dryden S, et al. Megestrol acetate stimulates food intake in the rat: effects on regional hypothalamic neuropeptide Y concentrations. *Eur. J. Pharmacol.* 1994; 265: 99–102.

27. Schwartz M.W. and Seeley R.J. Neuroendocrine responses to starvation and weight loss. *N. Engl. J. Med.* 1997; 336: 1802–1811.

28. Inui A. Feeding and body-weight regulation by hypothalamic neuropeptides — mediation of the actions of leptin. *Trends Neurosci.* 1999; 22: 62–67.

29. Friedman J.M. and Halaas J.L. Leptin and the regulation of body weight in mammals. *Nature* 1998; 395: 763–770.

30. Elmquist J.K., Maratos-Flier E., and Saper C.B. Unraveling the central nervous system pathways underlying responses to leptin. *Nat. Neurosci.* 1998; 1: 445–450.

31. Bray G.A. and York D.A. The MONA LISA hypothesis in the time of leptin. *Recent. Prog. Horm. Res.* 1998; 53: 95–117.

32. Kalra S.P., Dube M.G., and Pu S. Interacting appetite-regulating pathways in the hypothalamic regulation of body weight. *Endocr. Rev.* 1999; 20: 68–100.

33. Schwartz M.W., Woods S.C., and Porte D., Central nervous system control of food intake. *Nature* 2000; 404: 661–671.

34. Inui A. Transgenic approach to the study of body weight regulation. *Pharmacol-Rev.* 2000; 52: 35–61.

35. Inui A. Transgenic study of energy homeostasis equation: implications and confounding influences. *FASEB J.* 2000; 14: 2158–2170.

36. Plata-Salaman C.R. Central nervous system mechanisms contributing to the cachexia–anorexia syndrome. *Nutrition* 2000; 16: 1009–1012.

37. Costa A.M., Spence K.T., and Plata-Salaman C.R. Residual Ca2+ channel current modulation by megestrol acetate via a G-protein alpha s-subunit in rat hypothalamic neurons. *J. Physiol.* 1995; 487: 291–303.

38. McCarthy H.D., Crowder R.E., Dryden S., et al. Megestrol acetate stimulates food and water intake in the rat: effects on regional hypothalamic neuropeptide Y concentrations. *Eur. J. Pharmacol.* 1994; 265: 99–102.

39. Yeh S.S., Wu S.Y., Levine D.M., Parker T.S., Olson J.S., Stevens M.R., and Schuster M.W. The correlation of cytokine levels with body weight after megestrol acetate treatment in geriatric patients. *J. Gerontol. A Biol. Sci. Med. Sci.* 2001; 56: M48–M54.

40. Helle S.I., Lundgren S., Geisler S., Ekse D., Holly J.M., and Lonning P.E. Effects of treatment with megestrol acetate on the insulin-like growth factor system: time and dose dependency. *Eur. J. Cancer* 1999; 35: 1070–5.

41. Morley J.E. Orexigenic and anabolic agents. *Clin. Geriatr. Med.* 2002; 18: 853–866.

42. Castle S., Nguyen C., and Joaquin A. Megestrol acetate suspension therapy in the treatment of geriatric anorexia/cachexia in nursing home patients. *J. Am. Geriatr. Soc.* 1995; 43; 7: 835–836 (letter).

43. Graham N.M., Munoz A., Bacellar H., Kingsley L.A., Visscher B.R., and Phair J. Clinical factors associated with weight loss related to infection with the human immunodeficiency virus type 1 in the Multicenter AIDS Cohort Study. *Am. J. Epidemiol.* 1993; 137: 439–446.

44. Mantovani G., Maccio A., Esu S., et al. Medroxyprogesterone acetate reduces the in vitro production of cytokines and serotonin involved in anorexia/cachexia and emesis by peripheral blood mononuclear cells of cancer patients. *Eur. J. Cancer.* 1997; 33: 602–607.

45. Mantovani G., Maccio A., Bianchi, A., et al. Megestrol acetate in neoplasticss anorexia/cachexia: clinical evaluation and comparison with cytokine levels in patients with head and neck carcinoma treated with neoadjuvant chemotherapy. *Int. J. Clin. Lab. Res.* 1995; 25: 135–41.

46. McMillan D.C., O'Gorman P., Fearon K.C., and McArdle C.S. A pilot study of megestrol acetate and ibuprofen in the treatment of cachexia in gastrointestinal cancer patients. *Br. J. Cancer* 1997; 76: 788–790.

47. Lapp C.A., Thomas M.E., and Lewis J.B. Modulation by progesterone of IL-6 production by gingival fibroblasts. *J. Periodontol.* 1995; 66: 279–84.

48. Voth E.A. and Schwartz R.H. Medicinal applications of delta-9-tetrahydrocannabinol and marijuana. *Ann. Intern. Med.* 1997; 126: 791-798.

49. Mattes R.D., Engelman K., Shaw L.M., and Elsohly M.A. Cannabinoids and appetite stimulation. *Pharmacol. Biochem. Behav.* 1994; 49: 187–195.

50. Bredt B.M., Shade S.B., and Abrams D.I. Short-term effects of cannabinoids on immune phenotype and function in HIV-1-infected patients. *J. Clin. Pharmacol.* 2002; 42: 82S–89S.

51. Di Marzo V., Bifulco M., and Petrocellis L. The endocannabinoid system and its therapeutic exploitation. *Nat. Rev. Drug Discov.* 2004; 3: 771–784.

52. Henquet C., Krabbendam L., Spauwen J., Kaplan C., Lieb R., Wittchen van Os J. Prospective cohort study of cannabis use, predisposition for psychosis, and psychotic symptoms in young people. *BMJ* 2005; 330: 11–15.

53. Fattore L., Cossu G., Spano M.S., Deiana S., Fadda P., and Scherma M. Cannabinoids and reward: interactions with opioid system. *Crit. Rev. Neurobiol.* 2004; 16: 147–58.

54. Pillay S.S., Rogowska J., Kanayama G., Jon D.I., Gruber S., Simpson N., and Todd D.A. Neurophysiology of motor function following cannabis discontinuation in chronic cannabis smokers: an fMRI study. *Drug Alcohol Depend.* 2004; 76: 261–271.

55. Walsh D., Nelson K., and Mahmoud F.A. Established and potential therapeutic applications of cannabinoids in oncology. *Support Care Cancer* 2003; 11: 137–143.

56. Nelson K., Walsh D., Deeter P., and Sheehan F. A phase II study of delta-9-tetrahydrocannabinol for appetite stimulation in cancer-associated anorexia. *J. Palliat. Care* 1994; 10: 14–18.

57. Beal J.E., Olson R., Laubenstein L., Morales J.O., Bellman P., Yangco B., et al. Dronabinol as a treatment for anorexia associated with weight loss in patients with AIDS. *J. Pain. Symptom. Manage.* 1995; 10: 89–97.

58. Beal J.A. Appetite effect of dronabinol. *J. Clin. Oncol.* 1994; 12: 1524–1525.

59. Orr L.E., McKernan J.F., and Bloome B. Antiemetic effect of tetrahydrocannabinol. Compared with placebo and prochlorperazine in chemotherapy-associated nausea and emesis. *Arch. Intern. Med.* 1980; 140: 1431–1433.

60. Steele N., Gralla R.J., Braun D.W., Jr., and Young C.W. Double-blind comparison of the antiemetic effects of nabilone and prochlorperazine on chemotherapy-induced emesis. *Cancer Treat. Rep.* 1980; 64: 219–224.

61. Crawford S.M. and Buckman R. Nabilone and metoclopramide in the treatment of nausea and vomiting due to cisplatinum: a double blind study. *Med. Oncol. Tumor. Pharmacother.* 1986; 3: 39–42.

62. Cunningham D., Bradley C.J., and Forrest G.J. A randomized trial of oral nabilone and prochlorperazine compared to intravenous metoclopramide and dexamethasone in the treatment of nausea and vomiting induced by chemotherapy regimens containing cisplatin or cisplatin analogues. *Eur. J. Cancer. Clin. Oncol.* 1988; 24: 685–689.

63. Daviss W.B. and Scott J. A chart review of cyproheptadine for stimulant-induced weight loss. *J. Child Adolesc. Psychopharmacol.* 2004; 14: 65–73.

64. Genazzani A.D., Strucchi C., Malavasi B., Tortolani F., Vecci F., and Luisi P.F. Effects of cyproheptadine clorhydrate, a serotonin receptor antagonist, on endocrine parameters in weight-loss related amenorrhea. *Gynecol. Endocrinol.* 2001; 15: 279–285.

65. Kardinal C.G., Loprinzi C.L., and Schaid D.J. A controlled trial of cyproheptadine in cancer patients with anorexia and/or cachexia. *Cancer* 1990; 65: 2657–2664.

66. Kaegi, E. Unconventional therapies for cancer: 4. Hydrazine sulfate. Task Force on Alternative Therapies of the Canadian Breast Cancer Research Initiative. *CMAJ* 1998; 158: 1327–1330.

17 Cardiac Cachexia — Drugs in Clinical Use

Sabine Strassburg and Stefan D. Anker

CONTENTS

SUMMARY

Angiotensin converting enzyme (ACE)-inhibitors and beta-blockers are integral parts of standard therapy in chronic heart failure. Both of these prolong life and prevent weight loss. Whether these two effects are related still needs to be proven. The anticachectic effects of ACE-inhibitors and beta-blockers in chronic heart failure are thought to be mainly due to their effects against neurohormonal activation, oxidative stress, and endothelial dysfunction. As all these pathophysiological phenomena are proven or likely to be important in other chronic illness, ACE-inhibitors and beta-blockers may have anticachectic effects there as well.

17.1 INTRODUCTION

Cachexia has long been recognized as a late complication in patients with chronic heart failure (CHF). The onset of cachexia in CHF is associated with a poor prognosis in these patients.[1] Patients with cachexia due to heart failure (i.e., cardiac cachexia) suffer from an overall loss of fat, lean, and bone tissue.[2] The pathophysiology of cardiac cachexia is not fully clarified, but there is evidence that metabolic, neurohormonal, and immune abnormalities have an important role in the development of cardiac cachexia. The development of cachexia in other chronic diseases may be due to similar changes, and effects of drugs on cachexia in CHF may be of general relevance to other illnesses.

Cachectic heart failure patients show raised plasma levels of epinephrine, norepinephrine, and cortisol, and these neurohormonal abnormalities are strongly correlated with the degree of body wasting.[2,3] Other neurohormones, such as atrial and brain natriuretic peptide (ANP, BNP), endothelin-1 (ET-1), as well as aldosterone and renin, for instance, were also increased in adult patients with congenital heart disease who often suffer from cachexia.[4] Increased aldosterone levels and increased renin activity further indicate the importance of neuroendocrine activation in the development of cachexia in CHF since renin stimulates the production of angiotensin II (Renin–Angiotensin System). Considering these facts, it is conjecturable that a modulation of neurohormonal pathways might have a beneficial effect in cardiac cachexia patients. ACE-inhibitors and beta-blockers are part of the standard therapy of patients with heart failure — there is strong evidence that they modulate neurohormonal activity and that this is the major reason for their inherent mortality benefit.

In this chapter, we analyze ACE-inhibitors and beta-blockers in respect to their therapeutic potential in cachexia. This chapter focuses on data from clinical trials in humans, as cell culture and animal studies are discussed in other chapters. Additionally, it also focuses on data from heart failure patients, as this is where data on body weight changes is available.

17.2 Pathophysiologic Mechanisms

With the deterioration of cardiac function in CHF, a general stimulation of the sympathetic nervous system occurs, and this is combined with activation of the renin–angiotensin–aldosterone axis and the natriuretic peptide system.[2] Hence neurohormonal activation increases in CHF patients over time. Initially, these changes seem to have beneficial effects, for instance, they can contribute to an increase in the cardiac output. However, later they contribute to increased vascular resistance, increased afterload, ventricular dilatation, and myocardial remodeling. The neurohormones norepinephrine and epinephrine were shown to cause a catabolic metabolic shift and lead to an increase in resting energy expenditure in CHF patients resulting in further weight loss.[3,5,6]

It has been shown that increasing severity of chronic heart failure is associated with a stepwise increase in the concentrations of several neurohormones, for example, ANP, BNP, ET-1, and norepinephrine. This can be due to the impairment of the disease but could also suggest that neurohormones might boost the progress of the disease. In the latter case, inhibition of the neurohormonal activation would prevent further weight loss.

Both ACE inhibitors and beta-blockers can counteract neurohormonal activation and can thereby, in theory, have anticatabolic and hence also anticachectic effects.

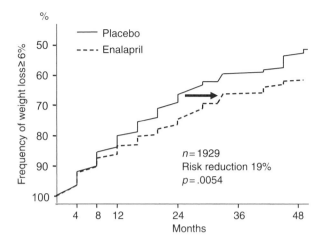

FIGURE 17.1 Weight loss in a reevaluation of the SOLVD trial. The risk of weight loss, ≥ 6% is reduced by elanapril (Adapted from Anker S.D., Negassa A., Coats A.J., Afzal R., Poole-Wilson P.A., Cohn J.N., and Yusuf S. *Lancet* 2003; 361: 1077–83. With Permission.).

17.3 HUMAN STUDIES — ACE INHIBITORS

Recent studies have proven the beneficial effect of ACE-inhibitors and beta-blockers on weight development in CHF. To start with the ACE-inhibitors, in a reanalysis of the studies of left ventricular dysfunction (SOLVD) trial, it has been found out that the ACE-inhibitor enalapril can prevent or delay the risk of weight loss in CHF patients[7] (Figure 17.1). In this study,[7] 817 (42%) out of 1929 CHF patients had a weight loss of 5% or more during the follow-up of this study. Several weight loss cutoff points of 5, 7.5, 10, and 15% (*a priori*) as well as 6% (post hoc) were used to define weight loss. Over an 8-month period, the weight loss in all of these cut points were significantly associated with reduced survival. Weight loss of 6% or more was the strongest predictor of impaired survival (adjusted hazard ratio 2.10; 95% CI, 1.77–2.49; $p < .0001$). In the SOLVD study, it was found that the ACE-inhibitor enalapril (given at a dose of 20 mg od) compared to placebo significantly reduced the risk of weight loss greater or equal to 6% (adjusted reduction 19%, $p = .0054$).[7]

In another study,[8] the ACE-inhibitor captopril was shown to have immune-modulatory effects *in vitro* in the blood from 74 CHF patients, especially in the cachectic subgroup; levels of tumor necrosis factor (TNF)-alpha were markedly increased in noncachectic CHF patients (366 ± 64 pg/ml) and even higher in cachectic CHF patients (622 ± 70 pg/ml) as compared with the healthy control subjects (221 ± 28 pg/ml). *In vitro* treatment with captopril reduced the inflammatory marker TNF-alpha in cultured peripheral blood mononuclear cells (PBMCs) from CHF patients (321 ± 39 pg/ml vs. 465 ± 52 pg/ml, $p < .05$), especially in the PBMCs from the cachectic subgroup (337 ± 38 pg/ml vs. 622 ± 70 pg/ml, $p < .05$).[8]

A combination therapy of the ACE-inhibitor enalapril with digoxin and furosemide was performed in Nigerian CHF patients and resulted in an increase in fat mass (four skinfold thickness) and muscle mass (mid-upper arm and thigh circumference).[9] The skinfold thickness increased from 27.6 ± 3.3 mm (40% of controls' thickness) to 30.1 ± 3.9 mm at 6 months, $p < .001$, 95% CI, 1.42–3.43 mm; the mid-upper arm circumference raised from 24.0 ± 2.4 cm (11% lower than controls) to 25.5 ± 3.1 cm at 6 months, $p < .001$, 95% CI, 0.87–2.1 cm; and the mid-thigh circumference showed a trend to increase from 39.3 ± 2.5 cm vs. 41.5 ± 3.5 cm at 6 month, $p = .14$, 95% CI, 0.93–5.3 cm. However, it cannot be clearly differentiated which of the used drugs in this combination was responsible for these effects.[9]

Angiotension converting enzyme inhibitors result in incomplete blockade of the production of angiotensin II. A further option to inhibit the activation of neurohormones would therefore be therapy with an angiotensin II type 1 receptor antagonist itself. Therapy with the angiotensin II type 1 receptor antagonist candesartan cilexetil resulted in reduced plasma levels of neurohormones and inflammatory cytokines.[10] However, so far there is no detailed analysis of the anticachectic effects of angiotensin II receptor antagonists in human subjects available.

17.4 HUMAN STUDIES — BETA-BLOCKERS

In patients with severe heart failure, the beta-blocker carvedilol decreased the risk of death by 35% vs. placebo.[11] In a reanalysis of this study, it could be shown that carvedilol (given at a target dose of 25 mg twice daily) reduced the risk of weight loss.[12] In this analysis, 2262 patients were studied out of which 14.1% of the placebo-treated patients and 10.2% of the carvedilol-treated patients developed cachexia during follow-up ($p = .005$). There was a highly significant decrease in mortality with increasing body mass index (BMI) ($p < .0001$). For each 1.0 unit in BMI, the risk of death decreased by 7.7%. Carvedilol-treated patients showed a significant increase in weight compared to the placebo-treated patients. Mean changes in weight in the carvedilol group vs. the placebo group after 4 months were +0.5 vs. −0.1 kg ($p = .0002$), after 8 months +0.9 vs. −0.1 kg ($p < .0001$), and after 12 months +1.1 vs. +0.2 kg ($p < .0001$).[12]

These results were confirmed by a study with a different beta-blocker, bisoprolol. A re-evaluation of a large-scale trial, in which the beta-blocker bisoprolol (given with a target dose of 10 mg od) vs. placebo was given on top of ACE-inhibitor therapy to CHF patients in New York Heart Association (NYHA) class III/IV, showed that bisoprolol prevents or delays the risk of weight loss in CHF patients (Figure 17.2).[13] In this study, a significant decrease in mortality with increasing BMI was shown as well (risk decrease of 3.5% per each 1.0 unit increase in BMI). CHF patients with a BMI greater than 25 kg/m^2 had a 25% lower mortality risk than patients with a BMI equal to or less than 25 kg/m^2 (95% CI, 8–39%). Furthermore, 11.4% (148 of 1296) of the placebo patients but only 8.6% (110 of 1281) of the bisoprolol-treated patients developed severe cachexia during

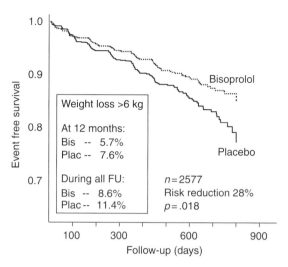

FIGURE 17.2   Weight loss >6 kg in CHF and treatment with a beta-blocker: The risk for a weight loss >6 kg was decreased in patients treated with bisoprolol compared to placebo group (Adapted from Anker, S.D., Lechat, P., and Dargie H.J. *J. Am. Coll. Cardiol.* 2003; 41: 156A–7A. With Permission.).

follow-up (hazard ratio for bisoprolol patients 0.72, 95% CI, 0.56–0.91, $p < .02$). Patients on bisoprolol showed a significant increase in weight (mean changes in weight in the bisoprolol group vs. placebo were +0.2 vs. −0.2 kg ($p = .0018$) after 15 weeks, +0.8 vs. −0.04 kg ($p < .0001$) after 12 months, and +1.2 vs. +0.03 kg ($p = .0041$) after 24 months of therapy.[13]

Therapy with carvedilol or metoprolol for 6 months caused a significantly greater weight gain in CHF patients with cachexia compared to the noncachectic group: 27 CHF patients received placebo or either carvedilol or metoprolol. After 6 months of therapy the beta-blocker groups had a significantly greater weight gain compared to the placebo group (+5.2±9.6 vs. +0.8±5.0), but there were no differences between the groups treated with carvedilol or metoprolol ($p > .2$).[14] Of note, both carvedilol[11] as well as metoprolol succinate[15] reduce mortality in CHF patients compared to placebo. In all studies mentioned above, patients with oedema at baseline were excluded to avoid bias.

17.5   OTHER CACHECTIC DISEASES

Neurohormonal activation is also seen in cachectic disease states other than heart failure, such as in cancers,[16] chronic obstructive pulmonary disease (COPD),[17] renal failure,[18,19] liver cirrhosis,[20,21] burn injuries[22], and acquired immunodeficiency syndrome (AIDS).[23] In these chronic disease states, inhibiting neurohormonal activation might also be useful to prevent and treat cachexia. This is a hypothesis that is unproven to date.

There are a few studies, however, which support this statement. A positive effect of the nonspecific beta-blocker propranolol was shown in cachectic states of different origins. For one, propranolol was shown to decrease energy expenditure, attenuate hypermetabolism, and even reverse muscle-protein catabolism in children after severe burns.[22] These burns were more than 40% of total body-surface area and propranolol was given to 13 out of 25 of these severely injured children via a nasogastral tube at a dose of 1.98 mg/kg body weight per day. Energy expenditure change at 2 weeks was -422 ± 197 kcal/day in the propranolol group vs. $+140 \pm 67$ kcal/day in the control group. Furthermore, the net muscle-protein balance increased by 82% over baseline values in the propranolol group ($p = .002$) and decreased nonsignificantly in the control group.[22]

Second, in a mouse cancer cachexia model, propranolol reduced the lipolysis induced by tumor lipid-mobilizing factor.[24] At a concentration of 10 μM, propranolol noncompetitively reduced the induction of lipolysis induced by lipid-mobilizing factor. In a third study, propranolol decreased resting energy expenditure in cancer cachexia patients.[25] Ten cancer patients with progressive weight loss were treated with the selective beta-1-blocker atenolol (50 mg/day) or the nonselective beta-1, 2-blocker propranolol (80 mg/day). Atenolol treatment reduced the resting energy expenditure by 77 ± 14 kcal/day and propranolol by 48 ± 13 kcal/day, respectively (both $p < .05$ vs. pretreatment values). However, the decrease in resting energy expenditure, attributable to the decline in heart rate, was significantly more pronounced following treatment with propanolol than with atenolol ($p < .05$).[25] We are not aware of published studies regarding the effect of ACE-inhibitors on cachexia disease states in other than cardiac.

17.6 Proven Effects with Anticachectic Potential

The available literature suggests that the positive effects of ACE-inhibitors and beta-blockers in cardiac cachexia are most probably due to an inhibition of the neurohormonal activation.[26] In 1994, it was found that ACE-inhibitors reduce the circulating levels of ANP and BNP.[27,28] It was also shown that ACE-inhibitors can restore levels of the circulating insulin-like growth factor-1 in CHF patients by reduction of the angiotensin II activity.[29] The ACE-inhibitor effects on neurohormones and catecholamines, as well as on endothelial function, are thought to improve nutritional status of tissues, reduce tissue ischemia, and local oxidative stress. All these effects may have anticachectic potential. However, many of this still needs to be proven *in vitro* and *in vivo*.

References

1. Anker, S.D., Ponikowski, P., Varney, S., Chua, T.P., Clark, A.L., Webb-Peploe, K.M., Harrington, D., Kox, W.J., Poole-Wilson, P.A., and Coats A.J. Wasting as independent risk factor for mortality in chronic heart failure. *Lancet.* 1997; 349: 1050–3.

2. Anker, S.D., Ponikowski, P.P., Clark, A.L., Leyva, F., Rauchhaus, M., Kemp, M., Teixeira M.M., Hellewell, P.G., Hooper, J., Poole-Wilson, P.A., and Coats A.J. Cytokines and neurohormones relating to body composition alterations in the wasting syndrome of chronic heart failure. *Eur. Heart J.* 1999; 20: 683–93.

3. Anker, S.D., Chua, T.P., Ponikowski, P., Harrington, D., Swan, J.W., Kox, W.J., Poole-Wilson P.A., and Coats A.J. Hormonal changes and catabolic/anabolic imbalance in chronic heart failure and their importance for cardiac cachexia. *Circulation* 1997; 96: 526–34.

4. Bolger, A.P., Sharma, R., Li, W., Leenarts, M., Kalra, P.R., Kemp, M., Coats, A.J., Anker S.D., and Gatzoulis M.A. Neurohormonal activation and the chronic heart failure syndrome in adults with congenital heart disease. *Circulation* 2002; 106: 92–9.

5. Lommi, J., Kupari, M., and Yki-Jarvinen H. Free fatty acid kinetics and oxidation in congestive heart failure. *Am. J. Cardiol.* 1998; 81: 45–50.

6. Poehlman, E.T. and Danforth, E., Jr. Endurance training increases metabolic rate and norepinephrine appearance rate in older individuals. *Am. J. Physiol.* 1991; 261: E233–E239.

7. Anker, S.D., Negassa, A., Coats, A.J., Afzal, R., Poole-Wilson, P.A., Cohn, J.N., and Yusuf S. Prognostic importance of weight loss in chronic heart failure and the effect of treatment with angiotensin-converting-enzyme inhibitors: an observational study. *Lancet* 2003; 361: 1077–83.

8. Zhao, S.P. and Xie X.M. Captopril inhibits the production of tumor necrosis factor-alpha by human mononuclear cells in patients with congestive heart failure. *Clin. Chim. Acta* 2001; 304: 85–90.

9. Adigun, A.Q. and Ajayi A.A. The effects of enalapril-digoxin-diuretic combination therapy on nutritional and anthropometric indices in chronic congestive heart failure: preliminary findings in cardiac cachexia. *Eur. J. Heart Fail.* 2001; 3: 359–63.

10. Tsutamoto, T., Wada, A., Maeda, K., Mabuchi, N., Hayashi, M., Tsutsui, T., Ohnishi M, Sawaki, M., Fujii, M., Matsumoto, T., and Kinoshita M. Angiotensin II type 1 receptor antagonist decreases plasma levels of tumor necrosis factor alpha, interleukin-6 and soluble adhesion molecules in patients with chronic heart failure. *J. Am. Coll. Cardiol.* 2000; 35: 714–21.

11. Packer, M., Coats, A.J., Fowler, M.B., Katus, H.A., Krum, H., Mohacsi, P., Rouleau, J.L., Tendera, M., Castaigne, A., Roecker, E.B., Schultz, M.K., and DeMets D.L. Effect of carvedilol on survival in severe chronic heart failure. *N. Engl. J. Med.* 2001; 344: 1651–8.

12. Anker, S.D., Coats, A.J.S., Roecker, E.B., Scherhag, A., and Packer M. Does carvedilol prevent and reverse cardiac cachexia in patients with severe heart failure? Results of the COPERNICUS study. *Eur. Heart J.* 2002; 23: 394.

13. Anker, S.D., Lechat, P., and Dargie H.J. Prevention and reversal of cachexia in patients with chronic heart failure by bisoprolol: results from the CIBIS-II study. *J. Am. Coll. Cardiol.* 2003; 41: 156A–157A.

14. Hryniewicz, K., Androne, A.S., Hudaihed, A., and Katz S.D. Partial reversal of cachexia by beta-adrenergic receptor blocker therapy in patients with chronic heart failure. *J. Card. Fail.* 2003; 9: 464–8.

15. Effect of metoprolol CR/XL in chronic heart failure: Metoprolol CR/XL Randomised Intervention Trial in Congestive Heart Failure (MERIT-HF). *Lancet* 1999; 353: 2001–7.

16. Yun, A.J., Lee, P.Y., and Bazar K.A. Clinical benefits of hydration and volume expansion in a wide range of illnesses may be attributable to reduction of sympathovagal ratio. *Med. Hypotheses* 2005; 64: 646–50.

17. Wouters, E.F., Creutzberg, E.C., and Schols A.M. Systemic effects in COPD. *Chest* 2002; 121: 127S–130S.

18. Smilde, T.D., Hillege, H.L., Navis, G., Boomsma, F., de Zeeuw, D., and van Veldhuisen D.J. Impaired renal function in patients with ischemic and non-ischemic chronic heart failure: association with neurohormonal activation and survival. *Am. Heart J.* 2004; 148: 165–72.

19. Augustyniak, R.A., Tuncel, M., Zhang, W., Toto, R.D., and Victor R.G. Sympathetic overactivity as a cause of hypertension in chronic renal failure. *J. Hypertens.* 2002; 20: 3–9.

20. Rector, W.G. and Robertson A.D. Prevalence and determinants of elevated plasma norepinephrine concentration in compensated cirrhosis. *Am. J. Gastroenterol.* 1994; 89: 2049–53.

21. Plauth, M. and Schutz E.T. Cachexia in liver cirrhosis. *Int. J. Cardiol.* 2002; 85: 83–7.

22. Herndon, D.N., Hart, D.W., Wolf, S.E., Chinkes, D.L., and Wolfe R.R. Reversal of catabolism by beta-blockade after severe burns. *N. Engl. J. Med.* 2001; 345: 1223–9.

23. Mulligan, K. and Schambelan M. Anabolic treatment with, G.H., IGF-I, or anabolic steroids in patients with HIV-associated wasting. *Int. J. Cardiol.* 2002; 85: 151–9.

24. Khan, S. and Tisdale M.J. Catabolism of adipose tissue by a tumour-produced lipid-mobilising factor. *Int. J. Cancer* 1999; 80: 444–7.

25. Hyltander, A., Daneryd, P., Sandstrom, R., Korner, U., and Lundholm K. Beta-adrenoceptor activity and resting energy metabolism in weight losing cancer patients. *Eur. J. Cancer* 2000; 36: 330–4.

26. Brink, M., Anwar, A., and Delafontaine P. Neurohormonal factors in the development of catabolic/anabolic imbalance and cachexia. *Int. J. Cardiol* 2002; 85: 111–21.

27. Sigurdsson, A., Swedberg, K., and Ullman B. Effects of ramipril on the neurohormonal response to exercise in patients with mild or moderate congestive heart failure. *Eur. Heart J.* 1994; 15: 247–54.

28. van Veldhuisen, D.J., Genth-Zotz, S., Brouwer, J., Boomsma, F., Netzer, T., Veld, T., Pinto, Y.M., Lie, K.I., and Crijns H.J. High- versus low-dose ACE inhibition in chronic heart failure: a double-blind, placebo-controlled study of imidapril. *J. Am. Coll. Cardiol.* 1998; 32: 1811–8.

29. Corbalan, R., Acevedo, M., Godoy, I., Jalil, J., Campusano, C., and Klassen J. Enalapril restores depressed circulating insulin-like growth factor 1 in patients with chronic heart failure. *J. Card. Fail.* 1998; 4: 115–9.

18 β₂-Adrenergic Agonists in the Treatment of Muscle Atrophy

Charles P. Lambert, Ergun Y. Uc, and William J. Evans

CONTENTS

SUMMARY

β₂-adrenergic agonists were first developed to treat reactive airway disease but are also known to have anabolic properties. They have recently been used in a number of human diseases to treat muscle wasting. Although the mechanism of action has not been elucidated in humans, in animal models there appears to be an increase in muscle protein synthesis and a reduction of muscle protein degradation. The increase in muscle protein synthesis may be due to an increase in insulin like growth factor-II (IGF-II) while the reduction in muscle protein degradation appears to be dependent on inhibition of the ATP-dependent ubiquitin proteolytic pathway as well as inhibition of the Ca^{2+} dependent calpain system for muscle protein breakdown. β₂-adrenergic agonist administration is associated with side effects and long-term administration has been associated with cardiac hypertrophy. Side effects of β_2-receptor agonists other than cardiovascular are headache, visual hallucinations, nervousness, irritability, restlessness, drowsiness, dizziness, insomnia, and tremors. The use of these agents in cachectic conditions appears to be warranted as they appear to be potently anabolic in a number of animal models and some human models; however, their widespread use in human clinical trials has been limited and thus the benefit to side effect ratio cannot be accurately determined at this time.

18.1 INTRODUCTION

Loss of skeletal muscle is a common feature of cachexia. Because it often accompanies weight loss in conditions such as HIV-associated wasting, cancer, heart failure, sepsis, and renal failure, it is often not recognized or treated. In addition, age-related loss of skeletal muscle, referred to as sarcopenia, is associated with reduced functional status and frailty. Although a number of anabolic agents are used to treat loss of muscle, such as testosterone, anabolic steroids, and growth hormone, the clinical use of β_2-adrenergic agonists for anabolic purposes remains very limited. β_2-adrenergic agonists were developed to treat patients with reactive airway disease. The short-acting selective β_2-adrenergic agonist albuterol was introduced in 1968; other short-acting β_2-adrenergic agonists include pirbuterol and terbutaline. More recently, long-acting β_2-adrenergic agonists were formulated, with salmeterol being introduced in the United Kingdom in 1990 and in the United States in 1994. Formoterol, another long-acting drug, was discovered in the mid-1980s and was approved in the United Kingdom in 1996 and in the United States in 2001. The evolution from nonspecific α- and β-agonists to β-specific and to β_2-specific agents originated with epinephrine (a.k.a. adrenaline), which is produced by the adrenal medulla in humans. The epinephrine molecule (stimulates α- and β-receptors) was discovered at the beginning of the 20th century.[1] β_2-adrenergic agonists are known to have an effect on skeletal muscle mass. While the precise mechanism of action remains unclear, the effects on increasing muscle size have been consistent. The purpose of this short review is to provide evidence of the anabolic effect of these agents and discuss their potential utility in treating cachectic conditions.

18.2 MECHANISM OF ACTION OF β_2-Adrenergic AGONISTS

The mechanism by which β_2-adrenergic agonists increase skeletal muscle protein content is not clear. However, a reduction in muscle proteolysis appears to be one mechanism.

18.2.1 Inhibition of Skeletal Muscle Proteolysis

There are three basic systems for skeletal muscle proteolysis, (1) the ATP-ubiquitin proteasome pathway, (2) the lysosomal pathway, and (3) the calpain pathway (which is calcium dependent). These pathways are not mutually exclusive as the proteasome is not able to degrade intact myofibrils. The calpains cannot degrade actin and myosin (reviewed in Jackman and Kandarin[2]). Studies performed in lambs and pigs suggest that β-agonist act to reduce skeletal muscle protein breakdown through the calpain system.[3,4] Calpastatin acts to inhibit the effects of calpains on skeletal muscle protein degradation. In pigs, Sensky et al.[4] reported that the infusion of the endogenous adrenergic agonist, epinephrine, increased skeletal muscle calpastatin activity by 92% and that the calpain:calpastatin ratio was significantly reduced. This would favor a reduction in muscle protein breakdown. Other studies have shown similar results.[3,5] There are other studies that support a decrease in the activation of the ATP-ubiquitin proteasome pathway. Viguerie et al.[6] examined the infusion of epinephrine at 0.04 μg/kg/min to nine healthy men. This corresponds to a plasma epinephrine concentration similar to maximal intensity physical exercise. Using microarray analysis these investigators found that the mRNA level of ubiquitin protein ligase E3A was decreased by epinephrine, suggesting reduced ubiquitin proteasome pathway activity. Costelli et al.[7] examined the effects of clenbuterol 1 mg/kg/day on tumor (Rat Ascites Hepatoma Yoshida AH-130) induced muscle wasting. They found that muscle proteolysis rates were reduced down close to control levels while there was no effect on muscle protein synthesis. They also reported reduced activation of the ATP-ubiquitin dependent protein degradative pathway.

Additionally, in the Lewis lung model of cancer[8] formoterol has been shown to reduce the gene expression of components of the ATP-ubiquitin proteolytic system induced by the tumor. Additionally, the tumors increased apoptosis in the tibialis anterior muscle and formoterol strongly inhibited this effect. Thus, these data support a role for inhibition of both the ATP-ubiquitin system and the calpain system in reducing muscle protein degradation as a result of β_2-agonist administration.

18.2.2 Stimulation of Skeletal Muscle Protein Synthesis

β_2-agonists have also been shown to stimulate muscle protein synthesis. Maclennan and Edwards[9] administered 0.250 mg/kg of clenbuterol to rats for 7 days. They found that muscle protein synthesis was increased by 43%. Previous studies[10,11] have demonstrated that the β_2-adrenergic agonist, clenbuterol, stimulates protein synthesis *in vivo* in rats. Also, β_2-adrenergic agonists act to increase α-actin mRNA,[12] α-actin,[3,12,13] and myosin light-chain[14] synthesis. The mechanism for this increase in muscle protein synthesis is less well established; however, a possible mechanism is via an increase in muscle blood flow and therefore an increase in amino acid delivery to the muscle as muscle amino acid uptake is the product of amino acid delivery and the arteriovenous difference for that amino acid. It is clear that acute administration of clenbuterol can increase muscle blood flow in rats[15]. Thus, it appears plausible that an increase in muscle blood flow may result in increased muscle protein synthesis. With regard to β_2-adrenergic agonist increase in growth factor induced production, Grant et al. [12] reported that the administration of the β_2-adrenergic agonist ractopamine did not increase the insulin-like growth factor-I (IGF-I) mRNA expression in the porcine longissimus muscle. However, another possible mechanism for the increase in muscle protein synthesis is an increase in IGF-II mRNA.[11] Clenbuterol has been shown to stimulate an $\sim$80% increase in IGF-II mRNA one day after administration and $\sim$50% increase after two days of administration in rats.[11]

18.3 β_2-Adrenergic Receptor Downregulation

One consideration when administering β_2-agonists is that chronic exposure to these agents will result in receptor downregulation. Huang et al.[16] reported a 35% reduction in the number of β_2-adrenergic receptors in rat skeletal muscle after 10 days of clenbuterol treatment (4 mg/kg of feed). Likewise, in rats, Rothwell et al.[15] reported that 18 days of treatment with clenbuterol (2 mg/kg/day) resulted in a reduction in the density of the β_2-adrenergic receptors in skeletal muscle by 65%. Along with the reduction in β_2-receptor density, there is a reduction of muscle blood flow with chronic clenbuterol administration.[15] This is believed to be due to a downregulation of the β_2-adrenergic receptors in the muscle vasculature. It is plausible to expect that if there is a downregulation of the β_2-receptors, the anabolic effect of the β_2-agonists would be reduced. Thus, from a practical standpoint with regard to maximizing efficacy, it may be advisable to have periodic reductions in the dose or removal of administration of β_2-agonists for periods of time to allow for the β_2-receptor number to return to the normal level and thus maximize the effects of these agents. To our knowledge, studies have not been conducted to test this hypothesis.

18.4 Effects of β_2-Agonists on Muscle Strength and Muscle Mass: Human Clinical Trials

Human clinical trials of β_2-adrenergic agonists have been conducted in facio-scapulohumeral, Duchenne, and Becker muscular dystrophies, in spinal muscular atrophy, lower body paraplegics, knee surgery patients, individuals with Parkinson's disease, chronic heart failure patients, individuals undergoing limb unloading, and normal healthy individuals. To our knowledge, no studies have examined the effects of β_2-adrenergic agonists in human cancer, chronic obstructive pulmonary disease, HIV infection, sepsis, renal failure, or in sarcopenia.

A number of studies have been conducted in humans on β_2-adrenergic agonists and the results of these studies are summarized in Table 18.1. In a pilot study, albuterol administration (16 mg/day for three months) in Facioscapulohumeral Muscular Dystrophy (FSHD) patients has been shown to increase lean body mass by 1.29 kg and improve strength by 12%.[17]

Kissel et al.[18] conducted a randomized, double-blind, placebo-controlled trial of albuterol in 90 patients with FSHD. They gave placebo, 8 mg of albuterol twice daily or 16 mg of albuterol twice daily over 1 year. They found no improvement in maximal voluntary isometric contraction or manual muscle testing but did see significant improvements in grip strength; lean body mass increased by 1.57 kg.

Van der Kooi et al.[19] examined the effects of resistance training and albuterol administration alone or in combination on muscle strength and muscle size in FSHD. They initiated progressive resistance training for 26 weeks prior to initiating albuterol administration. Albuterol administration was initiated at 8 mg/day for the first 2 weeks and then was increased to 16 mg/day. This occurred for an additional 26 weeks. Strength was increased in 7 out of the 12 muscle groups studied while muscle volume was increased by 1.5 l. Thus, it appears that albuterol administration with or without resistance training is effective in improving some indices of muscular strength and lean body mass in FSHD.

Fowler et al.[20] reported that administration of 8 mg/day of albuterol for 12 weeks resulted in a 2.8 fold improvement in the peak knee extensor moment (Nm) and a 6.9 fold improvement in the composite score for the maximal muscle test in 9 boys (age 5 to 9 years) with Becker or Duchenne muscular dystrophy.

Kinali et al.[21] examined the effects of 6 months of albuterol administration in children with spinal muscular atrophy (1 mg three times a day for children between the age of 5 and 6; if tolerated, dosage was increased to 2 mg 3 times per day; for older patients the dosage was 2 mg four times per day). Maximal voluntary isometric contraction was improved by 21.6%, forced vital capacity increased by 20%, and lean body mass increased by 4.2%.

Signorile et al.[22] reported that the administration of 80 mg of the β_2-adrenergic agonist metaproterenol for 4 weeks resulted in a significant increase in upper body muscle strength and forearm muscle size in lower body paraplegics.

Maltin et al.[23] examined the effects of clenbuterol (40 μg/day for 4 weeks) on patients undergoing medial meniscectomy of the knee. They found that the regain of strength in the operated leg was more rapid when clenbuterol was administered than when placebo was administered. Furthermore, in the unoperated leg, knee extensor strength improved above initial values at week 6 in the drug group but not in the placebo group.

We have reported[24] that albuterol administration (16 mg/day for 14 weeks) increased muscle mass by 5.3% (assessed by computed tomography scans) in elderly Parkinson's disease patients without resistance-exercise training (see Figure 18.1).

Harrington et al.[25] examined the effects of 16 mg of slow-release salbutamol or placebo for 3 weeks in 12 patients with chronic heart failure. There was no change in maximal voluntary isometric contraction strength, muscular fatigue, or muscle cross-sectional area. A potential reason for this negative finding is the short duration of administration of this β_2-adrenergic agent.

Caruso et al.[26] examined the effects of resistance training and albuterol administration (16 mg/day) to resistance training and placebo with regard to attenuating the decline in muscle strength and mass related to 40 days of unilateral limb suspension (limb unloading). They found no beneficial effect of albuterol in conjunction with resistance exercise compared to resistance exercise alone regarding the preservation of muscle mass. They found that concentric total work and eccentric

TABLE 18.1
Human Studies with β_2-Adrenergic Agonists Administration

Study	Patient population/mean age and range	Duration of study	Drug and dose	Resistance training	Δ Strength	Δ Lean body mass and/or muscle mass/method
Kissel et al.[17]	FSHD/36.2 (19–50)	12 weeks	Albuterol/ 16 mg/day	No	+12%	+3%/DEXA
Caruso et al.[28]	Healthy w/out disease/21.3 (18–27)	6 weeks	Albuterol/ 16 mg/day	Yes	+29.7%	NS/Circumferences with skinfolds
Martineau et al.[27]	Healthy w/out disease/27 (19–38)	3 weeks	Albuterol/ 16 mg/day	No	+12.0% Quadriceps; +22% hamstrings	NA
Maltin et al.[10]	Medial meniscal injury/31 (19–40)	4 weeks	Clenbuterol/ 40 μg/day	No	+4.5% Operated leg; +10.3% Unoperated leg	NS; CT scans
Fowler et al.[20]	Duchenne and Becker muscular dystrophy/7.4 (5–9)	12 weeks	Albuterol/ 8 mg/day	No	2.8 fold and 6.9 fold increases, for peak knee extensor moment and maximal muscle test, respectively	NA
Kissel et al.[18]	FSHD/35.4 (18–60)	52 weeks	Albuterol; 16 mg or 32 mg/day	No	Significant increase in grip strength in both treatment arms	Significant +1.57 kg; for 32 mg/day group via DEXA
Kinali et al.[21]	Spinal muscular atrophy; young children ~5–6	24 weeks	Albuterol; 3–8 mg/day	No	+21.6% in maximal voluntary contraction	Significant +4.2% at 6 months via DEXA

Reference	Disease/age	Duration	Drug; dose		Functional effect	Effect on muscle mass
Van der Kooi et al. (2002)	FSHD; 38 (18–60)	26 weeks of resistance training followed by 26 weeks of albuterol + resistance training	Albuterol 16 mg/day	Yes	Significant increase in maximal voluntary contraction in 7 of 12 muscle groups	Significant increase in muscle volume via skinfolds. Significant increase in muscle volume via placebo group; +1.5 L
Uc et al.[24]	Parkinson's disease; 59.8	14 weeks	Albuterol; 4 mg/day	No	No significant increase in leg extensor or chest-press strength	+5.3% increase in thigh muscle cross-sectional area; +9.5% increase in fat-free mass via whole body plethysmography combined with total body water measurement
Harrington et al.[25]	Chronic heart failure/61.9	3 weeks	Salbutamol; 16 mg/day	No	No effect on quadriceps maximal isometric strength or fatigability but significantly improved maximal expiratory mouth pressure	No significant increase in muscle size as measured by computed tomography.
Caruso et al.[28]	Young healthy/23.7	40 days	Albuterol; 16 mg/day	Yes	Total concentric and eccentric work over 8 maximal effort leg press contractions was improved with albuterol + RT vs. RT alone	No significant effect on muscle mass as measured by thigh circumference with skinfolds

FSHD is facioscapulohumeral muscular dystrophy; DEXA is dual x-ray absorptiometry; RT is resistance training; NS is nonsignificant; NA is not applicable; CT is computed tomography.

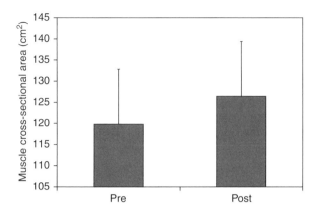

FIGURE **18.1**   Change in thigh muscle cross-sectional area in fluctuating Parkinson's disease patients as a result of ingestion of albuterol (16 mg/day) for 14 weeks.[24]

total work were improved during 8 maximal effort leg-press contractions in the albuterol + resistance exercise group when compared to the placebo + resistance exercise group. They suggested that the improvement in work capacity might have been due to heightened contractile protein sensitivity for calcium. The inability to detect changes in muscle mass may have been due to the use of leg circumferences with skinfolds for their cross-sectional area measurements rather than the use of computed tomography or magnetic resonance imaging scans, the latter two being more sensitive to changes in muscle mass.

Martineau et al.[27] administered 16 mg/day of albuterol orally for 3 weeks in a placebo controlled study of healthy young men and reported a 12.0% increase in the strength of the quadriceps after 14 days, which was still present at 21 days. Additionally, hamstring muscle strength increased by 22% at day 21. Regardless of the mechanism, it appears that albuterol administration and resistance training can counteract the effects of limb unloading in humans. In young healthy individuals free of disease, 9 weeks of albuterol administration (16 mg/day) combined with resistance training resulted in a significant 29.7% increase in muscle strength and the posttesting values were 23.5% higher for the albuterol and exercise group than the exercise-alone group.[28] Thus, based on these studies it would appear that albuterol is effective in stimulating significant strength improvements in individuals free of disease.

18.5   EFFECTS OF β_2-AGONISTS ON MUSCLE STRENGTH AND MUSCLE MASS, AND MYOSIN HEAVY-CHAIN CHANGES: ANIMAL STUDIES

Ryall et al.[29] treated groups of rats with fenoterol at doses of 0.025, 0.25, 0.5, 1.0, and 2.0 mg/kg/day for 4 weeks. They found progressive increases in muscle mass relative to body mass of the extensor digitorum longus (EDL) following the administration of 0.25, 0.5, 1.0 mg/kg/day. In the soleus there was an increase in muscle mass relative to body mass at the 0.25 and 0.5 mg/kg/day doses but not any further increase in muscle mass at the 1.0 or 2.0 mg/kg/day dose. Fenoterol treatment did not increase heart muscle mass relative to body mass at 0.025 mg/kg/day but did at the doses of 0.25, 0.5, and 1.0 mg/kg/day. Of particular interest is that fenoterol administration resulted in reduced fatigue resistance (increased fatigability) in both EDL and soleus. The authors suggested this was the result of reduced muscle oxidative capacity. This may be due to a progression of muscle fibers toward a faster phenotype.

Ryall et al.[30] administered clenbuterol (2 mg/kg/day) or fenoterol (2.8 mg/kg/day) at equimolar doses to rats. The authors chose this dose of clenbuterol because it appears to greatly stimulate the accretion of muscle with little desensitization of the β_2-adrenergic receptors. Fenoterol treatment for 4 weeks increased muscle mass of the EDL by 27% while clenbuterol administration increased muscle mass by 20%. Fenoterol increased soleus muscle mass by 26% vs. control and by 18% when compared to the clenbuterol treated rats. There was no significant effect of clenbuterol on soleus muscle mass. The force generating capacity of soleus and EDL muscles was significantly greater after the fenoterol treatment than the clenbuterol treatment. However, fenoterol had a greater effect on heart mass than clenbuterol with clenbuterol causing increased heart mass relative to controls. Of note is that fenoterol stimulates both β_1- and β_2-receptors while clenbuterol only acts at β_2-receptors.

Lynch et al.[31] reported that the administration of 1.5 to 2.0 mg/kg/day of clenbuterol for 1 year resulted in increases of muscle mass relative to body weight of 18% in the EDL, 18% in the soleus, and a 12% in the tibialis anterior muscle. The authors reported, however, that there was no improvement in maximal tetanic force production (Po), or normalized Po. Additionally, there was no effect of clenbuterol on normalized power output.

Based on a number of studies it appears that clenbuterol acts to decrease the effects of hindlimb unloading on muscle mass loss,[32–34] and decreases the effects of denervation on muscle atrophy.[35,36] The dosage range used in these studies was 1.5 to 3 mg/kg/day. Further, β_2-agonists appear to increase the recovery of muscle force production after injury.[37] Also, a dose as low as 10 μg/kg/day reduced the decline in isometric force production resulting from hindlimb suspension by 8%.[38]

Another, consistent finding in the animal literature is that in the soleus (a muscle with a very high percentage of myosin heavy chain-I) there is a $\sim$15–25% conversion to a myosin heavy chain type II phenotype with $\sim$2–8 weeks of administration of β_2-agonist administration.[34,39,40] If this fiber type change occurs with β_2-agonist administration in elderly humans this may be beneficial as typically there is a loss of type II fiber size and number with sarcopenia. This hypothesis warrants scientific inquiry.

Busquets et al.[8] examined the effects of the β_2-agonist formotorol on muscle wasting induced by the Rat Yoshida AH-130 Ascites Hepatoma and Mouse Lewis Lung Carcinoma lines. The tumor-bearing animals who received formotorol had muscles that were larger than those in the tumor-bearing rats that did not receive formoterol: 9% larger for gastrocnemius, 11% for the tibialis anterior, 16% for the EDL, and 11% for the heart. For the Lewis lung carcinoma model, formoterol with the tumor compared to tumor alone was 19% larger for the gastrocnemius, 13% for the tibialis anterior, and 18% for the soleus. The tumors also increased the expression of the calcium-dependent m-calpain in muscle but formoterol had no impact on expression of this gene.

18.6 SIDE EFFECTS OF β_2-AGONISTS

There are numerous reports from animal studies showing cardiac hypertrophy with β_2-agonist administration. There are many potential detrimental effects of cardiac hypertrophy on cardiac function. It is likely that similar effects occur in humans although to our knowledge this has not been substantiated by scientific data. In steer, the increase in heart rate seen with clenbuterol treatment appears to be a response at the β_2-receptor in the heart as a result of β_2-agonist-induced hypotension.[41] Another possibility is that clenbuterol is not strictly β_2-selective but also acts at β_1-receptors. Regardless of the exact mechanism it appears that β_2 stimulation is the cause for the increase in heart rate. With this in mind, Hoey et al.[41] suggested that the use of β_1-antagonists may be useful in limiting the effects of clenbuterol on the heart. This has yet to be experimentally determined.

The cardiac effects of β_2-agonists may not be detrimental in all cases, for example in cancer patients there is catabolism of cardiac muscle in addition to catabolism of skeletal muscle. Thus, the cardiotrophic effects of β_2-agonists on the heart may not be as detrimental in this situation.

Investigators from Greenhaff's laboratory have used another approach to minimize the effects of β_2-agonists on the heart.[42] They administered the prodrug BRL-47672 to rats. This prodrug is metabolized to clenbuterol after administration. BRL-47672 administration showed large and statistically significant blunted responses with regard to heart rate, the change in mean arterial pressure, and the change in vascular conductance when compared to the effects of clenbuterol. Thus, the use of prodrugs of commonly used β_2-agonists may be a way to minimize the deleterious effects of β_2-agonists on the heart.[42]

Duncan et al.[44] reported that the administration of clenbuterol to rats (2 mg/kg/day) combined with exercise training (3 groups; endurance swimming, treadmill sprinting, and voluntary wheel running for 16 to 20 weeks) resulted in reduced voluntary wheel running by ~57%, and the rats in the other two groups were unable to complete the sprinting and swimming protocols. These results were relative to animals that were trained but did not receive clenbuterol. Because there was a substantial increase in heart mass (~20%) these authors suggested that the increased heart mass may have been the cause of the reduced exercise capacity.

Side effects of β_2-receptor agonists other than cardiovascular are (specifically the Food and Drug Administration [FDA] approved albuterol) headache, visual hallucinations, nervousness, irritability, restlessness, drowsiness, dizziness, insomnia, and tremors.[24]

18.7 CONCLUSION

In comparison with androgens such as testosterone and anabolic steroids, it would appear that β_2-adrenergic agonists may be a safer alternative when considering that many cachectic states such as chronic obstructive pulmonary disease and chronic heart failure are associated with aging and aging may make one more susceptible to prostate carcinoma. Androgens, in turn, may exacerbate any preexisting prostate abnormalities. Exercise is another potential therapeutic option and would not appear to have apparent side effects. However, data are scant on the use of exercise in treating catabolic conditions. Nutritional support alone may be very safe but appears to be of limited efficacy in treating catabolic conditions. Acetylcholinesterase (ACE) inhibitors have been used successfully to treat weight loss associated with chronic heart failure; however, to our knowledge they have not been used in any other cachectic conditions. The use of β_2-adrenergic agonists to treat muscle wasting in human clinical trials of diseases traditionally considered cachectic such as chronic obstructive pulmonary disease (COPD), chronic heart failure, and cancer have received very limited attention. The fact that β_2-agonists appear to be effective in the treatment of various animal models of cancer and that they stimulate anabolic and inhibit catabolic pathways common to many muscle-wasting diseases suggests they may be effective in a number of cachectic disease states. Additional human clinical trials are necessary to substantiate this. Further, the potential improvements in muscle mass and strength in cachectic conditions must be weighed against the potentially detrimental effects on cardiac tissue mass and potential for detrimental alterations in cardiac function associated with β_2-adrenergic agonists.

REFERENCES

1. Hartung, W. Epinephrine and related compounds: Influence of structure on physiologic activity. *Chem. Rev.* 9: 389, 1931.
2. Jackman, R.W. and S.C. Kandarian. The molecular basis of skeletal muscle atrophy. *Am. J. Physiol. Cell Physiol.* 287: C834–C843, 2004.
3. Koohmaraie, M., S.D. Shackelford, N.E. Muggli-Cockett et al. Effect of the beta-adrenergic agonist L644,969 on muscle growth, endogenous proteinase activities, and postmortem proteolysis in wether lambs. *J. Anim. Sci.* 69: 4823–4835, 1991.
4. Sensky, P.L., T. Parr, R.G. Bardsley et al. The relationship between plasma epinephrine concentration and the activity of the calpain enzyme system in porcine longissimus muscle. *J. Anim. Sci.* 74: 380–387, 1996.
5. Killefer, J. and M. Koohmaraie. Bovine skeletal muscle calpastatin: Cloning, sequence analysis, and steady-state mRNA expression. *J. Anim. Sci.* 72: 606–614, 1994.
6. Viguerie, N., K. Clement, P. Barbe et al. In vivo epinephrine-mediated regulation of gene expression in human skeletal muscle. *J. Clin. Endocrinol. Metab.* 89: 2000–2014, 2004.
7. Costelli, P., C. Garcia-Martinez, M. Llovera et al. Muscle protein waste in tumor-bearing rats is effectively antagonized by a beta 2-adrenergic agonist

(clenbuterol). Role of the ATP-ubiquitin-dependent proteolytic pathway. *J. Clin. Invest.* 95: 2367–2372, 1995.

8. Busquets, S., M.T. Figueras, G. Fuster et al. Anticachectic effects of formoterol: A drug for potential treatment of muscle wasting. *Cancer Res.* 64: 6725–6731, 2004.

9. MacLennan, P.A. and R.H. Edwards. Effects of clenbuterol and propranolol on muscle mass. Evidence that clenbuterol stimulates muscle beta-adrenoceptors to induce hypertrophy. *Biochem. J.* 264: 573–579, 1989.

10. Maltin, C.A., S.M. Hay, M.I. Delday et al. The action of the beta-agonist clenbuterol on protein metabolism in innervated and denervated phasic muscles. *Biochem. J.* 261: 965–971, 1989.

11. Sneddon, A.A., M.I. Delday, J. Steven et al. Elevated IGF-II mRNA and phosphorylation of 4E-BP1 and p70(S6k) in muscle showing clenbuterol-induced anabolism. *Am. J. Physiol. Endocrinol. Metab.* 281: E676–E682, 2001.

12. Grant, A.L., D.M. Skjaerlund, W.G. Helferich et al. Skeletal muscle growth and expression of skeletal muscle alpha-actin mRNA and insulin-like growth factor I mRNA in pigs during feeding and withdrawal of ractopamine. *J. Anim. Sci.* 71: 3319–3326, 1993.

13. Helferich W.G., D.B. Jump, D.B. Anderson et al. Skeletal muscle alpha-actin synthesis is increased pretranslationally in pigs fed the phenethanolamine ractopamine. *Endocrinology* 126: 3096–3100, 1990.

14. Smith, S.B., D.K. Garcia, S.K. Davis et al. Elevation of a specific mRNA in longissimus muscle of steers fed ractopamine. *J. Anim. Sci.* 67: 3495–3502, 1989.

15. Rothwell, N.J., M.J. Stock, and D.K. Sudera. Changes in tissue blood flow and beta-receptor density of skeletal muscle in rats treated with the beta2-adrenoceptor agonist clenbuterol. *Br. J. Pharmacol.* 90: 601–607, 1987.

16. Huang, H., C. Gazzola, G.G. Pegg et al. Differential effects of dexamethasone and clenbuterol on rat growth and on beta2-adrenoceptors in lung and skeletal muscle. *J. Anim. Sci.* 78: 604–608, 2000.

17. Kissel, J.T., M.P. McDermott, R. Natarajan et al. Pilot trial of albuterol in facioscapulohumeral muscular dystrophy. FSH-DY Group. *Neurology* 50: 1402–1406, 1998.

18. Kissel, J.T., M.P. McDermott, J.R. Mendell et al. Randomized, double-blind, placebo-controlled trial of albuterol in facioscapulohumeral dystrophy. *Neurology* 57: 1434–1440, 2001.

19. van der Kooi, E.L., O.J. Vogels, R.J. van Asseldonk et al. Strength training and albuterol in facioscapulohumeral muscular dystrophy. *Neurology* 63: 702–708, 2004.

20. Fowler, E.G., M.C. Graves, G.T. Wetzel et al. Pilot trial of albuterol in Duchenne and Becker muscular dystrophy. *Neurology* 62: 1006–1008, 2004.

21. Kinali, M., E. Mercuri, M. Main et al. Pilot trial of albuterol in spinal muscular atrophy. *Neurology* 59: 609–610, 2002.

22. Signorile, J.F., K. Banovac, M. Gomez et al. Increased muscle strength in paralyzed patients after spinal cord injury: Effect of beta-2 adrenergic agonist. *Arch. Phys. Med. Rehabil.* 76: 55–58, 1995.

23. Maltin, C.A., M.I. Delday, J.S. Watson et al. Clenbuterol, a beta-adrenoceptor agonist, increases relative muscle strength in orthopaedic patients. *Clin. Sci. (Lond.)* 84: 651–654, 1993.

24. Uc, E.Y., C.P. Lambert, S.I. Harik et al. Albuterol improves response to levodopa and increases skeletal muscle mass in patients with fluctuating Parkinson disease. *Clin. Neuropharmacol.* 26: 207–212, 2003.

25. Harrington, D., T.P. Chua, and A.J. Coats. The effect of salbutamol on skeletal muscle in chronic heart failure. *Int. J. Cardiol.* 73: 257–265, 2000.

26. Caruso, J.F., J.L. Hamill, M. Yamauchi et al. Albuterol helps resistance exercise attenuate unloading-induced knee extensor losses. *Aviat. Space Environ. Med.* 75: 505–511, 2004.

27. Martineau, L., M.A. Horan, N.J. Rothwell et al. Salbutamol, a beta 2-adrenoceptor agonist, increases skeletal muscle strength in young men. *Clin. Sci. (Lond.)* 83: 615–621, 1992.

28. Caruso, J.F., J.F. Signorile, A.C. Perry et al. The effects of albuterol and isokinetic exercise on the quadriceps muscle group. *Med. Sci. Sports Exerc.* 27: 1471–1476, 1995.

29. Ryall, J.G., D.R. Plant, P. Gregorevic et al. Beta 2-agonist administration reverses muscle wasting and improves muscle function in aged rats. *J. Physiol.* 555: 175–188, 2004.

30. Ryall, J.G., P. Gregorevic, D.R. Plant et al. Beta 2-agonist fenoterol has greater effects on contractile function of rat skeletal muscles than clenbuterol. *Am. J. Physiol. Regul. Integr. Comp. Physiol.* 283: R1386–R1394, 2002.

31. Lynch, G.S., R.T. Hinkle, and J.A. Faulkner. Year-long clenbuterol treatment of mice increases mass, but not specific force or normalized power, of skeletal muscles. *Clin. Exp. Pharmacol. Physiol.* 26: 117–120, 1999.

32. Dodd, S.L. and T.J. Koesterer. Clenbuterol attenuates muscle atrophy and dysfunction in hindlimb-suspended rats. *Aviat. Space Environ. Med.* 73: 635–639, 2002.

33. Herrera, N.M., Jr., A.N. Zimmerman, D.D. Dykstra et al. Clenbuterol in the prevention of muscle atrophy: A study of hindlimb-unweighted rats. *Arch. Phys. Med. Rehabil.* 82: 930–934, 2001.

34. Ricart-Firinga, C., L. Stevens, M.H. Canu et al. Effects of beta(2)-agonist clenbuterol on biochemical and contractile properties of unloaded soleus fibers of rat. *Am. J. Physiol. Cell. Physiol.* 278: C582–C588, 2000.

35. Agrawal, S., P. Thakur, and S.S. Katoch. Beta adrenoceptor agonists, clenbuterol, and isoproterenol retard denervation atrophy in rat gastrocnemius muscle: Use of 3-methylhistidine as a marker of myofibrillar degeneration. *Japan. J. Physiol.* 53: 229–237, 2003.

36. Zeman, R.J., R. Ludemann, and J.D. Etlinger. Clenbuterol, a beta 2-agonist, retards atrophy in denervated muscles. *Am. J. Physiol.* 252: E152–E155, 1987.

37. Beitzel, F., P. Gregorevic, J.G. Ryall et al. Beta2-adrenoceptor agonist fenoterol enhances functional repair of regenerating rat skeletal muscle after injury. *J. Appl. Physiol.* 96: 1385–1392, 2004.

38. Chen, K.D. and S.E. Alway. Clenbuterol reduces soleus muscle fatigue during disuse in aged rats. *Muscle Nerve* 24: 211–222, 2001.

39. Jones, S.W., R.J. Hill, P.A. Krasney et al. Disuse atrophy and exercise rehabilitation in humans profoundly affects the expression of genes associated with the regulation of skeletal muscle mass. *FASEB J.* 18: 1025–1027, 2004.

40. Rajab P., J. Fox, S. Riaz et al. Skeletal muscle myosin heavy chain isoforms and energy metabolism after clenbuterol treatment in the rat. *Am. J. Physiol. Regul. Integr. Comp. Physiol.* 279: R1076–R1081, 2000.

41. Hoey, A.J., M.L. Matthews, T.W. Badran et al. Cardiovascular effects of clenbuterol are beta 2-adrenoceptor-mediated in steers. *J. Anim. Sci.* 73: 1754–1765, 1995.

42. Jones, S.W., D.J. Baker, S.M. Gardiner et al. The effect of the beta2-adrenoceptor agonist prodrug BRL-47672 on cardiovascular function, skeletal muscle myosin heavy chain, and MyoD expression in the rat. *J. Pharmacol. Exp. Ther.* 311: 1225–1231, 2004.

43. Duncan, N.D., D.A. Williams, and G.S. Lynch. Deleterious effects of chronic clenbuterol treatment on endurance and sprint exercise performance in rats. *Clin. Sci. (Lond.)* 98: 339–347, 2000.

44. van der Kooi, E.L., O.J. Vogels, R.J. Van Asseldonk et al. Strength training and albuterol in facioscapulohumeral muscular dystrophy. *Neurology* 63: 702–708, 2004.

19 Androgens, Anabolic Steroids, and Glucocorticoids

Lindsey Harle, Shehzad Basaria, M. D. and Adrian S. Dobs, M. D.

CONTENTS

SUMMARY

Patients suffering from cachexia are in a constant state of catabolism and suffer
from significant morbidity. In some conditions, wasting syndromes have also been
associated with increased mortality. Hence, there is a tremendous need to treat
this state in order to improve patient's independence, quality of life, and mortality.
Anabolic-androgenic agents and glucocorticoids have been studied for this purpose
in various patient populations suffering from cachexia. In certain diseases, they
have shown significant improvement in wasting. However, in some other states,
the evidence is insufficient to make any recommendations and more research is
needed. In this chapter, we summarize the efficacy of anabolic-androgenic steroids
and glucocorticoids in the treatment of various cachexia-syndromes.

19.1 INTRODUCTION

Patients who develop cachexia (of any etiology) experience significant loss of
lean body mass (LBM) and increased resting energy expenditure. This in part, is
due to the constant state of inflammation that is induced by chronic disease. In
addition to increased energy expenditure, many chronic diseases, including AIDS
and cancer, are frequently associated with hypogonadism. Hypogonadism, irrespective of its etiology, is itself associated with decreased LBM, reduced muscle
strength, and increased fat mass. Hence it is conceivable that hypogonadism that
occurs in wasting may at least in part be responsible for this unfavorable body composition. Since testosterone (and its derivatives) have anabolic effects, they have
been used in many clinical trials to increase LBM and muscle strength in patients
suffering from cachexia. Several testosterone derivatives have been developed,
such as nandrolone, oxymetholone, and oxandrolone, that have more anabolic and

less androgenic activity. In addition to androgens, glucocorticoids with their anti-inflammatory actions also have potential therapeutic benefit in cachectic patients. In this chapter, we briefly discuss first the normal endocrine physiology of gonadal and adrenal axes, the pharmacokinetics of androgens and glucocorticoids, and their mechanism of action. At the end of the chapter, we report the data on the efficacy of androgen and glucocorticoid administration in the treatment of wasting in various patient populations.

19.2 PHYSIOLOGY

19.2.1 Testosterone

In men, testosterone is produced by the Leydig cells of the testes. In women, testosterone is synthesized by the ovaries and adrenal glands. Cholesterol is the primary substrate for testosterone synthesis. The stimulus for testosterone production is luteinizing hormone (LH), which is released from the anterior pituitary in response to secretion of gonadotropin-releasing hormone (GnRH) from the hypothalamus.[1] Young healthy men secrete between 3–11 mg of testosterone daily, which translates into plasma concentrations of 300–1000 ng/dl.[2] Females have plasma testosterone level of 15–65 ng/dl.

Approximately 54% of plasma testosterone is loosely bound to albumin and easily dissociates from the protein to provide androgens to the tissues. Approximately 44% of testosterone is tightly bound to sex hormone binding globulin.[3] Hence, only 2% of testosterone circulates in the free form and is biologically active. The combination of free and albumin-bound testosterone is also called bioavailable testosterone. A small fraction of testosterone is converted to estradiol by the enzyme aromatase, which is abundant in adipose tissue. Similarly, in tissues like the prostate, seminal vesicles, and pubic skin, testosterone is converted to dihydrotestosterone (DHT) by the enzyme 5α-reductase. Hence, DHT is mainly responsible for the development of sexual organs. DHT has a 2- to 10-fold higher binding affinity to the androgen receptor.[4] Muscle tissue has minimal 5α-reductase enzyme, hence, testosterone itself is responsible for its anabolic effects on the muscle.[5] Synthetic testosterone derivatives like oxandrolone and nandrolone, etc. are more anabolic and less androgenic and have good oral bioavailability.

19.2.2 Glucocorticoids

Glucocorticoids are secreted by the zona fasciculata of the adrenal cortex. Cortisol comprises 95% of secreted glucocorticoids but corticosterone is secreted in a much smaller amount. Adrenocorticotropic hormone (ACTH) is the stimulus for glucocorticoid secretion. Adrenocorticotropic hormone is released from the anterior pituitary in response to corticotropin-releasing factor from the hypothalamus. Its secretion is inhibited by cortisol through negative feedback. The effect of ACTH on adrenocortical cells is mediated via production of cyclic AMP and activation of protein kinase A, resulting in the conversion of cholesterol to pregnenolone. A series of reactions results in the ultimate production of cortisol and corticosterone.[1]

In the absence of stress, a normal adult secretes 15 to 20 mg of cortisol daily, which follows a circadian rhythm (peaks early in the morning between 6–8 am). Cortisol exerts its effects via binding to intracellular receptors and directly regulating transcription of genes that contain glucocorticoid response elements (GRE).[6]

19.3 PHARMACOKINETICS

19.3.1 Testosterone and Other Anabolic Steroids

Testosterone in its endogenous form is rapidly degraded upon administration; therefore modified testosterone analogs have been developed to attain a longer half-life (Table 19.1). Class A analogs are produced through esterification of the 17β-hydroxyl group of the testosterone molecule; these include testosterone propionate, testosterone enanthate, and testosterone cypionate. The longer side chains of these molecules make them less polar, hence slowing their absorption and metabolism. Class B analogs have been alkylated at the 17α position, for example, methyltestosterone. 17α alkylation renders these molecules resistant to hepatic metabolism, thus making them orally active. Class C analogs are produced by modification of testosterone rings, for example, mesterolone (also orally active).

TABLE 19.1
Testosterone Derivatives and Anabolic Agents Currently Available

Drug	Route of delivery	Dosage
Testosterone		
Testosterone propionate	IM	10–25 mg 2–3x/week
Testosterone enanthate	IM	50–400 mg every 2–4 week
Testosterone cypionate	IM	50–400 mg every 2–4 weeks
Testosterone patch	Topical	5 mg/d
Testosterone gel	Topical	5 g/d (1% testosterone)
Anabolic Agents		
Nandrolone decanoate	IM	50–200 mg/week
Nandrolone phenpropionate	IM	50–200 mg/week
Methyltestosterone	PO	10–50 mg/d; 800 mg/d initially
Danazol	PO	400–600 mg/d
Fluoxymesterone	PO	5–20 mg/d
Methandrostenolone	PO	50–100 mg/d
Oxandrolone	PO	5–10 mg/d
Oxymetholone	PO	1–5 mg/kg/d
Stanozolol	PO	6 mg/d

Oral bioavailability of testosterone is poor due to extensive first-pass metabolism by the hepatic cytochrome P450 enzyme family. Intramuscular (IM) injection, implantable pellets, buccal, and transdermal systems of delivery are available. The less polar testosterone esters are absorbed and metabolized more slowly upon IM administration, giving them a longer half-life. These esters are metabolized in the liver and hydrolyzed to free testosterone. Injections typically have 2–4 weeks duration of action. Testosterone patches and gels result in continual absorption over a 24-h period, with the skin acting as a reservoir for sustained release. Implantable testosterone pellets last 3–4 months with a single administration providing sustained release. Buccal mucoadhesive systems of administration lead to testosterone absorption through the gum and cheek surfaces.

19.3.2 Glucocorticoids

Synthetic glucocorticoids are synthesized from bovine cholic acid or plant-based steroid sapogenins. They are rapidly and completely absorbed on oral administration. Chemical modification of the molecular structure can alter receptor binding affinities and half-life. Ninety percent of circulating cortisol is bound to corticosteroid-binding protein, the remainder being free or loosely bound to albumin and is biologically active. Cortisol has a half-life of 60 to 90 min; most of it is inactivated in the liver and secreted in the urine.

19.4 MECHANISM OF ACTION

19.4.1 Testosterone and Anabolic Steroids

The mechanisms through which anabolic steroids lead to net muscle synthesis are incompletely understood. Their primary effect is through transcriptional activation, leading to the production of various proteins. Other effects, such as induction of androgen receptor and growth hormone production, are the result of this transcriptional control. Testosterone also antagonizes the inflammatory action of physiologically produced glucocorticoids. Below is a brief review of these mechanisms.

19.4.1.1 Increased protein synthesis

The androgen receptor is ubiquitous throughout the body and its levels of expression are variable in various tissues.[7, 8] Testosterone exerts its effects by binding to a cytosolic receptor. Upon binding, the testosterone-receptor complex moves into the nucleus where it binds with a nuclear protein and induces DNA transcription and protein synthesis. The exact physiological effect of testosterone stimulation depends upon the tissue on which it acts; the common end-point is increased net protein synthesis. Various coactivators and corepressors of the androgen receptor are also present in different concentrations throughout the body, and these may contribute to the wide array of effects exerted by anabolic-androgenic steroids.

In muscle tissue, treatment with testosterone results in increased production of contractile and noncontractile proteins.[9–11] Similarly, short-term treatment with oxandrolone results in a 44% increase in muscle protein synthesis.[12]

19.4.1.2   Increased androgen receptor production

Supraphysiological doses of testosterone have been shown to result in increased muscle mass,[13] despite the fact that androgen receptors are saturated at physiological testosterone levels. One explanation for this may be the upregulation of synthesis of androgen receptor in response to exposure to anabolic steroids.[14–16] Oxandrolone, a synthetic testosterone analog, significantly increases mRNA levels of the skeletal muscle androgen receptor.[17] These findings suggest that supraphysiological doses of anabolic steroids could be effective due to their ability to increase the number of androgen receptors upon which they can act.

19.4.1.3   Stimulation of growth hormone secretion and insulin-like growth factor-I synthesis

Testosterone also exerts anabolic effects by increasing the release of growth hormone (GH) and increasing insulin-like growth factor-I (IGF-I) synthesis. GH is an anabolic hormone that leads to net protein synthesis. Hypogonadal patients have decreased levels of GH and androgen administration increases pituitary release of GH.[18,19] IGF-I (main mediator of GH action) also stimulates skeletal muscle protein synthesis.[20] Androgens also stimulate the production of IGF-I within the skeletal muscle since increased levels of IGF-I mRNA can be detected following androgen administration.[21,22] Testosterone also directly stimulates hepatic production of IGF-I.[18,19]

19.4.1.4   Antagonism of glucocorticoid effects

It has been proposed that androgens oppose the catabolic action of glucocorticoids. There is a high degree of homology between the androgen receptor and the glucocorticoid receptor.[23] High doses of testosterone may result in blockade of the glucocorticoid receptor, hence preventing glucocorticoid-induced protein degradation.[24–26] Testosterone also blocks the action of glucocorticoids in the induction of hepatic tyrosine amino transferase[27,28] and also antagonizes glucocorticoid hormone response elements at the level of the receptor.[26]

19.4.1.5   Decreased inflammatory cytokine production

The catabolic cytokines TNF-α and interleukin-1β (IL-1β) are elevated in many cachectic patients indicating an underlying inflammatory state. An inverse relationship between levels of these cytokines and testosterone has been demonstrated.[29]

This suggests that anabolic steroids may have an anticatabolic effect via interference with the production of these cytokines through a yet unknown mechanism. Several case reports and clinical trials have found testosterone therapy to be beneficial in patients with inflammatory diseases such as rheumatoid arthritis and systemic lupus erythematosus.[30–32] Testosterone has been shown to decrease production of TNF-α, IL-1β, and IL-6 *in vitro*.[33,34] One clinical trial found that testosterone administration in hypogonadal men results in a reduction in TNF-α and IL-1β and an increase in the anti-inflammatory cytokine IL-10.[35]

19.4.2 Glucocorticoids

Glucocorticoids have widespread effects, including stimulation of gluconeogenesis, protein catabolism, and lipolysis. It is the anti-inflammatory property of cortisol that is pertinent to the treatment of cachexia. Hence our discussion will be limited to these actions. Cortisol dramatically reduces the inflammatory response. Glucocorticoid administration leads to decreased serum concentrations of lymphocytes, monocytes, eosinophils, and basophils (due to their migration to lymphoid tissue) and increased serum concentration of neutrophils (due to decreased extravasation from the site of inflammation). Glucocorticoids also inhibit the function of macrophages and other antigen-presenting cells by stabilizing the lysosomal membrane, hence preventing the release of proteolytic enzymes.[1] Glucocorticoids inhibit the activation of phospholipase A_2 and cyclooxygenase II, thereby decreasing production of prostaglandins, leukotrienes, and platelet activating factor.[6] Since prostaglandins are potent anorexic agents (in animal models)[36], inhibition of their production by glucocorticoids may result in an improvement in appetite. Indeed, both dexamethasone and prednisone have been shown to increase appetite in cancer patients, although they had no effect on weight.[37]

Most of the anti-inflammatory effects of glucocorticoids are mediated via repression of inflammatory cytokine production. After binding to the glucocorticoid receptor, the receptor–ligand complex migrates into the cell nucleus where it activates transcription of numerous proteins. However, most of the inflammatory cytokines whose production is inhibited by glucocorticoids do not contain glucocorticoid response elements (GRE). Transcription factor NF-κB is activated by numerous stimuli, including lipopolysaccharides, viral infection, cytokines, mitogens, and reactive oxygen intermediates. It then dissociates from its cytoplasmic chaperone, IκB, and activates transcription of inflammatory cytokines.[38–41] Two studies have shown that glucocorticoid administration induces transcription of *IκB* gene. Increased levels of IκB trap NF-κB in inactive cytoplasmic complexes, hence inhibiting its ability to induce transcription of inflammatory cytokines.[42,43] Another study demonstrated the ability of the glucocorticoid-receptor complex to directly interact with NF-κB and prevent its binding to DNA.[44] It is currently thought that the inhibition of NF-κB is responsible for the immunosuppressive effects of glucocorticoids.

Other effects of glucocorticoids, including muscle proteolysis and lipolysis, make their use in cachectic patients questionable. Development of novel glucocorticoids with more immunosuppressive and fewer catabolic effects may lead to more efficacious treatment options.

19.4.2.1 Adverse effects

Since patient safety is of primary importance, the adverse effects and contraindications of androgen therapy must be reviewed before discussing their therapeutic applications. Though physiologic testosterone replacement is relatively safe, supraphysiologic doses may lead to adverse effects. In men, these include acne, increased body hair, aggressive behavior, irritability, polycythemia, worsening of existing benign prostatic hypertrophy, and in some cases worsening of sleep apnea. Women may experience acne and hirsutism. In some cases, virilization may develop (deepening of voice, clitoromegaly, male pattern baldness). Once these adverse effects develop, the androgen dose should be decreased or the treatment discontinued.

19.4.2.2 Contraindications

Absolute contraindications of androgen therapy include:

1. Hormone responsive cancer (prostate and breast cancer)
2. Allergy to androgen formulation
3. Hematocrit >55%
4. Female virilization

Relative contraindications of androgen therapy include:

1. Hematocrit >52%
2. Severe obstructive benign prostatic hyperplasia (BPH)
3. Advanced congestive heart failure
4. Severe sleep apnea

19.4.2.3 Therapeutic applications

19.4.2.3.1 HIV

Hypogonadism is a frequent finding in patients with human immunodeficiency virus (HIV) infection, with a prevalence of 20% in men and 33 to 66% in women.[45,46] This decrease in serum testosterone levels may be due to multiple etiologies.[47,48] Chronic illness and weight loss due to advanced AIDS can depress hypothalamic pituitary function, leading to secondary hypogonadism. The degree of weight loss in HIV-infected patients correlates with a reduction in testosterone levels.[49]

Opportunistic infections secondary to HIV can directly invade the testes and potentially impair hormone synthesis. Furthermore, many medications used in the treatment of secondary illnesses related to HIV also interfere with testosterone production, including antifungals, the appetite stimulant megesterol, and antiretrovirals.

19.4.3 Studies with Testosterone

Multiple studies have found that testosterone administration increases LBM and muscle strength in this population (Table 19.2).[50] Grinspoon et al.[51] showed that IM injection of 300 mg testosterone enanthate every 3 weeks improved LBM, muscle mass, and quality of life. In another study, patients with HIV gained an average of 2.3 kg over 12 weeks on 400 mg of IM testosterone cypionate every 2 weeks.[52] A trial by Bhasin of 34 hypogonadal, HIV-infected men treated with a nongenital, transdermal patch (5 mg daily) showed 1.34 kg increase in LBM, significant decrease in fat mass, and an increase in red cell count.[53] However, a multicenter study of 131 patients found no improvement in weight, body cell mass, or quality of life using a trans-scrotal testosterone system despite the fact that normal serum testosterone levels were achieved.[54] A meta-analysis of placebo-controlled trials highlights the fact that IM testosterone administration is more effective in the HIV/AIDS population. An average weight gain of 3.34 kg was seen when testosterone was administered IM compared to an average weight gain of 0.51 kg for all other administration routes.[55]

Two trials also looked at the additive effect of exercise and testosterone administration on muscle mass and strength. Grinspoon et al.[56] showed that both testosterone (200 mg/week) and resistance training (3 times/week) individually increased muscle mass; however, there was no additive benefit with combined intervention. Bhasin et al.[57] found similar results with both exercise and testosterone increasing muscle mass and strength individually, but without any synergistic effect.

One trial has studied the use of transdermal testosterone in HIV-positive hypogonadal women. Patients were randomized to three treatment groups, two placebo patches (PP), one active/one placebo patch (AP), or two active patches (AA). Each active patched contained 150 μg/d of testosterone. Patients in the AP group showed significant improvements in weight and quality of life, while those in the PP and AA groups did not, highlighting the fact that supraphysiological doses in these women did not lead to greater improvement in weight.[58]

19.4.4 Studies with Anabolic Steroids

Anabolic steroid derivatives of testosterone have also been studied in the treatment of HIV wasting. Recently, a 16-week double-blind, placebo-controlled trial of 63 patients with AIDS-associated malnutrition were randomized to two doses of oxandrolone (5 or 15 mg/d) or placebo.[59] The 15 mg/d dose showed a mean increase in body weight of 3.9 lb while the 5 mg/d dose showed a mean increase of 1.7 lb.

TABLE 19.2
Results of Trials Conducted on Testosterone and HIV-Associated Cachexia

Author, year	Participants	Product used	Duration of treatment	Effects on body weight	Other results
Fairfield, 2001	54 eugonadal men	Testosterone, 200 mg/week IM w/resistance training	12 weeks	↑ Lean body mass (LBM) ↓ Fat mass (FM)	No interaction found between testosterone and resistance training
Bhasin, 2000	61 hypogonadal men	Testosterone enanthate, 100 mg/week IM w/exercise	16 weeks	↑ Muscle strength	No interaction found between testosterone and resistance training
Grinspoon, 2000	54 eugonadal men	Testosterone enanthate, 200 mg/week IM w/resistance training	12 weeks	↑ LBM	↓ HDL ↓ Total cholesterol. No interaction found between testosterone and resistance training
Grinspoon, 1998	51 hypogonadal men	Testosterone enanthate, 300 mg/3 week IM	6 months	↑ LBM	↑ Quality of life
Bhasin, 1998	41 hypogonadal men	Testosterone transdermal patch, 5 mg/d	12 weeks	↑ LBM ↓ FM	↑ Red cell count ↑ Emotional stability
Dobs, 1999	133 hypogonadal men	Testosterone trans-scrotal patch, 6 mg/d	12 weeks	No change	↑ Serum testosterone
Miller, 1998	53 hypogonadal women	Testosterone transdermal patch, 150 μg or 300 μg/d	12 weeks	↑ Total body weight in 150 μg/24 h group	↑ Quality of life in 150 μg/24 h group

The 15 mg/d group had a significant increase in overall physical activity compared to placebo. Another trial of 24 eugonadal men with HIV-associated weight loss treated with oxandrolone (20 mg/d) and resistance exercise for 8 weeks showed a significant increase in nitrogen retention and LBM, muscle strength and body weight.[60] Hengge et al. [61] performed a placebo-controlled trial of 89 HIV-positive women and men treated with oxymetholone, 50 mg twice or thrice daily or placebo for 16 weeks. Oxymetholone led to significant increases in total body weight, body cell mass, appetite, and increased well-being. No difference was noted between groups dosed twice or thrice daily. However, both treatment groups experienced increases in serum alanine-aminotransferase levels, indicating potential liver toxicity. One study has found anabolic steroids to be effective in children with HIV. An unblinded prospective trial of oxandrolone 0.1 mg/kg/d orally in ten pediatric HIV-positive patients demonstrated an accelerated weight gain, increased body mass index, increased muscle mass, and decreased fat stores with therapy despite no change in calorie or protein intake.[62] Weight was maintained for 3 months after discontinuation of treatment. The results of studies evaluating the use of anabolic steroids in HIV patients are summarized in Table 19.3.

These studies show that anabolic-androgenic steroids do have beneficial effects in male patients with AIDS related wasting.

19.4.5   Studies with Glucocorticoids

The effect of glucocorticoids on HIV-associated wasting is incompletely understood. In a trial evaluating the effects of oral prednisone on viral load, an anticatabolic effect was noted among cachectic HIV-infected patients. The patients experienced a mean weight gain of 3.5 kg. There was a decrease in the concentration of receptors for tumor necrosis factor (TNF)-alpha (an inflammatory cytokine). Similarly, receptor levels for neopterin (a marker of progression of HIV infection) also decreased.[63] Further studies need to be conducted before the use of glucocorticoids in HIV wasting syndrome can be recommended.

19.5   CANCER

Cancer patients are in a chronic inflammatory state. The degree of weight loss in patients with cancer correlates with increased levels of soluble TNF receptor and decreased albumin and IGF-I levels. These patients also have significant hypogonadism.[64] Because of the cancer promoting potential of testosterone, its use in cancer patients has been questioned. Hence, trials evaluating its efficacy and safety are lacking.

19.5.1   Studies with Anabolic Steroids

The use of anabolic-androgenic therapy in the treatment of cancer cachexia is uncertain. Very few clinical trials have been performed to date. A preclinical study in rats did not support any positive effect on LBM, survival, or tumor growth.[65]

TABLE 19.3
Trials Conducted on Anabolic Steroids and HIV-Associated Cachexia

Author, year	Participants	Product used	Duration of treatment	Effects on body weight	Other results
Berger, 1996	63 eugonadal men	Oxandrolone, 5 or 15 mg/d	16 weeks	↑ Total body weight	↑ Physical activity in 15 mg/d group
Strawford, 1999	24 eugonadal men	Oxandralone, 20 mg/d with resistance exercise	8 weeks	↑ LBM ↑ Total body weight ↑ Muscle strength	↑ Nitrogen retention
Hengge, 2003	89 women and men	Oxymetholone 100 mg/d or 150 mg/d	16 weeks	↑ Total body weight	↑ Quality of life no benefit found with higher dosage
Fox-Wheeler, 1999	9 female and male children	Oxandrolone 0.1 mg/kg/d	3 months	↑ Total body weight ↑ LBM ↓ FM	Weight gain maintained after discontinuation of treatment

One placebo-controlled trial of 33 patients undergoing chemotherapy for metastatic cancer showed that IM administration of nandrolone decanoate significantly improved hemoglobin concentration and body weight and decreased the need for blood transfusions.[66] However, there was no improvement in survival rate. An open-label, 4-month trial of oxandrolone 20 mg/d in 131 cancer patients showed that 80% of patients maintained or gained weight on therapy, with an average increase in LBM of 4 lb.[67] There was also a significant improvement in functionality. One randomized, placebo-controlled trial of 37 patients with non-small cell lung cancer evaluated the efficacy and safety of IM nandrolone decanoate vs placebo.[68] Patients in the treatment group showed an increase in mean survival time (8.2 vs. 5.5 months) and decreased weight loss (0.21 vs. 0.84 kg), although neither of these achieved statistical significance.

At this time there is not enough data to support the use of anabolic/androgenic agents in the treatment of cancer cachexia; more research is warranted.

19.5.2   Studies with Glucocorticoids

Glucocorticoids have been shown to increase appetite and weight in cancer patients. Loprinzi et al.[69] found that dexamethasone at a dose of 0.75 mg four times daily led to improvement in appetite and increased nonfluid weight in patients with cancer cachexia. Another trial found prednisolone (10 mg twice daily) to maintain functional status in cancer patients and improvement in their Karnofsky and pain scores.[70] A 14 day randomized, double-blind trial of methylprednisolone 32 mg/d in terminally ill cancer patients showed improvements in appetite, activity level, and pain.[71] However, the use of glucocorticoids in wasting syndromes is questionable because of their association with TNF-α induced muscle proteolysis.[72]

19.6   PULMONARY DISEASE

19.6.1   Studies with Testosterone

Hypogonadism is prevalent among patients with chronic obstructive pulmonary disease (COPD). Patients with COPD and low serum testosterone levels have been shown to benefit from testosterone replacement and resistance exercise.[73] Intramuscular injection of 100 mg testosterone weekly led to increased LBM, decreased fat mass, and increased leg muscle strength. However, no change in pulmonary function was seen. Resistance training showed similar effects. Interestingly, testosterone combined with exercise showed the greatest improvement, although not statistically significant. Another study showed administration of 250 mg testosterone every 4 weeks decreased fat mass, increased lean body mass, and improved erectile function.[74] Pulmonary function was not affected. More research is needed to determine the efficacy, safety, and optimal dosage in this patient population before a recommendation can be made.

19.6.2 Studies with Anabolic Steroids

Oral anabolic steroids have been shown to be useful in the treatment of COPD-associated cachexia. An open-label trial with 10 mg twice a day of oxandralone showed an increase in body weight by 4.1 lb after 4 months of treatment.[75] Patients also showed increased endurance as measured by 6-min walking distance (6MWD). No changes in pulmonary function were detected. Since 6MWD correlates with overall subjective improvement in patients with COPD[76,77], the use of anabolic agents may be beneficial in improving overall quality of life in these patients. Recently a double-blind, placebo-controlled trial of 63 men with COPD with 50 mg nandrolone decanoate IM every 2 weeks showed an increase in LBM (mean 1.7 kg) in addition to improvements in muscle function.[78] Another trial by Ferriera et al.[79] showed that administration of 250 mg of IM testosterone at baseline followed by 12 mg oral stanozolol plus exercise improved total body weight, LBM, and muscle size in patients with COPD. No change in pulmonary function or 6MWD was seen. Another trial found IM nandrolone decanoate (25 mg in women, 50 mg in men) every two weeks combined with daily high caloric nutritional supplement increased body weight and LBM.[80]

These data suggest that anabolic-androgenic agents are beneficial in patients with wasting secondary to COPD. Long-term trials are needed to determine whether this benefit is maintained over a long period of time.

19.7 BURN INJURY

Severe burn injuries lead to a state of hypercortisolemia and hypoandrogenemia.[81,82] Both of these hormonal events lead the patient toward a state of catabolism.

19.7.1 Studies with Testosterone

Testosterone administration (200 mg/week IM) in severely burned patients leads to decreased protein breakdown while maintaining a normal protein synthesis rate, thus leading to net tissue anabolism.[25] More trials with testosterone in burn patients are needed to further understand its efficacy and safety in this catabolic population.

19.7.2 Studies with Anabolic Steroids

The use of anabolic steroids in burn patients has been more extensively studied. In a prospective, randomized trial of 13 severely burned patients, administration of 2 mg/kg/d of oxandrolone combined with a high-protein diet leads to significant increase in weight gain.[83] In another trial, serum protein kinetic studies showed an increase in net protein balance due to increased protein synthesis in 14 pediatric burn patients receiving oxandralone (0.1 mg/kg twice daily).[84] In a randomized trial of 40 severely burned patients, oxandrolone (10 mg twice daily for 4 weeks) increased the rate of restoration of total body weight (1.7 vs. 0.6 kg in placebo) and LBM and resulted in a 30% decrease in length of hospitalization.[85] This study also evaluated the effect of age on response to therapy and found that both young and elderly patients responded to oxandrolone administration. Another randomized trial showed that weight and LBM improvement in burn patients treated with oxandrolone (20 mg/d) were maintained even 6 months after cessation of therapy.[86] Patients on oxandrolone had a body composition of 76% LBM and 23% fat mass vs. 71% and 29%, respectively, in placebo-treated patients.

19.8 RENAL DISEASE

Patients with end-stage renal disease (ESRD) are frequently hypogonadal with two-thirds of men on dialysis having decreased serum testosterone levels.[87,88] Testosterone replacement, orally or intramuscularly, leads to increased serum androgen levels and decreased serum LH and follicle stimulating hormone (FSH) levels in hemodialysed men[89,90], but its effect on weight and muscle mass in this population has not been evaluated.

19.8.1 Studies with Anabolic Steroids

One trial evaluating the use of nandrolone decanoate (100 mg/week IM) showed a gain of 4.5 kg in LBM and a loss of 2.4 kg in fat mass (significantly greater than placebo). An increase in serum creatinine was also noted in the nandrolone group, signifying increased muscle mass.[91] A significant improvement in walking and stair-climbing was also seen. Based on these results, the use of anabolic steroids should be further studied in this patient population.

19.9 LIVER DISEASE

Like many other chronic disease states, end-stage liver disease is frequently associated with both cachexia and hypogonadism. The use of anabolic agents needs to be more extensively tested in this patient population.

19.9.1 Studies with Testosterone

Transdermal testosterone is thought to be the safe form of testosterone delivery in patients with end-stage liver disease since it eliminates first-pass hepatic metabolism, thus allowing patients with limited liver function to tolerate treatment. Recently a study was conducted on patients with end-stage liver disease who were randomized to treatment with 1% testosterone gel (5 g/d) compared to routine follow-up.[92] Patients on testosterone gel reported subjective improvement in muscle strength and overall well-being (although the study was not blinded). There was also an increase in serum albumin levels. The same group conducted another 6-month open-label study with testosterone gel on liver transplant recipients who had chronic allograft failure.[93] Patients on testosterone gel experienced improvement in muscle strength, subjective sense of well-being and Model for End-Stage Liver Disease (MELD) scores, an indicator of the severity of liver disease. The serum albumin levels improved in the testosterone group from 2.1 to 3.4 mg/dl while it declined in the untreated group.

Though these results are positive, these trials were, however, only conducted on very few patients. Future trials should include larger number of patients and should be double-blind in order to judge benefits of this therapy.

19.9.2 Studies with Anabolic Steroids

Patients with alcoholic hepatitis suffer from malnutrition. Mendenhall et al.[7] studied 273 male patients with alcoholic hepatitis who were malnourished and treated them with oxandrolone + nutrition supplement vs. placebo + nutrition supplement. Men on oxandrolone had significant improvements in liver injury scores and malnutrition compared to placebo. Furthermore, oxandrolone-treated men also had lower 1- and 6-month mortality compared to placebo. The authors recommended treatment with oxandrolone (along with nutrition therapy) in men with moderate malnutrition.

These results are positive but need to be duplicated in other trials. The main concern is the safety of anabolic agents in patients with existing liver disease. Transdermal testosterone may be more suitable in this patient population since it does not undergo first-pass metabolism.

REFERENCES

1. Guyton, A.C. and Hall, J.E., Endocrinology and Reproduction, in *Textbook of Medical Physiology*, Guyton, A.C. and Hall, J.E. (eds), 10th ed. WB Saunders Co, Philadelphia, 2000, chapter 74.
2. Rosenfield, R.L., Role of androgens in growth and development of the fetus, child, and adolescent, *Adv. Pediatr.*, 19, 172, 1972.
3. Pardridge, W.M., Serum bioavailability of sex steroid hormones, *Clin. Endocrinol. Metab.*, 15, 259, 1986.
4. Griffin, J.E. and Wilson, J.D., Disorders of the testes and the male reproductive tract. In: *Williams Textbook of Endocrinology*, Wilson, J.D., Foster, D.W., Kronenberg, H.M., and Larsen, P.R. (eds), 9th ed. Saunders, Philadelphia, 1998, pp. 819–876.
5. Wilson, J.D. and Gloyna, E., The intranuclear metabolism of testosterone in the accessory organs of male reproduction, *Recent Prog. Horm. Res.*, 26, 309, 1970.
6. Chrousos, G.P., Zoumakis, E.N., and Gravanis, A., The gonadal hormones and inhibitors, in *Basic and Clinical Pharmacology*, Katzung, B.G. (eds), 8th ed. Lange Medical Books, New York, 2001, chapter 40.

7. Kimura, N., Mizokami, M.A., Oonuma, T., et al., Immunocytochemical localization of androgen receptor with polyclonal antibody in paraffin-embedded human tissues, *J. Histochem. Cytochem.*, 41, 671, 1993.

8. Sar, M., Lubahn, D.B., French, F.S., et al., Immunohistochemical localization of the androgen receptor in rat and human tissues, *Endocrinology*, 127, 3180, 1990.

9. Mauras, N., Hayes, V., Welch, S., et al., Testosterone deficiency in young men: Marked alterations in whole body protein kinetics, strength, and adiposity, *J Clin, Endocrinol. Metab.*, 83, 1886, 1998.

10. Brodsky, I.G., Balagopal, P., and Nair, K.S., Effects of testosterone replacement on muscle mass and muscle protein synthesis in hypogonadal men — a clinical research center study, *J. Clin. Endocrinol. Metab.*, 81, 3469, 1996.

11. Wilson, J.D., Androgens, in *Goodman Gilman's Experimental Basis of Therapeutics*, Hardman, J.G., Limbird, L.E., Molinoff, P.B., Ruddon, R.W., and Goodman Gilman, A. (eds), McGraw-Hill, New York, 1996, pp. 1441–1457.

12. Sheffield-Moore, M., Urban, R.J., Wolf, S.E., et al., Short-term oxandrolone administration stimulates net muscle protein synthesis in young men, *J. Clin. Endo. Metab.*, 84, 2705, 1999.

13. Bhasin, S., Storer, T.W., Berman, N., et al., The effects of supraphysiological doses of testosterone on muscle size and strength in normal men, *N. Engl. J. Med.*, 335, 1, 1996.

14. Bricout, V., Germain, P., Serrurier, B., et al., Changes in testosterone muscle receptors: Effects of an androgen treatment on physically-trained rat, *Cell. Mol. Biol.*, 40, 291, 1994.

15. Doumit, M.E., Cook, D.R., and Merkel, R.A., Testosterone up-regulates androgen receptors and decreases differentiation of porcine myogenic satellite cells *in vitro*, *Endocrinology*, 137, 1385, 1996.

16. Kadi, F., Bonnerud, P., Eriksson, A., et al., The expression of androgen receptors in human neck and limb muscles: Effects of training and self-administration of androgenic-anabolic steroids, *Histochem. Cell. Biol.*, 113, 25, 2000.

17. Basaria, S., Wahlstrom, J.T., and Dobs, A.S., Anabolic-androgenic steroid therapy in the treatment of chronic diseases, *J. Clin. Endocrinol. Metab.*, 86, 5108, 2001.

18. Rosenfeld, R.G., Rosenbloom, A.L., and Guevara-Aguirre, J., Growth hormone (GH) insensitivity due to primary GH receptor deficiency, *Endocr. Rev.*, 15, 369, 1994.

19. Veldhuis, J.D. and Iranmanesh A., Physiological regulation of the human growth hormone (GH)-insulin-like-growth factor type I (IGF-I) axis: Predominant impact of age, obesity, gonadal function and sleep, *Sleep*, 19, S221, 1996.

20. Fryburg, D.A., Insulin-like growth factor-1 exerts growth hormone- and insulin-like actions on human muscle protein metabolism, *Am. J. Physiol.*, 267, E331, 1994.

21. Urban, R.J., Bodenburg, Y.H., Gilkison, C., et al., Testosterone administration to elderly men increases skeletal muscle strength and protein synthesis, *Am. J. Physiol.*, 269, E820, 1995.

22. Gayan-Ramirez, G., Rollier, H., Vanderhoydonc, F., et al., Nandrolone decanoate does not enhance training effects but increases IGF-I mRNA in rat diaphragm, *J. Appl. Physiol.*, 88, 26, 2000.

23. Hollenberg, S.M., Weinberg, C., Ong, E.S., et al., Primary structure and expression of a functional human glucocorticoid receptor cDNA, *Nature*, 318, 635, 1985.

24. Wu, F.C., Endocrine aspects of anabolic steroids, *Clin. Chem.*, 43, 1289, 1997.
25. Ferrando, A.A., Sheffield-Moore, M., Wolf, S.E., et al., Testosterone administration in severe burns ameliorates muscle catabolism, *Crit. Care Med.*, 29, 1936, 2001.
26. Hickson, R.C., Czerwinski, S.M., Falduto, M.T., et al., Glucocorticoid antagonism by exercise and androgenic-anabolic steroids, *Med. Sci. Sports Exerc.*, 22, 331, 1990.
27. Danhaive, P.A. and Rousseau, G.G., Binding of glucocorticoid antagonists to androgen and glucocorticoid hormone receptors in rat skeletal muscle, *J. Steroid. Biochem.*, 24, 481, 1986.
28. Danhaive, P.A. and Rousseau, G.G., Evidence for sex-dependent anabolic response to androgenic steroids mediated by muscle glucocorticoid receptors in the rat, *J. Steroid. Biochem.*, 29, 575, 1988.
29. Roubenoff, R., Grinspoon, S., Skolnik, P.R., et al., Role of cytokines and testosterone in regulating lean body mass and resting energy expenditure in HIV-infected men, *Am. J. Physiol. Endocrinol. Metab.*, 283, E138, 2002.
30. Cutolo, M., Balleari, E., Giusti, M., et al., Androgen replacement therapy in male patients with rheumatoid arthritis, *Arthritis Rheum.*, 34, 1, 1991.
31. Bizzarro, A., Valentini, G., Di Martino, G., et al., Influence of testosterone therapy on clinical and immunological features of autoimmune diseases associated with Klinefelter's syndrome, *J. Clin. Endorinol. Metab.*, 64, 32, 1987.
32. Olsen, N.J. and Kovacs, W.J., Case report: testosterone treatment of systemic lupus erythematosus in a patient with Klinefelter's syndrome, *Am. J. Med. Sci.*, 310, 158, 1995.
33. D'Agostino, P., Milano, S., Barbera, C., et al., Sex hormones modulate inflammatory mediators produced by macrophages, *Ann. N. Y. Acad. Sci.*, 876, 426, 1999.
34. Li, Z.G., Danis, V.A., and Brooks, P.M., Effects of gonadal steroids on the production of IL-1 and IL-6 by blood mononuclear cells *in vitro*, *Clin. Exp. Rheumatol.*, 11, 157, 1993.
35. Malkin, C.J., Pugh, P.J., Jones, R.D., et al., The effect of testosterone replacement on endogenous inflammatory cytokines and lipid profiles in hypogonadal men, *J. Clin. Endocrinol. Metab.*, 89, 3313, 2004.
36. Levine, A.S. and Morley, J.E., The effect of prostaglandins (PGE2 and PGF2 alpha) on food intake in rats, *Pharmacol. Biochem. Behav.*, 15, 735, 1981.
37. Schell, H.W., Adrenal corticosteroid therapy in far-advanced cancer, *Geriatrics*, 27, 131, 1972.
38. Baeuerle, P.A. and Baltimore, D., The physiology of the NF-κB transcription factor. Hormonal control regulation of gene transcription, *Mol. Aspects Cell Regul.*, 6, 409, 1991.
39. Baldwin, A.S., Azizkhan, J.C., Jensen, D.E., et al., Induction of NF-κB DNA-binding activity during the G_0-to-G_1 transition in mouse fibroblasts, *Mol. Cell. Biol.*, 11, 4943, 1991.
40. Lenardo, M.G., Fan, C.M., Maniatis, T., et al., The involvement of NF-κB in beta-interferon gene regulation reveals its role as widely inducible mediator of signal transduction, *Cell*, 57, 287, 1989.
41. Schreck, R., Rieber, P., and Baeuerle, P.A., Reactive oxygen intermediates as apparently widely used messengers in the activation of the NF-κB transcription factor and HIV-1, *EMBO J.*, 10, 2247, 1991.

42. Scheinman, F.I., Cogswell, P.C., Lofquist, A.K., et al., Role of transcriptional activation of IκBα in mediation of immunosuppression by glucocorticoids, *Science*, 270, 283, 1995.

43. Auphan, N., DiDonato, J.A., Rosette, C., et al., Immunosuppression by glucocorticoids: Inhibition of NF-κB activity through induction of IκB synthesis, *Science*, 270, 286, 1995.

44. Scheinman, R.I., Gualberto, A., Jewell, C.M., et al., Characterization of mechanisms involved in transrepression of NF-κB by activated glucocorticoid receptors, *Mol. Cell. Biol.*, 15, 943, 1995.

45. Rietschel, P., et al., Prevalence of hypogonadism among men with weight loss related to human immunodeficiency virus infection who were receiving highly active antiretroviral therapy, *Clin. Infect. Dis.*, 31, 1240, 2000.

46. Grinspoon, S., et al. Body composition and endocrine function in women with acquired immunodeficiency syndrome wasting, *J. Clin. Endocrinol. Metab.*, 82, 1332, 1997.

47. Dobs, A.S., Androgen therapy in AIDS wasting, *Baillieres Clin. Endrocrinol. Metab.*, 12, 379, 1998.

48. Wahlstrom, J.T. and Dobs, A.S., Acute and long-term effects of AIDS and injection drug use on gonadal function, *J. Acquir. Immune Defic. Syndr.*, 25, S27, 2000.

49. Coodley, G.O., Loveless, M.O., Nelson, H.D., et al., Endorcrine dysfunction in the HIV wasting syndrome, *J. Acquir. Immune Defic. Syndr.*, 7, 46, 1994.

50. Fairfield, W.P., Treat, M., Rosenthal, D.I., et al., Effects of testosterone and exercise on muscle leanness in eugonadal men with AIDS wasting. *J. Appl. Physiol.*, 90, 2166, 2001.

51. Grinspoon, S., Corcoran, C., Askari, A., et al., Effects of androgen administration in men with the AIDS wasting syndrome; a randomized, double-blind, placebo-controlled trial, *Ann. Intern. Med.*, 129, 18, 1998.

52. Wagner, G.J. and Rabkin, J.G., Testosterone therapy for clinical symptoms of hypogonadism in eugonadal men with AIDS, *Int. J. STD AIDS*, 9, 1, 1998.

53. Bhasin, S., Thomas, S., Asbel-Sethi, A., et al., Effects of testosterone replacement with a nongenital, transdermal system, Androderm, in Human Immunodeficiency Virus-infected men with low testosterone levels, *J. Clin. Endocrinol. Metab.*, 83, 3155, 1998.

54. Dobs, A.S., Cofrancesco, J., Nolten, W.E., et al., The use of a transscrotal testosterone delivery system in the treatment of weight loss in patients with HIV infection, *Am. J. Med.*, 107, 126, 1999.

55. Kong, A. and Edmonds, P., Testosterone therapy in HIV wasting syndrome: Systematic review and meta-analysis, *Lancet. Infect. Dis.*, 2, 692, 2002.

56. Grinspoon, S., Corcoran, C., Parlman, K., et al., Effects of testosterone and progressive resistance training in eugonadal men with AIDS wasting, *Ann. Intern. Med.*, 133, 348, 2000.

57. Bhasin, S., Storer, T., Javangakht, M., et al., Testosterone replacement and resistance exercise in HIV-infected men with weight loss and low testosterone levels, *JAMA*, 283, 763, 2000.

58. Miller, K., Corcoran, C., Armstrong, C., et al., Transdermal testosterone administration in women with acquired immunodeficiency syndrome wasting: A pilot study, *J. Clin. Endocrinol. Metab.*, 83, 2717, 1998.

59. Berger, J.R., Pall, L., Hall, C.D., et al., Oxandrolone in AIDS-wasting myopathy, *AIDS*, 10, 1657, 1996.

60. Strawford, A., Barbieri, T., Van Loan, M., et al., Resistance exercise and supra-physiologic androgen therapy in eugonadal men with HIV-related weight loss: A randomized controlled trial, *JAMA*, 281, 1282, 1999.

61. Hengge, U.R., Stocks, K., Wiehler, H., et al., Double-blind, randomized, placebo-controlled phase III trial of oxymetholone for the treatment of HIV wasting, *AIDS*, 17, 669, 2003.

62. Fox-Wheeler, S., Heller, L., Salata, C., et al. Evaluation of the effects of oxan-drolone on malnourished HIV-positive pediatric patients. *Pediatrics*, 104, e73, 1999.

63. Kilby, J.M., Tabereaux, P.B., Mulanovich, V., et al., Effects of tapering doses of oral prednisone on viral load among HIV-infected patients with unexplained weight loss, *AIDS Res. Hum. Retroviruses*, 13, 1533, 1997.

64. Simons, J.P., Schols, A.M., Buurman, W.A., et al., Weight loss and low body cell mass in males with lung cancer: Relationship with systemic inflammation, acute-phase response, resting energy expenditure, and catabolic and anabolic hormones, *Clin. Sci. (Lond.)*, 97, 215, 1999.

65. Lydén, E., Cvetkovska, E., Westin, T., et al. Effects of nandrolone propionate on experimental tumor growth and cancer cachexia, *Metabolism*, 44, 445, 1995.

66. Spiers, A.S., DeVita, S.F., Allar, M.J., et al., Beneficial effects of an anabolic steroid during cytotoxic chemotherapy for metastatic cancer, *J. Med.*, 12, 433, 1981.

67. Boughton, B. Drug increases lean tissue mass in patients with cancer, *Lancet Oncol.*, 4, 134, 2003.

68. Chlebowski, R.T., Herrold, J., Ali, I., et al., Influence of nandrolone decanoate on weight loss in advanced non-small cell lung cancer, *Cancer*, 58, 183, 1986.

69. Loprinzi, C.L., Kugler, J.W., Sloan, J.A., et al., Randomized comparison of megestrol acetate versus dexamethasone versus fluoxymesterone for the treatment of cancer anorexia/cachexia, *J. Clin. Oncol.*, 17, 3299, 1999.

70. Lundhom, K., Gelin, J., Hyltander, A., et al., Anti-inflammatory treatment may prolong survival in undernourished patients with metastatic solid tumors, *Cancer Res.*, 54, 5602, 1994.

71. Bruera, E., Roca, E., Cedaro, L., et al., Action of oral methylprednisolone in terminal cancer patients: A prospective randomized double-blind study, *Cancer Treat. Rep.*, 69, 751, 1985.

72. Hardin, T.C., Cytokine mediators of malnutrition: Clinical implications, *Nutr. Clin. Pract.*, 8, 55, 1993.

73. Casaburi, R., Bhasin, S., Cosentino, L., et al., Effects of testosterone and resistance training in men with chronic obstructive pulmonary disease, *Am. J. Respir. Crit. Care Med.*, 170, 870, 2004.

74. Svartberg, J., Aesebo, U., Hjalmarsen, A., et al., Testosterone treatment improves body composition and sexual function in men with COPD, in a 6-month randomized controlled trial, *Respir. Med.*, 98, 906, 2004.

75. Yeh, S., DeGuzman, B., and Kramer, T., Reversal of COPD-associated weight loss using the anabolic agent oxandrolone, *Chest*, 122, 421, 2002.

76. Efthimiou, J., Fleming, J., Gomes, C., et al., The effect of supplementary oral nutrition in poorly nourished patients with chronic obstructive pulmonary disease, *Am. Rev. Respir. Dis.*, 137, 1075, 1988.

77. Redelmeier, D.A., Bayoumi, A.M., Goldstein, R.S., et al., Interpreting small differ-ences in functional status: The six minute walk test in chronic lung disease patients, *Am. J. Respir. Crit. Care Med.*, 155, 1278, 1997.

78. Creutzberg, E.C., Wouters, E.F., Mostert, R., et al., A role for anabolic steroids in the rehabilitation of patients with COPD, *Chest*, 124, 1733, 2003.

79. Ferreira, I.M., Verreschi, I.T., Nery, L.E., et al., The influence of 6 months of oral anabolic steroids on body mass and respiratory muscles in undernourished COPD patients, *Chest*, 114, 19, 1998.

80. Schols, A.M., Soeters, P.B., Mostert, R., et al., Physiologic effects of nutritional support and anabolic steroids in patients with chronic obstructive pulmonary disease. A placebo-controlled randomized trial, *Am. J. Respir. Crit. Care Med.*, 152, 1268, 1995.

81. Vaughn, G.M., Becker, R.A., Allen, J.P., et al., Cortisol and corticotrophin in burned patients, *Adv. Wound Care*, 12, 1, 1982.

82. Lephart, E.D., Baxter, C.R., and Parker, C.R., Effects of burn trauma on adrenal and testicular steroid hormone production, *J. Clin. Endocrinol. Metab.*, 64, 842, 1987.

83. Demling, R.H. and DeSanti, L., Oxandrolone, and anabolic steroid, significantly increases the rate of weight gain in the recovery phase after major burns, *J. Trauma*, 43, 47, 1997.

84. Hart, D.W., Wolf, S.E., and Ramzy, P.I., Anabolic effects of oxandrolone after severe burn, *Ann. Surg.*, 233, 556, 2001.

85. Demling, R.H. and DeSanti, L., The rate of restoration of body weight after burn injury, using the anabolic agent oxandrolone, is not age dependent, *Burns*, 27, 46, 2001.

86. Demling, R.H. and DeSanti, L., Oxandrolone induced lean body mass gain during recovery from severe burns is maintained after discontinuation of anabolic steroid, *Burns*, 29, 793, 2003.

87. Handelsman, D., Hypothalamic-pituitary gonadal dysfunction in renal failure, dialysis and renal transplantation, *Endocr. Rev.*, 6, 151, 1985.

88. Singh, A., Norris, K., Modi, N., et al., Pharmacokinetics of a transdermal testosterone system in men with end stage renal disease receiving maintenance hemodialysis and healthy hypogonadal men, *J. Clin. Endocrinol. Metab.*, 86, 2437, 2001.

89. Van Coevorden, A., Stolear, J.C., Dhaene, M., et al., Effect of chronic oral testosterone undecanoate administration on the pituitary-testicular axes of hemodialyzed male patients, *Clin. Nephrol.*, 26, 48, 1986.

90. Barton, C.H., Mirahmadi, M.K., and Vaziri, N.D. Effects of long-term testosterone administration on pituitary-testicular axis in end-stage renal failure, *Nephron*, 31, 61, 1982.

91. Johansen, K.L., Mulligan, K., and Schambelan, M. Anabolic effects of nandrolone decanoate in patients receiving dialysis — a randomized controlled trial, *JAMA*, 281, 1275, 1999.

92. Neff, G.W., O'Brien, C.B., Montalbano, M., et al., Beneficial effects of topical testosterone replacement in patients with end-stage liver disease, *Dig. Dis. Sci.*, 49, 1186, 2004.

93. Neff, G.W., O'Brien, C.B., Shire, N.J., et al., Topical testosterone treatment for chronic allograft failure in liver transplant recipients with recurrent hepatitis C virus, *Transplant. Proc.*, 36, 3071, 2004.

94. Mendenhall, C.L., Moritz, T.E., Roselle, G.A., et al., A study of oral nutritional support with oxandrolone in malnourished patients with alcoholic hepatitis: results of a Department of Veterans Affairs Cooperative Study, *Hepatology*, 17, 564, 1993.

20 Growth Hormone

Paul V. Carroll

Contents

Summary

Advances in the understanding of the mechanisms responsible for cachexia in disease states have led to the development of therapeutic strategies to maintain skeletal mass and function in affected patients. Growth promoting agents have been the focus of considerable interest as potential treatments for cachexia. Growth hormone (GH) is a potent anabolic hormone and the potential benefits of therapeutic GH administration have been studied in patients with several conditions that result in skeletal muscle wasting. It is recognized that GH secretion and tissue sensitivity to the anabolic effects of GH are significantly altered in patients with chronic illness. The majority of studies investigating GH therapy have employed large doses of GH, have been small in size and short in duration. Most have investigated specific outcomes such as indices of protein metabolism or body composition. GH treatment has had potentially beneficial effects on protein metabolism, skeletal mass, and function in most of the conditions investigated. Studies in critically ill patients indicate that use of large doses of GH in patients with active critical illness results in increased mortality and morbidity, although the mechanisms responsible remain unclear. Further studies will determine whether the use of GH in patients with cachexia will result in substantial improvements in clinical well-being and reduce mortality in specific conditions.

20.1 INTRODUCTION

Disproportionate weight loss, particularly of skeletal muscle, leading to emaciation, weakness and morbidity is a feature common to a variety of medical disorders. Cachexia is a term used to define this syndrome in its most severe form. Increases in understanding of the pathophysiological and genetic basis of cachexia have been mirrored by the development of therapeutic strategies to attenuate the loss of healthy tissues and maintain function in patients at risk of cachexia. Treatment modalities include nutritional modification, exercise programmes, and nonpharmacological and specific drug therapies. Of the pharmacological treatments there has been considerable focus on agents known to have anabolic effects on protein metabolism. Evidence suggests that these anabolic strategies are similarly recognized and "abused" by athletes as part of performance enhancing methods. The availability of recombinant growth hormone (GH) has led to widespread interest in the potential of this potent anabolic peptide in the treatment of a variety of disorders resulting in cachexia. This chapter addresses the theoretical and practical basis for the use of GH in the treatment of these medical conditions.

20.2 GROWTH HORMONE: PHYSIOLOGY AND REGULATION IN HEALTH

Growth hormone the most abundant of the anterior pituitary hormones is a 191 amino acid, 22 KD single-chain peptide containing two intrachain disulphide bonds, which produce two loops. A 20 KD variant, with deletion of residues 32 to 46 constitutes 10% of total pituitary GH stores and approximately 5% of circulating GH.[1] Primate GH and its receptor have diverged considerably throughout evolution and until recently GH for therapeutic use in the human has only been available from human pituitary extracts.[2] Humans possess 5 linked *GH* genes, clustered on chromosome 17, but only one of these encodes for physiological GH.[3]

Growth hormone secretion in health is characterized by intermittent bursts of episodic secretion, with pulses ranging from 6 to 11/24-h period, with the majority occurring throughout the nocturnal sleeping period.[4] Several stimuli including exercise, hypoglycemia, stress, and sleep result in a GH secretory response. The synthesis and secretion of GH is regulated by three other hormones — GH-releasing hormone (GHRH), and ghrelin that are are stimulatory and somatostatin, which has has an inhibitory effect. Various inputs, including negative feedback from circulating GH and insulin-like growth factor-I (IGF-I), as well as central nervous system effects, modulate the controlling mechanisms of GH secretion. GH induces hepatic secretion of IGF-I, which in turn inhibits GH release both directly and via stimulating somatostatin activity.[5] Evidence indicates that GH also induces local synthesis of IGF-I (both paracrine and autocrine), which in turn acts to inhibit further GH release.[6]

20.3 REGULATION OF MUSCLE PROTEIN METABOLISM

20.3.1 Background

Skeletal muscle is the principal component of lean body mass (LBM) in humans[7] and is intimately involved in amino acid metabolism in health, during starvation, and throughout periods of illness.[8, 9] Muscle mass accounts for approximately 40% of total body weight in the healthy adult[10] but the total amount and composition of muscle declines with ageing, reflected by a selective decrease in type II muscle fibres.[11–13] Carbohydrate and protein metabolism are integral to muscle health and function and skeletal tissue is important in both the storage and release of amino acids and the synthesis of protein. Total free amino acids are present in higher concentrations in skeletal muscle than seen in the circulation[14] and muscle stores account for approximately 75% (or 850 g) of whole-body free amino acids. The concentrations of free amino acids in muscle are reduced during periods of net anabolism[15] and conversely are increased during periods of net catabolism. Loss of skeletal muscle (wasting) occurs when protein catabolism exceeds the rate of anabolism resulting in cachexia.[9] The precise mechanisms responsible for these changes are only partially understood and include alterations in the expression and circulating levels of pro-inflammatory cytokines and endocrine hormones.[16, 17]

20.3.2 Hormonal Effects

Insulin, GH, and IGF-I are the principal hormones regulating protein metabolism in the healthy human. Circulating concentrations of these agents vary with nutritional status and are altered in disease states. Several other systemic hormones (cortisol, glucagon, thyroid, androgens, and catecholamines) affect protein balance and thus skeletal muscle mass and function. Increasing evidences suggest that locally produced cytokines and inflammatory mediators including tumor necrosis factor-α (TNF-α) and several components of the interleukin (IL) superfamily also influence metabolic processes, including protein metabolism. It has been suggested that

Anabolic action	Catabolic action
Insulin	Cortisol
Growth hormone	Glucagon
Insulin-like growth factor-1	Thyroid hormones
Androgens	
Catecholamines	

FIGURE 20.1   Net effect of endogenous hormones on protein balance (Taken from De Feo. *European Journal of Endocrinology* 1996; 135: 7–18. With permission.).

these agents are important determinants of the skeletal muscle wasting seen in illness.[18] The net effects of the major endocrine factors on protein metabolism are summarized in Figure 20.1.

20.3.2.1   Insulin

In vitro studies indicate that insulin inhibits protein breakdown, stimulates amino acid uptake, and the synthesis of nucleic acids and protein.[19] In humans, the insulin deficiency of type 1 diabetes results in abnormal protein metabolism,[20, 21] characterized by increased plasma amino acid concentrations and protein flux. Insulin exerts a net anabolic effect on *in vivo* protein metabolism principally through reducing the rate of protein breakdown[22–24] with associated reductions in circulating amino acid levels. There is evidence that this action is enhanced by IGF-I.[25]

20.3.2.2   GH

As GH administration results in increases in IGF-I and insulin it is difficult to be certain in longer-term studies that measured effects are attributable to direct GH activity. GH has anabolic effects on protein metabolism, increasing both growth and muscle mass in GH-deficient children.[26, 27] GH administration in humans has been shown to reduce nitrogen excretion and increase skeletal muscle ribosome content, indicating an increase in muscle protein synthesis.[28] More recently GH treatment in GH-deficient adults has been shown to increase LBM and thigh mass.[29, 30] Studies using stable isotope tracers indicate that these effects are via GH-mediated increases in protein synthesis without effects on protein breakdown.[31] GH administration results in alterations in body composition characterized by reduced fat stores, particularly truncal fat.[32]

20.3.2.3   IGF-I

Initial isotopic studies indicated that an acute infusion of IGF-I has an insulin-like effect on protein metabolism, with a reduction in proteolysis without effects on protein synthesis.[33–35] These results were surprising, as IGF-I has been shown to mediate the anabolic actions of growth hormone *in vivo*.[36] However these studies were performed on fasted subjects with reduced circulating levels of amino acids. Subsequently Russell-Jones et al.[37] used an amino acid clamp protocol in which circulating levels of amino acids were maintained with an amino acid infusion to compare the actions of IGF-I and insulin on protein metabolism. IGF-I treatment resulted in an increase in protein synthesis, indicating that IGF-I has a direct effect on protein synthesis in the presence of an adequate supply of amino acids. The predominant action of insulin on protein metabolism in this study was suppression of protein breakdown with no effect on the rate of protein synthesis.

20.4 PROTEIN METABOLISM IN DISEASE STATES RESULTING IN CACHEXIA

Loss of protein as a result of trauma was first described in the 1930s by Cuthbertson who noted that healthy adults with a fracture of a long bone remained in a persistent net negative balance regardless of nutritional intake.[38] He postulated that the bulk of the nitrogen lost was derived from skeletal muscle.[39] It is now known that increased protein breakdown and increased nitrogen excretion are the hallmark of the hypermetabolic, catabolic state. The proteolysis associated with severe illness is thought to permit release of amino acids for high-priority use in threatened tissues.

A variety of techniques have shown that severe illness results in rapid and major changes in protein metabolism. Investigations have demonstrated decreased whole-body protein synthesis but an unchanged rate of protein breakdown in subjects following an elective surgical trauma,[40,41] but following more major surgery and in ICU patients whole-body proteolysis has been shown to be increased.[42–46] Depending on the nature, severity, and chronicity of illness these alterations may persist for periods varying from days to months and evidence suggests that protein catabolism influences outcome in critically ill patients and patients with cachexia as a result of chronic disease such as AIDS or chronic heart failure (CHF).[47,48]

20.5 THE GH/IGF-I AXIS IN CACHEXIA

Trauma, sepsis, surgery, and CHF are recognized to result in a state of acquired GH resistance, characterized by high GH levels, reduced levels of IGF-I, and a decreased anabolic response to GH administration. Studies assessing the secretion of GH in illness have yielded conflicting results. Elevated mean and interpulse (baseline) GH concentrations have been reported in critically ill subjects compared with matched healthy controls,[49] but these differences were not evident when the controls were in the fasting state. Similar findings were reported from a separate group of investigators[50] but decreased GH secretion in critically ill ICU patients has also been reported.[51] Less contentious is the observation that reduced circulating IGF-I is a characteristic finding in critically ill subjects and patients with CHF.[52,53] Changes in the clearance rate of IGF-I in catabolic states may be responsible for some of the observed decrease and IGF-I has been shown to have a shorter half-life when administered to patients following major surgery compared with healthy controls.[54] Evidence suggests that some of this decrease may be mediated by alterations in the levels of the IGF-binding proteins (IGF-BPs), in particular a reduction in the circulating level of the major constitutive binding protein IGFBP-3[55]. Protease activity associated with critical illness is thought to be responsible for this decrease.[56,57] More recent evidence indicates that nutritional deficiency per se results in decreased IGF-I expression and that alterations in cytokine and endotoxin may decrease hepatic GH receptor numbers leading to reduced levels of circulating IGF-I.[58] Several studies have addressed the metabolic responses to GH administration in catabolic illness. In septic patients treated with GH, IGF-I levels failed to rise compared with healthy controls and urinary urea excretion was not decreased to the same extent as in healthy subjects.[59] Similarly a study in AIDS patients indicated that the IGF-I response was blunted following GH therapy in comparison with the control population.[60]

Nutritional status may influence the pattern of GH secretion and the IGF-I generation response to exogenous GH. Fasting subjects have increased GH secretion characterized by increased pulse frequency and amplitude with the addition of raised interpulse levels.[61] In contrast total IGF-I levels fall during periods of fasting, particularly when dietary protein and total calories are restricted.[62] Treatment of fasting individuals with exogenous GH has been shown to result in an approximately 5-fold decrease in IGF-I response compared with well-nourished subjects.[63]

20.6 THERAPEUTIC USE OF GH IN CACHEXIA

The anabolic effects of GH, IGF-I, and to a lesser extent the androgens, testosterone, and dihydroepiandrosterone (DHEA) have been studied in a variety of clinical conditions that result in cachexia. These include surgery, burns, sepsis and trauma, AIDS associated wasting, cardiac failure, and elderly subjects. Table 20.1 summarizes results from studies assessing the effect of GH treatment in conditions associated with loss of LBM.

20.6.1 GH in Surgery

A large number of investigators, employing a variety of techniques, have assessed the effects of GH administration in subjects undergoing surgical procedures. In elective surgery GH treatment resulted in increased IGF-I levels, whole-body protein flux, and resting energy expenditure (REE).[64] When used in combination with hypocaloric nutritional support GH resulted in positive nitrogen balance.[65] Perioperative GH & total parenteral nutrition (TPN) therapy have been shown to improve glutamine retention, nitrogen balance, and polyribosome concentration.[66] Similarly unlike TPN alone, TPN with GH has resulted in a net positive nitrogen balance in humans undergoing upper GI resection for malignancy.[67] It has recently been demonstrated that the addition of combined GH + IGF-I to glutamine supplemented TPN, unlike nutritional support alone, results in net protein gain in critically ill

TABLE 20.1
Summary of the Controlled Studies Investigating the Effects of GH Administration on Protein Metabolism in Critically Ill Humans

Author	Year	n	Method	Patient category	Effect	Significance compared to placebo
Jiang[65]	1989	18	Nitrogen balance/body composition	Gastrectomy/colectomy	Increased nitrogen balance/maintained LBM	$p < .001$
Gore[79]	1991	10	Nitrogen balance/15N lysine isolated limb	Severe burns	Increased nitrogen balance/increased limb and WB protein synthesis	$p < .03$
Voerman[89]	1992	20	Nitrogen balance	Septic shock (+TPN)	Increased nitrogen balance	$p < .05$
Waters[106]	1996	60	Body composition	LBM in AIDS associated wasting	Increased LBM	$p < .001$
Carli[46]	1997	12	13C Leucine	Colonic resection (carcinoma)	Increased WB protein synthesis	$p < .03$
Jensen[120]	1998	24	DEXA/body composition	IAA	Increased nitrogen retention, LBM	Both <0.001
Berman[67]	1999	30	Nitrogen balance	Surgery (GI tract malignancy)	Increased nitrogen balance	$p < .05$
Barle[71]	1999	28	Ribosomal analysis	Elective cholecystectomy	Increased hepatic protein synthesis	$p < .01$
Kissmeyer-Nielsen[70]	1999	24	Body composition	ileo-anal J pouch surgery	Increased LBM	$p < .05$

GI, gastrointestinal; LBM, lean body mass; TPN, total parenteral nutrition; IAA, ileo-anal anastomosis; WB, whole-body.
Source: Carroll P.V. *Growth Hormone and IGF Research* 1999; 9: 400–413.

subjects, when administered soon after their admission to the ICU.[68] The majority of these patients were post emergency surgery cases.[69]

The effects of GH treatment have been assessed in adults undergoing surgery for ulcerative colitis.[70] Assessments of total muscle strength, fatigability, and body composition were made prior to operation and up to 90 days following surgery. GH treatment (from 2 days before to 7 days after surgery) resulted in preserved LBM, increased muscular strength, and decreased postoperative fatigue.

Barle and colleagues[71] assessed the effects of GH administration on liver protein synthesis in humans undergoing elective cholecystectomy in a randomized-controlled trial. Using a flooding dose of $[^2H_5]$-phenylalanine, GH treatment resulted in an increase in the fractional synthetic rate of liver proteins and maintained the hepatic polyribosome content, which decreased during surgery in those who received placebo. GH pretreatment in these patients decreased hepatic free amino acid concentrations and preserved liver protein synthesis. Further studies indicate that GH therapy alters substrate utilization in patients undergoing elective laparotomy, with increased fat oxidation, REE, and non-esterified fatty acid (NEFA) levels with reductions observed in carbohydrate oxidation.[72]

The effects of GH administration on cardiovascular performance were assessed by Barry and coworkers[73] in patients undergoing elective abdominal aortic aneurysm repair. Intraoperative cardiac index was assessed at induction, during and immediately following aortic clamping. A nonsignificant trend toward increased cardiac index and a significant increase in heart rate was observed in the GH treated patients compared with those randomized to receive placebo. In a similar patient group the effects of GH administration on hepatic acute phase protein responses and inflammatory cytokine production were assessed in a double-blind placebo-controlled trial.[74] GH (0.03 IU/kg by subcutaneous injection) or placebo was administered daily for 6 consecutive days prior to operation. Significant increases in the plasma levels of both GH and IGF-I were observed prior to surgery. No differences were recorded in the acute phase protein responses of the release of inflammatory cytokines following surgery.

These data suggest that the administration of GH ($\pm$IGF-I) can modify protein metabolism in critically ill subjects following both elective and emergency surgery. When used in conjunction with nutritional support this effect has translated into improved nitrogen balance in most but not all studies. Emerging evidence indicates that in addition to effects on protein metabolism GH administration in the surgical patient may modify cardiovascular dynamics and the expression of inflammatory mediators.

20.6.2   GH in Burns

In 1961, it was demonstrated that human GH derived from cadavers increased nitrogen balance in severely burned patients.[75] Subsequent studies confirmed this observation and indicated that GH administration increased oxygen consumption, decreased respiratory quotient,[76] while increasing insulin and glucose levels.[77] Using recombinant GH more recent studies have shown that GH treatment increased the rate of healing of graft donor sites[78] thereby reducing length of hospital stay in burns patients. Further investigations suggest that GH-mediated increases in protein synthesis may be responsible for this effect,[79] but provided evidence that GH administration induced insulin resistance in burns patients by decreasing glucose uptake and oxidation.[80] Double-blind placebo-controlled studies have demonstrated effective decreases in donor-site healing following GH treatment in severely burned children.[81,82] Evidence suggests that GH treatment maintains healthy growth in children with burns, who would otherwise expect to experience growth failure as a result of illness.[83] Recently GH treatment has been shown to increase left ventricular ejection fraction in children with burns,[84] but no major benefit in pulmonary function resulted from GH administration.[85]

20.6.3 GH in Sepsis and Trauma

In 1989, Okamura demonstrated that GH treatment increased nitrogen retention without affecting 3-methylhistidine excretion in a rat model of sepsis.[86] This experimental model provided evidence of an effect on protein synthesis with unaltered protein breakdown. A simultaneous study showed increased amino acid uptake into muscle protein following GH treatment in nutritionally deprived patients receiving TPN.[87] Further evidence of GH-mediated increases in protein synthesis without effects on proteolysis was provided by a stable isotope tracer study in trauma patients with sepsis.[88] In these studies fat oxidation was increased by GH treatment but leucine oxidation was decreased indicating a preferential use of fat rather than protein as a calorie source. A study of the effects of GH with the addition of TPN in sepsis demonstrated decreased nitrogen production during the treatment period.[89]

The systemic and splanchnic metabolic responses to GH were compared in patients exhibiting the systemic inflammatory response syndrome (SIRS), patients with SIRS receiving TPN, and normal subjects.[90] Fasting basal total IGF-I was reduced in both SIRS groups (by 75–83%) relative to the control subjects. Although minor increases in IGF-I were observed following 48 h GH treatment these values remained 74 to 76% lower in the sick patients. Urea excretion declined by a similar magnitude throughout the study in all three groups, and although hepatic blood flow was unaffected GH administration increased splanchnic oxygen consumption in all groups. The effects resulted in a decline in basal to end study hepatic venous oxygen saturation, suggesting that GH may induce hepatic hypoxia, which may in part account for the diminished IGF-I response.

Unneberg et al. studied the haemodynamic effects of GH and IGF-I in a piglet model of critical illness.[91] Haemodynamic and arterial blood gas measurements were made prior to and during sustained fatal haemorrhage in groups randomized to receive GH, IGF-I, and in control groups with either trauma and sepsis or trauma alone. Sustained volume loss before cardiac arrest was lower in the GH treated animals, which were characterized by metabolic acidosis. Visceral blood flow as a proportion of total cardiac output was lower in the GH group. The authors explained the findings as indicating that GH promoted a metabolic acidosis and impaired compensation to haemorrhage in traumatized sepsis. Overall the evidence indicates that GH treatment alters protein metabolism in sepsis and trauma via an increase in the rate of protein synthesis with largely unchanged rates of protein breakdown.

20.6.4   GH Treatment in Chronic Heart Failure

Cachexia is known to be a strong predictor of mortality in patients with chronic heart failure (CHF).[92] Fazio and colleagues were amongst the first to investigate the potential benefit of GH treatment in patients with CHF.[93] In this uncontrolled study GH at a pharmacological dose increased septal thickness, left ventricular mass, and cardiac output. These changes were associated with improvements in clinical well-being. Subsequently in a controlled study GH administration has been shown to increase left ventricular mass in patients with dilated cardiomyopathy (DCM)[94] but this change did not result in objective improvement in clinical status. The same group reported that in DCM patients the improvement in ejection fraction correlated with the increase in IGF-I following GH treatment.[95] Adamopoulos and colleagues[96] have recently reported improved contractile performance in DCM following GH administration and suggest that GH-mediated anti-inflammatory effects may be the mechanism behind these potential benefits.

Anker and coworkers investigated GH physiology in patients with CHF.[97] Comparisons were made between patients with cachexia and noncachectic patients as well as a healthy control group. Those with cachexia had lower IGF-I concentrations, despite higher GH levels, indicating the presence of GH resistance, whereas no differences were observed between the controls and noncachectic patients. Cachexia as a result of CHF was thus associated with GH resistance and a lack of response to exogenous GH. In a randomized trial, Van Thiel and colleagues studied the effects of GH treatment on left ventricular mass index and systolic, and diastolic function using cardiovascular magnetic resonance imaging.[98] The patients had heart failure attributable to ischaemic heart disease. No benefits on these variables were observed following 6 months GH treatment. Similar results were reported from an earlier small study of patients with CHF secondary to coronary artery disease.[99] Simultaneously unchanged left ventricular dimension was observed in a placebo-controlled trial of GH replacement in elderly GH-deficient patients. An increase in exercise capacity was recorded in this study following GH treatment but it is uncertain as to whether this was a result of improved cardiac performance.[100]

In a recent small study ghrelin was administered to patients with CHF for 3 weeks.[101] Left ventricular (LV) ejection fraction and LV mass increased in this uncontrolled study and was associated with increased LBM and exercise capacity. Ghrelin is a potent GH secretagogue, principally found in the intestinal tract. This early study raises the possibility that such secretagogue analogues may have a potential role for the treatment of cachexia related to CHF.

TABLE 20.2
Variables of the GH/IGF-I Axis in Healthy Control Subjects and Patients with CHF

	Control subjects	All patients with CHF	Noncachectic patients with CHF	Cachectic patients with CHF
n	26	72	51	21
Total GH (ng/ml)	1.22 ± 0.31	2.40 ± 0.47	1.20 ± 0.28	5.31 ± 1.29
IGF-I (ng/ml)	151 ± 9	142 ± 6	150 ± 8	124 ± 9
IGFBP-3 (μg/ml)	3.75 ± 0.14	3.49 ± 0.12	3.70 ± 0.13	3.13 ± 0.21
Log IGF-I/GH	2.72 ± 0.19	2.47 ± 0.12	2.80 ± 0.14	1.68 ± 0.16

The patients with cachexia related to CHF had lower IGF-I levels, despite increased GH concentrations indicating the presence of GH resistance.
GH; growth hormone, IGF-I; insulin-like growth factor-I, IGFBP-3; insulin-like growth factor binding protein-3.
Source: Anker S.D., Volterrani M., Pflaum C.D., et al. *Journal of the American College of Cardiology* 2001; 38: 443–452.

In totality, studies of GH administration in CHF have indicated improved left ventricular mass and function in some but not all investigations. Those with the most severe cachexia appear to be most resistant to GH, which may explain the lack of benefit in some investigations. It remains unknown whether differences in GH dosing schedules or the use of GH and IGF-I in combination may result in more positive benefits for patients with cachexia related to CHF (Table 20.2).[102]

20.6.5 GH in AIDS Wasting

AIDS is associated with protein wasting leading to loss of LBM and cachexia. In an early study reversal of pretreatment weight loss, increased IGF-I levels, and decreased fat mass and urinary nitrogen excretion were recorded in 10 AIDS patients treated with GH.[103] Virtually identical effects were seen in a subsequent study of HIV^{+ve} men with documented weight loss (19% average) treated with GH over 7 days.[104] Combined GH + IGF-I administration decreased fat mass (FM), and increased LBM after 6 weeks treatment in HIV infected males[105]; however these effects were not sustained over 12 weeks treatment. More recently a study of 60 patients with AIDS wasting indicated that both GH and IGF-I in isolation increased LBM and reduced FM, but these changes were greatest in patients who received these agents in combination. No changes were recorded in instruments used to assess quality of life in these patients.[106]

A simultaneous study reported on a larger cohort of 142 subjects, randomized to receive GH and IGF-I or placebo in a similar protocol lasting 12 weeks. Active treatment was associated with a transient increase in weight and fat-free mass but these effects were only observed after 3 weeks and not persistent throughout the 12 week duration, suggesting that combined GH + IGF-I had no major anabolic effect in HIV-associated wasting.[107] Overall the evidence suggests that GH treatment has effects on reducing weight loss and preserving LBM in patients with AIDS associated weight loss. It remains unclear whether these benefits would be sustained over protracted periods of treatment and it is not certain whether the altered protein metabolism would result in mortality benefits.

20.6.6 GH Treatment in Ageing

A number of studies have addressed the effects of GH administration in ageing humans. GH given to healthy older men for a period of 6 months resulted in a gain in LBM (3.7 kg) with an associated reduction in FM (2.4 kg) compared to controls.[108] These effects were sustained over 12 months.[109] In a similar placebo-controlled trial active treatment resulted in an increase in LBM (3.3 kg), together with significant gains in muscle strength for knee flexion and extension.[110] These studies have used doses of GH comparable, or even higher, to those used in the treatment of adult GH-deficiency resulting in a high incidence of GH related side effects leading to the withdrawal of large numbers of subjects.

In elderly subjects undergoing total hip replacement, perioperative GH administration resulted in preservation of thigh muscle area on CT and gain in hip abductor strength. This appeared to be largely due to a preoperative gain in muscle mass negating the catabolic effects of surgery.[111] Similarly, in older patients undergoing renal dialysis treated with GH, active treatment resulted in significantly greater increase in serum albumin, fat-free mass, and calf muscle area.[112] Much remains unclear with regard to the possible interaction between GH and gonadal steroids in both men and women. Administration of testosterone has been shown to increase IGF-I levels in normal men[113] and in older women taking oestrogen as hormone replacement therapy (HRT), oral estrogens have been shown to enhance GH secretion but reduce IGF-I generation.[114]

A recent study in middle aged and older men suggests that the subcutaneous administration of GHRH twice daily at high dosage can increase GH and IGF-I levels and over 3 months GHRH treatment resulted in an increase in LBM.[115] Overall the preliminary data in older subjects suggest that GH treatment may help maintain or increase skeletal muscle bulk and strength. Although considerable interest exists it remains uncertain whether such a treatment will have clinical application.

20.7 COMBINED ADMINISTRATION OF GH AND IGF-I IN CACHEXIA

As IGF-I is known to have a hypoglycemic effect, it has been suggested that coadministration of both GH and IGF-I may maintain blood glucose concentrations

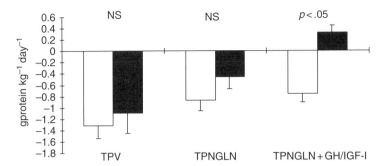

FIGURE 20.2 Net protein balance in critically ill subjects randomized to receive conventional TPN ($n = 7$), glutamine supplemented TPN (TPNGLN; $n = 7$) and TPNGLN with combined GH + IGF-I. Data were obtained using ^{13}C leucine and each group was studied initially in the fasting state (open bars) and subsequently following 3 days treatment (solid bars). The data indicate that following both TPN and TPNGLN the subjects remained in net negative protein balance but net protein gain was observed in these critically ill patients during treatment with TPNGLN + GH/IGF-I (Taken from Carroll P.V., Jackson N.C., Russell-Jones D.L., Treacher D.F., Sonksen P.H., and Umpleby A.M. *American Journal of Physiology* 2004; 286: E157–E157. With permission.).

and facilitate the anabolic actions of each agent. In a study of parenterally fed rats exposed to operative stress although both GH and IGF-I in isolation resulted in nitrogen retention and weight gain, combined treatment had a synergistic effect and increased fat oxidation.[116] In addition the combined treatment resulted in an additive rather than synergistic effect on weight gain. In these experiments TPN was provided and hypoglycemia was not observed. A recent study has investigated the effects of combined treatment with GH and IGF-I in addition to glutamine supplemented TPN in critically ill humans.[69] Using stable isotopes the findings indicate that this combined approach results in net protein gain in critically ill subjects without a major effect on whole-body glutamine plasma flux (see Figure 20.2). In the same study combined GH/IGF-I with glutamine supplemented TPN resulted in net whole-body protein anabolism (unlike nutritional support alone).[68] Future studies will determine whether use of IGF-I in combination with IGFBP-3 will be more effective than administration of IGF-I alone.[117]

20.8 SAFETY OF GH TREATMENT IN CACHEXIA

Pharmacological GH administration results in the adverse effects of fluid retention, arthralgia, jaw pain, and edema.[106, 107] These improve with dose reduction and do not usually require cessation of GH treatment. Hyperglycemia has been observed during GH therapy in a number of studies.[118] In severely ill patients with insulin resistance the requirement for exogenous insulin to maintain normoglycemia was increased by GH treatment.[119] These "adverse" effects are predictable based on the known effects of GH treatment, and in reported studies do not persist following cessation of GH administration.[118]

Few studies have been powered to assess safety issues of GH treatment at pharmacological doses, but clear evidence indicates that GH used as a treatment in critically ill, ICU dependent patients with diagnoses including cardiac and abdominal surgery, trauma, and respiratory failure results in increased mortality and morbidity.[119] The mechanism responsible for these deleterious effects on clinical outcome are unclear, although several possible factors have been proposed. These include enhancement of immune activation and worsening of the inflammatory response, prevention of amino acid mobilization (in particular glutamine), alterations in acid–base homeostasis, and disrupted cellular metabolism. In these ICU studies GH was administered early after the illness onset (in patients who had been in intensive care for 5 to 7 days), corresponding to the acute catabolic phase of illness. The majority of the excess deaths occurred within 10 days of commencement of GH treatment. It is likely that in both studies a variety of nutritional support regimens were provided and management with respect to the use of catecholamine and insulin infusions, as well as antibiotic usage, may have varied considerably. Nonetheless the results were homogenous within and across the participating countries, suggesting that the observed morbidity and mortality were directly related to GH treatment.[119, 120]

20.9 CONCLUSIONS

Advances in recombinant techniques leading to the widespread availability of peptide hormones, including GH and IGF-I, have resulted in studies assessing the potential of these agents in promoting anabolism in a variety of clinical conditions. Patients with catabolic skeletal muscle disorders resulting in cachexia have received particular attention and include those with AIDS, cardiac failure, sepsis, burns, trauma, and patients undergoing surgery.

Several well-conducted studies have been performed and these indicate that both GH and IGF-I, alone or in combination, result in positive nitrogen balance and net protein gain in different conditions using a variety of techniques. The majority of these trials have included selected patients and investigations have been performed over short periods. Not surprisingly, significant benefits in body composition, muscle function, and long-term clinical outcome are yet to be identified. Newer agents include oral GH secretagogues, which are being assessed in elderly subjects and patients with cardiac failure, and modulators of cytokine activity, which are currently undergoing preliminary assessment.

The studies of critically ill, ICU patients reported in 1999[118] provide clear evidence that GH, used in unselected ICU subjects, with acute illness results in major increases in morbidity and mortality. Although the mechanisms responsible remain poorly understood, these findings have had a significant impact on both clinical practice and research activities investigating the potential of anabolic agents.

It is likely that the anabolic benefit of GH is most pronounced when combined with appropriate provision of nutrition. Early evidence indicates that use of several anabolic treatments simultaneously may have synergistic effects on protein gain. Little information is available regarding the long-term use of these treatments and whether the use of growth promoting agents will result in improved clinical outcome is not clear. Importantly the widespread use of GH in acute illness results in increased mortality and current recommendations are that anabolic treatments should not be used in this setting. Ongoing research will address many of the outstanding issues relating to GH treatment for the promotion of protein gain. It remains to be seen whether GH will have a useful and safe therapeutic role in specific illnesses resulting in cachexia. Further well-designed, prospective, randomized, placebo-controlled studies are required to allow determination of whether GH treatment is safe, cost-effective, and, importantly, associated with clinical benefit in specific conditions characterized by catabolism.

REFERENCES

1. Bauman G., Stollar M.W., and Amburn K. Molecular forms of circulating GH during spontaneous secretory episodes and in the basal state. *Journal of Clinical Endocrinology and Metabolism* 1985; 60: 1216–1220.
2. Miller W.L. and Eberhardt N.L. Structure and evolution of the growth hormone gene family. *Endocrine Reviews* 1983; 4: 97–106.
3. Barsh G.S., Seeburg P.H., and Gelinas R.E. The human growth hormone gene family: Structure, evolution and allelic variations. *DNA* 1987; 6: 59–70.
4. Casaneuva, F.F. Physiology of GH secretion and action. *Endocrinology and Metabolism Clinics of North America* 1992; 21: 483–517.
5. Berlowitz M., Szabo M., Frohman L.A., Firestoke S., Chu L., and Hintz R.L. Somatomedin-C mediates growth hormone negative feedback by effects on both the hypothalamus and the pituitary. *Science* 1981; 212: 1279–1284.
6. Morita S., Yamashita S., and Melmed S. Insulin-like growth factor I action on rat anterior pituitary cell: Effects of intracellular messengers on growth hormone secretion and messenger ribonucleic acid levels. *Endocrinology* 1987; 121: 2000–2006.
7. Forbes G.B. *Human Body Composition: Growth, Aging, Nutrition and Activity*, 1987: p. 171. New York: Springer-Verlag.
8. Cahill G.F. Starvation in man. *New England Journal of Medicine* 1970; 282: 668–675.
9. Rennie M.J. Muscle protein turnover and the muscle wasting due to injury and disease. *British Medical Bulletin* 1985; 41: 257–264.
10. Cheek D.B. Muscle cell growth in normal children. In Cheek, D.B. (ed.) *Human Growth*, 1968; pp. 337–351. Philadelphia: Lea & Febiger.
11. Campbell M.J., McComas A.J., and Pepito F. Physiological changes in ageing muscles. *Journal of Neurology, Neurosurgery and Psychiatry* 1973; 36: 171–182.
12. Aniansson A., Grimby G., Hedberg M., et al. Muscle function in old age. *Scandinavian Journal of Rehabilitation Medicine* 1978; 6: 43–49.
13. Larsson L. Morphological and functional characteristics of ageing skeletal muscle in man. *Acta Physiologica Scandanavica* 1978; 457: 1–36.
14. Bergström J., Fürst P., Nòree L.-O., and Vinnars E. Intracellular free amino acid concentration in human muscle tissue. *Journal of Applied Physiology* 1974; 36: 693–697.

15. Waterlow J.C., Garlick P.J., and Millward D.J. *Protein Turnover in Mammalian Tissues and in the Whole Body.* 1978. Amsterdam: Elsevier-North Holland.

16. Chrysopoulo M.T., Jeschke M.G., Ramirez R.J., Barrow R.E., and Herndon D.N. Growth hormone attenuates tumor necrosis factor alpha in burned children. *Archives of Surgery* 1999; 134: 283–286.

17. Van den Berghe G. Endocrine evaluation of patients with critical illness. *Endocrinology and Metabolism Clinics North America* 2003; 32: 385–410.

18. Carroll P.V. Protein metabolism and the use of growth hormone and insulin-like growth factor-I in the critically ill patient. *Growth hormone and IGF Research* 1999; 9: 400–413.

19. Saltiel A.R. and Cuatrecasas P. In search of a second messenger for insulin. *American Journal of Physiology* 1988; 255: C1–C11.

20. Luzi L., Castellino P., Simonson D.C., Petrides A.S., and DeFronzo R.A. Leucine metabolism in IDDM. *Diabetes* 1990; 39: 38–48.

21. Nair K.S., Garrow J.S., Ford C., Mehler R.F., and Halliday D. Effect of poor diabetic control and obesity on whole body protein metabolism in man. *Diabetologia* 1993; 25: 400–403.

22. Umpleby A.M., Boroujerdi M.A., Brown P.M., Carson E.R., and Sönksen P.H. The effect of metabolic control on leucine metabolism in type-1 (insulin-dependent) diabetic patients. *Diabetologia* 1986; 29: 131–141.

23. Castellino P., Luzi L., Simonson D.C., Haymond M., and DeFronzo R.A. Effect of insulin and plasma amino acid concentrations on leucine metabolism in man. *Journal of Clinical Investigation* 1987; 80: 1784–1793.

24. Nair K.S., Ford G.C., and Halliday D. Effect of intravenous insulin treatment on in-vivo whole-body leucine kinetics and oxygen consumption in insulin-deprived type 1 diabetic patients. *Metabolism* 1987; 36: 491–495.

25. Carroll P.V., Christ E.R., Umpleby A.M., et al. IGF-I treatment in adults with type 1 diabetes: Effects on glucose and protein metabolism in the fasting state and during a hyperinsulinaemic-euglycaemic clamp. *Diabetes* 2000; 49: 789–796.

26. Tanner J.M., Hughes P.C.R., and Whitehouse R.H. Comparative rapidity of response of height, limb muscle and limb fat to treatment with human GH in patients with and without GH deficiency. *Acta Endocrinologica* 1977; 184: 681–696.

27. Hulthen L., Bengtsson B.A., Sunnerhagen K.S., et al. GH is needed for the maturation of muscle mass and strength in adolescents. *Journal of Clinical Endocrinology and Metabolism* 2001: 86: 4765–4770.

28. Kostyo, J.L. Separation of the effects of the effects of GH on muscle amino acid transport and protein synthesis. *Endocrinology* 1964; 75: 113–119.

29. Salomon F., Cuneo R.D., Hesp R., and Sönksen P.H. The effects of treatment with recombinant human growth hormone on body composition and metabolism in adults with growth hormone deficiency. *New England Journal of Medicine* 1989; 321: 1797–1803.

30. Jorgensen J.O.L., Pedersen S.A., Thuesen L., et al. Beneficial effects of growth hormone treatment in GH-deficient adults. *Lancet* 1989; i: 1221–1225.

31. Russell-Jones D.L., Weissberger A.J., Bowes S.B. et al. The effects of growth hormone on protein metabolism in adult growth hormone deficient patients. *Clinical Endocrinology* 1993; 38: 427–431.

32. Carroll P.V., Christ E.R., Bengtsson B.A., et al. Growth hormone deficiency in adulthood and the effects of growth hormone replacement: A review. *Journal of Clinical Endocrinology and Metabolism* 1998; 83: 382–395.

33. Boulware S.D., Tamborlane W.V., Rennert N.J., Gesundheit N., and Sherwin R.S. Comparison of the metabolic effects of recombinant human insulin-like growth factor-I and insulin: Dose–response relationships in healthy young and middle-aged adults. *Journal of Clinical Investigation* 1994; 93: 1131–1139.

34. Hussain M.A., Schmitz O., Mengel A., Keller A., Christiansen J.S., Zapf J., and Froesch E.R. Insulin-like growth factor-I stimulates lipid oxidation, reduces protein oxidation and enhances insulin sensitivity in humans. *Journal of Clinical Investigation* 1993; 95: 2249–2256.

35. Laager R., Ninnis R., and Keller U. Comparison of the effects of recombinant human insulin-like growth factor-I and insulin on glucose and leucine kinetics in humans. *Journal of Clinical Investigation* 1993; 92: 1903–1909.

36. Walker J.L., Ginalska-Malinowska M., Romer T.E., Pucilowska J., and Underwood L.E. Effects of infusion of IGF-I in a child with growth hormone insensitivity syndrome (Laron Dwarfism). *New England Journal of Medicine* 1991; 24: 1483–1488.

37. Russell-Jones D.L., Umpleby A.M., Hennessy T.R., et al. Use of a leucine clamp to demonstrate that IGF-I actively stimulates protein synthesis in normal humans. *American Journal of Physiology* 1994; 267: E591–E598.

38. Cutherbertson D.P. Disturbance of metabolism produced by bony and non-bony injury, with notes on certain abnormal conditions of bone. *Biochemical Journal* 1930; 24: 1244–1263.

39. Cutherbertson D.P. The distribution of nitrogen and sulphur in the urine during conditions of increased catabolism. *Biochemical Journal* 1931; 25: 236–244.

40. O Keefe S.J.D., Sender P.M., and James W.P.T. 'Catabolic' loss of body nitrogen in response to surgery. *Lancet* 1974; ii: 1035–1037.

41. Crane C.W., Picou D., Smith R., and Waterlow J.C. Protein turnover in patients before and after elective orthopaedic operations. *British Journal of Surgery* 1977; 64: 129–133.

42. Long C.L., Jeevanandam M., Kim B.M., and Kinney J.M. Whole body protein synthesis and catabolism in septic man. *American Journal of Clinical Nutrition* 1977; 30: 1340–1344.

43. Birkhahn R.H., Long C.L., Fitkin D., et al. Whole-body protein metabolism due to trauma in man as estimated by 1-(1, 14C)-leucine. *Surgery* 1980; 88: 294–300.

44. Tomkins A.M., Garlick P.J., Scholfield W.N., and Waterlow J.C. The combined effects of infection and malnutrition on protein metabolism in children. *Clinical Science* 1983; 65: 313–324.

45. Jahoor F., Desai M., Herndon D.N., and Wolfe R.R. Dynamics of the protein metabolic response to burn injury. *Metabolism* 1988; 37: 330–337.

46. Carli F., Webster J., Ramachandra V., et al. Aspects of protein metabolism after elective surgery in patients receiving constant nutritional support. *Clinical Science* 1990; 78: 621–628.

47. Windsor J.A. and Hill G.A. Risk factors of postoperative pneumonia: the importance of protein depletion. *Annals of Surgery* 1988; 208: 209–214.

48. Herrmann F.R., Safran C., Levkoff S.E., and Minaker K.L. Serum albumin level on admission as a predictor of death, length of stay, and readmission. *Archives of Internal Medicine* 1992; 152: 125–130.

49. Ross R.J.M., Miell J.P., Freeman E., et al. Critically ill patients have high basal growth hormone concentrations with attenuated oscillatory activity associated with

low concentrations of insulin-like growth factor-I. *Clinical Endocrinology* 1991; 35: 47–54.

50. Voerman B.J., Strack van Schijndel R.J., Groeneveld A.B., et al. Effects of human growth hormone in critically ill nonseptic patients: results from a prospective, randomized, placebo-controlled trial. *Critical Care Medicine* 1995; 23: 665–673.

51. Van den Berghe G. and de Zegher F. Anterior pituitary function during critical illness and dopamine treatment. *Critical Care Medicine* 1996; 24: 1580–1590.

52. Hawker F.H., Stewart P.M., Baxter R.C., et al. Relationship of somatomedin-C/ insulin-like growth factor I levels to conventional nutritional indices in critically ill patients. *Critical Care Medicine* 1987; 8: 732–739.

53. Anker S.D., Chua T.P., Poinkowski P., et al. Hormonal changes and anabolic/catabolic imbalance in chronic heart failure and their importance for cardiac cachexia. *Circulation* 1997; 15: 526–534.

54. Miell J.P., Taylor A.M., Jones J., et al. Administration of human recombinant insulin-like growth factor-I to patients following major gastrointestinal surgery. *Clinical Endocrinology* 1992; 37: 542–551.

55. Wolf S.E., Barrow R.E., and Herndon D.N. Growth hormone and IGF-I therapy in the hypercatabolic patient. *Baillieres Clinical Endocrinology and Metabolism* 1996; 10: 447–463.

56. Timmins A.C., Cotterill A.M., Cwyfan-Hughes S.C., et al. Critical illness is associated with low circulating concentrations of insulin-like growth factors-I and -II, alterations in insulin-like growth factor binding proteins, and induction of an insulin-like growth factor binding protein 3 protease. *Critical Care Medicine* 1996; 24: 1460–1466.

57. Davies S.C., Wass J.A., Ross R.J., et al. The induction of a specific protease for insulin-like growth factor binding protein-3 in the circulation during severe illness. *Journal of Endocrinology* 1991; 130; 469–473.

58. Defalque D., Brandt N., Ketelslegers J.-M., and Thissen J.-P. GH insensitivity induced by endotoxin injection is associated with decreased liver GH receptors. *American Journal of Physiology* 1999; 276: E565–E572.

59. Dahn M.S., Lange P., and Jacobs L.A. Insulin-like growth factor-I production is inhibited in human sepsis. *Archives of Surgery* 1988; 123: 1409–1413.

60. Mulligan K. and Schambelan M. Anabolic treatment with GH, IGF-I or anabolic steroids in patients with HIV-associated wasting. *International Journal of Cardiology* 2002; 85: 151–159.

61. Ho K.Y., Veldhuis J.D., Johnson M.L., et al. Fasting enhances growth hormone secretion and amplifies the complex rhythms of growth hormone secretion in man. *Journal of Clinical Endocrinology and Metabolism* 1988; 81: 968–975.

62. Isley W.L., Underwood L.E., and Clemmons D.R. Dietary components that regulate serum somatomedin-C concentrations in humans. *Journal of Clinical Investigation* 1983; 71: 175–182.

63. Merimee T.J., Zapf J., and Froesch E.R. Insulin-like growth factors in the fed and fasted states. *Journal of Clinical Endocrinology and Metabolism* 1982; 55: 999–1002.

64. Ponting G.A., Ward H.C., Halliday D., Teale J.D., and Sim A.J.W. Protein and energy metabolism with biosynthetic growth hormone in patients on full intravenous nutritional support. *Journal of Parenteral and Enteral Nutrition* 1990; 14: 437–441.

65. Jiang Z.-M., He G.-Z., Zhang S.-Y., et al. Low dose growth hormone and hypocaloric nutrition attenuate the protein-catabolic response after major operation. *Annals of Surgery* 1989; 210: 513–525.

66. Hammarqvist F., Strömberg C., von der Decken A., et al. Biosynthetic growth hormone preserves both muscle protein synthesis and the decrease in muscle-free glutamine, and improves whole-body nitrogen economy after operation. *Annals of Surgery* 1992; 216: 184–191.

67. Berman R.S., Harrison L.E., Pearlstone D.B., Burt M., and Brennan M.F. Growth hormone, alone and in combination with insulin, increases whole body and skeletal muscle protein kinetics in cancer patients after surgery. *Annals of Surgery* 1999; 229: 1–10.

68. Carroll P.V., Jackson N.C., Russell-Jones D.L., Treacher D.F., Sonksen P.H., and Umpleby A.M. Combined growth hormone/insulin-like growth factor in addition to glutamine-supplemented TPN results in net protein anabolism in critical illness. *American Journal of Physiology* 2004; 286: E157–E157.

69. Jackson N.C., Carroll P.V., Russell-Jones D.L., Sönksen P.H., Treacher D.F., and Umpleby A.M. Effects of glutamine supplementation, GH and IGF-I on glutamine metabolism in critically ill patients. *American Journal of Physiology* 2000; 278: E226–E233.

70. Kissmeyer-Nielsen P., Jensen M.B., and Laurberg S. Perioperative growth hormone treatment and functional outcome after major abdominal surgery: A randomized, double-blind, controlled study. *Annals of Surgery* 1999; 229: 298–302.

71. Barle H., Essen P., Nyberg B., Olivecrona H., Tally M., McNurlan M.A., Wernerman J., and Garlick P.J. Depression of liver protein synthesis during surgery is prevented by growth hormone. *American Journal of Physiology* 1999; 274: E620–E627.

72. Mjaaland M., Unneberg K., Larsson J., et al. Growth hormone after gastrointestinal surgery: attenuated forearm glutamine, alanine, alanine, 3-methylhistidine and total amino acid efflux in patients treated with total parenteral nutrition. *Annals of Surgery* 1993; 217: 413–422.

73. Barry M.C., Mealey K., Sheehan S.J., Burke P.E., Cunningham A.J., Leahy A, and Bouchier-Hayes D. The effects of recombinant human growth hormone on cardiopulmonary function in elective abdominal aortic aneurysm repair. *European Journal of Vascular and Endovascular Surgery* 1998; 16: 311–9.

74. Mealey K., Barry M., O Mahoney L., et al. Effects of recombinant human growth hormone (rhGH) on inflammatory responses in patients undergoing abdominal aortic aneurysm repair. *Intensive Care Medicine* 1998; 24: 128–131.

75. Liljedahl S., Gemzell C., Plantin L., and Birke G. Effect of human growth hormone in patients with severe burns. *Acta Chirurgica Scandinavica* 1961; 122: 1–14.

76. Soroff H.S., Rozin R.R., Mooty J.M., and Lister J. Role of human growth hormone in the response to trauma: 1. Metabolic effects following burns. *Annals of Surgery* 1967; 166: 739–752.

77. Wilmore D.W., Moylan J.A., Bristow B.F., Mason Jr A.D., and Pruitt Jr B.A. Anabolic effects of human growth hormone and high-caloric feedings following thermal injury. *Surgery Gynecology and Obstetrics* 1974; 138: 875–884.

78. Herndon D.N., Barrow R.E., Kunkle K.R., Broemeling L., and Rutan L. Effects of recombinant human growth hormone on donor site healing in severely burned children. *Annals of Surgery* 1990; 212: 424–429.

79. Gore D.C., Honeycutt D., Jahoor F., et al. Effect of exogenous growth hormone on whole body and isolated limb protein kinetics in burned patients. *Archives of Surgery* 1991; 126: 38–43.

80. Gore D.C., Honeycutt D., Jahoor F., et al. Effect of exogenous growth hormone on glucose utilisation in burn patients. *Journal of Surgical Research* 1991; 51: 518–523.

81. Gilpin D.A., Barrow R.E., Rutan R.L., Broemeling L., and Herndon D.N. Recombinant human growth hormone accelerates wound healing in children with large cutaneous burns. *Annals of Surgery* 1994; 220: 19–24.

82. Sherman S.K., Demling R.H., LaLonde C., et al. Growth hormone enhances re-epithelialization of human split skin graft donor sites. *Surgical Forum* 1989; 40: 37–39.

83. Aili Low J.F., Barrow R.E., Mittendorfer B., et al. The effect of short-term growth hormone treatment on growth and energy expenditure in burned children. *Burns* 2001; 27: 447–452.

84. Mlcak R.P., Suman O.E., Murphy K., and Herndon D.N. Effects of growth hormone on anthropometric measurements and cardiac function in children with thermal injury. *Burns* 2005; 31: 60–6.

85. Suman O.E., Mlcak R.P., and Herndon D.N. Effects of exogenous growth hormone on resting pulmonary function in children with thermal injury. *Journal of Burns Care and Rehabilitation* 2004; 25: 287–293.

86. Okamura K., Okuma T., Tabira Y., and Miyauchi Y. Effect of administered growth hormone on protein metabolism in septic rats. *Journal of Parenteral and Enteral Nutrition* 1989; 13: 450–454.

87. Fong Y., Rosenbaum M., Tracey K.J., et al. Recombinant growth hormone enhances muscle myosin heavy chain mRNA accumulation and amino acid accrual in humans. *Proceedings of the National Academy of Sciences of the USA* 1989; 86: 3371–3374.

88. Douglas R.G., Humberstone D.A., Haystead A., and Shaw J.H. Metabolic effects of recombinant human growth hormone: Isotopic studies in the postabsorptive state and during total parenteral nutrition. *British Journal of Surgery* 1990; 77: 785–790.

89. Voerman H.J., van Schijndel R.J., Groeneveld A.B., et al. Effects of recombinant human growth hormone in patients with severe sepsis. *Annals of Surgery* 1992; 216: 648–55.

90. Dahn M.S. and Lange M.P. Systemic and splanchnic metabolic response to exogenous human growth hormone. *Surgery* 1998; 123: 528–538.

91. Unneberg K., Balteskard L., Mjaaland M., and Revhaug A. Growth hormone impaired compensation of hemorrhagic shock after trauma and sepsis in swine. *Journal of Trauma-Injury Infection & Critical Care* 1996; 41: 775–780.

92. Anker S.D., Pnikowski P., Varney S., et al. Wasting as independent risk factor for mortality in chronic heart failure. *Lancet* 1997; 349: 1050–1053.

93. Fazio S., Sabatini D., Capaldo B., et al. A preliminary study of growth hormone in the treatment of dilated cardiomyopathy. *New England Journal of Medicine* 1996; 334: 809–814.

94. Osterziel K.J., Strohm O., Schuler J., et al. Randomised, double-blind, placebo-controlled trial of human recombinant growth hormone in patients with chronic heart failure due to dilated cardiomyopathy. *Lancet* 1998; 351: 1233–1237.

95. Perrot A., Ranke M.B., Dietz R., and Osterziel K.J. Growth hormone treatment in dilated cardiomyopathy. *Journal of Cardiac Surgery* 2001; 16: 127–131.

96. Adamopoulos S., Parissis J.T., Paraskevaidis I., et al. Effects of growth hormone on circulating cytokine network, and left ventricular contractile performance and geometry in patients with idiopathic dilated cardiomyopathy. *European Heart Journal* 2003; 24: 2186–2196.

97. Anker S.D., Volterrani M., Pflaum C.D., et al. Acquired growth hormone resistance in patients with chronic heart failure: Implications for therapy with growth hormone. *Journal of the American College of Cardiology* 2001; 38: 443–452.

98. van Thiel S.W., Smit J.W., de Roos A., et al. Six-months of recombinant human GH therapy in patients with ischaemic cardiac failure. *International Journal of Cardiovascular Imaging* 2004; 20: 53–60.

99. Spallarossa P., Rossettin P., Minuto F., et al. Evaluation of growth hormone administration in patients with chronic heart failure secondary to coronary artery disease. *American Journal of Cardiology* 1999; 84: 430–433.

100. Elgzyri T., Castenfors J., Haggg E., et al. The effects of GH replacement therapy on cardiac morphology and function, exercise capacity and serum lipids in elderly patients with GH deficiency. *Clinical Endocrinology* 2004; 61: 113–122.

101. Nagaya N., Moriya J., Yasumura Y., et al. Effects of ghrelin administration on left ventricular function, exercise capacity, and muscle wasting in patients with chronic heart failure. *Circulation* 2004; 110: 3674–3679.

102. Cicoira M., Kalra P.R., and Anker S.D. Growth hormone resistance in chronic heart failure and its therapeutic implications. *Journal of Cardiac Failure* 2003; 9: 219–226.

103. Krentz A.J., Koster F.T., Crist D.M. et al. Anthropometric, metabolic, and immunological effects of recombinant human growth hormone in AIDS and AIDS-related complex. *Journal of Acquired Immunodeficiency Syndromes* 1993; 6: 245–251.

104. Mulligan, K., Grunfeld, C., Hellerstein, M.K., Neise, R.A., and Schambelan, M. Anabolic effects of recombinant human growth hormone in patients associated with human immunodeficiency virus infection. *Journal of Clinical Endocrinology and Metabolism* 1993; 77: 956–962.

105. Schambelan M., Mulligan K., Grunfeld C., et al. Recombinant human growth hormone in patients with HIV wasting. A randomized, placebo-controlled trial. *Annals of Internal Medicine* 1996; 125: 873–882.

106. Waters D., Danska J., Hardy K., et al. Recombinant human growth hormone, insulin-like growth factor I., and combination therapy in AIDS associated wasting. A randomized, double-blind, placebo-controlled trial. *Annals of Internal Medicine* 1996; 125: 865–872.

107. Lee P.D., Pivarnik J.M., Bukar J.G., et al. A randomized, placebo-controlled trail of combined insulin-like growth factor I and low dose growth hormone therapy for wasting associated with human immunodeficiency virus infection. *Journal of Clinical Endocrinology and Metabolism* 1996; 81: 2968–2975.

108. Rudman D., Feller A.G., Nagraj H.S., et al. Effects of human growth hormone on men over 60 years old. *New England Journal of Medicine* 1990; 323: 1–6.

109. Rudman D., Feller A.G., Cohn L., Shetty K.R., Rudman I.W., and Draper M.W. Effects of human growth hormone on body composition in elderly men. *Hormone Research* 1991; 36: 73–81.

110. Welle S., Thornton C., Stott M., and McHenry B. Growth hormone increases muscle mass and strength but does not rejuvenate myofibrillar protein synthesis in healthy subjects over 60 years old. *Journal of Clinical Endocrinology and Metabolism* 1996; 83: 3239–3244.

111. Weissberger A.J., Anastasiadis A.D., Sturgess I., Martin F.C., Smith M.A., and Sonksen P.H. Recombinant human growth hormone treatment in elderly patients undergoing elective total hip replacement. *Clinical Endocrinology* 2003; 58: 99–107.

112. Johansson G., Bengtsson B.A., Ahlmen L. Double-blind, placebo-controlled study of growth hormone treatment in elderly patients undergoing chronic haemodialysis: anabolic effect and functional improvement. *American Journal of Kidney Disease* 1999; 33; 709–717.

113. Hobbs C.J., Plymate S.R., Rosen C.J., and Adler R.A. Testosterone administration increases IGF-I in normal men. *Journal of Clinical Endocrinology and Metabolism* 1993; 77: 776–779.

114. Friend K.E., Hartman M.L., Pezzoli S.S., Clasey J.L., and Thorner M.O. Both oral and transdermal oestrogen increase growth hormone release in post-menopausal women- a clinical research center study. *Journal of Clinical Endocrinology and Metabolism* 1996; 81: 2250–2256.

115. Veldhuis J.D., Patrie J.T., Frick K., Weltman J.Y., and Weltman A. Sustained growth hormone (GH) and insulin-like growth factor I responses to prolonged high-dose twice-daily GH-releasing hormone stimulation in middle-aged and older men. *Journal of Clinical Endocrinology and Metabolism* 2004; 89: 6325–6330.

116. Lo H.C., Hinton P.S., Peterson C.A., et al. Simultaneous treatment with IGF-I and growth hormone additively increases anabolism in parenterally fed rats. *American Journal of Physiology* 1995; 269: E368–E376.

117. O'Connell T. and Clemmons D.R. IGF-I/ IGF-binding protein-3 combination improves insulin resistance by GH-dependent and independent mechanisms. *Journal of Clinical Endocrinology and Metabolism* 2002; 87; 4356–4360.

118. Takala J., Ruokonen E., Webster N.R., et al. Increased mortality associated with growth hormone treatment in critically ill adults. *New England Journal of Medicine* 1999; 341: 785–92.

119. Carroll P.V. and van den Berghe G. Safety aspects of pharmacological GH therapy in adults. *Growth Hormone and IGF Research* 2001; 11: 166–72.

120. Jensen M.B., Kissmeyer-Nielsen P., and Laurberg S. Perioperative growth hormone treatment increases nitrogen and fluid balance and results in short-term and long-term conservation of lean tissue mass. *American Journal of Clinical Nutrition* 1998; 68: 840–846.

21 Anti-Inflammatory Agents in Cancer Cachexia

Mellar P. Davis

CONTENTS

SUMMARY

Corticosteroids, cyclopentenone prostaglandins, nonsteroidal anti-inflammatory drugs (NSAIDs), and eicosapentaenoic acid (EPA) have some rationale in the management of anorexia and cachexia.

The evidence is not strong enough to recommend the use of anti-inflammatory drugs as a routine practice. Nonsteroidal anti-inflammatory drugs have multiple anti-inflammatory targets, which are distinctly different depending upon

A World Health Organization project in palliative medicine.

the NSAID. Combination of anti-inflammatory drugs with nonoverlapping targets, would be an appropriate topic for future research. Such combinations could consist of NSAIDs with EPA, corticosteroids with EPA, and cyclopentenone prostaglandins with NSAIDs. Finally, combining anti-inflammatory drugs with appetite stimulants or other anticytokines would be a logical paradigm to pursue in randomized studies.

21.1 INTRODUCTION

The sickness behavior induced by inflammatory mediators such as lipopolysaccharides (LPSs) appears to be generated by proinflammatory cytokines such as interleukin 1B (IL_1B) and prostanoids. Such behavior in animals consists of anorexia, weight loss, depressed activity, and disappearance of body care activities.[1-3] The sickness response in these animal models is partially reversed by cyclooxygenase (COX) inhibitors such as indomethacin.[3]

Chronic cancer produces a similar sickness syndrome in humans marked by weight loss, anorexia, asthenia, anemia, and an acute phase response.[4-6] This sick syndrome has been termed the anorexia/cachexia syndrome, which is thought to be caused by the proinflammatory cytokines IL_1B, interleukin 6 (IL_6), tumor necrosis factor (TNF) alpha, and interferon (IFN) gamma.[4,7,8] Elevated serum concentrations of IL_1, IL_6, and TNF have been described in cancer patients.[8] Specific receptors for cytokines such as TNF have been found in both circulation and within the central nervous system (CNS) in cancer patients.[8-11] These inflammatory cytokines and prostanoids are generated by tumor and the host in response to the tumor in an interactive fashion.[12] However, cytokines are not always elevated in circulation in cancer but can still be a significant factor in cachexia and anorexia. Proinflammatory cytokines have diurnal variations, a short serum half-life, rapid clearance, and principally work as autocrine or paracrine mediators or regionally through peripheral mononuclear cells.[13,14] Prostanoids, particularly prostaglandin E2 (PGE_2) is a proximal mediator of the IL_1B and TNF metabolic effects.

The end result of PGE_2 upregulation is hypermetabolism, tumor-induced fever, elevated acute phase reactants, proteolysis, lipolysis, reduced lipogenesis, increased gluconeogenesis, and insulin resistance.[15] In this sense, cancer is a chronic inflammatory disorder. Such metabolic aberrations have a direct bearing on the degree of weight loss and survival in gastrointestinal cancer patients.[16,17]

Cyclooxygenase inhibitors and dietary omega-3 fatty acids have been used experimentally and clinically to treat cancer and reverse cancer cachexia and anorexia. The anti-inflammatory actions of nonsteroidal anti-inflammatory drugs (NSAIDs) and omega-3 fatty acids extend to more than one target. Salicylates, COX nonspecific inhibitors, COX-2 specific inhibitors, and omega-3 fatty acids activate certain nuclear transcription activator receptors, peroxisome proliferator activated receptors (PPARs) inhibit others, particularly nuclear factor kappa B (NF Kappa B), which are the principle means by which anti-inflammatory agents both reduce tumor growth and curtail cancer cachexia.[26] This chapter briefly reviews prostaglandin metabolism, COX enzymes, PPARs, and NF Kappa B as background to understanding NSAID and omega-3 fatty acid anti-inflammatory activity. The clinical experience in the treatment of cachexia with both classes of medications is also reviewed.

21.2 Prostaglandins

Prostaglandins are cyclized, oxygenated, and peroxidated derivates of arachidonic acid first extracted from seminal vesicles, prostate, and semen (hence the name) by Goldblatt and von Euler in the 1930s.[27] Arachidonic acid is released by phospholipase A_2, cyclized and oxygenated to prostaglandin G_2 (PGG$_2$) by COX (Figure 21.1). PGG$_2$ is reduced by a separate peroxidase on the COX enzyme to Prostaglandin H_2 (PGH$_2$). PGH$_2$ is isomerized to the various prostaglandins or exposed to oxireductases to form others.[27] Prostaglandin E_2 (PGE$_2$) is a major inflammatory mediator. Both PGE$_2$ and PGI$_2$ (also known as prostacyclin) cause pain by binding to prostaglandin receptors (EP$_1$) and activate sodium channels. In addition, PGE$_2$ promotes the translocation of NF Kappa B from cytoplasm to nuclear-binding sites, which activates transcription of proinflammatory cytokines.[15] Both PGE$_2$ and PGI$_2$ prevent IL$_2$ and alpha IFN production by T lymphocytes and thus are immune suppressive.

Prostaglandin E_2 is a potent stimulator of tumor angiogenesis, tumor proliferation, and vascular endothelial growth factor production. Tumor angiogenesis and upregulation of vascular endothelial growth factor occur through EP$_2$ and EP$_4$ receptors. These receptors provide a growth advantage to cancers that express COX-2.[27]

Prostanoids not only activate prostaglandin receptors (EP) but also activate nuclear receptors such as PPARs.[26,28] Cyclopentenone prostaglandins derived from PGD$_2$ (particularly 15dPGJ$_2$) stimulates PPAR gamma and alpha resulting in catabolism of prostaglandins as a negative feedback inhibitor and secondarily prevents the translocation of NF Kappa B.[27,29–31] The end result is the down-modulation of inflammation. Cyclopentenone prostaglandins inhibit the synthesis of TNF and IL$_1$B and IL$_6$.[15,30–34] Additionally, the cyclopentenone prostaglandin 15dPGJ$_2$ prevents COX-2 expression. This occurs through inhibition of the transcription-activating factor AP$_1$.[35] The other target of 15dPGJ$_2$ is IkB kinase (IKK), which phosphorylates IkB, the NF Kappa B inhibitor. IkB prevents the translocation of NF Kappa B to the nucleus but phosphorylation of IkB deactivates this inhibitor and allows NF Kappa B to be translocated.[32,36] The relative importance of 15dPGJ$_2$ activation of PPAR, inhibition of IKK, and direction inhibition of NF Kappa B and AP$_1$ will vary from cell to cell and is independent of COX enzyme levels.[32] Cyclopentenone prostaglandins act as autocrine and paracrine inhibitors of inflammation.[32,37,38] On the contrary, 15dPGJ$_2$ increases the expression of COX-2 by binding to the promoter site.[39]

Leukotrienes are produced from arachidonic acid via various lipoxygenases. Certain leukotrienes such as leukotriene B_4 and 8S-hydroxyeicosatetraenoic acid and 15-eicosatetraenoic acid are PPAR agonists.[15,31,40,41] Leukotrienes also stimulate COX-2 expression perhaps through activation of PPARs.[39]

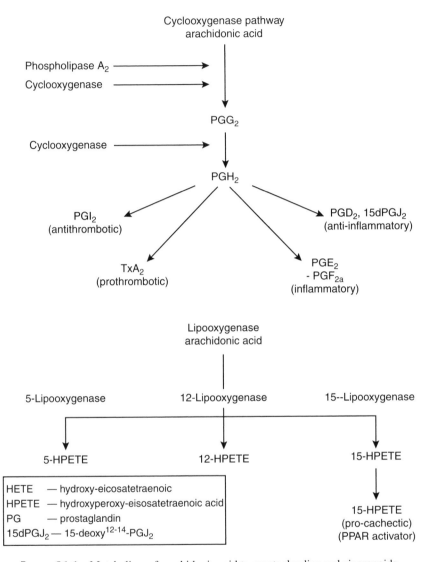

FIGURE 21.1   Metabolism of arachidonic acid to prostaglandins and eicosanoids.

Lipoxins derived from prostanoids down-modulate acute inflammation. Aspirin, which acetylates COX-2 does not entirely prevent prostaglandin formation but leads to the formation of a 15-epimer lipoxin from 15-hydroxyeicostatracroic acid, which is anti-inflammatory. Omega-3 fatty acids derived from membranes form leukotrienes of the five series and prostaglandins of the 3 series, which are much less inflammatory than the PGE_2.[42–44] The combination of a COX-2 inhibitor and omega-3 fatty acid forms a novel lipoxigenase generated lipoxin, which is a potent anti-inflammatory eicosanoid.[42]

21.3 OMEGA-3 FATTY ACIDS

A fatty acid structure is designated by a numerical notation indicating the number of carbons (C20) that is followed by the number of double bonds (C20:4 for arachidonic acid). The omega refers to the position of the first double bond from the methyl terminus.[43] Precursors for prostaglandins are 20 carbon tetraenoic fatty acids (C20:4 omega-6), which are part of the cell membrane and released by phospholipase A_2. The generation of prostaglandins depends as much on phospholipase A_2 as on COXs. Eicosapentaenoic acid (EPA) (C20:5 omega-3) is a 20 carbon polyunsaturated fatty acid containing five double bonds, the first located in the third position from the methyl terminus. Omega-3 fatty acids are derived from flaxseed (linseed) oil and fish oils. The linseed oil contains the omega-3 fatty acid, alpha linolenic acid is an 18 carbon fatty acid and must be lengthened to 20 carbons to be efficiently incorporated into cell membranes.[45–47] Fish oils (C20:5 omega-3) are easily incorporated into membranes in lieu of omega-6 fatty acids (arachidonic acid) resulting in a reduction in synthesis of the two series of prostaglandins (PGE_2) and the four series of leukotrienes (LTB_4) of which are proinflammatory and procoagulant. Omega-3 fatty acids stereotypically inhibit COX-2 by forming a strained conformation when bound to the active enzyme site, which significantly impairs enzyme kinetics.[48] In addition, EPA and other omega-3 fatty acids reduce NF Kappa B translocation to the nucleus and down-regulate IL_1B, TNF alpha, and IL_6 production.[43] This is done through decreasing IkB phosphorylation.[49,50] The EPA metabolite PGE_3 is much less efficient in initiating COX-2 expression than PGE_2. EPA also inhibits the second nuclear translocation factor AP_1 through inhibition of mitogen activated protein kinase (MAPK). Omega-3 fatty acids when given over months have produced responses in chronic inflammatory states such as inflammatory bowel disease, rheumatoid arthritis, psoriasis, and systemic lupus erythematosus.[43] Treatment however requires several months to be effective, presumably because it is necessary that EPA be incorporated into membranes, which is a gradual process.[43,46,51–54]

21.4 CYCLOOXYGENASE

Cyclooxygenase has been found in two isomeric forms. Cyclooxygenase 1 (COX-1) is constitutively expressed as a housekeeping COX and cyclooxygenase 2 (COX-2) is inducible under conditions of stress and inflammation.

COX-1 and COX-2 are products of two different genes. The enzyme proteins have 60% amino acid identity.[42] COXs do not require ATP for prostaglandin production.[27] Subtle differences in amino acid structure, membrane binding, and the active enzyme catalytic domains do not influence the enzyme kinetics, which are the same for COX-1 and COX-2.[27,30] Both COX-1 and COX-2 dimerize on the luminal side of microsomes, endoplasmic reticulum, and the nuclear envelope. Heterodimers between COX-1 and COX-2 do not normally occur.[27,30] COXs also carry a heme-binding site for peroxidase activity, which is unrelated to the COX site (and which is not inhibited by NSAIDs). Both are glycosylated at the time of translocation, which maintains the enzyme in a favorable conformation. COX-2 is much more efficient in metabolizing EPA than COX-1. Both have a short catalytic half-life (1 to 2 min) at high arachidonic acid concentrations.[27] COX-2 differs from COX-1 in that it has a TATA box and several inducible enhancer element-binding sites in a *cis* position next to the structural gene that act as promoter sites for transcription. These promoter sites are (1) NF Kappa B, (2) NF-IL$_6$, and (3) cyclic AMP response elements (CREs), which are involved in the expression and upregulation of COX-2 during inflammation.[30,55] It is important to note that there are exceptions to the usual understanding of COX-1 and COX-2. COX-1 is induced by phorbol esters in human monocyte leukemia cells and by stem cell factor and dexamethasone in mouse derived bone marrow mast cells.[30] COX-1 can be detected in an inflammatory response, which may reduce the benefits of COX-2 selective inhibitors. COX-2 is constitutively expressed within the spinal cord and kidney.[27,30]

During inflammation there are two peaks to COX-2 expression. The first peak is responsible for PGE$_2$. The second peak leads to down-modulation of inflammation and the production of PGD$_2$ and 15dPGJ$_2$. Prostaglandins of the J series activate PPAR, which catabolizes prostaglandins and down-modulates IL$_1$ and TNF.[30] Inhibition of COX-2 with selective COX-2 inhibitors can therefore paradoxically increase inflammation due to reduced 15dPDJ$_2$.[56-58] COXs are closely located next to phospholipase 2 on membranes and both are rate limiting to prostaglandin production. TNF alpha and IL$_1$ upregulate COX-2 expression through NF Kappa B.[32,59-61] Cytokines will also prevent COX-2 mRNA catabolism and prolong translation half-life of active enzymes.[27,62]

21.5 NF Kappa B

NF Kappa B is a generic term for a dimeric transcription factor formed by hetero- and homo-dimerization of five family proteins.[59] These five families of proteins are the RelA-p65, p50, p52, RelB, and C-Rel.[56] NF Kappa B hetero- and homo-dimers are responsible for binding to and activating transcriptions of over 150 genes at the NF Kappa B initiator site.[63] NF Kappa B, a nuclear transition activator factor, plays a major role in upregulating inflammatory gene expression. NF Kappa B induces expression of COX-2, nitric oxide synthase, TNF, IL$_1$, and IL$_6$.[32] The proinflammatory response occurs particularly with the formation of p65-p50 dimers, which act as the central control to an inflammatory response. NF Kappa B upregulation initiates the acute phase responses.[64]

Activation of NF Kappa B accelerates inflammation and increases cellular proliferation of tumors and prevents apoptosis.[64] Inhibition of NF Kappa B therefore is both antineoplastic and can reduce cachexia. Inhibition of NF Kappa B sensitizes tumors to chemotherapy and radiation, which is the subject of multiple research trials.[59]

Resolution of inflammation also requires NF Kappa B expression. The homo-dimer p50-p50 becomes the dominant form of NF Kappa B as inflammation evolves, which induces leukocyte apoptosis.[59] The NF Kappa B dimer C-ReL-p50 reduces nitric oxide synthase. Complete inhibition of NF Kappa B can lead to severe cellular apoptotic damage in critical illnesses.[65]

Nuclear factor Kappa B is kept in an inactive state by being bound to the inhibitor IkB.[59] NF Kappa B bound IkB cannot be translocated from cytoplasm to the nucleus due to masking of the nuclear translocation signal domain on NF Kappa B.[59,60] The proinflammatory cytokines IL_1 and TNF activate IKK, which phosphorylates and deactivates IkB. Phosphorylation of IkB is subsequently ubiquinated and catabolized through proteasomes.[59–61] TNF, IL_1B cause a sustained expression of IKK, which accelerates IkB degradation for which there is little tachyphylaxis.[66,67] As a positive feedback mechanism, PGE_2 promotes NF Kappa B by enhancing the formation of RelA-p65 subunits.[15]

The cyclopentenone prostaglandin 15dPGJ$_2$ derived from COX-2 down-regulates NF Kappa B by interfering with the p50 DNA binding and blocking IKK phosphorylation of IkB.[32,68,69]

21.6 PEROXISOME PROLIFERATOR-ACTIVATED RECEPTORS

Peroxisome proliferator-activated receptors are orphan nuclear receptors, similar to steroid receptors, which dimerize with the retinoid X receptor and then bind to various response elements within regulatory regions of genes.[31,39,70] Phosphorylation of PPAR by various kinases leads to degradation of PPAR by proteasomes, which is a regulator process similar to IkB.[71] PPARs are also central to the control of inflammation. Activation of PPAR leads to catabolism of PGE_2 by hydroxylation and oxidation and down-regulation of NF Kappa B and AP_1. The end result is reduced expression of TNF and IL_1.[71] Down-regulation of TNF and IL_1 by PPAR is at the level of transcription. There are three forms of PPAR (alpha, delta, and gamma); PPAR alpha is found predominantly in the liver and PPAR gamma in adipocytes. Fatty acids, 15dPGJ$_2$, EPA, certain NSAIDs, and leukotrienes activate PPAR.[29,31,39,72] Activation of PPAR also increases the expression of COX-2.[39,73] PPAR gamma causes maturation of adipocytes and increases insulin sensitivity and opposes the insulin resistance and lipolysis induced by TNF, IL_1, IL_6, and IFN gamma.[26,31] High concentrations of PPAR agonists prevent proinflammatory cytokine secretion, activation of AP_1, and NF Kappa B in macrophages.[74] Deficiencies of PPAR alpha lead to prolonged inflammatory responses in *PPAR* gene knockout mice.[75,76]

Skeletal muscle expresses little PPAR compared to adipocytes and macrophages. Hence PPAR agonists will influence lipid metabolism more than muscle proteolysis in cachexia.[30,31]

21.7 ANTI-INFLAMMATORY AGENTS IN CACHEXIA

Anti-inflammatory drugs, particularly NSAIDs, reduce animal tumors and experimental cachexia.[21–23,77] Nonsteroidal anti-inflammatory drugs have reversed some of the metabolic abnormalities associated with cancer cachexia in humans and are reported to prolong survival.[78,79] The anti-inflammatory EPA has been reported

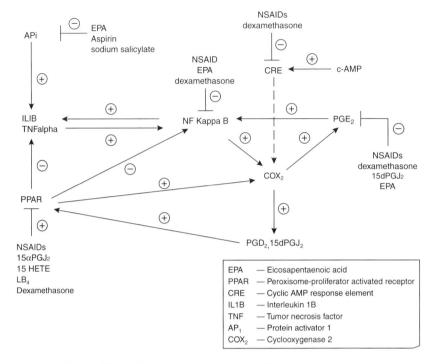

FIGURE 21.2   Site of action for various anti-inflammatory agents.

to improve cancer cachexia in cohort studies.[20,80,81] Randomized trials of EPA so far have been negative, though this may be related to study design.[82] This section reviews the sites of action and the clinical trials experience various anti-inflammatory drugs (Figure 21.2) (Table 21.1).

21.7.1 Corticosteroids

Corticosteroids are anti-inflammatory through several mechanisms.[83] Dexamethasone induces the transcription of IkB, which binds to NF Kappa B and prevents nuclear translocation.[84–87] Corticosteroids bind to the adjacent promoter regions of NF Kappa B and suppress transcription.[88,89] Dexamethasone destabilizes both TNF and COX-2 mRNA and prevents translation of the cytokines by monocytes.[55,73,90–92] Dexamethasone inhibits COX-2 transcription by inhibiting the p50 subunit of NF Kappa B. The mechanism by which dexamethasone inhibits NF Kappa B is distinctly different from that of salicylates but may be synergistic with salicylates.[63] Dexamethasone enhances the expression of PPAR alpha and delta through the CCAAT enhancer binding protein (which is probably the reason for enhanced lipogenesis with corticosteroids). Finally, corticosteroids do decrease AP$_1$ transcription activation.[33] Unfortunately, corticosteroids increase skeletal muscle proteolysis and muscle wasting.[29,93]

TABLE 21.1

The Mode of Action of Various Anti-Inflammatory Drugs

	AP$_1$ inhibitor	NF Kappa B inhibitor	PPAR agonist	COX inhibitor
Acetominophen	N/A	−	−	+
Aspirin	+	+	−	+
Sodium salicylate	+	+	−	±
Indomethacin	N/A	−	+	+
Ibuprofen	N/A	−	+	+
Diclofenac	N/A	−	Competitive antagonist	+
Sulindac	N/A	+	+	+
Dexamethasone	N/A	+	+	+
R-flurbiprofen	N/A	+	−	−
S-flurbiprofen	N/A	−	−	+
Ketorolac	N/A	+	−	+
Ketoprofen	N/A	+	−	+
Fenoprofen	N/A	−	+	+
Celecoxib	−	Activator at high doses	−	+
Rofecoxib	Activator at high doses	+	+	+
Naproxen	N/A	−	+	+
EPA	+	+	−	+

Note: API = Activator Protein 1; COX = cyclooxygenase; NF kappa B = nuclear factor kappa B; and PPAR = peroxisome proliferators activated receptor.

The clinical experience with corticosteroids in cancer cachexia is somewhat limited. Methylprednisolone has improved the quality of life in patients with advanced cancer.[94,95] A 14 day trial of methylprednisolone reported improvement in appetite, depression, and daily activity, as well as reduced analgesic consumption.[96] A randomized trial of dexamethasone 0.75 mg four times daily was compared to megestrol acetate 800 mg daily and fluoxymesterone 10 mg twice a day. Both dexamethasone and megestrol acetate improved nonfat weight and appetite to the same degree and both were superior to fluoxymesterone.[97] A prospective randomized trial compared indomethacin, prednisone, and placebo. Both anti-inflammatory drugs were superior to placebo in maintaining performance status and weight. Leukocytosis associated with cancer improved as well as pulse rate and liver function.[78]

21.7.2 Nonsteroidal Anti-Inflammatory Drugs

The NSAID anti-inflammatory activity has multiple targets including COX inhibition.[26] It is probably incorrect to classify NSAIDs based upon COX inhibition alone. COX expression is not necessary for NSAID anti-inflammatory or antineoplastic actions.[98] Sodium salicylate is a poor COX inhibitor and is easily overwhelmed by arachidonic acid, yet it is an anti-inflammatory drug, which reduces prostaglandin synthesis in intact cells.[99] Aspirin retains its anti-inflammatory activity in COX-2 gene knockout mice. The anti-inflammatory and analgesic potency of NSAIDs does not correlate with COX inhibition in animal models.[100] As an example, the R enantiomer of flurbiprofen does not block COX-2 but has analgesic and anti-inflammatory properties.[56] Celebrex, a COX-2 specific inhibitor, in high concentrations loses its anti-inflammatory effect and becomes proinflammatory.[58,101] Aspirin and other NSAIDs must be given in higher doses than those necessary to inhibit COX in order to reduce inflammation.

The COX-independent anti-inflammatory actions of NSAIDs are centered upon inactivation of NF Kappa B, activation by PPAR, and secondarily upon inhibition of AP_1 and CRE.[72,102,103] The COX-independent actions of NSAIDs may account for the reversal of cancer cachexia and tumor apoptosis. Several NSAIDs inhibit NF Kappa B activation. Aspirin and sodium salicylate prevent NF Kappa B activation by stabilizing IkB and preventing its phosphorylation by IKK.[69,102,104–110] As a result, TNF, IL_1 and COX-2 are down-modulated at the level of transcription.[39,111] The induction of NF Kappa B by TNF is inhibited since TNF is a major upregulator of IKK expression.[105] Other NSAIDs, which inhibit NF Kappa B, are rofecoxib, flufenamic acid, sulindac, and R-flurbiprofen.[56,64,72,112] Certain COX-2 selective inhibitors (SC 236) directly inhibit the RelA/p65 subunits of NF Kappa B and thus prevent NF Kappa B translocation independent of IKK.[113] Celecoxib at high concentrations promotes the degradation of IkB and activates NF Kappa B promoting the expression of COX-2 and TNF.[58] This may be one of the reasons why celecoxib has a poor anti-inflammatory response compared to nonselective COX inhibitors in certain animal models and why there is a ceiling dose.[114] Rofecoxib upregulates AP_1 transcription factor and can be proinflammatory as well.[115] Acetaminophen, indomethacin, S-flurbiprofen, ketoprofen, and diclofenac do not inhibit NF Kappa B activation.[56,72,103,105,116] NSAIDs, which do not block NF Kappa B, will not reduce the expression of proinflammatory cytokines such as IL_6 nor prevent the upregulation of COX-2.[72,117] The antineoplastic activity of NSAIDs also appears to be largely COX independent due to suppression of transcription factors necessary for cell proliferation and neoangiogenesis.[118,119]

Certain NSAIDs selectively activate PPAR alpha and gamma in the liver and monocytes resulting in a reduction PGE_2 generated NF Kappa B and AP_1 activation.[120–122] This is done in a stereospecific fashion with certain enantiomers more effective than others. The NSAID order for activation of PPAR alpha is S-ibuprofen > R-ibuprofen > indomethacin > S-naproxen. The NSAID order of activity for PPAR gamma is S-naproxen > indomethacin > S-ibuprofen > R-ibuprofen.[26,123] Nonsteroidal anti-inflammatory doses need to be higher in order to activate PPAR than to inhibit COX-1.[26,124] Sodium salicylate and aspirin do not activate PPAR.[29] Diclofenac binds to PPAR but fails to activate PPAR and thus acts as a competitive antagonist.[40,120] Other NSAIDs, which are PPAR activators, are meclofenamate, mefenamic acid, sulindac, nimesulide, and rofecoxib.[39,125,126]

21.7.2.1 Additional anti-inflammatory mechanisms of NSAIDs

Aspirin, but not sodium salicylate, indomethacin, or peroxicam induces the production of an anti-inflammatory lipoxin through 15-HETE.[99, 127, 128] Certain NSAIDs, including sulindac, activate heat shock transcription factor 1 (HSF-1), which inhibits monocyte production of IL_1, IL_6, and TNF.[129] Selective NSAIDs and aspirin inhibit cellular kinases other than IKK such as ribosomal S6 kinase 2, which is responsible for activating CRE.[130] CREs stimulate the expression of COX-2.[131] Aspirin also causes the accumulation of adenosine in inflammatory exudates, which is anti-inflammatory.[100]

21.7.2.2 Proinflammatory mechanisms of NSAIDs

Nonsteroidal anti-inflammatory drugs can become proinflammatory by reducing cyclopentenone prostaglandins and through activation of PPAR, which increases COX-2 expression. Celecoxib activates NF Kappa B at high concentrations and rofecoxib AP_1 both of which can induce proinflammatory cytokines and COX-2 expression.[56, 58, 72, 73, 101]

21.7.2.3 Animal studies with NSAIDs

Cyclooxygenase-2 inhibitors have both anticachectic and antitumor activity in several animal models.[21-23],[132-134] The benefits appear partly related to inhibition of COX-2, since COX-1 agents are ineffective. Both PGE_2 and proinflammatory cytokines are reduced, which correlates with a reversal of cachexia and antitumor responses. In certain tumor models, COX-2 expression is not necessary for tumor suppression by NSAIDs.[98]

A combination of celecoxib plus a PPAR agonist have potent antitumor activity and improve carcass weight in experimental animals.[135] In other animal models, however (Morris 777 hepatoma model and the Walker 256 carcinosarcoma), NSAIDs neither improve carcass weight nor reduce tumor.[12]

It is difficult to determine whether the upregulated cytokines are derived from tumor or are the result of a host response to tumor. By using an IL_6 gene knockout mouse model, indomethacin did not reduce IL_6 derived from tumor. Tumor production of cytokines can be a more important determinant to tumor growth and weight loss than host derived cytokines.[136] Tumors incubated with monocytes stimulate production of PGE_2 from monocytes, which in turn increases monocyte production of COX-2, TNF, and IL_6. Prostaglandins derived from tumor-infiltrating monocytes can be crucial to tumor growth in animal models. Tumor derived NF Kappa B becomes activated by stromal monocyte generated PGE_2. COX-2 inhbitors, such as indomethacin can also reduce tumor derived PGE_2, TNF, and IL_6, which inhibits NF Kappa B.[122, 137] In most animal tumor models, it is difficult to separate the antineoplastic and anticachectic benefits of NSAIDs since they appear together. It may be that by reducing tumor growth, carcass weight is maintained or improved rather than as a result of direct inhibition of the catabolic processes.

21.7.2.4 Clinical studies

The earliest published experience of NSAIDs in cancer cachexia used indomethacin and did not find any alteration in resting energy expenditures, lipolysis, or acute phase reactants (such as C-reactive protein). The beta-blocker, propanolol, improved resting energy expenditure, reduced lipolysis, and C-reactive protein indicating that perhaps sympathetic activity rather than inflammation was responsible for much of the abnormal metabolic activity in cancer cachexia.[138]

A prospective randomized trial published 1 year later demonstrated survival benefits of indomethacin. In the trial, 135 weight-losing cancer patients were randomized between indomethacin 50 mg twice daily, prednisone and placebo. Patients were matched by gender, tumor, stage, nutritional status, biochemical, and physiologic and functional state.[78] Indomethacin doubled survival compared to placebo (510 ± 28 days vs. 250 ± 28 days). Both anti-inflammatory drugs (indomethacin and prednisone) improved survival over placebo when combined and compared to placebo. The nutritional response was not the primary aim of this study; however, it was noted that leukocytosis improved, pulse rate normalized, and liver function improved. Performance status did not deteriorate as rapidly on anti-inflammatory drugs compared to placebo.

Ibuprofen was used in a small group of colon cancer patients with liver metastases and elevated C-reactive protein prior to surgery. Patients were randomized between no drug and ibuprofen three times daily. Ibuprofen reduced IL_6 levels, serum C-reactive protein, and whole body protein synthesis (presumably related to increased production of acute phase reactants).[24]

A large study involving nearly 300 patients over 16 years compared indomethacin with matched controls. All patients had lost a significant amount of body weight, had advanced stage solid tumors, and were not on corticosteroids or chemotherapy. Indomethacin reduced resting energy expenditure and improved food intake compared to matched controls. C-reactive protein was reduced. Indomethacin improved fat weight.[18]

A combination of medroxyprogesterone 500 mg twice daily, celecoxib 200 mg twice daily plus nutritional supplement was used to treat a small number of advanced nonsmall cell lung cancer patients. All had weight loss >10% of baseline weight and all had elevated C-reactive protein, fatigue, and reduced performance score (>2 by an Eastern Cooperative Oncology Group Performance Scale). After 6 weeks of treatment, weight loss was curtailed, body mass index and lean body mass slightly improved, and mid-arm circumference (a measure of lean body mass) improved. Caloric intake, performance score, nausea, fatigue, and early satiety were better with medroxyprogesterone and celecoxib.[139] Reduction in C-reactive protein correlated with improvement in body mass index and lean body mass.

A combination of indomethacin and erythropoietin was used in approximately 300 patients with gastrointestinal cancers. This randomized trial was analyzed by treatment analysis and intention to treat analysis. The intention to treat analysis was negative but the treatment analysis demonstrated improved survival, body weight, food intake, and metabolic efficiency but not lean body mass with the combination.[79]

21.7.2.5 Summary of studies

Nonsteroidal anti-inflammatory drugs improve certain parameters associated with cachexia. Survival may be improved but additional studies will be necessary to validate this point. Combinations with progesterone and erythropoietin have demonstrated benefit but it is difficult to determine if the combination is an improvement over NSAIDs, progesterone, or erythropoietin alone. Outcomes among the various trials have been quite diverse; hence, making comparisons is difficult at best. The use of other NSAIDs, which block NF Kappa B such as sulindac, R-flurbiprofen, aspirin, and sodium salicylate or combinations may be more appropriate. Combinations of NSAIDs with cyclopentenone prostaglandins or EPA would be of interest. There are nonoverlapping anti-inflammatory actions with particular drug combinations. Certain NSAIDs should be avoided due to upregulation of proinflammatory transcription factors at certain doses or due to competitive antagonism of PPAR (rofecoxib, celecoxib, and diclofenac).

21.7.3 Omega-3 Fatty Acids

The anti-inflammatory effects of EPA include (1) replacement of N-6 fatty acids within cellular membranes reducing arachidonic acid substrates, (2) conversion to less inflammatory prostaglandins and leukotrienes, (3) inhibition of COX-2, (4) inhibition of NF Kappa B translocation, (5) inhibition of AP_1 translocation, and (6) reduction in expression of IL_1 and TNF.[15,42,43,47–50]

21.7.3.1 Animal studies

Eicosapentaenoic acid inhibits tumor growth in animals.[140] *In vitro*, commercial omega-3 fatty acids reduce TNF expression in macrophages stimulated by LPS.[141]

21.7.3.2 Clinical studies

Six months of dietary omega-3 fatty acids reduces IL_1 and TNF expression in peripheral blood monocytes in multiple sclerosis.[51] Other studies have demonstrated improvement in autoimmune diseases, but prolonged treatment was necessary.[46,52,53,142] These findings concur with studies done in animals.[54]

Clinically tolerable doses of omega-3 fatty acids are 0.3 g/kg. Side effects such as diarrhea or gastrointestinal intolerance limit further dose escalation.[143,144] Unfortunately, most trials in cancer patients have used two or less grams in a nutritional supplement, which limited dose escalation. Most trials are 4 weeks or less and all randomized trials have used low doses.[82,145,146] The failure to demonstrate significant weight gain in randomized trials may reflect the dose and duration of treatment rather than the failure of omega-3 fatty acids to improve cancer cachexia.

Single-arm trials have demonstrated reduction in IL_6, reduction in insulin resistance, improvement in physical activity, stabilization of body weight, reduction in resting energy expenditure, and improvement in energy intake.[20,80,81,147,148] Additional randomized trials using higher doses over longer periods of time are needed in order to either confirm or disprove the benefits of EPA in the treatment of cancer cachexia.

21.8 STATUS OF SPECIFIC THERAPIES

Nonsteroidal anti-inflammatory agents are increasingly approved for arthritis and pain but not cachexia. Corticosteroids have multiple indications and have been used as appetite stimulants but are neither licensed for such purposes nor for the treatment of cachexia. Omega-3 fatty acids are available in a nutritional supplement and licensed as a nutritional product and not a medication for cancer patients. Omega-3 fatty acids are also available as a noncontrolled over-the-counter product but are not licensed or approved as a treatment for cachexia.

ACNOWLEDGMENT

The author would like to thank Becky Phillips for her superb manuscript skills in preparing this chapter.

REFERENCES

1. Kent, S., Bluthe, R.M., Kelley, K.W., and Dantzer, R. Sickness behavior as a new target for drug development. *Trends Pharmacol. Sci.* 1992;**13**:24–28.
2. Spadaro, F. and Dunn, A.J. Intracerebroventricular administration of interleukin-1 to mice alters investigation of stimuli in a novel environment. *Brain Behav. Immun.* 1990;**4**:308–322.
3. Johnson, R.W., Curtis, S.E., Dantzer, R., and Kelley, K.W. Central and peripheral prostaglandins are involved in sickness behavior in birds. *Physiol. Behav.* 1993;**53**:127–131.
4. Argiles, J.M., Moore-Carrasco, R., Fuster, G., Busquets, S., and Lopez-Soriano, F.J. Cancer cachexia: The molecular mechanisms. *Int. J. Biochem. Cell. Biol.* 2003;**35**:405–409.
5. Bosaeus, I., Daneryd, P., Svanberg, E., and Lundholm, K. Dietary intake and resting energy expenditure in relation to weight loss in unselected cancer patients. *Int. J. Cancer* 2001;**93**:380–383.
6. Bosaeus, I., Daneryd, P., and Lundholm, K. Dietary intake, resting energy expenditure, weight loss and survival in cancer patients. *J. Nutr.* 2002;**132**(11 Suppl.):3465S–3466S.
7. Barber, M.D., Ross, J.A., and Fearon, K.C. Disordered metabolic response with cancer and its management. *World J. Surg.* 2000;**24**:681–689.
8. Ramos, E.J., Suzuki, S., Marks, D., Inui, A., Asakawa, A., and Meguid, M.M. Cancer anorexia-cachexia syndrome: Cytokines and neuropeptides. *Curr. Opin. Clin. Nutr. Metab. Care* 2004;**7**:427–434.
9. Simons, J.P., Schols, A.M., Buurman, W.A., and Wouters, E.F. Weight loss and low body cell mass in males with lung cancer: relationship with systemic inflammation, acute-phase response, resting energy expenditure, and catabolic and anabolic hormones. *Clin. Sci. (Lond.)* 1999;**97**:215–223.

10. Barton, B.E. IL-6-like cytokines and cancer cachexia: consequences of chronic inflammation. *Immunol. Res.* 2001;**23**:41–58.

11. O'Riordain, M.G., Falconer, J.S., Maingay, J., Fearon, K.C., and Ross, J.A. Peripheral blood cells from weight-losing cancer patients control the hepatic acute phase response by a primarily interleukin-6 dependent mechanism. *Int. J. Oncol.* 1999;**15**:823–827.

12. McCarthy, D.O. Inhibitors of prostaglandin synthesis do not improve food intake or body weight of tumor-bearing rats. *Res. Nurs. Health* 1999;**22**:380–387.

13. Noguchi, Y., Yoshikawa, T., Matsumoto, A., Svaninger, G., and Gelin, J. Are cytokines possible mediators of cancer cachexia? *Surg. Today* 1996;**26**:467–475.

14. Dayer, J.M., Beutler, B., and Cerami, A. Cachectin/tumor necrosis factor stimulates collagenase and prostaglandin E2 production by human synovial cells and dermal fibroblasts. *J. Exp. Med.* 1985;**162**:2163–2168.

15. Ross, J.A., and Fearon, K.C. Eicosanoid-dependent cancer cachexia and wasting. *Curr. Opin. Clin. Nutr. Metab. Care* 2002;**5**:241-248.

16. Scott, H.R., McMillan, D.C., Watson, W.S., Milroy, R., and McArdle, C.S. Longitudinal study of resting energy expenditure, body cell mass and the inflammatory response in male patients with non-small cell lung cancer. *Lung Cancer* 2001;**32**:307–312.

17. McMillan, D.C., Canna, K., and McArdle, C.S. Systemic inflammatory response predicts survival following curative resection of colorectal cancer. *Br. J. Surg.* 2003;**90**:215–219.

18. Lundholm, K., Daneryd, P., Korner, U., Hyltander, A., and Bosaeus, I. Evidence that long-term COX-treatment improves energy homeostasis and body composition in cancer patients with progressive cachexia. *Int. J. Oncol.* 2004;**24**:505–512.

19. Wigmore, S.J., Falconer, J.S., Plester, C.E., Ross, J.A., Maingay, J.P., Carter, D.C., et al. Ibuprofen reduces energy expenditure and acute-phase protein production compared with placebo in pancreatic cancer patients. *Br. J. Cancer* 1995;**72**:185–188.

20. Wigmore, S.J., Barber, M.D., Ross, J.A., Tisdale, M.J., and Fearon, K.C. Effect of oral eicosapentaenoic acid on weight loss in patients with pancreatic cancer. *Nutr. Cancer* 2000;**36**:177–184.

21. Hussey, H.J., and Tisdale, M.J. Effect of the specific cyclooxygenase-2 inhibitor meloxicam on tumour growth and cachexia in a murine model. *Int. J. Cancer* 2000;**87**:95–100.

22. Peluffo, G.D., Stillitani, I., Rodriguez, V.A., Diament, M.J., and Klein, S.M. Reduction of tumor progression and paraneoplastic syndrome development in murine lung adenocarcinoma by nonsteroidal antiinflammatory drugs. *Int. J. Cancer* 2004;**110**:825–830.

23. Davis, T.W., Zweifel, B.S., O'Neal, J.M., Heuvelman, D.M., Abegg, A.L., Hendrich, T.O., et al. Inhibition of cyclooxygenase-2 by celecoxib reverses tumor-induced wasting. *J. Pharmacol. Exp. Ther.* 2004;**308**:929–934.

24. Preston, T., Fearon, K.C., McMillan, D.C., Winstanley, F.P., Slater, C., Shenkin, A., et al. Effect of ibuprofen on the acute-phase response and protein metabolism in patients with cancer and weight loss. *Br. J. Surg.* 1995;**82**:229–234.

25. Dannenberg, A.J., and Subbaramaiah, K. Targeting cyclooxygenase-2 in human neoplasia: Rationale and promise. *Cancer Cell* 2003;**4**:431–436.

26. Jaradat, M.S., Wongsud, B., Phornchirasilp, S., Rangwala, S.M., Shams, G., Sutton, M., et al. Activation of peroxisome proliferator-activated receptor isoforms

and inhibition of prostaglandin H(2) synthases by ibuprofen, naproxen, and indomethacin. *Biochem. Pharmacol.* 2001;**62**:1587–1595.

27. Simmons, D.L., Botting, R.M., and Hla, T. Cyclooxygenase isozymes: The biology of prostaglandin synthesis and inhibition. *Pharmacol. Rev.* 2004;**56**:387–437.

28. Hinz, B., Brune, K., and Pahl, A. 15-Deoxy-Delta(12,14)-prostaglandin J2 inhibits the expression of proinflammatory genes in human blood monocytes via a PPAR-gamma-independent mechanism. *Biochem. Biophys. Res. Commun.* 2003;**302**:415–420.

29. Lehmann, J.M., Lenhard, J.M., Oliver, B.B., Ringold, G.M., and Kliewer, S.A. Peroxisome proliferator-activated receptors alpha and gamma are activated by indomethacin and other non-steroidal anti-inflammatory drugs. *J. Biol. Chem.* 1997;**272**:3406–3410.

30. Gilroy, D.W. and Colville-Nash, P.R. New insights into the role of COX 2 in inflammation. *J. Mol. Med.* 2000;**78**:121–129.

31. Gelman, L., Fruchart, J.C., and Auwerx, J. An update on the mechanisms of action of the peroxisome proliferator-activated receptors (PPARs) and their roles in inflammation and cancer. *Cell. Mol. Life Sci.* 1999;**55**:932–943.

32. Straus, D.S., Pascual, G., Li, M., Welch, J.S., Ricote, M., Hsiang, C.H., et al. 15-deoxy-delta 12,14-prostaglandin J2 inhibits multiple steps in the NF-kappa B signaling pathway. *Proc. Natl. Acad. Sci. USA* 2000;**97**:4844–4849.

33. Jiang, C., Ting, A.T., and Seed, B. PPAR-gamma agonists inhibit production of monocyte inflammatory cytokines. *Nature* 1998;**391**:82–86.

34. Koppal, T., Petrova, T.V., and Van Eldik, L.J. Cyclopentenone prostaglandin 15-deoxy-Delta(12,14)-prostaglandin J(2) acts as a general inhibitor of inflammatory responses in activated BV-2 microglial cells. *Brain Res.* 2000;**867**: 115–121.

35. Sawano, H., Haneda, M., Sugimoto, T., Inoki, K., Koya, D., and Kikkawa, R. 15-Deoxy-Delta12,14-prostaglandin J2 inhibits IL-1beta-induced cyclooxygenase-2 expression in mesangial cells. *Kidney Int.* 2002;**61**:1957–1967.

36. Boyault, S., Bianchi, A., Moulin, D., Morin, S., Francois, M., Netter, P., et al. 15-Deoxy-delta(12,14)-prostaglandin J(2) inhibits IL-1beta-induced IKK enzymatic activity and IkappaBalpha degradation in rat chondrocytes through a PPARgamma-independent pathway. *FEBS Lett.* 2004;**572**:33–40.

37. Shibata, T., Kondo, M., Osawa, T., Shibata, N., Kobayashi, M., and Uchida, K. 15-deoxy-delta 12,14-prostaglandin J2. A prostaglandin D2 metabolite generated during inflammatory processes. *J. Biol. Chem.* 2002;**277**:10459–10466.

38. Cuzzocrea, S., Wayman, N.S., Mazzon, E., Dugo, L., Di Paola, R., Serraino, I., et al. The cyclopentenone prostaglandin 15-deoxy-Delta(12,14)-prostaglandin J(2) attenuates the development of acute and chronic inflammation. *Mol. Pharmacol.* 2002;**61**:997–1007.

39. Meade, E.A., McIntyre, T.M., Zimmerman, G.A., and Prescott, S.M. Peroxisome proliferators enhance cyclooxygenase-2 expression in epithelial cells. *J. Biol. Chem.* 1999;**274**:8328–8334.

40. Wood, I.C. Pro-inflammatory mechanisms of a non-steroidal anti-inflammatory drug. *Trends Pharmacol. Sci.* 2002;**23**:109.

41. Kliewer, S.A., Sundseth, S.S., Jones, S.A., Brown, P.J., Wisely, G.B., Koble, C.S., et al. Fatty acids and eicosanoids regulate gene expression through direct interactions with peroxisome proliferator-activated receptors alpha and gamma. *Proc. Natl. Acad. Sci. USA* 1997;**94**:4318–4323.

42. Serhan, C.N., Clish, C.B., Brannon, J., Colgan, S.P., Chiang, N., and Gronert, K. Novel functional sets of lipid-derived mediators with anti-inflammatory actions generated from omega-3 fatty acids via cyclooxygenase 2-nonsteroidal antiinflammatory drugs and transcellular processing. *J. Exp. Med.* 2000;**192**:1197–1204.

43. Jho, D.H., Cole, S.M., Lee, E.M., and Espat, N.J. Role of omega-3 fatty acid supplementation in inflammation and malignancy. *Integr. Cancer Ther.* 2004;**3**:98–111.

44. Endres, S., and von Schacky, C. n-3 polyunsaturated fatty acids and human cytokine synthesis. *Curr. Opin. Lipidol.* 1996;**7**:48–52.

45. Sinclair, A.J., Murphy, K.J., and Li, D. Marine lipids: Overview "news insights and lipid composition of Lyprinol." *Allerg Immunol. (Paris)* 2000;**32**:261–271.

46. Robinson, D.R., Tateno, S., Knoell, C., Olesiak, W., Xu, L., Hirai, A., et al. Dietary marine lipids suppress murine autoimmune disease. *J. Intern. Med. Suppl.* 1989;**225**:211–216.

47. Carrick, J.B., Schnellmann, R.G., and Moore, J.N. Dietary source of omega-3 fatty acids affects endotoxin-induced peritoneal macrophage tumor necrosis factor and eicosanoid synthesis. *Shock* 1994;**2**:421–426.

48. Malkowski, M.G., Thuresson, E.D., Lakkides, K.M., Rieke, C.J., Micielli, R., Smith, W.L., et al. Structure of eicosapentaenoic and linoleic acids in the cyclooxygenase site of prostaglandin endoperoxide H synthase-1. *J. Biol. Chem.* 2001;**276**:37547–37555.

49. Babcock, T., Helton, W.S., and Espat, N.J. Eicosapentaenoic acid (EPA): An anti-inflammatory omega-3 fat with potential clinical applications. *Nutrition* 2000;**16**:1116–1118.

50. Bagga, D., Wang, L., Farias-Eisner, R., Glaspy, J.A., and Reddy, S.T. Differential effects of prostaglandin derived from omega-6 and omega-3 polyunsaturated fatty acids on COX-2 expression and IL-6 secretion. *Proc. Natl. Acad. Sci. USA* 2003;**100**:1751–1756.

51. Gallai, V., Sarchielli, P., Trequattrini, A., Franceschini, M., Floridi, A., Firenze, C., et al. Cytokine secretion and eicosanoid production in the peripheral blood mononuclear cells of MS patients undergoing dietary supplementation with n-3 polyunsaturated fatty acids. *J. Neuroimmunol.* 1995;**56**:143–153.

52. Kremer, J.M. and Robinson, D.R. Studies of dietary supplementation with omega 3 fatty acids in patients with rheumatoid arthritis. *World Rev. Nutr. Diet.* 1991;**66**:367–382.

53. Singer, P., Wirth, M., Berger, I., Heinrich, B., Godicke, W., Voigt, S., et al. Long-chain omega 3 fatty acids are the most effective polyunsaturated fatty acids for dietary prevention and treatment of cardiovascular risk factors. Conclusions from clinical studies. *World Rev. Nutr. Diet.* 1992;**69**:74–112.

54. Robinson, D.R., Xu, L.L., Tateno, S., Guo, M., and Colvin, R.B. Suppression of autoimmune disease by dietary n-3 fatty acids. *J. Lipid Res.* 1993;**34**:1435–1444.

55. Inoue, H., Tanabe, T., and Umesono, K. Feedback control of cyclooxygenase-2 expression through PPARgamma. *J. Biol. Chem.* 2000;**275**:28028–28032.

56. Tegeder, I., Niederberger, E., Israr, E., Guhring, H., Brune, K., Euchenhofer, C., et al. Inhibition of NF-kappaB and AP-1 activation by R- and S-flurbiprofen. *FASEB J.* 2001;**15**:2–4.

57. Kanekura, T., Goorha, S., Kirtikara, K., and Ballou, L.R. The involvement of NF-kappa B in the constitutive overexpression of cyclooxygenase-2 in cyclooxygenase-1 null cells. *Biochem. Biophys. Acta* 2002;**1542**:14–22.

58. Niederberger, E., Tegeder, I., Vetter, G., Schmidtko, A., Schmidt, H., Euchenhofer, C., et al. Celecoxib loses its anti-inflammatory efficacy at high doses through activation of NF-kappaB. *FASEB J.* 2001;**15**:1622–1624.

59. Lawrence, T., Gilroy, D.W., Colville-Nash, P.R., and Willoughby, D.A. Possible new role for NF-kappaB in the resolution of inflammation. *Nat. Med.* 2001;**7**:1291–1297.

60. Zhou, W., Jiang, Z.W., Tian, J., Jiang, J., Li, N., and Li, J.S. Role of NF-kappaB and cytokine in experimental cancer cachexia. *World J. Gastroenterol.* 2003;**9**:1567–1570.

61. Zingarelli, B., Sheehan, M., Hake, P.W., O'Connor, M., Denenberg, A., and Cook, J.A. Peroxisome proliferator activator receptor-gamma ligands, 15-deoxy-Delta(12,14)-prostaglandin J2 and ciglitazone, reduce systemic inflammation in polymicrobial sepsis by modulation of signal transduction pathways. *J. Immunol.* 2003;**171**:6827–6837.

62. Ristimaki, A., Garfinkel, S., Wessendorf, J., Maciag, T., and Hla, T. Induction of cyclooxygenase-2 by interleukin-1 alpha. Evidence for post-transcriptional regulation. *J. Biol. Chem.* 1994;**269**:11769–11775.

63. Tegeder, I., Pfeilschifter, J., and Geisslinger, G. Cyclooxygenase-independent actions of cyclooxygenase inhibitors. *FASEB J.* 2001;**15**:2057–2072.

64. Schwartz, S.A., Hernandez, A., and Mark Evers, B. The role of NF-kappa B/kappa B proteins in cancer: Implications for novel treatment strategies. *Surg. Oncol.* 1999;**8**:143–153.

65. Groesdonk, H.V., and Senftleben, U. Modulation of inhibitor kappaB kinase/nuclear factor kappaB signaling during critical illness: A double-edged sword. *Crit. Care Med.* 2004;**32**:1239–1240.

66. Poppers, D.M., Schwenger, P., and Vilcek, J. Persistent tumor necrosis factor signaling in normal human fibroblasts prevents the complete resynthesis of I kappa B-alpha. *J. Biol. Chem.* 2000;**275**:29587–29593.

67. Yamamoto, Y., Yin, M.J., and Gaynor, R.B. IkappaB kinase alpha (IKKalpha) regulation of IKKbeta kinase activity. *Mol. Cell. Biol.* 2000;**20**:3655–3666.

68. Cernuda-Morollon, E., Pineda-Molina, E., Canada, F.J., and Perez-Sala, D. 15-Deoxy-Delta 12,14-prostaglandin J2 inhibition of NF-kappa B-DNA binding through covalent modification of the p50 subunit. *J. Biol. Chem.* 2001;**276**:35530–35536.

69. Hinz, B., Kraus, V., Pahl, A., and Brune, K. Salicylate metabolites inhibit cyclooxygenase-2-dependent prostaglandin E(2) synthesis in murine macrophages. *Biochem. Biophys. Res. Commun.* 2000;**274**:197–202.

70. Berger, J. and Moller, D.E. The mechanisms of action of PPARs. *Annu. Rev. Med.* 2002;**53**:409–435.

71. Blanquart, C., Barbier, O., Fruchart, J.C., Staels, B., and Glineur, C. Peroxisome proliferator-activated receptors: Regulation of transcriptional activities and roles in inflammation. *J. Steroid Biochem. Mol. Biol.* 2003;**85**:267–273.

72. Paik, J.H., Ju, J.H., Lee, J.Y., Boudreau, M.D., and Hwang, D.H. Two opposing effects of non-steroidal anti-inflammatory drugs on the expression of the inducible cyclooxygenase. Mediation through different signaling pathways. *J. Biol. Chem.* 2000;**275**:28173–28179.

73. Pang, L., Nie, M., Corbett, L., and Knox, A.J. Cyclooxygenase-2 expression by nonsteroidal anti-inflammatory drugs in human airway smooth muscle cells: Role of peroxisome proliferator-activated receptors. *J. Immunol.* 2003;**170**:1043–1051.

74. Spiegelman, B.M. and Heinrich, R. Biological control through regulated transcriptional coactivators. *Cell* 2004;**119**:157–167.
75. Cabrero, A., Laguna, J.C., and Vazquez, M. Peroxisome proliferator-activated receptors and the control of inflammation. *Curr. Drug Targets Inflamm. Allergy* 2002;**1**:243–248.
76. Chinetti, G., Fruchart, J.C., and Staels, B. Peroxisome proliferator-activated receptors (PPARs): Nuclear receptors at the crossroads between lipid metabolism and inflammation. *Inflamm. Res.* 2000;**49**:497–505.
77. Wang, W., Lonnroth, C., Svanberg, E., and Lundholm, K. Cytokine and cyclooxygenase-2 protein in brain areas of tumor-bearing mice with prostanoid-related anorexia. *Cancer Res.* 2001;**61**:4707–4715.
78. Lundholm, K., Gelin, J., Hyltander, A., Lonnroth, C., Sandstrom, R., Svaninger, G., et al. Anti-inflammatory treatment may prolong survival in undernourished patients with metastatic solid tumors. *Cancer Res.* 1994;**54**:5602–5606.
79. Lundholm, K., Daneryd, P., Bosaeus, I., Korner, U., and Lindholm, E. Palliative nutritional intervention in addition to cyclooxygenase and erythropoietin treatment for patients with malignant disease: Effects on survival, metabolism, and function. *Cancer* 2004;**100**:1967–1977.
80. Barber, M.D., Ross, J.A., Voss, A.C., Tisdale, M.J., and Fearon, K.C. The effect of an oral nutritional supplement enriched with fish oil on weight-loss in patients with pancreatic cancer. *Br. J. Cancer* 1999;**81**:80–86.
81. Barber, M.D., McMillan, D.C., Preston, T., Ross, J.A., and Fearon, K.C. Metabolic response to feeding in weight-losing pancreatic cancer patients and its modulation by a fish-oil-enriched nutritional supplement. *Clin. Sci. (Lond.)* 2000;**98**:389–399.
82. Jatoi, A., Rowland, K., Loprinzi, C.L., Sloan, J.A., Dakhil, S.R., MacDonald, N., et al. An eicosapentaenoic acid supplement versus megestrol acetate versus both for patients with cancer-associated wasting: a North Central Cancer Treatment Group and National Cancer Institute of Canada collaborative effort. *J. Clin. Oncol.* 2004;**22**:2469–2476.
83. Rowland, T.L., McHugh, S.M., Deighton, J., Ewan, P.W., Dearman, R.J., and Kimber, I. Differential effect of thalidomide and dexamethasone on the transcription factor NF-kappa B. *Int. Immunopharmacol.* 2001;**1**:49–61.
84. Scheinman, R.I., Cogswell, P.C., Lofquist, A.K., and Baldwin, A.S., Jr. Role of transcriptional activation of I kappa B alpha in mediation of immunosuppression by glucocorticoids [comment]. *Science* 1995;**270**:283–286.
85. Crinelli, R., Antonelli, A., Bianchi, M., Gentilini, L., Scaramucci, S., and Magnani, M. Selective inhibition of NF-kB activation and TNF-alpha production in macrophages by red blood cell-mediated delivery of dexamethasone. *Blood Cells Mol. Dis.* 2000;**26**:211–222.
86. Zhang, Y., Xu, J., and Zhong, N. [Proinflammatory cytokine stimulated NF-kappa B activation and the effect of dexamethasone in the human airway epithelial cells]. *Zhonghua Jie He He Hu Xi Za Zhi* 2000;**23**:296–299.
87. Kurokouchi, K., Kambe, F., Kikumori, T., Sakai, T., Sarkar, D., Ishiguro, N., et al. Effects of glucocorticoids on tumor necrosis factor alpha-dependent activation of nuclear factor kappaB and expression of the intercellular adhesion molecule 1 gene in osteoblast-like ROS17/2.8 cells. *J. Bone Miner. Res.* 2000;**15**:1707–1715.
88. Joyce, D.A., Gimblett, G., and Steer, J.H. Targets of glucocorticoid action on TNF-alpha release by macrophages. *Inflamm. Res.* 2001;**50**: 337–340.

89. Quan, N., He, L., Lai, W., Shen, T., and Herkenham, M. Induction of IkappaB alpha mRNA expression in the brain by glucocorticoids: A negative feedback mechanism for immune-to-brain signaling. *J. Neurosci.* 2000;**20**:6473–6477.

90. Han, J., Thompson, P., and Beutler, B. Dexamethasone and pentoxifylline inhibit endotoxin-induced cachectin/tumor necrosis factor synthesis at separate points in the signaling pathway. *J. Exp. Med.* 1990;**172**:391–394.

91. Ramsay, R.G., Ciznadija, D., Vanevski, M., and Mantamadiotis, T. Transcriptional regulation of cyclo-oxygenase expression: three pillars of control. *Int. J. Immunopathol. Pharmacol.* 2003;**16**(2 Suppl.):59–67.

92. Nishimori, T., Inoue, H., and Hirata, Y. Involvement of the 3′-untranslated region of cyclooxygenase-2 gene in its post-transcriptional regulation through the glucocorticoid receptor. *Life Sci.* 2004;**74**:2505–2513.

93. Du, J., Mitch, W.E., Wang, X., and Price, S.R. Glucocorticoids induce proteasome C3 subunit expression in L6 muscle cells by opposing the suppression of its transcription by NF-kappa B. *J. Biol. Chem.* 2000;**275**: 19661–19666.

94. Della Cuna, G.R., Pellegrini, A., and Piazzi, M. Effect of methylprednisolone sodium succinate on quality of life in preterminal cancer patients: a placebo-controlled, multicenter study. The Methylprednisolone Preterminal Cancer Study Group. *Eur. J. Cancer Clin. Oncol.* 1989;**25**:1817–1821.

95. Popiela, T., Lucchi, R., and Giongo, F. Methylprednisolone as palliative therapy for female terminal cancer patients. The Methylprednisolone Female Preterminal Cancer Study Group. *Eur. J. Cancer Clin. Oncol.* 1989;**25**: 1823–1829.

96. Bruera, E., Roca, E., Cedaro, L., Carraro, S., and Chacon, R. Action of oral methylprednisolone in terminal cancer patients: a prospective randomized double-blind study. *Cancer Treat. Rep.* 1985;**69**:751–754.

97. Loprinzi, C.L., Kugler, J.W., Sloan, J.A., Mailliard, J.A., Krook, J.E., Wilwerding, M.B., et al. Randomized comparison of megestrol acetate versus dexamethasone versus fluoxymesterone for the treatment of cancer anorexia/cachexia. *J. Clin. Oncol.* 1999;**17**:3299–3306.

98. Hawcroft, G., Gardner, S.H., and Hull, M.A. Activation of peroxisome proliferator-activated receptor gamma does not explain the antiproliferative activity of the nonsteroidal anti-inflammatory drug indomethacin on human colorectal cancer cells. *J. Pharmacol. Exp. Ther.* 2003;**305**:632–637.

99. Amann, R. and Peskar, B.A. Anti-inflammatory effects of aspirin and sodium salicylate. *Eur. J. Pharmacol.* 2002;**447**:1–9.

100. Cronstein, B.N., Montesinos, M.C., and Weissmann, G. Sites of action for future therapy: An adenosine-dependent mechanism by which aspirin retains its antiinflammatory activity in cyclooxygenase-2 and NFkappaB knockout mice. *Osteoarthritis Cartilage* 1999;**7**:361–363.

101. Ajuebor, M.N., Singh, A., and Wallace, J.L. Cyclooxygenase-2-derived prostaglandin D(2) is an early anti-inflammatory signal in experimental colitis. *Am. J. Physiol. Gastrointest. Liver Physiol.* 2000;**279**:G238–244.

102. Shackelford, R.E., Alford, P.B., Xue, Y., Thai, S.F., Adams, D.O., and Pizzo, S. Aspirin inhibits tumor necrosis factoralpha gene expression in murine tissue macrophages. *Mol. Pharmacol.* 1997;**52**:421–429.

103. Kopp, E. and Ghosh, S. Inhibition of NF-kappa B by sodium salicylate and aspirin. *Science* 1994;**265**:956–959.

104. Yan, F., and Polk, D.B. Aminosalicylic acid inhibits IkappaB kinase alpha phosphorylation of IkappaBalpha in mouse intestinal epithelial cells. *J Biol. Chem.* 1999;**274**:36631–36636.

105. Schreiber, S. Activation of nuclear factor KB as a target for anti-inflammatory therapy. *Gut* 1999;**44**:309–310.

106. Xu, X.M., Sansores-Garcia, L., Chen, X.M., Matijevic-Aleksic, N., Du, M., and Wu, K.K. Suppression of inducible cyclooxygenase 2 gene transcription by aspirin and sodium salicylate. *Proc. Natl. Acad. Sci. USA* 1999;**96**:5292–5297.

107. Fernandez de Arriba, A., Cavalcanti, F., Miralles, A., Bayon, Y., Alonso, A., Merlos, M., et al. Inhibition of cyclooxygenase-2 expression by 4- trifluoromethyl derivatives of salicylate, triflusal, and its deacetylated metabolite, 2-hydroxy-4-trifluoromethylbenzoic acid. *Mol. Pharmacol.* 1999;**55**:753–760.

108. Catania, A., Arnold, J., Macaluso, A., Hiltz, M.E., and Lipton, J.M. Inhibition of acute inflammation in the periphery by central action of salicylates. *Proc. Natl. Acad. Sci. USA* 1991;**88**:8544–8547.

109. Yin, M.J., Yamamoto, Y., and Gaynor, R.B. The anti-inflammatory agents aspirin and salicylate inhibit the activity of I(kappa)B kinase-beta. *Nature* 1998;**396**:77–80.

110. Schwenger, P., Alpert, D., Skolnik, E.Y., and Vilcek, J. Activation of p38 mitogen-activated protein kinase by sodium salicylate leads to inhibition of tumor necrosis factor-induced IkappaB alpha phosphorylation and degradation. *Mol. Cell. Biol.* 1998;**18**:78–84.

111. Lo, C.J., Cryer, H.G., Fu, M., and Lo, F.R. Regulation of macrophage eicosanoid generation is dependent on nuclear factor kappa B. *J. Trauma* 1998;**45**:19–23; discussion 23–24.

112. Callejas, N.A., Fernandez-Martinez, A., Castrillo, A., Bosca, L., and Martin-Sanz, P. Selective inhibitors of cyclooxygenase-2 delay the activation of nuclear factor kappa B and attenuate the expression of inflammatory genes in murine macrophages treated with lipopolysaccharide. *Mol. Pharmacol.* 2003;**63**:671–677.

113. Wong, B.C., Jiang, X., Fan, X.M., Lin, M.C., Jiang, S.H., Lam, S.K., et al. Suppression of RelA/p65 nuclear translocation independent of IkappaB-alpha degradation by cyclooxygenase-2 inhibitor in gastric cancer. *Oncogene* 2003;**22**:1189–1197.

114. Pinheiro, R.M., and Calixto, J.B. Effect of the selective COX-2 inhibitors, celecoxib and rofecoxib in rat acute models of inflammation. *Inflamm. Res.* 2002;**51**:603–610.

115. Niederberger, E., Tegeder, I., Schafer, C., Seegel, M., Grosch, S., Geisslinger, G. Opposite effects of rofecoxib on nuclear factor-kappaB and activating protein-1 activation. *J. Pharmacol. Exp. Ther.* 2003;**304**:1153–1160.

116. Pierce, J.W., Read, M.A., Ding, H., Luscinskas, F.W., and Collins, T. Salicylates inhibit I kappa B-alpha phosphorylation, endothelial-leukocyte adhesion molecule expression, and neutrophil transmigration. *J. Immunol.* 1996;**156**: 3961–3969.

117. Hinson, R.M., Williams, J.A., and Shacter, E. Elevated interleukin 6 is induced by prostaglandin E2 in a murine model of inflammation: possible role of cyclooxygenase-2. *Proc. Natl. Acad. Sci. USA* 1996;**93**:4885–4890.

118. Lonnroth, C., Andersson, M., and Lundholm, K. Indomethacin and telomerase activity in tumor growth retardation. *Int. J. Oncol.* 2001;**18**:929–937.

119. Altinoz, M.A. and Korkmaz, R. NF-kappaB, macrophage migration inhibitory factor and cyclooxygenase-inhibitions as likely mechanisms behind the acetaminophen- and NSAID-prevention of the ovarian cancer. *Neoplasma* 2004;**51**:239–247.

120. Adamson, D.J., Frew, D., Tatoud, R., Wolf, C.R., and Palmer, C.N. Diclofenac antagonizes peroxisome proliferator-activated receptor-gamma signaling. *Mol. Pharmacol.* 2002;**61**:7–12.

121. Han, S., Inoue, H., Flowers, L.C., and Sidell, N. Control of COX-2 gene expression through peroxisome proliferator-activated receptor gamma in human cervical cancer cells. *Clin. Cancer. Res.* 2003;**9**:4627–4635.

122. Eisengart, C.A., Mestre, J.R., Naama, H.A., Mackrell, P.J., Rivadeneira, D.E., Murphy, E.M., et al. Prostaglandins regulate melanoma-induced cytokine production in macrophages. *Cell. Immunol.* 2000;**204**:143–149.

123. Kojo, H., Fukagawa, M., Tajima, K., Suzuki, A., Fujimura, T., Aramori, I., et al. Evaluation of human peroxisome proliferator-activated receptor (PPAR) subtype selectivity of a variety of anti-inflammatory drugs based on a novel assay for PPAR delta(beta). *J. Pharmacol. Sci.* 2003;**93**:347–355.

124. Delerive, P., Fruchart, J.C., and Staels, B. Peroxisome proliferator-activated receptors in inflammation control. *J. Endocrinol.* 2001;**169**:453–459.

125. Shaik, M.S., Chatterjee, A., and Singh, M. Effect of a selective cyclooxygenase-2 inhibitor, nimesulide, on the growth of lung tumors and their expression of cyclooxygenase-2 and peroxisome proliferator- activated receptor-gamma. *Clin. Cancer Res.* 2004;**10**:1521–1529.

126. Konturek, P.C., Konturek, S.J., Bielanski, W., Kania, J., Zuchowicz, M., Hartwich, A., et al. Influence of COX-2 inhibition by rofecoxib on serum and tumor progastrin and gastrin levels and expression of PPARgamma and apoptosis-related proteins in gastric cancer patients. *Dig. Dis. Sci.* 2003;**48**:2005–2017.

127. Serhan, C.N. and Oliw, E. Unorthodox routes to prostanoid formation: New twists in cyclooxygenase-initiated pathways. *J. Clin. Invest.* 2001;**107**:1481–1489.

128. Paul-Clark, M.J., Van Cao, T., Moradi-Bidhendi, N., Cooper, D., and Gilroy, D.W. 15-epi-lipoxin A4-mediated induction of nitric oxide explains how aspirin inhibits acute inflammation. *J. Exp. Med.* 2004;**200**:69–78.

129. Housby, J.N., Cahill, C.M., Chu, B., Prevelige, R., Bickford, K., Stevenson, M.A., et al. Non-steroidal anti-inflammatory drugs inhibit the expression of cytokines and induce HSP70 in human monocytes. *Cytokine* 1999;**11**:347–358.

130. Stevenson, M.A., Zhao, M.J., Asea, A., Coleman, C.N., and Calderwood, S.K. Salicylic acid and aspirin inhibit the activity of RSK2 kinase and repress RSK2-dependent transcription of cyclic AMP response element binding protein- and NF-kappa B-responsive genes. *J. Immunol.* 1999;**163**:5608–5616.

131. Wu, K.K. Aspirin and other cyclooxygenase inhibitors: new therapeutic insights. *Semin. Vasc. Med.* 2003;**3**:107–112.

132. Sandstrom, R., Gelin, J., and Lundholm, K. The effect of indomethacin on food and water intake, motor activity and survival in tumor-bearing rats. *Eur. J. Cancer* 1990;**26**:811–814.

133. Okamoto, T. NSAID zaltoprofen improves the decrease in body weight in rodent sickness behavior models: proposed new applications of NSAIDs (Review). *Int. J. Mol. Med.* 2002;**9**:369–372.

134. Roe, S.Y., Cooper, A.L., Morris, I.D., and Rothwell, N.J. Involvement of prostaglandins in cachexia induced by T-cell leukemia in the rat. *Metabolism* 1997;**46**:359-365.

135. Badawi, A.F., Eldeen, M.B., Liu, Y., Ross, E.A., and Badr, M.Z. Inhibition of rat mammary gland carcinogenesis by simultaneous targeting of

cyclooxygenase-2 and peroxisome proliferator-activated receptor gamma. *Cancer Res.* 2004;**64**:1181–1189.

136. Cahlin, C., Gelin, J., Delbro, D., Lonnroth, C., Doi, C., and Lundholm, K. Effect of cyclooxygenase and nitric oxide synthase inhibitors on tumor growth in mouse tumor models with and without cancer cachexia related to prostanoids. *Cancer Res.* 2000;**60**:1742–1749.

137. Kimura, T., Iwase, M., Kondo, G., Watanabe, H., Ohashi, M., Ito, D., et al. Suppressive effect of selective cyclooxygenase-2 inhibitor on cytokine release in human neutrophils. *Int. Immunopharmacol.* 2003;**3**:1519–1528.

138. Hyltander, A., Korner, U., and Lundholm, K.G. Evaluation of mechanisms behind elevated energy expenditure in cancer patients with solid tumours. *Eur. J. Clin. Invest.* 1993;**23**:46–52.

139. Cerchietti, L.C., Navigante, A.H., Peluffo, G.D., Diament, M.J., Stillitani, I., Klein, S.A., et al. Effects of celecoxib, medroxyprogesterone, and dietary intervention on systemic syndromes in patients with advanced lung adenocarcinoma: a pilot study. *J. Pain Symptom Manage.* 2004;**27**:85–95.

140. Sauer, L.A., Dauchy, R.T., Blask, D.E. Mechanism for the antitumor and anticachectic effects of n-3 fatty acids. *Cancer Res.* 2000;**60**:5289–5295.

141. Babcock, T.A., Helton, W.S., Hong, D., and Espat, N.J. Omega-3 fatty acid lipid emulsion reduces LPS-stimulated macrophage TNF-alpha production. *Surg. Infect. (Larchmt)* 2002;**3**:145–149.

142. Robinson, D.R., Knoell, C.T., Urakaze, M., Huang, R., Taki, H., Sugiyama, E., et al. Suppression of autoimmune disease by omega-3 fatty acids. *Biochem. Soc. Trans.* 1995;**23**:287–291.

143. Burns, C.P., Halabi, S., Clamon, G.H., Hars, V., Wagner, B.A., Hohl, R.J., et al. Phase I clinical study of fish oil fatty acid capsules for patients with cancer cachexia: Cancer and leukemia group B study 9473. *Clin. Cancer Res.* 1999;**5**:3942–3947.

144. Barber, M.D. and Fearon, K.C. Tolerance and incorporation of a high-dose eicosapentaenoic acid diester emulsion by patients with pancreatic cancer cachexia. *Lipids* 2001;**36**:347–351.

145. Moses, A.W., Slater, C., Preston, T., Barber, M.D., and Fearon, K.C. Reduced total energy expenditure and physical activity in cachectic patients with pancreatic cancer can be modulated by an energy and protein dense oral supplement enriched with n-3 fatty acids. *Br. J. Cancer* 2004;**90**:996–1002.

146. Fearon, K.C., Von Meyenfeldt, M.F., Moses, A.G., Van Geenen, R., Roy, A., Gouma. D.J., et al. Effect of a protein and energy dense N-3 fatty acid enriched oral supplement on loss of weight and lean tissue in cancer cachexia: A randomised double blind trial. *Gut* 2003;**52**:1479–1486.

147. Barber, M.D., Fearon, K.C., Tisdale, M.J., McMillan, D.C., and Ross, J.A. Effect of a fish oil-enriched nutritional supplement on metabolic mediators in patients with pancreatic cancer cachexia. *Nutr. Cancer* 2001;**40**:118–124.

148. Gogos, C.A., Ginopoulos, P., Salsa, B., Apostolidou, E., Zoumbos, N.C., and Kalfarentzos, F. Dietary omega-3 polyunsaturated fatty acids plus vitamin E restore immunodeficiency and prolong survival for severely ill patients with generalized malignancy: a randomized control trial. *Cancer* 1998;**82**:395–402.

V

Drugs in Research and Development

22 The Yin and Yang of IGF-1 and Ghrelin

Ross G. Clark

CONTENTS

SUMMARY

Normal whole body growth in young, growing mammals requires the anabolic hormones GH and IGF-1 acting in concert; the maintenance of optimal tissue structure and function in adult mammals likewise requires an intact GH/IGF-1 axis. High blood levels of IGF-1 are predictive of an ability to survive serious acute illnesses in humans. In contrast, reduced signaling through the GH/IGF-1 axis due to IGF-1 deficiency and GH deficiency has been shown to lead to an increased lifespan in experimental animals maintained under laboratory conditions. This difference may be an artifact of the benign environment under which experimental animals are maintained. GH is an approved anabolic therapy in GH deficiency and

AIDS wasting. In other severe catabolic states, such as cancer cachexia, the use of ghrelin to stimulate the GH and IGF-1 axis is being explored. Inhibitors of the GH/IGF-1 axis, such as GH receptor antagonists or IGF-1 receptor antagonists, are being tested as inhibitors of tumor growth. In cachexic cancer patients it is possible that, although inhibiting the GH/IGF-1 axis might be of benefit to slow tumor growth, it might also worsen tissue wasting and reduce survival. Endocrinologists therefore point to potential benefits of agonists of ghrelin, GH, and IGF-1 as therapeutics, whereas oncologists point to their potential risks and propose the use of antagonists. The weight of evidence suggests that maintaining GH and IGF-1 levels within age- and sex-adjusted normal ranges is beneficial in most situations and will likely be of benefit to cachexic patients.

22.1 INTRODUCTION

The endocrinology and physiology of the regulation of the growth hormone (GH) and insulin-like growth factor-1 (IGF-1) axis is now well understood. The endocrine mechanisms mediating GH activity were discovered by Daughaday and Salmon in 1956 when they[1] proposed the somatomedin hypothesis; that the effects of GH were indirect and mediated by a second hormonal factor, which they called somatomedin.[2] Somatomedin was subsequently purified and re-named IGF-1.[3] It is now known that GH acts as an endocrine hormone to generate IGF-1 in the liver, but that GH also acts directly on tissues to generate local IGF-1. Both sources of IGF-1 are important for normal body growth and anabolism (Figure 22.1). The purification of the IGF-1 and GH receptors in the 1980s and the elucidation of their intracellular signaling pathways has firmly established the mechanism of action of these hormones.

In the 1970s and 1980s the regulation of the synthesis and release of pituitary GH by the hypothalamic hormones, GH releasing hormone (GHRH) and somatostatin, was elucidated as was their effect on body growth.[4] The late 1990s brought the discovery of a third hormone in the stomach, ghrelin, which also modulates GH secretion and feeding.[5] The physiological actions of ghrelin on the regulation of the GH axis and GH secretion are controversial as the gene knockout of ghrelin in mice does not significantly impact growth or IGF-1 levels in mice.[6] By comparison, GHRH deficiency or mutations in the GHRH receptor, such as in the lit/lit mouse or in humans, produce profound GH and IGF-1 deficiency and dramatically affect body growth.[4]

Loss of weight in humans is associated with an increase in mortality in many epidemiology studies[7] with unintentional weight loss in some studies being a greater risk.[8] Weight loss due to a loss of fat can decrease subsequent mortality whereas weight loss due to a loss of lean body mass can increase mortality.[9] Weight loss was associated with low IGF-1 levels in the Framingham Heart Study;[10] higher IGF-1 levels and a decreased loss of fat-free mass were also associated with reduced mortality during the next 2 years.[11] Such data sets the stage for the use of therapies that increase blood GH or IGF-1 levels as a means to prevent weight loss and reduce mortality in cachexic patients.

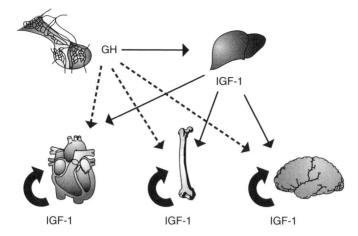

FIGURE 22.1 GH released from the pituitary gland acts on the liver to release large amounts of IGF-1 into the blood; this IGF-1, in turn, acts hormonally on many tissues. GH also acts directly on many tissues, including heart, bone, cartilage and brain, resulting in production of local IGF-1 that acts as a paracrine and autocrine growth factor. In many tissues, local and systemic IGF-1 act together to produce an anabolic effect.

Growth hormone and IGF-1 have been tested as anabolic therapies in catabolic states while small molecule ghrelin receptor agonists, the so called GH releasing peptide (GHRP) or GH secretagogue molecules, some of them oral agents, have also been tested as pharmacological agents.[4,12] Administration of ghrelin agonists could increase GH secretion, increase IGF-1 levels, and have anabolic actions in animals and humans (Figure 22.1). The actions of ghrelin are complex as it also increases food intake[13] and the combination of a stimulation of the GH/IGF-1 axis and food intake suggests that ghrelin agonists might be useful in catabolic states such as that in the frail elderly or in cachexia.[12] Several compounds that affect the GH/IGF-1 axis have been or are being tested as anabolic agents in catabolic humans. This review will focus on the potential risks and benefits of IGF-1 and ghrelin or ghrelin agonists as therapeutic agents in humans.

22.2 REPLACEMENT THERAPY FOR GH DEFICIENCY, IGF-1 DEFICIENCY, OR GHRELIN DEFICIENCY

A basic tenet of endocrinology is that when the blood levels of a hormone are low there are adverse consequences and so the deficient hormone should be replaced. For GH or IGF-1 deficiency, this concept has been challenged by recent animal data suggesting that a low blood level of GH or IGF-1 might be beneficial in terms of prolonging lifespan.[14] The suggestion that these deficiencies favor longevity challenges the well-established medical use of GH replacement as an anabolic therapy in children and adults.[15] The importance of this issue deserves close examination since GH is an approved anabolic therapy and IGF-1 is now approved (as Increlex™) by the United States Food and Drug Administration for the long-term treatment of short stature in children with severe primary IGF-1 deficiency. An antagonist to the GH receptor is now an approved therapeutic agent[16] and antagonists of the IGF-1 receptor are being developed[17] and tested as therapeutics for the inhibition of tumor growth. In contrast, agonists of the GH/IGF-1 axis such as ghrelin are also being tested as therapeutics for the treatment of cancer cachexia.[12] The overall benefit to cancer patients of either inhibiting or stimulating the GH/IGF-1 axis is therefore in question. Obviously, it is very important to understand the role of the GH/IGF-1 axis, especially in catabolic states such as cancer cachexia, so that patient care can be optimized.

Aside from the effects of a genetic deficiency of GH or IGF-1, the effects on organ function and/or on lifespan of a low blood IGF-1 level due to the chronic decline in GH and IGF-1 levels with aging or in catabolic states is also a matter of current controversy. In animal models of GH and IGF-1 deficiency, some studies indicate that lifespan is extended, whereas other studies have shown that GH and IGF-1 replacement therapy can reverse the age-related decline of many functional measures such as lean body mass, bone density, immune function, cardiac function, learning, memory, and adiposity. There are therefore data supporting two opposing concepts which appear to be mutually incompatible. These are: (1) GH and IGF-1 inhibit the age-related functional decline of many organ systems and therefore increase lifespan; or (2) GH and IGF-1 accelerate the aging of cells and tissues and therefore decrease lifespan. The availability of therapeutics that either stimulate or inhibit the GH/IGF-1 axis compels a discussion of the apparent paradox. The ultimate logic for using agents to combat cachexia is to improve lifespan, so the impact of manipulating the GH/IGF-1 axis on lifespan is of obvious interest.

22.3 EFFECTS OF GH/IGF-1 DEFICIENCY AND BENEFITS OF STIMULATING THE GH/IGF-1 AXIS

The anabolic actions of ghrelin, GH, or IGF-1 are not restricted to an increase in bone and muscle mass, but extend to structural changes in most cell types, tissues, and organs. These anabolic actions are also associated with effects on cell, tissue, and organ function. The beneficial structural and functional effects of these hormones likely counter their possible direct, deleterious, and nonspecific effects on cell lifespan and therefore on longevity. In catabolic states several specific beneficial effects of stimulating the GH/IGF-1 axis have been established and some of these are summarized below.

22.3.1 Short Stature

It has long been held that short stature is undesirable in both children and adults, irrespective of its cause. Short children can be subject to juvenilization, teasing, bullying, exclusion from activities and peer groups, and impairment of the normal progression toward independence.[18] In adulthood there can be social isolation, reduced likelihood of marriage, perceptions of lower competence, ineligibility for occupations such as the military or police due to minimum height requirements for employees, and height-related aspects of daily living can provide additional challenges to individual with short stature. Overall, normalizing GH or IGF-1 levels in childhood or in a young animal appears to be beneficial. Early short-term treatment with GH had the long-term effect of extending lifespan in a recent animal study in GH and IGF-1 deficient dw/dw rats treated with GH replacement therapy from 4 to 14 weeks of age.[19]

22.3.2 Obesity and Diabetes

The first rat with an isolated GH deficiency was the dwarf (dw/dw) rat discovered and characterized in the 1980s by the author.[20] The dwarf rat was not obese, unlike humans with adult GH deficiency. We discovered that this was likely an artifact of the rodents' laboratory diet, which consisted of a low-fat grain chow. When the dwarf rat was fed a diet with the fat content of that eaten by most humans, extreme obesity resulted. In addition we showed that this obesity was accompanied by insulin resistance and could be reversed by combined treatment with GH and IGF-1.[21] Therefore GH deficiency and IGF-1 deficiency in this situation led to obesity and most likely, in the long-term, to an adverse effect on lifespan. This illustrates the artifacts that can be produced by the experimental conditions under which experimental laboratory animals are maintained and the danger of extrapolating from data in GH or IGF-1 deficient experimental animals to data in humans.[22]

22.3.3 Immune System

A wealth of studies show that the GH/IGF-1 axis affects the immune system, including a series of studies by Kelley showing that the susceptibility of animals to bacterial infections and toxins was increased by GH deficiency.[23] Direct beneficial effects of IGF-1 on the immune system were shown to include an enhancement of B-cell number and function, including the ability of IGF-1 to increase an antibody response *in vivo*.[24] The detrimental effects of IGF-1 deficiency on the immune system are normally not seen as animals in modern animal houses are usually not exposed to pathogens.[22] In contrast, humans are exposed to a variety of immune challenges and so deficiencies of GH and IGF-1 are likely to adversely affect immune function and therefore longevity.

In 1997, based on this immunology work, the author proposed that a key function of the IGF-1 system was to oppose the actions of adrenal steroids.[24] This proposal was later confirmed by experimental studies in catabolic animals treated with IGF-1 and dexamethasone[25] and expanded to include prolactin and GH as natural counters to the effects of adrenal steroids.[22] These authors added support to the proposal[24] that animals with a compromised GH/IGF-1 system are sensitive to and adversely affected by stress to a greater extent than animals with a normal

GH/IGF-1 axis. Similar beneficial effects on the immune system have been shown following ghrelin treatment of old animals.[12]

Laboratory animals generally have a compromised adrenal system as the adrenal gland of laboratory mice and rats is greatly reduced in size compared to that of rats and mice caught in the wild. It would be of interest to study the longevity of rats and mice with "normal adrenal function" and a compromised GH/IGF-1 axis. The counterbalancing effects of the stress hormones and the anabolic hormones must be considered when interpreting the effect of GH and IGF-1 deficiency on longevity and which have implications for the use of GH and IGF-1 as anabolic therapies.

22.3.4 Effect of GH and IGF-1 on Cardiac Structure and Function

GH and IGF-1 directly affect cardiovascular (CV) structure and function and have beneficial effects in pathological situations such as maintaining the structure and function of a damaged or failing heart.[26] In mice with a liver-specific IGF-1 knock-out there was a decreased stroke volume and cardiac output, which was likely secondary to the increased peripheral resistance, indicating that liver-derived IGF-1 is involved in the regulation of blood pressure suggesting a mechanism for the link between blood IGF-1 levels and CV disease or mortality.[27] These discoveries in mice were followed by human studies showing that IGF-1 deficiency is linked to the pathogenesis of atherosclerosis and of ischaemic heart disease (IHD).[28] Subjects with IGF-1 deficiency but no IHD have, during a 15-year follow-up period, a significantly increased risk of developing IHD and there is a strong epidemiological link between IGF-1 deficiency and the risk of developing CV disease.[29] Elderly individuals in the Framingham Heart Study with serum IGF-1 levels at or above the median value had half the risk of heart failure of those with serum IGF-1 levels below the median.[30] Higher IGF-1 levels and a decreased loss of fat-free mass were also associated with reduced mortality during the next 2 years.[11] Low IGF-1 levels in the Framingham Heart Study were associated with poor nutritional status.[10] IGF-1 deficiency likely increases the risk of CV disease and mortality, and normalizing IGF-1 levels will likely confer a long-term health benefit by maintaining fat-free mass and reducing the risk of CV disease and CV mortality.

22.3.5 Beneficial Effects of GH/IGF-1 on the Brain

The dramatic neural effects of extreme IGF-1 deficiency in experimental animals due to IGF-1 gene deletions[31] are mirrored in the microcephaly and mental retardation seen in humans with IGF-1 gene defects.[32,33] There is also evidence that the decline in GH/IGF-1 activity with aging has adverse effects on the brain in animals and in humans.[34] Therefore, the long-term effects of manipulations of the GH/IGF-1 axis on the brain need to be considered, especially if GH or IGF-1 antagonists are used long-term.

22.4 Evidence that GH and IGF-1 Deficiency can Affect Lifespan

The prime reason for using agents to combat cachexia is obviously to improve lifespan so data showing how manipulating the GH/IGF-1 axis might affect lifespan is of clear importance in this paradigm.

When recombinant GH became available and long-term studies of GH therapy in GH-deficient adults were performed in the late 1980s, beneficial effects on body composition, bone mass, cardiac functions, and some measures of psychological function were seen.[35] Such efficacy data was accompanied by epidemiological assessments of the long-term effects of GH deficiency in adult humans. Adult GH deficiency was associated with marked visceral obesity, a risk factor for mortality due to CV disease. Not surprisingly, adult GH deficiency led to an increased mortality due to CV disease.[36] Such data led regulatory authorities to approve the use of GH as replacement therapy in adult GH deficiency.[15]

Deleting the IGF-1 gene in the mouse made the adverse effect of a complete absence of IGF-1 on lifespan very clear in the early 1990s. Such mice usually died at birth due to the muscles of the chest and diaphragm being so severely compromised that they could not breathe.[37] The decreased lifespan of these IGF-1 null mice puts into perspective subsequent observations of an increased lifespan in animal models of less severe IGF-1 deficiency. Therefore, in patients with cachexia and lowered blood IGF-1 levels, it appears that blood IGF-1 levels should be maintained in the normal range to reduce CV mortality, perhaps by administering IGF-1 or ghrelin.

Insulin administration to patients in intensive care dramatically reduces mortality,[38] while high dose GH administration increases mortality,[39] which is not seen when IGF-1 is administered to critically ill patients[40] or to patients with HIV.[41] The adverse effect of GH could be due to the well-known insulin resistance caused by GH while the lack of an adverse effect of IGF-1 alone is likely due to its insulin-like effect. Coadministering IGF-1 and GH to counter the diabetogenic effect of GH, would also likely have a beneficial insulin-like effect in cachexic patients and an anabolic effect.[4,40,41] Administering ghrelin acutely increased GH levels and food intake in patients with cancer cachexia[42] so the results of chronic studies are awaited with interest.[12]

The epidemiological data and studies in cachexic patients suggest that, in subjects at high risk of suffering a severe illness or who are in a catabolic state, low IGF-1 levels increase mortality and so using a therapeutic agent to increase IGF-1 levels would likely reduce mortality.

22.5 The Use of IGF-1 Receptor Antagonists in Cancer

An epidemiological association between high blood IGF-1 levels and the risk of developing cancer, along with data showing that IGF-1 and the IGF-1 receptor are important to the growth of tumor cells, has lead to the development of IGF-1 receptor antagonists as treatments for cancer.[43,44] It is hoped that such therapies, by inhibiting tumor growth or inducing cell apoptosis, will reduce tumor size and improve survival of cancer patients.[45] However, it is also possible that IGF-1 receptor antagonists, despite having possible beneficial effects on tumor growth, might worsen the catabolic state and cause a loss of lean tissue mass and an increased susceptibility to adverse cardiac events. Therefore, these antagonists may have an overall detrimental effect by inducing a state of relative IGF-1 deficiency and a catabolic state. It might be advisable to test IGF-1 receptor antagonists in patients who are not catabolic or at least closely monitor cachexia in treated patients.

22.6   THE USE OF GH AND IGF-1 IN CATABOLIC AND CACHEXIC STATES

The likely utility of ghrelin, GH and/or IGF-1 as anabolic therapies is supported by evidence that a deficiency of GH or IGF-1 can shorten lifespan in humans. This evidence has led to GH replacement therapy in AIDS patients, which improves lean body mass, and to the regulatory approval of GH therapy for AIDS wasting. It has also been proposed that ghrelin agonists might be of benefit in cancer cachexia by increasing GH and IGF-1 exposure and increasing appetite thereby ameliorating the catabolic state.[12] It might be prudent to test anabolic therapies in cancer patients with tumors that have been shown to be nonresponsive to GH or IGF-1. These anabolic therapies may prove more promising in cardiac cachexia, where there would seem to be fewer issues surrounding the use of stimulants of the GH/IGF-1 system than in oncology.

22.7   SUMMARY

There is controversy regarding the therapeutic benefit of manipulating the GH/IGF-1 axis. Endocrinologists point to the potential benefits of GH and IGF-1 therapy and the evidence that maintaining GH and IGF-1 levels within age- and sex-adjusted normal ranges is beneficial in patients with cachexia. By contrast, oncologists point to potential risks of increasing IGF-1 levels and suggest that antagonizing GH and IGF-1 might be beneficial in patients with cancer. New therapies based on manipulating the GH/IGF-1 axis are now being tested in the laboratory and the clinic and the results of these studies are awaited with interest as they should help resolve these controversial issues. The weight of evidence suggests that maintaining GH and IGF-1 levels within age- and sex-adjusted normal ranges is beneficial in most situations and will likely be of benefit to cachexic patients.

REFERENCES

1. Daughaday W.H. A personal history of the origin of the somatomedin hypothesis and recent challenges to its validity. *Perspect. Biol. Med.* 1989; **32**: 194–211.
2. Salmon W.D., Jr. and Daughaday W.H. A hormonally controlled serum factor which stimulates sulfate incorporation by cartilage *in vitro*. 1956. *J. Lab. Clin. Med.* 1990; **116**: 408–19.

3. Rinderknecht E. and Humbel R.E. The amino acid sequence of human insulin-like growth factor I and its structural homology with proinsulin. *J. Biol. Chem.* 1978; **253**: 2769–76.

4. Clark R.G. and Robinson I.C. Up and down the growth hormone cascade. *Cytokine Growth Factor Rev.* 1996; **7**: 65–80.

5. Kojima M., Hosoda H., Matsuo H., and Kangawa K. Ghrelin: Discovery of the natural endogenous ligand for the growth hormone secretagogue receptor. *Trends Endocrinol. Metab.* 2001; **12**: 118–22.

6. Sun Y., Ahmed S., and Smith R.G. Deletion of ghrelin impairs neither growth nor appetite. *Mol. Cell. Biol.* 2003; **23**: 7973–81.

7. Kuller L. and Wing R. Weight loss and mortality. *Ann. Intern. Med.* 1993; **119**: 630–2.

8. Gregg E.W., Gerzoff R.B., Thompson T.J., and Williamson D.F. Intentional weight loss and death in overweight and obese U.S. adults 35 years of age and older. *Ann. Intern. Med.* 2003; **138**: 383–9.

9. Allison D.B., Zannolli R., Faith M.S., et al. Weight loss increases and fat loss decreases all-cause mortality rate: Results from two independent cohort studies. *Int. J. Obes. Relat. Metab. Disord.* 1999; **23**: 603–11.

10. Harris T.B., Kiel D., Roubenoff R., et al. Association of insulin-like growth factor-I with body composition, weight history, and past health behaviors in the very old: The Framingham Heart Study. *J. Am. Geriatr. Soc.* 1997; **45**: 133–9.

11. Roubenoff R., Parise H., Payette H.A., et al. Cytokines, insulin-like growth factor 1, sarcopenia, and mortality in very old community-dwelling men and women: The Framingham Heart Study. *Am. J. Med.* 2003; **115**: 429–35.

12. Smith R.G. Development of Growth Hormone Secretagogues. *Endocr. Rev.* 2005; **26**: 346–60.

13. Kojima M., Hosoda H., and Kangawa K. Clinical endocrinology and metabolism. Ghrelin, a novel growth-hormone-releasing and appetite-stimulating peptide from stomach. *Best Pract. Res. Clin. Endocrinol. Metab.* 2004; **18**: 517–30.

14. Kenyon C. The plasticity of aging: Insights from long-lived mutants. *Cell* 2005; **120**: 449–60.

15. Vance M.L. and Mauras N. Growth hormone therapy in adults and children. *N. Engl. J. Med.* 1999; **341**: 1206–16.

16. Trainer P.J., Drake W.M., Katznelson L., et al. Treatment of acromegaly with the growth hormone-receptor antagonist pegvisomant. *N. Engl. J. Med.* 2000; **342**: 1171–7.

17. Cohen B.D., Baker D.A., Soderstrom C., et al. Combination therapy enhances the inhibition of tumor growth with the fully human anti-type 1 insulin-like growth factor receptor monoclonal antibody CP-751, 871. *Clin. Cancer. Res.* 2005; **11**: 2063–73.

18. Voss L.D. and Mulligan J. Bullying in school: Are short pupils at risk? Questionnaire study in a cohort. *BMJ* 2000; **320**: 612–3.

19. Sonntag W.E., Carter C.S., Ikeno Y., et al. Adult-onset growth hormone and IGF-1 deficiency reduces neoplastic disease, modifies age-related pathology and increases lifespan. *Endocrinology* 2005; **146**: 2920–32.

20. Charlton H.M., Clark R.G., Robinson I.C., et al. Growth hormone-deficient dwarfism in the rat: A new mutation. *J. Endocrinol.* 1988; **119**: 51–8.

21. Clark R.G., Mortensen D.L., Carlsson L.M., Carlsson B., Carmignac D., and Robinson I.C. The obese growth hormone (GH)-deficient dwarf rat: Body fat responses

to patterned delivery of GH and insulin-like growth factor-I. *Endocrinology* 1996; **137**: 1904–12.

22. Dorshkind K. and Horseman N.D. The roles of prolactin, growth hormone, insulin-like growth factor-I, and thyroid hormones in lymphocyte development and function: Insights from genetic models of hormone and hormone receptor deficiency. *Endocr. Rev.* 2000; **21**: 292–312.

23. Kelley K.W. Growth hormone in immunobiology. In: Ader R. (ed.) *Psychoneuroimmunology*. New York: Academic Press, 1991: pp. 377–402.

24. Clark R. The somatogenic hormones and insulin-like growth factor-1: Stimulators of lymphopoiesis and immune function. *Endocr. Rev.* 1997; **18**: 157–79.

25. Hinton P.S., Peterson C.A., Dahly E.M., and Ney D.M. IGF-I alters lymphocyte survival and regeneration in thymus and spleen after dexamethasone treatment. *Am. J. Physiol.* 1998; **274**: R912–20.

26. Duerr R.L., McKirnan M.D., Gim R.D., Clark R.G., Chien K.R., and Ross J., Jr. Cardiovascular effects of insulin-like growth factor-1 and growth hormone in chronic left ventricular failure in the rat. *Circulation* 1996; **93**: 2188–96.

27. Tivesten A., Bollano E., Andersson I., et al. Liver-derived insulin-like growth factor-I is involved in the regulation of blood pressure in mice. *Endocrinology* 2002; **143**: 4235–42.

28. Juul A., Scheike T., Davidsen M., Gyllenborg J., and Jorgensen T. Low serum insulin-like growth factor I is associated with increased risk of ischemic heart disease: A population-based case-control study. *Circulation* 2002; **106**: 939–44.

29. Laughlin G.A., Barrett-Connor E., Criqui M.H., and Kritz-Silverstein D. The prospective association of serum insulin-like growth factor I (IGF-I) and IGF-binding protein-1 levels with all cause and cardiovascular disease mortality in older adults: The Rancho Bernardo Study. *J. Clin. Endocrinol. Metab.* 2004; **89**: 114–20.

30. Vasan R.S., Sullivan L.M., D'Agostino R.B., et al. Serum insulin-like growth factor I and risk for heart failure in elderly individuals without a previous myocardial infarction: The Framingham Heart Study. *Ann. Intern. Med.* 2003; **139**: 642–8.

31. Beck K.D., Powell-Braxton L., Widmer H.R., Valverde J., and Hefti F. Igf-1 gene disruption results in reduced brain size, CNS hypomyelination, and loss of hippocampal granule and striatal parvalbumin-containing neurons. *Neuron* 1995; **14**: 717–30.

32. Woods K.A., Camacho-Hubner C., Savage M.O., and Clark A.J. Intrauterine growth retardation and postnatal growth failure associated with deletion of the insulin-like growth factor I gene. *N. Engl. J. Med.* 1996; **335**: 1363–7.

33. Walenkamp M.J., Karperien M., Pereira A.M., et al. Homozygous and Heterozygous Expression of a Novel Igf-I Mutation. *J. Clin. Endocrinol. Metab.* 2005; **90**: 2855–64.

34. Smith R.G., Betancourt L., and Sun Y. Molecular endocrinology and physiology of the aging central nervous system. *Endocr. Rev.* 2005; **26**: 203–50.

35. Hartman M.L. The Growth Hormone Research Society consensus guidelines for the diagnosis and treatment of adult GH deficiency. *Growth Horm. IGF Res.* 1998; **8**: 25–9.

36. Rosen T. and Bengtsson B.A. Premature mortality due to cardiovascular disease in hypopituitarism. *Lancet* 1990; **336**: 285–8.

37. Powell-Braxton L., Hollingshead P., Warburton C., et al. IGF-I is required for normal embryonic growth in mice. *Genes Dev.* 1993; **7**: 2609–17.

38. van den Berghe G., Wouters P., Weekers F., et al. Intensive insulin therapy in the critically ill patients. *N. Engl. J. Med.* 2001; **345**: 1359–67.

39. Takala J., Ruokonen E., Webster N.R., et al. Increased mortality associated with growth hormone treatment in critically ill adults. *N. Engl. J. Med.* 1999; **341**: 785–92.

40. Hatton J., Rapp R.P., Kudsk K.A., et al. Intravenous insulin-like growth factor-I (IGF-I) in moderate-to-severe head injury: A phase II safety and efficacy trial. *J. Neurosurg.* 1997; **86**: 779–86.

41. Waters D., Danska J., Hardy K., et al. Recombinant human growth hormone, insulin-like growth factor 1, and combination therapy in AIDS-associated wasting. A randomized, double-blind, placebo-controlled trial. *Ann. Intern. Med.* 1996; **125**: 865–72.

42. Neary N.M., Small C.J., Wren A.M., et al. Ghrelin increases energy intake in cancer patients with impaired appetite: Acute, randomized, placebo-controlled trial. *J. Clin. Endocrinol. Metab.* 2004; **89**: 2832–6.

43. Chan J.M., Stampfer M.J., Giovannucci E., et al. Plasma insulin-like growth factor-I and prostate cancer risk: A prospective study. *Science* 1998; **279**: 563–66.

44. Burtrum D., Zhu Z., Lu D., et al. A fully human monoclonal antibody to the insulin-like growth factor I receptor blocks ligand-dependent signaling and inhibits human tumor growth *in vivo*. *Cancer Res.* 2003; **63**: 8912–21.

45. Ibrahim Y.H. and Yee D. Insulin-like growth factor-I and breast cancer therapy. *Clin. Cancer Res.* 2005; **11**: 944s–50s.

23 Erythropoietin — A Novel Therapeutic Option for Cachectic Patients

Piotr Ponikowski, Ewa A. Jankowska, and Waldemar Banasiak

CONTENTS

SUMMARY

Cachexia and anemia often coexist with similar mechanisms underlying both conditions. Erythropoietin (EPO) therapy has already been well accepted as a part of anemia correction in cachectic patients. Only recently, in the experimental and clinical studies numerous favorable biological properties of EPO beyond the hematopoietic system have been reported including antiapoptotic, anti-inflammatory, and anabolic effects of EPO in the peripheral tissues. It forms a solid basis to expect that EPO therapy may be extended also to nonanemic cachectic patients. Nevertheless, the potential side effects (increased hematocrit, increased incidence of thrombosis, and hypertension, undefined effect on neoplasmatic processes) should be carefully taken into the consideration when administering EPO.

23.1 INTRODUCTION

Cachexia, defined as progressive and marked loss of tissue in all compartments and organs of the body, is a common consequence of many chronic diseases (e.g., chronic heart failure [CHF], chronic obstructive pulmonary disease [COPD], cancer, chronic infectious diseases — AIDS and tuberculosis, liver cirrhosis, end-stage renal disease [ESRD], chronic inflammatory diseases — rheumatoid arthritis, autoimmunological syndromes). Cachexia usually develops at a late stage of the chronic illness, is related to very poor quality of life, and predicts a miserable prognosis. Irrespective of the underlying etiology of body wasting, the clinical picture remains fairly similar and it is hypothesized that analogous pathophysiological mechanisms are directly involved in the development of the cachectic syndrome across the entire spectrum of chronic diseases. It is therefore reasonable to expect that similar therapeutic options may well be applicable and efficacious in the management of body wasting. In this chapter, we will discuss the potential usefulness of erythropoietin (EPO) as a novel therapy for patients with cachexia.

EPO is a glycoprotein hormone, synthesized mainly by peritubular cells in the cortex-medullary border of the kidney, which promotes the survival, proliferation, and differentiation of erythrocytic progenitors in hemopoietic tissues.[1,2] The major stimulus for EPO production is tissue hypoxia, resulting in activation of the hypoxia-inducible factor 1 pathway, which upregulates the expression of EPO.[3] EPO and its downstream signaling plays a pivotal, protective role against tissue hypoxia by maintaining adequate tissue oxygen concentrations through an adjustment in the number of red blood cells, using a hormonal feedback-control system that involves the kidney and the bone marrow.

Recombinant human EPO, and recently also novel recombinant erythropoiesis stimulating protein — darbepoetin alfa, have been widely used in the treatment of anemia from a variety of etiologies including ESRD, cancer, myelodysplastic syndromes, HIV-infection, rheumatoid arthritis, and CHF.[4–8] The recognition of the numerous, nonhematopoietic effects of EPO has formed the background for a much broader clinical application of this hormone. There is growing interest in EPO as a novel cytoprotective agent in cardiovascular and neuronal systems.[2,3] It is conceivable that all the biological properties of EPO may translate into favorable effects in patients with cachexia.

23.2 ANEMIA — AN ESTABLISHED INDICATION FOR EPO IN CACHEXIA

Anemia is a common pathology, occurring as a result of underlying disease or an effect of concomitant treatment in the majority of chronic diseases.[9] In particular, patients with cachectic syndrome as a complication of chronic illness may easily become anemic, with the prevalence of anemia increasing with disease severity.

This observation is mainly based on clinical experience from studies in cancer and ESRD but has been recently extended to the population with CHF. Sharma et al.[10] reported that among 3044 elderly patients recruited in the

Evaluation of Losartan In The Elderly (ELITE II) trial, lower hemoglobin level coexisted with lower body mass index (BMI). In our own study comprising 205 male CHF subjects with advanced CHF (New York Heart Association [NYHA] class I to IV) who participated in the program of metabolic assessment at our institution, anemia (defined as a serum hemoglobin level <12.5 g/dl) was present in 13%, and those with anemia had significantly lower BMI as compared with nonanemic patients (24.9 vs. 27.4 kg/m², respectively, $p = .005$). Additionally, anemia was more prevalent among patients with cachexia, and this relationship was independent of CHF severity expressed by NYHA class and left ventricular function (Jankowska E.A. et al., unpublished).

The origin of cachexia-related anemia may be multifactorial, with the following underlying mechanisms: insufficient nutritional status, blood loss, side effects of therapy, renal dysfunction, exaggerated immune activation, limited anabolic stimulation, tumor infiltration of the bone marrow (in cancer patients), and opportunistic infections (in HIV-infected patients). In cardiac cachexia, hemodilution and decreased perfusion to the bone marrow may also contribute. In most cases, it represents the well-described anemia of chronic disease, a form of impaired erythropoiesis, secondary to augmented production of the cytokines that mediate the immune or inflammatory response (such as tumor necrosis factor [TNF]-alpha, interleukins [IL]-1 and 6, and interferons), and are characterized by erythroid bone marrow hypoplasia, reduced life-time of circulating red blood cells, and disturbed iron homeostasis (an impaired mobilization of reticuloendothelial iron stores).[9,11,12] Serum level of EPO may significantly vary in patients with anemia of chronic disease (in some cases being abnormally elevated), but blunted hemopoietic response to EPO with subsequent peripheral EPO resistance is common. This phenomenon has been recently well-characterized in patients with ESRD undergoing dialysis.[4,9] Kalantar-Zadeh et al. demonstrated that the so-called malnutrition–inflammation complex syndrome often complicated the clinical course in dialysis patients with the following possible causes — associated illnesses, oxidative and carbonyl stress, nutrient loss through dialysis, anorexia and low nutrient intake, uremic toxins, decreased clearance of inflammatory cytokines, volume overload, and dialysis-related factors.[13,14] All the elements of this syndrome may blunt the responsiveness of anemia of ESRD to EPO.

Metabolic (disturbed energy balance), immune (chronic inflammatory processes), and hormonal alterations (anabolic/catabolic imbalance in favor of catabolism) seen in nonanemic cachectic patients are also present in subjects with anemia of chronic disease. So far, it remains unclear and difficult to discriminate whether in the natural history of the cachectic syndrome, anemia is a secondary pathology naturally occurring at the advanced stages, or rather constitutes another element of generalized body wasting, with similar mechanisms promoting the development of both body wasting and anemia.

Cachexia is always linked to severely impaired anabolic drive. On the other hand, there is limited evidence on pathophysiological links between hematopoiesis and anabolic/catabolic metabolism. According to Ellegala et al.,[15] among men with pituitary adenoma resulting in hypogonadism, hemoglobin level was positively related to serum testosterone concentration; almost 50% of hypogonadal men were anemic, whereas the pharmacological substitution of gonadal andropenia resulted in an increase in hematocrit. In men with CHF, there are positive correlations between hemoglobin levels, and parameters of both body composition and anabolic status (serum total testosterone level) (own unpublished data). In the already mentioned study, in male CHF patients with gonadal andropenia (total testosterone <3 ng/ml), anemia is present in 24% of patients as compared to 7% in those with normal total testosterone level. On the other hand, in anemic men with CHF, the prevalence of gonadal andropenia reaches up to 54%, whereas in nonanemic men with CHF, only 21% (Jankowska E.A. et al., unpublished).

It is commonly known that anabolic steroids, particularly when overdosed by athletes, can strongly stimulate erythropoiesis. Also in men with COPD, a therapy with anabolic steroids results in an increase in fat free mass, which is accompanied by an increase in both hemoglobin and EPO levels.[16] In these patients, the change in peak workload during an exercise test correlates positively with the change in hemoglobin levels, whereas the change in isokinetic legwork is positively related to the change in EPO levels, both correlations remaining significant even when adjusted for changes in fat free mass.[16]

However, what seems more interesting is that there are few experimental and clinical data revealing the stimulatory effects of EPO on male gonadal steroidogenesis.[17–20] It has been demonstrated that EPO interacting specifically with its receptors can stimulate testosterone production in isolated adult rat Leydig cells.[17,18] Moreover, Foresta et al.[19] revealed that venous injections of recombinant human EPO increased the testicular steroidogenesis and testosterone production in young adult men, acting independently of gonadotrophin secretion. It has also been revealed in patients with chronic renal failure (CRF) that both those undergoing and not undergoing hemodialysis, recombinant human EPO significantly increased both hematocrit and serum testosterone levels.[20]

Anemia itself can additionally exaggerate the unfavorable catabolic changes in the peripheral tissues, mainly as the effect of acute and chronic lower oxygenation. This mechanism is particularly important in actively working skeletal muscles, as insufficient oxygen supply impairs exercise capacity and ultimately causes symptoms of fatigue and dyspnoea.[11,21] In fact, cachectic patients with concomitant anemia frequently complain of severe symptoms of exercise intolerance occurring even during normal daily activities, which dramatically impairs quality of life. Additionally, anemia complicating the syndrome of the cachexia is a well-recognized predictor of high morbidity and mortality, also lessening the effects of therapy.

The high prevalence of anemia in patients with cachectic syndrome, with markedly deranged EPO signaling in hematopoietic tissues constitute crucial premises in favor of the application of EPO therapy in this group of patients. There are numerous reports of the beneficial effects of human-recombinant EPO and erythropoiesis stimulating protein — darbepoetin alfa in the treatment of anemic patients with cancer and ESRD, most of which are cachectic.[22–24] The use of EPO has broadened to include HIV-infected adults with anemia, patients with rheumatoid arthritis, and recently also those with CHF-related anemia.[5,25-28]

23.3 PLEIOTROPIC EFFECTS OF ERYTHROPOIETIN

Erythropoietin (described also as a hematopoietic cytokine) is no longer considered only as a regulator of hematopoiesis.[3,29,30] Currently, pleiotropic effects of EPO are becoming of a special interest, in particular, in the context of potential novel therapeutic approaches for cachectic patients that may arise from other cellular mechanisms of EPO.[3,29]

EPO has an analogous molecular structure and acts via similar intracellular signaling pathways as the family of type 1 cytokines.[1,3,31] Interacting with its specific receptors (which belong to the cytokine receptor superfamily), EPO can activate selected transcriptional factors, and subsequently suppress apoptosis in various cell types.[1,29,32]

Erythropoietin is produced not only by the kidneys, but also by several other tissues, for example, liver, endothelial cells, enterocytes, skeletal and cardiac myocytes, and smooth muscle cells.[3,29] Specific EPO receptors are present in a wide range of tissues, not only within the hematopoietic system, but also on vascular smooth muscle cells, endothelial cells, skeletal myoblasts, and neonatal cardiomyocytes.[1,3,29,33] Only recently, has it been demonstrated that EPO receptors are localized also within the adult rat heart tissue on endothelial cells, fibroblasts, and differentiated cardiomyocytes, suggesting that also these cells are responsive to stimulation with EPO.[34] The EPO signaling for heart morphogenesis seems to be of crucial significance, as mice deficient in genes for either EPO or EPO receptor develop severe ventricular hypoplasia with a reduced number of proliferative cardiomyocytes and significant pathologies within cardiac vasculature.[35]

The intracellular mechanisms of EPO are not fully understood. At least some of the biological functions of EPO are independent of the hematopoietic system, and linked to autocrine–paracrine mechanisms.[29] The fundamental mechanism is related to the activation of pathways leading to the activation of transcriptional factors, involved in the regulation of cell growth and apoptosis.[1,3,29,32]

In *in vitro* models, recombinant human EPO stimulates the neonatal rat myocyte proliferation, inhibits their apoptosis, increases the Na^+/K^+ATP-ase activity,[36] mobilizes intracellular calcium through interacting with phospholipase C pathways,[1,37] and induces a broad range of other cellular responses, including mitogenesis, chemotaxis, and angiogenesis in various peripheral tissues (also in the skeletal muscle and the myocardium).[1,29] There is also evidence of anti-inflammatory properties of EPO, as its administration can reduce the number of inflammatory cells in injured brain tissue in rats.[38] In *in vitro* models, EPO antagonizes the effects of pro-inflammatory cytokines (such as IL-6, TNF-α, monocyte chemotactic protein 1)[39] and protects against lipopolysaccharide-mediated apoptosis.[40]

Among the effects of EPO beyond the hematopoietic system, the following may be of clinical importance: an increase in the number of circulating endothelial progenitor cells, a stimulation of several endothelial-derived modulators of vascular tone, and cardio- and neuroprotective properties.[3,29]

Therefore, in view of the numerous biological properties of EPO, it may well be expected that EPO therapy by far surpasses reversal of anemia, and the other cellular mechanisms are of a major clinical significance, which form the basis for broader EPO application.[29]

23.4 ERYTHROPOIETIN FOR THE IMPROVEMENT OF EXERCISE INTOLERANCE IN CACHECTIC PATIENTS

Patients with cachexia often complain of very poor exercise tolerance manifested as severe fatigue, weakness, and shortness of breath on minimal exertion. Exercise intolerance significantly impairs quality of life and predicts increased mortality in patients with cachexia. There are several pathophysiological mechanisms already operating in cachexia that may cause or promote the development of exercise intolerance and among them:

1. Reduced supply of oxygen and energy substrates to the working skeletal muscles (resulting from anemia or impaired peripheral blood flow)
2. Impaired energy metabolism in the peripheral tissues, including abnormal utilization of glucose, amino acids, and fatty acids
3. Profound abnormalities in the skeletal musculature including reduction in muscle mass, and deranged muscle structure and metabolism (e.g., deficiency in type 1 muscle fibers, excessive degradation of skeletal muscle protein and glycogen, and mitochondrial oxidative dysfunction)
4. Impaired reflex control within the cardiorespiratory system (mainly affecting the reflex-based regulation of generalized response to exercise, with augmented response from peripheral and central chemoreceptors and overactive ergoreceptors in the skeletal muscle)
5. Myocardial dysfunction and subsequently deranged hemodynamic response to exercise (primarily in patients with CHF-related cachexia but very often also in those with ESRD)
6. Abnormally elevated ventilatory response to exercise (well-characterized in CHF, but also present in cancer patients)
7. Various other reasons — psychological changes (e.g., depression), long-term physical inactivity, side effects of therapy

In addition to disease-specific therapies to improve the underlying cause of cachexia, there is a need for therapies to alleviate the symptoms of exercise intolerance. In this aspect, EPO may appear to be an interesting option.

As learned from doping strategies, a fundamental factor optimizing muscular activity and exercise performance is the degree of blood oxygenation, which can be improved for example, by using EPO. There are numerous papers demonstrating that in healthy subjects therapy with EPO significantly enhances exercise capacity (as evidenced by an improvement in virtually all indices characterizing ability of the body to perform various forms of exercise).[41,42] Whether this can be extended to patients with chronic disorders complicated by cachexia and suffering from exercise limitation is currently being investigated.

In patients with CHF, across a wide range of severity of the disease, hemoglobin levels strongly predict exercise capacity expressed as peak oxygen consumption, which gives the rationale for interventions aiming at the correction of anemia in patients with CHF.[43] This relationship is particularly strong in patients with CHF and hemoglobin level below 14.5 g/dl, whereas there is no correlation between these variables in subjects with hemoglobin level equal to or greater than 14.5 g/dl, suggesting that in this subpopulation oxygen transport capacity seems to be crucial for exercise capacity.[43]

According to Silverberg et al.,[44,45] the administration of recombinant human EPO in anemic patients with severe CHF (NYHA class III–IV) resulted in significant improvement of clinical status, myocardial function, and finally a reduction in hospitalization rate. These effects were seen in diabetic and nondiabetic patients with severe drug-resistant CHF.[46] In all these studies, the authors clearly demonstrated that long-term therapy with EPO significantly reduced symptoms of exercise intolerance and improved quality of life. In the study of Mancini et al.,[47] the 3-month treatment with recombinant human EPO in patients with moderate-to-severe CHF improved exercise capacity as evidenced by increased peak oxygen consumption, increased exercise duration time, and improved quality of life. Moreover, the change in peak oxygen consumption correlated positively with the change in hemoglobin level, confirming the significance of hemoglobin as an oxygen carrier as a determinant of exercise capacity.[47]

In patients with ESRD, impaired exercise capacity is partially reversed during the therapy with EPO.[48–51] Moreover, in these patients, EPO treatment can improve skeletal muscle function and oxygen utilization.[49–52] Nevertheless, the increase in peak oxygen consumption in ESRD patients after increasing hemoglobin up to almost normal levels due to EPO therapy is lower than expected.[53,54] This phenomenon is explained by mechanisms that directly limit the transport and uptake of oxygen by exercising muscles.[53,54] Firstly, the correction of anemia is followed by the elimination of an hyperdynamic cardiac status and a significant reduction in peak blood flow to skeletal muscles.[54] Secondly, in ESRD patients the oxygen conductance from muscle microcirculation to mitochondria is markedly reduced,[54] mainly due to microvascular rarification and capillary-microfiber dissociation.[55,56]

In anemic cancer patients, therapy with recombinant human EPO or darbepoetin alfa results in a correction of anemia and is accompanied by a marked improvement in quality of life; there is a positive correlation between the quantitative hematopoietic response to EPO treatment and an increase in scores describing quality of life of cancer patients.[21,57–59] In unselected weight-losing patients with malignant disease (mainly solid gastrointestinal tumors), therapy with recombinant human EPO prevented the development of cachexia- and cancer-related anemia, and this intervention was associated with a better preserved exercise capacity, explained in part by improved whole-body metabolic and energy efficiency during work load (even when adjusted for lean mass) in EPO-treated vs. nontreated patients.[60] However, controversies still exist. Only recently, the same group reported the study on the relationships between objectively assessed exercise capacity and subjectively assessed scoring of physical functioning and well-being after EPO treatment in cancer patients with cachexia on palliative care. Although they confirmed that recombinant human EPO therapy on top of anti-inflammatory treatment with indomethacin was effective as evidenced by time course changes of biochemical, physiological, and nutritional parameters, individual self-reported scoring of physical functioning and general health did not indicate a clear-cut effectiveness, particularly at moderately subnormal hemoglobin levels (the patients' own subjective scoring was insufficient to sense such improvements).

EPO can also exert some favorable effects on skeletal musculature. There is some evidence suggesting the significance of EPO for the development, differentiation, repair, and degradation of skeletal muscle tissue.[61]

There is no doubt that at least some beneficial effects of EPO on skeletal muscles are due to an increased hematocrit and subsequently augmented oxygen supply to the working muscles. EPO can also stimulate angiogenesis in skeletal muscle tissue, and therefore additionally and directly enhance oxygen supply to these cells.[62] Finally, mouse primary satellite cells and cultured myoblasts have EPO receptors, and it has been demonstrated that the exposure to EPO results in augmented proliferation and reduced differentiation of these cells.[33]

There are clinical studies showing that in hemodialysis patients with ESRD, therapy with recombinant human EPO can result in an increase in the average glycogen content in skeletal muscle tissue, an improvement in the mean diameter of the type 1 muscle fibers, a reduction in cytoarchitectural abnormalities,[63] a partial normalization of mitochondrial abnormalities evidenced as diminished lactic acid accumulation, and an increase in skeletal muscle strength.[64] Recently, Sarikaya et al. using 99Tcm-sestamibi leg scintigraphy demonstrated an improvement in muscle metabolism among uremic patients after EPO therapy.[65]

However, it needs to be remembered that most of the studies documenting favorable effects of EPO on exercise capacity have been performed among anemic patients. Therefore, it still needs to be established whether these promising results can be extended to nonanemic patients with cachexia who experience symptoms of exercise intolerance.

23.5 EFFECTS OF EPO ON BODY COMPOSITION AND NUTRITIONAL STATUS

Despite its unquestionable effects on erythropoiesis, it has been shown in ESRD patients that EPO treatment can also improve nutritional status,[66] as it amends amino acid metabolism,[67] glucose metabolism (insulin sensitivity),[68] and reduces generation of free radicals and subsequently lipid peroxidation.[69] Experimental data suggest that EPO can also stimulate the process of wound healing in rats.[70]

In the study of Barany et al., the correction of anemia with EPO treatment in anemic hemodialysis patients with underweight was accompanied by an increase in BMI, subcutaneous fat tissue mass, and muscle protein content.[71]

As it has been established in both an experimental model and in the clinic, treatment with EPO can markedly reduce the magnitude of weight loss in cachectic subjects.[60, 72, 73] Mice with adenocarcinoma treated with recombinant human EPO demonstrated a significant reduction in weight loss as compared to nontreated animals; there was a negative correlation between the change in hemoglobin concentration and the extent of weight loss during the study period.[72] In weight-losing nonanemic cancer patients, treatment with EPO prevented the development of anemia, and simultaneously attenuated weight loss as compared to nontreated patients.[60] Moreover, children undergoing chemotherapy for solid tumors who were treated with recombinant human EPO demonstrated a better clinical status and reduced weight loss as compared to nontreated subjects.[73]

Vaisman et al. have recently evaluated the effect of the correction of anemia on resting energy expenditure in subjects with CHF.[74] After anemia correction with EPO, patients tended to increase weight, daily caloric intake increased significantly but no significant changes were observed in body composition. Most importantly, resting energy expenditure was also increased significantly at the end of therapy.

23.6 ANTI-APOPTOTIC AND ISCHEMIA-PROTECTIVE EFFECTS OF EPO WITHIN MYOCARDIUM

There is increasing evidence of a protective role of EPO in experimental models of both myocardial infarction and ischemia/reperfusion injury.[34, 75–78] According to Parsa et al., an administration of EPO reveals antiapoptotic effects both *in vitro* on embryonic rat myoblasts and *in vivo* on cardiomyocytes in rabbits exposed to either ischemia/reperfusion injury or experimental myocardial infarction.[76, 77] In the study of Calvillo et al., EPO prevents apoptosis of cultured adult rat cardiomyocytes when exposed to hypoxia.[78] Additionally, EPO-treated rabbits demonstrate smaller infarct size and improved inotropic reserve when compared with untreated animals.[76] In isolated rat hearts, perfusion with EPO during ischemia/reperfusion injury reduces cellular damage by 56%, diminishes apoptosis and improves the recovery of left ventricular pressure and blood flow when compared to nonperfused organs.[34] A single dose of recombinant human EPO administered immediately after the artificial induction of myocardial infarction in rats results in a 50% reduction of apoptosis within ischemic area, diminishes the infarct size by 75 to 85%, and markedly attenuates the functional decline of injured myocardium, as compared to untreated animals.

There are also studies in ESRD patients demonstrating that EPO therapy can reduce exercise-induced myocardial ischemia in subjects with coronary artery disease,[79] and can ameliorate left ventricular hyperthrophy.[80]

EPO signaling is also involved in the recruitment and activation of circulating bone marrow-derived endothelial progenitor cells (EPCs, CD34+), which are involved in regenerative processes within the cardiovascular system, being responsible mainly for endothelial and vascular repair.[81,82] In patients with renal anemia, treatment with recombinant human EPO or darbepoietin alfa results in an increase in the number of circulating EPCs and increases in the number of functionally active EPCs.[83,84] The stimulatory effects on EPCs can be observed even at subtherapeutic EPO doses with respect to the correction of renal anemia.[84] Moreover, activating some pro-survival cellular pathways, EPO makes EPCs more resistant to ischemic stimuli.[85]

23.7 THERAPY WITH EPO — REMAINING CONCERNS

Before advocating to broaden clinical application of EPO, potential side effects of such therapy should be always considered. They may be divided into two major groups — risk related to increased hematocrit as a natural result of EPO substitution and hematocrit-independent side effects.

In the former group, clinically the major issue is the increased incidence of thrombotic events, including deep venous thrombosis, pulmonary thrombosis, ischemic stroke, and myocardial infarction, in the course of EPO substitution.[1,86–89] Apart from increased hematocrit, they are secondary to well-described cardiovascular effects of EPO: thrombotic/fibrynolytic imbalance in favor of thrombotic factors, enhanced endothelial activation and platelet hyperreactivity.[1,90–92] Therapy with EPO may also induce or worsen arterial hypertension in part due to the following mechanisms: increased blood viscosity, enhanced vascular reactivity to angiotensin II, endothelin and norepinephrine, increased endothelin production and release, elevated intracellular calcium concentration, ability to upregulate tissue renin–angiotensin system.[1] In anemic cancer patients pooled results from several randomized, controlled trials revealed a 1.55-fold elevated risk of thromboembolic events and 1.25-fold elevated risk of hypertension (both statistically significant) with recombinant human EPO therapy compared with controls.[87]

There are still undefined effects of EPO treatment on the progression of neoplasmatic processes.[87,93] Only recently, Henke et al. reported that in patients with anemia in the course of head and neck cancer, long-term therapy with epoetin beta (recombinant EPO) significantly impaired disease control and decreased survival.[31]

It must be also remembered that even in the area of anemia due to cancer or ESRD as well-established indication for EPO, there is a debate regarding the optimal strategy of EPO substitution in order to achieve the best possible balance between benefits and risks. In this aspect, the following questions still remain to be answered: what is the hemoglobin threshold for initiation of EPO therapy, what is the optimal target for hemoglobin concentration, what are the optimal doses particularly in those who are EPO resistant, how to identify patients at high risk for thromboembolic events and hypertension?

Finally, when starting the long-term EPO treatment, the cost-effectiveness should be taken into consideration.[94]

TABLE 23.1
Properties of EPO of Potential Benefit to Cachectic Patients

Dependent on the correction of anemia
Increase in oxygen supply to working skeletal muscles (the correction of anemia and hypoxia)
Independent of the correction of anemia
Suppression of apoptosis in various cell types[1, 29, 32, 40, 76–78]
Partial normalization of mitochondrial oxidative abnormalities in skeletal muscles[95]
Increase in Na^+/K^+ATP-ase activity[36]
Mobilization of intracellular calcium[37]
Anti-inflammatory properties[38, 39]
Improvement in energy expenditure in peripheral tissues, including optimalization of glucose (increase in insulin sensitivity),[68] amino acid,[67] and fatty acid metabolism[69]
Induction of mitogenesis, chemotaxis, and angiogenesis in various tissue types[1, 29, 32]
Augmented proliferation and reduced differentiation of skeletal myoblasts[33]
Recruitment and activation of circulating bone marrow-derived EPCs[81–84]
Stimulation of gonadal steroidogenesis (testosterone synthesis)[17–20]
Protective role against myocardial ischaemia[34, 75–78]
Increase in subcutaneous fat tissue mass,[71] muscle protein content,[71] glycogen content,[63] the type 1 muscle fibres[63] and a reduction in the rate of weight loss[60, 72, 73]
Increase in skeletal muscle strength[64]
Improvement in hemodynamic, systolic function of the left ventricle[44–46]
Improvement in exercise capacity (increased peak oxygen consumption)[47–51]

23.8 CONCLUSIONS

Cachexia and anemia often coexist with similar mechanisms underlying both conditions. EPO therapy has already been widely accepted as a part of anemia correction in patients with the syndrome of cachexia. There are many potential mechanisms explaining the beneficial effects of EPO (Table 23.1), but interestingly, most of EPO properties are beyond the hematopoietic system. It still remains unclear whether the therapy with EPO should be restricted only to cachectic patients with an accompanying anemia. Recent experimental and clinical data are very promising and form a solid basis to expect that EPO therapy may be extended also to nonanemic cachectic patients. Nevertheless, the potential side effects (increased hematocrit, increased incidence of thrombosis and hypertension, undefined effect on neoplasmatic processes) should be carefully taken into consideration when administering EPO.

REFERENCES

1. Smith, K.J., Bleyer, A.J., Little, W.C., and Sane, D.C. The cardiovascular effects of erythropoietin. *Cardiovasc. Res.* 2003;59:538–548.

2. Jelkmann, W. and Wagner, K. Beneficial and ominous aspects of the pleiotropic action of erythropoietin. *Ann. Hematol.* 2004;83:673–686.
3. Maiese, K., Li, F., and Chong, Z.Z. New avenues of exploration for erythropoietin. *JAMA* 2005;293:90–95.
4. Ferrario, E., Ferrari, L., Bidoli, P., De Candis, D., Del Vecchio, M., De Dosso, S., Buzzoni, R., and Bajetta, E. Treatment of cancer-related anemia with epoetin alfa: A review. *Cancer Treat. Rev.* 2004;30:563–575.
5. Fangman, J.J. and Scadden, D.T. Anemia in HIV-infected adults: Epidemiology, pathogenesis, and clinical management. *Curr. Hematol. Rep.* 2005;4:95–102.
6. Felker, G.M., Adams, K.F., Gattis, W.A., and O'Connor, C.M. Anemia as a risk factor and therapeutic target in heart failure. *J. Am. Coll. Cardiol.* 2004;44: 959–966.
7. Wilson, A., Yu, H.T., Goodnough, L.T., and Nissenson, A.R. Prevalence and outcomes of anemia in rheumatoid arthritis: A systematic review of the literature. *Am. J. Med.* 2004;116(Suppl 7A):S50–S57.
8. Locatelli, F., Aljama, P., Barany, P., Canaud, B., Carrera, F., Eckardt, K.U., Horl, W.H., Macdougal, I.C., Macleod, A., and Wiecek, A., Cameron S; and European Best Practice Guidelines Working Group. Revised European best practice guidelines for the management of anaemia in patients with chronic renal failure. *Nephrol. Dial. Transplant.* 2004;19(Suppl 2):ii1–47.
9. Silverberg, D.S., Iaina, A., Wexler, D., and Blum, M. The pathological consequences of anaemia. *Clin. Lab. Haematol.* 2001;23:1–6.
10. Sharma, R., Francis, D.P., Pitt, B., Poole-Wilson, P.A., Coats, A.J., and Anker, S.D. Haemoglobin predicts survival in patients with chronic heart failure: A substudy of the ELITE II trial. *Eur. Heart J.* 2004;25:1021–1028.
11. Ferrario, E., Ferrari, L., Bidoli, P., De Candis, D., Del Vecchio M, De Dosso, S., Buzzoni, R., and Bajetta, E. Treatment of cancer-related anemia with epoetin alfa: A review. *Cancer Treat. Rev.* 2004;30:563–575.
12. Kalmanti, M. and Kalmantis, T. Committed erythroid progenitors and erythropoietin levels in anemic children with lymphomas and tumors. *Pediatr. Hematol. Oncol.* 1989;6:85–93.
13. Kalantar-Zadeh, K., McAllister, C.J., Lehn, R.S., Lee, G.H., Nissenson, A.R., and Kopple, J.D. Effect of malnutrition-inflammation complex syndrome on EPO hyporesponsiveness in maintenance hemodialysis patients. *Am. J. Kidney Dis.* 2003;42:761–773.
14. Kalantar-Zadeh, K., Ikizler, T.A., Block, G., Avram, M.M., and Kopple J.D. Malnutrition-inflammation complex syndrome in dialysis patients: Causes and consequences. *Am. J. Kidney Dis.* 2003;42:864–881.
15. Ellegala, D.B., Alden, T.D., Couture, D.E., Vance, M.L., Maartens, N.F., and Laws, E.R. Jr. Anemia, testosterone, and pituitary adenoma in men. *J. Neurosurg.* 2003;98:974–977.
16. Creutzberg, E.C., Wouters, E.F., Mostert, R., Pluymers, R.J., and Schols, A.M. A role for anabolic steroids in the rehabilitation of patients with COPD? A double-blind, placebo-controlled, randomized trial. *Chest* 2003;124:1733–1742.
17. Mioni, R., Gottardello, F., Bordon, P., Montini, G., and Foresta, C. Evidence for specific binding and stimulatory effects of recombinant human erythropoietin on isolated adult rat Leydig cells. *Acta. Endocrinol.* 1992;127:459–465.

18. Foresta, C., Mioni, R., Bordon, P., Gottardello, F., Nogara, A., and Rossato, M. Erythropoietin and testicular steroidogenesis: The role of second messengers. *Eur. J. Endocrinol.* 1995;132:103–108.

19. Foresta, C., Mioni, R., Bordon, P., Miotto, D., Montini, G., and Varotto, A. Erythropoietin stimulates testosterone production in man. *J. Clin. Endocrinol. Metab.* 1994;78:753–756.

20. Wu, S.C., Lin, S.L., and Jeng, F.R. Influence of erythropoietin treatment on gonadotropic hormone levels and sexual function in male uremic patients. *Scand. J. Urol. Nephrol.* 2001;35:136–140.

21. Libretto, S.E., Barrett-Lee, P.J., Branson, K., Gorst, D.W., Kaczmarski, R., McAdam, K., Stevenson, P., and Thomas, R. Improvement in quality of life for cancer patients treated with epoetin alfa. *Eur. J. Cancer Care* 2001;10:183–191.

22. Justice, G., Kessler, J.F., Jadeja, J., Campos, L., Weick, J., Chen, C.F., Heatherington, A.C., and Amado, R.G. A randomized, multicenter study of subcutaneous and intravenous darbepoetin alfa for the treatment of chemotherapy-induced anemia. *Ann. Oncol.* 2005; 16: 1192–1198.

23. Vansteenkiste, J. and Wauters, I. The use of darbepoetin alfa for the treatment of chemotherapy-induced anaemia. *Expert. Opin. Pharmacother.* 2005;6:429–440.

24. Toto, R.D., Pichette, V., Navarro, J., Brenner, R., Carroll, W., Liu, W., and Roger, S. Darbepoetin alfa effectively treats anemia in patients with chronic kidney disease with de novo every-other-week administration. *Am. J. Nephrol.* 2004;24:453–460.

25. Deicher, R. and Horl, W.H. Differentiating factors between erythropoiesis-stimulating agents: A guide to selection for anaemia of chronic kidney disease. *Drugs* 2004;64:499–509.

26. John, M., Hoernig, S., Doehner, W., Okonko, D.D., Witt, C., and Anker, S.D. Anemia and inflammation in COPD. *Chest* 2005;127:825–829.

27. Means, R.T. Jr. Advances in the anemia of chronic disease. *Int. J. Hematol.* 1999;70:7–12.

28. Lindholm, E., Daneryd, P., Korner, U., Hyltander, A., Fouladiun, M., and Lundholm, K. Effects of recombinant erythropoietin in palliative treatment of unselected cancer patients. *Clin. Cancer Res.* 2004;10:6855–6864.

29. Lappin, T.R., Maxwell, A.P., and Johnston, P.G. EPO's alter ego: Erythropoietin has multiple actions. *Stem Cells* 2002;20:485–492.

30. Jelkmann, W. and Wagner, K. Beneficial and ominous aspects of the pleiotropic action of erythropoietin. *Ann. Hematol.* 2004;83:673–686.

31. Ozaki, K. and Leonard, W.J. Cytokine and cytokine receptor pleiotropy and redundancy. *J. Biol. Chem.* 2002;277:29355–29358.

32. Farrell, F. and Lee, A. The erythropoietin receptor and its expression in tumor cells and other tissues. *Oncologist* 2004;9(Suppl 5):18–30.

33. Ogilvie, M., Yu, X., Nicolas-Metral, V., Pulido, S.M., Liu, C., Ruegg, U.T., and Noguchi, C.T. Erythropoietin stimulates proliferation and interferes with differentiation of myoblasts. *J. Biol. Chem.* 2000;275:39754–39761.

34. van der Meer, P., Lipsic, E., Henning, R.H., de Boer, R.A., Suurmeijer, A.J., van Veldhuisen, D.J., and van Gilst, W.H. Erythropoietin improves left ventricular function and coronary flow in an experimental model of ischemia-reperfusion injury. *Eur. J. Heart Fail.* 2004;6:853–859.

35. Wu, H., Lee, S.H., Gao, J., Liu, X., and Iruela-Arispe, M.L. Inactivation of erythropoietin leads to defects in cardiac morphogenesis. *Development* 1999;126:3597–3605.

36. Wald, M., Gutnisky, A., Borda, E., and Sterin-Borda, L. Erythropoietin modified the cardiac action of ouabain in chronically anaemic-uraemic rats. *Nephron* 1995;71:190–196.

37. Marrero, M.B., Venema, R.C., Ma, H., Ling, B.N., and Eaton, D.C. Erythropoietin receptor-operated Ca2+ channels: Activation by phospholipase C-gamma 1. *Kidney Int.* 1998;53:1259–1268.

38. Brines, M.L., Ghezzi, P., Keenan, S., Agnello, D., de Lanerolle, N.C., Cerami, C., Itri, L.M., and Cerami, A. Erythropoietin crosses the blood-brain barrier to protect against experimental brain injury. *Proc. Natl Acad. Sci. USA* 2000;97:10526–10531.

39. Chong, Z.Z., Kang, J.Q., and Maiese, K. Hematopoietic factor erythropoietin fosters neuroprotection through novel signal transduction cascades. *J. Cereb. Blood Flow Metab.* 2002;22:503–514.

40. Carlini, R.G., Alonzo, E.J., Dominguez, J., Blanca, I., Weisinger, J.R., Rothstein, M., and Bellorin-Font, E. Effect of recombinant human erythropoietin on endothelial cell apoptosis. *Kidney Int.* 1999;55:546–553.

41. Clyne, N., Berglund, B., and Egberg, N. Treatment with recombinant human erythropoietin induces a moderate rise in hematocrit and thrombin antithrombin in healthy subjects. *Thromb. Res.* 1995;79:125–129.

42. Shaskey, D.J. and Green, G.A. Sports haematology. *Sports Med.* 2000;29:27–38.

43. Kalra, P.R., Bolger, A.P., Francis, D.P., Genth-Zotz, S., Sharma R, Ponikowski, P.P., Poole-Wilson, P.A., Coats, A.J., and Anker, S.D. Effect of anemia on exercise tolerance in chronic heart failure in men. *Am. J. Cardiol.* 2003;91:888–891.

44. Silverberg, D.S., Wexler, D., Blum, M., Keren, G., Sheps, D., Leibovitch, E., Brosh, D., Laniado, S., Schwartz, D., Yachnin, T., Shapira, I., Gavish, D., Baruch, R., Koifman, B., Kaplan, C., Steinbruch, S., and Iaina, A. The use of subcutaneous erythropoietin and intravenous iron for the treatment of the anemia of severe, resistant congestive heart failure improves cardiac and renal function and functional cardiac class, and markedly reduces hospitalizations. *J. Am. Coll. Cardiol.* 2000;35:1737–1744.

45. Silverberg, D.S., Wexler, D., Sheps, D., Blum, M., Keren, G., Baruch, R., Schwartz, D., Yachnin, T., Steinbruch, S., Shapira, I., Laniado, S., and Iaina, A. The effect of correction of mild anemia in severe, resistant congestive heart failure using subcutaneous erythropoietin and intravenous iron: A randomized controlled study. *J. Am. Coll. Cardiol.* 2001;37:1775–1780.

46. Silverberg, D.S., Wexler, D., Blum, M., Tchebiner, J.Z., Sheps, D., Keren, G., Schwartz, D., Baruch, R., Yachnin, T., Shaked, M., Schwartz, I., Steinbruch, S., and Iaina, A. The effect of correction of anaemia in diabetics and non-diabetics with severe resistant congestive heart failure and chronic renal failure by subcutaneous erythropoietin and intravenous iron. *Nephrol. Dial. Transplant.* 2003;18:141–146.

47. Mancini, D.M., Katz, S.D., Lang, C.C., LaManca, J., Hudaihed, A., and Androne, A.S. Effect of erythropoietin on exercise capacity in patients with moderate to severe chronic heart failure. *Circulation* 2003;107:294–299.

48. Macdougall, I.C., Lewis, N.P., Saunders, M.J., Cochlin, D.L., Davies, M.E., Hutton, R.D., Fox, K.A., Coles, G.A., and Williams, J.D. Long-term cardiorespiratory effects of amelioration of renal anaemia by erythropoietin. *Lancet* 1990;335:489–493.

49. Marrades, R.M., Roca, J., Campistol, J.M., Diaz, O., Barbera, J.A., Torregrosa, J.V., Masclans, J.R., Cobos, A., Rodriguez-Roisin, R., and Wagner, P.D.

Effects of erythropoietin on muscle O2 transport during exercise in patients with chronic renal failure. *J. Clin. Invest.* 1996;97:2092–2100.

50. Metra, M., Cannella, G., La Canna, G., Guaini, T., Sandrini, M., Gaggiotti, M., Movilli, E., and Dei Cas, L. Improvement in exercise capacity after correction of anemia in patients with end-stage renal failure. *Am. J. Cardiol.* 1991;68:1060–1066.

51. McMahon, L.P., McKenna, M.J., Sangkabutra, T., Mason, K., Sostaric, S., Skinner, S.L., Burge, C., Murphy, B., and Crankshaw, D. Physical performance and associated electrolyte changes after haemoglobin normalization: A comparative study in haemodialysis patients. *Nephrol. Dial. Transplant.* 1999;14:1182–1187.

52. Sarikaya, A., Sen, S., Cermik, T.F., Birtane, M., and Berkarda, S. Evaluation of skeletal muscle metabolism and response to erythropoietin treatment in patients with chronic renal failure using 99Tcm-sestamibi leg scintigraphy. *Nucl. Med. Commun.* 2000;21:83–87.

53. Marrades, R.M., Alonso, J., Roca, J., Gonzalez de Suso, J.M., Campistol, J.M., Barbera, J.A., Diaz, O., Torregrosa, J.V., Masclans, J.R., Rodriguez-Roisin, R., and Wagner, P.D. Cellular bioenergetics after erythropoietin therapy in chronic renal failure. *J. Clin. Invest.* 1996;97:2101–2110.

54. Marrades, R.M., Roca, J., Campistol, J.M., Diaz, O., Barbera, J.A., Torregrosa, J.V., Masclans, J.R., Cobos, A., Rodriguez-Roisin, R., and Wagner, P.D. Effects of erythropoietin on muscle O2 transport during exercise in patients with chronic renal failure. *J. Clin. Invest.* 1996;97:2092–2100.

55. Bradley, J.R., Anderson, J.R., Evans, D.B., and Cowley, A.J. Impaired nutritive skeletal muscle blood flow in patients with chronic renal failure. *Clin. Sci.* 1990;79:239–245.

56. Diesel, W., Emms, M., Knight, B.K., Noakes, T.D., Swanepoel, C.R., van Zyl Smit, R., Kaschula, R.O., and Sinclair-Smith, C.C. Morphologic features of the myopathy associated with chronic renal failure. *Am. J. Kidney Dis.* 1993;22:677–684.

57. Demetri, G.D., Kris, M., Wade, J., Degos, L., and Cella, D. Quality-of-life benefit in chemotherapy patients treated with epoetin alfa is independent of disease response or tumor type: results from a prospective community oncology study. Procrit Study Group. *J. Clin. Oncol.* 1998;16:3412–3425.

58. Glaspy, J., Bukowski, R., Steinberg, D., Taylor, C., Tchekmedyian, S., and Vadhan-Raj, S. Impact of therapy with epoetin alfa on clinical outcomes in patients with nonmyeloid malignancies during cancer chemotherapy in community oncology practice. Procrit Study Group. *J. Clin. Oncol.* 1997;15:1218–1234.

59. Littlewood, T.J., Bajetta, E., Nortier, J.W., Vercammen, E., and Rapoport B; and Epoetin Alfa Study Group. Effects of epoetin alfa on hematologic parameters and quality of life in cancer patients receiving nonplatinum chemotherapy: Results of a randomized, double-blind, placebo-controlled trial. *J. Clin. Oncol.* 2001;19:2865–2874.

60. Daneryd, P., Svanberg, E., Korner, U., Lindholm, E., Sandstrom, R., Brevinge, H., Pettersson, C., Bosaeus, I., and Lundholm, K. Protection of metabolic and exercise capacity in unselected weight-losing cancer patients following treatment with recombinant erythropoietin: A randomized prospective study. *Cancer Res.* 1998;58:5374–5379.

61. Scoppetta, C. and Grassi, F. Erythropoietin: A new tool for muscle disorders? *Med. Hypotheses* 2004;63:73–75.

62. Vaziri, N.D. Cardiovascular effects of erythropoietin and anemia correction. *Curr. Opin. Nephrol. Hypertens.* 2001 Sep;10(5):633–637.

63. Davenport, A., King, R.F., Ironside, J.W., Will, E.J., and Davison, A.M. The effect of treatment with recombinant human erythropoietin on the histological appearance and glycogen content of skeletal muscle in patients with chronic renal failure treated by regular hospital haemodialysis. *Nephron* 1993;64:89–94.

64. Davenport, A. The effect of treatment with recombinant human erythropoietin on skeletal muscle function in patients with end-stage renal failure treated with regular hospital hemodialysis. *Am. J. Kidney Dis.* 1993;22:685–690.

65. Sarikaya, A., Sen, S., Cermik, T.F., Birtane, M., and Berkarda, S. Evaluation of skeletal muscle metabolism and response to erythropoietin treatment in patients with chronic renal failure using 99Tcm-sestamibi leg scintigraphy. *Nucl. Med. Commun.* 2000;21:83–87.

66. Kaupke, C.J. and Vaziri, N.D. Effect of recombinant erythropoietin on electrolytes and nutrition in end-stage renal disease patients. *Int. J. Artif. Organs* 1993;16:59–62.

67. Riedel, E., Nundel, M., Wendel, G., and Hampl, H. Amino acid and alpha-keto acid metabolism depends on oxygen availability in chronic hemodialysis patients. *Clin. Nephrol.* 2000;53(1 Suppl):S56–S60.

68. Spaia, S., Pangalos, M., Askepidis, N., Pazarloglou, M., Mavropoulou, E., Theodoridis, S., Dimitrakopoulos, K., Milionis, A., and Vayonas, G. Effect of short-term rHuEPO treatment on insulin resistance in haemodialysis patients. *Nephron* 2000;84:320–325.

69. Sommerburg, O., Grune, T., Hampl, H., Riedel, E., Ehrich, J.H., and Siems, W.G. Does treatment of renal anemia with recombinant erythropoietin influence oxidative stress in hemodialysis patients? *Clin. Nephrol.* 2000;53(1 Suppl):S23–S29.

70. Fatouros, M.S., Vekinis, G., Bourantas, K.L., Mylonakis, E.P., Scopelitou, A.S., Malamou-Mitsis, V.D., and Kappas, A.M. Influence of growth factors erythropoietin and granulocyte macrophage colony stimulating factor on mechanical strength and healing of colonic anastomoses in rats. *Eur. J. Surg.* 1999;165:986–992.

71. Barany, P., Pettersson, E., Ahlberg, M., Hultman, E., and Bergstrom, J. Nutritional assessment in anemic hemodialysis patients treated with recombinant human erythropoietin. *Clin. Nephrol.* 1991;35:270–279.

72. van Halteren, H.K., Bongaerts, G.P., Verhagen, C.A., Kamm, Y.J., Willems, J.L., Grutters, G.J., Koopman, J.P., and Wagener, D.J. Recombinant human erythropoietin attenuates weight loss in a murine cancer cachexia model. *J. Cancer Res. Clin. Oncol.* 2004;130:211–216.

73. Csaki, C., Ferencz, T., Schuler, D., and Borsi, J.D. Recombinant human erythropoietin in the prevention of chemotherapy-induced anaemia in children with malignant solid tumours. *Eur. J. Cancer* 1998;34:364–367.

74. Vaisman, N., Silverberg, D.S., Wexler, D., Niv, E., Blum, M., Keren, G., Soroka, N., and Iaina, A. Correction of anemia in patients with congestive heart failure increases resting energy expenditure. *Clin. Nutr.* 2004;23:355–361.

75. Moon, C., Krawczyk, M., Ahn, D., Ahmet, I., Paik, D., Lakatta, E.G., and Talan, M.I. Erythropoietin reduces myocardial infarction and left ventricular functional decline after coronary artery ligation in rats. *Proc. Natl Acad. Sci. USA* 2003;100:11612–11617.

76. Parsa, C.J., Matsumoto, A., Kim, J., Riel, R.U., Pascal, L.S., Walton, G.B., Thompson, R.B., Petrofski, J.A., Annex, B.H., Stamler, J.S., and Koch, W.J.

A novel protective effect of erythropoietin in the infarcted heart. *J. Clin. Invest.* 2003;112:999–1007.

77. Parsa, C.J., Kim, J., Riel, R.U., Pascal, L.S., Thompson, R.B., Petrofski, J.A., Matsumoto, A., Stamler, J.S., and Koch, W.J. Cardioprotective effects of erythropoietin in the reperfused ischemic heart: A potential role for cardiac fibroblasts. *J. Biol. Chem.* 2004;279:20655–20662.

78. Calvillo, L., Latini, R., Kajstura, J., Leri, A., Anversa, P., Ghezzi, P., Salio, M., Cerami, A., and Brines, M. Recombinant human erythropoietin protects the myocardium from ischemia-reperfusion injury and promotes beneficial remodeling. *Proc. Natl Acad. Sci. USA* 2003;100:4802–4806.

79. Wizemann, V., Kaufmann, J., and Kramer, W. Effect of erythropoietin on ischemia tolerance in anemic hemodialysis patients with confirmed coronary artery disease. *Nephron* 1992;62:161–165.

80. Portoles, J., Torralbo, A., Martin, P., Rodrigo, J., Herrero, J.A., and Barrientos, A. Cardiovascular effects of recombinant human erythropoietin in predialysis patients. *Am. J. Kidney Dis.* 1997;29:541–548.

81. Fliser, D., de Groot, K., Bahlmann, F.H., and Haller, H. Cardiovascular disease in renal patients–a matter of stem cells? *Nephrol. Dial. Transplant.* 2004;19:2952–2954.

82. Bahlmann, F.H., de Groot, K., Haller, H., and Fliser, D. Erythropoietin: Is it more than correcting anaemia? *Nephrol. Dial. Transplant.* 2004;19:20–22.

83. Bahlmann, F.H., DeGroot, K., Duckert, T., Niemczyk, E., Bahlmann, E., Boehm, S.M., Haller, H., and Fliser, D. Endothelial progenitor cell proliferation and differentiation is regulated by erythropoietin. *Kidney Int.* 2003;64:1648–1652.

84. Bahlmann, F.H., De Groot, K., Spandau, J.M., Landry, A.L., Hertel, B., Duckert, T., Boehm, S.M., Menne, J., Haller, H., and Fliser, D. Erythropoietin regulates endothelial progenitor cells. *Blood* 2004;103:921–926.

85. Mangi, A.A., Noiseux, N., Kong, D., He, H., Rezvani, M., Ingwall, J.S., and Dzau, V.J. Mesenchymal stem cells modified with Akt prevent remodeling and restore performance of infarcted hearts. *Nat. Med.* 2003;9:1195–1201.

86. Strippoli, G.F., Craig, J.C., Manno, C., and Schena, F.P. Hemoglobin targets for the anemia of chronic kidney disease: A meta-analysis of randomized, controlled trials. *J. Am. Soc. Nephrol.* 2004;15:3154–3165.

87. Bokemeyer, C., Aapro, M.S., Courdi, A., Foubert, J., Link, H., Osterborg, A., Repetto, L., and Soubeyran, P. EORTC guidelines for the use of erythropoietic proteins in anaemic patients with cancer. *Eur. J. Cancer* 2004;40:2201–2216.

88. Muirhead, N., Laupacis, A., and Wong, C. Erythropoietin for anaemia in haemodialysis patients: Results of a maintenance study (the Canadian Erythropoietin Study Group). *Nephrol. Dial. Transplant.* 1992;7:811–816.

89. Besarab, A., Bolton, W.K., Browne, J.K., Egrie, J.C., Nissenson, A.R., Okamoto, D.M., Schwab, S.J., and Goodkin, D.A. The effects of normal as compared with low hematocrit values in patients with cardiac disease who are receiving hemodialysis and epoetin. *N. Engl. J. Med.* 1998;339:584–590.

90. Tobu, M., Iqbal, O., Fareed, D., Chatha, M., Hoppensteadt, D., Bansal, V., and Fareed, J. Erythropoietin-induced thrombosis as a result of increased inflammation and thrombin activatable fibrinolytic inhibitor. *Clin. Appl. Thromb. Hemost.* 2004;10:225–232.

91. Malyszko, J., Suchowierska, E., Malyszko, J.S., and Mysliwiec, M. Some aspects of hemostasis in CAPD patients treated with erythropoietin. *Kidney Blood Press. Res.* 2002;25:240–244.

92. Stohlawetz, P.J., Dzirlo, L., Hergovich, N., Lackner, E., Mensik, C., Eichler, H.G., Kabrna, E., Geissler, K., and Jilma, B. Effects of erythropoietin on platelet reactivity and thrombopoiesis in humans. *Blood* 2000;95:2983–2989.

93. Henke, M., Laszig, R., Rube, C., Schafer, U., Haase, K.D., Schilcher, B., Mose, S., Beer, K.T., Burger, U., Dougherty, C., and Frommhold, H. Erythropoietin to treat head and neck cancer patients with anaemia undergoing radiotherapy: Randomised, double-blind, placebo-controlled trial. *Lancet* 2003;362:1255–1260.

94. Tonelli, M., Winkelmayer, W.C., Jindal, K.K., Owen, W.F., and Manns, B.J. The cost-effectiveness of maintaining higher hemoglobin targets with erythropoietin in hemodialysis patients. *Kidney Int.* 2003;64:295–304.

95. chronic uraemia on skeletal muscle metabolism in man. *Nephrol. Dial. Transplant.* 1993;8:218–222.

24 Statins

Stephan von Haehling and Stefan D. Anker

Summary

Statins have revolutionized the treatment of hypercholesterolemia. These drugs are usually well tolerated and generally safe. Effects beyond mere cholesterol reduction, the so-called pleiotropic effects, have recently received tremendous attention. These effects include improvement of endothelial dysfunction, release of endothelial progenitor cells, anti-inflammatory properties, and a number of antitumour activities. The first three are likely to be of benefit in patients with cardiac and possibly other forms of cachexia. The antitumour activities of statins, which include the induction of apoptosis and growth arrest in malignant cells, are likely to slow down the progression of cancer cachexia. While the improvement of endothelial dysfunction and a number of anti-inflammatory properties have already been proven in clinical settings, the potential anti-cancer effects of these drugs are still awaiting clinical verification.

24.1 INTRODUCTION

Statins, also known as 3-hydroxy-3-methylglutaryl-coenzyme A (HMG-CoA) reductase inhibitors, were originally designed to lower plasma cholesterol levels. Several studies have shown that statins have beneficial effects beyond mere cholesterol reduction, and that they may provide a means to treat diseases other than cardiovascular illnesses. Recent work on statins has therefore largely focused on their cholesterol-independent, so-called pleiotropic effects. Indeed, promising results have been published to suggest beneficial effects in patients with multiple sclerosis,[1,2] Alzheimer's disease,[3] osteoporosis,[4] age-related macular degeneration (with mixed results),[5,6] some types of cancer,[7] and heart failure.[8,9] Particularly cancer and heart failure are frequently accompanied by cachexia in advanced stages of the disease. It is therefore tempting to speculate that statin administration can prevent the development of cachexia and improve the clinical status of patients already diagnosed as being cachectic.

The introduction of lovastatin to the market in 1987 has not only revolutionized the treatment of hypercholesterolemia, but it also triggered an avalanche of research in recent years. This research has led to the approval of five more statins: pravastatin, simvastatin, fluvastatin, atorvastatin, and rosuvastatin (Figure 24.1). The latter was approved by the FDA in August 2003. The most recent addition to the class is pitavastatin, which is currently approved only in Japan.[10,11] The pleiotropic effects of these substances include the improvement of endothelial function, anti-inflammatory properties, the mobilization of bone marrow-derived endothelial progenitor cells, and pro-apoptotic effects. These effects seem to vary according to drug-specific properties and according to the pharmacological profile (Table 24.1). This chapter will discuss the basic mechanisms of statin action, their pharmacological profiles, and, most importantly, their pleiotropic effects that are likely to be of benefit to cachectic patients.

24.2 PHARMACOLOGIC PROFILE AND OVERVIEW

24.2.1 Statin Development

In 1971, the Japanese biochemist Akira Endo and his colleagues set about screening more than 6000 microbial strains for their ability to block cholesterol biosynthesis.[12] Their idea was that certain microorganisms would produce such compounds in order to fight back microbes that require sterols or other isoprenoids for growth. In 1973 they eventually extracted 23 mg of a substance they termed ML-236B from 600 L of culture filtrate from the mold *Penicillium citrinum*. This

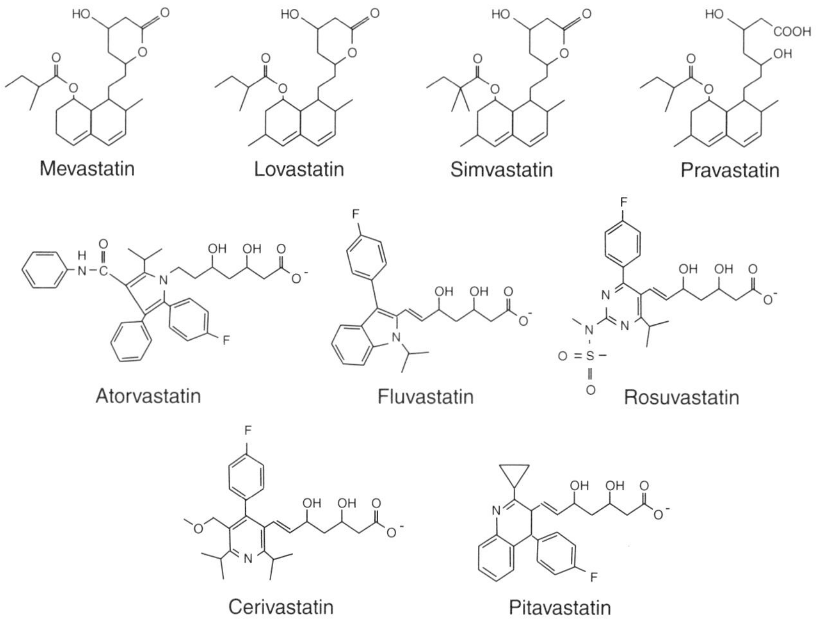

FIGURE 24.1 Chemical structures of various statins. Mevastatin is not in clinical use, but it is largely used in *in vitro* studies. Pitavastatin is currently approved only in Japan.

substance was subsequently termed mevastatin (Figure 24.1), and it was found to be highly effective at reducing plasma cholesterol levels.[13] The same compound was later independently isolated by Brown and colleagues who termed it compactin.[12] In 1980, a mevastatin analogue was isolated from *Aspergillus terreus*.[14] It was initially named mevinolin and later marketed as lovastatin.

The first patient to be treated with a statin was a 17-year old girl who suffered from homozygous familial hypercholesterolemia.[13] She had presented with a total cholesterol of 1000 mg/dL and had experienced repeated episodes of angina. Mevastatin was therefore administered at a dose of 500 mg/day, which led to a reduction in plasma cholesterol levels by approximately 20%. Unfortunately, creatine kinase and transaminases were elevated after two weeks of treatment, and the young patient developed muscular weakness.[13] Later studies in dogs revealed that mevastatin produced toxic effects.[12] Thus, the drug never reached the market. The reasons were never made entirely clear.

24.2.2 Basic Mechanisms of Action

Cholesterol synthesis is regulated by its actual levels. Indeed, an increased food intake leads to a decrease in endogenous cholesterol production. Intracellular cholesterol pools in the endoplasmic reticulum provide the main regulatory parameter by which the need for an increase or a decrease in cholesterol synthesis is sensed.

TABLE 24.1
Pharmacological Data of Various Statins

	Atorvastatin	Fluvastatin	Lovastatin	Pitavastatin	Pravastatin	Rosuvastatin	Simvastatin
Synonyms	CI-981	XU 62-320	Mevinolin, Monacolin K	NK-104, itavastatin	CS-514, SQ 31000	ZD4522	MK-733, synvinolin
FDA approval	Dec 1996	Dec 1993	Aug 1987	Not approved	Oct 1991	Aug 2003	Dec 1991
Available doses	10, 20, 40, 80	20, 40, (80ª)	10, 20, 40	Not approved	10, 20, 40, 80	5, 10, 20, 40	5, 10, 20, 40, 80
Administered form	Active form	Active form	Inactive lactone prodrug	Active form	Active form	Active form	Inactive lactone prodrug
Source	Synthetic	Synthetic	Fungal fermentation (*Aspergillus terreus*)	Synthetic	Fungal fermentation (chemical modification of lovastatin)	Synthetic	Fungal fermentation (chemical modification of lovastatin)
Solubility	Lipophilic	Ambiphilic	Lipophilic	Lipophilic	Hydrophilic	Hydrophilic	Lipophilic
Cytochrome P450 metabolism	CYP3A4	CYP2C9	CYP3A4 CYP3A5 CYP3A7	CYP2C9	None	CYP2C9 CYP2C19	CYP3A4
Metabolites	Yes (active)	Yes (mainly inactive)	Yes (active)	Negligible metabolism	Yes (mainly inactive)	Negligible metabolism	Yes (active)

Excretion	Feces: >90% Urine: <1%	Feces: 93% Urine: 6%	Feces: 83% Urine: 13%	Feces: >90% Urine: <2%	Feces: 70% Urine: 20%	Feces: 90% Urine: 10%	Feces: 60% Urine: 13%
Hepatic excretion of absorbed dose	>70%	68%	>70%	N/A	50%	90%	78–87%
Absolute bioavailability	12–14%	20–30%	<5%	60%	17%	20%	5%
Absorption rate from gastrointestinal tract	30%	98%	30%	80%	34%	50%	60–80%
Plasma protein binding	≥98%	99%	≥95%	96%	50%	90%	95%
Elimination half-life	14.0 h	0.5–2.3 h	3.0 h	11 h	1.3–2.6 h	20.8 h	1.9–3.0 h
Time to peak plasma concentration (T_{max})	2.0–4.0 h	0.5 h	2.8 h	0.5–0.8 h	0.9–1.6 h	3.0 h	1.1–3.0 h
Peak plasma concentration (C_{max})[b]	1.95–252 μg/L	269 μg/L	4.0 μg/L	26.1 μg/L	9.1–45.8 μg/L	37 μg/L	3.65–22.15 μg/L

Pitavastatin is currently approved only in Japan

[a] The 80 mg dose of fluvastatin is available as an extended release formulation only.

[b] C_{max} may vary according to the administered dose and the patient population investigated.

The regulatory protein has been termed sterol regulatory element binding protein (SREBP).[15] Its activated form acts as a transcription factor for the LDL receptor and HMG-CoA reductase. While LDL receptors scavenge circulating LDL from the plasma, HMG-CoA reductase represents the rate-limiting step in cholesterol biosynthesis in the so-called mevalonate pathway. A large part of these mechanisms has been clarified by the landmark studies of Michael S. Brown and Joseph L. Goldstein.[16, 17] In 1985, they received the Nobel Prize in Physiology or Medicine for their work.

Statin application yields two responses. By increasing the amount of HMG-CoA, the cell compensates for the inhibition of the enzyme. Thus, the direct reduction in circulating cholesterol remains small. The other response to HMG-CoA reductase inhibition is an upregulation in the number of LDL receptors on hepatocytes.[18] This leads to an increased LDL uptake and eventually to a dramatic fall in plasma cholesterol levels. These mechanisms explain why statins are ineffective in patients with homozygous familial hypercholesterolemia. These patients present an inherited recessive and ineffective allele for LDL receptor production from both parents. Thus, an increase in the number of LDL receptors, an effect statins are highly dependent on, cannot be achieved. Atorvastatin seems to be an exception to that rule.[19]

The intracellular mechanism by which statins act is the (very selective) inhibition of the rate-limiting step in cholesterol biosynthesis. The dihydroxy heptenoic acid side chain of all statins inhibits HMG-CoA reductase by binding to its active site. This enzyme catalyzes the conversion of HMG-CoA to mevalonate (Figure 24.2). Several important intermediates from this so-called mevalonate pathway supply other pathways with their substrates. One such by-product is farnesyl pyrophosphate, a precursor of not only cholesterol, but also of heme A, dolichols, and ubiquinones. Moreover, geranylgeranyl pyrophosphate is also synthetisized from farnesyl pyrophosphate (Figure 24.2). Both these substrates are important for the activation of various intracellular products, a process referred to as (iso)prenylation. Therefore, statins inhibit not only cholesterol synthesis but also a number of other intracellular pathways, including the activation of Ras, nuclear lamins, transducin γ, rhodopsin kinase, and Rho. Proteins of the Rho family (e.g., Rho, Rac1, Cdc42) are involved in the regulation of cell morphology, cell adhesion, cell motility, cell growth, and cancer cell metastasis.[20] Ras proteins, on the other hand, are associated with mitogenic signal transduction in response to growth factor stimulation.[7] Other pleiotropic effects are less well understood, and most of them are probably independent of the mevalonate pathway.

24.2.3   Statin Classification

A number of different statin classifications have been suggested. In essence, statins are currently subdivided according to their chemical structure (open-ring vs. closed-ring structure), their origin (natural vs. synthetic), and their solubility (hydrophilic vs. lipophilic) (Table 24.1). It is interesting to note that only

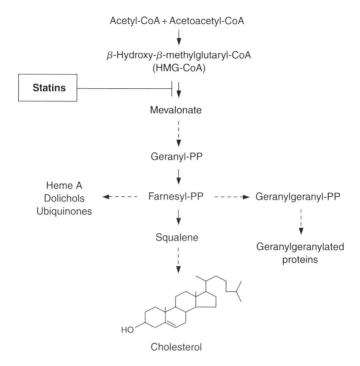

FIGURE 24.2   Pathway of cholesterol biosynthesis. Statins block the rate-limiting step in the cascade by inhibition of HMG-CoA reductase. The inhibition can be overcome by addition of mevalonate in experimental models. CoA, coenzyme A; HMG, 3-hydroxy-3-methylglutaryl; PP, pyrophosphate.

compounds with an open-ring structure inhibit HMG-CoA reductase. Thus, closed-ring statins undergo metabolic changes that eventually yield an open-ring structure. This has to be kept in mind for *in vitro* studies.

Natural statins are, like mevastatin or lovastatin, derived from fungal fermentation. Replacing the highly functionalized decalin ring of the fungal products with a variety of aromatic and heteroaromatic nuclei has yielded synthetic compounds, such as atorvastatin and fluvastatin. Lipophilic statins were originally thought to penetrate cell membranes more effectively than hydrophilic ones[21]; however, the evidence suggests otherwise.

24.2.4   Side Effects

Statins are generally considered safe and well tolerated. Indeed, a number of large-scale trials have largely dispelled doubt about the safety and tolerability of this class of drugs. Most adverse events are relatively mild and often transient, and they include gastrointestinal symptoms, headaches, and rash.[22] However, these mild side effects may occur in up to 15% of all patients. Liver transaminase elevations may also occur within the first three months of treatment. These require monitoring.

A more important and in some cases life threatening adverse reaction is myopathy. This term refers to any noninherited disorder of the skeletal muscle that causes proximal muscle weakness.[23] The clinical spectrum is broad and varies from mild muscle aches to severe pain and restriction in mobility with grossly elevated creatine kinase levels.[23] The onset of myopathy varies from a few weeks to as long as two years after the initiation of statin therapy.[24] The symptoms progress to rhabdomyolysis as long as the patient remains on statin therapy. Rhabdomyolysis describes a clinical condition that results from severe skeletal muscle injury and lysis, causing the widespread release of myoglobin with dark brown urine secondary to myoglobinuria.[22] In clinical trials the incidence of myopathy was low in the order of 0.1 to 0.2%.[23] However, the testing of creatine kinase values in the plasma may still fail at detecting statin-associated myopathy, because a few cases of myopathies with normal values of the enzyme have been reported.[25] Overall, the occurrence of fatal rhabdomyolysis has been reported at rates of approximately only 0.15 deaths per one million prescriptions for all currently available statins.[22] It has recently been speculated that these side effects might be caused by the interference of statins with isopentenylation of selenocysteine-tRNA, thus leading to a fall in available selenoproteins.[26]

Taken together, the safety profile of statins is much better with a much lower likelihood of adverse events than that of other lipid-lowering substances. However, patients receiving a combination of any statin with other lipid-lowering substances or certain other drugs (e.g., erythromycin, itraconazole, cyclosporine) should be monitored carefully. In fact, cerivastatin was withdrawn from the market in 2001 as it was implicated in a total of 52 deaths worldwide.[27] Most patients who died from severe rhabdomyolysis had received a combination of cerivastatin and gemfibrozil, a fibric acid derivative that is particularly effective in reducing total plasma triglyceride concentrations.[28] The majority of those on cerivastatin monotherapy had received the highest dose of the drug (0.8 mg/day).

Data from the early 1990s suggested that low total plasma cholesterol levels could be associated with an increased risk of cancer.[29] Moreover, experimental data suggested that statins could increase the occurrence of several types of cancer in rodents.[30]

Although the debate is far from finished, very recent work has contributed to dispelling this concern. In fact, an analysis of the Scandinavian Simvastatin Survival Study (4S) showed that there was no difference in mortality from incidence of cancer between the simvastatin and the placebo group 10 years after study termination.[31] The data are in line with earlier reports from smaller studies[32] and a recent meta-analysis.[33]

24.3 POTENTIAL PLEIOTROPIC EFFECTS ON THE CARDIOVASCULAR SYSTEM

Statin treatment consistently reduces cardiovascular risk.[34, 35] However, a reduction in recurrent coronary events was observed as early as 16 weeks after the initiation of therapy.[36] This timeframe is far too short to be ascribed to the positive effects of cholesterol reduction alone.[37] Moreover, it has been reported that statins can ameliorate morbidity and mortality in coronary artery disease (CAD) irrespective of serum cholesterol levels. These findings sparked an avalanche of research into the pleiotropic effects of these drugs.

24.3.1 Endothelial Dysfunction, Nitric Oxide, and Oxidative Stress

The endothelium is the largest autocrine, paracrine, and endocrine organ.[38] It covers approximately 700 m^2 and weighs 1.5 kg. The endothelium releases the potent vasodilator nitric oxide (NO), which increases soluble guanylate cyclase activity in smooth muscle cells, thereby inducing vasodilatation. Moreover, NO inhibits the expression of adhesion molecules both on endothelial cells and neutrophils, thus preventing the adherence of leukocytes to the vascular wall.[39] NO also has antithrombotic effects by inhibiting platelet aggregation and directly influencing the synthesis of different factors involved in the coagulation cascade.[39] NO appears to have some antiproliferative properties as well. Endothelial dysfunction, mainly characterized by lack of NO, is a common feature of chronic heart failure, hypercholesterinemia, atherosclerosis, hypertension, and certain inflammatory diseases. In fact, systemic inflammation, as reflected by elevated levels of tumor necrosis factor-α (TNFα) or C-reactive protein (CRP), may yield expression of cell-surface adhesion molecules and induction of pro-coagulant activity.[40]

NO is produced by two constitutive isoforms of nitric oxide synthase (NOS) from the amino acid L-arginine. The endothelial isoform (eNOS) is primarily located within specific microdomains of the plasma membrane, termed caveolae.[41] Caveolae express a specific marker protein that directly interacts with eNOS and thereby inhibits the activity of the enzyme.[42] Thus, a reduction in caveolin abundance would be expected to improve endothelial function.

Statins have been found to improve endothelial function by several mechanisms. A recent study by Feron and colleagues has shown that atorvastatin reduces caveolin expression in endothelial cells.[43] Treatment of such cells with increasing doses of LDL cholesterol induces caveolin expression, which leads to an impairment in NO release due to stabilization of the inhibitory caveolin/eNOS complex. Atorvastatin, despite decreasing net caveolin levels, increases the amount of unbound eNOS for activation and NO production.

However, statins do not only improve endothelial function by altering caveolin abundance. It has been demonstrated that both lovastatin and simvastatin directly induce eNOS gene transcription in human endothelial cells.[44] Additionally, it is interesting to note that pravastatin at a dose that does not reduce LDL levels improves endothelial function in monkeys.[45] While coronary arteries of pravastatin-fed monkeys dilated significantly as quantified using angiography compared to the status before treatment, those of the control group constricted.

Superoxide anion inactivates NO to a great extent. Such radicals are, for example, derived from mitochondria, immune cells, and especially from purine breakdown via xanthine oxidase. Hyperuricemia, a frequent finding in patients with chronic heart failure, reflects an upregulated xanthine oxidase activity and an overall increased production of oxygen free radicals. It is also an independent marker of a poor prognosis in this condition.[46] An increased amount of such oxygen free radicals contributes to the development of endothelial dysfunction. A diversity of antioxidant systems, such as superoxide dismutase and catalase, counteract the continuous generation of reactive oxygen species. It is therefore noteworthy that statins have been proposed to reduce oxidative stress.[47] In this study, atorvastatin reduced vascular production of reactive oxygen species to 6 ± 12% of control in spontaneously hypertensive rats.[47] The authors, however, believe that this effect might be mediated via statin-induced downregulation of vascular angiotensin receptors, because stimulation of this receptor leads to an increase of free radical production. Atorvastatin has also been shown to upregulate the expression of catalase at the mRNA and protein level in cultured rat aortic vascular smooth muscle cells.[48] In another animal model, vascular superoxide anion generation was unaffected by statin treatment, but it was increased during withdrawal.[49]

24.3.2 Endothelial Progenitor Cells

Hematopoiesis is a lifelong process responsible for replenishing both hematopoietic progenitor cells and mature blood cells from a pool of pluripotent, long-term reconstituting stem cells.[50] A number of studies have recently shown that statins are able to enhance the mobilization of bone marrow-derived endothelial progenitor cells. Mobilization of endothelial progenitor cells is a complex process that is initiated following the activation of matrix metalloproteinase-9 (MMP-9).[51] This enzyme catalyzes the transformation of membrane bound c-*kit* ligand to soluble c-*kit* ligand, which results in the detachment of early c-*kit*-positive progenitor cells from the bone marrow stromal niche and their movement to the vascular zone.[52] Endothelial progenitor cells can therefore be isolated from both bone marrow aspirate and from peripheral blood. Bone marrow-derived cells, however, represent more immature cells, because they express CD133, an early hematopoietic marker that is lost gradually in the peripheral circulation as they differentiate into mature endothelial cells.[51] Interestingly, endothelial progenitor cells have recently been found to be able to trans-differentiate into beating cardiomyocytes when cocultured with neonatal rat cardiomyocytes or when injected into post-ischemic adult mouse heart.[53]

In a recent study, wild-type mice were treated with atorvastatin at a dose of 50 mg/kg four times daily for 4 weeks after extensive anterior myocardial infarction.[54] The number of endothelial progenitor cells was elevated in all wild-type mice 4 weeks after myocardial infarction (by >100% as compared to sham animals, $p < .05$), however, atorvastatin significantly augmented the mobilization of endothelial progenitor cells (by > 200%, $p < .05$). Atorvastatin also increased the survival rate during 4 weeks after myocardial infarction in these animals (atorvastatin: 80%, placebo: 46%, $p < .01$, $n = 75$). No effect was observed in eNOS knockout mice in this study.[54] This report is in line with an earlier study which demonstrated that simvastatin augments the circulating population of endothelial progenitor cells in simvastatin-treated mice.[55] This effect was potentially mediated by Akt protein kinase in endothelial progenitor cells, which was rapidly activated following simvastatin treatment.[55]

These findings may also hold true in a clinical setting. In fact, a prospective trial in 15 patients with angiographically documented stable coronary artery disease found recently, that treatment with atorvastatin 40 mg once daily for 4 weeks increased the number of endothelial progenitor cells.[56] In this study, endothelial progenitor cells were isolated from peripheral blood and counted. Atorvastatin treatment was associated with a one and a half-fold increase in the number of circulating endothelial progenitor cells after 1 week of treatment. This was followed by a sustained three-fold increase in the levels of endothelial progenitor cells throughout the 4 week study period.[56]

It is not known if all statins share the capability of mobilizing endothelial progenitor cells. Finally, this effect might be restricted to this type of progenitor cells. Indeed, mevastatin was found to inhibit human bone marrow granulocyte progenitor cell proliferation, an effect that was entirely reversed by the addition of mevalonate.[57]

24.4 POTENTIAL PLEIOTROPIC EFFECTS ON THE IMMUNE SYSTEM

Although a final common pathway of all forms of cachexia has as yet not been established, it is thought that pro-inflammatory cytokines play a crucial role in this sense. For example, TNFα, interleukin (IL)-1, and IL-6 are known to stimulate the ubiquitin–proteasome pathway in skeletal muscle.[58–60] Among these substances, TNFα has been considerably researched during the last several years. TNFα plays a considerable role in chronic heart failure and its progression to cardiac cachexia, as this cytokine is involved in endothelial dysfunction and left ventricular impairment.[61] High plasma levels of TNFα and its soluble receptors TNFR-1 and TNFR-2 predict a poor survival.[62, 63]

Several studies have shown that statins may inhibit the production of such pro-inflammatory substances. Lovastatin, for example, was found to inhibit the induction of TNFα, IL-1, and IL-6 in certain rat cell lines.[64] In a study in 40 hypercholesterolemic patients, pravastatin 40 mg once daily led to a significant reduction in TNFα plasma levels after eight weeks of treatment as compared to placebo (pravastatin 1.10, placebo 1.33 pg/ml, $p = .032$).[65] This effect was most pronounced in the subgroup of smokers. IL-6 plasma values were left unaffected. Another study found a significant decrease in TNFα release into the supernatant of phorbol myristate acetate (PMA) stimulated human monocytes after the addition of increasing doses of pravastatin as compared to PMA treated samples ($p < .01$).[66] Pravastatin had no effect on PMA induced IL-6 production. Data from large-scale clinical studies are currently not available.

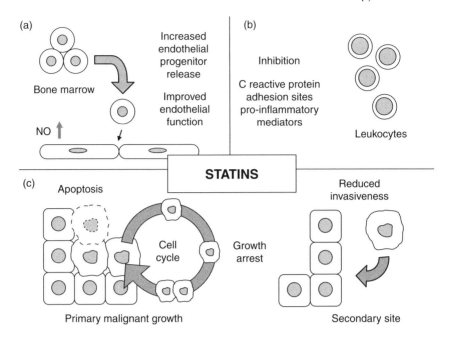

FIGURE 24.3   Overview of the principal pleiotropic effects of statins on (a) the endothelium, (b) the immune system, and (c) tumor development and metastasis. Several statins have been shown to improve endothelial function by increasing nitric oxide (NO) production and release. Statins also increase the number of endothelial progenitor cells in the bloodstream. Production of pro-inflammatory mediators (e.g., TNF-α, IL-1, IL-6) and adhesion molecule expression (e.g., ICAM-1) are all reduced with statin treatment in experimental models. The release of C-reactive protein (CRP) has been proven in several clinical settings. Antitumor activities of statins comprise the induction of apoptosis, growth arrest by inhibiting the cell cycle of malignant cells, and the reduction of the invasiveness of malignant cells at secondary sites.

C-reactive protein (CRP) levels are an independent predictor of future coronary events.[67] Its plasma values are positively correlated with weight loss, the occurrence of cachexia, and recurrence in advanced cancer.[68] Its role as a predictor of survival has been shown in multiple myeloma, melanoma, lymphoma, ovarian, renal, pancreatic, and gastrointestinal tumors.[68] Recent studies suggest that CRP is much more than a mere marker of the body's inflammatory load. In cultured human umbilical vein endothelial cells, CRP was shown to activate endothelial cells, which, in turn, express Intracellular Adhesion Molecule-1 (ICAM-1).[69] CRP also induces other adhesion molecules in endothelial cells such as vascular-cell adhesion molecule-1 (VCAM-1) and E-selectin. These molecules are involved in leukocyte-binding to the endothelial layer. CRP also activates the expression of monocyte chemotactic protein-1 (MCP-1).[70]

Several recent large-scale studies have shown that plasma CRP levels can be reduced with statin therapy. The Pravastatin Inflammation/CRP Evaluation

(PRINCE) study demonstrated a significant CRP reduction in 1182 patients with a history of myocardial infarction, stroke, or arterial revascularization procedure using pravastatin 40 mg once daily.[71] The reduction was 13.1% as compared to baseline. A reanalysis of the Pravastatin or Atorvastatin Evaluation and Infection Therapy-Thrombolysis in Myocardial Infarction 22 (PROVE IT-TIMI 22) database found recently that both drugs reduce CRP levels within 30 days of initiation significantly among 3745 patients with acute coronary syndromes.[72] Atorvastatin 80 mg once daily was more effective than pravastatin 40 mg once daily at 30 days (median CRP level: 1.6 vs. 2.3 mg/L, $p < .001$) and at 4 months (1.3 vs. 2.1 mg/L, $p < .001$). It is interesting to note that the achieved CRP and LDL levels were independent of each other.[72]

Statins also appear to reduce the number of inflammatory cells in atherosclerotic plaques. This observation is in line with reports that showed a statin-mediated reduction in the number of adhesion sites on the cell surfaces of leukocytes. A study in isolated human monocytes recently found that both pravastatin and fluvastatin inhibited the expression of ICAM-1 on these cells.[73] In this study, the release of TNFα and interferon-γ (IFN-γ) was inhibited by both statins only in the presence of IL-18 (formerly known as IFN-γ-inducing factor). Another study described a statin-mediated reduction of the expression of a counterreceptor of ICAM-1 on leukocytes.[74] This downregulation reduced their adhesion to endothelial cells.[74] These results are consistent with reports in which simvastatin pretreatment inhibited *Staphylococcus aureus*-induced leukocyte rolling and adherence in the rat mesenteric circulation as assessed using intravital microscopy.[75] Leukocyte transmigration was also significantly decreased by such treatment.

Some of these statin-mediated effects on immunological responses might be regulated by Rho proteins. These form a family of proteins that activate various kinases. Extracellular stimuli convert the inactive form of Rho (GDP-Rho) to its active form (GTP-Rho). In fact, efficient leukocyte extravasation requires Rho signaling not only within the migrating leukocytes but also within the endothelial lining of the vessel wall.[76] Moreover, Rho mediates dynamic reorganization of cytoskeletal proteins, such as stress fiber and focal adhesion formation.[77] Thus, Rho is believed to play an important role in local inflammatory responses.

24.5 Effects on Skeletal Muscle and the Proteasome

24.5.1 Statin-Mediated Effects on Skeletal Muscle

As mentioned earlier, the occurrence of myopathy is one of the few potentially serious side effects of statin treatment. However, little is known about how statins produce such muscle injury. One theory holds that blocking cholesterol synthesis reduces the cholesterol content of skeletal muscle cell membranes, which makes them unstable.[78] It has also been argued that the lack of ubiquinone, an essential compound for mitochondrial energy production, is responsible for the development of myopathy.[78] Indeed, statins block the production of farnesyl pyrophosphate, an intermediary for the production of ubiquinone (Figure 24.2). Serum levels of ubiquinone thus decrease with statin therapy.[79] On the other hand, intramuscular levels of ubiquinone are not reduced by statin treatment,[80] which argues against the hypothesis that decreases in ubiquinone metabolism cause muscle complaints with statin therapy.[78] Interestingly, the ratio of lactate to pyruvate is higher in statin-treated patients than in untreated hypercholesterolemic patients or in healthy controls ($p < .05$).[81] This finding suggests a shift toward anaerobic metabolism and possibly mitochondrial dysfunction. It remains uncertain to what extent low serum levels of ubiquinone could explain the mitochondrial dysfunction, but biopsy studies have recently confirmed mitochondrial dysfunction in patients with muscle complaints without creatine kinase activation.[82] More recent work indicates that Rho family proteins might be involved in the myotoxicity of statins. Their inactivity could be the reason for an enhancement of apoptosis in vascular smooth muscle cells that was recently reported after atorvastatin, lovastatin, and simvastatin treatment *in vitro*.[83] This effect was reversed by mevalonate, farnesyl pyrophosphate, and geranylgeranyl pyrophosphate application. These results thus document that statins may induce apoptosis, at least in vascular smooth muscle cells. However, this effect was independent of ubiquinone. The precise actions of statins on skeletal muscle still remain controversial.

24.5.2 Statin-Mediated Effects on the Proteasome

The proteasome represents the predominant pathway of protein turnover in all eukaryotic cell types.[84] It requires adenosine triphosphate. The proteasome complex forms a multisubunit protease that specifically degrades ubiquitin-labeled proteins. It is the main mediator of muscle wasting in man. Few studies have investigated the effects of statins on the proteasome. However, a recent study has challenged the view that only open-ring statins can exert pleiotropic effects. In this study, proteasome activity in cell lysates of a human breast cancer cell line was inhibited only by closed-ring lovastatin.[85] This lactone pro-drug form failed to inhibit HMG-CoA reductase. A study in terminally differentiated murine neuroblastoma cells showed that mevastatin reduced viability of these cells by inhibiting proteasome activity.[86] High-performance liquid chromatography analysis of the extract obtained from mevastatin-treated growth medium and differentiated cells revealed that differentiated cells did not convert any portion of mevastatin into an open-ring structure and accumulated only mevastatin with a closed-ring structure. These data are noteworthy, as it was commonly assumed that conversion to an open-ring structure is an important prerequisite for statin action. The addition of mevalonate failed to prevent mevastatin-induced inhibition of the proteasome, suggesting a HMG-CoA reductase independent pathway.[86] The closed-ring structure of both lovastatin and simvastatin stimulated the chymotrypsin-like activity and inhibited the peptidylglutamylpeptide hydrolyzing activity of the proteasome in purified bovine pituitary cells, whereas the open-ring structure of both drugs failed to produce such effects.[87] These effects could not be prevented by mevalonate. Importantly, statins are structurally similar to the proteasome inhibitor lactacystin. A study to investigate osteoblast proteasome activity during statin treatment recently failed to show an inhibition of the complex.[88] Taken together, it remains rather speculative if statins can inhibit proteasome activity, although some studies indicate that only the closed-ring structure of certain statins is able to do so. Furthermore, it remains a matter of speculation if these effects can be clinically implemented.

24.6 POTENTIAL ANTITUMOR PROPERTIES

Over the last several years a number of studies have suggested that statins may have potent antitumor properties. These mechanisms are not yet fully understood. Inhibition of protein prenylation has been generally accepted as a key mechanism for statins to exert their antitumor activities. However, even cholesterol reduction itself might be beneficial in the treatment of certain malignant tumors that require increased levels of cholesterol and cholesterol precursors for growth.[89] Downregulation of the levels of cholesterol by diet or drugs resulted in decreased tumor growth in some studies.[7,90] In animals injected subcutaneously with Novikoff ascites tumor cells, survival was significantly better in animals with low cholesterol plasma values than in those with higher levels.[91]

24.6.1 Growth Arrest and Apoptosis

A number of studies have shown that statins can inhibit cell growth by two main mechanisms. Indeed, statins were found to induce growth arrest predominantly in the G1 phase of the cell cycle. For example, treatment of NIH3T3 cells, a murine fibroblast cell line, with lovastatin led to a G1 arrest that was released by both mevalonate and geranylgeraniol, but not farnesol treatment.[92] This suggests that geranylgeranylated proteins are required for cells to proceed from the G1 to the S phase of the cell cycle. This finding is in line with similar studies. In prostate cancer cells from transgenic mice with adenocarcinoma of the prostate, lovastatin treatment led to actin fiber disassembly, cell rounding, growth arrest in G1, cell detachment, and apoptosis.[93] In this study, addition of geranylgerandiol yielded activation of Rho, which was followed by stress fibre assembly, flat cell morphology, and cell growth. Indeed, inactivation of Rho family proteins is known to cause cell rounding and actin filament disassembly.[94] The effects of lovastatin on murine prostate tumor cells could be prevented by the addition of geranylgeraniol.[94]

An earlier study showed that lovastatin at a concentration of 2 to 10 μM, arrested human bladder carcinoma T24 cells in the G1 phase of the cell cycle.[95] A minor fraction was also arrested in the G2 phase. Lovastatin removal led to progression of the cells through the cycle, and they entered phase S within approximately 6 h.[95] The cytostatic effects of lovastatin were, as expected, reversed by the addition of mevalonate. However, a concentration of 50 μM of lovastatin was cytotoxic.[95] The studies available to date indicate that growth arrest and apoptosis occur *in vitro* at lovastatin concentrations ranging from 0.1 to 100 μM, depending on the cell line investigated.[7]

In addition to interference with the cell cycle, the induction of apoptosis might be another important mechanism by which statins exert antitumor activity. An *in vitro* study in malignant mesothelioma cells found that lovastatin at doses between 5 and 30 μM decreased cell viability in a dose-dependent fashion.[96] Morphological changes, histological evidence of nuclear condensation and degradation, and flow-cytometric analysis of DNA content led these authors to the conclusion that lovastatin induced apoptosis in these cells.[96] These findings are supported by a study in the prostate cancer cell line LNCaP in which lovastatin at a dose of 30 μM effectively induced apoptosis.[97] The effect was prevented by simultaneous addition of 300 μM mevalonate.[97] In this context, it is interesting to note that 1 μM lovastatin induced apoptosis in a primitive neuroectodermal tumor cell line (Ewing's sarcoma cell line CHP-100) but failed to induce growth arrest of the cell cycle.[98] It remains unknown if these features are specific for lovastatin or if antitumor features are a class effect of the statins. It might well be that certain statins are more potent than others at inducing antitumor activity. One study demonstrated that cerivastatin is at least ten times more potent than fluvastatin and atorvastatin at triggering tumor-specific apoptosis in various acute myeloid leukemia cell lines.[99] These cells were only weakly sensitive to lovastatin.[99] Interestingly, the addition of 1.5% serum from fluvastatin-treated patients to human vascular smooth muscle cells *in vitro* had an antiproliferative effect.[100] A proapoptotic effect was achieved only at higher serum concentrations.[100]

Few clinical trials to study the effects of lovastatin in cancer patients are available so far. One study investigated lovastatin administration in 88 patients with solid tumors.[101] In this study, doses ranging from 2 to 45 mg/kg of body weight per day during monthly seven-day courses were generally well tolerated and yielded lovastatin plasma concentrations ranging from 0.1 to 3.9 μM.[101] Myopathy was the dose-limiting factor. Its occurrence could be prevented by ubiquinone supplementation.[101] These data suggest that the achievable plasma concentrations of lovastatin are sufficient to treat at least some tumors *in vivo*.[7] Therefore, a phase II study was conducted in 16 patients with advanced measurable gastric adenocarcinoma.[102] High doses of lovastatin (35 mg/kg per day during monthly seven-day cycles for 1 to 4 months) were administered together with ubiquinone to prevent myopathy.[102] Unfortunately, no patient responded. The only available randomized, placebo controlled trial to investigate the impact of pravastatin treatment in 91 patients with unresectable hepatocellular carcinoma yielded promising results.[103] All patients received standard treatment. Patients on additional treatment with 40 mg of pravastatin once daily ($n = 41$) had a median survival of 18 months, compared to 9 months in the placebo group ($n = 42$, survival $p = .006$). In summary, it appears that these results will encourage future randomized trials of statins in patients with cancer.

24.6.2 Anti-Metastatic Properties

The metastatic process includes cell detachment from the primary tumor, their spreading through the circulation, and finally their growth at a secondary site. A number of different mechanisms support the metastatic process. These include invasion of the basement membrane, cell migration, angiogenesis, and attachment to and detachment from the extracellular matrix. Statins have been shown to interfere with this process by various mechanisms. Lovastatin, cerivastatin, and simvastatin were found to reduce the expression of MMP-9.[104–106] This might be part of the explanation for why statins have been found to reduce the invasiveness of several cancer cell lines.[7] Lovastatin, for example, reduced the invasiveness of lymphoma cells,[107] human glioma cells,[108] and melanoma cells.[109] Fluvastatin was effective at blocking the invasiveness of pancreatic cancer cells,[110] colon and breast cancer cells.[7] Some of these studies also describe a reduced ability of the investigated cells to attach to the extracellular matrix after statin treatment.[107, 109]

These results were partly confirmed by *in vivo* studies. In a study in female BALB/c mice, the animals were inoculated subcutaneously with F3II cells, a highly invasive and metastatic murine tumor cell line.[111] The animals were then treated intraperitoneally with 10 mg/kg of lovastatin or administered orally at a level corresponding to the human dosage of 1–2 mg/kg/day. Lovastatin treatment was associated with significantly prolonged tumor latency and reduced metastatic dissemination to the lungs from established mammary tumors.[111] In a study in mice injected with pancreatic tumor cells, treatment with fluvastatin resulted in a marked reduction in the number and size of metastatic nodules in the liver.[112] In fact, the median number of liver nodules after 44 days of fluvastatin treatment (0.6 mg/kg/day) was 6 (range: 0–14) as compared to the control group with 25 (16–35) nodules ($p < .01$).[112]

24.6.3 Angiogenesis

The effects of different statins on angiogenesis are somewhat confusing and appear to be dose-related. Using cerivastatin and atorvastatin *in vitro*, a recent study found that endothelial cell proliferation, migration, and differentiation were enhanced at low statin concentrations (0.005–0.01 μM), but inhibited at high concentrations (0.05–1 μM).[113] The latter concentrations were associated with decreased endothelial release of vascular endothelial growth factor and increased endothelial apoptosis. The same study described that inflammation-induced angiogenesis in mice was enhanced with low-dose statin therapy (0.5 mg/kg/day), but inhibited with high concentrations of both cerivastatin and atorvastatin (2.5 mg/kg/day).[113] A study in ischemic limbs of normocholesterolemic rabbits found an increase in revascularization with simvastatin treatment.[114] Based on these findings, a number of authors concluded that long-term treatment with simvastatin and other statins may promote vascularization of developing tumors.[115, 116] Various other reports, on the other hand, have advocated antiangiogenic effects of statins. Cerivastatin, for example, was shown to decrease endothelial cell locomotion, which was mainly related to delocalization of RhoA from the cell membrane to the cytoplasm.[117] The net effect was an inhibition of angiogenesis observed in this study. A study of capillary growth in both vascular endothelial growth factor-stimulated chick chorioallantoic membranes and basic fibroblast growth factor-stimulated mouse corneas had a similar outcome.[118] In fact, simvastatin inhibited capillary growth in both these systems. Since simvastatin was additionally found to inhibit membrane localization of RhoA in a concentration-dependent fashion similar to its effect on tube formation, it is tempting to speculate that the effect is RhoA-mediated.[118]

24.7 CONCLUSION

Statins have revolutionized the treatment of hypercholesterolemia. They have been widely hailed as the aspirin of the new millennium. Their efficacy in the treatment of various cardiovascular diseases is beyond doubt, although their usefulness in the treatment of chronic heart failure has not been shown in prospective trials as yet. Some studies, such as the Controlled Rosuvastatin Multinational Trial in Heart Failure (CORONA) and the GISSI-HF trial (Gruppo Italiano per lo Studio della Sopravvivenza nell'Insufficienza Cardiaca) are underway.[119] CORONA aims at studying rosuvastatin (10 mg once daily) or placebo in more than 4500 patients with chronic heart failure of ischemic origin. All patients will be followed-up for approximately 3 years, and cardiovascular death, nonfatal myocardial infarction, and nonfatal stroke are the primary end-points. In GISSI-HF, a prospective, multicenter, randomized, double blind study, patients will be randomized in two steps to (i) n-3 PUFA (1 g once daily) or placebo and (ii) rosuvastatin (10 mg once daily) or placebo. No study has as yet specifically investigated cachectic patients. This is also true for cancer patients, although, as discussed above, a randomized, placebo controlled trial to investigate the impact of pravastatin treatment in 91 patients with unresectable hepatocellular carcinoma has yielded promising results.

The available data indicate that statins may improve a number of perturbations associated with the cachectic syndrome. Cachectic patients may benefit from improving endothelial function per se. Moreover, the inhibition of proinflammatory mediators is likely to slowdown the progression of the disease and possibly improve the clinical status of the patients. The effects of statins on proteasome activity remain speculative. We are far from having established the right dose, the right statin, or the right patient who may benefit from such treatment. The only study in an animal model of cancer cachexia, however, ended in disappointment.[120] In essence, simvastatin (20 mg/kg/day) negatively affected the wasting pattern induced by Yoshida AH-130 ascites hepatoma in rats. The organ weights of liver, spleen, heart, gastrocnemius, and soleus decreased by 4 days of simvastatin treatment. The administration of simvastatin to control animals (without hepatoma) yielded no effects with the exception of a significant reduction in heart weight (controls without simvastatin: 501 ± 58 mg, controls with simvastatin: 426 ± 65 mg, $p < .05$). These findings once again underscore the fact that we do not know if pleiotropic effects are a class effect or if they are specific to certain substances. Moreover, we do not know if it is possible to achieve pleiotropic effects without lowering plasma cholesterol concentrations, which per se is a potentially harmful effect. This might be possible with very low statin doses, but it remains to be established. This chapter of cachexia research is far from finished.

REFERENCES

1. Sena A., Pedrosa R., and Graca Morais M. Therapeutic potential of lovastatin in multiple sclerosis. *J. Neurol.* 2003; 250: 754–755.
2. Vollmer T., Key L., Durkalski V., Tyor W., Corboy J., Markovic-Plese S., Preiningerova J., Rizzo M., and Singh I. Oral simvastatin treatment in relapsing-remitting multiple sclerosis. *Lancet* 2004; 363: 1607–1608.
3. Jick H., Zornberg G.L., Jick S.S., Seshadri S., and Drachman D.A. Statins and the risk of dementia. *Lancet* 2000; 356: 1627–1631.
4. Pasco J.A., Kotowicz M.A., Henry M.J., Sanders K.M., and Nicholson G.C. Statin use, bone mineral density, and fracture risk: Geelong Osteoporosis Study. *Arch. Intern. Med.* 2002; 162: 537–540.
5. McGwin G. Jr., Owsley C., Curcio C.A., and Crain R.J. The association between statin use and age related maculopathy. *Br. J. Ophthalmol.* 2003; 87: 1121–1125.
6. Klein R., Klein B.E., Tomany S.C., Danforth L.G., and Cruickshanks K.J. Relation of statin use to the 5-year incidence and progression of age-related maculopathy. *Arch. Ophthalmol.* 2003; 121: 1151–1155.
7. Graaf M.R., Richel D.J., van Noorden C.J., and Guchelaar H.J. Effects of statins and farnesyltransferase inhibitors on the development and progression of cancer. *Cancer Treat. Rev.* 2004; 30: 609–641.
8. Horwich T.B., MacLellan W.R., and Fonarow G.C. Statin therapy is associated with improved survival in ischemic and non-ischemic heart failure. *J. Am. Coll. Cardiol.* 2004; 43: 642–648.
9. Anker S.D., Clark A.L., Kilkowski C., Zugck C., Ponikowski P., Davos C.H., Banasiak W., Zardini P., Haass M., Senges J., and Coats A.J. Statins and survival in 2068 CHF patients with ischemic and non-ischemic etiology. *Circulation* 2002; 106 Suppl II: 2535 (Abstract).
10. Mukhtar R.Y., Reid J., and Reckless J.P. Pitavastatin. *Int. J. Clin. Pract.* 2005; 59: 239–252.
11. von Haehling S., Okonko D.O., and Anker S.D. Statins: a treatment option for chronic heart failure? *Heart Fail. Monit.* 2004; 4: 90–97.
12. Endo A. The discovery and development of HMG-CoA reductase inhibitors. *J. Lipid Res.* 1992; 33: 1569–1582.
13. Yamamoto A., Sudo H., and Endo A. Therapeutic effects of ML-236B in primary hypercholesterolemia. *Atherosclerosis* 1980; 35: 259–266.
14. Alberts A.W., Chen J., Kuron G., Hunt V., Huff J., Hoffman C., Rothrock J., Lopez M., Joshua H., Harris E., Patchett A., Monaghan R., Currie S, Stapley E., Albers-Schonberg G., Hensens O., Hirshfield J., Hoogsteen K., Liesch J., and Springer J. Mevinolin: a highly potent competitive inhibitor of hydroxymethylglutaryl-coenzyme A reductase and a cholesterol-lowering agent. *Proc. Natl. Acad. Sci. USA* 1980; 77: 3957–3961.
15. Rawson R.B. The SREBP pathway – insights from Insigs and insects. *Nat. Rev. Mol. Cell. Biol.* 2003; 4: 631–640.
16. Brown M.S. and Goldstein J.L. A receptor-mediated pathway for cholesterol homeostasis. *Science* 1986; 232: 34–47.
17. Goldstein J.L. and Brown M.S. Regulation of the mevalonate pathway. *Nature* 1990; 343: 425–430.
18. Page C.P., Curtis M.J., Sutter M.C., Walker M.J., and Hoffman B.B. Integrated Pharmacology. *Mosby* 1997: 267–270.

19. Yamamoto A., Harada-Shiba M., Kawaguchi A., Oi K., Kubo H., Sakai S., Mikami Y., Imai T., Ito T., Kato H., Endo M., Sato I., Suzuki Y., and Hori H. The effect of atorvastatin on serum lipids and lipoproteins in patients with homozygous familial hypercholesterolemia undergoing LDL-apheresis therapy. *Atherosclerosis* 2000; 153: 89–98.

20. Aznar S. and Lacal J.C. Rho signals to cell growth and apoptosis. *Cancer Lett.* 2001; 165: 1–10.

21. Liao J.K. Isoprenoids as mediators of the biological effects of statins. *J. Clin. Invest.* 2002; 110: 285–288.

22. Bellosta S., Paoletti R., and Corsini A. Safety of statins: focus on clinical pharmacokinetics and drug interactions. *Circulation* 2004; 109: III50–III57.

23. Hamilton-Craig I. Statin-associated myopathy. *Med. J. Aust.* 2001; 175: 486–489.

24. Tobert J.A. Efficacy and long-term adverse effect pattern of lovastatin. *Am. J. Cardiol.* 1988; 62: 28J–34J.

25. Phillips P.S., Haas R.H., Bannykh S., Hathaway S., Gray N.L., Kimura B.J., Vladutiu G.D., and England J.D. Statin-associated myopathy with normal creatine kinase levels. *Ann. Intern. Med.* 2002; 137: 581–585.

26. Moosmann B. and Behl C. Selenoprotein synthesis and side-effects of statins. *Lancet* 2004; 363: 892–894.

27. Rosenson R.S. Current overview of statin-induced myopathy. *Am. J. Med.* 2004; 116: 408–416.

28. Todd P.A. and Ward A. Gemfibrozil. A review of its pharmacodynamic and pharmacokinetic properties, and therapeutic use in dyslipidaemia. *Drugs* 1988; 36: 314–339.

29. Jacobs D., Blackburn H., Higgins M., Reed D., Iso H., McMillan G., Neaton J., Nelson J., Potter J., Rifkind B., et al. Report of the Conference on Low Blood Cholesterol: Mortality Associations. *Circulation* 1992; 86: 1046–1060.

30. Newman T.B. and Hulley S.B. Carcinogenicity of lipid-lowering drugs. *JAMA* 1996; 275: 55–60.

31. Strandberg T.E., Pyorala K., Cook T.J., Wilhelmsen L., Faergeman O., Thorgeirsson G., Pedersen T.R., and Kjekshus J. for the 4S Group. Mortality and incidence of cancer during 10-year follow-up of the Scandinavian Simvastatin Survival Study (4S). *Lancet* 2004; 364: 771–777.

32. Kaye J.A. and Jick H. Statin use and cancer risk in the General Practice Research Database. *Br. J. Cancer* 2004; 90: 635–637.

33. Bjerre L.M. and LeLorier J. Do statins cause cancer? A meta-analysis of large randomized clinical trials. *Am. J. Med.* 2001; 110: 716–723.

34. Shepherd J., Cobbe S.M., Ford I., Isles C.G., Lorimer A.R., MacFarlane P.W., McKillop J.H., and Packard C.J. for the West of Scotland Coronary Prevention Study Group. Prevention of coronary heart disease with pravastatin in men with hypercholesterolemia. *N. Engl. J. Med.* 1995; 333: 1301–1307.

35. The Long-Term Intervention with Pravastatin in Ischaemic Disease (LIPID) Study Group. Prevention of cardiovascular events and death with pravastatin in patients with coronary heart disease and a broad range of initial cholesterol levels. *N. Engl. J. Med.* 1998; 339: 1349–1357.

36. Schwartz G.G., Olsson A.G., Ezekowitz M.D., Ganz P., Oliver M.F., Waters D., Zeiher A., Chaitman B.R., Leslie S., and Stern T. Effects of atorvastatin on early recurrent ischemic events in acute coronary syndromes: the MIRACL study: a randomised controlled trial. *JAMA* 2001; 285: 1711–1718.

37. Vaughan C.J., Gotto A.M., and Basson C.T. The evolving role of statins in the management of atherosclerosis. *J. Am. Coll. Cardiol.* 2000; 35: 1–10.

38. Vogel R.A. Cholesterol lowering and endothelial dysfunction. *Am. J. Med.* 1999; 107: 479–487.

39. Tiefenbacher C.P. and Kreuzer J. Nitric oxide-mediated endothelial dysfunction — is there need to treat? *Curr. Vasc. Pharmacol.* 2003; 1: 123–133.

40. Vallance P., Collier J., and Bhagat K. Infection, inflammation, and infarction: does acute endothelial dysfunction provide a link? *Lancet* 1997; 349: 1391–1392.

41. Hecker M., Mulsch A., Bassenge E., Forstermann U., and Busse R. Subcellular localization and characterization of nitric oxide synthase(s) in endothelial cells: physiological implications. *Biochem. J.* 1994; 299: 247–252.

42. von Haehling S., Anker S.D., and Bassenge E. Statins and the role of nitric oxide in chronic heart failure. *Heart Fail. Rev.* 2003; 8: 99–106.

43. Feron O., Dessy C., Desager J.P., and Balligand J.L. Hydroxy-methylglutaryl-coenzyme A reductase inhibition promotes endothelial nitric oxide synthase activation through a decrease in caveolin abundance. *Circulation* 2001; 103: 113–118.

44. Laufs U., Fata V.L., Plutzky J., and Liao J.K. Upregulation of endothelial nitric oxide synthase by HMG CoA reductase inhibitors. *Circulation* 1998; 97: 1129–1135.

45. Williams J.K., Sukhova G.K., Herrington D.M., and Libby P. Pravastatin has cholesterol-lowering independent effects on the artery wall of atherosclerotic monkeys. *J. Am. Coll. Cardiol.* 1998; 31: 684–691.

46. Anker S.D., Doehner W., Rauchhaus M., Sharma R., Francis D., Knosalla C., Davos C.H., Cicoira M., Shamim W., Kemp M., Segal R., Osterziel K.J., Leyva F., Hetzer R., Ponikowski P., and Coats A.J. Uric acid and survival in chronic heart failure: validation and application in metabolic, functional, and hemodynamic staging. *Circulation* 2003; 107: 1991–1997.

47. Wassmann S., Laufs U., Baumer A.T., Muller K., Ahlbory K., Linz W., Itter G., Rosen R., Bohm M., and Nickenig G. HMG-CoA reductase inhibitors improve endothelial dysfunction in normocholesterolemic hypertension via reduced production of reactive oxygen species. *Hypertension* 2001; 37: 1450–1457.

48. Wassmann S., Laufs U., Muller K., Konkol C., Ahlbory K., Baumer A.T., Linz W., Bohm M., and Nickenig G. Cellular antioxidant effects of atorvastatin *in vitro* and *in vivo*. *Arterioscler. Thromb. Vasc. Biol.* 2002; 22: 300–305.

49. Vecchione C. and Brandes R.P. Withdrawal of 3-hydroxy-3-methylglutaryl coenzyme A reductase inhibitors elicits oxidative stress and induces endothelial dysfunction in mice. *Circ. Res.* 2002; 91: 173–179.

50. Lyman S.D. and Jacobsen S.E. c-kit ligand and Flt3 ligand: stem/progenitor cell factors with overlapping yet distinct activities. *Blood* 1998; 91: 1101–1134.

51. Hristov M. and Weber C. Endothelial progenitor cells: characterization, pathophysiology, and possible clinical relevance. *J. Cell. Mol. Med.* 2004; 8: 498–508.

52. Heissig B., Hattori K., Dias S., Friedrich M., Ferris B., Hackett N.R., Crystal R.G., Besmer P., Lyden D., Moore M.A., Werb Z., and Rafii S. Recruitment of stem and progenitor cells from the bone marrow niche requires MMP-9 mediated release of kit-ligand. *Cell* 2002; 109: 625–637.

53. Condorelli G., Borello U., De Angelis L., Latronico M., Sirabella D., Coletta M., Galli R., Balconi G., Follenzi A., Frati G, Cusella De Angelis M.G., Gioglio L.,

Amuchastegui S., Adorini L., Naldini L., Vescovi A., Dejana E., and Cossu G. Cardiomyocytes induce endothelial cells to trans-differentiate into cardiac muscle: implications for myocardium regeneration. *Proc. Natl. Acad. Sci. USA* 2001; 98: 10733–10738.

54. Landmesser U., Engberding N., Bahlmann F.H., Schaefer A., Wiencke A., Heineke A., Spiekermann S., Hilfiker-Kleiner D., Templin C., Kotlarz D., Mueller M., Fuchs M., Hornig B., Haller H., and Drexler H. Statin-induced improvement of endothelial progenitor cell mobilization, myocardial neovascularization, left ventricular function, and survival after experimental myocardial infarction requires endothelial nitric oxide synthase. *Circulation* 2004; 110: 1933–1939.

55. Llevadot J., Murasawa S., Kureishi Y., Uchida S., Masuda H., Kawamoto A., Walsh K., Isner J.M., and Asahara T. HMG-CoA reductase inhibitor mobilizes bone marrow-derived endothelial progenitor cells. *J. Clin. Invest.* 2001; 108: 399–405.

56. Vasa M., Fichtlscherer S., Adler K., Aicher A., Martin H., Zeiher A.M., and Dimmeler S. Increase in circulating endothelial progenitor cells by statin therapy in patients with stable coronary artery disease. *Circulation* 2001; 103: 2885–2890.

57. Hoffman P.C., Richman C.M., Larson R.A., and Yachnin S. Cholesterol and mevalonic acid are independent requirements for the *in vitro* proliferation of human bone marrow granulocyte progenitor cells: studies using ML-236B. *Blood* 1983; 61: 667–671.

58. Zamir O., Hasselgren P.O., Kunkel S.L., Frederick J., Higashiguchi T., and Fischer J.E. Evidence that tumor necrosis factor participates in the regulation of muscle proteolysis during sepsis. *Arch. Surg.* 1992; 127: 170–174.

59. Zamir O., Hasselgren P.O., von Allmen D., and Fischer J.E. *In vivo* administration of interleukin-1 alpha induces muscle proteolysis in normal and adrenalectomized rats. *Metabolism* 1993; 42: 204–208.

60. Goodman M.N. Interleukin-6 induces skeletal muscle protein breakdown in rats. *Proc. Soc. Exp. Biol. Med.* 1994; 205: 182–185.

61. von Haehling S., Jankowska E.A., and Anker S.D. Tumour necrosis factor-alpha and the failing heart: pathophysiology and therapeutic implications. *Basic Res. Cardiol.* 2004; 99: 18–28.

62. Deswal A., Petersen N.J., Feldman A.M., Young J.B., White B.G., and Mann D.L. Cytokines and cytokine receptors in advanced heart failure: an analysis of the cytokine database from the Vesnarinone trial (VEST). *Circulation* 2001; 103: 2055–2059.

63. Rauchhaus M., Doehner W., Francis D.P., Davos C., Kemp M., Liebenthal C., Niebauer J., Hooper J., Volk H.D., Coats A.J., and Anker S.D. Plasma cytokine parameters and mortality in patients with chronic heart failure. *Circulation* 2000; 19: 3060–3067.

64. Pahan K., Sheikh F.G., Namboodiri A.M., and Singh I. Lovastatin and phenyl-acetate inhibit the induction of nitric oxide synthase and cytokines in rat primary astrocytes, microglia, and macrophages. *J. Clin. Invest.* 1997; 100: 2671–2679.

65. Solheim S., Seljeflot I., Arnesen H., Eritsland J., and Eikvar L. Reduced levels of TNF alpha in hypercholesterolemic individuals after treatment with pravastatin for 8 weeks. *Atherosclerosis* 2001; 157: 411–415.

66. Grip O., Janciauskiene S., and Lindgren S. Pravastatin down-regulates inflammatory mediators in human monocytes *in vitro. Eur. J. Pharmacol.* 2000; 410: 83–92.
67. Ridker P.M., Cushman M., Stampfer M.J., Tracy R., and Hennekens C.H. Inflammation, aspirin, and the risk of cardiovascular disease in apparently healthy men. *N. Engl. J. Med.* 1997; 336: 973–979.
68. Mahmoud F.A. and Rivera N.I. The role of C-reactive protein as a prognostic indicator in advanced cancer. *Curr. Oncol. Rep.* 2002; 4: 250–255.
69. Pasceri V., Willerson J.T., and Yeh E.T.H. Direct proinflammatory effect of C-reactive protein on human endothelial cells. *Circulation* 2000; 102: 2165–2168.
70. Pasceri V., Chang J.S., Willerson J.T., et al. Modulation of C-reactive protein-mediated monocyte chemoattractant protein-1 induction in human endothelial cells by anti-atherosclerosis drugs. *Circulation* 2001; 103: 2531–2534.
71. Albert M.A., Danielson E., Rifai N., and Ridker P.M. PRINCE Investigators. Effect of statin therapy on C-reactive protein levels: the pravastatin inflammation/CRP evaluation (PRINCE): a randomized trial and cohort study. *JAMA* 2001; 286: 64–70.
72. Ridker P.M., Cannon C.P., Morrow D., Rifai N., Rose L.M., McCabe C.H., Pfeffer M.A., and Braunwald E. on behalf of the Pravastatin or Atorvastatin Evaluation and Infection Therapy-Thrombolysis in Myocardial Infarction 22 (PROVE IT-TIMI 22) Investigators. C-reactive protein levels and outcomes after statin therapy. *N. Engl. J. Med.* 2005; 352: 20–28.
73. Takahashi H.K., Mori S., Iwagaki H., Yoshino T., Tanaka N., Weitz-Schmidt G., and Nishibori M. Differential effect of LFA703, pravastatin, and fluvastatin on production of IL-18 and expression of ICAM-1 and CD40 in human monocytes. *J. Leukoc. Biol.* 2005; 77: 400–407.
74. Weitz-Schmidt G., Welzenbach K., Brinkmann V., Kamata T., Kallen J., Bruns C., Cottens S., Takada Y., and Hommel U. Statins selectively inhibit leukocyte function antigen-1 by binding to a novel regulatory integrin site. *Nat. Med.* 2001; 7: 687–692.
75. Pruefer D., Makowski J., Schnell M., Buerke U., Dahm M., Oelert H., Sibelius U., Grandel U., Grimminger F., Seeger W., Meyer J., Darius H., and Buerke M. Simvastatin inhibits inflammatory properties of *Staphylococcus aureus* alpha-toxin. *Circulation* 2002; 106: 2104–2110.
76. Strey A., Janning A., Barth H., and Gerke V. Endothelial Rho signaling is required for monocyte transendothelial migration. *FEBS Lett.* 2002; 517: 261–266.
77. Amano M., Fukata Y., and Kaibuchi K. Regulation and functions of Rho-associated kinase. *Exp. Cell Res.* 2000; 261: 44–51.
78. Thompson P.D., Clarkson P., and Karas R.H. Statin-associated myopathy. *JAMA* 2003; 289: 1681–1690.
79. Ghirlanda G., Oradei A., Manto A., Lippa S., Uccioli L., Caputo S., Greco A.V., and Littarru G.P. Evidence of plasma CoQ10-lowering effect by HMG-CoA reductase inhibitors: a double-blind, placebo-controlled study. *J. Clin. Pharmacol.* 1993; 33: 226–229.
80. Laaksonen R., Jokelainen K., Sahi T., Tikkanen M.J., and Himberg J.J. Decreases in serum ubiquinone concentrations do not result in reduced levels in muscle tissue during short-term simvastatin treatment in humans. *Clin. Pharmacol. Ther.* 1995; 57: 62–66.
81. De Pinieux G., Chariot P., Ammi-Said M., Louarn F., Lejonc J.L., Astier A., Jacotot B., and Gherardi R. Lipid-lowering drugs and mitochondrial

function: effects of HMG-CoA reductase inhibitors on serum ubiquinone and blood lactate/pyruvate ratio. *Br. J. Clin. Pharmacol.* 1996; 42: 333–337.

82. Phillips P.S., Haas R.H., Bannykh S., Hathaway S., Gray N.L., Kimura B.J., Vladutiu G.D., and England J.D. Scripps Mercy Clinical Research Center. Statin-associated myopathy with normal creatine kinase levels. *Ann. Intern. Med.* 2002; 137: 581–585.

83. Guijarro C., Blanco-Colio L.M., Ortego M., Alonso C., Ortiz A., Plaza J.J., Diaz C., Hernandez G., and Egido J. 3-Hydroxy-3-methylglutaryl coenzyme a reductase and isoprenylation inhibitors induce apoptosis of vascular smooth muscle cells in culture. *Circ. Res.* 1998; 83: 490–500.

84. von Haehling S., Genth-Zotz S., Anker S.D., and Volk H.D. Cachexia: a therapeutic approach beyond cytokine antagonism. *Int. J. Cardiol.* 2002; 85: 173–183.

85. Rao S., Porter D.C., Chen X., Herliczek T., Lowe M., and Keyomarsi K. Lovastatin-mediated G1 arrest is through inhibition of the proteasome, independent of hydroxymethyl glutaryl-CoA reductase. *Proc. Natl. Acad. Sci. USA* 1999; 96: 7797–7802.

86. Kumar B., Andreatta C., Koustas W.T., Cole W.C., Edwards-Prasad J., and Prasad K.N. Mevastatin induces degeneration and decreases viability of cAMP-induced differentiated neuroblastoma cells in culture by inhibiting proteasome activity, and mevalonic acid lactone prevents these effects. *J. Neurosci. Res.* 2002; 68: 627–635.

87. Wojcik C., Bury M., Stoklosa T., Giermasz A., Feleszko W., Mlynarczuk I., Pleban E., Basak G., Omura S., and Jakobisiak M. Lovastatin and simvastatin are modulators of the proteasome. *Int. J. Biochem. Cell Biol.* 2000; 32: 957–965.

88. Murray S.S., Tu K.N., Young K.L., and Murray E.J. The effects of lovastatin on proteasome activities in highly purified rabbit 20 S proteasome preparations and mouse MC3T3-E1 osteoblastic cells. *Metabolism* 2002; 51: 1153–1160.

89. Buchwald H. Cholesterol inhibition, cancer, and chemotherapy. *Lancet* 1992; 339: 1154–1156.

90. Littman M.L., Taguchi T., and Mosbach E.H. Effect of cholesterol-free, fat-free diet and hypocholesteremic agents on growth of transplantable animal tumors. *Cancer Chemother. Rep.* 1966; 50: 25–45.

91. Schneider P.D., Chan E.K., Guzman I.J., Rucker R.D., Varco R.L., and Buchwald H. Retarding Novikoff tumor growth by altering host rat cholesterol metabolism. *Surgery* 1980; 87: 409–416.

92. Vogt A., Qian Y., McGuire T.F., Hamilton A.D., and Sebti S.M. Protein geranylgeranylation, not farnesylation, is required for the G1 to S phase transition in mouse fibroblasts. *Oncogene* 1996; 13: 1991–1999.

93. Ghosh P.M., Ghosh-Choudhury N., Moyer M.L., Mott G.E., Thomas C.A., Foster B.A., Greenberg N.M., and Kreisberg J.I. Role of RhoA activation in the growth and morphology of a murine prostate tumor cell line. *Oncogene* 1999; 18: 4120–4130.

94. Fenton R.G., Kung H.F., Longo D.L., and Smith M.R. Regulation of intracellular actin polymerization by prenylated cellular proteins. *J. Cell. Biol.* 1992; 117: 347–356.

95. Jakobisiak M., Bruno S., Skierski J.S., and Darzynkiewicz Z. Cell cycle-specific effects of lovastatin. *Proc. Natl. Acad. Sci. USA* 1991; 88: 3628–3632.

96. Rubins J.B., Greatens T., Kratzke R.A., Tan A.T., Polunovsky V.A., and Bitterman P. Lovastatin induces apoptosis in malignant mesothelioma cells. *Am. J. Respir. Crit. Care Med.* 1998; 157: 1616–1622.

97. Marcelli M., Cunningham G.R., Haidacher S.J., Padayatty S.J., Sturgis L., Kagan C., and Denner L. Caspase-7 is activated during lovastatin-induced apoptosis of the prostate cancer cell line LNCaP. *Cancer Res.* 1998; 58: 76–83.

98. Kim J.S., Pirnia F., Choi Y.H., Nguyen P.M., Knepper B., Tsokos M., Schulte T.W., Birrer M.J., Blagosklonny M.V., Schaefer O., Mushinski J.F., and Trepel J.B. Lovastatin induces apoptosis in a primitive neuroectodermal tumor cell line in association with RB down-regulation and loss of the G1 checkpoint. *Oncogene* 2000; 19: 6082–6090.

99. Wong W.W., Tan M.M., Xia Z., Dimitroulakos J., Minden M.D., and Penn L.Z. Cerivastatin triggers tumor-specific apoptosis with higher efficacy than lovastatin. *Clin. Cancer Res.* 2001; 7: 2067–2075.

100. Buemi M., Allegra A., Senatore M., Marino D., Medici M.A., Aloisi C., Di Pasquale G., and Corica F. Pro-apoptotic effect of fluvastatin on human smooth muscle cells. *Eur. J. Pharmacol.* 1999; 370: 201–203.

101. Thibault A., Samid D., Tompkins A.C., Figg W.D., Cooper M.R., Hohl R.J., Trepel J., Liang B., Patronas N., Venzon D.J., Reed E., and Myers C.E. Phase I study of lovastatin, an inhibitor of the mevalonate pathway, in patients with cancer. *Clin. Cancer Res.* 1996; 2: 483–491.

102. Kim W.S., Kim M.M., Choi H.J., Yoon S.S., Lee M.H., Park K., Park C.H., and Kang W.K. Phase II study of high-dose lovastatin in patients with advanced gastric adenocarcinoma. *Invest. New Drugs* 2001; 19: 81–83.

103. Kawata S., Yamasaki E., Nagase T., Inui Y., Ito N., Matsuda Y., Inada M., Tamura S., Noda S., Imai Y., and Matsuzawa Y. Effect of pravastatin on survival in patients with advanced hepatocellular carcinoma. A randomized controlled trial. *Br. J. Cancer* 2001; 84: 886–891.

104. Wang I.K., Lin-Shiau S.Y., and Lin J.K. Suppression of invasion and MMP-9 expression in NIH 3T3 and v-H-Ras 3T3 fibroblasts by lovastatin through inhibition of ras isoprenylation. *Oncology* 2000; 59: 245–254.

105. Ganne F., Vasse M., Beaudeux J.L., Peynet J., Francois A., Mishal Z., Chartier A., Tobelem G., Vannier J.P., Soria J., and Soria C. Cerivastatin, an inhibitor of HMG-CoA reductase, inhibits urokinase/urokinase-receptor expression and MMP-9 secretion by peripheral blood monocytes — a possible protective mechanism against atherothrombosis. *Thromb. Haemost.* 2000; 84: 680–688.

106. Wong B., Lumma W.C., Smith A.M., Sisko J.T., Wright S.D., and Cai T.Q. Statins suppress THP-1 cell migration and secretion of matrix metalloproteinase 9 by inhibiting geranylgeranylation. *J. Leukoc. Biol.* 2001; 69: 959–962.

107. Matar P., Rozados V.R., Binda M.M., Roggero E.A., Bonfil R.D., and Scharovsky O.G. Inhibitory effect of Lovastatin on spontaneous metastases derived from a rat lymphoma. *Clin. Exp. Metastasis* 1999; 17: 19–25.

108. Prasanna P., Thibault A., Liu L., and Samid D. Lipid metabolism as a target for brain cancer therapy: synergistic activity of lovastatin and sodium phenylacetate against human glioma cells. *J. Neurochem.* 1996; 66: 710–716.

109. Jani J.P., Specht S., Stemmler N., Blanock K., Singh S.V., Gupta V., and Katoh A. Metastasis of B16F10 mouse melanoma inhibited by lovastatin, an inhibitor of cholesterol biosynthesis. *Invasion Metastasis* 1993; 13: 314–324.

110. Kusama T., Mukai M., Iwasaki T., Tatsuta M., Matsumoto Y., Akedo H., and Nakamura H. Inhibition of epidermal growth factor-induced RhoA translocation and invasion of human pancreatic cancer cells by 3-hydroxy-3-methylglutaryl-coenzyme a reductase inhibitors. *Cancer Res.* 2001; 61: 4885–4891.

111. Alonso D.F., Farina H.G., Skilton G., Gabri M.R., De Lorenzo M.S., and Gomez D.E. Reduction of mouse mammary tumor formation and metastasis by lovastatin, an inhibitor of the mevalonate pathway of cholesterol synthesis. *Breast Cancer Res. Treat.* 1998; 50: 83–93.

112. Kusama T., Mukai M., Iwasaki T., Tatsuta M., Matsumoto Y., Akedo H., Inoue M., and Nakamura H. 3-hydroxy-3-methylglutaryl-coenzyme a reductase inhibitors reduce human pancreatic cancer cell invasion and metastasis. *Gastroenterology* 2002; 122: 308–317.

113. Weis M., Heeschen C., Glassford A.J., and Cooke J.P. Statins have biphasic effects on angiogenesis. *Circulation* 2002; 105: 739–745.

114. Kureishi Y., Luo Z., Shiojima I., Bialik A., Fulton D., Lefer D.J., Sessa W.C., and Walsh K. The HMG-CoA reductase inhibitor simvastatin activates the protein kinase Akt and promotes angiogenesis in normocholesterolemic animals. *Nat. Med.* 2000; 6: 1004–1010.

115. Ungvari Z., Pacher P., and Csiszar A. Can simvastatin promote tumor growth by inducing angiogenesis similar to VEGF? *Med. Hypotheses* 2002; 58: 85–86.

116. Simons M. Molecular multitasking: statins lead to more arteries, less plaque. *Nat. Med.* 2000; 6: 965–966.

117. Vincent L., Chen W., Hong L., Mirshahi F., Mishal Z., Mirshahi-Khorassani T., Vannier J.P., Soria J., and Soria C. Inhibition of endothelial cell migration by cerivastatin, an HMG-CoA reductase inhibitor: contribution to its anti-angiogenic effect. *FEBS Lett.* 2001; 495: 159–166.

118. Park H.J., Kong D., Iruela-Arispe L., Begley U., Tang D., and Galper J.B. 3-hydroxy-3-methylglutaryl coenzyme A reductase inhibitors interfere with angiogenesis by inhibiting the geranylgeranylation of RhoA. *Circ. Res.* 2002; 91: 143–150.

119. von Haehling S. and Anker S.D. Statins for heart failure: at the crossroads between cholesterol reduction and pleiotropism? *Heart* 2005; 91: 1–2.

120. Muscaritoli M., Costelli P., Bossola M., Grieco G., Bonelli G., Bellantone R., Doglietto G.B., Rossi-Fanelli F., and Baccino F.M. Effects of simvastatin administration in an experimental model of cancer cachexia. *Nutrition* 2003; 19: 936–939.

25 Xanthine Oxidase Inhibitors and Insulin Sensitizers

Wolfram Doehner

CONTENTS

25.1 INTRODUCTION

Our understanding of the complex pathophysiological mechanisms underlying chronic diseases increasingly acknowledge metabolic aspects significantly contributing to both morbidity and progression of disease processes. Modern techniques shed more light on the cellular and subcellular processes of functional and structural metabolic pathways and new pharmacological agents that aim at metabolic intervention are increasingly investigated. An imbalance between anabolic and catabolic pathways that occurs in the course of chronic disease underlies a process of initially slow but measurable tissue wasting. If not stopped or even better reversed, this may ultimately lead to clinically apparent body wasting, that is, cachexia. Although a multitude of factors are involved in the complex regulation of metabolic balance, certain pathways may be identified that enable targeted metabolic intervention. The following chapter is focused on two metabolic aspects, namely insulin resistance and upregulated xanthine oxidase activity that have been identified as contributors to the pathophysiology of chronic diseases, such as chronic heart failure, and that are observed in the context of cachexia. The potential of novel treatment strategies aimed at these metabolic pathways will be explored.

25.1.1 Historic Milestones in Metabolic Studies

Already in the 19th century the clinical observation that an elevated metabolic rate is associated with weight losses that occur in the course of chronic diseases was made. This is in contrast to uncomplicated starvation, where energy expenditure is decreased to compensate for reduced nutrient supply. Several symptoms such as tachycardia, hyperpnea, sweating, and a rise in body temperature indicated an increase of the metabolic rate.[1] As early as 1916, the increase in the basal metabolic rate was specifically documented.[2] Increased metabolic demands of several specific tissues were discussed such as decreased efficiency of the respiratory system,[3] or of the hypertrophied myocardium in heart failure.[4]

Signs of subclinical chronic pro-inflammatory immune activation in the context of chronic disease were systematically studied in the 1930s when Cohn and Steel discussed the possibility of a yet unidentified pyrogen[5] causing the elevated temperature that was consistently found in heart failure patients.[6] The accurate observations and thoughtful interpretation of these old studies is stunning in light of today's knowledge of immune activation affecting imbalanced weight control and changes in body composition.

The observation of constant "air hunger" lead researchers to explore the role of reduced oxygen supply and cellular hypoxia became the focus of metabolic research at the beginning of the 20th century. First, blood gas analyses were reported as early as 1919[7] and in 1923 it was observed that at high altitudes study subjects involuntarily lost weight.[8] Increased lactate levels were recognized as a measure of an imbalance between aerobic and anaerobic metabolism[9] and in 1958, Huckabee introduced the concept of "excess lactate" production as a measure of tissue hypoxia.[10] Those observations suggested that tissue hypoxia also accounted for impaired protein biosynthesis leading to the downregulation of anabolic and the increase of catabolic pathways. Based on observations of tachycardia, vasoconstriction with reduced cutaneous and renal blood flow, an overactivity of the sympathetic nervous system was recognized, further contributing to the increased caloric turnover. As a result, catabolic drive may chronically dominate the anabolic pathways. The constant drain of the body's energy reserves may eventually lead to pathological tissue degradation.

Based on these pioneering studies, modern research has developed fundamental insight both at a clinical and molecular level into mechanisms of metabolic regulation and pathophysiology. With regard to therapeutic options, however, weight loss in the course of chronic diseases is still a mostly unconquered territory and specific metabolic treatment options are rare. In the following sections the potential significance of two specific metabolic aspects, namely insulin resistance and xanthine oxidase activation are discussed.

25.2 INSULIN RESISTANCE

25.2.1 Introduction

Chronic diseases such as chronic heart failure,[11] chronic renal failure[12] and chronic inflammatory diseases are in general associated with impaired sensitivity of tissue to insulin-mediated uptake and utilization of glucose (i.e., insulin resistance). This insulin resistance is characterized by both fasting and stimulated hyperinsulinemia and a reduced rate of insulin stimulated glucose uptake by the peripheral tissues. Importantly, it has been shown that in chronic heart failure, insulin resistance occurs secondarily to the chronic disease and independently of underlying etiological factors. Moreover, this abnormal metabolic regulation is not restricted to the myocardium[13] but, in the course of the chronic disease, it affects the body in general, specifically the skeletal muscle, the main tissue of insulin-mediated glucose utilization.[14] Beside its crucial role in energy homeostasis, insulin is one of the strongest anabolic hormones of the human body. It is therefore centrally involved in both structural and functional metabolic regulation. The imbalance in these metabolic pathways, that results from insulin resistance, may importantly contribute to both morbidity (i.e., symptomatic impairment) and mortality (i.e., progression) in chronic disease.

25.2.2 Insulin Resistance is Associated with Myocardial and Skeletal Muscle Functional Impairment

Skeletal muscle functional capacity is impaired if insulin dependent glucose utilization is reduced. Lower exercise capacity, early dyspnea and muscle fatigue are common symptoms in chronic disease and are specifically pronounced in cachectic patients, which usually comprise such patients with advanced disease state. From the central role of insulin in the regulation of energy balance, a direct impact of insulin resistance on impaired functional capacity of skeletal muscle seems obvious. It has indeed been shown that insulin resistance and impaired glucose utilization are directly related to reduced muscle strength, peak oxygen uptake[15], and exercise capacity (Figure 25.1).[14, 16]

Insulin resistance has also been described to imply direct adverse effects on myocardial function. After myocardial infarction the normal compensatory

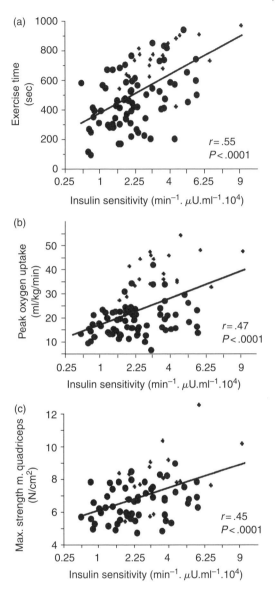

FIGURE 25.1   Impaired insulin sensitivity relates to reduced exercise capacity: (a) exercise time in a symptom limited treadmill exercise test, (b) peak oxygen uptake during symptom limited treadmill exercise test, and (c) maximum quadriceps isometric muscle strength. •, chronic heart failure patient, +, healthy control (Adapted from Doehler, W., Rachhaus, M., Davos, C.H., Sharma, R., Bolger, A., and Anker, S.D. *Circulation* 2000;102(Suppl.):II 719, With permission.).

hyperkinetic response in the noninfarcted myocardial area is blunted among insulin resistant patients with diabetes.[17] Diabetic patients have a higher incidence of heart failure after myocardial infarction despite smaller infarct sizes and an ejection fraction similar to those in subjects without diabetes.[18] Further, it has been shown that impaired glucose metabolism is related to an abnormal diastolic filling pattern,[19,20] that is, diastolic dysfunction, which is a characteristic feature of diabetes-associated myocardial disease.[21,22]

25.2.3 Insulin Resistance and Anabolic Capacity

The definition of insulin resistance as a "state in which a normal amount of insulin produces a subnormal biological response"[23] conventionally refers to the glucoregulatory action of insulin.[24] Insulin is, however, a central metabolic hormone and its pleiotropic effects reach far beyond the regulation of glucose homeostasis. Notably, insulin is one of the most powerful anabolic hormones. If insulin resistance affects structural metabolic pathways to a similar degree as energy metabolism, this may contribute to body composition alterations that are often found in the process of advancing chronic disease. A catabolic/anabolic imbalance has been described for chronic heart failure[25] that underlies the development of tissue wasting and ultimately leads to cardiac cachexia. Accordingly, insulin resistance has been observed in patients suffering from cardiac cachexia. Moreover, the insulin action index, the product of insulin sensitivity and the measured plasma insulin concentration is particularly low in cachectic patients (Figure 25.2).

25.2.4 Peripheral Hemodynamic Aspects of Insulin Resistance in CHF

Beyond its effects to regulate glucose metabolism, insulin has a direct effect on endothelial function in the peripheral[26] and coronary[27] vascular beds. Insulin resistance is directly associated with impaired endothelium-dependent vasodilation.[28] Endothelial dysfunction leads to increased vascular resistance and reduced perfusion of skeletal muscle[29] and is closely related to impaired exercise capacity and muscular fatigue. This effect in insulin resistance may further contribute to the functional impairment of these tissues (see below).

25.2.5 Insulin Sensitivity and Prognosis

There is increasing evidence to suggest that insulin resistance is indeed pathophysiologically linked with chronic diseases such as chronic heart failure. Insulin resistance progresses parallel to the severity of the disease (Figure 25.3).[14,16] Based on the latter it could be hypothesized that insulin resistance may contribute to impaired prognosis in these patients. Indeed in prospective trials it has been shown that insulin resistance is a significant prognostic marker in chronic heart failure

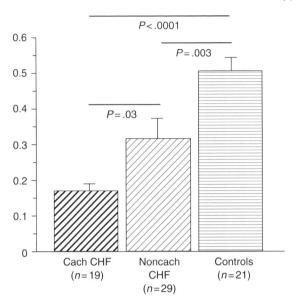

FIGURE 25.2 Insulin action index (mean ± SD) in cachectic patients and noncachectic patients with chronic heart failure. Insulin action index is the product of insulin sensitivity and plasma insulin concentration (dimensionless). This is particularly low in cachectic patients with chronic heart failure (CHF) suggesting a role of impaired insulin efficiency in the development of tissue wasting and ultimately cachexia.

patients independently of, and additionally to, other established prognostic markers such as age, New York Heart Association (NYHA) functional class, exercise capacity, or left ventricular ejection fraction.[30,31]

25.2.6 Mechanisms of Insulin Resistance in Chronic Disease

The underlying mechanisms of insulin resistance in chronic heart failure and other chronic diseases are complex and far from being fully understood. As energy metabolism in general and glucose homeostasis is regulated within a complex web of metabolic interrelations, a combination of a number of factors is likely to account for the net effect of insulin resistance. On the basis of age and genetic predisposition, factors such as immune activation (elevated cytokine levels such as tumor necrosis factor-α) neurohormonal activation (elevated catecholamine levels), impairment of other hormonal regulatory mechanisms (e.g., hyperleptinemia, growth hormone resistance), endothelium dysfunction, and impaired tissue perfusion (tissue hypoxia and reduced substrate supply) are discussed in this context. Furthermore, specific changes have been observed in skeletal muscle structure (fiber type shift) and in its metabolic apparatus (impaired oxidative capacity) as well as direct impairment to intracellular insulin signaling and reduced

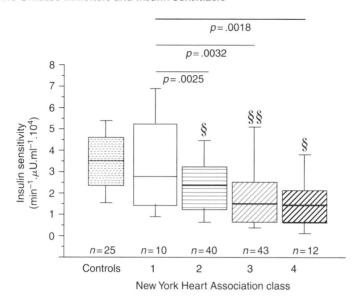

FIGURE 25.3 Insulin sensitivity relates to disease severity in patients with chronic heart failure according to New York Heart Association (NYHA) functional class suggesting a pathophysiological link to disease progression (mean ± SD, § $P < .05$, §§ $P < .01$ compared with healthy control subjects of similar age) (Adapted from Doehner, W., Rauchhaus, M., Ponikowski, P., Godsland, von Haehling, S., Okonko, D.A., Leyva, F., Proudler, T., Coats, A.J., and Anker, S.D. *J. Am. Coll. Cardiol.* 2005 [in press]. With permission.).

amount or translocation of the GLUT4, the predominant glucose transport protein in skeletal muscle.[32] Specific disease-related secondary conditions such as atrophy due to reduced physical activity, changes in eating behavior and diet, and treatment-related side-effects (diuretics, nonsteroidal anti-inflammatory drugs) may further contribute to insulin resistance.

25.2.7 Insulin Resistance as a Potential Novel Therapeutic Target

On the basis of the above it seems intriguing to test whether therapeutically targeting and restoring insulin sensitivity may exert additional beneficial effects on the symptomatic status as well as potentially on disease progression and hence prognosis in chronic disease, and especially in cachectic patients. Theoretically, improved energy utilization and increased anabolic drive may imply substantial benefit for patients with cachexia due to advanced chronic disease. Evaluating the potential impact on the metabolic profile of currently available treatments in chronic disease may provide indirect support of the metabolic approach.

Beside insulin, the second major stimulus to regulate glucose uptake, via increased GLUT4 translocation from an intracellular pool to the plasma membrane, is muscle fiber contraction. Accordingly, reduced exercise, which regularly occurs in chronic patients, results in a decrease in GLUT4. The well-known positive impact of physical exercise on insulin sensitivity[33] may be involved in the physiological mechanisms that underlie the beneficial effects of exercise training on morbidity and mortality in chronic diseases such as chronic heart failure.[34] Neurohormonal activation is discussed to contribute to insulin resistance in chronic diseases and, accordingly, treatment with angiotensin converting enzyme inhibitors and angiotensin 1 receptor antagonists have been shown to improve glucose utilization. Beta receptor blockers provide, however, an inconclusive picture as positive and negative effects on insulin sensitivity have been described. Several studies report that carvedilol, for example, has an improving effect on metabolic profile and insulin.[35, 36] This is paralleled by the observation of a moderate weight gain following carvedilol treatment. Whether this positive metabolic profile of this agent substantially contributes to the treatment effect that has been demonstrated for neurohormonal blocking therapy in chronic heart failure and whether this effect can be translated to other chronic diseases that result in cachexia needs to be studied in more detail.

Only recently have new treatment options emerged that directly target insulin sensitivity. Thiazolidinediones (glitazones), or insulin sensitizers, are selective agonists for the peroxisome proliferator activated receptor-γ (PPAR-γ). Activation of PPAR-γ has been found to modulate the transcription of a number of insulin responsive genes involved in the control of glucose and lipid metabolism. This leads to improved insulin sensitivity, predominantly by stimulating insulin-mediated glucose utilization by the skeletal muscle. The exact mechanisms of action of glitazones are still unclear, but it has been reported in animal studies that glitazone administration leads to stimulation of GLUT4 gene expression,[37] and to reduced leptin (ob gene product) and TNF-α expression.[38] Other favorable effects of glitazones have been described such as anti-inflammatory, antithrombotic,[39] and antiatherosclerotic effects,[40] attenuation of endothelial dysfunction[41] and improvement of lipid profiles by reducing triglycerides and free fatty acids, and increasing HDL cholesterol.[42] As the above findings are characteristic in various chronic diseases, glitazones may have a beneficial impact in these conditions beyond restoring insulin sensitivity. In a retrospective study, a decreased incidence of chronic heart failure has been observed in diabetic patients receiving glitazone therapy.[43] This has recently been confirmed in a large observational study.[44]

Most importantly, weight increase has been confirmed to accompany the metabolic effect of glitazones, presumably due to improved insulin action as an anabolic hormone. This weight gain is considered as an unwanted effect in diabetic patients; however in cachetic patients; it might be viewed as a desirable result from glitazone treatment. The increase in body mass results partly from fluid retention but more importantly also from enhanced adipogenesis, that is, increased energy storage which would be a distinct treatment goal in cachectic patients.

It has to be pointed out, however, that glitazones are currently to be used in chronic heart failure with great caution as increased fluid retention has been described which in heart failure patients potentially results in a worsening of symptoms.[45] Ongoing studies could, however, not confirm adverse effects on cardiac structure or function. In a randomized open label echocardiography study in 203 type 2 diabetic patients the effect of long-term treatment with rosiglitazone 8 mg/d vs. glyburide on cardiac structure has been evaluated.[46] The two-year follow-up data showed no significant changes from baseline for any of the echo-cardiographic parameters tested (left ventricular mass index, ejection fraction, left ventricular end diastolic volume).[47]

Intervention studies designed to target impaired insulin sensitivity may be of benefit in patients with chronic diseases such as chronic heart failure and specific-ally in cachectic patients. Those studies are clearly warranted, but need to be done carefully and with great attention to potential adverse effects.

25.3 XANTHINE OXIDASE

25.3.1 Introduction

The upregulated xanthine oxidase (XO) metabolic pathway may be of specific importance in the pathophysiology of chronic diseases and specifically in cachexia. In humans, uric acid forms the metabolic endpoint of purine degradation. The last metabolic steps in this process (from hypoxanthine to xanthine and from xanthine to uric acid) are promoted by the enzyme xanthine oxidoreductase (EC1.1.3.22). This enzyme is a flavoprotein that contains both iron and molybdenum and uses NAD^+ as an electron acceptor. It exists in two interconvertible forms, xanthine dehydrogenase and XO. Only in its oxidase form, this enzyme transfers the redu-cing equivalent generated by the oxidation of substrates to molecular oxygen with the resultant production of superoxide anion and hydrogen peroxide (Figure 25.4). Hydrogen peroxide can be converted to free hydroxyl radicals, starting a cascade of free radical generation.

There is increasing evidence that the xanthine oxidase metabolic pathway is not merely the final step in the purine degradation with the formation of uric acid as a metabolically inert waste product. The generation of free oxygen radicals by XO may have an important pathophysiological role in tissues and in vascular regulation in chronic diseases with upregulated XO activity. In 1968 the cytosolic XO was the first documented putative biological generator of oxygen derived free radicals.[48] Since then it has been established that XO is a major source of free oxygen radical production in the human body.[49] In humans the organs with the highest XO activity are the intestine and the liver, with low or undetectable levels in the brain, kidney, lung, and muscle.[50] The localization of XO primarily in the endothelial cells of the capillaries suggests that XO has a specific function in the vascular system.[51] Given the capacity to generate free oxygen radicals, this enzyme may have a role in bactericidal defense mechanisms,[52] especially at the barrier between the intestinal lumen and the body tissues. One could hypothesize that, analogous to other regulatory systems in chronic diseases, long-term stimulation of XO may result in chronic activation of this mechanism leading to maladaptive processes and eventually harmful effects. The latter provides the pathophysiological link

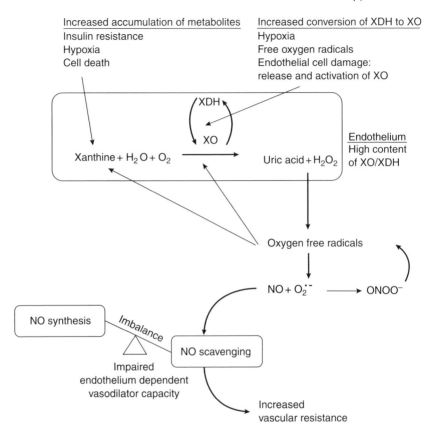

Increased accumulation of metabolites   Increased conversion of XDH to XO
Insulin resistance   Hypoxia
Hypoxia   Free oxygen radicals
Cell death   Endothelial cell damage:
                                           release and activation of XO

FIGURE 25.4 Role of xanthine oxidase (XO)-derived free oxygen radicals in endothelium dysfunction. Increased substrate supply and XO/XH imbalance lead to upregulated XO activity resulting in increased free oxygen radical production. Superoxide anions react with endothlium derived NO, causing impaired NO dependent regulation of vascular tone and hence increased peripheral vascular resistance (Adapted from Doehner, W. and Anker, S.D. *Semin. Nephrol.* 2005;25:61–66. With permission.).

between elevated uric acid and a large variety of detrimental processes, including increased cytokine production,[53] cell apoptosis,[54] and endothelial dysfunction[55] all of which occur in chronic diseases.

This metabolic pathway is of particular significance in conditions of tissue hypoxia and ischemia/reperfusion[56] as increased degradation of ATP to adenosine leads to increased substrate load for XO.[57] Accordingly, elevation of serum uric acid has been observed in hypoxic states, such as obstructive pulmonary disease,[58] neonatal hypoxia,[59] cyanotic heart disease,[60] and acute heart failure.[61] Of note, in ischemia/hypoxia xanthine dehydrogenase is increasingly converted to XO, which further adds to accelerated radical production.[62]

In cachexia elevated uric acid levels result partly from increased substrate supply due to an increase in catabolic processes. Accordingly, uric acid levels

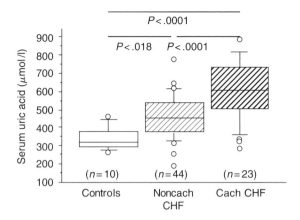

FIGURE 25.5 Serum uric acid levels in controls and in patients with chronic heart failure subgrouped as noncachectic (ncCHF) and cachectic (cCHF) chronic heart failure patients. Box plot displaying the 10th, 25th, 50th, 75th, and 90th percentiles (Adapted from Doehner, W., Rauchhaus, M., Florea, V.G., Sharma, R., Bolger, A.P., Davos, C.H., Coats, A.J.S., and Anker, S.D. *Am. Heart J.* 2001;141:792–799. With permission.).

are elevated in chronic heart failure patients but are highest in those with cardiac cachexia (Figure 25.5). Although impaired renal excretion might contribute to the accumulation of uric acid in chronic diseases, xanthine oxidase activation may be the dominant factor accounting for hyperuricemia. In chronic heart failure, high serum uric acid levels have been confirmed to indicate XO activation.[63,64] This may lead to an increase of free radical oxygen load in patients with chronic disease such as chronic heart failure, which has indeed been observed.[65,66]

25.3.2 The XO Metabolic Pathway as One Part of a Complex Metabolic Web

Elevated plasma uric acid levels are closely associated with insulin resistance.[67] In fact, hyperuricemia has been added to the expanded definition of the insulin resistance syndrome, a cluster of metabolic abnormalities that represents an increased risk for cardiovascular and metabolic diseases.[68]

The pathophysiological connection of insulin resistance and hyperuricemia following upregulated XO activity has been hypothesized to result from the accumulation and diversion of glycolytic intermediates toward the pentose phosphate pathway.[69] One branch point for diversion of glycolytic intermediates toward the pentose phosphate pathway is controlled by the enzyme glycerylaldehyde-3-phosphate dehydrogenase (GA3PDH). This enzyme catalyzes the only oxidative step in glycolysis and its activity is regulated by insulin.[70,71] Increased metabolic activity via this pathway results in accumulation of phosphoribosylpyrophosphate (PPRP),[72] in itself a key precursor of the purine de-novo synthesis. Increased substrate supply is followed by upregulated purine synthesis and metabolic turnover

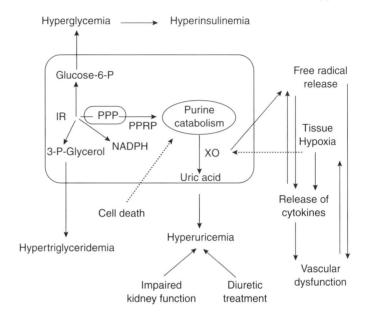

FIGURE 25.6 The inter-relationship between hyperuricemia, xanthine oxidase (XO), cell metabolism and insulin resistance (IR), tissue hypoxia, vascular dysfunction, cytokines, oxygen free radicals, and adhesion molecules in chronic heart failure. Insulin resistance, tissue wasting (cell death), and tissue hypoxia can cause accumulation of purine bodies and activation of xanthine oxidase, finally causing hyperuricemia. Impaired kidney function and diuretic treatment can also contribute to hyperuricemia. PPP — pentose phosphate pathway, PPRP — phosphoribosylpyrophosphate (Adapted from Anker, S.D., Doehner, W., Rauchhaus, M., Sharma, R., Francis, D., Knosalla, C., Davos, C.H., Cicoira, M., Shamim, W., Kemp, M., Segal, R., Osterziel, K.J., Leyva, F., Hetzer, R., Ponikowski, P., and Coats, A.J. *Circulation* 2003;107:1991–1997. With permission.).

toward uric acid. Factors such as hypoxia, oxygen radical accumulation cytokine exposure, cell death and others further contribute to the complex regulation of the XO pathway (Figure 25.6).

25.3.3 Cardiovascular Effect of XO-Derived Free Oxygen Radicals

XO-derived oxygen free radicals lead to impaired endothelium-dependent vasodilator capacity due to its effect to diminish vasoactive NO.[73,74] XO generated free oxygen radicals interact with endothelium derived NO to form peroxynitrate ($ONOO^-$, in itself a highly active oxygen radical), starting a cascade of detrimental oxygen radical effects (Figure 25.4). In chronic heart failure hyperuricemia has been shown to be predictive of impaired peripheral vasodilator capacity, and this association was especially pronounced in cachectic patients (Figure 25.7).[75] Impaired vasodilator capacity, the hallmark of endothelium dysfunction, leads to

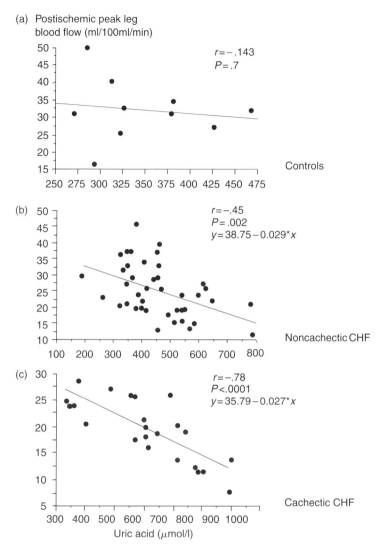

FIGURE 25.7 Correlation of serum uric acid with postischemic peak leg blood flow in healthy controls (a), noncachectic CHF patients (b), and cachectic CHF patients (c).

impaired tissue perfusion and thereby constitutes an important aspect of the pathophysiology of chronic diseases such as diabetes, hypertension, arteriosclerosis, and chronic heart failure. Endothelium dysfunction, in turn, is closely related to prominent clinical symptoms of chronic disease such as reduced exercise capacity and early muscle fatigue.[76,77]

Consequently, blocking the XO metabolic pathway emerged as an intriguing novel treatment option to prevent oxygen radical accumulation and its adverse effects. Recently, several clinical studies have been initiated to test whether this therapeutic approach results in improvement of clinical status or pathophysiological surrogate disease markers. In aortocoronary bypass surgery, treatment with allopurinol has been shown to counteract effects of ischemia-reperfusion injury[78] leading to a reduction in the need for inotropic support[79] and to increased immediate and late (after 6 months) left ventricular contractility.[80]

Beside the regulation of peripheral vascular tone, inhibition of XO appears to exert direct myocardial effects. In animal models, allopurinol reduces myocardial oxygen consumption[81] and improves systolic function,[82, 83] resulting in increased myocardial energetic efficiency. Recently, this has been confirmed in patients with chronic heart failure.[84] Although the underlying mechanism is not yet fully uncovered, some authors have suggested a specific effect of allopurinol to sensitize cardiac myofilaments to Ca^{2+}.[85]

The XO inhibiting approach has been further tested in chronic heart failure patients in a number of studies. In patients with hyperuricemia, treatment with allopurinol improved endothelial function and peripheral blood flow,[86–88] while markers of free oxygen radical generation were reduced.[86] In contrast to these studies a recent placebo-controlled study by Gavin and Struthers did not find an improvement in exercise capacity on maximum exercise level nor on submaximum level.[89] To explain the neutral result of this study, it should be noted, that a pronounced effect of XO inhibition by allopurinol may be observed only if XO activity is upregulated. Accordingly, in previous treatment trials, the improvement in endothelium dysfunction was tested particularly in hyperuricemic patients.[86–88] In contrast, no effect on vasodilator capacity was observed following allopurinol treatment in patients with normal uric acid levels (Figure 25.8).[86, 88] It could be concluded that targeted XO inhibition as a novel treatment approach might be suitable only in those patients where increased uric acid levels are an indicator of upregulated XO activity. This would include the majority of cachectic patients. Extending this approach to patients with normal uric acid levels might increase the number of nonresponders.

25.3.4 Hyperuricemia as a Novel Prognostic Marker

In addition to the involvement in the regulation of vascular tone, increasing data suggests prognostic significance for the XO metabolic pathway in chronic disease. Recently, it has been shown that in chronic heart failure patients, high uric acid levels are a predictor of impaired survival, independently of established parameters such as the clinical status, exercise capacity, parameters of kidney function, and diuretic therapy.[90] This is further supported by the results from a recent retrospective study that examined the effect of allopurinol in chronic heart failure patients on mortality and hospitalization.[91] It was observed that in these patients long-term high dose of allopurinol (≥ 300 mg/d) may be associated with a lower all-cause mortality (adjusted RR 0.59, 95%CI 0.37 to 0.95, $P < .05$) than low dose (<300 mg/d) assuming a dose-related effect of allopurinol. Whether this effect can be observed as well in other chronically diseased

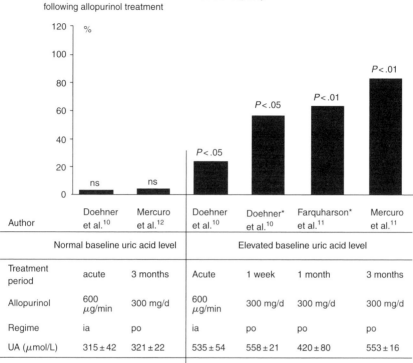

Change of endothelium dependent vasodilator capacity following allopurinol treatment

Author	Doehner et al.[10]	Mercuro et al.[12]	Doehner et al.[10]	Doehner* et al.[10]	Farquharson* et al.[11]	Mercuro et al.[11]
	Normal baseline uric acid level		Elevated baseline uric acid level			
Treatment period	acute	3 months	Acute	1 week	1 month	3 months
Allopurinol	600 μg/min	300 mg/d	600 μg/min	300 mg/d	300 mg/d	300 mg/d
Regime	ia	po	ia	po	po	po
UA (μmol/L)	315±42	321±22	535±54	558±21	420±80	553±16

FIGURE 25.8 Allopurinol treatment in patients with chronic heart failure and its effect on vasodilator capacity. Allopurinol improved endothelium dependent vasodilator capacity in hyperuricemic patients but not in patients with normal uric acid levels (sorted by normal vs. elevated uric acid levels and duration of treatment, data in % change vs. baseline, *% change vs. placebo) (Adapted from Doehner, W., Anker, S.D. *Heart* 2005 [in press]. With permission.).

patients needs to be tested. It should be noted, however, that treatment with allopurinol is not free of problems. It can induce gout attacks, kidney dysfunction, or skin reactions and further work is required to confirm the potential benefit of this treatment.

Studies are underway using allopurinol and also other XO inhibitors like oxopurinol.[92] Those studies will show whether the therapeutic approach of XO inhibition may translate into meaningful clinical benefit in chronic disease and specifically in cachectic patients.

REFERENCES

1. Silver, S., Proto, P., and Crohn, E.B. Hypermetabolic states without hyperthyroidism (non-thyroidism hypermetabolism). *Arch. Int. Med.* 1950;85: 479–482.

2. Peabody, F.W., Meyer, A.L., and DuBois, E.F. Clinical calorimetry. VII. The basal metabolism of patients with cardiac and renal disease. *Arch. Int. Med.* 1916;17:980–1009.

3. Christie, R.V. and Meakins, J.C. The intra-pleural pressure in congestive heart failure and its clinical significance. *J. Clin. Invest.* 1934;13:323–345.

4. Blain, J.M., Schafer, H., Siegel, A.L., and Bing, R.J. Studies on myocardial metabolism. IV Myocardial metabolism in congestive failure. *Am. J. Med.* 1956;20:820–833.

5. Cohn, A.E. and Steel, J.M. Unexplained fever in heart failure. *J. Clin. Invest.* 1934;13:853–868.

6. Kinsey, D. and White, P.D. Fever in congestive heart failure. *Arch. Int. Med.* 1940;65:163–170.

7. Harrop, G.A. The oxygen and carbon dioxide content of arterial and venous blood in normal individuals and in patients with anaemia and heart disease. *J. Exp. Med.* 1919;30:241–257.

8. Barcroft, J. Binger, C.A., Bock, A.V., et al. Observation upon the effect of high altitude on physiological processes of the human body carried out in the Peruvian Andes, chiefly at Cerroda Pasca. *Phil. Trans. R. Soc.* 1923;B211:351.

9. Weiss, S. and Ellis, L.B. Oxygen utilisation and lactic acid production in the extremities during rest and exercise in subjects with normal and in those with diseased cardiovascular systems. *Arch. Int. Med.* 1935;55:665–680.

10. Huckabee, W.E. Relationships of pyruvate and lactated during anaerobic metabolism. I. Effects of infusion of pyruvate or glucose and of hyperventilation. *J. Clin. Invest.* 1958;37:244–254.

11. Swan, J.W., Walton, C., Godsland, I.F., Clark, A.L., Coats, A.J.S., and Oliver, M.F. Insulin resistance in chronic heart failure. *Eur. Heart J.* 1994;15:1528–1532.

12. Serum leptin concentrations correlate to plasma insulin concentrations independent of body fat content in chronic renal failure. *Nephrol. Dial. Transplant.* 1997;12:1321–1325.

13. Paternostro, G., Camici, P.G., Lammerstma, A.A., Marinho, N., Baliga, R.R., Kooner, J.S., Radda, G.K., and Ferrannini, E. Cardiac and skeletal muscle insulin resistance in patients with coronary heart disease. A study with positron emission tomography. *J. Clin. Invest.* 1996;98:2094–2099.

14. Swan, J.W., Anker, S.D., Walton, C., Godsland, I.F., Clark, A.L., Leyva, F., Stevenson, J.C., and Coats, A.J. Insulin resistance in chronic heart failure: Relation to severity and etiology of heart failure. *J. Am. Coll. Cardiol.* 1997;30:527–532.

15. Doehner, W., Rauchhaus, M., Davos, C.H., Sharma, R., Bolger, A., and Anker, S.D. Clinical significance of impaired insulin sensitivity in chronic heart failure. *Circulation* 2000;102 (Suppl.):II 719.

16. Suskin, N., McKelvie, R.S., Burns, R.J., et al. Glucose and insulin abnormalities relate to functional capacity in patients with congestive heart failure. *Eur. Heart J.* 2000;21:1368–1375.

17. Woodfield, S.L., Lundergan, C.F., Reiner, J.S., Greenhouse, S.W., Thompson, M.A., Rohrbeck, S.C., Deychak, Y., Simoons, M.L., Califf, R.M., Topol, E.J., and Ross, A.M. Angiographic findings and outcome in diabetic patients treated with thrombolytic therapy for acute myocardial infarction: The GUSTO-I experience. *J. Am. Coll. Cardiol.* 1996;28:1661–1669.

18. Stone, P.H., Muller, J.E., Hartwell, T., York, B.J., Rutherford, J.D., Parker, C.B., Turi, Z.G., Strauss, H.W., Willerson, J.T., Robertson, T., et al. The effect of

diabetes mellitus on prognosis and serial left ventricular function after acute myocardial infarction: Contribution of both coronary disease and diastolic left ventricular dysfunction to the adverse prognosis. The MILIS Study Group. *J. Am. Coll. Cardiol.* 1989;14:49–57.

19. Danielsen, R., Nordrehaug, J.E., Lien, E., and Vik-Mo, H. Subclinical left ventricular abnormalities in young subjects with long-term type 1 diabetes mellitus detected by digitized M-mode echocardiography. *Am. J. Cardiol.* 1987;60:143–146.

20. Perez, J.E., McGill, J.B., Santiago, J.V., Schechtman, K.B., Waggoner, A.D., Miller, J.G., and Sobel, B.E. Abnormal myocardial acoustic properties in diabetic patients and their correlation with the severity of disease. *J. Am. Coll. Cardiol.* 1992;19:1154–1162.

21. Okura, H., Inoue, H., Tomon, M., Nishiyama, S., Yoshikawa, T., Yoshida, K., and Yoshikawa, J. Impact of Doppler-derived left ventricular diastolic performance on exercise capacity in normal individuals. *Am. Heart J.* 2000;139:716–722.

22. Fein, F.S. and Sonnenblick, E.H. Diabetic cardiomyopathy. *Cardiovasc. Drugs Ther.* 1994;8:65–73.

23. Kahn, C.R. Insulin resistance, insulin insensitivity, and insulin unresponsiveness: A necessary distinction. *Metabolism* 1978;27:1893–1902.

24. Beck-Nielsen, H. Insulin resistance: A scientific and clinical challenge. *Diabetes Metab. Rev.* 1989;5:529–530.

25. Anker, S.D., Chua, T.P., Ponikowski, P., Harrington, D., Swan, J.W., Kox, W.J., Poole-Wilson, P.A., and Coats, A.J. Hormonal changes and catabolic/anabolic imbalance in chronic heart failure and their importance for cardiac cachexia. *Circulation* 1997;96:526–534.

26. Petrie, J.R., Ueda, S., Webb, D.J., Elliott, H.L., and Connell, J.M. Endothelial nitric oxide production and insulin sensitivity. A physiological link with implications for pathogenesis of cardiovascular disease. *Circulation* 1996;93:1331–1333.

27. Inoue, T., Matsunaga, R., Sakai, Y., Yaguchi, I., Takayanagi, K., and Morooka, S. Insulin resistance affects endothelium-dependent acetylcholine-induced coronary artery response. *Eur. Heart J.* 2000;21:895–900.

28. Jonkers, I.J., van de Ree, M.A., Smelt, A.H., de Man, F.H., Jansen, H., Meinders, A.E., van der Larse, A., and Blauw, G.J. Insulin resistance but not hypertriglyceridemia per se is associated with endothelial dysfunction in chronic hypertriglyceridemia. *Cardiovasc. Res.* 2002;53:496–501.

29. Drexler, H., Hayoz, D., Munzel, T., Hornig, B., Just, H., Brunner, H.R., and Zelis, R. Endothelial function in chronic congestive heart failure. *Am. J. Cardiol.* 1992;69:1596–1601.

30. Paolisso, G., Tagliamonte, M.R., Rizzo, M.R., et al. Prognostic importance of insulin-mediated glucose uptake in aged patients with congestive heart failure secondary to mitral and/or aortic valve disease. *Am. J. Cardiol.* 1999;83:1338–1344.

31. Doehner, W., Rauchhaus, M., Ponikowski, P., Godsland, von Haehling, S., Okonko, D.A., Leyva, F., Proudler, T., Coats, A.J., and Anker, S.D. Impaired insulin sensitivity as an independent risk factor for mortality in patients with stable chronic heart failure. *J. Am. Coll. Cardiol.* 2005;46:1019–1026.

32. Doehner, W., Gathercole, D.V., Cicoira, M., et al. Glucose transport protein GLUT4 levels in skeletal muscle and its relationship to insulin resistance in chronic heart failure. *Circulation* 2002;101 (Suppl.):II-569 (abstr.).

33. Lund, S., Holman, G.D., Schmitz, O., and Pedersen, O. Contraction stimulates translocation of glucose transporter GLUT4 in skeletal muscle through a mechanism distinct from that of insulin. *Proc. Natl. Acad. Sci. USA* 1995;92: 5817–5821.

34. Piepoli, M.F., Davos, C., Francis, D.P., Coats, A.J., and ExTraMATCH Collaborative. Exercise training meta-analysis of trials in patients with chronic heart failure (ExTraMATCH). *BMJ* 2004;328:189–196.

35. Giugliano, D., Acampora, R., Marfella, R., De Rosa, N., Ziccardi, P., Ragone, R., De Angelis, L., and D'Onofrio, F. Metabolic and cardiovascular effects of carvedilol and atenolol in non-insulin-dependent diabetes mellitus and hypertension. A randomized, controlled trial. *Ann. Intern. Med.* 1997;126: 955–959.

36. Bakris, G.L., Fonseca, V., Katholi, R.E., McGill, J.B., Messerli, F.H., Phillips, R.A., Raskin, P., Wright, J.T., Jr., Oakes, R., Lukas, M.A., Anderson, K.M., Bell, D.S., and GEMINI Investigators. Metabolic effects of carvedilol vs. metoprolol in patients with type 2 diabetes mellitus and hypertension: A randomized controlled trial. *JAMA* 2004;292:2227–2236.

37. Bahr, M., Spelleken, M., Bock, M., von Holtey, M., Kiehn, R., and Eckel, J. Acute and chronic effects of troglitazone (CS-045) on isolated rat ventricular cardiomyocytes. *Diabetologia* 1996;39:766–774.

38. De Vos, P., Lefebvre, A.M., Miller, S.G., Guerre-Millo, M., Wong, K., Saladin, R., Hamann, L.G., Staels, B., Briggs, M.R., and Auwerx, J. Thiazolidinediones repress ob gene expression in rodents via activation of peroxisome proliferator-activated receptor gamma. *J. Clin. Invest.* 1996;98:1004–1009.

39. Marx, N., Sukhova, G.K., Collins, T., Libby, P., and Plutzky, J. PPAR alpha activators inhibit cytokine-induced vascular cell adhesion molecule-1 expression in human endothelial cells. *Circulation* 1999;99:3125–3131.

40. Flavell, D.M., Jamshidi, Y., Hawe, E., Pineda Torra, I., Taskinen, M.R., Frick, M.H., Nieminen, M.S., Kesaniemi, Y.A., Pasternack, A., Staels, B., Miller, G., Humphries, S.E., Talmud, P.J., and Syvanne, M. Peroxisome proliferator-activated receptor alpha gene variants influence progression of coronary atherosclerosis and risk of coronary artery disease. *Circulation* 2002;105:1440–1445.

41. Yue, T., Chen, J., Bao, W., Narayanan, P.K., Bril, A., Jiang, W., Lysko, P.G., Gu, J.L., Boyce, R., Zimmerman, D.M., Hart, T.K., Buckingham, R.E., and Ohlstein, E.H. In vivo myocardial protection from ischemia/reperfusion injury by the peroxisome proliferator-activated receptor-agonist rosiglitazone. *Circulation* 2001;104:2588–2594.

42. Patel, J., Miller, E., and Patwardhan, R. Rosiglitazone improves glycaemic control when used as a monotherapy in type 2 diabetic patients. *Diabetic Med.* 1998;15 (Suppl. 2):37–38.

43. Rajagopalan, R., Rosenson, R.S., Fernandes, A.W., Khan, M., and Murray, F.T. Association between congestive heart failure and hospitalization in patients with type 2 diabetes mellitus receiving treatment with insulin or pioglitazone: A retrospective data analysis. *Clin. Ther.* 2004;26:1400–1410.

44. Masoudi, F.A., Inzucchi, S.E., Wang, Y., Havranek, E.P., Foody, J.M., and Krumholz, H.M. Thiazolidinediones, metformin, and outcomes in older patients with diabetes and heart failure: An observational study. *Circulation* 2005;111:583–590.

45. Marceille, J.R., Goins, J.A., Soni, R., Biery, J.C., and Lee, T.A. Chronic heart failure-related interventions after starting rosiglitazone in patients receiving insulin. *Pharmacotherapy* 2004;24:1317–1322.
46. St John Sutton, M., Rendell, M., Dandona, P., Dole, J.F., Murphy, K., Patwardhan, R., Patel, J., and Freed, M. A comparison of the effects of rosiglitazone and glyburide on cardiovascular function and glycemic control in patients with type 2 diabetes. *Diabetes Care* 2002;25:2058–2064.
47. Salzman, A. and Murphy, K. *Diabetes Res. Clin. Pract.* 2001;50 (Suppl. 1):P309.
48. McCord, J.M. and Fridovich, I. The reduction of cytochrome c by milk xanthine oxidase. *J. Biol. Chem.* 1968;243:5753–5760.
49. Terada, L.S., Guidot, D.M., Leff, J.A., Willingham, I.R., Hanley, M.E., Piermattei, D., and Repine, J.E. Hypoxia injures endothelial cells by increasing endogenous xanthine oxidase activity. *Proc. Natl. Acad. Sci. USA* 1992;89:3362–3366.
50. Sarnesto, A., Linder, N., and Raivio, K.O. Organ distribution and molecular forms of human xanthine dehydrogenase/xanthine oxidase protein. *Lab. Invest.* 1996;74:48–56.
51. Jarasch, E.D., Grund, C., Bruder, G., Heid, H.W., Keenan, T.W., and Franke, W.W. Localization of xanthine oxidase in mammary-gland epithelium and capillary endothelium. *Cell* 1981;25:67–82.
52. Tubaro, E., Lotti, B., Cavallo, G., Croce, C., and Borelli, G. Liver xanthine oxidase increase in mice in three patholgoical models. A possible defence mechanism. *Biochem. Pharmacol.* 1980;29:1939–1943.
53. Bolger, A.P. and Anker, S.D. Tumour necrosis factor in chronic heart failure: A peripheral view on pathogenesis, clinical manifestations and therapeutic implications. *Drugs* 2000;60:1245–1257.
54. Adams, V., Jiang, H., Yu, J., et al. Apoptosis in skeletal myocytes of patients with chronic heart failure is associated with exercise intolerance. *J. Am. Coll. Cardiol.* 1999;33:959–965.
55. Hornig, B., Maier, V., and Drexler, H. Physical training improves endothelial function in patients with chronic heart failure. *Circulation* 1996;93:210–214.
56. Zweier, J.L., Kuppusamy, P., and Lutty, G.A. Measurement of endothelial cell free radical generation: Evidence for a central mechanism of free radical injury in postischemic tissues. *Proc. Natl. Acad. Sci. USA* 1988;85:4046–4050.
57. McCord, J.M. and Roy, R.S. The pathophysiology of superoxide: Roles in inflammation and ischemia. *Can. J. Physiol. Pharmacol.* 1982;60:1346–1352.
58. Braghiroli, A., Sacco, C., Erbetta, M., Ruga, V., and Donner, C.F. Overnight urinary uric acid: Creatinine ratio for detection of sleep hypoxemia. Validation study in chronic obstructive pulmonary disease and obstructive sleep apnea before and after treatment with nasal continuous positive airway pressure. *Am. Rev. Respir. Dis.* 1993;148:173–178.
59. Porter, K.B., O'Brien, W.F., and Benoit, R. Comparison of cord purine metabolites to maternal and neonatal variables of hypoxia. *Obstet. Gynecol.* 1992;79:394–397.
60. Hayabuchi, Y., Matsuoka, S., Akita, H., and Kuroda, Y. Hyperuricaemia in cyanotic congenital heart disease. *Eur. J. Pediatr.* 1993;152:873–876.
61. Woolliscroft, J.O., Colfer, H., and Fox, I.H. Hyperuricemia in acute illness: A poor prognostic sign. *Am. J. Med.* 1982;72:58–62.

62. Ashraf, M. and Samra, Z.Q. Subcellular distribution of xanthine oxidase during cardiac ischemia and reperfusion: An immunocytochemical study. *J. Submicrosc. Cytol. Pathol.* 1993;25:193–201.

63. Bakhtiiarov, Z.A. [Changes in xanthine oxidase activity in patients with circulatory failure.] *Ter Arkh* 1989;61:68–69. (Article in Russian.)

64. Landmesser, U., Spiekermann, S., Dikalov, S., Tatge, H., Wilke, R., Kohler, C., Harrison, D.G., Hornig, B., and Drexler, H. Vascular oxidative stress and endothelial dysfunction in patients with chronic heart failure: Role of xanthine-oxidase and extracellular superoxide dismutase. *Circulation* 2002;106: 3073–3078.

65. Belch, J.J., Bridges, A.B., Scott, N., and Chopra, M. Oxygen free radicals and congestive heart failure. *Br. Heart J.* 1991;65:245–248.

66. Keith, M., Geranmayegan, A., Sole, M.J., Kurian, R., Robinson, A., Omran, A.S., and Jeejeebhoy, K.N. Increased oxidative stress in patients with congestive heart failure. *J. Am. Coll. Cardiol.* 1998;31:1352–1356.

67. Vuorinen-Markkola, H. and Yki-Jarvinen, H. Hyperuricemia and insulin resistance. *J. Clin. Endocrinol. Metab.* 1994;78:25–29.

68. Reaven, G.M. Role of insulin resistance in human disease (syndrome X): An expanded definition. *Annu. Rev. Med.* 1993;44:121–131.

69. Leyva, F., Wingrove, C.S., and Godsland, I.F., and Stevenson, J.C. The glycolytic pathway to coronary heart disease: A hypothesis. *Metabolism* 1998;47: 657–662.

70. Alexander, M., Curtis, G., Avruch, J., and Goodman, H.M. Insulin regulation of protein biosynthesis in differentiated 3T3 adipocytes. Regulation of glyceraldehyde-3-phosphate dehydrogenase. *J. Biol. Chem.* 1985;260:11978–11985.

71. Alexander, M.C., Lomanto, M., Nasrin, N., and Ramaika, C. Insulin stimulates glyceraldehyde-3-phosphate dehydrogenase gene expression through cis-acting DNA sequences. *Proc. Natl. Acad. Sci. USA* 1988;85:5092–5096.

72. Kunjara, S., Sochor, M., Ali, S.A., Greenbaum, A.L., and McLean, P. Hepatic phosphoribosyl pyrophosphate concentration. Regulation by the oxidative pentose phosphate pathway and cellular energy status. *Biochem. J.* 1987;244: 101–108.

73. Keaney, J.F. Jr., and Vita, J.A. Atherosclerosis, oxidative stress, and antioxidant protection in endothelium-derived relaxing factor action. *Prog. Cardiovasc. Dis.* 1995;38:129–154.

74. Indik, J.H., Goldman, S., and Gaballa, M.A. Oxidative stress contributes to vascular endothelial dysfunction in heart failure. *Am. J. Physiol. Heart Circ. Physiol.* 2001;281:H1767–H1770.

75. Doehner, W., Rauchhaus, M., Florea, V.G., Sharma, R., Bolger, A.P., Davos, C.H., Coats, A.J.S., and Anker, S.D. Uric acid in cachectic and non-cachectic CHF patients — relation to leg vascular resistance. *Am. Heart J.* 2001;141:792–799.

76. Zelis, R. and Flaim, S.F. Alterations in vasomotor tone in congestive heart failure. *Prog. Cardiovasc. Dis.* 1982;24:437–459.

77. Anker, S.D., Swan, J.W., Volterrani, M., et al. The influence of muscle mass, strength, fatiguability and blood flow on exercise capacity in cachectic and non-cachectic patients with chronic heart failure. *Eur. Heart J.* 1997;18: 259–269.

78. Gimpel, J.A., Lahpor, J.R., van der Molen, A.J., Damen, J., and Hitchcock, J.F. Reduction of reperfusion injury of human myocardium by allopurinol: A clinical study. *Free Radic. Biol. Med.* 1995;19:251–255.

79. Coghlan, J.G., Flitter, W.D., Clutton, S.M., et al. Allopurinol pretreatment improves postoperative recovery and reduces lipid peroxidation in patients undergoing coronary artery bypass grafting. *J. Thorac. Cardiovasc. Surg.* 1994;107:248–256.

80. Guan, W., Osanai, T., Kamada, T., Hanada, H., Ishizaka, H., Onodera, H., Iwasa, A., Fujita, N., Kudo, S., Ohkubo, T., and Okumura, K. Effect of allopurinol pretreatment on free radical generation after primary coronary angioplasty for acute myocardial infarction. *J. Cardiovasc. Pharmacol.* 2003;41: 699–705.

81. Ekelund, U.E., Harrison, R.W., Shokek, O., et al. Intravenous allopurinol decreases myocardial oxygen consumption and increases mechanical efficiency in dogs with pacing-induced heart failure. *Circ. Res.* 1999;85:437–445.

82. Ukai, T., Cheng, C.P., Tachibana, H., et al. Allopurinol enhances the contractile response to dobutamine and exercise in dogs with pacing-induced heart failure. *Circulation* 2001;103:750–755.

83. Saavedra, W.F., Paolocci, N., St John, M.E., Skaf, M.W., Stewart, G.C., Xie, J.S., Harrison, R.W., Zeichner, J., Mudrick, D., Marban, E., Kass, D.A., and Hare, J.M. Imbalance between xanthine oxidase and nitric oxide synthase signaling pathways underlies mechanoenergetic uncoupling in the failing heart. *Circ. Res.* 2002;90:297–304.

84. Cappola, T.P., Kass, D.A., Nelson, G.S., et al. Allopurinol improves myocardial efficiency in patients with idiopathic dilated cardiomyopathy. *Circulation* 2001;104:2407–2411.

85. Perez, N.G., Gao, W.D., and Marban, E. Novel myofilament Ca2+-sensitizing property of xanthine oxidase inhibitors. *Circ. Res.* 1998;83:423–430.

86. Doehner, W., Schoene, N., Rauchhaus, M., Leyva-Leon, F., Pavitt, D.V., Reaveley, D.A., Schuler, G., Coats, A.J.S., Anker, S.D., and Rainer Hambrecht. The effects of xanthine oxidase inhibition with allopurinol on endothelial function and peripheral blood flow in hyperuricemic patients with chronic heart failure — results from two placebo controlled studies. *Circulation* 2002;105: 2619–2624.

87. Farquharson, C.A., Butler, R., Hill, A., Belch, J.J., and Struthers, A.D. Allopurinol improves endothelial dysfunction in chronic heart failure. *Circulation* 2002;106:221–226.

88. Mercuro, G., Vitale, C., Cerquetani, E., Zoncu, S., Deidda, M., Fini, M., and Rosano, G.M. Effect of hyperuricemia upon endothelial function in patients at increased cardiovascular risk. *Am. J. Cardiol.* 2004;94:932–935.

89. Gavin, A.D. and Struthers, A.D. Allopurinol reduces β-type natriuretic peptide concentrations and haemoglobin but does not alter exercise capacity in chronic heart failure. *Heart* 2005;91:749–753.

90. Anker, S.D., Doehner, W., Rauchhaus, M., Sharma, R., Francis, D., Knosalla, C., Davos, C.H., Cicoira, M., Shamim, W., Kemp, M., Segal, R., Osterziel, K.J., Leyva, F., Hetzer, R., Ponikowski, P., and Coats, A.J. Uric acid and survival in chronic heart failure: Validation and application in metabolic, functional, and hemodynamic staging. *Circulation* 2003;107:1991–1997.

91. Struthers, A.D., Donnan, P.T., Lindsay, P., McNaughton, D., Broomhall, J., and MacDonald, T.M. Effect of allopurinol on mortality and hospitalisations in chronic heart failure: A retrospective cohort study. *Heart* 2002;87:229–234.

92. Freudenberger, R.S., Schwarz, R.P., Jr., Brown, J., Moore, A., Mann, D., Givertz, M.M., Colucci, W.S., and Hare, J.M. Rationale, design and organisation of an efficacy and safety study of oxypurinol added to standard therapy in patients with NYHA class III–IV congestive heart failure. *Expert Opin. Investig. Drugs* 2004;13:1509–1516.

93. Doehner, W. and Anker, S.D. Uric acid in chronic heart failure. *Semin. Nephrol.* 2005;25:61–66.

94. Doehner, W. and Anker, S.D. Xanthine oxidase inhibition for chronic heart failure. Is allopurinol the next therapeutic advance in heart. *Heart* 2005;91:707–709.

26 Tumor Necrosis Factor Inhibitors in Cancer-Associated Weight Loss Syndrome

Smitha Patiyil and Aminah Jatoi

Contents

Summary

The vast majority of patients with advanced cancer suffer from cancer-associated weight loss syndrome, which adversely affects prognosis and quality of life. This syndrome is characterized by loss of appetite, erosion of lean and adipose tissue and causes progressive weight loss, which is refractory to caloric supplementation. Tumor necrosis factor (TNF) plays a crucial role in the pathogenesis of this syndrome. Various strategies to inhibit TNF and thus reverse the cancer-associated weight loss syndrome are being explored. A randomized controlled study on the use of Pentoxifylline had not shown any benefit. Other inhibitors of TNF, namely Thalidomide and Melatonin, have shown some promising results. Currently, there is interest in testing novel inhibitors of TNF including infliximab and etanercept, which are approved for the treatment of Rheumatoid Arthritis and other inflammatory conditions. Ongoing clinical trials should define the role of these agents in the management of the cancer-associated weight loss syndrome.

26.1 INTRODUCTION

Tumor necrosis factor (TNF) is a key molecular mediator of cancer-associated weight loss. Discovered in the 1980s, this cytokine was very early on implicated in inflammation and sepsis and was therefore labeled "cachectin." Only later was "cachectin" found to be identical to another molecule known as TNF alpha, which had been implicated in the apoptosis of tumor cells.[1] The term "cachectin" is now virtually obsolete — typically this molecule is now referred to as TNF.

Since the discovery of TNF, three different lines of investigation have converged to suggest a rationale for focusing on this molecule as a therapeutic target for cancer-associated weight loss. First, over several decades, mounting evidence has shown that the cancer-associated weight loss syndrome carries with it grave consequences relevant to prognosis and quality of life in patients with advanced cancer. There is no question that the well-established, negative clinical impact of this syndrome — coupled with limited available therapeutic options — mandate the study of promising therapeutic strategies. Second, several recent studies suggest that TNF is a mediator of cancer-associated weight loss. In effect, these studies have identified TNF as a potential therapeutic target for the cancer-associated weight loss syndrome. Third, various inhibitors of TNF have emerged and now allow for the testing of inhibitors of this cytokine in the clinical setting. Among these relatively new agents are infliximab and etanercept. This chapter reviews some of the information relevant to all three of these lines of investigation.

26.2 OVERVIEW OF THE CANCER ANOREXIA/WEIGHT LOSS SYNDROME

The vast majority of patients with advanced solid tumors suffer from the cancer-associated weight loss syndrome. This syndrome is characterized by involuntary weight loss, a loss of appetite (or anorexia), a predominant wasting of lean tissue, a decline in functional status, and a shortened survival[2] (see Figure 26.1).

Highly distressing, this syndrome is also highly prevalent. Tchekmedyian and others evaluated 644 consecutive, ambulatory cancer patients who were receiving care in a cancer clinic only to find that 54% weighed less than their anticipated weight.[3] The perception among many oncologists is that among patients with advanced incurable malignancies, the prevalence of weight loss is likely to be even higher.

As noted above, the implications of this loss of weight are grave to say the least. Functional status — defined here as a patient's ability to perform various activities essential for daily living and enjoyment of life — declines with weight

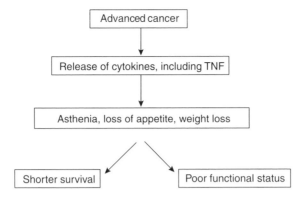

FIGURE 26.1   An overview of the cancer associated weight loss syndrome.

loss. Dewys et al.[4] have reported that the percentage of cancer patients with weight loss was significantly lower among patients with an Eastern Cooperative Oncology Group performance status of 0–1 vs. 2–4 ($p < .01$ among all nine tumor types). In other words, weight loss and a decline in performance score, or functional status, are intimately intertwined. Similarly, Finkelstein and others provided data that made this very same point.[5] With the use of questionnaires that assessed patients' activity level, these investigators observed that weight loss of >5% was associated with a poor functional status ($p = .004$).[5] Again, this clear association between loss of weight and decline in functional status suggests a potential cause and effect relationship between these two entities.

Weight loss is also associated with a poor prognosis. In the same study referenced above, Dewys et al.[4] and others found that patients who reported a loss of greater than 5% of premorbid weight over the preceding 6 months, died sooner than patients who reported weight maintenance. This prognostic effect occurred independently of tumor stage, tumor histology, and performance status, observations that underscore the powerful and independent predictive capability of this red-flag sign.

From a mechanistic standpoint, an erosion of lean tissue may account for much of what characterizes this syndrome and of what is described earlier. Indeed, an erosion of lean tissue appears to be a very real phenomenon in these weight-losing cancer patients. In a matched case control study, Cohn et al.[6] assessed body composition in 10 healthy individuals and 17 weight-losing cancer patients. Cancer patients appeared to waste muscle excessively compared to other body compartments such as adipose tissue. Although, to our knowledge, the direct prognostic effect of this erosion of lean tissue has not been clearly demonstrated in cancer patients, such evidence is forthcoming in patients with other diseases. For example, Kotler et al.[7] analyzed data from 32 AIDS patients who were in retrospect within 100 days of death. These investigators measured whole body 40 K. This naturally occurring isotope is particularly relevant here because over 97% of potassium is found in nonadipose tissue, thereby allowing for a reasonable estimate of body cell mass with the detection of 80 K. These investigators found a direct association between total body potassium, which approximates body cell mass, and survival ($r = 0.48$; $p = .0013$). This finding underscores the importance of lean tissue in contributing to survival and provides further evidence of a link between erosion of this body compartment and early death. This observation also begs the following two questions: (1) What is happening at the muscle level to explain this pattern of wasting? and (2) What sorts of mechanism-based interventions can be tested to attenuate this erosion of lean tissue and to perhaps result in an improvement in survival and quality of life? Attempts to answer these two questions are made in the following sections.

26.3 MECHANISMS AND MEDIATORS: RECENT DATA ON TUMOR NECROSIS FACTOR

There is much that remains unknown about lean tissue wasting in cancer patients. However, several studies have begun to focus on TNF. A host of other inflammatory cytokines, such as IL-1β and IL-6, has also been implicated, but a growing body of evidence continues to point to TNF as a particularly key mediator. It is important to point out that several previous studies suggest serum or plasma concentrations of TNF do not correlate with the presence or degree of weight loss or even with the presence or degree of anorexia. For example, a recent North Central Cancer Treatment Group study found that among 118 cancer patients with notable loss of appetite and weight, 54% had undetectable TNF concentrations in their serum.[8] However, others studies suggest a direct relationship between concentrations of this cytokine within peripheral blood mononuclear cells and cancer-associated weight loss.[9] The detection of this cytokine within peripheral blood mononuclear cells suggests that TNF may in fact be present, detectable, and active at the cell/tissue level, as opposed to at the serum level. In short, previous studies that failed to reveal TNF elevation in the serum or plasma of weight-losing cancer patients should not be invoked as part of a counterargument that this cytokine is at best only an effete mediator of muscle wasting.

To further complete this picture, Llovera et al.[10] others found that implantation of the Lewis lung carcinoma in gene knockout mice deficient in TNF demonstrated a different pattern of wasting compared to tumor-implanted wild-type mice. The mice that were deficient in TNF manifested rates of protein degradation that were much lower than the tumor-bearing wild-type mice. Moreover, the ubiquitin-proteasome system was activated to a lesser degree. In another study from this same group, Llovera et al.[11] also found that TNF led to an increase in expression of ubiquitin genes in skeletal muscle. This latter point is an important one because the ubiquitin-proteasome system appears to account for 80% of muscle wasting in cancer,[12] as shown by Baracos et al. who utilized an animal model in formulating these conclusions. Overall, these findings suggest that TNF plays an active role in the erosion of lean tissue and therefore appears to carry a major negative clinical impact among patients with advanced malignancies.

More recently, two other lines of evidence provide further insight into what occurs at the muscle level in patients with cancer-associated weight loss. There appears to be selective targeting of myosin heavy chains. Tumor necrosis factor in combination with interferon-gamma downregulates myosin heavy chains.[13] This effect appears to occur only with this combination of cytokines and does not occur with exposure to TNF exclusively. Second, nuclear factor kappa B (NF kappa B) has been implicated in muscle wasting as well.[14] Although studies have not specifically focused on cancer-associated weight loss, NF kappa B is upregulated by TNF, and it suppresses MyoD, a transcription factor that is essential for muscle differentiation and repair of damaged tissue. These two pieces of evidence provide further understanding of the potentially critical role of TNF in cancer-associated weight loss.

Some of the most provocative data to suggest the importance of TNF inhibition and the potential success of this approach in treating cancer-associated weight loss comes from Torelli et al.[15] These investigators examined changes in food intake and body weight in tumor-bearing rodents. In a placebo-controlled trial, they tested a dimeric, pegylated 55-kDa TNF receptor construct that acted as a TNF inhibitor. A total of 16 mice were studied. Mice that received the TNF inhibitor ate far better in an ad libitum setting and overall achieved much higher levels of caloric intake over a 1-week period compared to animals that received only vehicle or placebo. Mice that received the TNF inhibitor also gained weight, whereas the vehicle-only-exposed group did not. This study did not evaluate muscle mass or survival, but these findings are nonetheless highly provocative. The results of this study raise the possibility of testing TNF inhibition in a clinical setting with the hope of achieving similar favorable effects.

26.4 THERAPEUTICS: WHAT DOES NOT WORK — AND WHAT MIGHT WORK

Practicing oncologists know that apart from highly effective antineoplastic treatment, there is no successful therapy that reverses all aspects of the cancer-associated weight loss syndrome. Practicing oncologists also know that the majority of advanced solid tumor cancer patients suffer from chemotherapy-refractory malignancies. Hence, as stated earlier, the majority of patients with advanced cancer suffer from unremitting loss of weight, loss of appetite, and a decline in functional status.

What does work to reverse this syndrome? No therapy has been proven to reverse all aspects of this syndrome. Hormonal agents such as progestational agents and corticosteroids are used in clinical practice, but, at best, they provide only limited benefits. As reviewed elsewhere,[16] these agents improve appetite, but they provide no positive impact on global quality of life or survival. Moreover, their appetite-stimulatory effects are short-lived at best. These limited benefits, coupled with occasional side effects, lend credence to the contention that other promising agents deserve further investigation. Hence, the prospect of targeting

TNF in a clinical trial setting among weight-losing cancer patients is clearly justified.

Are there clinical data to suggest that TNF inhibition can result in reversal of the cancer-associated weight loss syndrome? In an effort to provide a balanced answer to this question, it is important to first cite data that impart a negative answer. Pentoxifylline is a xanthine derivative, and it also appears to prevent the synthesis of certain inflammatory cytokines such as TNF. To illustrate its effects on TNF, Noel et al.[17] studied patients who had received cadaveric kidney transplants. Ninety-six patients were randomly assigned to receive either pentoxifylline 1200 mg/day when their serum creatinine was acceptable vs. placebo. Over time, monthly circulating TNF concentrations were consistently lower among the pentoxifylline-treated patients as compared to the placebo-exposed patients. This finding suggests that pentoxifylline is in fact an effective agent for lowering circulating concentrations of TNF.

Despite this suppressive effect of pentoxifylline on TNF, this agent proved to be inactive in the treatment of the cancer-associated weight loss syndrome. In a North Central Cancer Treatment Group study, Goldberg et al.[18] conducted a trial that closed prematurely because of an absolute lack of efficacy of pentoxifylline in this setting. This placebo-controlled trial tested pentoxifylline 400 mg orally three times a day and found this agent did nothing to improve weight or appetite within a group of advanced cancer patients who were suffering from weight loss and loss of appetite. The investigators concluded that this particular TNF inhibitor was ineffective in the treatment of the cancer-associated weight loss syndrome.

Nonetheless, this study should not dampen enthusiasm for studying other inhibitors of TNF. The reasons to justify further study of TNF inhibitors are threefold. First, the study by Goldberg et al.[18] represents an isolated example of one negative endeavor. Second, whether the dose of pentoxifylline used in this study was achieving the desired threshold of TNF suppression was not assessed and, in fact, is unknown. It is true that the previously mentioned trial by Noel et al.[17] utilized the exact same dose of pentoxifylline as the Goldberg study and achieved a decline in circulating TNF concentrations. The fact remains, however, that it is totally unknown what degree of cytokine suppression is safe and potentially effective in treating cancer-associated weight loss. In the setting of cancer, tissue concentrations of TNF may be greater and higher doses of pentoxifylline might have been required to achieve favorable clinical effects at the tissue level. Third, several other inhibitors of TNF appear to be preliminarily effective in the treatment of cancer-associated weight loss. Two of these agents are discussed below: thalidomide and melatonin.

Thalidomide is a promising agent (see Table 26.1). It has been associated with weight gain, an increase in lean body mass, and possibly even maintenance of functional status. Studies have invoked downregulation of tumor necrosis production by mononuclear cells to explain the effects of thalidomide on lean tissue. Previous investigations suggest that thalidomide shortens the half-life of TNF mRNA.[19] As an immunomodulator, however, thalidomide has also demonstrated dose-dependent, bidirectional regulation of TNF.

TABLE 26.1
Preliminary Benefits of Thalidomide in Weight-Losing Cancer Patients

Better appetite[a]
Weight gain[b]
Less nausea[a]
Improved sense of well being[a]
Better sleep[a]

[a] Bruera E., Neumann C.M., Pituskin E., et al. *Ann. Oncol.* 10: 857–9, 1999.
[b] Khan Z.H., Simpson E.J., Cole A.T., Holt M., MacDonald I., Pye D., Austin A., and Freeman J.G. *Aliment. Pharmacol. Ther.* 17: 677–82, 2003.

It is therefore not surprising that at least four studies in cancer patients have been reported and all suggest that thalidomide may be effective in the treatment of cancer-associated weight loss. First, in a preliminary investigation, Mahmoud et al.[20] administered thalidomide to a small 15-patient cohort. All patients were begun on thalidomide 50 mg/day and were titrated up to 100 mg/day if no improvements in appetite were noted at two weeks. All patients had to have received at least two weeks of treatment to be considered evaluable, thus defining the 15-patient cohort described above. Nine of these patients noted an improvement in appetite and as many as five actually gained weight with median weight being 48.4 kg at baseline and increasing to 54.9 kg on day 21. It is important to point out that this study did not allow for concomitant cancer treatment or for the use of other cancer agents to boost appetite or weight, and therefore the favorable effects described above are more likely attributable to thalidomide, as opposed to another concomitant pharmacological intervention. Thus, these preliminary findings suggest this approach of TNF inhibition is promising.

Similarly, in a second report, Boasberg et al.[21] observed weight stability in a cohort of 15 cancer patients with a dose of 100 to 200 mg per day of thalidomide. All members of this cohort reported a weight loss of 5% of body weight over the preceding 6 months, and all had metastatic disease. Thalidomide was well tolerated and over a 6-week period all patients managed to maintain their weight. Although quality of life and adverse event data are not reported in detail, one minor point is that patients overall reported sleeping better with thalidomide. Again, this study provides further data on the promise of thalidomide.

Third, Khan et al.[22] evaluated 11 patients with incurable esophageal cancer but without notable dysphagia. In this phase II study, these cancer patients were treated with an isocaloric diet for 2 weeks followed by thalidomide 200 mg orally

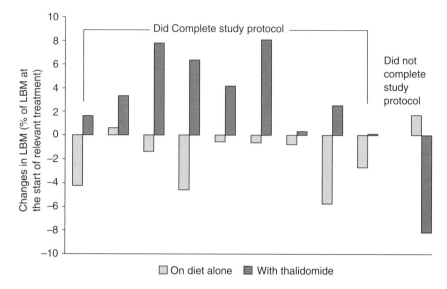

FIGURE 26.2   (Taken from Khan Z.H., Simpson E.J., Cole A.T., Holt, M., Macdonald, I., Pye, D., Austin, A., and Freeman, J.G. *Aliment. Pharmacol. Ther.* 17: 677–82, 2003. With permission.).

daily. Again, these patients had no severe dysphagia that precluded compliance. Ten patients were ultimately evaluable. During the first 2 weeks, nine of ten patients lost weight with diet alone. During the subsequent two weeks after the initiation of thalidomide, a mean weight gain of 1.29 kg was observed (see Figure 26.2 and Figure 26.3). A total of eight out of ten of these patients gained weight, and dual x-ray absorptiometry suggested that this weight gain consisted of lean tissue. Small and exploratory in nature, this phase II study is certainly not definitive in its conclusions but it suggests that thalidomide perhaps by means of its suppressive effects on TNF might play a role in the treatment of the cancer-associated weight loss syndrome.

Finally, in one of the earlier studies on thalidomide for the treatment of cancer-associated weight loss, Bruera et al.[23] tested this drug in 72 cancer patients. This cohort was treated with thalidomide 100 mg/night for a period of 10 days. As often occurs with such end-of-life studies, 35 patients dropped out of the study because of cancer-related morbidity. Of the remaining patients, sizable percentages reported improvements in various symptoms. Improvement in insomnia occurred in 69 %, improvement in nausea in 44%, and improvement in appetite in 63%. Finally, 53% of patients reported an improvement in their overall sense of well-being. As part of this preliminary investigation, these findings were compared to historical data from 28 cancer patients who had been treated with megestrol acetate alone. Overall treatment with thalidomide appeared to provide more favorable effects: mean difference $\pm$ standard deviation: -1.09 ± 2.67 vs. 0.04 ± 1.71 for nausea ($p = .05$); -2.21 ± 2.83 vs. -1.03 ± 2.49 for appetite ($p = .073$); and -1.65 ± 3.19

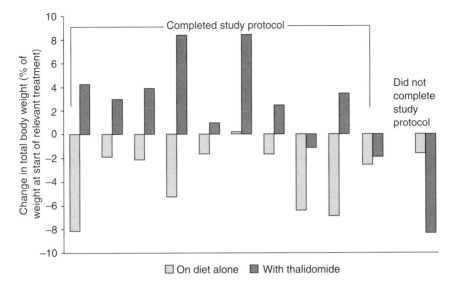

FIGURE 26.3 (Taken from Khan Z.H., Simpson E.J., Cole A.T., Holt, M., Macdonald, I., Pye, D., Austin, A., and Freeman, J.G. *Aliment. Pharmacol. Ther.* 17: 677–82, 2003. With permission.).

vs. -0.61 ± 1.42 for sense of well being ($p = .033$), for thalidomide- vs. megestrol acetate-treated patients, respectively. Although this study was neither randomized nor double-blinded, it nonetheless suggests that thalidomide may play a role in the supportive care of weight-losing cancer patients with advanced disease. Moreover, when reviewed in context with the three other studies described earlier, this study provides a powerful impetus for the further investigation of TNF blockade.

Although thalidomide appears as if it might be a promising therapy for patients with the cancer-associated weight loss syndrome, it does carry with it side effects that make its testing in cancer patients potentially problematic. Common major side effects consist of somnolence, rash, sensorimotor-peripheral neuropathy, and constipation. Other more infrequent side effects include mood changes, dry mouth, headache, nausea, edema, dry skin, pruritis, bradycardia, thyroid dysfunction, and alterations in serum glucose levels. Teratogenicity is another risk but in some respects is perhaps less of a concern among weight-losing patients with advanced malignancies. However, because of this drug's notorious history relevant to birth defects, careful screening and counseling are mandatory before prescribing this drug even to patients with incurable malignancies. Hence a less toxic, simpler intervention that specifically blocks tumor necrosis would be preferable.

In addition to thalidomide, another agent that appears promising and that may work by means of inhibition of TNF is melatonin. This hormone is secreted by the pineal gland and plays a role in neuroendocrine regulation. Preliminary clinical data also suggest that TNF appears to drop in cancer patients who receive

melatonin. Lissoni et al.[24] studied this hormone in 100 cancer patients with metastatic disease. In a comparative trial, an oral dose of melatonin 20 mg/day appeared
to stabilize weight whereas patients who were receiving supportive care exclusively had high rates of weight loss (10% or more of baseline weight loss) ($p < .01$).
In a second study from this same group, seventy patients with advanced non-small
cell lung cancer were randomly assigned to receive cisplatin and etoposide vs.
this same chemotherapy combination along with melatonin 20 mg/day orally.[25]
Once again, a larger percentage of melatonin-treated patients manifested weight
stability. These two studies strongly suggest that melatonin merits further study in
weight-losing cancer patients with advanced disease. Although it is likely that this
hormone exerts these promising effects on weight by means of multiple different
mechanisms, preliminary data that suggest suppression of TNF provide further
justification for focusing on this cytokine as a potential therapeutic target.

26.5 TUMOR NECROSIS FACTOR BLOCKADE WITH INFLIXIMAB AND ETANERCEPT: A POTENTIAL TREATMENT STRATEGY

A few different anti-TNF agents are available, but, to our knowledge, ongoing clinical trials in patients with cancer-associated weight loss have focused on infliximab
and etanercept. Infliximab is a highly specific monoclonal antibody that targets
TNF. This agent has demonstrated efficacy in both rheumatoid arthritis and Crohn's
disease. It is highly specific in blocking the binding of TNF to its p55 and p75
receptors. Yet, at the same time and in contrast to etanercept, infliximab does not
counteract the effects of lymphotoxin, a cytokine that shares some of the same
receptors as TNF. Infliximab is an IgG monoclonal antibody, formulated as a result of DNA recombinant methodology and following a single injection this agent
demonstrates a predictable, dose-dependent increase in the serum after administration of doses of 1 to 20 mg/kg of body weight. Liver or kidney dysfunction does
not alter this pharmacokinetic profile and pharmacokinetics remains consistent
among patients from different demographic groups.

In addition, infliximab also holds a reasonable safety profile, having been
administered to thousands of patients thus far. Hundreds of these patients have
participated in placebo-controlled trials, and based on toxicity data from these
trials, infliximab has been approved by the Food and Drug Administration for
the treatment of rheumatoid arthritis and Crohn's disease. However, whenever an
agent is being tested in a new clinical setting prudent safety monitoring is indicated. Three particular concerns remain germane to cancer patients. First, because
TNF is an important mediator of inflammation, there have been concerns that
its suppression may alter the host's ability to fight infections. Reports of activation of dormant tuberculosis have been recently forthcoming.[26] Other infections
also remain of concern, especially among patients who are already somewhat
immunosuppressed from cancer therapy. At the same time, the fact that many
patients with Crohn's disease suffer with infected fistulas but have not experienced increased rates of infection at these sites might provide reassurance. Perhaps

with prudent monitoring including halting drug administration in the setting of an infection, infliximab might be tested safely in weight-losing cancer patients or in cancer patients at risk for weight loss. Second, another concern that has perhaps not been totally explained revolves around congestive heart failure. Earlier trials with infliximab in heart failure patients resulted in a higher death rate among patients with particularly severe heart failure.[27] The reasons behind these adverse outcomes remain unexplained, but it appears judicious to refrain from exposing cancer patients with severe heart failure to infliximab and to remain vigilant for cardiac-related adverse events that might occur in other cancer patients. Third, another area of concern involves the development of new malignancies, such as lymphomas. Previous studies suggest perhaps a slight increase in risk of lymphomas in patients who had received anti-TNF therapy for other indications, but there is no clear consensus on a cause and effect relationship.[28] Although this issue is less of a concern among patients with incurable cancer with a limited life expectancy and a great deal of other morbidity, close monitoring for second malignancies is nonetheless indicated.

Another potential issue revolves around the administration of infliximab to cancer patients with weight loss in the absence of data that have previously demonstrated safety with concomitant chemotherapy administration. Certainly, earlier studies have examined the concomitant use of infliximab and methotrexate in patients with rheumatoid arthritis. An increase in adverse events was not observed with concomitant therapy, and specifically there was no increased risk of myelosuppression and infection. However, infliximab serum concentrations tended to be higher among patients who were receiving methotrexate. In the setting of cancer, the Mayo Clinic recently published anecdotal experience with infliximab 5 mg/kg/day given intravenously on a weekly schedule on weeks 1, 3, 5 with docetaxel given on weeks 1, 2, 3, 4, 5, and 6 of an 8-week treatment cycle.[29] Using Common Toxicity Criteria from the National Cancer Institute, these investigators observed no grade 4 or 5 events among 4 patients who enrolled on this lead-in portion of a clinical trial. Moreover, when the investigative team evaluated pharmacokinetics of infliximab, they found no evidence that docetaxel was increasing the bioavailability of the former. These data are highly preliminary, but they do suggest that infliximab and docetaxel can be administered together on the above schedule with some degree of safety and they do put in place the groundwork for further clinical investigation with this combination.

Etanercept is a similar compound, representing a dimeric soluble recombinant form of the extracellular domain of the p75 TNF receptor. This agent is used for the treatment of rheumatoid arthritis, ankylosing spondylitis, and psoriatic arthritis. Therapy usually consists of 25 mg subcutaneous injections twice a week. Many of the same concerns for adverse events — most particularly risk of infection — pertain to the use of this drug as well and have recently been reviewed by Scheinfeld.[30]

Preliminary data also suggest that etanercept can be safely given to cancer patients. In a preliminary 13-patient, as yet unpublished, placebo-controlled study, our group did not find any notable adverse events that could be directly attributable to etanercept when this agent was given to patients at the doses described above. Many of these advanced cancer patients were receiving concomitant chemotherapy. These safety data provide preliminary justification for moving forward with a larger trial that tests etanercept for cancer-associated weight loss.

Is anti-TNF therapy potentially detrimental because of risk of tumor growth in patients with an underlying malignancy? A growing body of evidence has suggested that tumor growth is unlikely to be a concern with either of these agents, and some data suggest the opposite may occur. In fact, suppression of TNF might play a role in preventing cancer and might even carry some anticancer effects. As a case in point, Moore et al.[31] found that mice deficient in TNF were resistant to skin carcinogenesis. As noted earlier, data are sparse in denoting outcomes with these cytokine inhibitors in patients with metastatic cancer, but to date, there is no definitive evidence that low doses of anti-TNF treatment causes cancers to grow or metastasize at greater rates. Admittedly, early work with TNF had suggested that this cytokine might play a role in cancer treatment — hence the term "tumor necrosis factor." Subsequent trials failed to demonstrate that this cytokine is a useful antineoplastic agent, but to date no studies have shown that suppression of TNF causes tumor growth.

26.6 A Brief Overview of Ongoing Clinical Trials from the North Central Cancer Treatment Group

Given the foregoing, the North Central Cancer Treatment Group is embarking on two clinical trials to examine the role of TNF blockade in patients with advanced malignancies. The first study focuses on patients with non-small cell lung cancer and does allow for the inclusion of patients with no prior weight loss in an effort to explore whether this treatment will prevent such weight loss. All patients will receive chemotherapy with docetaxel and then will be randomly assigned to receive either infliximab or placebo. This 220-patient trial includes weight as its primary endpoint, but a variety of other endpoints such as quality of life, survival, body composition, and TNF polymorphisms will also be assessed. The latter will be assessed for whether they correlate with weight change at the time of study enrollment and with weight maintenance or weight gain after enrollment. The study is anticipated to complete its accrual in 3 years and results are expected to shed light on whether there is a role for an agent such as infliximab in treating or preventing weight loss. The second study includes patients who are suffering from loss of weight or loss of appetite at the time of study enrollment. Patients can have a variety of malignancies, but all must have incurable cancer. All patients will be randomly assigned to etanercept vs. placebo. Patients are allowed to be receiving chemotherapy at the same time they are receiving etanercept or placebo. Again, the primary endpoint of this study focuses on weight, but other parameters such as appetite, global quality of life, and survival will also be assessed.

Although these studies are unlikely to provide the final solution to this troubling syndrome of cancer-associated weight loss, it is hoped that they will generate

TABLE 26.2

Comparison of purported TNF Inhibitors

Agent	Mechanism of TNF inhibition	Increase in weight	Increase in appetite	Effect on sleep	Side effects
Pentoxifylline	Inhibits TNF transcription	No[18]	No[18]	NR	No toxicities observed
Thalidomide	Degrades TNF m-RNA	Yes[20]	Yes[20]	Yes[21]	Constipation, rash, sedation, teratogen
Melatonin	Decreases circulating TNF level	No*[25]	NR	NR	*Reduced incidence of weight loss. Few reports of headache and paradoxical insomnia
Etanercept/ Infliximab	Endogenous inhibition	?	?	?	

preliminary data to clarify whether further study of highly specific TNF inhibitors is warranted. In summary, the hypothesis that anti-TNF can help weight-losing cancer patients appears plausible, and opportunities to begin to test this hypothesis have now come to fruition (see Table 26.2 for a summary of various TNF antagonists). Over the next few years, preliminary answers on the role of TNF inhibitors in treating cancer-associated weight loss will likely become available.

REFERENCES

1. Beutler B. and Cerami A. Cachectin: more than a tumor necrosis factor. *N. Engl. J. Med.* 316: 379–85, 1987.
2. Strasser F. and Bruera E.D. Update on anorexia and cachexia. *Hematol. Oncol. Clin. North. Am.* 16: 589–617, 2002.
3. Tchekmedyian N.S. Costs and benefits of nutrition support in cancer. *Oncology* 9: 79–84, 1995.
4. Dewys W.D., Begg C., Lavin P.T., et al. Prognostic effect of weight loss prior to chemotherapy in cancer patients. Eastern Cooperative Oncology Group. *Am. J. Med.* 69: 491–7, 1980.
5. Finkelstein D.M., Cassileth B.R., Bonomi P.D., et al. A pilot study of the Functional Living Index-Cancer (FLIC) scale for the assessment of quality of life for metastatic lung cancer patients. An Eastern Cooperative Oncology Group study. *Am. J. Clin. Oncol.* 11: 630–3, 1988.
6. Cohn S.H., Gartenhaus W., Sawitsky A., et al. Compartmental body composition of cancer patients by measurement of total body nitrogen, potassium, and water. *Metabolism* 30: 222–9, 1981.

7. Kotler D.P., Tierney A.R., Wang J., and Pierson R.N. Magnitude of body cell mass depletion and the timing of death from wasting in AIDS. *Am. J. Clin. Nutr.* 50: 444–7, 1989.

8. Jatoi A., Egner J., Loprinzi C.L., et al. Investigating the utility of serum cytokine measurements in a multi-institutional cancer anorexia/weight loss trial. *Support. Care Cancer* 12: 640–4, 2004.

9. Mantovani G., Maccio A., Lai P., et al. Cytokine activity in cancer-related anorexia/cachexia: role of megestrol acetate and medroxyprogesterone acetate. *Semin. Oncol.* 25: 45–52, 1998.

10. Llovera M., Garcia-Martinez C., Lopez-Soriano J., et al. Role of TNF receptor 1 in protein turnover during cancer cachexia using gene knockout mice. *Mol. Cell. Endocrinol.* 142: 183–9, 1998.

11. Llovera M., Garcia-Martinez C., Lopez-Soriano J., et al. Protein turnover in skeletal muscle of tumor-bearing transgenic mice overexpressing the soluble TNF receptor-1. *Cancer Lett.* 130: 19–27, 1998.

12. Baracos V.E., DeVivo C., Hoyle D.H., and Goldberg A.L. Activation of the ATP-ubiquitin-proteasome pathway in skeletal muscle of cachectic rats bearing hepatoma. *Am. J. Physiol.* 268: E996–E1006, 1995.

13. Acharyya S., Ladner K.J., Nelsen L.L., et al. Cancer cachexia is regulated by selective targeting of skeletal muscle gene products. *J. Clin. Invest.* 114: 370–8, 2004.

14. Langen R.C., Van Der Velden J.L., et al. Tumor necrosis factor-alpha inhibits myogenic differentiation through MyoD protein destabilization. *FASEB J.* 18: 227–37, 2004.

15. Torelli G.F., Meguid M.M., Moldawer L.L., et al. Use of recombinant soluble TNF receptor in anorectic tumor-bearing rats. *Am. J. Physiol.* 277: R850–5, 1999.

16. Jatoi A., Kumar S., Sloan J.A., and Nguyen P.L. On appetite and its loss. *J. Clin. Oncol.* 18: 2930–2, 2000.

17. Noel C., Copin M.-C., Hazzan M., et al. Immunomodulatory effect of pentoxifylline during human allograft refection: involvement of tumor necrosis factor and adhesion molecules. *Transplantation* 69: 1102–7, 2000.

18. Goldberg R.M., Loprinzi C.L., Mailliard J.A., et al. Pentoxifylline for treatment of cancer anorexia and cachexia? A randomized, double-blind, placebo-controlled trial. *J. Clin. Oncol.* 13: 2856–9, 1995.

19. Makonkawkeyoon S., Limson-Pobre R.N., Moreira A.L., et al. Thalidomide inhibits the replication of human immunodeficiency virus type 1. *Proc. Natl. Acad. Sci. USA* 90: 5974–8, 1993.

20. Mahmoud F.A., Walsh D., Davis S., et al. A dose titration study of thalidomide in cancer anorexia. *Proc. Am. Soc. Clin. Oncol.* 22: 789, 2003.

21. Boasberg P., O'Day S., Weisberg M., et al. Thalidomide induced cessation of weight loss and improved sleep in advanced cancer patients with cachexia. *Proc. Am. Soc. Clin. Oncol.* #2396, 2000.

22. Khan Z.H., Simpson E.J., Cole A.T., et al. Oesophageal cancer and cachexia: the effect of short-term treatment with thalidomide on weight loss and lean body mass. *Aliment. Pharmacol. Ther.* 17: 677–82, 2003.

23. Bruera E., Neumann C.M., Pituskin E., et al. Thalidomide in patients with cachexia due to terminal cancer: preliminary report. *Ann. Oncol.* 10: 857–9, 1999.

24. Lissoni P., Paolorossi F., Tancini G., et al. Is there a role for melatonin in the treatment of neoplastic cachexia? *Eur. J. Cancer* 32A: 1340–3, 1996.

25. Lissoni P., Chilelli M., Villa S., et al. Five years survival in metastatic non-small cell lung cancer patients treated with chemotherapy alone or chemotherapy and melatonin: a randomized trial. *J. Pineal. Res.* 35: 12–15, 2003.

26. Keane J., Gershon S., Wise R.P., et al. Tuberculosis associated with infliximab, a tumor necrosis factor alpha-neutralizing agent. *N. Engl. J. Med.* 345: 1098–104, 2001.

27. Chung E.S., Packer M., Lo K.H., et al. Randomized, double-blind, placebo-controlled, pilot trial of infliximab, a chimeric monoclonal antibody to tumor necrosis factor-alpha, in patients with moderate-to-severe heart failure: results of the anti-TNF Therapy Against Congestive Heart Failure (ATTACH) trial. *Circulation* 107: 3133–40, 2003.

28. Wolfe F. and Michaud K. Lymphoma in rheumatoid arthritis: the effect of methotrexate and anti-tumor necrosis factor therapy in 18, 572 patients. *Arthritis Rheum.* 50: 1740–51, 2004.

29. Jatoi A., Jett J.R., Sloan J., et al. A pilot study on safety and pharmacokinetics of infliximab for the cancer anorexia/weight loss syndrome in non-small cell lung cancer patients. *Support Care Cancer* 12: 859–63, 2004.

30. Scheinfeld N. A comprehensive review and evaluation of the side effects of the tumor necrosis factor alpha blockers etanercept, infliximab, and adalimumab. *J. Dermatolog. Treat.* 15: 280–94, 2004.

31. Moore R.J., Owens D.M., Stamp G., et al. Mice deficient in tumor necrosis factor-alpha are resistant to skin carcinogenesis. *Nat. Med.* 5: 828–31, 1999.

VI

Future Drug Targets

27 Central Pharmacological Targets

Daniel L. Marks

CONTENTS

27.1 INTRODUCTION: THE CENTRAL CONTROL OF BODY WEIGHT

Under normal circumstances, body weight is regulated with remarkable precision. Indeed, body mass and fat mass are regulated to within 0.5 to 1% per year under basal conditions, even in the face of disturbances in energy balance brought about by changes in food intake or exercise. The mechanism whereby this fidelity is maintained has been the subject of an intense research effort over the last decade. The investigation into the mechanisms of peripheral feedback control of food intake and body mass intensified dramatically with the discovery of the peptide hormone leptin. Leptin is secreted by adipocytes and regulates adiposity and metabolic rate by reducing food intake and increasing energy expenditure.[1-4] Remarkably, leptin is a member of the IL-6 (interleukin-6) superfamily of proteins and has many biochemical features of a cytokine molecule.[1,5] Experimental elevation of leptin within the physiological range produces weight loss and hypophagia, while decreased leptin levels lead to the complex neuroendocrine response that occurs during starvation.[6,7] Leptin secretion is increased by both central and systemic immunological challenge and has therefore been proposed as a potential mediator of inflammation-induced anorexia.[8-11] Thus, the data regarding the mechanisms whereby leptin regulates body weight may provide us with important clues for understanding body weight regulation during chronic illness. For example, in addition to peripheral mechanisms, we now understand that the responsibility for maintaining energy homeostasis is shared by several regions of the brain, including the brainstem, hypothalamus, limbic structures, and the cortex (for recent reviews, see References 12 and 13. Indeed, leptin is thought to exert its effects on feeding and metabolism primarily via regulation of hypothalamic neurons.[14–19] The majority of research has focused on understanding feeding regulatory centers in the hypothalamus and brainstem, and we now have a basic understanding of the neuroanatomical organization of these regions. This chapter will first discuss the central anatomy involved in the regulation of energy homeostasis and then highlight what is known about the role of the various neuronal systems in the development of cachexia. Because research in this area remains in its infancy, much of this discussion will, by necessity, be speculative, particularly as it relates to human disease. Nonetheless, research in this field has already provided exciting possibilities for the pharmacotherapy of cachexia and an understanding of the central control of body weight and muscle mass will continue to be essential for medical professionals working to combat this devastating disorder of nutrient balance.

27.2 THE ROLE OF THE HYPOTHALAMUS IN THE REGULATION OF ENERGY BALANCE

Much of what we currently understand regarding the function of the hypothalamus in body weight regulation has its roots in classical neuroanatomical and ablation studies. The idea that there were "feeding centers" and "satiety centers" in the hypothalamus was established by lesioning studies performed more than 50 years ago.[20–22] The basic layout of the hypothalamus that was defined during this period remains central to our current understanding of hypothalamic feeding circuits, and is shown in Figure 27.1. One of the structures that has received a great deal of attention from researchers is the arcuate nucleus (ARC). The ARC occupies a unique position between the floor of the third ventricle and the median eminence. Because neurons in the ARC respond to a large variety of hormones, cytokines, and nutrients, this nucleus is thought to be functionally outside of the blood brain barrier.[23] The ARC also has extensive reciprocal connections with other hypothalamic nuclei including the periventricular nucleus (PVN), the ventromedial hypothalamic nucleus (VMH), the dorsomedial hypothalamic nucleus (DMH), and the lateral hypothalamic area (LHA). Collectively, these hypothalamic nuclei are thought to play a critical role in the physiological response to illness, including the production of fever, the activation of the hypothalamic-pituitary-adrenal (HPA) axis, decreased locomotion, anorexia, and loss of lean body mass. Thus, the ARC is in an ideal anatomical position to relay peripheral signals of illness (e.g., cytokines) to higher brain centers involved in the production of illness behaviors and cachexia. It is particularly interesting that this area of the brain is known to express receptors for

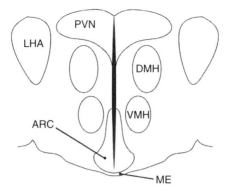

FIGURE 27.1 A coronal section of the rat hypothalamus showing the major hypothalamic anatomy. ARC, arcuate nucleus; ME, median eminence; VMH, ventromedial nucleus; DMH, dorsomedial nucleus; PVN, paraventricular nucleus; LHA, lateral hypothalamic area. Note the position of the ARC in close proximity to both the highly vascular median eminence and the third ventricle. The symbols represent the general anatomical location of the neuropeptide systems described in this text: ^Corticotropin Releasing Hormone (CRH), # Neuropeptide Y (NPY), and Agouti Related Peptide (AgRP), * Proopiomelano-cortin (POMC) and Cocaine and Amphetamine-Regulated Transcript (CART), + Orexin, @ Melanin Concentrating Hormone (MCH).

cachexigenic cytokines (including interleukin-1 beta, tumor necrosis factor-alpha, and leukemia inhibitory factor).[24–27] Furthermore, it is known that peripheral inflammation causes rapid and extensive activation of the immediate early gene cFos within this nucleus.[28] As discussed below, there are two primary neuronal systems within the arcuate nucleus that are thought to have reciprocal effects on energy balance. The primary feeding stimulatory population is comprised of neurons that express both neuropeptide Y (NPY) and agouti related peptide (AgRP), whereas the feeding inhibitory pathway is made up of neurons co-expressing proopiomelanocortin (POMC) and cocaine- and amphetamine-regulated transcript (CART).[29–31]

One of the most important hypothalamic nuclei innervated by the arcuate nucleus is the VMH. Early lesioning studies established this nucleus as an important satiety center, although it is now clear that this is neither an entirely specific or exclusive role for this nucleus. Although neurons in this nucleus are known to express the leptin receptor and to be sensitive to glucose, relatively little is known about the types of neurons found in this nucleus. Steroidogenic factor-1 (SF-1) is an orphan receptor that is also expressed in the VMH where it is known to play a role in organizing the architecture of this nucleus.[32] It is not currently known if this receptor plays a significant physiological role in body weight regulation and at this point has no obvious role in the treatment of cachexia. VMH neurons also express histamine receptor subtypes and corticotropin-releasing factor 2 receptors (CRF-R2), both of which have been implicated in the control of body weight, as discussed below.[33,34] Finally, neurons in the VMH are known to be responsive to circulating glucose and may play a role in terminating feeding as blood glucose rises.[35,36]

Signals from the ARC are also highly integrated with signals from the classical "feeding center" in the LHA.[37] Lesions of this area of the hypothalamus lead to many features typical of cachexia including anorexia, changes in resting metabolic rate, and loss of lean body mass whereas stimulation of this area leads to opposite effects.[38] In contrast to the VMH, much more is known about the identity of the appetite regulating neurons in the LHA. Neurons which express the orexigenic peptides melanin-concentrating hormone (MCH) or orexin are found in this nucleus, as are the type 5 NPY receptors (NPY Y_5) which are thought to be responsible for the orexigenic actions of NPY.[39] The potential role of each of these neuropeptides in the regulation of body weight is considered separately, below.

Other hypothalamic areas thought to be involved in body weight homeostasis in the hypothalamus include the DMH and the PVN. The DMH is thought to integrate a number of signals from other hypothalamic and brainstem nuclei and expresses both insulin and leptin receptors. The PVN receives dense innervation from the ARC and the LHA as well as from feeding centers in the brainstem. Neurons in this region of the hypothalamus also express CRF, a neuropeptide known to be involved in the regulation of feeding.

27.3 THE ROLE OF THE BRAINSTEM IN ENERGY BALANCE

While most of the recent research into the central regulation of body weight has focused on the role of the hypothalamus, it is clear that nuclei in the brainstem play a critical role as well. Several of the neuropeptide systems present in the hypothalamus are also found in the brainstem (e.g., POMC, CART, and neuromedin U), and the brainstem is known to respond to cytokine signals from the periphery. In particular, the nucleus of the solitary tract (NTS) appears to play a unique and important role in the regulation of food intake. This nucleus contains cell bodies that express POMC, as well as cells expressing CART (for review, see Reference 12). Furthermore, there is a well-developed network of POMC fibers and melanocortin receptors in the brainstem, and stimulation of these receptors is known to inhibit food intake.[40] The fact that the NTS is in close anatomical proximity to a circumventricular organ (the area postrema) raises the intriguing possibility that this area may have direct access to cytokines and other circulating compounds elaborated during chronic disease.

27.3.1 Hypothalamic Systems Involved in Energy Balance and their Potential Role in Cachexia

There are a large number of neuropeptides and "classical" neurotransmitters in the hypothalamus and brainstem. Many of these have been shown to play a role in the regulation of food intake, illness behavior, and energy balance, at least in experimental systems. Thus, any attempt at a comprehensive review of the neural control of these processes and their potential derangement in chronic illness will by necessity be a simplification. Nonetheless, it is possible to build a basic framework for the understanding of these neural systems and to suggest the potential role of these systems in designing treatments for cachexia.

27.4 NEUROPEPTIDE SYSTEMS IN THE ARCUATE NUCLEUS

As mentioned above, neurons in the ARC play a key role in the regulation of food intake, metabolic rate, and nutrient partitioning. One of the most important hypothalamic systems is the melanocortin system, here defined as the hypothalamic and brainstem neurons expressing POMC, the hypothalamic neurons expressing NPY and the melanocortin antagonist AgRP, and the neurons downstream of these systems. POMC neurons are thought to provide an important tonic inhibition of food intake and energy storage, primarily via production and release of alpha-melanocyte stimulating hormone (alpha-MSH) from the POMC precursor. Alpha-MSH binds to central melanocortin receptors (including the type 4 melanocortin receptor, MC4-R) where it acts to inhibit food intake and stimulate metabolic rate.[41] POMC neurons in the arcuate nucleus express the leptin receptor and MC4-R KO mice are leptin resistant, leading several investigators to propose that melanocortin neurons mediate many of the anorexic effects of elevated leptin.[42,43] Further investigation of melanocortin signaling led to the cloning and characterization of AgRP, an endogenous hypothalamic melanocortin receptor antagonist.[44] One of the more intriguing phenotypes of the MC4-R KO mouse is a tremendous increase in lean body mass relative to WT littermates.[45] This has also been observed in humans with defective melanocortin signaling.[46,47] Given that maintenance of lean body mass is perhaps the most critical therapeutic goal of cachexia therapy, this phenotype is of particular relevance in this condition. Indeed, as discussed below, therapy based on antagonism of the MC4-R has already shown benefit in the preservation of lean body mass in experimental models of cachexia.

POMC neurons are leptin responsive and may mediate many of its feedback effects.[48] In addition, the melanocortin system has effects on weight and metabolism that are independent of leptin feedback and leptin-deficient animals with disrupted melanocortin signaling remain sensitive to leptin administration.[49,50] In general, POMC neurons represent a logical target for leptin and other cytokine-mediated feedback on feeding behavior and metabolic control, and recent data suggest a role in anorexia and cachexia as well.[51–54] Specifically, it has been shown that genetic deletion of the MC4-R in mice ameliorates the cachexia that normally accompanies both acute illness (due to injection of lipopolysaccharide, LPS), and more prolonged illness (tumor growth).[52,55] Furthermore, intracerebroventricular (icv) injection of peptide melanocortin antagonists has also been shown to reverse many features of cachexia brought about by injections of LPS or cytokines in mice, or brought about by tumor growth in mice or rats.[52–55] In each of these studies, benefit has been shown in multiple aspects of cachexia, including increasing food intake, decreasing basal metabolic rate, and prevention or reversal of the loss of lean body mass. These results have led to the development of small molecule antagonists of the MC4-R as potential therapeutic agents for cachexia. These compounds appear to be effective in preventing the anorexia and loss of lean body mass that normally accompanies tumor growth in animal models, paving the way for trials in patient populations.[56,57]

Although AgRP clearly functions as an antagonist at the MC4-R both *in vitro* and *in vivo*, there is evidence that it may have other effects that are not due to receptor antagonism. Perhaps the most compelling evidence in this regard is the extremely long duration of the orexigenic effect of AgRP when injected icv. Hagan et al.[58] demonstrated that a single injection of AgRP could induce hyperphagia for as long as one week in the rat. Furthermore, while the animals were hyperphagic on the second day after the injection, they were normally sensitive to further injections of melanocortin agonists. These data are not consistent with a role as a competitive antagonist, and suggest that there may be genetic effects that depend on AGRP signaling, perhaps by inverse agonism at the MC4-R.[59] Thus, another rational therapeutic strategy would be to create compounds that not only function as melanocortin antagonists, but rather as AgRP mimetics. These compounds would have the ability not only to block the binding of alpha-MSH to the MC4-R, but would also be able to decrease the constitutive cyclic AMP generation that this receptor is known to have.[59]

In addition to the melanocortin system, other neuronal pathways have also been identified as targets of leptin and other cytokine action within the hypothalamus. In particular, neuropeptide Y (NPY) is produced in hypothalamic nuclei known to regulate appetite and metabolism, and NPY is a potent orexigen when injected centrally.[39,60,61] Leptin deficiency results in a significant upregulation of NPY expression while leptin administration causes a decrease in expression.[62,63] Furthermore, genetic deletion of NPY in leptin-deficient animals results in a significant decrease in their degree of obesity.[64] In general, it appears that NPY synthesis and secretion are upregulated in most models of energy deficiency or increased metabolic demand.[65,66] Thus, a decrease in the level of NPY expression was proposed as one potential mechanism for the anorexic effects of leptin and potentially for the cytokine-mediated anorexia seen in animal models of inflammation and chronic disease. However, the observations that there was no change, or even an increase in NPY mRNA and peptide expression in tumor-bearing rats, rats with dehydration-induced anorexia, and rats treated with LPS, indicates that downregulation of NPY expression is not a primary stimulus for anorexia in these models.[67–70] Furthermore in anorectic, tumor-bearing animals, hypothalamic content of NPY is increased, and animals respond to NPY injections with *worsened* anorexia.[71,72] Thus, the utility of nonsubtype specific NPY analogs in the treatment of cachexia appears doubtful. However, it remains possible that analogs that are specific for individual NPY receptor subtypes may be useful in this condition.

Unlike the extensive body of literature on the regulation of energy balance by the melanocortin system and NPY, relatively little is known about the physiological role of CART. Certainly the fact that CART is colocalized with POMC provides a clue that this peptide is likely to play an important role in weight regulation.

Indeed, it is known that icv administration of CART potently inhibits feeding, and that the transcription of CART is regulated by leptin.[73] Furthermore, CART is robustly expressed in brainstem nuclei known to play a role in the regulation of feeding and gastric motility (e.g., the NTS).[74] Thus, a CART receptor antagonist is a plausible therapeutic target for cachexia. At this point, no such compound exists, and the development of these compounds will likely be very difficult until the CART receptor is cloned.

Overall, it appears that the ARC plays a key integrative role in the reception, amplification, and processing of chemical signals from the periphery. During chronic illness, changes in circulating cytokines, insulin, leptin, and other factors are likely to be sensed by neurons in the ARC, with a resulting output that produces or exacerbates the clinical features of cachexia. A hypothetical model that contrasts the starvation response in the ARC with that found during cachexia is shown in Figure 27.2.

27.5 NEUROPEPTIDE SYSTEMS IN THE LHA

As discussed above, the LHA has extensive reciprocal innervation with the ARC, and damage to this area of the hypothalamus leads to severe anorexia. At this point, there are two primary neuropeptide systems that have been described in the LHA that are known to play a role in food intake and energy balance. Melanin concentrating hormone (MCH) is a 19 amino acid peptide found in neurons in the zona incerta and LHA, with widespread connections throughout the brain and posterior pituitary.[75–78] MCH has the properties of an energy conserving, anabolic neuropeptide. Chronic administration or over-expression of MCH is known to stimulate feeding and increase body weight, whereas lack of MCH is known to produce hypophagia, an increased metabolic rate, and decreased fat mass.[79–83] The mechanism of the anabolic effects of MCH are not yet fully known, but it is known that MCH-KO mice have an increased basal metabolic rate, perhaps due to direct projections from MCH neurons to brown adipose tissue (BAT).[84,85]

MCH exerts its effects by binding to one of two specific MCH receptors. The MCH-R1 is widely distributed throughout the brain, with relatively high concentrations found in the ARC and the VMH.[86–88] The MCH-R2 is also found in the brain, although at lower expression levels and primarily in cortical areas.[89] MCH-R1 KO mice are lean, hypermetabolic, and have a tendency to develop osteoporosis even in the face of increased physical activity.[84,85,90] Overall, the physiology of the MCH system would suggest that this is an ideal target for the development of drugs for the treatment of cachexia. Specific and potent MCH-R1 antagonists have already been developed and shown to reduce food intake in rodents.[91] Thus, MCH-R1 agonists have the potential to be effective in the pharmacotherapy of cachexia, although no studies have yet been published on this subject.

A second neuron group in the LHA are the neurons that synthesize the orexins (orexin A and orexin B; also known as the hypocretins). These neuropeptides

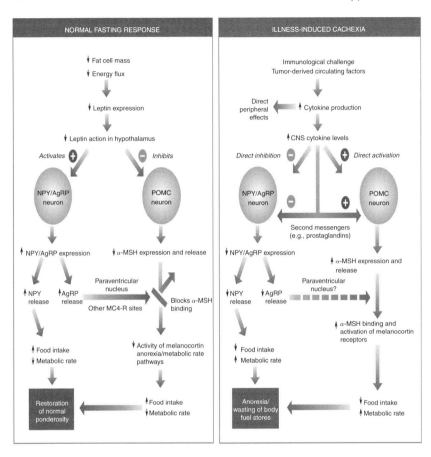

FIGURE 27.2   Hypothetical models for the regulation of feeding and metabolic rate during normal fasting (left panel), and illness-induced cachexia. During illness, cytokines and other circulating factors result in activation of central anorexic/metabolic pathways, resulting in decreased energy intake and pathological wasting of body fuels, even in a relatively starved state (right panel) (From Marks, D.L. and Cone, R.D. *Recent Prog. Horm. Res.* 56:359–375. With permission. Copyright 2001, The Endocrine Society).

are cleaved from a common precursor peptide, preproorexin, and have somewhat different pharmacology and physiological effects. Orexin neurons project very widely within the CNS including connections to MCH neurons within the LHA and POMC/CART and NPY/AGRP neurons within the ARC.[92–94] Orexin A can stimulate an acute increase in feeding, but does not appear to increase overall 24 hour food intake or lead to weight gain with chronic experimental administration.[95,96] Orexin A also has been shown to increase metabolic rate and raise body temperature when injected centrally.[97,98] The effects of the orexins are mediated by the orexin receptors OX-1 and OX-2 which are widely distributed in the brain.[99–101] Most of the feeding effects are thought to be due to the binding of orexin A to

OX-1, a receptor that is particularly abundant in the VMH.[100, 101] Although the orexins may play a role in the short term regulation of feeding, their primary function appears to be regulating arousal and the sleep-wake cycle.[102–105] Indeed, genetic deletion of orexin or its receptor results in severe narcolepsy, and the observed decrease in food intake in these models is likely to result from an over-all decrease in feeding activity rather than a change in appetite per se.[103] Thus, the role of the orexins in the disordered appetite and loss of lean body mass that accompanies chronic disease appears to be limited. However, illness com-monly leads to sleep disruption, lethargy, and decreased movement. The role of the orexins in regulating these processes may therefore indicate that orexin agon-ists may be useful in treating this aspect of the cachexia syndrome. The recent development of selective nonpeptide orexin receptor antagonists with biological activity indicates that the development of agonists should be possible in the near future.[106, 107]

27.6 NEUROPEPTIDES IN THE PVN

Corticotropin-releasing hormone (CRH, also know as corticotropin releasing factor, CRF) is synthesized in neurons in the VMH where it is well known to play a critical role in activation of the hypothalamo-pitutory axis (HPA) axis via neural connections to the portal vasculature. CRH has been shown to play an important role in anorexia resulting from a variety of stressors, including injections of cach-exigenic cytokines,[108–112] but deletion of CRH does not abolish the anorexia in response to chronic stress.[113] Overall, the data suggest that CRF may be one of the effectors downstream from a number of other neuropeptide systems, including the melanocortin system, and may be of particular importance in the inhibition of feeding brought about by increases in interleukin-1 in the brain.[111, 112, 114] CRH action is mediated largely by two central CRH receptors, CRF-1 and CRF-2, which are widely distributed in the CNS.[115, 116] The majority of the effects of CRH on feeding are thought to be due to activation of the CRF-2 receptor, although the CRF-1 receptor clearly plays a role in the overall stress response as well.[117, 118] Thus, specific antagonism of the CRF-2 may be a plausible therapy for cachexia. However, the complexity and importance of the stress response, and the anxiety and fearfulness observed in the CRF-2 knockout mouse indicate that a therapy based on this mechanism may be of limited clinical utility.[119] For a further discus-sion of the role of the CRF receptors and their ligands in stress and feeding, refer to an excellent recent review by Bale et al.[120]

27.7 OTHER NEUROPEPTIDES AND NONPEPTIDE
   NEUROTRANSMITTERS

One of the common complaints of individuals with chronic disease is the loss of enjoyment of eating. The hedonic aspects of food consumption are well known, particularly in the case of consumption of palatable high-fat and sweet tasting foods. The endogenous opioids are a group of peptide agonists that act as central opioid receptors where they are thought to be involved in the reward aspects of food consumption. The role of opioids in the physiological regulation of food intake and body weight is quite complex. For example, chronic administration of potent opioid compounds does not lead to sustained changes in food intake or body weight, suggesting that tolerance quickly develops to the hyperphagic activity of these compounds.[121] On the other hand, genetic deletion of beta-endorphin (an endogenous opioid derived from cleavage of POMC) leads to hyperphagia and obesity in mice.[122] Ultimately, the cognitive effects and addictive potential of the existing opioid agonists limit their usefulness in the treatment of cachexia. Nonetheless, these compounds will continue to be important in future research related to food intake and quality of life in chronic disease.

Another neurotransmitter that is thought to play a role in regulating the hedonic properties of food is the endocannabinoid system. To date, three endocannabinoid agonists have been shown to be produced in the brain; anandamide, sn-2-arachidonylglycerol (2-AG), and nolodin ether.[123, 124] In the brain, the primary cannabinoid receptor (CB-1) is widely distributed and known to play a role in food intake, particularly in response to starvation.[125] Both endogenous and synthetic cannabinoids have been shown to have a potent stimulatory effect on food intake, whereas CB-1 receptor antagonists are known to inhibit feeding.[126, 127] Collectively, the evidence suggests that endocannabinoids play a role in driving food intake by altering the response to a rewarding palatable diet, perhaps by interacting with the known opiodergic reward pathways.[128] Early clinical trials aimed at antagonizing the effects of the endogenous cannabinoid system as a treatment for obesity are already underway. Research into the effective use of cannabinoid agonists as drugs for the treatment of cachexia have been somewhat limited, due in large part to the psychotropic activity of these drugs. As we begin to understand the physiology and pharmacology of these compounds, it is likely that drugs that show beneficial effects on food intake and quality of life without undesirable psychological side effects can be created.

Abundant evidence exists regarding the role of monoaminergic neurotransmitters (dopamine, histamine, serotonin, and noradrenaline) in the regulation of food intake in experimental models. In particular, the serotonin (5-HT) system is known to inhibit feeding and stimulate basal metabolic rate, and agonists of this system (e.g., sibutramine) are known to lead to weight loss in experimental animals and humans.[129–131] Although the 5-HT receptor system is extraordinarily complex, it is known that the 5-HT_{2c} subtype plays a significant role in body weight regulation, and therefore represents an important target for future development of therapeutics for cachexia. For example, genetic deletion of the 5-HT_{2c} receptor leads to hyperphagia and obesity, indicating that antagonists of this receptor may be useful in the treatment of cachexia.[132] There is also evidence to suggest that serotonergic drugs may exert their effects by activating the central melanocortin system.[133] Thus, drugs designed to block activation of the 5-HT_{2c} receptor may be particularly effective when combined with a melanocortin antagonist.

27.8 SUMMARY

Although numerous central targets for drug therapy for cachexia have been discussed in this section, this list is by no means comprehensive. Indeed, there exists suggestive evidence for the role of central ghrelin, glucagon-like peptide-1 (GLP-1), dopamine, and other neurotransmitter systems in the regulation of body weight, and each of these systems may prove to be an important target for drug therapy for cachexia. Furthermore, there are many peripheral hormones that will undoubtedly be considered as therapeutics for this disorder, many of which exert their activity, at least in part, via activation or inhibition of the central body weight pathways described in the preceding sections. At this point, the data suggests that neurons located in parts of the brain with an attenuated blood brain barrier (e.g., the ARC and the NTS) may play a primary role in transducing signals elaborated during illness into the complex physiological and psychological behaviors typical of chronic diseases. The fact that many of these systems appear to be redundant indicates that therapies directed at the central regulation of body weight in chronic diseases will likely involve the use of a multimodal therapy strategy designed to regulate several central neurotransmitter systems simultaneously.

REFERENCES

1. Zhang, Y., Proenca, R., Maffei, M., Barone, M., Leopold, L., and Friedman, J.M. 1994. Positional cloning of the mouse obese gene and its human homologue. *Nature* 372:425–432.
2. Campfield, L.A., Smith, F.J., Guisez, Y., Devos, R., and Burn, P. 1995. Recombinant mouse OB protein: Evidence for a peripheral signal linking adiposity and central neural networks. *Science* 269:546–549.
3. Halaas, J., Gajiwala, K., Maffei, M., Cohen, S., Chait, B., Rabinowitz, D., Lallone, R., Burley, S., and Friedman, J.M. 1995. Weight-reducing effects of the plasma protein encoded by the *obese* gene. *Science* 269:543–546.
4. Pelleymounter, M., Cullen, M., Baker, M., Hecht, R., Winters, D., Boone, T., and Collins, F. 1995. Effects of the obese gene product on body weight regulation in *ob/ob* mice. *Science* 269:540–543.
5. Zhang, F.M., Basinski, M.B., and Beals, J.M. 1997. Crystal structure of the obese protein leptin E-100. *Nature* 387:206–209.
6. Halaas, J.L., Boozer, C., Blair-West, J., Fidahusein, N., Denton, D.A., and Friedman, J.M. 1997. Physiological response to long-term peripheral and central leptin infusion in lean and obese mice. *Proc. Natl. Acad. Sci. USA* 94:8878–8883.
7. Ahima, R., Prabakaran, D., Mantzoros, C., Qu, D., Lowell, B., Maratos-Flier, E., and Flier, S. 1996. Role of leptin in the neuroendocrine response to fasting. *Nature* 382:250–252.
8. Finck, B.N. and Johnson, R.W. 1999. Intracerebroventricular injection of lipopolysaccharide increases plasma leptin levels. *NeuroReport* 10:153–156.
9. Finck, B.N., Kelly, K.W., Dantzer, R., and Johnson, R.W. 1998. In vivo and in vitro evidence for the involvement of tumor necrosis factor-alpha in the induction of leptin by lipopolysaccharide. *Endocrinology* 139:2278–2283.

10. Grunfeld, C., Zhao, C., Fuller, J., Pollock, A., Moser, A., Friedman, J., and Feingold, K.R. 1996. Endotoxin and cytokine induce expression of leptin, the ob gene product, in hamsters. *J. Clin. Invest.* 97:2152–2157.

11. Sarraf, P., Frederich, R.C., Turner, E.M., Ma, G., Jaskowiak, N.T., Rivet, D.J.I., Filer, J.S., Lowell, B.B., Fraker, D.L., and Alexander, H.R. 1997. Multiple cytokines and acute inflammation raise mouse leptin levels: Potential role in inflammatory anorexia. *J. Exp. Med.* 185:171–175.

12. Ellacott, K.L. and Cone, R.D. 2004. The central melanocortin system and the integration of short- and long-term regulators of energy homeostasis. *Recent Prog. Horm. Res.* 59:395–408.

13. Harrold, J.A. 2004. Hypothalamic control of energy balance. *Curr. Drug Targets* 5:207–219.

14. Elmquist, J.K., Ahima, R.S., Maratos Flier, E., Flier, J.S., and Saper, C.B. 1997. Leptin activates neurons in ventrobasal hypothalamus and brainstem. *Endocrinology* 138:839–842.

15. Jacob, R.J., Dziura, J., Medwick, M., Leone, P., Caprio, S., During, M., Shulman, G.I., and Sherwin, R.S. 1997. The effect of leptin is enhanced by microinjection into the ventromedial hypothalamus. *Diabetes* 46:150–152.

16. Woods, A.J. and Stock, M.J. 1996. Leptin activation in hypothalamus. *Nature* 381:745.

17. Cheung, C.C., Clifton, D.K., and Steiner, R.A. 1997. Proopiomelanocortin neurons are direct targets for leptin in the hypothalamus. *Endocrinology* 138:4489–4492.

18. Baskin, D.G., Breininger, J.F., and Schwartz, M.W. 1999. Leptin receptor mRNA identifies a subpopulation of neuropeptide Y neurons activated by fasting in rat hypothalamus. *Diabetes* 48:828–833.

19. Glaum, S.R., Hara, M., Bindokas, V.P., Lee, C.C., Polonsky, K.S., Bell, G.I., and Miller, R.J. 1997. Leptin, the obese gene product, rapidly modulates synaptic transmission in the hypothalamus. *Mol. Pharmacol.* 50:230–235.

20. Anand, B.K. and Broback, J.R. 1951. Hypothalamic control of food intake in rats and cats. *Yale J. Biol. Med.* 24:123–140.

21. Broback, J.R., Tepperman, J., and Long, C.N.H. 1943. Experimental hypothalamic hyperphagia in the albino rat. *Yale J. Biol. Med.* 15:831–853.

22. Hotheriogtoa, A.W. and Ranson, S.W. 1940. Hypothalamic lesions and adiposity in the rat. *Anat. Rec.* 78:149–172.

23. Cone, R.D., Cowley, M.A., Butler, A.A., Fan, W., Marks, D.L., and Low, M.J. 2001. The arcuate nucleus as a conduit for diverse signals relevant to energy homeostasis. *Int. J. Obes. Relat. Metab. Disord.* 25(Suppl 5):S63–S67.

24. Nadeau, S. and Rivest, S. 1999. Effects of circulating tumor necrosis factor on the neuronal activity and expression of the genes encoding the tumor necrosis factor receptors (p55 and p75) in the rat brain: A view from the blood-brain barrier. *Neuroscience* 93:1449–1464.

25. Rizk, N.M., Joost, H.G., and Eckel, J. 2001. Increased hypothalamic expression of the p75 tumor necrosis factor receptor in New Zealand obese mice. *Horm. Metab. Res.* 33:520–524.

26. Yamakuni, H., Minami, M., and Satoh, M. 1996. Localization of mRNA for leukemia inhibitory factor receptor in the adult rat brain. *J. Neuroimmunol.* 70:45–53.

27. Ericsson, A., C., L., Hart, R.P., and Sawchenko, P.E. 1995. Type 1 interleukin-1 receptor in the rat brain: Distribution, regulation, and relationship to sites of IL-1-induced cellular activation. *J. Comp. Neurol.* 361:681–698.

28. Elmquist, J.K., Scammell, T.E., Jacobsen, C.D., and Saper, C.B. 1996. Distribution of Fos-like immunoreactivity in the rat brain following intravenous lipopolysaccharide administration. *J. Comp. Neurol.* 371:85–103.

29. Elias, C.F., Saper, C.B., Maratos-Flier, E., Tritos, N.A., Lee, C., Kelly, J., Tatro, J.B., Hoffman, G.E., Ollmann, M.M., Barsh, G.S. et al. 1998. Chemically defined projections linking the mediobasal hypothalamus and the lateral hypothalamic area. *J. Comp. Neurol.* 402:442–459.

30. Broberger, C., Johansen, J., Johansson, C., Schalling, M., and Hokfelt, T. 1998. The neuropeptide Y/agouti gene-related protein (AGRP) brain circuitry in normal, anorectic, and monosodium glutamate-treated mice. *Proc. Natl. Acad. Sci. USA* 95:15043–15048.

31. Jacobowitz, D.M. and O'Donohue, T.L. 1978. α-Melanocyte-stimulating hormone: Immunohistochemical identification and mapping in neurons of rat brain. *Proc. Natl. Acad. Sci. USA* 75:6300–6304.

32. Davis, A.M., Seney, M.L., Stallings, N.R., Zhao, L., Parker, K.L., and Tobet, S.A. 2004. Loss of steroidogenic factor 1 alters cellular topography in the mouse ventromedial nucleus of the hypothalamus. *J. Neurobiol.* 60:424–436.

33. Hashimoto, K., Nishiyama, M., Tanaka, Y., Noguchi, T., Asaba, K., Hossein, P.N., Nishioka, T., and Makino, S. 2004. Urocortins and corticotropin releasing factor type 2 receptors in the hypothalamus and the cardiovascular system. *Peptides* 25:1711–1721.

34. Magrani, J., de Castro e Silva, E., Varjao, B., Duarte, G., Ramos, A.C., Athanazio, R., Barbetta, M., Luz, P., and Fregoneze, J.B. 2004. Histaminergic H1 and H2 receptors located within the ventromedial hypothalamus regulate food and water intake in rats. *Pharmacol. Biochem. Behav.* 79:189–198.

35. Kang, L., Routh, V.H., Kuzhikandathil, E.V., Gaspers, L.D., and Levin, B.E. 2004. Physiological and molecular characteristics of rat hypothalamic ventromedial nucleus glucosensing neurons. *Diabetes* 53:549–559.

36. Levin, B.E., Dunn-Meynell, A.A., and Routh, V.H. 1999. Brain glucose sensing and body energy homeostasis: Role in obesity and diabetes. *Am. J. Physiol.* 276:R1223–R1231.

37. Aaronson, S.A. 1991. Growth factors and cancer. *Science* 254:1146–1153.

38. Shimazu, T., Fukuda, A., and Ban, T. 1966. Reciprocal influences of the ventromedial and lateral hypothalamic nuclei on blood glucose level and liver glycogen content. *Nature* 210:1178–1179.

39. Campbell, R.E., ffrench-Mullen, J.M., Cowley, M.A., Smith, M.S., and Grove, K.L. 2001. Hypothalamic circuitry of neuropeptide Y regulation of neuroendocrine function and food intake via the Y5 receptor subtype. *Neuroendocrinology* 74:106–119.

40. Grill, H.J., Ginsberg, A.B., Seeley, R.J., and Kaplan, J.M. 1998. Brainstem application of melanocortin receptor ligands produces long-lasting effects on feeding and body weight. *J. Neurosci.* 18:10128–10135.

41. Fan, W., Boston, B.A., Kesterson, R.A., Hruby, V.J., and Cone, R.D. 1997. Role of melanocortinergic neurons in feeding and the agouti obesity syndrome. *Nature* 385:165–168.

42. Friedman, J.M. 1997. The alphabet of weight control. *Nature* 385:119–120.

43. Gura, T. 1997. Obesity sheds its secrets. *Science* 275:751–753.

44. Ollmann, M.M., Wilson, B.D., Yang, Y.-K., Kerns, J.A., Chen, Y., Gantz, I., and Barsh, G.S. 1997. Antagonism of central melanocortin receptors in vitro and in vivo by agouti-related protein. *Science* 278:135–137.

45. Huszar, D., Lynch, C.A., Fairchild-Huntress, V., Dunmore, J.H., Fang, Q., Berkemeier, L.R., Gu, W., Kesterson, R.A., Boston, B.A., Cone, R.D. et al. 1997. Targeted disruption of the melanocortin-4 receptor results in obesity in mice. *Cell* 88:131–141.

46. Farooqi, I.S., Keogh, J.M., Yeo, G.S., Lank, E.J., Cheetham, T., and O'Rahilly, S. 2003. Clinical spectrum of obesity and mutations in the melanocortin 4 receptor gene. *N. Engl. J. Med.* 348:1085–1095.

47. Farooqi, I.S. and O'Rahilly, S. 2005. Monogenic obesity in humans. *Annu. Rev. Med.* 56:443–458.

48. Cowley, M.A., Smart, J.L., Cerdan, M., Rubinstein, M., Diano, C., Horvath, T.L., Cone, R.D., and Low, M.J. 2001. Leptin activates anorexigenic POMC neurons through a neural network in the arcuate nucleus. *Nature* 24:155–163.

49. Boston, B.A., Blaydon, K.M., Varnerin, J., and Cone, R.D. 1997. Independent and additive effects of central POMC and leptin pathways on murine obesity. *Science* 278:1641–1644.

50. Marsh, D.J., Hollopeter, G., Huszar, D., Laufer, R., Yagaloff, K.A., Fisher, S.L., Burn, P., and Palmiter, R.D. 1999. Response of melanocortin-4 receptor-deficient mice to anorectic and orexigenic peptides. *Nature Genetics* 21:119–122.

51. Marks, D.L., Butler, A.A., Turner, R., Brookhart, G.B., and Cone, R.D. 2003. Differential role of melanocortin receptor subtypes in cachexia. *Endocrinology* 144:1513–1523.

52. Marks, D.L., Ling, N., and Cone, R.D. 2001. Role of the central melanocortin system in cachexia. *Cancer Res.* 61:1432–1438.

53. Huang, Q., Hruby, V.J., and Tatro, J.B. 1999. Role of central melanocortins in endotoxin-induced anorexia. *Am. J. Physiol.* 276:R864–R871.

54. Wisse, B.E., Frayo, R.S., Schwartz, M.W., and Cummings, D.E. 2001. Reversal of cancer anorexia by blockade of central melanocortin receptors in rats. *Endocrinology* 142:3292–3301.

55. Marks, D.L. and Cone, R.D. 2001. Central melanocortins and the regulation of weight during acute and chronic disease. *Recent Prog. Horm. Res.* 56:359–375.

56. Foster, A., Markinson, S., Brookhart, G.B., and Marks, D.L. 2005. The regulation of feeding and reversal of cachexia with peripheral administration of a selective melanocortin-4 antagonist. *Endocrinology* 146: 2766–2773.

57. Vos, T.J., Caracoti, A., Che, J.L., Dai, M., Farrer, C.A., Forsyth, N.E., Drabic, S.V., Horlick, R.A., Lamppu, D., Yowe, D.L. et al. 2004. Identification of 2-[2-[2-(5-bromo-2- methoxyphenyl)-ethyl]-3-fluorophenyl]-4,5-dihydro-1H-imidazole (ML00253764), a small molecule melanocortin 4 receptor antagonist that effectively reduces tumor-induced weight loss in a mouse model. *J. Med. Chem.* 47:1602–1604.

58. Hagan, M.M., Rushing, P.A., Pritchard, L.M., Schwartz, M.W., Strack, A.M., Van Der Ploeg, L.H., Woods, S.C., and Seeley, R.J. 2000. Long-term orexigenic effects of AgRP-(83-132) involve mechanisms other than melanocortin receptor blockade. *Am J. Physiol. Regul. Integr. Comp. Physiol.* 279:R47–R52.

59. Nijenhuis, W.A., Oosterom, J., and Adan, R.A. 2001. AGRP (83-132) acts as an inverse agonist on the human melanocortin-4 receptor. *Mol. Endocrinol.* 15:164–171.

60. Stanley, B.G., Kyrkouli, S.E., Lampert, S., and Leibowitz, S.F. 1986. Neuropeptide Y chronically injected into the hypothalamus: A powerful neurochemical inducer of hyperphagia and obesity. *Peptides* 7:1189–1192.

61. Sahu, A. 1998. Leptin decreases food intake induced by melanin-concentrating hormone (MCH), galanin (GAL) and neuropeptide Y (NPY) in the rat. *Endocrinology* 139:4739–4742.

62. Wilding, J.P.H., Gilbey, S.G., Bailey, C.J., Batt, R.A.L., Williams, G., Ghatei, M.A., and Bloom, S.R. 1993. Increased neuropeptide-Y messenger ribonucleic acid (mRNA) and decreased neurotensin mRNA in the hypothalamus of the obese (ob/ob) mouse. *Endocrinology* 132:1939–1944.

63. Stephens, T.W., Basinsky, M., Bristow, P.K., Bue-Valleskey, J.M., Burgett, S.G., Craft, L., Hale, J., Hoffmann, J., Hsiung, H.M., Kriauciunas, A. et al. 1995. The role of neuropeptide Y in the antiobesity action of the *obese* gene product. *Nature* 377:530–532.

64. Erickson, J., Hollopeter, G., and Palmiter, J.D. 1996. Attenuation of the obesity syndrome of ob/ob mice by the loss of neuropeptide Y. *Science* 274:1704–1707.

65. Smith, M.S. 1993. Lactation alters neuropeptide-Y and proopiomelanocortin gene expression in the arcuate nucleus of the rat. *Endocrinology* 133:1258–1265.

66. Inui, A. 1999. Feeding and body-weight regulation by hypothalamic neuropeptides — mediation of the actions of leptin. *Trends Neurosci.* 22:62–67.

67. Gayle, D., Ilyin, S.E., Flynn, M.C., and Plata-Salaman, C.R. 1998. Lipopolysaccharide (LPS)- and muramyl dipeptide (MDP)-induced anorexia during refeeding following acute fasting: Characterization of brain cytokine and neuropeptide systems mRNAs. *Brain Res.* 795:77–86.

68. Watts, A.G., Sanchez-Watts, G., and Kelly, A.B. 1999. Distinct patterns of neuropeptide gene expression in the lateral hypothalamic area and arcuate nucleus are associated with dehydration-induced anorexia. *J. Neurosci.* 19:6111–6121.

69. Plata-Salaman, C.R., Ilyin, S.E., and Gayle, D. 1998. Brain cytokine mRNAs in anorectic rats bearing prostate adenocarcinoma tumor cells. *Am. J. Physiol.* 275:R566–R573.

70. Jensen, P.B., Blume, N., Mikkelsen, J.D., Larsen, P.J., Jensen, H.I., Holst, J.J., and Madsen, O.D. 1998. Transplantable rat glucagonomas cause acute onset of severe anorexia and adipsia despite highly elevated NPY mRNA levels in the hypothalamic arcuate nucleus. *J. Clin. Invest.* 101:503–510.

71. Chance, W.T., Balasubramaniam, A., Thompson, H., Mohapatra, B., Ramo, J., and Fischer, J.E. 1996. Assessment of feeding response of tumor-bearing rats to hypothalamic injection and infusion of neuropeptide Y. *Peptides* 17:797–801.

72. Chance, W.T., Sheriff, S., Kasckow, J.W., Regmi, A., and Balasubramaniam, A. 1998. NPY messenger RNA is increased in medial hypothalamus of anorectic tumor-bearing rats. *Regul. Pept.* 25:347–353.

73. Kristensen, P., Judge, M.E., Thim, L., Ribel, U., Christjansen, K.N., Wulff, B.S., Clausen, J.T., Jensen, P.B., Madsen, O.D., Vrang, N. et al. 1998. Hypothalamic CART is a new anorectic peptide regulated by leptin. *Nature* 393:72–76.

74. Zheng, H., Patterson, L.M., and Berthoud, H.R. 2002. CART in the dorsal vagal complex: Sources of immunoreactivity and effects on Fos expression and food intake. *Brain Res.* 957:298–310.

75. Naito, N., Kawazoe, I., Nakai, Y., and Kawauchi, H. 1988. Melanin-concentrating hormone-like immunoreactive material in the rat hypothalamus; characterization and subcellular localization. *Cell Tissue Res.* 253:291–295.

76. Bittencourt, J.C., Presse, F., Arias, C., Peto, C., Vaughan, J., Nahon, J.L., Vale, W., and Sawchenko, P.E. 1992. The melanin-concentrating hormone system of the rat brain: An immuno- and hybridization histochemical characterization. *J. Comp. Neurol.* 319:218–245.

77. Skofitsch, G., Jacobowitz, D.M., and Zamir, N. 1985. Immunohistochemical localization of a melanin concentrating hormone-like peptide in the rat brain. *Brain Res. Bull.* 15:635–649.

78. Zamir, N., Skofitsch, G., and Jacobowitz, D.M. 1986. Distribution of immunoreactive melanin-concentrating hormone in the central nervous system of the rat. *Brain Res.* 373:240–245.

79. Mahieu, S., del Carmen Contini, M., Gonzalez, M., Millen, N., and Elias, M.M. 2000. Aluminum toxicity. Hematological effects. *Toxicol. Lett.* 111:235–242.

80. Gonzalez, M.I., Vaziri, S., and Wilson, C.A. 1996. Behavioral effects of alpha-MSH and MCH after central administration in the female rat. *Peptides* 17:171–177.

81. Gomori, A., Ishihara, A., Ito, M., Mashiko, S., Matsushita, H., Yumoto, M., Tanaka, T., Tokita, S., Moriya, M., Iwaasa, H. et al. 2003. Chronic intracerebroventricular infusion of MCH causes obesity in mice. Melanin-concentrating hormone. *Am. J. Physiol. Endocrinol. Metab.* 284:E583–E588.

82. Ito, M., Gomori, A., Ishihara, A., Oda, Z., Mashiko, S., Matsushita, H., Yumoto, M., Sano, H., Tokita, S., Moriya, M. et al. 2003. Characterization of MCH-mediated obesity in mice. *Am. J. Physiol. Endocrinol. Metab.* 284:E940–E945.

83. Shimada, M., Tritos, N.A., Lowell, B.B., Flier, J.S., and Maratos-Flier, E. 1998. Mice lacking melanin-concentrating hormone are hypophagic and lean. *Nature* 396:670–674.

84. Marsh, D.J., Weingarth, D.T., Novi, D.E., Chen, H.Y., Trumbauer, M.E., Chen, A.S., Guan, X.M., Jiang, M.M., Feng, Y., Camacho, R.E. et al. 2002. Melanin-concentrating hormone 1 receptor-deficient mice are lean, hyperactive, and hyperphagic and have altered metabolism. *Proc. Natl. Acad. Sci. USA* 99:3240–3245.

85. Chen, Y., Hu, C., Hsu, C.K., Zhang, Q., Bi, C., Asnicar, M., Hsiung, H.M., Fox, N., Slieker, L.J., Yang, D.D. et al. 2002. Targeted disruption of the melanin-concentrating hormone receptor-1 results in hyperphagia and resistance to diet-induced obesity. *Endocrinology* 143:2469–2477.

86. Hervieu, G. and Nahon, J.L. 1995. Pro-melanin concentrating hormone messenger ribonucleic acid and peptides expression in peripheral tissues of the rat. *Neuroendocrinology* 61:348–364.

87. Chambers, J., Ames, R.S., Bergsma, D., Muir, A., Fitzgerald, L.R., Hervieu, G., Dytko, G.M., Foley, J.J., Martin, J., Liu, W.S. et al. 1999. Melanin-concentrating hormone is the cognate ligand for the orphan G-protein-coupled receptor SLC-1. *Nature* 400:261–265

88. Hervieu, G.J., Cluderay, J.E., Harrison, D., Meakin, J., Maycox, P., Nasir, S., and Leslie, R.A. 2000. The distribution of the mRNA and protein products of the melanin-concentrating hormone (MCH) receptor gene, slc-1, in the central nervous system of the rat. *Eur. J. Neurosci.* 12:1194–1216.

89. Rodriguez, M., Beauverger, P., Naime, I., Rique, H., Ouvry, C., Souchaud, S., Dromaint, S., Nagel, N., Suply, T., Audinot, V. et al. 2001. Cloning and

molecular characterization of the novel human melanin-concentrating hormone receptor MCH2. *Mol. Pharmacol.* 60:632–639.

90. Bohlooly, Y.M., Mahlapuu, M., Andersen, H., Astrand, A., Hjorth, S., Svensson, L., Tornell, J., Snaith, M.R., Morgan, D.G., and Ohlsson, C. 2004. Osteoporosis in MCHR1-deficient mice. *Biochem. Biophys. Res. Commun.* 318:964–969.

91. Kowalski, T.J., Farley, C., Cohen-Williams, M.E., Varty, G., and Spar, B.D. 2004. Melanin-concentrating hormone-1 receptor antagonism decreases feeding by reducing meal size. *Eur. J. Pharmacol.* 497:41–47.

92. Dube, M.G., Horvath, T.L., Kalra, P.S., and Kalra, S.P. 2000. Evidence of NPY Y5 receptor involvement in food intake elicited by orexin A in sated rats. *Peptides* 21:1557–1560.

93. Elias, C.F., Lee, C.E., Kelly, J.F., Ahima, R.S., Kuhar, M., Saper, C.B., and Elmquist, J.K. 2001. Characterization of CART neurons in the rat and human hypothalamus. *J. Comp. Neurol.* 432:1–19.

94. Jain, M.R., Horvath, T.L., Kalra, P.S., and Kalra, S.P. 2000. Evidence that NPY Y1 receptors are involved in stimulation of feeding by orexins (hypocretins) in sated rats. *Regul. Pept.* 87:19–24.

95. Edwards, C.M., Abusnana, S., Sunter, D., Murphy, K.G., Ghatei, M.A., and Bloom, S.R. 1999. The effect of the orexins on food intake: Comparison with neuropeptide Y, melanin-concentrating hormone and galanin. *J. Endocrinol.* 160:R7–R12.

96. Haynes, A.C., Jackson, B., Overend, P., Buckingham, R.E., Wilson, S., Tadayyon, M., and Arch, J.R. 1999. Effects of single and chronic intracerebroventricular administration of the orexins on feeding in the rat. *Peptides* 20:1099–1105.

97. Lubkin, M. and Stricker-Krongrad, A. 1998. Independent feeding and metabolic actions of orexins in mice. *Biochem. Biophys. Res. Commun.* 253:241–245.

98. Yoshimichi, G., Yoshimatsu, H., Masaki, T., and Sakata, T. 2001. Orexin-A regulates body temperature in coordination with arousal status. *Exp. Biol. Med. (Maywood)* 226:468–476.

99. Marcus, J.N., Aschkenasi, C.J., Lee, C.E., Chemelli, R.M., Saper, C.B., Yanagisawa, M., and Elmquist, J.K. 2001. Differential expression of orexin receptors 1 and 2 in the rat brain. *J. Comp. Neurol.* 435:6–25.

100. Lu, X.Y., Bagnol, D., Burke, S., Akil, H., and Watson, S.J. 2000. Differential distribution and regulation of OX1 and OX2 orexin/hypocretin receptor messenger RNA in the brain upon fasting. *Horm. Behav.* 37:335–344.

101. Trivedi, P., Yu, H., MacNeil, D.J., Van der Ploeg, L.H., and Guan, X.M. 1998. Distribution of orexin receptor mRNA in the rat brain. *FEBS Lett.* 438:71–75.

102. Willie, J.T., Chemelli, R.M., Sinton, C.M., and Yanagisawa, M. 2001. To eat or to sleep? Orexin in the regulation of feeding and wakefulness. *Annu. Rev. Neurosci.* 24:429–458.

103. Chemelli, R.M., Willie, J.T., Sinton, C.M., Elmquist, J.K., Scammell, T., Lee, C., Richardson, J.A., Williams, S.C., Xiong, Y., Kisanuki, Y. et al. 1999. Narcolepsy in orexin knockout mice: Molecular genetics of sleep regulation. *Cell* 98:437–451.

104. Hara, J., Beuckmann, C.T., Nambu, T., Willie, J.T., Chemelli, R.M., Sinton, C.M., Sugiyama, F., Yagami, K., Goto, K., Yanagisawa, M. et al. 2001. Genetic ablation of orexin neurons in mice results in narcolepsy, hypophagia, and obesity. *Neuron* 30:345–354.

105. Willie, J.T., Chemelli, R.M., Sinton, C.M., Tokita, S., Williams, S.C., Kisanuki, Y.Y., Marcus, J.N., Lee, C., Elmquist, J.K., Kohlmeier, K.A. et al. 2003. Distinct narcolepsy syndromes in Orexin receptor-2 and Orexin null mice: Molecular genetic dissection of non-REM and REM sleep regulatory processes. *Neuron* 38:715–730.

106. Langmead, C.J., Jerman, J.C., Brough, S.J., Scott, C., Porter, R.A., and Herdon, H.J. 2004. Characterisation of the binding of [3H]-SB-674042, a novel nonpeptide antagonist, to the human orexin-1 receptor. *Br. J. Pharmacol.* 141:340–346.

107. Hirose, M., Egashira, S., Goto, Y., Hashihayata, T., Ohtake, N., Iwaasa, H., Hata, M., Fukami, T., Kanatani, A., and Yamada, K. 2003. N-acyl 6,7-dimethoxy-1,2,3,4-tetrahydroisoquinoline: The first orexin-2 receptor selective non-peptidic antagonist. *Bioorg. Med. Chem. Lett.* 13:4497–4499.

108. Samarghandian, S., Ohata, H., Yamauchi, N., and Shibasaki, T. 2003. Corticotropin-releasing factor as well as opioid and dopamine are involved in tail-pinch-induced food intake of rats. *Neuroscience* 116:519–524.

109. Okamoto, S., Kimura, K., and Saito, M. 2001. Anorectic effect of leptin is mediated by hypothalamic corticotropin-releasing hormone, but not by urocortin, in rats. *Neurosci. Lett.* 307:179–182.

110. Koob, G.F. and Heinrichs, S.C. 1999. A role for corticotropin releasing factor and urocortin in behavioral responses to stressors. *Brain Res.* 848:141–152.

111. Vergoni, A.V., Bertolini, A., Wikberg, J.E.S., and Schioth, H.B. 1999. Corticotropin-releasing factor (CRF) induced anorexia is not influenced by a melanocortin 4 receptor blockage. *Peptides* 20:509–513.

112. Suto, G., Kiraly, A., Plourde, V., and Tache, Y. 1996. Intravenous interleukin-1-beta-induced inhibition of gastric emptying: Involvement of central corticotrophin-releasing factor and prostaglandin pathways in rats. *Digestion* 57:135–140.

113. Weninger, S.C., Muglia, L.J., Jacobson, L., and Majzoub, J.A. 1999. CRH-deficient mice have a normal anorectic response to chronic stress. *Regul. Pept.* 84:69–74.

114. Suto, G., Kiraly, A., and Tache, Y. 1994. Interleukin 1 beta inhibits gastric emptying in rats: Mediation through prostaglandin and corticotropin-releasing factor. *Gastroenterology* 106:1568–1575.

115. Richard, D., Rivest, R., Naimi, N., Timofeeva, E., and Rivest, S. 1996. Expression of corticotropin-releasing factor and its receptors in the brain of lean and obese Zucker rats. *Endocrinology* 137:4786–4795.

116. Richard, D., Lin, Q., and Timofeeva, E. 2002. The corticotropin-releasing factor family of peptides and CRF receptors: Their roles in the regulation of energy balance. *Eur. J. Pharmacol.* 440:189–197.

117. Smagin, G.N., Howell, L.A., Ryan, D.H., De Souza, E.B., and Harris, R.B. 1998. The role of CRF2 receptors in corticotropin-releasing factor- and urocortin-induced anorexia. *Neuroreport* 9:1601–1606.

118. Contarino, A., Dellu, F., Koob, G.F., Smith, G.W., Lee, K.F., Vale, W.W., and Gold, L.H. 2000. Dissociation of locomotor activation and suppression of food intake induced by CRF in CRFR1-deficient mice. *Endocrinology* 141:2698–2702.

119. Bale, T.L., Contarino, A., Smith, G.W., Chan, R., Gold, L.H., Sawchenko, P.E., Koob, G.F., Vale, W.W., and Lee, K.F. 2000. Mice deficient for corticotropin-releasing hormone receptor-2 display anxiety-like behaviour and are hypersensitive to stress. *Nat. Genet.* 24:410–414.

120. Bale, T.L. and Vale, W.W. 2004. CRF and CRF receptors: Role in stress responsivity and other behaviors. *Annu. Rev. Pharmacol. Toxicol.* 44:525–557.

121. Smith, S.L., Harrold, J.A., and Williams, G. 2002. Diet-induced obesity increases mu opioid receptor binding in specific regions of the rat brain. *Brain Res.* 953:215–222.

122. Appleyard, S.M., Hayward, M., Young, J.I., Butler, A.A., Cone, R.D., Rubinstein, M., and Low, M.J. 2003. A role for the endogenous opioid beta-endorphin in energy homeostasis. *Endocrinology* 144:1753–1760.

123. Mechoulam, R., Shabat, S.B., Hanus, L., Fride, E., Bayewitch, M., and Vogel, Z. 1996. Endogenous cannabinoid ligands. *Adv. Exp. Med. Biol.* 402:95–101.

124. Hanus, L., Abu-Lafi, S., Fride, E., Breuer, A., Vogel, Z., Shalev, D.E., Kustanovich, I., and Mechoulam, R. 2001. 2-arachidonyl glyceryl ether, an endogenous agonist of the cannabinoid CB1 receptor. *Proc. Natl. Acad. Sci. USA* 98:3662–3665.

125. Di Marzo, V., Goparaju, S.K., Wang, L., Liu, J., Batkai, S., Jarai, Z., Fezza, F., Miura, G.I., Palmiter, R.D., Sugiura, T. et al. 2001. Leptin-regulated endocannabinoids are involved in maintaining food intake. *Nature* 410:822–825.

126. Williams, C.M. and Kirkham, T.C. 1999. Anandamide induces overeating: Mediation by central cannabinoid (CB1) receptors. *Psychopharmacology (Berl.)* 143:315–317.

127. Harrold, J.A. and Williams, G. 2003. The cannabinoid system: a role in both the homeostatic and hedonic control of eating? *Br. J. Nutr.* 90:729–734.

128. Welch, S.P. and Eads, M. 1999. Synergistic interactions of endogenous opioids and cannabinoid systems. *Brain Res.* 848:183–190.

129. Leibowitz, S.F. 1986. Brain monoamines and peptides: Role in the control of eating behavior. *Fed. Proc.* 45:1396–1403.

130. Leibowitz, S.F. and Alexander, J.T. 1998. Hypothalamic serotonin in control of eating behavior, meal size, and body weight. *Biol. Psychiatry* 44:851–864.

131. Sramek, J.J., Leibowitz, M.T., Weinstein, S.P., Rowe, E.D., Mendel, C.M., Levy, B., McMahon, F.G., Mullican, W.S., Toth, P.D., and Cutler, N.R. 2002. Efficacy and safety of sibutramine for weight loss in obese patients with hypertension well controlled by beta-adrenergic blocking agents: A placebo-controlled, double-blind, randomised trial. *J. Hum. Hypertens.* 16:13–19.

132. Tecott, L.H., Sun, L.M., Akana, S.F., Strack, A.M., Lowenstein, D.H., Dallman, M.F., and Julius, D. 1995. Eating disorder and epilepsy in mice lacking 5-HT2c serotonin receptors. *Nature* 374:542–546.

133. Heisler, L.K., Cowley, M.A., Tecott, L.H., Fan, W., Low, M.J., Smart, J.L., Rubinstein, M., Tatro, J.B., Marcus, J.N., Holstege, H. et al. 2002. Activation of central melanocortin pathways by fenfluramine. *Science* 297:609–611.

28 The Ubiquitin-Proteasome Pathway

Marijke Brink

Contents

Summary

Regardless of its cause, skeletal muscle wasting is the net outcome of a changed balance between the rates of protein synthesis and protein degradation. Suppressed rates of protein synthesis, elevated rates of protein degradation, or a combination of the two cause a rapid loss of muscle protein. The concept has evolved that specific activation of one of the intracellular protease systems, the ubiquitin-proteasome system, is instrumental in atrophy-related protein degradation. This system is better known for its involvement in the specific degradation of short-lived regulatory proteins, but now also appears to be implicated in the final steps of degradation of long-lived proteins such as the contractile proteins that make up muscle cells. Drugs have been developed that specifically block the activity of the proteasome and these are now in clinical trials for diseases such as cancer and inflammation, in which the short-lived regulatory proteins are targeted. In animal and cell culture models of atrophy, use of similar proteasome inhibitors provided evidence for the involvement of this system in cachexia; however, their therapeutic potential for this disease remains to be tested. One of the problems faced is to define a therapeutic window that will be effective in wasting. The existing inhibitors of total proteasome activity target many cellular regulatory proteins, which very likely will result in undesirable side effects. Recently, new components of the ubiquitin-proteasome system have been identified that are expressed only in striated muscle, are involved in regulating muscle mass, and moreover are strongly increased in several wasting diseases. These proteins, Muscle RING (really interesting new gene) Finger-1 (MuRF-1) and atrogin-1, also called Muscle Atrophy F-box (MAFbx), are members of the large family of ubiquitin E3 ligases. The E3 enzymes, together with their E2 counterparts, are involved in the selective ubiquitination of proteins that will subsequently be digested in the proteasome, thereby determining specificity of the ubiquitin-proteasome system. Given this advantage of high selectivity, the challenge is now to define protein targets for as well as inhibitors of MuRF-1 and atrogin-1.[1,2] A more detailed understanding of ubiquitin-proteasome mechanisms and the signaling pathways that activate this system in wasting will help to design these and other new therapeutic strategies. This chapter summarizes our current knowledge on the ubiquitin-proteasome system in wasting.

28.1 INTRODUCTION

The proteolytic activities of each cell can be grouped into those of (1) the vacuolar pathways, including endosomes, the endoplasmic reticulum, and lysosomes, where lysosomal proteases such as the cathepsins degrade mainly membrane proteins and extracellular proteins, and (2) the cytosolic pathways, which include several cysteine proteases (the Ca^{2+}-activated ATP-independent calpains, and the caspases) and the enzymes of the ATP-dependent ubiquitin-proteasome system. Not long ago, the lysosomes with their considerable complement of broad-specificity proteases were thought to be the principal locus of cellular protein degradation. The discovery of the ubiquitin-proteasome pathway of degradation, however, led to major progress in understanding mechanisms by which a cell distinguishes between its nonlysosomal constituent cytosolic proteins and is able to decide which proteins remain intact and which are degraded within minutes.[3] More recently, the concept has developed that the majority of protein degradation in muscle wasting takes place in the proteasome (for reviews, see References 4 to 11). While the focus of this chapter is on the ubiquitin-proteasome system in muscle wasting, we will also summarize the signaling molecules that activate this system and the other enzyme activities including those of the calpains and caspases that are involved in targeting proteins for specific degradation.

28.2 THE UBIQUITIN-PROTEASOME PATHWAY OF PROTEIN DEGRADATION

28.2.1 The Ubiquitin-Proteasome System as Central Regulator of Cellular Functions

Initially, the proteasome was thought of as the waste bin of a cell: the place where misfolded or damaged polypeptides are disposed of. This function of the proteasome ensures quality of intracellular proteins and prevents the potentially toxic accumulation of damaged cellular proteins.[12] In the last decade, however, evidence has accumulated that ubiquitin-mediated mechanisms also target a wide range of functional cellular proteins for degradation, thereby modulating a long list of basic cellular processes, including the cell cycle and division, differentiation and development, response to stress and extracellular effectors, morphogenesis, modulation of cell surface receptors, ion channels and the secretory pathway, DNA repair, transcriptional regulation and silencing, long-term memory, circadian rhythms, regulation of immune and inflammatory responses, and biogenesis of organelles. Thus, the proteasome, which makes up 1% of a cell's total protein, degrades countless cellular proteins with a multitude of cellular functions. It is not surprising then that the targeting of these regulatory proteins for degradation requires very precise mechanisms and indeed is tightly controlled in a complex multistep pathway. For extensive discussion of the molecular biology of the ubiquitin-proteasome pathway the following reviews are recommended.[13-17]

28.2.2 Enzyme Activities Reside Inside the 20S Core Particle of the Proteasome

The 2.4-MDa proteasome is composed of a 20S core particle in which proteins are digested to short peptides, and one or two 19S regulatory particles, responsible for substrate recognition and transport into the core particle.[13] The 20S particle consists of four stacked rings: two outer α-rings on either side of two inner β-rings and each ring contains seven different subunits. Substrates can enter the 20S particle only through a gated channel in the center of the upper α-ring, which is normally maintained in a closed state. This channel, while being quite narrow even in its open state, allows entry of unfolded proteins only. The recognition, unfolding, and linearization that must happen to substrates before their injection into the 20S particle, requires complex mechanisms. Access to the 20S particle is controlled by ATPases in the 19S particle. The two inner β-rings of the 20S proteasome form a central chamber containing the proteolytic sites, thereby ensuring that protease activities are not exposed to the surrounding cytosol, which would lead to nonspecific degradation of cytosolic proteins. Proteasomes can cut most types of peptide bonds, because they contain two sites that cleave after hydrophobic residues ("chymotrypsin-like"), two after acidic residues ("post-glutamyl peptide hydrolase," or "caspase-like") and two after basic residues ("trypsin-like").[13-17]

28.2.3 Ubiquitinating Enzymes Ensure Selectivity and Specificity

Whereas the specific architecture of the proteasome ensures that only those proteins that have entered the proteasome are degraded, the addition of a polyubiquitin chain is a prerequisite for entry of these proteins into the proteasome. Certain proteins are extremely short-lived in the cell, whereas others are far more stable, and a major question is as to how the system achieves this selectivity.[15, 18] It appears to be controlled by the E3s and the ancillary proteins. The E3s, generally referred to as "ubiquitin–protein ligases," are one of three types of ubiquitin-conjugating enzymes that act together to link chains of ubiquitin onto target proteins to mark them for degradation by the proteasome. The others are the E1, the ubiquitin-activating enzyme, and the E2s, the ubiquitin-carrier proteins, which prepare ubiquitin for conjugation onto proteins. The E3 ligases are key enzymes, because the appropriate E3 is required for each target protein substrate to selectively couple ubiquitin to lysine residues, thereby acting as the substrate recognition component of the ubiquitin conjugation machinery. As a consequence, the E3 ubiquitin ligases turn out to be one of the largest functional families of proteins in mammals. Certain structural features or degradation signals must be present on the target protein for recognition by an E3 and, in some cases E3 ligases themselves must undergo posttranslational modification before they become active. These specific modifications, or association with ancillary proteins such as molecular chaperones, create a link to the appropriate ligase thereby yielding a degradable form of the target protein. For example, phosphorylation triggers ubiquitination and subsequent degradation of inhibitory factor-κB (I-κB) required for activation of the transcription factor nuclear factor κB (NF-κB). In other cases, such as those of certain transcription factors, the target protein has to dissociate from the specific DNA sequence that it binds to, to be recognized by the system. Thus, in addition to the E3 ligases themselves, modifying enzymes such as kinases, ancillary proteins, or DNA sequences to which substrates bind, all play an important role in the recognition process and determine whether a protein gets degraded in the 20S core particle of the proteasome.[18]

28.2.4   Postproteasomal Mechanisms

It is of note that degradation is not yet complete after the proteasome, that is, the proteasome does not lead to degradation into single amino acids. The 26S proteasome complex generates peptides of diverse sizes, most of which are subsequently rapidly degraded to amino acids; some however are translocated to the cell surface to serve in antigen presentation.[19] Rapid clearance of free peptides in the cytosol by several specific peptidases provides amino acids for use in the synthesis of new proteins, important under conditions where the supply of exogenous amino acids is limiting.[20, 21]

28.2.5   The Ubiquitin-Proteasome System in Disease

As already mentioned earlier, the ubiquitin pathway targets many substrates and processes, and aberrations of the system underlie the pathogenesis of several diseases. The system is involved in malignancies, Liddle's syndrome, Angelman's syndrome, neurodegenerative diseases, cystic fibrosis, immune and inflammatory responses, and has also been implicated in disease associated wasting.[4, 22] What is the experimental evidence underlying the conclusion that this system is involved in the degradation of long-lived proteins such as muscle proteins and validates the view that it is the predominant system active in muscle wasting? These are questions we address in Section 28.3.

28.3   The Ubiquitin-Proteasome System in Muscle Wasting

28.3.1   Multiple Methods Provide Evidence in Rodent Models

The evidence, which led to the concept that the ubiquitin-proteasome pathway plays a primary role in the enhanced protein degradation that occurs in atrophying muscle, is derived from a wide variety of rodent models (for review, see References 4,5,7,10,23,24). Table 28.1 summarizes the animal models mimicking

TABLE 28.1

Animal Models that Show Involvement of the Ubiquitin-Proteasome System

Model	Increased mRNA levels	Reference
Fasting	Ubiquitin, PolyUb, C-1, C-3, C-5, C-8, C-9, $E2_{14k}$, atrogin-1	33,37,42,46,98
Denervation	PolyUb, C-1, C-3, C-5, C-8, C-9	29,32,33,42
Disuse atrophy	Ub, C-2, C-9, $E2_{14k}$, MuRF-1, atrogin-1	49,70,92,64
Diabetes	C-3, Ub, $E3\alpha$, $E2_{14k}$	31,34,36,184,185
Renal failure/acidosis	Ub, C-2, C-3, C-9, $E2_{14k}$	25,26,34,186
Sepsis	C-3, C-7, C-9, $E3\alpha$, $E2_{14k}$, Cathepsin B, m-calpain	32,39,74,187–190
Burn injury	Ub, C-2, $E2_{14k}$	191
Endotoxemia	Ub, C-2, $E2_{14k}$	192
Cancer	Ub, C-8, C-9, $E2_{14k}$	30,38,193
Hyperthyroidism	Ub	32
Glucocorticoid excess	PolyUb, C-8, Ub-E2G, $E2_{14k}$	35,40

human diseases that have been used to analyze the role of the ubiquitin-proteasome system in muscle wasting. In general, in these studies the results obtained with diseased animals were compared with pair-fed control animals, which were fed an amount of food identical to that eaten by the diseased animals. This allows the distinction of disease-related effects from diet-related effects.

In early studies, intact muscles from diseased animals and pair-fed controls were incubated in organ chambers under conditions that block the different proteolytic systems. Blocking of lysosomal and calcium-dependent proteases was performed by adding specific inhibitors of these enzymes to the incubation medium and blockers of ATP synthesis were used to deplete ATP.[25, 26] The latter approach in fact led to the initial discovery of the existence of an ATP-dependent pathway of protein degradation.[27, 28] In most models, ATP depletion indeed resulted in significant decreases in protein degradation, proving that the process requires ATP and thus supporting a role for the ubiquitin-proteasome system.[25, 26, 29] Further evidence was obtained after specific inhibitors of proteasome activity had been developed such as MG132 or methylamine.[25, 30–32] Blocking of lysosomal or calcium-dependent calpain proteases generally led to only up to 20% reduction in protein degradation, whereas blocking the proteasome led to 40 to 100% inhibition.

A complementary approach consisted of the measurement of levels of messenger RNA (mRNA) transcripts for members of the ubiquitin-proteasome pathway relative to total muscle RNA and mRNA content in several muscles, including fast and slow fiber muscles, for example, the mixed-fiber gastrocnemius, the white-fiber extensor digitorus longus, and the red-fiber soleus. Increased transcription (generally 2- to 3-fold increases) of pathway members has been interpreted as a sign for activation of ubiquitin-mediated proteolysis in wasting muscle. These include (1) mRNAs of multiple proteasome subunits, such as C-1, C-3, C-5, C-8, and C-9.[2,33] In insulinopenic and renal failure rats, it was demonstrated that C-3 mRNA is regulated at the level of gene transcription.[25,34] (2) Poly-ubiquitin mRNA was increased typically 2- to 5-fold in muscles during fasting, after denervation,[33] or after application of catabolic doses of glucocorticoids.[35] (3) In models of diabetes and renal failure, ubiquitin mRNA was increased, and molecular evidence was provided to substantiate that changes in transcript levels were due to enhanced gene transcription of ubiquitin gene rather than, for example, changes in mRNA stability.[25,31,34] (4) Transcripts of certain ubiqui-tinating enzymes, including the $E2_{14k}$ and $E3\alpha$ were increased typically 2- to 3-fold in diabetes,[34,36] fasting,[37] cancer,[38] sepsis,[39] or after treatment with the glucocorticoid analogue dexamethasone.[40] (5) Finally, specific deubiquitination enzymes exist and provide one of the ways by which a cell can regulate levels of protein-ubiquitination. In response to fasting, streptozotocin-induced diabetes, dexamethasone treatment and cancer, expression of the mRNA of the deubiqui-tinating enzyme USP19 was increased in skeletal muscle by $\sim$30 to 200% and was always inversely correlated with muscle mass. Deubiquitinating enzymes can thus be added to the list of ubiquitin-proteasome system-related enzymes regulated in the process of atrophy.[41]

A further important step to support the importance of the ubiquitin pathway was the demonstration of the presence of ubiquitin–protein conjugates,[35,42] although in these studies it was unclear which proteins were ubiquitinated.

Most recently, the availability of cDNA microarray techniques allowed a more global analysis of transcriptional changes in single models and led to the identification of new proteins involved in wasting caused by disuse,[43] sepsis,[44,45] food deprivation,[46] or diabetes.[47] The fact that similarities existed in several models,[48,49] prompted Goldberg and colleagues to propose that in all types of atrophying muscle, a common set of transcriptional adaptations and biochemical changes develop that enhance the cell's capacity for protein breakdown.[1] A recent comparison of the changes in content of specific mRNAs in muscle atrophying from fasting, cancer, streptozotocin-induced diabetes mellitus, or uremia, confirmed that a common set of genes was induced.[2] Many of the strongly induced genes in muscles in these four catabolic states coded for proteins involved in degradation, including polyubiquitins, ubiquitin fusion proteins, the ubiquitin ligases atrogin-1/MAFbx and MuRF-1, and the multiple but not all subunits of the 20S proteasome and its 19S regulator, but interestingly, the lysosomal enzyme cathepsin L as well. Indeed, involvement of nonproteasomal protein degradation pathways had been assessed in several earlier studies by measuring mRNA for cathepsin L, cathepsin D, and calpain 1.[50,51] In particular, cathepsin L was confirmed to be important in sepsis.[44] Atrogin-1/MAFbx and MuRF-1 are discussed in more detail later, because their muscle-specific expression pattern as well as the strong increases measured in various models of atrophy makes them potential new targets in muscle wasting.

Taken together, the evidence for involvement of the ubiquitin-proteasome in muscle wasting *in vivo* has initially been based to a large extent on the upregulation of ubiquitin and of components of the proteasome and on measurements performed in the isolated organ. In addition, proteins targeted for degradation were shown to be of myofibrillar origin mainly.[52,53] Particular combinations of the ubiquitinating enzymes E2 and E3 play a pivotal role in distinguishing which proteins are degraded by the proteasome. The list of discovered E3s keeps growing and it has become a challenge to identify the E3s that specifically recognize the myofibrillar proteins and mediate or cause their degradation during atrophy. We next describe the progress in this field.

28.3.2 Involvement of the E3α/E2$_{14k}$ Enzymes in Muscle Wasting

E3α (also called Ubr1) acts together with E2$_{14k}$ in the N-end rule pathway, and, based on the fact that mRNA for one or both of these two factors increased, this was the first enzyme implicated in muscle atrophy.[54–56] In the N-end rule pathway of protein degradation, substrates that begin with unusual NH_2-terminal residues, such as unblocked hydrophobic or basic amino acids are recognized, which serve as destabilizing amino acids[57,58] Distinct binding sites on these NH_2-terminal residues are known and can be blocked by selective dipeptide inhibitors or amino acid esters. Evidence that the E2$_{14k}$/E3α combination is important in muscle atrophy was provided in fasted,[37] diabetic,[36] hyperthyroid,[56] septic,[39,59] and tumor-bearing[38] animals, as well as after glucocorticoid administration.[40]

Using cell extracts, Lecker and colleagues[36] provided evidence that the upregulation of these enzymes indeed resulted in enhanced ubiquitination: a specific substrate of the N-end rule pathway, α-lactalbumin, was ubiquitinated faster in extracts of diabetic rats and a dominant negative form of E2$_{14k}$ inhibited this increase in ubiquitination rates. Furthermore, these authors showed that both E2$_{14k}$ and E3α were rate limiting for ubiquitin conjugation, because adding small amounts of the purified proteins to extracts stimulated ubiquitin conjugation. Consistently, in muscle extracts from sepsis and tumor-bearing animals, selective inhibitors of E3α markedly inhibited ^{125}I-ubiquitin conjugation and ATP-dependent degradation of endogenous proteins. Importantly, in muscle extracts the N-end rule pathway catalyzed most ubiquitin conjugation, but in extracts from non-muscle origin such as those from HeLa cells, it made only a minor contribution to overall protein ubiquitination. Thus, these data suggested that protein degradation in cachectic muscle involves this E2/E3 combination.[55,56]

Recently, the functional significance of this mechanism in degrading muscle proteins has been questioned. First, it is still unclear which protein substrates are ubiquitinated by E3α in muscle. It has been suggested that for ubiquitination by this pathway, the target proteins have to be "clipped" by another proteolytic enzyme to generate the free amino-terminal residues recognized by the E3α. Moreover, in sepsis the changes in E2$_{14k}$ expression paralleled the effects of sepsis on protein breakdown in muscle, but the increased message levels were not associated with increased E2$_{14k}$ protein levels.[39] Another reason for doubt came from mice lacking E2$_{14k}$ or E3α, which show no defects in muscle development or growth but undergo atrophy just like control animals.[60,61] However, multiple isoforms of these E2 and E3s exist and redundancy makes it difficult to draw conclusions on the role of the N-end rule pathway in muscle wasting.[1] Lecker's recent results obtained with cDNA microarrays in various models of atrophy demonstrated that the vast majority of ubiquitin-conjugating enzymes does not change, including $E2_{14k}$ and $E3\alpha$, suggesting that the changes in $E2_{14k}$ and $E3\alpha$ are not general features of atrophying muscle; however, they still may be important in individual models.[2] Those recent studies in which transcriptional profiles in several models of atrophy were compared; however, identified two E3s, of which the mRNAs were induced far more dramatically than those coding for $E2_{14k}$ and $E3\alpha$, namely atrogin-1/MAFbx and MuRF-1, as being a common feature of various models of wasting.[2]

28.3.3 Involvement of Atrogin-1/MAFbx and MuRF-1

These recently discovered muscle-specific ubiquitin-protein ligases are now thought to be central in the development of muscle atrophy.[48,49] The gene most strongly induced is the ubiquitin-ligase atrogin-1. Cloning and further analysis of this gene demonstrated that it contains an F-box domain, which is characteristic of a particular type of E3s, namely the SCF family of E3 ubiquitin ligases.[48,49] The SCF complex name comes from the fact that it involves stable interactions between Skp1, Cullin1, and one of many F-box containing proteins. F-box proteins are the key components that link the target protein with the rest of the E3 and ubiquitination machinery. Atrogin-1 was therefore also designated as Muscle Atrophy F-box (MAFbx).[49]

Atrogin-1/MAFbx is induced 8- to 40-fold in atrophy during fasting, diabetes, cancer, renal failure,[46,48] denervation,[49] and in sepsis.[45] The expression of atrogin-1 increases at least 12 h before muscle weight loss, and its mRNA is high when net proteolysis is rapid.[48] Mice lacking atrogin-1 have reduced rates of denervation atrophy and therefore it was concluded that this E3 plays a role in the loss of muscle protein.[49] The motifs that recognize specific proteins in muscle, that is, a substrate recognition site remains to be identified on atrogin-1, and may involve protein–protein interactions for direct muscle protein degradation via ubiquitination of myofibrillar components, or regulation of a nuclear regulatory protein or transcription factor, as suggested by the presence of a nuclear localization sequence.[1]

The other E3 that was identified by the genomic approach, MuRF-1,[49] belongs to the RING finger E3 ligase subfamily. It was initially found in association with the myofibril, in particular with its constituent titin,[62,63] and therefore thought to play a role in the breakdown of filamentous proteins. It will be discussed in more detail later in this chapter.

28.3.4 Nedd4 and KIAA10

In a similar study, global analysis of gene expression patterns during disuse led to the identification of another E3, Nedd4, which was found upregulated to a similar extent as MuRF-1 and atrogin-1.[64,65] Nedd4 has been implicated in targeting membrane proteins for degradation by the proteasome, and, colocated with substrates such as the IGF-I receptor,[66] it has been shown to be involved in the regulation of the IGF-I receptor. Thus, Nedd4 could influence myotube hypertrophy/atrophy by affecting IGF-I receptor-mediated effects on growth. Finally, KIAA10, a HECT-domain ubiquitin-protein ligase, is also highly expressed in muscle,[67,68] and its role in wasting is currently being studied.[1]

Further study, in particular of MAFbx/atrogin-1 and MuRF-1, is needed to better understand their role in muscle atrophy because, at present knowledge on the substrates and mechanisms of action of these E3s remains rudimentary.

28.4 MAKING MYOFIBRILLAR PROTEINS AVAILABLE FOR DEGRADATION BY THE UBIQUITIN-PROTEASOME PATHWAY: PREPROTEASOMAL PROTEOLYTIC MECHANISMS

Being the most abundant protein of a muscle cell, myofibrillar protein is the main source of protein expected to be broken down and to contribute to changes in protein turnover. This expectation was confirmed in early studies in which total and myofibrillar protein breakdown was measured as the release of tyrosine and 3-methylhistidine into incubation medium, respectively. 3-Methylhistidine is an amino acid that is mainly incorporated in the contractile proteins actin and myosin, and not recycled by the cell,[6] and it could therefore be concluded that it is mostly myofibrillar proteins that are degraded in wasting muscle.[52,53] The predominant release of 3-methylhistidine in these studies was not just due to the high abundance of these proteins in muscle cells, but also to the fact that myofibrillar proteins were degraded at higher rates than other cellular proteins. It is of note that transcripts of myofibrillar or cytoskeletal genes were found not to be changed under conditions of muscle atrophy induced by fasting, diabetes, uremia, and cancer, as demonstrated by microarray analysis[2] and therefore the contractile protein loss is likely mainly due to a change in rates of protein turnover.

To better understand how cytoskeletal proteins that are part of a complex structure are prepared for degradation, a brief description follows of the sarcomere, which is the structural unit of the myofibrils in striated muscle responsible for contraction (for review, see Reference 69). The sarcomere consists of myofilaments, including the thin actin filaments (chains of actin monomers) and the thick myosin filaments (chains of myosin monomers), and a number of associated proteins necessary to regulate contraction and to build and maintain the sarcomeric structure. Each sarcomere has two Z-disks at either end, which anchor thin filaments and link them from one sarcomere to the next all along the myofibril. The thick myosin filaments are located in the middle of the sarcomere and held in place by the M-band, another transverse structure. In addition, the very thin and elastic titin filaments serve as a template for myofibril assembly and have an important anchoring function by positioning the thick filaments during contraction. Contractile proteins that are present in these highly organized sarcomere structures cannot enter the proteasome as such, but have to dissociate in order to become readily degradable by the ubiquitin-proteolytic pathway.[54] Release of myofilaments therefore may be an initial and perhaps rate-limiting element in the process of muscle breakdown. Cytosolic endopeptidases such as caspases and calpains may each in their own way contribute to the steps that need to take place before degradation of contractile proteins in the proteasome.

28.4.1 Calcium-Dependent Release of Contractile Proteins by Calpains

Whereas animal studies led to a consensus on the view that the ubiquitin-proteasome pathway mediates the majority of protein degradation that occurs in the muscle wasting response, in several studies the use of specific inhibitors demonstrated that multiple pathways are involved in the breakdown of cellular proteins, including the calpains.[7] Involvement of the calpains was shown to take place as one of the early events during unloading dystrophy and the presence of inhibitors of calcium-activated proteases attenuated degradation rates to a small extent.[70] Evidence for involvement of calpains was obtained in other models of muscle wasting, such as Duchenne muscular dystrophy,[71] fasting,[72] cancer,[73] and sepsis.[74, 75] Moreover, overexpression in mice of calpastatin, an endogenous inhibitor of the calpains, led to diminished reduction of muscle fiber cross-sectional area in a model of unloading.[76]

Calpains are concentrated in the Z-disk,[77] the site where assembly of the sarcomere is mediated[78] but also where disassembly may begin. In cachectic muscle from septic rats, increased calpain expression as well as morphologic evidence for disintegration of the Z-disk was found.[79] This was associated with the release of myofilaments from the myofibrils, and this release was inhibited with dantrolene, a substance that inhibits the release of calcium from intracellular stores. This study therefore suggests that calcium-dependent release of myofilaments from myofibrils is part of the mechanism that leads to contractile protein degradation.[79]

In vitro, calpains can initiate degradation of individual myofibrillar proteins, including actin, myosin, desmin, filamin, tropomyosin, troponin T and I, vinculin, C-protein, nebulin, gelsolin, and titin.[80–82] By overexpression of dominant-negative m-calpain or of the calpastatin inhibitory domain in L8 myoblast cells, Huang demonstrated that inhibition of calpain stabilized the calpain substrates fodrin and nebulin and reduced protein degradation, indicating that calpains may play a role in the disassembly of sarcomeric proteins.[82] In another study, the cytoskeletal proteins desmin and dystrophin increased markedly when calpain, proteasome, and lysosome activities were inhibited, but sarcomeric proteins such as α-actinin, tropomyosin, and filamin were relatively insensitive to the addition of protease inhibitors.[83] Notably, while having specific cleavage sites on several contractile proteins, the calpains cannot fully degrade these proteins by themselves[80] and it is therefore suggested that the calpain-mediated cleavage of myofibrillar proteins may have a destabilizing effect and precede ubiquitination and further degradation by the proteasome. This notion was supported in a study in which treatment of L6 myotubes with the calcium ionophore A23187 or thapsigargin, substances that increase intracellular calcium levels through different mechanisms, resulted in an increase in proteasome activity. Further data suggested that this was at least in part regulated by calmodulin, calcium calmodulin-dependent kinase II, caspases, and calpains. These observations support a role of calcium in the regulation of proteasome-dependent protein breakdown in skeletal muscle.[84]

28.4.2 Caspases

Interesting new protein–protein interactions were identified in cardiac muscle cells.[85] A yeast two-hybrid system was used to screen for caspase-3 interacting proteins of the cardiac cytoskeleton and a specific cleavage site for caspase-3 was mapped to ventricular essential myosin light chain (vMLC1). It was demonstrated that vMLC1 cleavage in failing myocardium *in vivo* was associated with a morphological disruption of the organized vMLC1 staining of sarcomeres and with a reduction in myocyte contractile performance. Adenoviral gene transfer of the caspase inhibitor p35 *in vivo* prevented caspase-3 activation and vMLC1 cleavage, with a positive impact on contractility. Caspase-3 is well known as one of the caspases involved in apoptosis, however in the context of this paper, its activation may lead to contractile dysfunction before cell death.[85] It is speculated that in skeletal muscle, similar caspase-mediated cleavage may be the step before degradation of contractile proteins by the proteasome.[86] The following study has provided some evidence for this.

Besides the above-mentioned new site on MLC, caspases have specific cleavage sites on actin, and recombinant caspase-3 was shown to cleave actomyosin complexes *in vitro*.[86] In cultured cells, cleavage resulted in accumulation of a small actin proteolytic fragment of 14-kDa, which was visible only in the presence of proteasome blockers indicating that these fragments are subsequently degraded by the ubiquitin-proteasome system. The characteristic 14-kDa fragment was abundant in rat muscles undergoing atrophy due to diabetes, as well as in cultured muscle cells deprived of serum.[86] Activation of caspases has also been implicated in the inhibition of myogenesis caused by tumor necrosis factor (TNF)-α.[87] Caspase activity was found increased in diabetic rats, in rats with cancer, and in cachectic animals due to treatment with angiotensin II.[86,88–90] In the study on angiotensin II-induced wasting, caspase activation was shown to be related to enhanced apoptosis.[89] For a review on the role of apoptosis in cachexia, see Reference 91. In conclusion, cleavage by caspases may destabilize myofibrillar proteins and trigger subsequent proteasomal degradation. In some models, this can be followed by apoptosis of the muscle cells, but it may very well be that the process stops after disassembly and digestion of myofibrils, thus only affecting protein turnover and not cell death.

28.4.3 Cathepsins

Various isoforms of the cathepsins are increased in atrophy,[44,64,70,92] however, they do not seem to be involved in myofibrillar protein degradation, because when their activity is blocked, rates of protein degradation are not changed significantly.[44,50,70,74,92,93] A minor involvement of the cathepsins was demonstrated in some *in vivo* studies and confirms that lysosomal degradation mainly concerns degradation of membrane proteins including receptors, ligands, channels, and transporters and that these lysosomal enzymes do not degrade contractile proteins themselves (reviewed in Reference 94). Instead, by degrading specific membrane associated proteins, they may contribute to the trigger that causes the atrophy to occur.[65] In this respect it is interesting to note that the more recent gene array studies on several models of cachexia identified cathepsin L as one of the genes increased in all models.[2]

28.4.4 E3 Ligases

We have introduced MuRF-1 as one of the genes strongly induced in various forms of muscle atrophy. MuRF-1 was discovered some years ago, using titin-based yeast two-hybrid screens, as a novel, muscle-specific protein that interacts with titin. MuRF-1, which contains all features of RING-domain-containing monomeric ubiquitin ligases, is located in close proximity to titin's catalytic kinase domain and has therefore been suggested to modulate titin's kinase activity.[62] In addition it has been suggested that, since RING domains mediate ubiquitin transfer, the association of MuRF-1 with the M- and Z-line regions could link ubiquitin-conjugation pathways to the titin filament system.[62] Indeed, a recent publication demonstrates that expression of MuRF-1 severely disrupts the integrity of titin's M-line region, and causes a perturbation of thick filament components, but not of the NH_2-terminal or I-band regions of titin, the Z-lines, or the thin filaments, indicating that the interaction between titin and MuRF-1 is important for the stability of the sarcomeric M-line region.[95] The authors speculate that the titin–MuRF-1 interaction functions in the assembly of M-line and thick filament components during myofibrillogenesis, as well as during myofibril turnover.[95] Together with the studies by Bodine,[96] in which it was shown that MuRF-1 knockout mice were resistant to muscle atrophy, this suggests that MuRF-1 is involved in a pathway responsible for titin turnover and that degradation of titin could facilitate further disassembly of the sarcomere and allow degradation of its other components, thus providing a mechanism as to how MuRF-1 might mediate skeletal muscle atrophy.

28.5 SIGNALING MOLECULES INVOLVED IN ACTIVATION OF THE UBIQUITIN-PROTEASOME SYSTEM

In different disease states, distinct signals may lead to weight loss, for example in diabetes it involves elevated levels of glucocorticoids and low insulin, in tumor cachexia and sepsis it may be TNF-α, whereas in renal failure it could be metabolic acidosis. Still, it appears that a common set of molecules from the ubiquitin-proteasome system is induced,[2] and the question is whether there is also a common intracellular signaling pathway induced by the different triggers, which ultimately leads to activation of molecules of the protein degradation machinery. Identification of transcriptional factors that regulate expression levels of calpains, caspases, MuRF-1, atrogin-1, and ubiquitin components of the proteasome will be an important step toward understanding the atrophy response. We discuss current knowledge on the two best known catabolic factors, glucocorticoids and TNF-α, and the way by which both affect the ubiquitin-proteasome pathway by distinct mechanisms. These factors have in common their ability to modulate insulin-like growth factor (IGF)-I IGF-I/insulin specific signal transduction pathways, which are directly involved metabolism, protein synthesis, and hypertrophy of the cell, and they, therefore, are also discussed.

28.5.1 The Role of Glucocorticoids

In animal models of acidosis, fasting, diabetes, uremia, and sepsis, physiological levels of glucocorticoids were shown to be required for stimulation of the ubiquitin-proteasome pathway in a permissive way.[34,97–100] It was shown in adrenalectomy experiments that abolishing glucocorticoid production resulted in decreased rates of ATP-dependent protein degradation. Administration of low doses of dexamethasone to the adrenalectomized animals provided evidence that the measured changes were due to glucocorticoids. More recently, glucocorticoids were also shown to participate in the upregulation of MuRF-1 and atrogin-1/MAFbx: treatment of rats with the glucocorticoid receptor antagonist RU-486 prevented sepsis-induced increases in mRNA levels for MuRF-1 and atrogin-1/MAFbx.[45]

Injection of catabolic doses of glucocorticoids in the rat directly activates the ubiquitin pathway of protein degradation, and induces loss of body weight.[35,100–108] Cell culture experiments elucidated part of the mechanism, while showing that glucocorticoids directly control transcription of components of the ubiquitin system in the cell.[109] In addition, administration of corticosteroids decreases IGF-I gene expression in muscle and liver, resulting in significantly reduced serum levels of IGF-I.[110] Endotoxin-treatment also reduced circulating, liver, and muscle IGF-I, which appeared to be due to increased corticosteroid levels measured in these rats, because RU-486 abolished the decrease in IGF-I and reversed body weight loss.[111] Similarly, angiotensin II-treated animals displayed low muscle mass and reduced muscle and systemic IGF-I levels, which were related to high glucocorticoids.[89,112] The latter studies are relevant to the wasting syndrome observed in patients with severe heart failure, in which angiotensin II levels are often increased.[113]

Low IGF-I does not only occur in steroid-induced cachexia, but is a common feature of many models of wasting. It has, for example, also been described in hindlimb unloading[114] and in a model of heart failure, induced by ligation of the left coronary artery.[115] The reduction in IGF-I in the heart failure model correlated with a decreased muscle fiber cross-sectional area and inversely with the local expression of interleukin-1β (IL-1β), suggesting that both factors may contribute to catabolic metabolism that finally results in skeletal muscle atrophy and cardiac cachexia.[115] Notably, both circulating[116] and muscle[117] IGF-I is reduced in cachectic heart failure patients. These and other studies suggest a role for IGF-I in cachexia and recent studies show that it is one of the common signaling molecules that modulate transcription of the ubiquitin-proteasome system.

28.5.2 IGF-I Prevents Induction of Components of the Ubiquitin-Proteasome System

As described above, glucocorticoids are required for, or at higher concentrations, induce muscle atrophy *in vivo*. The analogue dexamethasone was shown to cause strong increases in transcripts of ubiquitin, $E2_{14k}$, Ub-E2G, and the C-2, -3, and -8 proteasome subunits. Injections of IGF-I in dexamethasone-treated animals caused a significant suppression of the same transcripts and provided support to the notion that IGF-I regulates the expression of mRNAs encoding components of the ubiquitin pathway during catabolism.[40,118] Similarly, IGF-I significantly inhibited lysosomal and ubiquitin-proteasome-dependent protein breakdown in skeletal muscle from burned and septic rats, but did not influence calcium/calpain-dependent protein breakdown.[119–122]

Dexamethasone was also shown to induce atrogin-1 and MuRF-1,[49] and IGF-I inhibited the increases in mRNA levels of these ubiquitin ligases.[123] The dexamethasone- and IGF-I-induced changes in atrogin-1 correlated well with changes in the actual breakdown of long-lived cell proteins, especially of myofibrillar proteins, as measured by 3-methylhistidine release, whereas a correlation with MuRF-1 mRNA was not present. The effect of IGF-I on atrogin-1 expression was achieved by rapid reduction of synthesis of its mRNA, without affecting mRNA degradation, indicating that IGF-I is involved in the rapid modulation of atrogin-1.[123] The effect of IGF-I on MuRF-1 mRNA was much slower.

Insulin also suppresses proteolysis, and inversely, muscle protein wasting in insulinopenia includes activation of the ubiquitin-proteasome pathway with increased expression of the ubiquitin gene.[8,31,34] The lack of insulin may have a similar downstream effect as the lack of IGF-I, as insulin in part uses the same intracellular pathways as IGF-I. Consistent with this idea, IGF-I treatment was able to blunt diabetes-induced increases in the expression of atrogin-1 and ubiquitin.[124,125]

Cell culture studies helped to identify molecular mechanisms by which IGF-I affect protein turnover in the cell. Studies on rat L6 myotubes showed that insulin and IGF-I lowered levels of $E2_{14k}$ mRNA.[37,126] In these studies, IGF-I stimulated degradation of the mRNA transcript encoding for this ubiquitin-conjugating enzyme, thus modulating levels of this enzyme at the posttranscriptional level.[126] Other components of the ubiquitin-proteasome system appeared to be regulated via transcriptional mechanisms. In L8 cells, IGF-I reduced increases in transcripts of the proteasome subunit C-2, had no effect on cathepsin B and D gene expression but slightly increased m-calpain, indicating that levels of calpain are not determining the myofibrils for degradation.[127] This is in contrast to a more recent study, in which the use of different inhibitors of the individual proteolytic pathways demonstrated that IGF-I inhibited dexamethasone-induced proteasome-dependent proteolysis, but also blunted increases in cathepsin B and calpain activities.[107] In conclusion, several *in vivo* and *in vitro* studies show that IGF-I plays a role in the molecular mechanisms responsible for induction of muscle wasting such as that induced by glucocorticoids, by modulating transcription of muscle specific components of the proteasome pathway, and some studies indicate that it interferes with multiple cellular degradation pathways.

28.5.3 IGF-I-Mediated Regulation of Atrogin-1 Involves PI3K/Akt/Foxo Signaling

IGF-I contributes to multiple functions in muscle, including activation of satellite cells, proliferation and differentiation of myoblasts, enhanced protein synthesis, and diminished protein degradation in myotubes. Thus, the muscle wasting observed in models with low muscle and systemic IGF-I levels can be explained by multiple mechanisms. Well-known are the IGF-I and insulin signaling mechanisms important for protein synthesis in differentiated muscle cells (Figure 28.1), which involve a stimulation of phosphatidylinositol-3-kinase (PI3K) and its downstream effector, a serine/threonine kinase called Akt (or PKB, protein kinase B). Further downstream, protein translation is enhanced through activation of the mammalian target of rapamycin (mTOR), because it phosphorylates the translation initiation factor 4E-binding protein and the 70-kDa ribosomal protein S6 kinase.[96, 128–130] In addition, the PI3K/Akt pathway targets glycogen synthase kinase 3, another regulatory protein involved in protein translation and synthesis. Besides this PI3K/Akt pathway, intracellular cascades normally induced by IGF-I include calcineurin, and the mitogen-activated protein kinase pathways, consisting of a sequence of successively acting kinases that ultimately result in activation of terminal kinases such as extracellular-signal regulated kinases (ERKs). The question now is whether the same or different pathways are used to modulate transcription of components of the ubiquitin-proteasome system. This question has recently been addressed in dexamethasone-induced wasting, where the protein degradation pathway that changed atrogin-1 mRNA levels involved only the PI3K/Akt pathway.[124, 131] The mechanism downstream of PI3K/Akt leading to the inhibition of atrogin-1 and MuRF-1 transcription involves inhibition of the Foxo family of transcription factors (Figure 28.1, Foxo = forkhead box-containing protein, O-subfamily).[131–133] Phosphorylated Foxo proteins are unable to translocate to the nucleus. Reduced IGF-I/insulin signaling decreases the phosphorylation of Foxo, promotes its nuclear translocation, and increases transcription of atrogin-1 and MuRF-1.[131, 132] IGF-I treatment or Akt overexpression inhibited Foxo translocation and atrogin-1 expression, whereas constitutively active Foxo acted on the atrogin-1 promoter to cause atrogin-1 transcription and dramatic atrophy of myotubes and muscle fibers. When Foxo activation was blocked by a dominant-negative construct in myotubes or by RNAi in mouse muscles *in vivo*, atrogin-1 induction during starvation and atrophy of myotubes induced by glucocorticoids was prevented. Thus, this forkhead factor plays a critical role in the development of muscle atrophy.[132] Consistently, a mutant form of Foxo prevented Akt-mediated

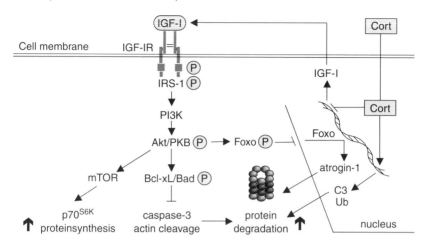

FIGURE 28.1   Schematic representation showing that the IGF-I/PI3K/Akt pathway is involved in protein synthesis as well as protein degradation and that glucocorticoids modulate this pathway. One of the proteins targeted by Akt is mTOR, which in turn phosphorylates p70^{S6K}, a kinase involved in the translation of mRNA into protein. Two mechanisms are depicted by which Akt may prevent protein degradation. Akt reduces caspase-3-mediated actin cleavage by phosphorylating and inactivating the proapoptotic protein Bad, and Akt prevents translocation of the transcription factor Foxo to the nucleus, thereby diminishing transcription of atrogin-1, an E3 ubiquitin ligase involved in protein degradation. In addition, this scheme depicts the role of corticosteroids on transcription of genes coding for IGF-I and members of the proteasome pathway. In particular the diminished IGF-I transcription will result in reduced PI3K/Akt signaling, and consequently a shift of the delicate balance between protein synthesis and degradation, which ultimately leads to muscle wasting (IGF-I = insulin-like growth factor-I; IGF-IR = IGF-I receptor; P indicates protein phosphorylation; IRS-1 = insulin receptor substrate-1; PI3K = phosphatidylinositol-3 kinase; Akt = PKB = protein kinase B; mTOR = mammalian target of rapamycin; p70^{S6K} = 70-kDa ribosomal protein S6 kinase; Bcl-xL = antiapoptotic protein of the Bcl family; Bad = proapoptotic protein of the Bcl family; Foxo = forkhead transcription factor; C-3 = subunit of the proteasome core particle; Ub = ubiquitin; Cort = corticosterone).

inhibition of MuRF-1 and atrogin-1 upregulation.[131] These findings were confirmed *in vivo* in diabetic muscle: mRNA of the forkhead transcription factor, its nuclear translocation, and binding to the atrogin-1 promoter were increased in diabetes.[133] In conclusion, IGF-I is, via PI3K/Akt, not only capable of activating protein synthesis pathways, but is simultaneously suppressing catabolic pathways, allowing it to prevent glucocorticoid and denervation-induced muscle atrophy, and these effects are mediated by Foxo.

28.5.4 IGF-I Regulation of Caspase and Calpain Activity

IGF-I is known to prevent apoptosis by blocking activation of the intrinsic apoptotic caspase cascade. Part of the mechanism involves phosphorylation and inactivation of a proapoptotic protein called Bad, which suppresses activity of the downstream enzyme caspase-3. As described above, caspase-3 may be one of the initial steps leading to myofibril breakdown. The beneficial effect of IGF-I on muscle wasting may therefore in part consist of the inhibition of this initial step.[86] Calpain activity, which was also suggested as one of the initiating steps in muscle wasting, was not affected by IGF-I *in vivo*,[120] but in a cell culture system IGF-I blocked dexamethasone-induced activation of calpain activity.[107] Further studies are required to analyze the role of IGF-I in triggering protein degradation via activation of either enzyme.

28.5.5 The Role of TNF-α and NF-κB

Cytokines and tumor factors are thought to mediate muscle wasting in cancer and other diseases such as sepsis and heart failure. Induction of ubiquitin and E3 ligase has been shown in several models of wasting that are associated with high cytokine levels, and studies in tumor-bearing or septic rats demonstrated that anticytokine treatment prevented the increased ATP-ubiquitin- and Ca^{2+}-dependent proteolysis, suggesting a role for both systems in this type of wasting.[134–136] Several studies have analyzed the mechanism by which TNF-α inhibits myogenesis[128,137] and here we summarize the role of the proteasome in this process.

Like the glucocorticoids, one mechanism by which TNF-α may block myogenesis is by changing the activation state of the PI3K/Akt pathway thus interfering with insulin/IGF signaling. It has been shown that infusion of TNF-α reduces circulating concentrations of IGF-I in fasted rats,[138] and this has an effect on protein synthesis[139] as well as on specific protein degradation pathways (see the previous sections). Several *in vitro* studies confirm that there is extensive cross talk between TNF-α and IGF signaling and analyze its role in proliferation, protein synthesis, and differentiation from myoblast to myotube.

The other mechanism by which TNF-α, either by itself or together with other cytokines, may enhance or induce the breakdown of muscle fibers is by activating nuclear factor (NF)-κB, one of its most important downstream mediators (Figure 28.2). The proteasome is involved in NF-κB activation as follows. In its inactive form, NF-κB is bound to the inhibitory protein, I-κB, which maintains it in the cytosol. To function, NF-κB must enter the nucleus, which happens upon activation by phosphorylation for example by cellular stress or inflammatory factors such as TNF-α. These stimuli lead to rapid I-κB ubiquitination, followed by its degradation in the proteasome. Once NF-κB is released from I-κB, it translocates into the nucleus where it can activate various genes, for example, the inflammatory cytokines. It is expected that inhibition of proteasome activity will blunt the inflammatory response, and therefore maybe also the atrophy response.

A direct relation between NF-κB signaling and the ubiquitin-proteasome system was shown in mice with muscle-specific transgenic expression of activated I-κB kinase beta (MIKK), which results in activation of NF-κB. Profound muscle wasting was detected in these mice, the muscle loss due to accelerated protein

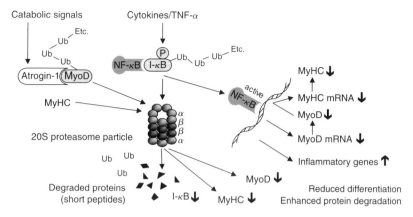

Figure 28.2 Potential mechanisms by which the ubiquitin-proteasome pathway may influence muscle mass. Catabolic signals enhance atrogin-1 transcription. Atrogin-1 interacts directly with MyoD, causing its ubiquitination and subsequent degradation by the proteasome. MyoD normally causes transcription of the MyHC gene and myotube differentiation, a response that is blunted under catabolic conditions. One of the catabolic mediators is TNF-α, which causes NF-κB translocation to the nucleus. This process requires degradation of the inhibitory I-κB by the proteasome. Active NF-κB blunts MyoD transcription and, as a consequence, MyHC gene transcription. In addition to these transcriptional mechanisms, MyHC protein in the cell may diminish by its direct degradation via the ubiquitin-proteasome pathway (Ub = ubiquitin; MyoD = myogenic transcription factor; MyHC = myosin heavy chain; TNF = tumor necrosis factor; NF-κB = nuclear factor-κB; I-κB = inhibitory factor-κB).

breakdown through ubiquitin-dependent proteolysis. Expression of the E3 ligase MuRF-1 was increased in the MIKK mice and pharmacological or genetic inhibition of the I-κB kinase/NF-κB/MuRF-1 pathway reversed muscle atrophy. On the other hand, denervation- and tumor-induced muscle loss were substantially reduced and survival rates improved by NF-κB inhibition in mice, in which NF-κB is inhibited through expression of the I-κBα superrepressor, consistent with a critical role for NF-κB in the pathology of muscle wasting.[140] Several other studies confirm that inhibition of NF-κB decreases activation of the proteasome's chymotrypsin-like activity.[141–143]

Using C2C12 cell cultures, Guttridge and colleagues analyzed the proteins targeted by this pathway, while aiming to further characterize the cellular mechanism of muscle loss. The TNF-α-induced activation of NF-κB was shown to reduce predominantly myosin heavy chain (MyHC) in the cell.[143–145] This contractile protein was the preferred target of procachectic factors such as TNF-α, because mRNA and protein levels of other contractile proteins, including troponin T, tropomyosin, and actin, were not reduced.[146] Normal expression of MyHC requires binding of the transcription factor MyoD to the MyHC IIb promotor. In addition to this role in muscle-specific gene expression, MyoD is known to control the exit from the cell cycle.[147] Activation of NF-κB represses synthesis of MyoD by reducing its transcription in myoblasts undergoing differentiation.[145, 148, 149] Taken together, these studies have demonstrated that the mechanism by which MyHC expression is reduced involves activation of NF-κB, which leads to reduced MyoD transcription, thus lower MyoD protein levels, and consequently results in reduced transcription of MyHC. Inhibition of proteasome activity inhibits NF-κB activation, thereby providing one of the mechanisms for blunted loss in muscle mass that has been observed in the presence of these inhibitors, and suggesting that the mechanism involves the cellular differentiation process.

Recent data suggest that MyoD levels can also be reduced by enhanced degradation of the MyoD protein. Direct biological and biochemical evidence has been provided that atrogin-1 interacts with MyoD and thereby induces its ubiquitin-mediated proteolysis. A core LXXLL motif sequence in MyoD was identified, which is necessary for binding to atrogin-1, and a mutation in this LXXLL core motif repressed ubiquitination and degradation of MyoD. Overexpression of atrogin-1 suppressed MyoD-induced differentiation and inhibited myotube formation and therefore a role was attributed to this E3 enzyme in the course of muscle differentiation by determining the abundance of MyoD via this posttranslational protein degradation mechanism.[150] Taken together, blocking proteasome activity may prevent or reverse disease-induced muscle loss by enhancing myogenic differentiation in two ways. One is by blocking NF-κB activity, thereby preventing the TNF-α-induced reduction in cellular transcripts of MyoD, whereas the other is by diminishing the degradation of MyoD protein in the proteasome.

A recent study suggests that not only protein levels of the transcription factor MyoD, but also those of MyHC itself are determined by proteasomal degradation. In an *in vivo* tumor model, a selective reduction of this myofibrillar protein was found, as well as an induction of ubiquitin and E3 ligase genes. Together with the presence of ubiquitin-MyHC conjugates, these results suggested that the ubiquitin-proteasome system is involved in the regulation of MyHC at the protein level, implying that MyHC is one of the substrates of ubiquitin ligases.[146]

28.6 TARGETING THE UBIQUITIN-PROTEASOME SYSTEM IN DISEASE

28.6.1 Evidence for a Role of the Ubiquitin-Proteasome System in Patients

For some time, it was not known whether the ubiquitin-proteasome system is implicated in human muscle wasting. Above, we described the various means by which it was demonstrated in animal models that an important part of wasting-related protein degradation is mediated by the proteasome. Similar approaches have recently been used to confirm relevance of these mechanisms in patients. In biopsies from septic patients compared with nonseptic controls, 3- to 4-fold higher mRNA levels for ubiquitin and certain 20S proteasome subunits were detected.[151] Also in muscle biopsies from severe head trauma patients, increased muscle mRNA levels for critical components of the ubiquitin proteolytic pathway, including ubiquitin, E2$_{14k}$, and various proteasome subunits, as well as for cathepsin D and m-calpain were detected, and these paralleled metabolic adaptations that were analyzed in those patients.[152] Similarly, in patients with AIDS or cancer, increased expression of genes encoding for the ubiquitin-proteasome pathway were detectable, thus confirming experimental data.[153–156] On the other hand, a study performed on lung cancer patients did not confirm this conclusion,[157] but postulates that cathepsin B may have a role in inducing muscle wasting in the early stages of lung cancer. Two other studies report no alterations in gene expression of components of the ubiquitin-proteasome proteolytic pathway, the one being in dystrophin-deficient muscles of Duchenne patients[158] and the other in Cushings syndrome.[159] The lack of regulation of proteolytic genes in such patients was suggested to represent an adaptive regulatory mechanism, preventing sustained increased protein breakdown and avoiding rapid muscle wasting.

28.6.2 Cancer and Inflammatory Disease

Based on the unique mechanism of protein degradation in the proteasome, specific inhibitors have been developed that do not interfere with other proteolytic enzymes of the cell. Currently, several inhibitors are available which are divided into four classes,[160] most of which are analogs of the substrates of the chymotryptic site.[161–163] One compound, PS-341, later called Bortezomib, was selected for intensive study based on its low IC_{50} and other advantageous properties. Obviously, inhibition of these core enzymes of the pathway that reside in the proteasome itself may have undesirable effects on some of the many cellular processes that the ubiquitin-proteasome system regulates. Existing drugs such as Bortezomib preserve the wide range of targets digested by the proteasome. Nevertheless, it has appeared possible to define a narrow window between beneficial effects and toxicity for short-term treatment of certain diseases.[22] Experimental and clinical studies, in particular in the field of cancer, asthma, brain infarct, and autoimmune encephalomyelitis, show that positive effects can be achieved. In fact, the highly potent inhibitor Bortezomib is in phase II and phase III trials against various cancers,[160] including lung cancer,[164] breast cancer,[165] prostate cancer,[166] and multiple myeloma.[160, 164, 167–170]

In malignancies, one of the main underlying mechanisms of the beneficial effects is thought to be via inhibition of NF-κB activation, which leads to inhibition of the antiapoptotic action of this transcription factor. The mechanism of NF-κB activation involves ubiquitin-proteasome mediated degradation of the inhibitory protein, I-κB, which is required to allow NF-κB to enter the nucleus, where, in this case, it activates an antiapoptotic mechanism. This happens for example in response to cellular stress. One of the reasons why chemotherapy is not always efficient is because NF-κB is activated and protects the cells from apoptosis and confers resistance. In cancer therapy, inhibition of its activation by inhibition of ubiquitin-proteasome system is due to stabilization of I-κB and may sensitize cells to irradiation and chemotherapy-induced apoptosis.[170] As the proteasome has a direct role in cell cycle control by degrading cell cycle regulatory proteins, inhibition of proliferation is a second mechanism by which proteasome inhibitors may act. Finally, Bortezomib may also inhibit angiogenesis.[171–174]

NF-κB is stimulated under various conditions of inflammation, and stimulates expression of most of the key components of inflammatory responses. Based on their inhibitory effect on NF-κB and therefore anti-inflammatory effects, proteasome inhibitors have been tested *in vivo* in several animal models of inflammatory disease, including asthma,[175,176] arthritis,[177] cardiovascular disease,[178] and autoimmune encephalomyelitis.[179] The synthetic derivative PS-519 of the natural product lactacystin is in clinical trials against stroke, where the mechanism probably involves blocking the inflammation that occurs after reperfusion of ischemic tissue.[180,181] In the heart, proteasome inhibition was shown to ablate activation of NF-κB in myocardial reperfusion and to reduce reperfusion injury.[182] Its effectiveness has also been demonstrated in psoriasis, for which topical application of the drug is possible, which offers advantages, because the systemic toxicity of proteasome inhibition can be avoided.[183] Similarly, in neuroprotection, the effects of proteasome inhibitors are thought to be due to inhibition of the local inflammatory response.

28.6.3 Potential Use of Ubiquitin-Proteasome System Inhibitors in Cachexia

Experimental evidence together with more recent data obtained in cachectic patients have led many authors to suggest the proteasome as an attractive target for cachexia therapy. However, the ubiquitin-proteasome system has a central role in a wide range of basic cellular processes and thus modulation or interference with its activity by means of drugs may be complicated or result in many side effects. For example, its blocking activity on angiogenesis described as one of the beneficial effects in cancer therapy will probably have adverse effects in muscle wasting. Moreover, the existing drugs are very efficient at inhibiting the short-lived regulatory proteins of the cell, but drugs that selectively block degradation of the long-lived contractile proteins remain to be developed. Targeting the muscle specific MuRF-1 and atrogin-1 would be the approach of choice, since effects will be restricted to striated muscle only. Once the substrates of these enzymes are identified, molecules may be developed that either bind to these substrates or their ancillary proteins or to the E3 proteins themselves.[22] Selective inhibition of these E3 targets may potentially lead to a selective inhibition of the degradation process in muscle.

REFERENCES

1. Lecker S.H., Ubiquitin-protein ligases in muscle wasting: multiple parallel pathways?, *Curr. Opin. Clin. Nutr. Metab. Care* 6, 271, 2003.
2. Lecker S.H. et al., Multiple types of skeletal muscle atrophy involve a common program of changes in gene expression, *FASEB J.* 18, 39, 2004.

3. Ciechanover A. et al., Ubiquitin-mediated proteolysis: biological regulation via destruction, *Bioessays* 22, 442, 2000.
4. Mitch W.E. and Goldberg A.L., Mechanisms of muscle wasting. The role of the ubiquitin-proteasome pathway, *N. Engl. J. Med.* 335, 1897, 1996.
5. Lecker S.H. et al., Muscle protein breakdown and the critical role of the ubiquitin-proteasome pathway in normal and disease states, *J. Nutr.* 129, 227S, 1999.
6. Jagoe R.T. and Goldberg A.L., What do we really know about the ubiquitin-proteasome pathway in muscle atrophy?, *Curr. Opin. Clin. Nutr. Metab. Care* 4, 183, 2001.
7. Hasselgren P.O. and Fischer J.E., Muscle cachexia: current concepts of intracellular mechanisms and molecular regulation, *Ann. Surg.* 233, 9, 2001.
8. Price S.R. et al., Molecular mechanisms regulating protein turnover in muscle, *Am. J. Kidney Dis.* 37, S112, 2001.
9. Combaret L. et al., Nutritional and hormonal control of protein breakdown, *Am. J. Kidney Dis.* 37, S108, 2001.
10. Mitch W.E. and Price S.R., Mechanisms activating proteolysis to cause muscle atrophy in catabolic conditions, *J. Ren. Nutr.* 13, 149, 2003.
11. Jackman R.W. and Kandarian S.C., The molecular basis of skeletal muscle atrophy, *Am. J. Physiol. Cell Physiol.* 287, C834, 2004.
12. Goldberg A.L., Protein degradation and protection against misfolded or damaged proteins, *Nature* 426, 895, 2003.
13. Voges D. et al., The 26S proteasome: A molecular machine designed for controlled proteolysis, *Ann. Rev. Biochem.* 68, 1015, 1999.
14. Weissman A.M., Themes and variations on ubiquitilation, *Nat. Rev. Mol. Cell Biol.* 2, 169, 2001.
15. Glickman M.H. and Ciechanover A., The ubiquitin-proteasome proteolytic pathway: destruction for the sake of construction, *Physiol. Rev.* 82, 373, 2002.
16. Pickart C.M. and Cohen R.E., Proteasomes and their kin: proteases in the machine age, *Nat. Rev. Mol. Cell Biol.* 5, 177, 2004.
17. Ciechanover A. and Ben-Saadon R., N-terminal ubiquitination: more protein substrates join in, *Trends Cell Biol.* 14, 103, 2004.
18. Pickart C.M., Mechanisms underlying ubiquitination, *Ann. Rev. Biochem.* 70, 503, 2001.
19. Rock K.L. et al., Post-proteasomal antigen processing for major histocompatibility complex class I presentation, *Nat. Immunol.* 5, 670, 2004.
20. Saric T. et al., Pathway for degradation of peptides generated by proteasomes: a key role for thimet oligopeptidase and other metallopeptidases, *J. Biol. Chem.* 279, 46723, 2004.
21. Hasselgren P.O. et al., Molecular regulation of muscle cachexia: it may be more than the proteasome, *Biochem. Biophys. Res. Commun.* 290, 1, 2002.
22. Ciechanover A., The ubiquitin proteolytic system and pathogenesis of human diseases: a novel platform for mechanism-based drug targeting, *Biochem. Soc. Trans.* 31, 474, 2003.
23. Attaix D. et al., Adaptation of the ubiquitin-proteasome proteolytic pathway in cancer cachexia, *Mol. Biol. Rep.* 26, 77, 1999.
24. Baracos V.E., Regulation of skeletal-muscle-protein turnover in cancer-associated cachexia, *Nutrition* 16, 1015, 2000.

25. Bailey J.L. et al., The acidosis of chronic renal failure activates muscle proteolysis in rats by augmenting transcription of genes encoding proteins of the ATP- dependent ubiquitin-proteasome pathway, *J. Clin. Invest.* 97, 1447, 1996.

26. Mitch W.E. et al., Metabolic acidosis stimulates muscle protein degradation by activating the adenosine triphosphate-dependent pathway involving ubiquitin and proteasomes, *J. Clin. Invest.* 93, 2127, 1994.

27. Etlinger J.D. and Goldberg A.L., A soluble ATP-dependent proteolytic system responsible for the degradation of abnormal proteins in reticulocytes, *Proc. Natl Acad. Sci. USA* 74, 54, 1977.

28. Hershko A. and Ciechanover A., The ubiquitin system for protein degradation, *Ann. Rev. Biochem.* 61, 761, 1992.

29. Medina R. et al., Activation of the ubiquitin-ATP-dependent proteolytic system in skeletal muscle during fasting and denervation atrophy, *Biomed. Biochim. Acta.* 50, 347, 1991.

30. Baracos V.E. et al., Activation of the ATP-ubiquitin-proteasome pathway in skeletal muscle of cachectic rats bearing a hepatoma, *Am. J. Physiol.* 268, E996, 1995.

31. Price S.R. et al., Muscle wasting in insulinopenic rats results from activation of the ATP-dependent, ubiquitin-proteasome proteolytic pathway by a mechanism including gene transcription, *J. Clin. Invest.* 98, 1703, 1996.

32. Tawa N.E., Jr. et al., Inhibitors of the proteasome reduce the accelerated proteolysis in atrophying rat skeletal muscles, *J. Clin. Invest.* 100, 197, 1997.

33. Medina R. et al., Increase in levels of polyubiquitin and proteasome mRNA in skeletal muscle during starvation and denervation atrophy, *Biochem. J.* 307 (Pt 3), 631, 1995.

34. Mitch W.E. et al., Evaluation of signals activating ubiquitin-proteasome proteolysis in a model of muscle wasting, *Am. J. Physiol.* 276, C1132, 1999.

35. Auclair D. et al., Activation of the ubiquitin pathway in rat skeletal muscle by catabolic doses of glucocorticoids, *Am. J. Physiol.* 272, C1007, 1997.

36. Lecker S.H. et al., Ubiquitin conjugation by the N-end rule pathway and mRNAs for its components increase in muscles of diabetic rats, *J. Clin. Invest.* 104, 1411, 1999.

37. Wing S.S. and Banville D., 14-kDa ubiquitin-conjugating enzyme: structure of the rat gene and regulation upon fasting and by insulin, *Am. J. Physiol.* 267, E39, 1994.

38. Temparis S. et al., Increased ATP-ubiquitin-dependent proteolysis in skeletal muscles of tumor-bearing rats, *Cancer Res.* 54, 5568, 1994.

39. Hobler S.C. et al., Sepsis is associated with increased ubiquitin-conjugating enzyme E214k mRNA in skeletal muscle, *Am. J. Physiol.* 276, R468, 1999.

40. Chrysis D. and Underwood L.E., Regulation of components of the ubiquitin system by insulin-like growth factor I and growth hormone in skeletal muscle of rats made catabolic with dexamethasone, *Endocrinology* 140, 5635, 1999.

41. Combaret L. et al., USP19 is a ubiquitin specific protease regulated in rat skeletal muscle during catabolic states, *Am. J. Physiol. Endocrinol. Metab.* 2004.

42. Wing S.S. et al., Increase in ubiquitin-protein conjugates concomitant with the increase in proteolysis in rat skeletal muscle during starvation and atrophy denervation, *Biochem. J.* 307 (Pt 3), 639, 1995.

43. Cros N. et al., Analysis of altered gene expression in rat soleus muscle atrophied by disuse, *J. Cell Biochem.* 83, 508, 2001.

44. Deval C. et al., Identification of cathepsin L as a differentially expressed message associated with skeletal muscle wasting, *Biochem. J.* 360, 143, 2001.

45. Wray C.J. et al., Sepsis upregulates the gene expression of multiple ubiquitin ligases in skeletal muscle, *Int. J. Biochem. Cell Biol.* 35, 698, 2003.

46. Jagoe R.T. et al., Patterns of gene expression in atrophying skeletal muscles: response to food deprivation, *FASEB J.* 16, 1697, 2002.

47. Yechoor V.K. et al., Coordinated patterns of gene expression for substrate and energy metabolism in skeletal muscle of diabetic mice, *Proc. Natl Acad. Sci. USA* 99, 10587, 2002.

48. Gomes M.D. et al., Atrogin-1, a muscle-specific F-box protein highly expressed during muscle atrophy, *Proc. Natl Acad. Sci. USA* 98, 14440, 2001.

49. Bodine S.C. et al., Identification of ubiquitin ligases required for skeletal muscle atrophy, *Science* 294, 1704, 2001.

50. Furuno K. and Goldberg A.L., The activation of protein degradation in muscle by Ca2+ or muscle injury does not involve a lysosomal mechanism, *Biochem. J.* 237, 859, 1986.

51. Lowell B.B. et al., Evidence that lysosomes are not involved in the degradation of myofibrillar proteins in rat skeletal muscle, *Biochem. J.* 234, 237, 1986.

52. Long C.L. et al., Urinary excretion of 3-methylhistidine: An assessment of muscle protein catabolism in adult normal subjects and during malnutrition, sepsis, and skeletal trauma, *Metabolism* 30, 765, 1981.

53. Hasselgren P.O. et al., Total and myofibrillar protein breakdown in different types of rat skeletal muscle: effects of sepsis and regulation by insulin, *Metabolism* 38, 634, 1989.

54. Solomon V. and Goldberg A.L., Importance of the ATP-ubiquitin-proteasome pathway in the degradation of soluble and myofibrillar proteins in rabbit muscle extracts, *J. Biol. Chem.* 271, 26690, 1996.

55. Solomon V. et al., The N-end rule pathway catalyzes a major fraction of the protein degradation in skeletal muscle, *J. Biol. Chem.* 273, 25216, 1998.

56. Solomon V. et al., Rates of ubiquitin conjugation increase when muscles atrophy, largely through activation of the N-end rule pathway, *Proc. Natl Acad. Sci. USA* 95, 12602, 1998.

57. Varshavsky A., The N-end rule: functions, mysteries, uses, *Proc. Natl Acad. Sci. USA* 93, 12142, 1996.

58. Varshavsky A., Recent studies of the ubiquitin system and the N-end rule pathway, *Harvey Lect.* 96, 93, 2000.

59. Fischer D. et al., The gene expression of ubiquitin ligase E3alpha is upregulated in skeletal muscle during sepsis in rats-potential role of glucocorticoids, *Biochem. Biophys. Res. Commun.* 267, 504, 2000.

60. Adegoke O.A. et al., Ubiquitin-conjugating enzyme E214k/HR6B is dispensable for increased protein catabolism in muscle of fasted mice, *Am. J. Physiol. Endocrinol. Metab.* 283, E482, 2002.

61. Kwon Y.T. et al., Construction and analysis of mouse strains lacking the ubiquitin ligase UBR1 (E3alpha) of the N-end rule pathway, *Mol. Cell Biol.* 21, 8007, 2001.

62. Centner T. et al., Identification of muscle specific ring finger proteins as potential regulators of the titin kinase domain, *J. Mol. Biol.* 306, 717, 2001.

63. McElhinny A.S. et al., Muscle-specific RING finger-1 interacts with titin to regulate sarcomeric M-line and thick filament structure and may have nuclear functions via

its interaction with glucocorticoid modulatory element binding protein-1, *J. Cell Biol.* 157, 125, 2002.

64. Stevenson E.J. et al., Global analysis of gene expression patterns during disuse atrophy in rat skeletal muscle, *J. Physiol.* 551, 33, 2003.

65. Kandarian S.C. and Stevenson E.J., Molecular events in skeletal muscle during disuse atrophy, *Exerc. Sport Sci. Rev.* 30, 111, 2002.

66. Vecchione A. et al., The Grb10/Nedd4 complex regulates ligand-induced ubiquitination and stability of the insulin-like growth factor I receptor, *Mol. Cell Biol.* 23, 3363, 2003.

67. You J. and Pickart C.M., A HECT domain E3 enzyme assembles novel polyubiquitin chains, *J. Biol. Chem.* 276, 19871, 2001.

68. You J. et al., Proteolytic targeting of transcriptional regulator TIP120B by a HECT domain E3 ligase, *J. Biol. Chem.* 278, 23369, 2003.

69. Clark K.A. et al., Striated muscle cytoarchitecture: an intricate web of form and function, *Ann. Rev. Cell Dev. Biol.* 18, 637, 2002.

70. Taillandier D. et al., Coordinate activation of lysosomal, Ca 2+-activated and ATP-ubiquitin-dependent proteinases in the unweighted rat soleus muscle, *Biochem. J.* 316 (Pt 1), 65, 1996.

71. Arahata K. et al., Dystrophin digest, *Nature* 337, 606, 1989.

72. Ilian M.A. and Forsberg N.E., Gene expression of calpains and their specific endogenous inhibitor, calpastatin, in skeletal muscle of fed and fasted rabbits, *Biochem. J.* 287 (Pt 1), 163, 1992.

73. Costelli P. et al., Activation of Ca(2+)-dependent proteolysis in skeletal muscle and heart in cancer cachexia, *Br. J. Cancer* 84, 946, 2001.

74. Voisin L. et al., Muscle wasting in a rat model of long-lasting sepsis results from the activation of lysosomal, Ca2+-activated, and ubiquitin-proteasome proteolytic pathways, *J. Clin. Invest.* 97, 1610, 1996.

75. Hotchkiss R.S. and Karl I.E., Dantrolene ameliorates the metabolic hallmarks of sepsis in rats and improves survival in a mouse model of endotoxemia, *Proc. Natl Acad. Sci. USA* 91, 3039, 1994.

76. Tidball J.G. and Spencer M.J., Expression of a calpastatin transgene slows muscle wasting and obviates changes in myosin isoform expression during murine muscle disuse, *J. Physiol.* 545, 819, 2002.

77. Kumamoto T. et al., Localization of the Ca(2+)-dependent proteinases and their inhibitor in normal, fasted, and denervated rat skeletal muscle, *Anat. Rec.* 232, 60, 1992.

78. Agarkova I. et al., M-band: a safeguard for sarcomere stability?, *J. Muscle Res. Cell Motil.* 24, 191, 2003.

79. Williams A.B. et al., Sepsis stimulates release of myofilaments in skeletal muscle by a calcium-dependent mechanism, *FASEB J.* 13, 1435, 1999.

80. Thompson M.G. and Palmer R.M., Signalling pathways regulating protein turnover in skeletal muscle, *Cell Signal* 10, 1, 1998.

81. Sorimachi H. et al., Skeletal muscle-specific calpain, p94, and connectin/titin: their physiological functions and relationship to limb-girdle muscular dystrophy type 2A, *Adv. Exp. Med. Biol.* 481, 383, 2000.

82. Huang J. and Forsberg N.E., Role of calpain in skeletal-muscle protein degradation, *Proc. Natl Acad. Sci. USA* 95, 12100, 1998.

83. Purintrapiban J. et al., Degradation of sarcomeric and cytoskeletal proteins in cultured skeletal muscle cells, *Comp. Biochem. Physiol. B, Biochem. Mol. Biol.* 136, 393, 2003.

84. Menconi M.J. et al., Treatment of cultured myotubes with the calcium ionophore A23187 increases proteasome activity via a CaMK II-caspase-calpain-dependent mechanism, *Surgery* 136, 135, 2004.

85. Moretti A. et al., Essential myosin light chain as a target for caspase-3 in failing myocardium, *Proc. Natl Acad. Sci. USA* 99, 11860, 2002.

86. Du J. et al., Activation of caspase-3 is an initial step triggering accelerated muscle proteolysis in catabolic conditions, *J. Clin. Invest.* 113, 115, 2004.

87. Coletti D. et al., TNFalpha inhibits skeletal myogenesis through a PW1-dependent pathway by recruitment of caspase pathways, *EMBO J.* 21, 631, 2002.

88. Belizario J.E. et al., Cleavage of caspases-1, -3, -6, -8 and -9 substrates by proteases in skeletal muscles from mice undergoing cancer cachexia, *Br. J. Cancer* 84, 1135, 2001.

89. Song Y.H. et al., Muscle-specific expression of IGF-1 blocks angiotensin II-induced skeletal muscle wasting, *J. Clin. Invest.* 2005.

90. Smith H.J. and Tisdale M.J., Induction of apoptosis by a cachectic-factor in murine myotubes and inhibition by eicosapentaenoic acid, *Apoptosis* 8, 161, 2003.

91. Sandri M., Apoptotic signaling in skeletal muscle fibers during atrophy, *Curr. Opin. Clin. Nutr. Metab. Care* 5, 249, 2002.

92. Ikemoto M. et al., Space shuttle flight (STS-90) enhances degradation of rat myosin heavy chain in association with activation of ubiquitin-proteasome pathway, *FASEB J.* 15, 1279, 2001.

93. Tischler M.E. et al., Different mechanisms of increased proteolysis in atrophy induced by denervation or unweighting of rat soleus muscle, *Metabolism* 39, 756, 1990.

94. Mayer R.J., The meteoric rise of regulated intracellular proteolysis, *Nat. Rev. Mol. Cell Biol.* 1, 145, 2000.

95. McElhinny A.S. et al., Muscle-specific RING finger-2 (MURF-2) is important for microtubule, intermediate filament and sarcomeric M-line maintenance in striated muscle development, *J. Cell Sci.* 117, 3175, 2004.

96. Bodine S.C. et al., Akt/mTOR pathway is a crucial regulator of skeletal muscle hypertrophy and can prevent muscle atrophy *in vivo*, *Nat. Cell Biol.* 3, 1014, 2001.

97. May R.C. et al., Metabolic acidosis stimulates protein degradation in rat muscle by a glucocorticoid-dependent mechanism, *J. Clin. Invest.* 77, 614, 1986.

98. Wing S.S. and Goldberg A.L., Glucocorticoids activate the ATP-ubiquitin-dependent proteolytic system in skeletal muscle during fasting, *Am. J. Physiol.* 264, E668, 1993.

99. Price S.R. et al., Acidosis and glucocorticoids concomitantly increase ubiquitin and proteasome subunit mRNAs in rat muscle, *Am. J. Physiol.* 267, C955, 1994.

100. Tiao G. et al., Energy-ubiquitin-dependent muscle proteolysis during sepsis in rats is regulated by glucocorticoids, *J. Clin. Invest.* 97, 339, 1996.

101. Isozaki U. et al., Protein degradation and increased mRNAs encoding proteins of the ubiquitin-proteasome proteolytic pathway in BC3H1 myocytes require an interaction between glucocorticoids and acidification, *Proc. Natl Acad. Sci. USA* 93, 1967, 1996.

102. Wang L. et al., Dexamethasone stimulates proteasome- and calcium-dependent proteolysis in cultured L6 myotubes, *Shock* 10, 298, 1998.

103. Luo G. et al., Heat shock protects L6 myotubes from catabolic effects of dexamethasone and prevents downregulation of NF-kappaB, *Am. J. Physiol. Regul. Integr. Comp. Physiol.* 281, R1193, 2001.

104. Sun X. et al., Expression and binding activity of the glucocorticoid receptor are upregulated in septic muscle, *Am. J. Physiol. Regul. Integr. Comp. Physiol.* 282, R509, 2002.

105. Penner C.G. et al., The transcription factors NF-kappab and AP-1 are differentially regulated in skeletal muscle during sepsis, *Biochem. Biophys. Res. Commun.* 281, 1331, 2001.

106. Penner G. et al., C/EBP DNA-binding activity is upregulated by a glucocorticoid-dependent mechanism in septic muscle, *Am. J. Physiol. Regul. Integr. Comp. Physiol.* 282, R439, 2002.

107. Li B.G. et al., Insulin-like growth factor-I blocks dexamethasone-induced protein degradation in cultured myotubes by inhibiting multiple proteolytic pathways: 2002 ABA paper, *J. Burn. Care Rehabil.* 25, 112, 2004.

108. Combaret L. et al., Glucocorticoids regulate mRNA levels for subunits of the 19 S regulatory complex of the 26 S proteasome in fast-twitch skeletal muscles, *Biochem. J.* 378, 239, 2004.

109. Du J. et al., Glucocorticoids induce proteasome C3 subunit expression in L6 muscle cells by opposing the suppression of its transcription by NF-kappa B, *J. Biol. Chem.* 275, 19661, 2000.

110. Gayan-Ramirez G. et al., Acute treatment with corticosteroids decreases IGF-1 and IGF-2 expression in the rat diaphragm and gastrocnemius, *Am. J. Respir. Crit. Care Med.* 159, 283, 1999.

111. Li Y.H. et al., Differential role of glucocorticoids in mediating endotoxin-induced changes in IGF-I and IGFBP-1, *Am. J. Physiol.* 272, R1990, 1997.

112. Brink M. et al., Angiotensin II induces skeletal muscle wasting through enhanced protein degradation and down-regulates autocrine insulin-like growth factor I, *Endocrinology* 142, 1489, 2001.

113. Brink M. et al., Neurohormonal factors in the development of catabolic/anabolic imbalance and cachexia, *Int. J. Cardiol.* 85, 111., 2002.

114. Criswell D.S. et al., Overexpression of IGF-I in skeletal muscle of transgenic mice does not prevent unloading-induced atrophy, *Am. J. Physiol.* 275, E373, 1998.

115. Schulze P.C. et al., Muscular levels of proinflammatory cytokines correlate with a reduced expression of insulin-like growth factor-I in chronic heart failure, *Basic Res. Cardiol.* 98, 267, 2003.

116. Anwar A. et al., Effect of congestive heart failure on the insulin-like growth factor-1 system, *Am. J. Cardiol.* 90, 1402, 2002.

117. Hambrecht R. et al., Reduction of insulin-like growth factor-I expression in the skeletal muscle of noncachectic patients with chronic heart failure, *J. Am. Coll. Cardiol.* 39, 1175, 2002.

118. Chrysis D. et al., Divergent regulation of proteasomes by insulin-like growth factor I and growth hormone in skeletal muscle of rats made catabolic with dexamethasone, *Growth Horm. IGF Res.* 12, 434, 2002.

119. Fang C.H. et al., Burn injuries in rats upregulate the gene expression of the ubiquitin-conjugating enzyme E2(14k) in skeletal muscle, *J. Burn. Care Rehabil.* 21, 528, 2000.

120. Fang C.H. et al., Insulin-like growth factor-I inhibits lysosomal and proteasome-dependent proteolysis in skeletal muscle after burn injury, *J. Burn. Care Rehabil.* 23, 318, 2002.

121. Fang C.H. et al., Insulin-like growth factor I reduces ubiquitin and ubiquitin-conjugating enzyme gene expression but does not inhibit muscle proteolysis in septic rats, *Endocrinology* 141, 2743, 2000.

122. Fang C.H. et al., Treatment of burned rats with insulin-like growth factor I inhibits the catabolic response in skeletal muscle, *Am. J. Physiol.* 275, R1091, 1998.

123. Sacheck J.M. et al., IGF-I stimulates muscle growth by suppressing protein breakdown and expression of atrophy-related ubiquitin ligases, atrogin-1 and MuRF1, *Am. J. Physiol. Endocrinol. Metab.* 287, E591, 2004.

124. Dehoux M. et al., Role of the insulin-like growth factor I decline in the induction of atrogin-1/MAFbx during fasting and diabetes, *Endocrinology* 145, 4806, 2004.

125. Heszele M.F. and Price S.R., Insulin-like growth factor I: the yin and yang of muscle atrophy, *Endocrinology* 145, 4803, 2004.

126. Wing S.S. and Bedard N., Insulin-like growth factor I stimulates degradation of an mRNA transcript encoding the 14 kDa ubiquitin-conjugating enzyme, *Biochem. J.* 319, 455, 1996.

127. Hong D. and Forsberg N.E., Effects of serum and insulin-like growth factor I on protein degradation and protease gene expression in rat L8 myotubes, *J. Anim. Sci.* 72, 2279, 1994.

128. Guttridge D.C., Signaling pathways weigh in on decisions to make or break skeletal muscle, *Curr. Opin. Clin. Nutr. Metab. Care* 7, 443, 2004.

129. Rommel C. et al., Mediation of IGF-1-induced skeletal myotube hypertrophy by PI(3)K/Akt/mTOR and PI(3)K/Akt/GSK3 pathways, *Nat. Cell Biol.* 3, 1009, 2001.

130. Glass D.J., Molecular mechanisms modulating muscle mass, *Trends Mol. Med.* 9, 344, 2003.

131. Stitt T.N. et al., The IGF-1/PI3K/Akt pathway prevents expression of muscle atrophy-induced ubiquitin ligases by inhibiting FOXO transcription factors, *Mol Cell* 14, 395, 2004.

132. Sandri M. et al., Foxo transcription factors induce the atrophy-related ubiquitin ligase atrogin-1 and cause skeletal muscle atrophy, *Cell* 117, 399, 2004.

133. Lee S.W. et al., Regulation of muscle protein degradation: coordinated control of apoptotic and ubiquitin-proteasome systems by phosphatidylinositol 3 kinase, *J. Am. Soc. Nephrol.* 15, 1537, 2004.

134. Costelli P. et al., Anticytokine treatment prevents the increase in the activity of ATP-ubiquitin- and Ca(2+)-dependent proteolytic systems in the muscle of tumour-bearing rats, *Cytokine* 19, 1, 2002.

135. Combaret L. et al., Torbafylline (HWA 448) inhibits enhanced skeletal muscle ubiquitin-proteasome-dependent proteolysis in cancer and septic rats, *Biochem. J.* 361, 185, 2002.

136. Reid M.B. and Li Y.P., Tumor necrosis factor-alpha and muscle wasting: a cellular perspective, *Respir. Res.* 2, 269, 2001.

137. Langen R.C. et al., Tumor necrosis factor-alpha inhibits myogenic differentiation through MyoD protein destabilization, *FASEB J.* 18, 227, 2004.

138. Fan J. et al., Regulation of insulin-like growth factor-I (IGF-I) and IGF-binding proteins by tumor necrosis factor, *Am. J. Physiol.* 269, R1204, 1995.

139. Frost R.A. et al., Transient exposure of human myoblasts to tumor necrosis factor-alpha inhibits serum and insulin-like growth factor-I stimulated protein synthesis, *Endocrinology* 138, 4153, 1997.

140. Cai D. et al., IKKbeta/NF-kappaB activation causes severe muscle wasting in mice, *Cell* 119, 285, 2004.

141. Wyke S.M. et al., Induction of proteasome expression in skeletal muscle is attenuated by inhibitors of NF-KappaB activation, *Br. J. Cancer* 91, 1742, 2004.

142. Whitehouse A.S. and Tisdale M.J., Increased expression of the ubiquitin-proteasome pathway in murine myotubes by proteolysis-inducing factor (PIF) is associated with activation of the transcription factor NF-kappaB, *Br. J. Cancer* 89, 1116, 2003.

143. Ladner K.J. et al., Tumor necrosis factor-regulated biphasic activation of NF-kappa B is required for cytokine-induced loss of skeletal muscle gene products, *J. Biol. Chem.* 278, 2294, 2003.

144. Guttridge D.C. et al., NF-kappaB-induced loss of MyoD messenger RNA: possible role in muscle decay and cachexia, *Science* 289, 2363, 2000.

145. Li Y.P. and Reid M.B., NF-kappaB mediates the protein loss induced by TNF-alpha in differentiated skeletal muscle myotubes, *Am. J. Physiol. Regul. Integr. Comp. Physiol.* 279, R1165, 2000.

146. Acharyya S. et al., Cancer cachexia is regulated by selective targeting of skeletal muscle gene products, *J. Clin. Invest.* 114, 370, 2004.

147. Tintignac L.A. et al., Cyclin E-cdk2 phosphorylation promotes late G1-phase degradation of MyoD in muscle cells, *Exp. Cell. Res.* 259, 300, 2000.

148. Langen R.C. et al., Inflammatory cytokines inhibit myogenic differentiation through activation of nuclear factor-kappaB, *FASEB J.* 15, 1169, 2001.

149. Li Y.P. et al., Skeletal muscle myocytes undergo protein loss and reactive oxygen-mediated NF-kappaB activation in response to tumor necrosis factor alpha, *FASEB J.* 12, 871, 1998.

150. Tintignac L.A. et al., Degradation of MyoD Mediated by the SCF (MAFbx) Ubiquitin Ligase, *J. Biol. Chem.* 280, 2847, 2005.

151. Tiao G. et al., Sepsis is associated with increased mRNAs of the ubiquitin-proteasome proteolytic pathway in human skeletal muscle, *J. Clin. Invest.* 99, 163, 1997.

152. Mansoor O. et al., Increased mRNA levels for components of the lysosomal, Ca^{2+}-activated, and ATP-ubiquitin-dependent proteolytic pathways in skeletal muscle from head trauma patients, *Proc. Natl Acad. Sci. USA* 93, 2714, 1996.

153. Llovera M. et al., Ubiquitin and proteasome gene expression is increased in skeletal muscle of slim AIDS patients, *Int. J. Mol. Med.* 2, 69, 1998.

154. Williams A. et al., The expression of genes in the ubiquitin-proteasome proteolytic pathway is increased in skeletal muscle from patients with cancer, *Surgery* 126, 744, 1999.

155. Bossola M. et al., Increased muscle ubiquitin mRNA levels in gastric cancer patients, *Am. J. Physiol. Regul. Integr. Comp. Physiol.* 280, R1518, 2001.

156. Bossola M. et al., Increased muscle proteasome activity correlates with disease severity in gastric cancer patients, *Ann. Surg.* 237, 384, 2003.

157. Jagoe R.T. et al., Skeletal muscle mRNA levels for cathepsin B, but not components of the ubiquitin-proteasome pathway, are increased in patients with lung cancer referred for thoracotomy, *Clin. Sci. (Lond.)* 102, 353, 2002.

158. Combaret L. et al., No alteration in gene expression of components of the ubiquitin-proteasome proteolytic pathway in dystrophin-deficient muscles, *FEBS Lett.* 393, 292, 1996.

159. Ralliere C. et al., Glucocorticoids do not regulate the expression of proteolytic genes in skeletal muscle from Cushing's syndrome patients, *J. Clin. Endocrinol. Metab.* 82, 3161, 1997.

160. Adams J., The development of proteasome inhibitors as anticancer drugs, *Cancer Cell* 5, 417, 2004.

161. Kisselev A.F. and Goldberg A.L., Proteasome inhibitors: from research tools to drug candidates, *Chem. Biol.* 8, 739, 2001.

162. Kisselev A.F. et al., The caspase-like sites of proteasomes, their substrate specificity, new inhibitors and substrates, and allosteric interactions with the trypsin-like sites, *J. Biol. Chem.* 278, 35869, 2003.

163. Goldberg A.L. and Rock K., Not just research tools–proteasome inhibitors offer therapeutic promise, *Nat. Med.* 8, 338, 2002.

164. Lara P.N., Jr. et al., Proteasome inhibition with PS-341 (bortezomib) in lung cancer therapy, *Semin Oncol.* 31, 40, 2004.

165. Orlowski R.Z. and Dees E.C., The role of the ubiquitination-proteasome pathway in breast cancer: applying drugs that affect the ubiquitin-proteasome pathway to the therapy of breast cancer, *Breast Cancer Res.* 5, 1, 2003.

166. Papandreou C.N. and Logothetis C.J., Bortezomib as a potential treatment for prostate cancer, *Cancer Res.* 64, 5036, 2004.

167. Stanford B.L. and Zondor S.D., Bortezomib treatment for multiple myeloma, *Ann. Pharmacother.* 37, 1825, 2003.

168. Richardson P.G. et al., A phase 2 study of bortezomib in relapsed, refractory myeloma, *N. Engl. J. Med.* 348, 2609, 2003.

169. Voorhees P.M. et al., The proteasome as a target for cancer therapy, *Clin. Cancer Res.* 9, 6316, 2003.

170. Adams J., The proteasome: Structure, function, and role in the cell, *Cancer Treat. Rev.* 29 Suppl 1, 3, 2003.

171. Williams S. et al., Differential effects of the proteasome inhibitor bortezomib on apoptosis and angiogenesis in human prostate tumor xenografts, *Mol. Cancer Ther.* 2, 835, 2003.

172. Nawrocki S.T. et al., Effects of the proteasome inhibitor PS-341 on apoptosis and angiogenesis in orthotopic human pancreatic tumor xenografts, *Mol. Cancer Ther.* 1, 1243, 2002.

173. Sunwoo J.B. et al., Novel proteasome inhibitor PS-341 inhibits activation of nuclear factor-kappa B, cell survival, tumor growth, and angiogenesis in squamous cell carcinoma, *Clin. Cancer Res.* 7, 1419, 2001.

174. Drexler H.C. et al., Inhibition of proteasome function induces programmed cell death in proliferating endothelial cells, *FASEB J.* 14, 65, 2000.

175. Elliott P.J. et al., Proteasome inhibition: A novel mechanism to combat asthma, *J. Allergy Clin. Immunol.* 104, 294, 1999.

176. Elliott P.J. et al., Proteasome inhibition: a new anti-inflammatory strategy, *J. Mol. Med.* 81, 235, 2003.

177. Palombella V.J. et al., Role of the proteasome and NF-kappaB in streptococcal cell wall-induced polyarthritis, *Proc. Natl Acad. Sci. USA* 95, 15671, 1998.

178. Herrmann J. et al., The ubiquitin-proteasome system in cardiovascular diseases-a hypothesis extended, *Cardiovasc. Res.* 61, 11, 2004.

179. Vanderlugt C.L. et al., Treatment of established relapsing experimental autoimmune encephalomyelitis with the proteasome inhibitor PS-519, *J Autoimmun.* 14, 205, 2000.

180. Phillips J.B. et al., Proteasome inhibitor PS519 reduces infarction and attenuates leukocyte infiltration in a rat model of focal cerebral ischemia, *Stroke* 31, 1686, 2000.

181. Wojcik C. and Di Napoli M., Ubiquitin-proteasome system and proteasome inhibition: new strategies in stroke therapy, *Stroke* 35, 1506, 2004.

182. Pye J. et al., Proteasome inhibition ablates activation of NF-kappa B in myocardial reperfusion and reduces reperfusion injury, *Am. J. Physiol. Heart Circ. Physiol.* 284, H919, 2003.

183. Zollner T.M. et al., Proteasome inhibition reduces superantigen-mediated T cell activation and the severity of psoriasis in a SCID-hu model, *J. Clin. Invest.* 109, 671, 2002.

184. Pepato M.T. et al., Role of different proteolytic pathways in degradation of muscle protein from streptozotocin-diabetic rats, *Am. J. Physiol.* 271, E340, 1996.

185. Merforth S. et al., Alterations of proteasome activities in skeletal muscle tissue of diabetic rats, *Mol. Biol. Rep.* 26, 83, 1999.

186. Bailey J.L. and Mitch W.E., Twice-told tales of metabolic acidosis, glucocorticoids, and protein wasting: what do results from rats tell us about patients with kidney disease?, *Semin. Dial.* 13, 227, 2000.

187. Hobler S.C. et al., Activity and expression of the 20S proteasome are increased in skeletal muscle during sepsis, *Am. J. Physiol.* 277, R434, 1999.

188. Garcia-Martinez C. et al., Ubiquitin gene expression in skeletal muscle is increased during sepsis: involvement of TNF-alpha but not IL-1, *Biochem. Biophys. Res. Commun.* 217, 839, 1995.

189. Tiao G. et al., Sepsis stimulates nonlysosomal, energy-dependent proteolysis and increases ubiquitin mRNA levels in rat skeletal muscle, *J. Clin. Invest.* 94, 2255, 1994.

190. Wray C.J. et al., Dantrolene downregulates the gene expression and activity of the ubiquitin-proteasome proteolytic pathway in septic skeletal muscle, *J. Surg. Res.* 104, 82, 2002.

191. Chai J. et al., The relationship between skeletal muscle proteolysis and ubiquitin-proteasome proteolytic pathway in burned rats, *Burns* 28, 527, 2002.

192. Chai J. et al., Role of ubiquitin-proteasome pathway in skeletal muscle wasting in rats with endotoxemia, *Crit. Care Med.* 31, 1802, 2003.

193. Llovera M. et al., Anti-TNF treatment reverts increased muscle ubiquitin gene expression in tumour-bearing rats, *Biochem. Biophys. Res. Commun.* 221, 653, 1996.

29 Peripheral Signaling Pathways Involved in Muscle Loss

Stefanie Possekel, Thomas Meier, and Markus A. Ruegg

CONTENTS

SUMMARY

Muscle wasting observed in cachexia is thought to be based on the misbalance between anabolic and catabolic pathways. Here, we discuss selected peripheral signaling pathways relevant for muscle growth and maintenance. In the first section, we focus on positive regulators of muscle growth including the calcineurin pathway and pathways activated by insulin-like growth factor (IGF-1). This is followed by a discussion of the NF-κB and the myostatin signaling pathway, both of which have a negative influence on muscle growth. Although the pathways are discussed separately they have several points of interaction, which may offer promising entry points for the treatment of muscle loss. As several of these pathways are also deregulated in situations of uncontrolled cell growth, such as cancer, current drug development efforts are aimed in fact to suppress anabolic pathways and to increase catabolic pathways, which is just the opposite of what would be desirable to overcome conditions of muscle wasting. Although this is certainly a challenge, the known chemical templates that may interact with components of these pathways may offer a unique opportunity for the development of pharmacological entities to treat muscle loss. Besides cachexia, the pathways described are equally relevant for other pathological conditions such as muscular dystrophies. However, the complexity of the described pathways as well as their contribution to a variety of essential cellular processes will be challenging for the development of efficacious treatments with the required safety profile.

29.1 INTRODUCTION

Muscle wasting as observed in cachexia is due to an imbalance between anabolic and catabolic processes and, therefore, both principal pathways can be modified to counteract this wasting. While inhibition of catabolic pathways (Figure 29.1B; right) is likely to be a promising strategy for treatment of muscle loss, equally important are strategies that stimulate anabolic pathways to enhance muscle growth (Figure 29.1A). Several lines of evidence strongly suggest that muscle hypertrophy induced by exercise, overload, or during recovery from atrophy is based on two concomitantly occurring processes (1) increase in the cytoplasmic volume within individual muscle fibers and (2) increase in the number of muscle fibers per muscle due to an altered fusion pattern of satellite cells with existing muscle fibers.

In this chapter, we discuss selected signaling pathways that have recently been shown to be involved in the growth of muscle. These pathways represent possible entry points that might allow the maintenance of muscle mass in cachectic situations where catabolic pathways are hyperactive and thus prevail. As other chapters already cover the targets that affect appetite or hormonal regulation, we only specifically discuss peripheral signaling pathways and potential drug targets suitable for pharmacological interventions.

It has become evident that the molecular mechanisms involved in the wasting of muscle mass are similar irrespective of the underlying cause of this wasting process (see for example Lecker et al., 2004). Conditions that can trigger muscle wasting

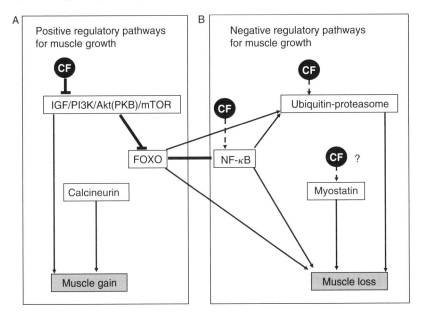

FIGURE 29.1 Schematic drawing of the peripheral signaling pathways discussed in this chapter. Anabolic pathways such as the IGF/PI3K/Akt(PKB)/mTOR and the calcineurin pathway cause muscle hypertrophy in animal models upon activation. In contrast, activation of the ubiquitin-proteasome or the NF-κB pathway by cachectic factors (CF), such as cytokines, hormones or tumor-derived factors, or genetic manipulation in animal models, cause muscle atrophy. Both sets of pathways are linked by forkhead transcription factors (FOXO).

include immobilization or denervation and some pathological conditions, such as muscle dystrophy or cachexia. Thus, although for most examples discussed in this chapter, the primary evidence may in fact be derived from work on experimental paradigms other than cachexia (e.g., hypertrophy, dystrophy), it is highly probable that these mechanisms also play a role in cachexia.

29.2 THE CALCINEURIN/NF-AT SIGNALING PATHWAY

Calcium-dependent processes are known to regulate many aspects of muscle growth, in particular myoblast fusion. The most prominent pathway involved in this fusion process, which also has been implicated in hypertrophy of skeletal and heart muscle, activates the "nuclear factor of activated T-cells" (NF-AT) family of transcription factors (Figure 29.2). Muscle expresses several NF-AT isoforms including NF-ATc1 and NF-ATc2. The analysis of mouse mutants deficient in NF-AT expression has provided indirect evidence that activation of the calcineurin/NF-AT pathway correlates with gain of muscle mass. For example, mice deficient in NF-ATc2 have smaller skeletal muscle fibers and show a deficit in myoblast fusion (Horsley et al., 2001). This effect is, at least partially, based on

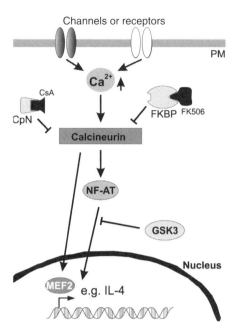

FIGURE 29.2   Schematic representation of the calcineurin signaling pathway. Multiple signals can cause an increase in intracellular calcium. Rise in calcium concentration, by entry through channels or by activation of receptors in the plasma membrane (PM), activates calcineurin that dephosphorylates the nuclear factor of activated T-cells (NF-AT) transcription factors, which then translocate into the nucleus. Activated NF-AT induces expression of target genes such as interleukin-4 (IL-4) and others. Myocyte enhance factor (MEF)2 can also be directly activated by calcineurin. The immunosuppressants CsA, by binding to CpN, and FK 506, by binding to FKBP, are selective inhibitors of calcineurin.

the increase of expression of interleukin-4 (IL-4) by NF-ATc2. IL-4 in turn acts in an autocrine fashion to promote myoblast fusion. Consistent with the concept that NF-ATc2 is mainly involved in the regulation of IL-4, NF-ATc2-deficient myoblasts form large myotubes in the presence of IL-4. Moreover, muscle from both IL-4- and IL-4 receptor-deficient mice is considerably smaller and contains fewer myonuclei than that of wild-type mice (Horsley et al., 2003). In these animals, skeletal muscle can regenerate to some degree after freeze damage; however, the newly formed myofibers remain small and contain fewer myonuclei than their wild-type controls.

The NF-AT transcription factors are well-known downstream targets of the calcineurin pathway. Calcineurin, a serine/threonine phosphatase controlled by calcium, was originally identified in the mammalian brain. The discovery that the clinically used immunosuppressants cyclosporin A (CsA) and tacrolimus (formerly FK 506) inhibit the phosphatase activity of calcineurin has greatly helped to elucidate the function of calcineurin. Interestingly, both drugs exert their effect on calcineurin not directly but via a complex of CsA and cyclophilin (CpN) or FK 506 and FK506-binding protein (FKBP) (Figure 29.2). In skeletal muscle, calcineurin is involved in determining the muscle fiber type and in heart it is involved in controlling its size (reviewed in Wilkins and Molkentin 2004). Thus, the finding that (1) NF-ATc2 is involved in determining muscle fiber size and that (2) CsA prevents overload-induced hypertrophy (Dunn et al., 1999; Semsarian et al., 1999; Musaro et al., 1999) has led to the hypothesis that calcineurin signaling is also involved in muscle hypertrophy in the adult. There is, however, still some controversy as to whether the calcineurin pathway is indeed involved in this process (see e.g., Dunn et al., 2002; Parsons et al., 2004; reviewed in Schiaffino and Serrano 2002).

Recently it has been demonstrated that activation of the calcineurin/NF-AT pathway may hold the potential to ameliorate muscular dystrophy. c-Jun amino-terminal kinase (JNK)1 is constitutively active in dystrophic muscle and causes the phosphorylation of NF-ATc1 leading to its nuclear export and loss of appropriate target gene activation. The hypothesis that restoration of NF-ATcl activity could ameliorate muscle dystrophy was tested in dystrophin-deficient mice that expressed constitutively active forms of calcineurin to counteract the JNK1 kinase activity. Indeed, activated calcineurin in these *mdx* mice caused significant amelioration of muscle histology including the reconstitution of the dystrophin-associated protein complex at the sarcolemma, reduction in the extent of central nucleation and amelioration of membrane integrity (Chakkalakal et al., 2004). Interestingly, it has also been shown recently that glucocorticoids, currently the only pharmacological treatment option routinely used in patients with Duchenne muscular dystrophy (DMD), can upregulate the activity of calcineurin in dystrophic muscle, an effect that could be completely abolished by the calcineurin inhibitor CsA (St-Pierre et al., 2004). Taken together, this set of experiments indicates that the calcineurin/NF-AT pathway might offer pharmacological entry points for treatment of muscle loss caused by dystrophin deficiencies (dystrophinopathies) and also other cachectic conditions.

29.3 THE IGF-1/PI3K/Akt(PKB)/mTOR SIGNALING PATHWAYS

In the past few years convincing evidence has accumulated that pathways activated by insulin-like growth factor-1 (IGF-1) are critically involved in regulating muscle mass. Exercise or overload-induced hypertrophy of skeletal muscle leads to direct upregulation of IGF-1 and of a muscle-specific splice variant thereof, called mechano-growth factor (MGF; Yang et al., 1996). Recent evidence also indicates that the amount of IGF-binding proteins is altered in experimental paradigms that affect muscle growth (Lecker et al., 2004). All this rather circumstantial evidence for a role of IGF-1 in muscle growth is strongly supported by the fact that transgenic overexpression of IGF-1 in skeletal muscle induces muscle hypertrophy *in vivo* (Semsarian et al., 1999; Musaro et al., 1999). The activity of IGF-1 on muscle growth is based on both, an increase in the number of myonuclei per muscle fiber and an increase in protein synthesis. IGF-1 has also been shown to be efficacious in delaying onset of symptoms in animal models for DMD (Barton et al., 2002).

29.3.1 The Central Elements Akt(PKB) and mTOR are Positive Regulators of Muscle Size

While earlier reports suggested that the binding of IGF-1 to its cognate receptor activates the calcineurin pathway (Semsarian et al., 1999; Musaro et al., 1999) recent work *in vitro* and *in vivo* clearly indicates that it rather acts via activation of phosphatidylinositol-3-kinase (PI3K), Akt (also called protein kinase B [PKB]) and mammalian target of rapamycin (mTOR) (Bodine et al., 2001; Rommel et al., 2001). IGF-1, through activation of the IGF-1 receptor tyrosine kinase, induces autophosphorylation of the receptor that creates docking sites for scaffolding proteins such as the insulin receptor substrate (IRS). These substrates, in turn, become phosphorylated by the IGF receptor and activate the p85 regulatory subunit of class I PI3K. As shown in Figure 29.3A, activated PI3K converts phosphatidylinositol-4,5-biphosphate (PIP2) into phosphatidylinositol-3,4,5-triphosphate (PIP3). This process is controlled by the phosphatase and tensin homologue PTEN, a widely acting tumor suppressor that is mutated in multiple advanced cancers. The protein kinase Akt (PKB) and 3-phosphoinositide-dependent protein kinase (PDK)1 are recruited and activated by PIP3. There are three Akt (PKB) genes in mammals (Akt1, 2, 3 or PKBα, β, γ). Consistent with the idea that Akt1/(PKBα) is mainly involved in muscle biology, Akt1/(PKBα)-, but not Akt2/(PKBβ)-deficient mice, have smaller organs including smaller muscle fibers (Chen et al., 2001) and constitutively active forms of Akt1/(PKBα) induce hypertrophy in myotubes *in vitro* (Rommel et al., 1999) and *in vivo* (Rommel et al., 2001; Lai Gonzalez et al., 2004). Moreover, overexpression of the SH2-domain-containing inositol 5′-phosphatase (SHIP)2, which inactivates Akt/(PKB), inhibits hypertrophy while dominant-negative forms of SHIP2 cause hypertrophy (Rommel et al., 2001). One of the central components of the IGF/Akt (PKB) pathway is TOR. TOR proteins were initially identified in yeast based on genetic screens that confer resistance to the growth-inhibitory properties of the immunophilin-immunosuppressant complex between FKBP and rapamycin (Heitman et al., 1991). Like CsA and FK506, which act via calcineurin, rapamycin has been approved as an immunosuppressant and is successfully marketed.

The two mutants characterized that make yeast resistant to rapamycin encode the two TOR1 and TOR2 proteins (Kunz et al., 1993). Homologues of TOR1 but not of TOR2 are found in all species including man where they are known as mTOR (Brown et al., 1994). TOR in yeast controls an unusually abundant and diverse set of readouts including activation of translation, transcription, protein kinase C signaling and actin organization as well as inhibition of autophagy or nutrient uptake (reviewed in Schmelzle and Hall 2000; Raught et al., 2001). Thus, TOR in yeast is involved in many cellular processes. In mammals, mTOR can be activated by growth factors including IGF-1 or by availability of nutrients, such as amino acids or glucose (Figure 29.3A). Moreover, the energy state of a cell sensed

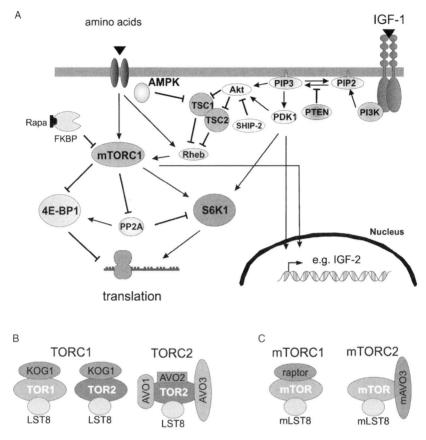

FIGURE 29.3 (A) Signaling pathway involving mTOR activation in mammalian cells include growth factors and the availability of nutrients (adapted from Jacinto and Hall 2003). The pathway depicted here is rapamycin-sensitive and represents a key pathway involved in regulation of protein translation. In addition, the rapamycin-sensitive mTOR pathway can also induce transcription of some genes including IGF-2. Note that the immunosuppressant rapamycin, by its binding to FKBP, is an inhibitor of the rapamycin-sensitive mTOR pathway (mTORC1, see Section 29.3.2). (B,C) TOR complexes formed in yeast and mammals. (B) In yeast, TOR complex1 (TORC1) is rapamycin-sensitive and contains TOR1 or TOR2 and its associated components KOG1 and LST8. TOR complex 2 (TORC2) is rapamycin-insensitive and contains TOR2 and its associated components AVO1, AVO2, AVO3 and LST8. (C) The mammalian counterparts of TORC1 and TORC2 can be distinguished by the selective binding of raptor (equivalent to KOG1) and mAVO3, respectively.

by the 5′ AMP-activated protein kinase (AMPK) regulates activation of mTOR. The current view predicts that high energy levels and thus inhibition of AMPK inhibits tuberous sclerosis complex (TSC) gene products TSC1 and TSC2, which in turn inhibit the small G protein Rheb. Rheb is an important activator of the rapamycin-sensitive component of mTOR, called mTORC1 (see Section 29.3.2). mTOR acts in a rapamycin-sensitive manner to control protein translation. This is mainly achieved by its activation of ribosomal S6 kinase (S6K)1 and the inhibition of eukaryote initiation factor 4E- binding protein (4E-BP)1, which in turn is an inhibitor of protein translation. Thus, the rapamycin-sensitive function of mTOR is mainly the control of protein translation. However, mTOR can also alter gene transcription; for example in cultured muscle cells the mTOR pathway is involved in the transcription of IGF-2 (Erbay et al., 2003).

29.3.2 mTOR Assembles into Two Distinct Complexes

In yeast, two distinct TOR complexes have been identified (Loewith et al., 2002). One of these complexes, whose function is inhibited by rapamycin, is called TOR complex (TORC)1 and consists of TOR1 or TOR2 and two associated molecules, called Kontroller of growth (KOG)1 and lethal with sec-thirteen (LST)8 (Figure 29.3B). The second complex is rapamycin insensitive and is composed of TOR2, LST8, and additional associated proteins. Because the latter adhere voraciously (AVO) to TOR, they were termed AVO1, AVO2, AVO3. In yeast, this second complex, called TORC2 is involved in regulating the actin cytoskeleton. As mentioned above, mammals have only one mTOR gene but mTOR also assembles into two distinct complexes that differ by their interaction with the mammalian homologues of KOG1 (called raptor; Kim et al., 2002), LST8 (called mLST8; Loewith et al., 2002) or GβL (Kim et al., 2003), and AVO3 (called mAVO3 or rictor; Sarbassov et al., 2004). TORC1 in mammals consists of mTOR, mLST8, and raptor while TORC2 contains mTOR, mLST8, and mAVO3 (Figure 29.3C). Like in yeast, the two complexes can be distinguished by (1) the rapamycin sensitivity of the binding of individual interactors to mTOR, (2) the capability of individual complexes to induce phosphorylation of S6K1, and (3) by the mutually exclusive association of some of the components (Jacinto et al., 2004; Sarbassov et al., 2004). Like the TORC1 complex in yeast, mTORC1 appears to be responsible for aspects involving cell growth, while mTORC2 affects the actin cytoskeleton. Only little is known about the signaling components up- and downstream of mTORC2. It is of course possible that the same components that activate mTORC1 also influence the activity of mTORC2. Downstream signals of mTORC2 appear to influence the cytoskeleton by regulating the phosphorylation of protein kinase C (PKC)α and paxillin (Jacinto et al., 2004; Sarbassov et al., 2004). This is corroborated by the fact that RNAi-induced depletion of mAVO3 in cultured HEK293 cells causes a decrease in the phosphorylation of PKCα and of paxillin, an adaptor molecule that associates with integrin-dependent focal adhesion sites and is known to influence the actin cytoskeleton (reviewed by Turner 2000). Moreover, mAVO3- but not raptor-depleted cells show alterations in cell spreading and the assembly of F-actin fibers (Jacinto et al., 2004). Similarly, localization of paxillin to peripheral stress fiber-like structures is abrogated and becomes localized to cytoplasmic assemblies of the F-actin (Sarbassov et al., 2004).

Thus, the finding that the TOR pathway also regulates cell growth in mammals may have strong implications for the biology of skeletal muscle. In particular, the mTORC2 pathway may contribute also in the organization of the actin cytoskeleton, which is of key importance to warrant muscle integrity and whose maintenance is disturbed in pathological situations that cause muscle atrophy and wasting. These new findings also question the experimental paradigms that only use rapamycin, which interferes only with TORC1, to address the general role of mTOR in muscle biology. Clearly, new approaches including tissue-selective inactivation of mTOR and its binding partners are needed to fully understand the role of this signaling pathway in muscle biology.

29.3.3 Potential Targets in the TOR Signaling Pathway

The pathways discussed above seem to be promising entry points to treat muscle loss. There are several potential ways for the development of inhibitors or activators that could interfere with muscle loss in pathological conditions including cachexia. Among the most upstream entry points are IGF-1 and its receptor, as well as Akt(PKB). For all these components, there is strong evidence that an increase in their activity would lead to an increase in muscle mass. For example, constitutively active forms of Akt(PKB) lead to rapid hypertrophy in mice (Lai et al., 2004) and transgenic mice overexpressing IGF-1 in muscle have an increased muscle mass (Barton-Davis et al., 1998). From a therapeutic point of view, the identification of small-molecule antagonists for inhibitors of the different signaling pathways could be more attractive option. We have recently found that inhibition of either TSC1 or TSC2 by RNA interference *in vivo* causes a tremendous increase in muscle fiber size in denervated muscle and that this increase can be inhibited by rapamycin (S. Lin and MAR; unpublished observation). Thus, synthetic antagonists of TSCs or of other endogenous inhibitors of the pathway, such as SHIP2, might be interesting drugs.

29.4 THE NF-κB SIGNALING PATHWAY

Transcription factor nuclear factor kappa B (NF-κB) is expressed in a variety of cell types, including mature muscle, where it participates in the regulation of a broad range of physiological and pathological processes. Among its functions are the inhibition of apoptosis and the control of the innate and adaptive immune response. NF-κB transcription factors are evolutionarily highly conserved from insects to humans. They belong to the Rel family that in mammalian cells consists of five proteins: RelA (also known as p65 and NF-κB3), c-Rel, RelB, p50 (also known as NF-κB1), and p52 (also known as NF-κB2). All these proteins share a Rel homology domain, which is required for dimerization, DNA binding, and regulatory function (Baldwin 1996; Karin et al., 2004). NF-κB proteins are dimers of variable subunits; the most common form is the heterodimer of the p50 and p65 subunits. In their inactive state, NF-κB dimers are bound to a specific inhibitor protein IκB and are predominately located in the cytoplasm. Only recently they were also shown to be located in mitochondria (Cogswell et al., 2003). The IκB proteins are masking the nuclear localization sequence within the Rel domain of NF-κB.

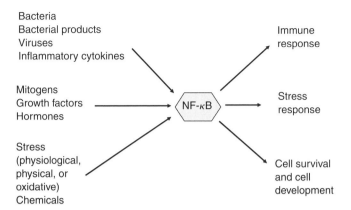

Bacteria
Bacterial products
Viruses
Inflammatory cytokines

Mitogens
Growth factors
Hormones

Stress
(physiological,
physical, or
oxidative)
Chemicals

NF-κB

Immune
response

Stress
response

Cell survival
and cell
development

FIGURE 29.4   NF-κB and its associated signaling pathways. NF-κB is a central mediator for a variety of cellular responses that can also modulate growth of muscle mass.

29.4.1 Activation of NF-κB and Control of Gene Transcription

TNFα is a highly potent activator of NF-κB but there is a wide range of other intra- and extracellular stimuli. Among them are inflammatory cytokines, growth factors, viral and bacterial products, and prooncogenic signals (see Figure 29.4). Activation of NF-κB is mediated through the IκB kinase (IKK) complex, which functions to phosphorylate two serine residues on IκB proteins. This phosphorylation causes the ubiquitination and subsequent degradation of IκB proteins by the 26S proteasome complex (Guttridge 2004). The IKK complex itself is composed of two catalytically active subunits: IKKα and IKKβ, which can directly phosphorylate IκB and a regulatory subunit IKKγ. Only IKKβ is required for cytokine-dependent activation of NF-κB (Karin and Lin 2002). The free NF-κB translocates to the nucleus where it binds to the promoters of specific target genes and thus regulates inflammatory, proliferative, and apoptotic processes. NF-κB can induce transcription of certain target genes within minutes following exposure to the relevant stimulus (Baldwin 1996). Among many other genes, NF-κB also induces the gene expression of its own inhibitor IκBα. This is why the activation of NF-κB is a transient response that reaches its peak usually within 30 min followed by a return to basal levels within 1 to 4 h. However, Ladner et al. (2003) described a biphasic activation of NF-κB through TNFα in skeletal muscle cells. Here, the first transient phase is followed by a second persistent phase of NF-κB activity.

29.4.2 THE Role of NF- κB in Muscle Growth

While the role of the NF-κB signaling pathway in innate and adaptive immune responses, inflammation, and apoptosis has long been recognized, its importance for muscle growth and differentiation has only been discovered in the last few

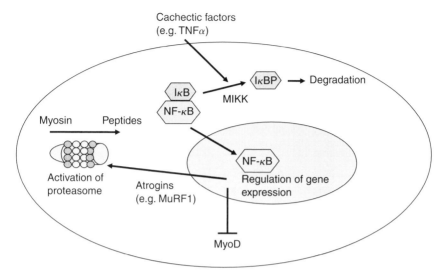

FIGURE 29.5   NF-κB as a target for the treatment of cachexia. In the cytosol, NF-κB is bound to its inhibitor protein IκB. Upon activation by cachectic factors such as tumor necrosis factor (TNF)α the catalytic β subunit of the muscle specific IκB kinase complex (MIKK) induces phosphorylation and ubiquitin-dependent degradation of IκB. The liberated NF-κB translocates to the nucleus where it becomes transcriptionally active by (1) downregulation of MyoD expression and (2) induction of proteasome expression and protein degradation.

years and is still far from being understood in detail (Guttridge 2004). Figure 29.5 illustrates the role of NF-κB as a negative regulator of muscle growth by acting as a repressor of MyoD and as an enhancer of the ubiquitin-proteasome system. However the role of NF-κB is not limited to these two functions. NF-κB is a strong antiapoptotic factor, assuring survival of the differentiated muscle cells. Furthermore, it was postulated that NF-κB enhances the transformation of myoblasts to myotubes via nitric acid production (Kaliman et al., 1999).

There seems to be a consensus that NF-κB is a key mediator in muscle damage and wasting and also in disuse atrophy and dystrophy. Studies in *mdx* mice, a mouse model of DMD, suggest that the NF-κB pathway is predominantly active in this severe neuromuscular disease. The basal level of NF-κB activity was higher in the diaphragm in *mdx* mice as compared to control mice, and it was further enhanced through mechanical stress (Kumar and Boriek 2003). There is evidence that disuse muscle atrophy is distinct from atrophy seen in cachexia. In contrast to the TNFα/NF-κB pathway seen in cachexia the alternative NF-κB pathway used in disuse atrophy cachexia is not activated by cytokines and does not involve activation of p65 (Hunter et al., 2002). Instead, a protooncogene and so-called unusual member of the IκB family that acts as a transcriptional coactivator is involved (Hunter and Kandarian 2004).

29.4.2.1 NF-κB influences skeletal muscle differentiation through MyoD

The differentiation of undifferentiated cells to skeletal muscle cells is under the control of the transcriptional factor MyoD. It was postulated that NF-κB suppresses MyoD mRNA level and consequently inhibits skeletal muscle differentiation. By showing that cells devoid of NF-κB activity retained MyoD mRNA in the presence of TNFα, it was further demonstrated that NF-κB is a downstream effector of tumor necrosis factor (TNF) that regulates MyoD expression (Guttridge et al., 2000). These authors were also able to show that the p65 subunit alone can suppress MyoD expression. There is also evidence suggesting that a second inflammatory cytokine, IFNγ, in addition to TNFα is required to induce a NF-κB-regulated suppression of MyoD synthesis (Guttridge et al., 2000; Acharyya et al., 2004). In adult skeletal muscle, MyoD is expressed at relatively low levels but in response to injury its expression is induced from satellite cells. MyoD is required for satellite cells to proliferate and reinitiate skeletal muscle differentiation and can therefore be considered as essential for the repair of damaged muscle tissue. It was shown that NF-κB is activated by injury and also by sepsis and might play a role in the pathogenesis of these conditions via the inhibition of MyoD expression.

29.4.2.2 NF-κB enhances muscle degradation through activation of the ubiquitin-proteasome system

It was postulated that direct inhibition of NF-κB could reduce muscle wasting in cancer patients and also in other muscle wasting conditions. For example, pharmacological inhibition of NF-κB was shown to block cachexia in animal models. Wyke et al. (2004) demonstrated that resveratrol, a natural phytoalexin found in red wine and an inhibitor of IKK, significantly attenuated weight loss that, at least in part, resulted from a reduction in protein degradation. Other researchers have also found the ubiquitin-proteasome system to be responsible for NF-κB-enhanced muscle degradation. Cai et al. (2004) showed that activation of NF-κB through muscle-specific transgenic expression of constitutively activated IkB kinase β causes profound muscle wasting. This phenotype could be reversed by muscle-specific inhibition of NF-κB through expression of IκBα super-repressor and by high doses of salicylates (Cai et al., 2004). They also showed that selective NF-κB blockade in muscle decreased muscle wasting and prolonged survival in a Lewis lung carcinoma mouse model of cancer cachexia. It was postulated that the muscle wasting upon *in vivo* activation of IKKβ and NF-κB is mediated by the ubiquitin-proteasome mechanism via the E3 ubiquitin ligase MuRF1 (muscle RING finger protein 1). This view was supported by experimental data that showed that MG132, a pharmacological inhibitor of the proteasome, could prevent an increase in protein catabolism in mice expressing constitutively activated NF-κB (Cai et al., 2004).

29.4.2.3 Interaction of NF-κB with other pathways

The action of NF-κB is clearly context-dependent or according to Karin and Lin (2002) NF-κB "cannot function alone." The NF-κB pathway interacts with various other pathways and these connections are far from being fully understood. For example, the IGF/PI3K/Akt(PKB)/mTOR pathways can influence the NF-κB pathway through FOXO1, a member of the forkhead family of transcription factors (see Figure 29.1). FOXO1 can therefore be considered as a link between the muscle anabolic and catabolic pathways. Akt phosphorylation keeps FOXO1 in its inactive state. In contrast, reduced Akt phosphorylation leaves FOXO1 in its underphosphorylated active form and induces not only the expression of the ubiquitin ligase atrogin-1/MAFbx but also of MuRF1 that is equally induced by NF-κB (Cai et al., 2004; Stitt et al., 2004).

29.4.3 NF-κB as a Potential Drug Target for the Treatment of Cancer Cachexia

NF-κB and the IKK system have long been recognized as highly attractive targets for drug development, mainly for indications such as cancer and chronic and acute inflammatory diseases. Since the central role of NF-κB for muscle growth and differentiation has been recognized, it has also become a highly promising pharmacological target for the treatment of muscle atrophy as seen in cachexia. Various strategies for inhibiting NF-κB have been explored. Molecular therapies such as RelA antisense oligonucleotides, RNA interference (RNAi), and NF-κB decoy oligo-DNA are in development to interfere with the binding of NF-κB to DNA. For the latter, synthetic double-stranded DNA with high affinity for a target transcription factor is introduced into target cells as "decoy" cis element to bind the transcriptional factor and alter gene transcription. NF-κB decoy oligodeoxynucleotides have already been tested in animal models of cancer cachexia and resulted in attenuation of the reduction in body weight, epididymal fat, gastrocnemius muscle mass, and food intake (Morishita et al., 2004). Also the dimerization of NF-κB proteins was targeted. A special focus has been on strategies to interfere with the activation of NF-κB. Therefore, inhibition of IKK activity has to be considered the most effective and selective approach. The pharmaceutical industry has undertaken major efforts to identify small-molecule inhibitors of IKK-α and/or IKK-β catalytic activity (for a review see Karin et al., 2004). Among these compounds are imidazoquinoxaline derivatives (Burke et al., 2003), beta-carbolines (Castro et al., 2003), and diarylpyridine derivatives (Murata et al., 2003) that all have an IC$_{50}$ for IKK-2 in the nanomolar range. However, these inhibitors still need to prove successful in clinical trials. Also, NF-κB nuclear translocation inhibitors and inhibitors of NF-κB transcriptional activity (such as glucocorticoids) have been investigated. Furthermore, some naturally occurring (e.g., lactacystin) and synthetic inhibitors (e.g., bortezomib, marketed as Velcade$^{®}$) of the ubiquitin-proteasome that can block NF-κB activation by preventing IκB degradation have been identified. These proteasome inhibitors that were initially developed as anticancer drugs are increasingly considered promising drug candidates for the treatment of muscle wasting conditions, although their toxicology profile might be of concern.

Although inhibition of NF-κB is effective in reducing muscle catabolism, it might have undesirable side effects. For instance the general inhibition of NF-κB over longer time periods is likely to lead to immunodeficiency. In order to prevent such a generalized effect, it might be an option to interfere with muscle-specific factors in the pathway. For this reason, the E3 ligase MuRF1 was proposed as a novel drug target (Cai et al., 2004). MuRF1 belongs to the so-called atrogins, mediators of muscle atrophy. MuRF1 was shown to be upregulated in various settings of muscle atrophy. The increasing knowledge of the function of these muscle specific factors will inspire the search for pharmacological approaches to NF-κB regulation. Without any doubt, drugs targeting the NF-κB system will play an important role for future pharmacological treatment of cachexia.

29.5 THE MYOSTATIN SIGNALING PATHWAY

Since the discovery of myostatin (previously termed growth and differentiation factor 8, GDF8) as a negative regulator of skeletal muscle growth it has been postulated that this protein may be useful in the treatment of musculodegenerative states such as muscular dystrophy, neuromuscular diseases, or cancer cachexia (McPherron et al., 1997). This postulate was initially motivated by the observation that mutations in the gene encoding for myostatin (*MSTN*) were associated with the "double-muscle" appearance of certain naturally occurring cattle breeds and mice carrying a deletion of the C-terminal domain of myostatin. In general, the observed marked increase in muscle mass appears to be the result of both hyperplasia (increase in the number of muscle fibers) and hypertrophy (increased size of muscle fibers) of virtually all muscles in animals with myostatin mutations. Myostatin's function in the regulation of muscle mass is evolutionarily conserved and applies to human muscle as well. Specifically, it has been shown that a child with a mutation in the myostatin gene displays marked muscle hypertrophy (Schuelke et al., 2004) substantiating the hypothesis that reducing levels of myostatin might be of therapeutic use for muscle diseases.

29.5.1 Myostatin-Mediated Regulation of Muscle Growth

Myostatin primarily limits muscle growth by suppressing the proliferation of myoblasts, the pool of satellite cells associated with differentiated muscle fibers that upon cell division fuse with one another and preexisting muscle fibers to increase muscle mass. Myostatin appears to inhibit cell proliferation by preventing satellite cells' transition from the G0 or G1 cell cycle phase to the S phase, thereby initiating the withdrawal of satellite cells from the cell cycle. Specifically, cyclin-dependent kinases (e.g., Cdk2) are downregulated indirectly by upregulating cyclin-dependent kinase inhibitors (e.g., p21, p27, and p53). Consequently, myostatin functions to maintain the quiescent state of muscle satellite cells (McCroskery et al., 2003). In addition to the repression of satellite cell division, myostatin also inhibits differentiation of myotubes by decreasing expression of muscle-specific basic helix-loop-helix (bHLH) transcription factors including MyoD, myf5, and myogenin that have a pivotal role in the regulation of muscle fiber differentiation.

29.5.2 Processing of Myostatin Protein

Myostatin is a member of the transforming growth factor beta (TGF-β) family of proteins expressed predominantly in muscle tissue and, interestingly, to some extent in adipose tissue as well. The protein is characterized by a signal sequence at the amino terminal followed by a dibasic proteolytic processing site and nine distinctly spaced cysteine residues within the carboxyterminal portion of the protein. The precursor protein, which forms homodimers via a disulfide link in the C-terminal region of the protein, is cleaved proteolytically. The first cleavage removes a short signal peptide of each of the monomer units while the second cleavage generates two products, a 27.6 kDa propeptide and the 12.4 kDa C-terminal fragment. The latter maintains its dimeric state via a disulfide bridge located in the C-terminal fragment and carries the biological function of the protein. Binding to the propeptide region in a latent complex inhibits the formation of the functional C-terminal dimer of myostatin as for instance determined by receptor binding capacity. Finally, additional proteolytic cleavage of the propeptide by members of the bone morphogenetic protein (BMP)-1/tolloid family of metalloproteinases activates the C-terminal dimer (Wolfman et al., 2003).

The importance of the propeptide as a regulatory unit of the myostatin complex has been demonstrated in several experimental settings. For example, overexpression of the propeptide in mice blocks all myostatin C-terminal peptide function and results in a phenotype comparable to myostatin knock-out mice with respect to increased muscle mass (Yang et al., 2001). Likewise, mice injected weekly with a human propeptide carrying a mutation that does not allow inactivation by BMP-1/tolloid protease cleavage also show increased muscle mass. In addition to the propeptide itself, several other proteins are capable of inhibiting the activity of the C-terminal myostatin dimer (Figure 29.6). For instance, biochemical experiments demonstrate that follistatin can block myostatin activity *in vitro* (Zimmers et al., 2002). Consistent with such an antagonistic function, overexpression of follistatin leads to a dramatic increase in muscle growth (Lee and McPherron 2001) while deficiency of follistatin results in reduced muscle mass (Matzuk et al., 1995). Myostatin activity is further regulated by two proteins, FLRG (product of the follistatin related gene) and GASP-1 (growth and differentiation factor associated serum protein-1), which bind with high affinity to the C-terminal dimer thereby blocking its activity. Interestingly, it appears that myostatin circulating in the blood is complexed with these proteins (Hill et al., 2002; Hill et al., 2003). In addition it has been postulated that GASP-1 might regulate myostatin function by inhibiting

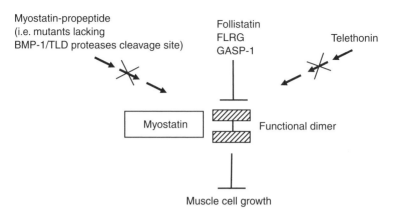

FIGURE 29.6 Myostatin is a negative regulator of skeletal muscle growth. Myostatin (i.e., its C-terminal region) can be inhibited by rendering attached to propeptide, which is mutated in the cleavage site used by members of the bone morphogenetic protein-1/tolloid (BMP-1/TLD) metalloproteases. Follistatin as well as FLRG and GASP-1 can directly interact with the C-terminal myostatin dimer thereby inhibiting its function. The titin-cap protein (telethonin) can bind to mature myostatin in the cytoplasm or Golgi interfering with the secretion of myostatin and thereby preventing the formation of the latent complex.

proteinase activity required for propeptide cleavage. It is possible that binding of FLRG, GASP-1, and propeptide to the C-terminal dimer is mutually exclusive and that the circulating form of myostatin may be a mixture of these complexes. Finally, titin-cap protein (telethonin) limits the secretion of myostatin by binding to mature myostatin in the cytoplasm or Golgi and thereby prevents the formation of the latent complex (Nicholas et al., 2002). In summary, the functional activity of myostatin is regulated by interaction with several possible proteins forming inactive complexes with the C-terminal fragment of myostatin.

29.5.3 Therapeutic Relevance of Myostatin Signaling

Resembling the situation found with other TGF-β related ligands, the myostatin C-terminal dimer acts through a heteromeric complex of type I and type II serine/threonine kinase receptors. Specifically, myostatin binds to the activin type II receptors (ActRIIA and ActRIIB) *in vitro*. It is assumed that myostatin binds to ActRIIA and ActRIIB and that this ligand/type II receptor complex then recruits a type I receptor kinase. Initial direct evidence supporting that ActRIIB is the functional receptor for the myostatin C-terminal dimer came from experiments with transgenic mice expressing a dominant-negative form of ActRIIB. These mice lacking the kinase domain of the ActRIIB receptor appear to be phenocopies of the myostatin knock-out mice and show a dramatic increase in muscle mass as a combination of hyperplasia and hypertrophy (Lee and McPherron 2001). Given the large body of evidence supporting that myostatin is a key regulator of muscle mass affecting both cell number and myotube differentiation it is tempting to speculate that inhibition of myostatin could be used therapeutically to ameliorate muscle loss in conditions of atrophy, dystrophy, or cachexia. In the following we summarize the evidence that supports the view that myostatin signaling indeed may have therapeutic value.

Initial observations that point toward a contribution of myostatin signaling mediating muscle atrophy were based on measurements of myostatin levels. For example, expression was increased in patients experiencing loss of muscle mass as a consequence of prolonged bed rest or after disuse atrophy irrespective of age. Likewise, myostatin protein was elevated in elderly patients suffering from muscle wasting (Zachwieja et al., 1999; Reardon et al., 2001; Yarasheski et al., 2002). These results support the view that myostatin is the mediator of the wasting observed. However, not all experimental conditions are in agreement with this hypothesis. For instance, atrophy induced by hind-limb suspension in myostatin-deficient mice was more pronounced than in normal controls (McMahon et al., 2003). Moreover, in rodents myostatin expression was increased only several days after the onset of denervation-induced muscle atrophy. This is consistent with the observation that modification of the myostatin signaling pathway appears not to be an early response in experimentally induced atrophic conditions. It has been speculated, therefore, that the pattern of myostatin function is correlated with the rate of occurring muscle loss; however this postulation needs further experimental investigation.

More direct evidence for the potential of myostatin as therapeutic target has been collected in dystrophic conditions. The *mdx* mouse is a widely used animal model for dystrophinopathies in general and for DMD in particular. *Mdx* mice that carry a null-mutation in myostatin showed increased muscle mass and improved muscle strength comparable to what was observed with myostatin null but dystrophin positive mice (Wagner et al., 2002). Similarly, applying myostatin-neutralizing antibodies to *mdx* mice resulted in increased muscle mass and improved force production (Bogdanovich et al., 2002). In addition to an increase in muscle mass, myostatin inhibition in both experimental settings also lead to improved histology paralleled by decreased fibrosis lending support to the interpretation that myostatin inhibition may have had enhanced the regenerative capacity of the dystrophin-deficient muscle, presumably by accelerating the regeneration of muscle fibers from activated satellite cells. In patients with DMD myostatin repression may actually be more harmful than beneficial since it may accelerate the recruitment of satellite cells to replenish lost muscle mass but, consequently, the limited pool of satellite cells might be exhausted even quicker under this condition. Although it has been reported that increased muscle mass and reduced fibrosis was seen even in 9 month old myostatin-deficient *mdx* mice it is still too early to judge whether myostatin repression in patients with DMD may be beneficial in the long run.

Whether myostatin inhibition also holds a potential for the treatment of cachexia in chronic diseases such as AIDS, cancer, or in severe sepsis is still under investigation. In support of this view are experiments in which myostatin was overexpressed in mice leading to loss of muscle and fat mass even though caloric intake was not restricted, thereby resembling to a certain extent some forms of human cachexia (Zimmers et al., 2002). Likewise, serum and intramuscular levels of myostatin-immunoreactive protein was increased in HIV-infected men with weight loss. This observation would indicate that the myostatin signaling pathway is a mediator of muscle loss in this condition (Gonzalez-Cadavid et al., 1998). However, it is based on experiments using an antibody to myostatin-immunoreactive protein, which has raised concerns about its specificity to myostatin. Therefore, additional experimental support is needed in order to judge whether reduction of myostatin protein or suppression of the myostatin signaling pathway would eventually slow down the muscle loss induced in cachectic conditions.

29.5.4 Myostatin as a Potential Drug Target to Prevent Muscle Loss

Myostatin is currently considered as an attractive drug target, despite the fact that it is still not entirely clear whether interfering with the myostatin signaling pathway could be an effective strategy for treating muscle loss in conditions of atrophy, dystrophy, or cachexia.

It certainly is advantageous that effects caused by repression of myostatin protein function appear to be confined to muscle and adipose tissue; other tissues seem not to be affected by myostatin blockade. Moreover, circulating myostatin in human serum is attractive in light of possible pharmacological interventions including the use of proteins and peptides that bind to and interfere with the function of myostatin. Finally, myostatin acts in a dose-dependent manner, as has been shown first in mice carrying null-mutations in the myostatin gene. Heterozygous mice showed an increase in muscle mass that was considerably less than in homozygous myostatin null animals. Therefore, it appears that a partial but nevertheless therapeutically relevant increase in muscle mass might be accomplished by partial inhibition of myostatin.

So far myostatin blockade is limited to proteins and peptides; small molecule inhibitors have not been identified yet. Antibodies as well as propeptide lacking the BMP-1/tolloid metalloproteinase cleavage site (Wolfman et al., 2003) may be used therapeutically by direct binding to myostatin. By analogy, FLRG product also has been identified as a negative regulator of myostatin (Hill et al., 2002) while the potential of follistatin to bind to and antagonize the effect of myostatin is still not clear. In support of follistatin as a possible myostatin inhibitor are experiments demonstrating its inhibitory effect in receptor binding and reporter gene assays as well as in transgenic mice showing increased muscle mass upon its overexpression (Lee and McPherron 2001; Zimmers et al., 2002). Interestingly, one recent study (Hill et al., 2002) could not confirm the presence of follistatin bound in a latent complex with myostatin in serum, which may argue against this approach to inhibit myostatin. The therapeutic use of follistatin to regulate muscle mass is questionable since follistatin, by interacting with a myostatin-related protein, GDF-11, may deform the axial skeleton. This stems from experiments in developing animals where deletion of GDF-11 caused homeotic transformations of the axial skeleton

(Matzuk et al., 1995; McPherron et al., 1999). Whether skeletal deformation upon GDF-11 inhibition will be observed after maturation of the skeletal apparatus is still unresolved. However, this may raise concerns for the safety of using follistatin as an inhibitor of myostatin. Finally, although theoretically possible, it is too early to say whether GASP-1 via its interaction with mature myostatin protein or myostatin propeptide in human serum can be of therapeutic use.

Myostatin is now a well established key regulator of muscle growth and has been identified as a potential therapeutic target for normalizing muscle mass in conditions of atrophy, dystrophy, or cachexia. The strongest evidence that inhibition of myostatin protein (or interfering with myostatin signaling) may be of therapeutic relevance was established in conditions of dystrophy with the caveat that the dystrophic condition in the *mdx* mouse clearly has a different time-course than that seen in patients with DMD. It is not clear at this point whether the effects observed in this animal model can be safely transferred to the human condition. Whether myostatin could serve as a peripheral target to prevent muscle loss in conditions of atrophy or cachexia still needs further work in appropriate animal models.

REFERENCES

Acharyya, S., Ladner, K.J. et al. (2004). "Cancer cachexia is regulated by selective targeting of skeletal muscle gene products." *J. Clin. Invest.* 114: 370–8.

Baldwin, A.S. Jr. (1996). "The NF-kappa B and I kappa B proteins: new discoveries and insights." *Annu. Rev. Immunol.* 14: 649–83.

Barton, E.R., Morris, L. et al. (2002). "Muscle-specific expression of insulin-like growth factor I counters muscle decline in mdx mice." *J. Cell. Biol.* 157: 137–48.

Barton-Davis, E.R., Shoturma, D.I. et al. (1998). "Viral mediated expression of insulin-like growth factor I blocks the aging-related loss of skeletal muscle function." *Proc. Natl Acad. Sci. USA* 95: 15603–7.

Bodine, S.C., Stitt, T.N. et al. (2001). "Akt/mTOR pathway is a crucial regulator of skeletal muscle hypertrophy and can prevent muscle atrophy in vivo." *Nat. Cell Biol.* 3: 1014–9.

Bogdanovich, S., Krag, T.O. et al. (2002). "Functional improvement of dystrophic muscle by myostatin blockade." *Nature* 420: 418–21.

Brown, E.J., Albers, M.W. et al. (1994). "A mammalian protein targeted by G1-arresting rapamycin-receptor complex." *Nature* 369: 756–8.

Burke, J.R., Pattoli, M.A. et al. (2003). "BMS-345541 is a highly selective inhibitor of I kappa B kinase that binds at an allosteric site of the enzyme and blocks NF-kappa B-dependent transcription in mice." *J. Biol. Chem.* 278: 1450–6.

Cai, D., Frantz, J.D. et al. (2004). "IKKbeta/NF-kappaB activation causes severe muscle wasting in mice." *Cell* 119: 285–98.

Castro, A.C., Dang, L.C. et al. (2003). "Novel IKK inhibitors: beta-carbolines." *Bioorg. Med. Chem. Lett.* 13: 2419–22.

Chakkalakal, J.V., Harrison, M.A. et al. (2004). "Stimulation of calcineurin signaling attenuates the dystrophic pathology in mdx mice." *Hum. Mol. Genet.* 13: 379–88.

Chen, W.S., Xu, P.Z. et al. (2001). "Growth retardation and increased apoptosis in mice with homozygous disruption of the Akt1 gene." *Genes Dev.* 15: 2203–8.

Cogswell, P.C., Kashatus, D.F. et al. (2003). "NF-kappa B and I kappa B alpha are found in the mitochondria. Evidence for regulation of mitochondrial gene expression by NF-kappa B." *J. Biol. Chem.* 278: 2963–8.

Dunn, S.E., Burns, J.L. et al. (1999). "Calcineurin is required for skeletal muscle hypertrophy." *J. Biol. Chem.* 274: 21908–12.

Dunn, S.E., Simard, A.R. et al. (2002). "Calcineurin and skeletal muscle growth." *Nat. Cell Biol.* 4: E46–7.

Erbay, E., Park, I.H. et al. (2003). "IGF-II transcription in skeletal myogenesis is controlled by mTOR and nutrients." *J. Cell. Biol.* 163: 931–6.

Gonzalez-Cadavid, N.F., Taylor, W.E. et al. (1998). "Organization of the human myostatin gene and expression in healthy men and HIV-infected men with muscle wasting." *Proc. Natl Acad. Sci. USA* 95: 14938–43.

Guttridge, D.C. (2004). "Signaling pathways weigh in on decisions to make or break skeletal muscle." *Curr. Opin. Clin. Nutr. Metab. Care* 7: 443–50.

Guttridge, D.C., Mayo, M.W. et al. (2000). "NF-kappaB-induced loss of MyoD messenger RNA: possible role in muscle decay and cachexia." *Science* 289: 2363–6.

Heitman, J., Movva, N.R. et al. (1991). "Targets for cell cycle arrest by the immunosuppressant rapamycin in yeast." *Science* 253: 905–9.

Hill, J.J., Davies, M.V. et al. (2002). "The myostatin propeptide and the follistatin-related gene are inhibitory binding proteins of myostatin in normal serum." *J. Biol. Chem.* 277: 40735–41.

Hill, J.J., Qiu, Y. et al. (2003). "Regulation of myostatin *in vivo* by growth and differentiation factor-associated serum protein-1: a novel protein with protease inhibitor and follistatin domains." *Mol. Endocrinol.* 17: 1144–54.

Horsley, V., Friday, B.B. et al. (2001). "Regulation of the growth of multinucleated muscle cells by an NFATC2-dependent pathway." *J. Cell. Biol.* 153: 329–38.

Horsley, V., Jansen, K.M. et al. (2003). "IL-4 acts as a myoblast recruitment factor during mammalian muscle growth." *Cell* 113: 483–94.

Hunter, R.B. and Kandarian, S.C. (2004). "Disruption of either the Nfkb1 or the Bcl3 gene inhibits skeletal muscle atrophy." *J. Clin. Invest.* 114: 1504–11.

Hunter, R.B., Stevenson, E. et al. (2002). "Activation of an alternative NF-kappaB pathway in skeletal muscle during disuse atrophy." *FASEB J.* 16: 529–38.

Jacinto, E. and Hall, M.N. (2003). "Tor signalling in bugs, brain and brawn." *Nat. Rev. Mol. Cell Biol.* 4: 117–26.

Jacinto, E., Loewith, R. et al. (2004). "Mammalian TOR complex 2 controls the actin cytoskeleton and is rapamycin insensitive." *Nat. Cell Biol.* 6: 1122–8.

Kaliman, P., Canicio, J. et al. (1999). "Insulin-like growth factor-II, phosphatidylinositol 3-kinase, nuclear factor-kappaB and inducible nitric-oxide synthase define a common myogenic signaling pathway." *J. Biol. Chem.* 274: 17437–44.

Karin, M. and Lin, A. (2002). "NF-kappaB at the crossroads of life and death." *Nat. Immunol.* 3: 221–7.

Karin, M., Yamamoto, Y. et al. (2004). "The IKK NF-kappa B system: a treasure trove for drug development." *Nat. Rev. Drug Discov.* 3: 17–26.

Kim, D.H., Sarbassov, D.D. et al. (2002). "mTOR interacts with raptor to form a nutrient-sensitive complex that signals to the cell growth machinery." *Cell* 110: 163–75.

Kim, D.H., Sarbassov dos, D. et al. (2003). "GbetaL, a positive regulator of the rapamycin-sensitive pathway required for the nutrient-sensitive interaction between raptor and mTOR." *Mol. Cell.* 11: 895–904.

Kumar, A. and Boriek, A.M. (2003). "Mechanical stress activates the nuclear factor-kappaB pathway in skeletal muscle fibers: a possible role in Duchenne muscular dystrophy." *FASEB J.* 17: 386–96.

Kunz, J., Henriquez, R. et al. (1993). "Target of rapamycin in yeast, TOR2, is an essential phosphatidylinositol kinase homolog required for G1 progression." *Cell* 73: 585–96.

Ladner, K.J., Caligiuri, M.A. et al. (2003). "Tumor necrosis factor-regulated biphasic activation of NF-kappa B is required for cytokine-induced loss of skeletal muscle gene products." *J. Biol. Chem.* 278: 2294–303.

Lai, K.M., Gonzalez, M. et al. (2004). "Conditional activation of akt in adult skeletal muscle induces rapid hypertrophy." *Mol. Cell. Biol.* 24: 9295–304.

Lecker, S.H., Jagoe, R.T. et al. (2004). "Multiple types of skeletal muscle atrophy involve a common program of changes in gene expression." *FASEB J.* 18: 39–51.

Lee, S.J. and McPherron, A.C. (2001). "Regulation of myostatin activity and muscle growth." *Proc. Natl Acad. Sci. USA* 98: 9306–11.

Loewith, R., Jacinto, E. et al. (2002). "Two TOR complexes, only one of which is rapamycin sensitive, have distinct roles in cell growth control." *Mol. Cell* 10: 457–68.

Matzuk, M.M., Lu, N. et al. (1995). "Multiple defects and perinatal death in mice deficient in follistatin." *Nature* 374: 360–3.

McCroskery, S., Thomas, M. et al. (2003). "Myostatin negatively regulates satellite cell activation and self-renewal." *J. Cell Biol.* 162: 1135–47.

McMahon, C.D., Popovic, L. et al. (2003). "Myostatin-deficient mice lose more skeletal muscle mass than wild-type controls during hindlimb suspension." *Am. J. Physiol. Endocrinol. Metab.* 285: E82–7.

McPherron, A.C., Lawler, A.M. et al. (1997). "Regulation of skeletal muscle mass in mice by a new TGF-beta superfamily member." *Nature* 387: 83–90.

McPherron, A.C., Lawler, A.M. et al. (1999). "Regulation of anterior/posterior patterning of the axial skeleton by growth/differentiation factor 11." *Nat. Genet.* 22: 260–4.

Morishita, R., Tomita, N. et al. (2004). "Molecular therapy to inhibit NFkappaB activation by transcription factor decoy oligonucleotides." *Curr. Opin. Pharmacol.* 4: 139–46.

Murata, T., Shimada, M. et al. (2003). "Discovery of novel and selective IKK-beta serine-threonine protein kinase inhibitors. Part 1." *Bioorg. Med. Chem. Lett.* 13: 913–8.

Musaro, A., McCullagh, K.J. et al. (1999). "IGF-1 induces skeletal myocyte hypertrophy through calcineurin in association with GATA-2 and NF-ATc1." *Nature* 400: 581–5.

Nicholas, G., Thomas, M. et al. (2002). "Titin-cap associates with, and regulates secretion of, Myostatin." *J. Cell. Physiol.* 193: 120–31.

Parsons, S.A., Millay, D.P. et al. (2004). "Genetic loss of calcineurin blocks mechanical overload-induced skeletal muscle fiber type switching but not hypertrophy." *J. Biol. Chem.* 279: 26192–200.

Raught, B., Gingras, A.C. et al. (2001). "The target of rapamycin (TOR) proteins." *Proc. Natl Acad. Sci. USA* 98: 7037–44.

Reardon, K.A., Davis, J. et al. (2001). "Myostatin, insulin-like growth factor-1, and leukemia inhibitory factor mRNAs are upregulated in chronic human disuse muscle atrophy." *Muscle Nerve* 24: 893–9.

Rommel, C., Bodine, S.C. et al. (2001). "Mediation of IGF-1-induced skeletal myotube hypertrophy by PI(3)K/Akt/mTOR and PI(3)K/Akt/GSK3 pathways." *Nat. Cell Biol.* 3: 1009–13.

Rommel, C., Clarke, B.A. et al. (1999). "Differentiation stage-specific inhibition of the Raf-MEK-ERK pathway by Akt." *Science* 286: 1738–41.

Sarbassov, D.D., Ali, S.M. et al. (2004). "Rictor, a novel binding partner of mTOR, defines a rapamycin-insensitive and raptor-independent pathway that regulates the cytoskeleton." *Curr. Biol.* 14: 1296–302.

Schiaffino, S. and Serrano, A. (2002). "Calcineurin signaling and neural control of skeletal muscle fiber type and size." *Trends Pharmacol. Sci.* 23: 569–75.

Schmelzle, T. and Hall, M.N. (2000). "TOR, a central controller of cell growth." *Cell* 103: 253–62.

Schuelke, M., Wagner, K.R. et al. (2004). "Myostatin mutation associated with gross muscle hypertrophy in a child." *N. Engl. J. Med.* 350: 2682–8.

Semsarian, C., Wu, M.J. et al. (1999). "Skeletal muscle hypertrophy is mediated by a Ca2+-dependent calcineurin signalling pathway." *Nature* 400: 576–81.

Stitt, T.N., Drujan, D. et al. (2004). "The IGF-1/PI3K/Akt pathway prevents expression of muscle atrophy-induced ubiquitin ligases by inhibiting FOXO transcription factors." *Mol. Cell* 14: 395–403.

St-Pierre, S.J., Chakkalakal, J.V. et al. (2004). "Glucocorticoid treatment alleviates dystrophic myofiber pathology by activation of the calcineurin/NF-AT pathway." *Faseb J.* 18: 1937–9.

Turner, C.E. (2000). "Paxillin and focal adhesion signalling." *Nat. Cell Biol.* 2: E231–6.

Wagner, K.R., McPherron, A.C. et al. (2002). "Loss of myostatin attenuates severity of muscular dystrophy in mdx mice." *Ann. Neurol.* 52: 832–6.

Wilkins, B.J. and Molkentin, J.D. (2004). "Calcium-calcineurin signaling in the regulation of cardiac hypertrophy." *Biochem. Biophys. Res. Commun.* 322: 1178–91.

Wolfman, N.M., McPherron, A.C. et al. (2003). "Activation of latent myostatin by the BMP-1/tolloid family of metalloproteinases." *Proc. Natl Acad. Sci. USA* 100: 15842–6.

Wyke, S.M., Russell, S.T. et al. (2004). "Induction of proteasome expression in skeletal muscle is attenuated by inhibitors of NF-kappaB activation." *Br. J. Cancer.* 91: 1742–1750.

Yang, J., Ratovitski, T. et al. (2001). "Expression of myostatin pro domain results in muscular transgenic mice." *Mol. Reprod. Dev.* 60: 351–61.

Yang, S., Alnaqeeb, M. et al. (1996). "Cloning and characterization of an IGF-1 isoform expressed in skeletal muscle subjected to stretch." *J. Muscle Res. Cell Motil.* 17: 487–95.

Yarasheski, K.E., Bhasin, S. et al. (2002). "Serum myostatin-immunoreactive protein is increased in 60–92 year old women and men with muscle wasting." *J. Nutr. Health Aging* 6: 343–8.

Zachwieja, J.J., Smith, S.R. et al. (1999). "Plasma myostatin-immunoreactive protein is increased after prolonged bed rest with low-dose T3 administration." *J. Gravit Physiol.* 6: 11–5.

Zimmers, T.A., Davies, M.V. et al. (2002). "Induction of cachexia in mice by systemically administered myostatin." *Science* 296: 1486–8.

30 Perspectives and Outlook

Karl G. Hofbauer, Stefan D. Anker, Akio Inui, Janet R. Nicholson

Contents

Summary

Cachexia is a term without a generally approved definition. This precludes the establishment of formal diagnostic and therapeutic criteria and also makes it difficult to precisely estimate its prevalence. This problem is at least in part related to the fact that cachexia presents itself in highly diverse clinical pictures as a consequence of different underlying diseases. However, it appears that there are a limited number of final common pathways, which might represent common denominators in the pathophysiology of cachexia regardless of the initiating event. Such mechanisms would be promising targets for new drugs that should show better efficacy and tolerability than the currently available agents.

30.1 Definition

Cachexia is a syndrome without an internationally recognized medical definition.[1] It is therefore difficult to estimate its prevalence and to establish criteria for its diagnosis and treatment. Numerous terms describe some aspects of cachexia but do not encompass the full complexity of the clinical picture (see Figure 30.1). A comprehensive definition of cachexia would have to include at least the following elements[2]:

> First, cachexia is usually associated with a chronic disease such as cancer. Whether it may also occur in the context of a physiological process such as aging is a matter of debate.

565

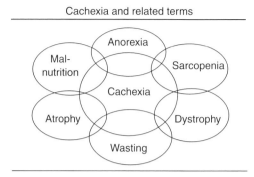

FIGURE 30.1 Various terms describing symptoms observed in patients with cachexia. Despite multiple overlaps none of them is sufficient to define the full clinical picture of this syndrome.

Second, the most prominent clinical feature of cachexia is weight loss. This results from a negative energy balance, which is due to reduced food intake and increased resting energy expenditure.

Third, the pathophysiology of cachexia is complex. It can for instance be accompanied by inflammatory symptoms, pronounced catabolism, neurohormonal activation, and host defense reactions such as an immune response.

Fourth, the final result of cachexia is a reduction in fat and lean body mass, that is, the loss of both lipids and proteins, which eventually leads to impaired body functions and death.

The criteria for the diagnosis of cachexia are variable but according to a widely accepted recommendation a loss of more than 5% of body weight over a period of 6 months should be regarded as cachexia.[1] However, it becomes immediately apparent that such a definition neither makes a distinction between fat and lean body mass nor takes into account the patient's condition, for example, the possible existence of obesity before the development of cachexia. Moreover, this cutoff point has never been validated in any of the clinical syndromes of cachexia. From these considerations it is obvious that a clinically validated definition of cachexia in the context of the various underlying diseases is urgently needed. This issue may best be addressed by organizing an international consensus conference.

30.2 CLINICAL ASPECTS

The cachexia syndrome may develop in association with diseases such as cancer, infections, or chronic heart failure. However, it appears that despite the different initiating events there is a common pattern in the body's reaction to disease, which includes immune responses and neuro hormonal activation.[1, 3–6] If the final responses leading to cachexia are mediated by the same effector systems despite different pathogenic stimuli then the cachexia syndrome should show some common features. This indeed appears to be the case with regard to the stimulation of a class of mediators, the cytokines, which are peptides derived from lymphocytes, and inflammatory cells. The main representatives of this class are tumor necrosis factor α, interferon γ, and interleukin 1 and 6.[7] Their effects on central and peripheral functions, including appetite and energy expenditure, explain many of the symptoms associated with cachexia. Other pathogenic factors include stress hormones such as cortisol, and neuropeptides in the brain.[1, 5]

30.3 PATHOPHYSIOLOGICAL ASPECTS

It is interesting to note that a few central and peripheral systems seem to be of particular importance for mediating the effects of cytokines. In the central nervous system for instance, cytokines appear to act on the pro-opiomelanocortin system in the hypothalamus and finally induce the activation of melanocortin-4 receptors.[8] In the periphery the targets for cachexia-induced proteolysis may be more selective than previously assumed. In skeletal muscle for instance, it seems that myosin heavy chain is more prone to degradation than the other constituents of the contractile machinery.[9] This is not only interesting from a pathophysiological point of view but may also be important for the design of new and selective anticachexia drugs.[10]

Despite possible common patterns in its pathophysiology cachexia presents itself in highly diverse clinical pictures. This is mainly due to the nature of the underlying disease, which usually dominates the clinical situation. Due to the progressive nature of the underlying diseases cachexia is also a highly dynamic syndrome and the symptoms and their severity keep changing over time. Nevertheless, several characteristic symptoms such as weakness and increased fatigability, anorexia, depression, and a general lack of thrive dominate the clinical picture. Finally, increased morbidity and mortality are common features.

30.4 THERAPEUTIC ASPECTS

For the treatment of patients with cachexia several factors have to be taken into consideration.[11] Obviously the most important of them would be the adequate treatment of the underlying disease, for example, cancer, which is sometimes difficult or impossible. However, it has been shown that the successful therapy of cachexia in cancer patients also improves their tolerance vis-à-vis radio- and chemotherapy. This enhancement of the efficacy of causal treatment indicates an obvious synergy between the treatment of cachexia and that of the underlying disease. Nutritional interventions are currently a cornerstone of anticachexia therapy. Although they are not sufficient as the sole measure against cachexia they provide the basis for pharmacological interventions to stimulate appetite and caloric intake. Physical therapy is important to accomplish the desired nutrient partitioning, that is, to achieve an increase not only of body fat but also of skeletal muscle. Behavioral therapy may provide benefit for patients who need psychological support and motivation for continuing therapy.

The pharmacotherapy of cachexia is probably one of the most diverse areas of clinical pharmacology, which is also reflected by the various different chapters on drugs in the present book. Centrally acting drugs are given to stimulate appetite, hormones or analogues thereof are prescribed to substitute for the decreased production of the endogenous substances, and a variety of other drugs with different mechanisms of action are applied to support body functions that are impaired in cachectic patients. Most, if not all, of these drugs provide symptomatic relief but no causal treatment. Moreover, many of these agents are used although their clinical efficacy has never been convincingly demonstrated. In particular, mortality trials are lacking for most of the available compounds and some of the double-blind, placebo controlled studies with the newer agents, for example, anticytokine drugs, unfortunately had negative results.[11]

30.5 POSSIBLE FUTURE DEVELOPMENTS

There are several ways to improve the unsatisfactory situation of current drug treatment of cachexia. The establishment of optimal treatment schedules for existing drugs could enhance their therapeutic ratio. The combination of these drugs or their combination with supportive therapy should be systematically studied to increase their therapeutic efficacy while reducing unwanted effects. The careful definition of criteria for or against the prescription of individual drugs including an internationally recognized description of the stages of cachexia would be another improvement. All these activities are entirely feasible but carefully planned clinical trials with sufficient patient numbers are needed. Individualized treatment may be another promising possibility. However, it requires the identification of patient subpopulations or individual patients with a certain clinical or genetic profile, which precludes short-term success.

Progress in the pharmacotherapy of cachexia may be expected from two sources. On the one hand, existing drugs that have not been used for the treatment of cachexia but have shown promising effects in other trials could be evaluated in various forms of cachexia. To find such candidates available databases could be systematically explored for effects on body weight, an approach that has been successfully applied for angiotensin-converting enzyme inhibitors and β-blockers. On the other hand, an improvement in the pharmacotherapy of cachexia may originate from the development of new drugs with novel mechanisms of action. However, it is difficult to select the best strategy for drug discovery and development. Should anticachexia agents show a broad spectrum of activity to be widely efficacious in as many patients as possible or should they rather be highly specific and thereby convey a maximum of benefit to a well-defined but small subpopulation of patients? The most promising approach would be to design drugs that could be used soon after the initial diagnosis of a serious disease in order to prevent the development of cachexia.

REFERENCES

1. Inui, A., Cancer anorexia–cachexia syndrome: current issues in research and management, *CA: A Cancer Journal for Clinicians*, 52, 72–91, 2002.
2. Kotler, D.P., Cachexia, *Annals of Internal Medicine*, 133, 622–634, 2000.
3. Brink, M., Anwar, A., and Delafontaine, P., Neurohormonal factors in the development of catabolic/anabolic imbalance and cachexia, *International Journal of Cardiology*, 85, 111–121, 2002.
4. Kalra, P.R. and Tigas, S., Regulation of lipolysis: natriuretic peptides and the development of cachexia, *International Journal of Cardiology*, 85, 125–132, 2002.
5. Ramos, E.J.B., et al., Cancer anorexia-cachexia syndrome: cytokines and neuropeptides, *Current Opinion in Clinical Nutrition and Metabolic Care*, 7, 427–434, 2004.
6. Sharma, R. and Anker, S.D., Cytokines, apoptosis and cachexia: the potential for TNF antagonism, *International Journal of Cardiology*, 85, 161–171, 2002.
7. Argilés, J.M., Busquets, S., and Lopez-Soriano, F.J., Cytokines in the pathogenesis of cancer cachexia, *Current Opinion in Clinical Nutrition and Metabolic Care*, 6, 401–406, 2003.
8. Marks, D.L., Ling, N., and Cone, R.D., Role of the central melanocortin system in cachexia, *Cancer Research*, 61, 1432–1438, 2001.
9. Acharyya, S., et al., Cancer cachexia is regulated by selective targeting of skeletal muscle gene products, *Journal of Clinical Investigation*, 114, 370–378, 2004.
10. Glass, D., Signalling pathways that mediate skeletal muscle hypertrophy and atrophy, *Nature Cell Biology*, 5, 87–90, 2003.
11. Argilés, J.M., et al., The pharmacological treatment of cachexia, *Current Drug Targets*, 5, 265–277, 2004.

Index

A

ACE inhibitors, *see* Angiotensin converting enzyme inhibitors
Acquired immunodeficiency syndrome (AIDS), *see* HIV/AIDS
Adaptor proteins, 54
Addison's disease, and steroid hormones, 75–76
Adenocarcinoma, EPO therapy, 415
Adrenal steroids, IGF-1 opposition to, 399
Adrenergic beta agonists, *see* β_2-Adrenergic agonists
Adrenocorticotrophic hormone deficiency, 77
Advanced glycation end-products (AGEs), dietary restriction of, 202–203
Aerobic exercise, effect on skeletal muscle apoptosis, 62
AGEs, *see* Advanced glycation end-products
Aging
  see also Geriatrics
  GH therapy, 358
  skeletal muscle wasting, 32
    IL-6 role in, 191
    in normal aging, 60–62
Agouti related peptides (AgRP), 493, 496
AgRP, *see* Agouti related peptides
AIDS, *see* HIV/AIDS
Akt
  inhibition of MURF1 and MAFbx, 39
  regulation of muscle size, 548–550
Akt/GSK3/eIF2B pathway, 35
Akt1/mTOR/p70S6K pathway, genetic evidence for, 34–35
Albuterol
  effect on FSHD, 314–315
  effects on spinal muscular atrophy, 315
  and resistance training
    effect on FSHD, 315

effect on muscle strength and mass of unilateral limb suspension, 315–316
Alcoholic hepatitis
  malnutrition in
    effect of anabolic steroids, 340
Allopurinol, 464
  dose-related effect, 464–465
α-Melanocyte-stimulating hormone (α-MSH), 5, 24, 128, 129
α-MSH, *see* α-Melanocyte-stimulating hormone
ALT-711, 203
Amino acid-based peritoneal dialysis fluids, 198–199
Amino acids
  and muscle wasting in COPD, 239
  nutritional status in ESRD, 187
  requirements in cachexia, 274
Anabolic androgenic steroids, 204
Anabolic myotrophic cytokines, 108–109
Anabolic steroids, 328
  effect on burn injury-associated cachexia, 338–339
  effect on cancer cachexia, 335, 337
  effect on COPD-associated cachexia, 338
  effect on ESRD-associated cachexia, 339
  effect on HIV/AIDS-associated cachexia, 333, 335, *336*, 156
  mechanism of action, 329–330
  for PEM/wasting in ESRD, 201
  stimulation of erythropoiesis, 410
Androgen receptors
  effect of glucocorticoids, 330
  effect of testosterone, 330
Androgen therapy, contraindications, 332
Anemia, as indication for EPO in cachexia, 408–411
Anemic cancer
  EPO therapy, 413–414
    side effects, 416

O